Atomic Numbers and Atomic Weights

Element	Symbol	Atomic Number	Atomic Weight
Actinium	Ac	89	227.0278
Aluminum	Al	13	26.98154
Americium	Am	95	(243)
Antimony	Sb	51	121.75
Argon	Ar	18	39.948
Arsenic	As	33	74.9216
Astatine	At	85	(210)
Barium	Ba	56	137.33
Berkelium	Bk	97	(247)
Beryllium	Be	4	9.01218
Bismuth	Bi	83	208.9804
Boron	B	5	10.81
Bromine	Br	35	79.904
Cadmium	Cd	48	112.41
Calcium	Ca	20	40.08
Californium	Cf	98	(251)
Carbon	C	6	12.011
Cerium	Ce	58	140.12
Cesium	Cs	55	132.9054
Chlorine	Cl	17	35.453
Chromium	Cr	24	51.996
Cobalt	Co	27	58.9332
Copper	Cu	29	63.546
Curium	Cm	96	(247)
Dysprosium	Dy	66	162.50
Einsteinium	Es	99	(252)
Erbium	Er	68	167.26
Europium	Eu	63	151.96
Fermium	Fm	100	(257)
Fluorine	F	9	18.998403
Francium	Fr	87	(223)
Gadolinium	Gd	64	157.25
Gallium	Ga	31	69.72
Germanium	Ge	32	72.59
Gold	Au	79	196.9665
Hafnium	Hf	72	178.49
Helium	He	2	4.00260
Holmium	Ho	67	164.9304
Hydrogen	H	1	1.0079
Indium	In	49	114.82
Iodine	I	53	126.9045
Iridium	Ir	77	192.22
Iron	Fe	26	55.847
Krypton	Kr	36	83.80
Lanthanum	La	57	138.9055
Lawrencium	Lr	103	(260)
Lead	Pb	82	207.2
Lithium	Li	3	6.941
Lutetium	Lu	71	174.967
Magnesium	Mg	12	24.305
Manganese	Mn	25	54.9380
Mendelevium	Md	101	(258)
Mercury	Hg		
Molybdenum	Mo	42	95.94
Neodymium	Nd	60	144.24
Neon	Ne	10	20.179
Neptunium	Np	93	237.0482
Nickel	Ni	28	58.69
Niobium	Nb	41	92.9064
Nitrogen	N	7	14.0067
Nobelium	No	102	(259)
Osmium	Os	76	190.2
Oxygen	O	8	15.9994
Palladium	Pd	46	106.42
Phosphorous	P	15	30.97376
Platinum	Pt	78	195.08
Plutonium	Pu	94	(244)
Polonium	Po	84	(209)
Potassium	K	19	39.0983
Praseodymium	Pr	59	140.9077
Promethium	Pm	61	(145)
Protactinium	Pa	91	231.0359
Radium	Ra	88	226.0254
Radon	Rn	86	(222)
Rhenium	Re	75	186.207
Rhodium	Rh	45	102.9055
Rubidium	Rb	37	85.4678
Ruthenium	Ru	44	101.07
Samarium	Sm	62	150.36
Scandium	Sc	21	44.9559
Selenium	Se	34	78.96
Silicon	Si	14	28.0855
Silver	Ag	47	107.868
Sodium	Na	11	22.98977
Strontium	Sr	38	87.62
Sulfur	S	16	32.06
Tantalum	Ta	73	180.9479
Technetium	Tc	43	(98)
Tellurium	Te	52	127.60
Terbium	Tb	65	158.9254
Thallium	Tl	81	204.383
Thorium	Th	90	232.0381
Thulium	Tm	69	168.9342
Tin	Sn	50	118.69
Titanium	Ti	22	47.88
Tungsten	W	74	183.85
Uranium	U	92	238.0289
Vanadium	V	23	50.9415
Xenon	Xe	54	131.29
Ytterbium	Yb	70	173.04
Yttrium	Y	39	88.9059
Zinc	Zn	30	65.38
Zirconium	Zr	40	91.22

Aquatic Chemistry Concepts

Aquatic Chemistry Concepts

James F. Pankow

Department of Environmental
Science and Engineering
Oregon Graduate Institute
Beaverton, Oregon

Library of Congress Cataloging-in-Publication Data

Pankow, James F.
Aquatic chemistry concepts / James F. Pankow
p. cm.
Includes bibliographical references and index.
ISBN 0-87371-150-5
1. Water chemistry. I. Title.
GB855.P36 1991
551.46—dc20 91-18535
CIP

Lewis Publishers is an imprint of CRC Press LLC

No claim to original U.S. Government works
International Standard Book Number 0-87371-150-5
Library of Congress Card Number 91-18535
Printed in the United States of America 1 2 3 4 5 6 7 8 9 0
Printed on acid-free paper

In appreciation for their many contributions
to their fields of science
and to their students,
this book is dedicated to
Gilbert E. Janauer,
James J. Morgan,
and
Werner Stumm

Contents

Preface

In the last decade, the field of natural water chemistry has matured from an infancy born of L. G. Sillén to an acknowledged field of science as nurtured by W. Stumm, J. J. Morgan, J. D. Hem, R. Garrels, C. L. Christ, J. N. Butler, F. M. M. Morel, R. M. Pytkowicz, and many others. As a result, there now are several textbooks available which are very useful to the natural water chemist. Why then has this author spent so many hours creating another text? One answer to this question lies in the fact that I perceive that the field of aquatic chemistry needs a book that is very *didactic* in nature. That is, a book which spends considerable time going into the details behind some of the complicated equations and principles of aquatic chemistry. Thus, it is my hope that interested scientists and engineers will be able to use this text in a self-instructional manner. Of course, I also hope that the text sees some classroom use.

Anyone who is familiar with Stumm and Morgan's (1981) text *Aquatic Chemistry* will note certain similarities between that book and this one. The reason for this is two-fold. Firstly, given the natural order present in *Aquatic Chemistry*, it would be difficult to find a more logical sequence with which to present the various important principles. Secondly, so that these two books can reinforce one another, the specific chemical examples of some of those principles together with their accompanying figures have been intentionally structured in a similar manner.

One is finally brought to the question as to why an environmental researcher and teacher such as myself should put so much energy into collecting and structuring "old" science into a text. After all, conducting research and publishing "new" science sometimes seems to be so much more exciting. (Also, the simultaneous jobs of research and teaching seem to leave little time for anything else.) For me, the answer to this question is clear. Indeed, if one is lucky, perhaps ten or so articles written during a lifetime of research will have some significant impact; most of the articles written will be much less important. Thus, placing established science into a text that helps others to learn, to solve important practical environmental problems, as well as to do research just may provide a contribution that far outweighs the importance of one's own original research.

The scope of local, regional, and global environmental problems seems to grow with each passing day. We are thus in a race which we do not wish to lose. My hope is that this text will help to move us all along in that race by at least a few steps.

Portland, Oregon
May 1991

James F. Pankow

Acknowledgements

Firstly, I thank the many aquatic chemistry students at my institution for providing motivation to deepen my knowledge of this subject, as well as for providing me much input on how early drafts of this book might be improved. I also gratefully acknowledge Dr. James J. Morgan for the initial idea for the redox problem considered in Chapter 24. I thank Dr. Richard L. Johnson and Mr. Donald A. Buchholz for assistance with the computer software and hardware used to prepare this book. The helpful comments of Dr. David R. Boone on a portion of the text are also appreciated. For conscientious work in the preparation and assembly of the manuscript, I thank Ms. Judi Irvine, Ms. Dorothy Malek, and Ms. Edie Taylor. For exacting help in the preparation of many of the figures, I thank Ms. Barbara Ryall. For careful layout and typesetting preparation of the manuscript, I am grateful to the staff of ETP Services (Portland, Oregon). Finally, for providing overall confidence and support, I would like to thank my family, and also my publisher, Mr. Brian A. Lewis.

Acknowledgements

PART I

Introduction

1

Overview

1.1 GENERAL IMPORTANCE OF AQUATIC CHEMISTRY

A knowledge of water chemistry is important in understanding the processes governing a wide variety of aquatic situations. The specific environmental settings in which aquatic chemistry is important include:

a. lakes
b. rivers
c. groundwaters
d. soil/water/air systems (vadose zone)
e. estuaries
f. oceans
g. sediments
h. water treatment
i. aerosols and hydrosols
j. rain

1.2 IMPORTANT TYPES OF CHEMICAL REACTIONS IN NATURAL WATERS

Most reactions of interest in aquatic systems can be classified into one of the following five different categories:

a. acid/base
b. complexation
c. dissolution/precipitation
d. redox (i.e., reduction/oxidation)
e. adsorption

Examples of the five above reaction types are given below:

a. acid/base, e.g., dissociation of bicarbonate:

$$HCO_3^- = H^+ + CO_3^{2-} \tag{1.1}$$

b. complexation, e.g., reaction of Fe^{3+} with OH^-:

$$Fe^{3+} + OH^- = FeOH^{2+} \tag{1.2}$$

c. dissolution/precipitation, e.g., dissolution of calcite:

$$CaCO_{3(s)} = CO_3^{2-} + Ca^{2+} \tag{1.3}$$

d. redox, e.g., reduction of oxygen with electrons (e^-):

$$O_2 + 4e^- + 4H^+ = 2H_2O \tag{1.4}$$

e. adsorption, e.g., adsorption of Pb^{2+} on a generic surface:

$$\equiv S + Pb^{2+} = \equiv S - Pb \tag{1.5}$$

1.3 CONCENTRATION SCALES

A variety of units can be used to express the concentration of a given material in an aqueous solution. In one way or another, all of these units are expressed in terms of the mole concept. The unit of "one mole" is fundamentally no more complicated than the unit of "one dozen." A mole is merely much bigger than a dozen. Obviously, if you have one dozen of something, you have 12 items. If you have one mole, you have 6.023×10^{23} items. The number 6.023×10^{23} is often referred to as Avogadro's Number. Except when used in the expression "mole fraction," the term "mole" is often abbreviated simply as "mol." When a solution contains 1.0 mols of a particular species per liter of aqueous system, we say that the solution is 1.0 *molar* (1.0 *M*) in that species.

Before going on, it will also be useful to define the concept of **formula weight** (FW). One FW of a given material is often thought of as being equivalent to *one mol* of that material. While this is true for species that exist as identifiable atoms or molecules, the FW concept is also very useful for materials that do not really exist as molecules per se in their natural forms. For example, in the case of the solid $NaCl_{(s)}$, each Na^+ is not associated with a particular Cl^-, but rather exists in a crystal lattice surrounded by a certain number of Cl^- ions, and vice versa. Thus, one does not speak of a "one mol" of $NaCl_{(s)}$, since in $NaCl_{(s)}$ there is no NaCl *molecule*. Rather, we say we have *one FW* of $NaCl_{(s)}$.

One FW of NaCl is the weight of one formula's worth of $NaCl_{(s)}$. Let us consider the chemistry of a solution obtained by adding 1.0 FW of $NaCl_{(s)}$ (58.4428 g) to enough water to obtain a 1 liter (L) system. When $NaCl_{(s)}$ dissolves, it *dissociates* to form Na^+ and Cl^- ions. Thus, just as we do not have an NaCl molecule in the $NaCl_{(s)}$ lattice, the solution under consideration is not a 1.0 *M* solution of dissolved NaCl. The 1.0 FW/L (i.e., 1.0 *formal*, or simply 1 *F*) solution is, however, 1.0 *M* in both Na^+ and Cl^-.

The FW concept is also useful when describing aqueous solutions of materials that may exist largely as molecules in their pure form, but can *change* when added to water. Acids and bases are good examples. For a 0.001 *F* solution of the acid HA, while 0.001 mols of the acid molecule HA might have been added to the water, some of the HA will certainly dissociate, forming a certain amount of A^- and leaving a solution that is something less than 0.001 M in HA. Regardless of how much of the HA dissociates, however, the system remains a 0.001 F solution of HA.

Molarity and formality are not the only concentration units that come up in the field of natural water chemistry. The common concentration units in aquatic chemistry and some corresponding examples are summarized as follows:

1. Molarity, M (mol/L): the number of mols of a *specific species* per liter of solution. Example: The atomic weight of oxygen is 15.9994 g/mol, and the atomic weight of hydrogen is 1.0079 g/mol. Therefore, if there is (15.9994 + 1.0079) g = 17.0073 g of OH^- present for each liter of solution, then the molarity of OH^- is 1.0000 M.

2. Molality, m (mols/kg): the number of mols of a *specific species* per kilogram of solvent. Units of molality are particularly convenient when one does not want to have to worry about changes in volume that might occur due to changes in temperature, pressure, and/or composition. (Situations involving seawater are good cases in point.) Example: If there is (15.9994 + 1.0079) g = 17.0073 g of OH^- present for each kg of water, the molality of OH^- is 1.0000 m.

3. Formality, F (formula weights/L): the number of formula weights per liter of solution. The atomic weight of Na is 22.98977 g/mol. If (22.98977 + 17.0073) g = 39.9971 of $NaOH_{(s)}$ is added to enough water to make a liter of solution, then that solution is 1.0000 F in NaOH.

4. Normality, N (equivalents/L): the number of equivalents of acid, base, or redox-active species per liter of solution. Examples: a solution that is 0.01 F in HCl is 0.01 N in H^+. A solution that is 0.01 F in H_2SO_4 is 0.02 N in acid.

5. Mole fraction, X (dimensionless): the fractional concentration of a *specific species* where the fraction is computed based on a knowledge of the numbers of mols of *all* of the species present. The mole fraction concentration scale can be used equally well for gases, liquids, and solids. It is important to note that for any given phase, when all concentrations are expressed as mole fractions, the sum of those mole fractions must be identically equal to 1.0. If the number of mols of species i in a given phase is denoted n_i, then

$$X_i = \frac{n_i}{\sum_j n_j} \tag{1.6}$$

where $\sum_j n_j$ represents the sum of the numbers of mols of all species in the system (including all ions). The mole fraction of a *truly* pure material is always exactly 1.0. However, the mole fraction of H_2O in a system produced from just water alone would not be exactly 1.0 since there would be trace amounts of H^+ and OH^- present. Example: if an aqueous solution contains 30.0 g of water and 0.100 FW of $NaCl_{(s)}$, the mole fractions of the various species are computed as follows. We first note that 30.0 g of H_2O = 30.0 g/(18.0152 g/mol) = 1.665 mols of H_2O. Therefore, since the $NaCl_{(s)}$ dissolves to form equal amounts of Na^+ and Cl^-, if we neglect the presence of H^+ and OH^-, we have

$$X_{H_2O} = \frac{1.665}{1.665 + 0.100 + 0.100} = 0.893 \tag{1.7}$$

$$X_{Na^+} = \frac{0.100}{1.665 + 0.100 + 0.100} = 0.054 \quad (1.8)$$

$$X_{Cl^-} = \frac{0.100}{1.665 + 0.100 + 0.100} = 0.054. \quad (1.9)$$

In this discussion, we stated that the mole fraction concentration scale can be used for solids. Actually, any concentration scale can be used for solids, it is just that the mole fraction scale is the scale most often applied. When more than one solid is present, those solids will always tend to dissolve in one another, to at least some small extent. Metal alloys are examples of solid solutions with which we are all familiar. An example of an ionic solid solution is the substitution of chloride for iodide in AgI. This leads to a solid solution of AgCl in AgI.

1.4 ACTIVITY AND ACTIVITY COEFFICIENTS

The concentration of species i is often symbolized using brackets, that is, as $[i]$. For reasons that are discussed later, sometimes a given chemical does not *act* as though it has the concentration that it does. It might act less concentrated or more concentrated than it is. How it *acts* is called its *act*ivity. It is almost a truism to note that the degree to which a given chemical species re*acts* and inter*acts* with other species is determined by its *act*ivity, not its concentration.

Whenever the activity of the species does not equal its concentration, the species is said to be behaving **non-ideally**. The correction factor which interrelates the activity and concentration scales is called the activity coefficient. If the activity of species i is represented as $\{i\}$ or a_i, then it may be defined that

$$\{i\} = a_i = \gamma_i[i]/[i]^{\circ} \quad (1.10)$$

where γ_i is the activity coefficient (dimensionless) of species i in the solution of interest, and $[i]^{\circ}$ has a numerical value equal to 1.0 and has the same dimensions as $[i]$. With this definition, $\{i\}$ is dimensionless. Division by $[i]^{\circ}$ in Eq. (1.10) is just a convenience adopted by thermodynamicists to make $\{i\}$ dimensionless. For example, when we first discuss chemical potentials in Chapter 2, we will see that we need to evaluate $\log a_i$, and the mathematical side of physical chemists tells them that they should only be taking logarithms of pure numbers, that is, numbers with no units associated with them. For the purposes of this text, the reader need not worry any further about the convention that all activities are dimensionless. Indeed, when expressing a concentration on the molarity or molality scale, no harm will be done if units of molarity or molality are given to the corresponding activity values. As will be discussed in detail in Chapter 2, the specific numerical value of a given activity coefficient will depend on the scale (units) used to express the concentration.

Activity coefficients of ions are frequently less than 1.0. We also note that the values of activity coefficients are both species and matrix dependent. An example of species dependence is the fact that in a 0.1 F solution of $MgCl_2$, the activity coefficient of Mg^{2+}($\gamma \simeq 0.2$) is significantly smaller than the activity coefficient of Cl^-($\gamma \simeq 0.8$). An example of matrix dependence is the fact that the activity coefficient of Mg^{2+} is significantly smaller in a 0.1 F $MgCl_2$ solution ($\gamma \simeq 0.2$) than it is in a 0.01 F $MgCl_2$ solution ($\gamma \simeq 0.5$).

When species i acts less concentrated than it really is, then $\gamma_i < 1.0$. When species i acts more concentrated than it really is, then $\gamma_i > 1.0$. When molality is used to express $[i]$, then $\gamma_i = 1.0$ in an ideal solution, and the concentration of species i is equal to its activity:

$$[i] = \{i\} \qquad \text{ideal behavior.} \tag{1.11}$$

The above discussion for aqueous solutions applies equally well to gases and solids. The importance of liquid phase chemistry in aquatic chemistry should be obvious. Gases and solids are also of great interest since both gases and solids can dissolve in, and thereby interact and alter the properties of an aqueous system. For example, CO_2 in the atmosphere and $CaCO_{3(s)}$ (limestone) often dissolve in and affect the pH of natural waters.

If species i is present in more than one phase in a given system, it is important to realize that it can act more ideally in one of the phases than in another. In other words, the activity coefficient for species i will not generally be the same in all phases of the system unless it is behaving ideally in all of the phases.

In aqueous solutions, it is when we are dealing with ionic species that we most often need to consider the possibility of non-unity activity coefficients. Consider for example the acid dissociation reaction

$$HA = H^+ + A^-. \tag{1.12}$$

If a given solution is 10^{-5} F in HA and also contains some NaCl (say about 0.005 F), the H^+ ions that form by dissocation of the HA will tend to be surrounded by Cl^- ions, and the A^- ions will tend to be surrounded by Na^+ ions (see Figure 1.1). This will lead to shielding effects, and neither H^+ nor A^- will be as active as they would be in the absence of the NaCl: the activity coefficients for both H^+ and A^- will be less than 1.0.

It should also be noted that H^+ ions can also be shielded by A^- ions, and vice-versa, so increasing non-ideality will set in for H^+ and A^- as the formality of a solution of HA is increased even when there is no NaCl in the system. When the total electrolyte concentration becomes large, γ_i values can actually become greater than 1.0. In this

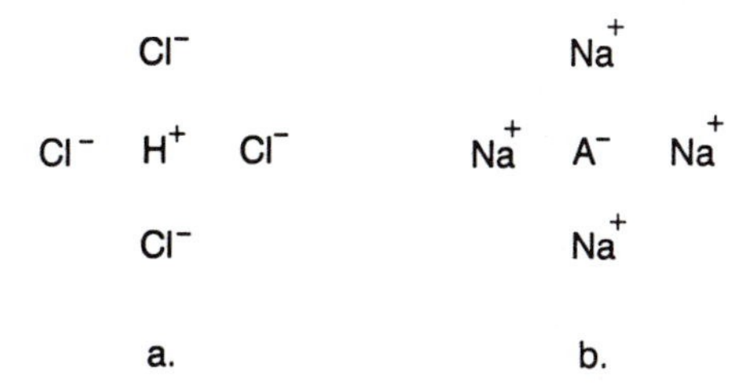

Figure 1.1 Cross sections of ion clouds surrounding H^+ and A^- ions in a solution in which most of the negative ions are Cl^-, and most of the positive ions are Na^+. (a) Shielding of H^+ by Cl^- ions. (b) Shielding of A^- by Na^+ ions.

case, the aqueous system does not have enough water to effectively solvate all of the ions, and this leads to a higher activities for all of the ions (see Chapter 2).

The most common *gas phase* concentration/activity scale is not molarity, molality, or mole fraction, but rather *atmospheres* of pressure. Strictly speaking, gaseous pressure is not a *concentration*. However, it can be related to the concentration scale of mols per L of gas phase if the temperature is known (e.g., for an ideal gas, the relation needed is the ideal gas law, i.e., $p_i V = n_i RT$ which yields $n_i/V\,(\text{mols/L}) = p_i/RT$ where p_i is the pressure of species i (atm), V (L) is the volume of the gaseous portion of the system, and R is the gas constant (= 0.082 L-atm/mol-deg) and T is the temperature (K), respectively). When one wants to emphasize the *possibility* of gas phase non-ideality, a special name is used for that activity, that is, the "fugacity" (f). In a manner that is analogous to the case of liquid and solid solutions, strictly speaking, a fugacity is dimensionless. Commonly, however, the units are the same as used for pressure, that is, atm. Except for when gas pressures are tens to hundreds of atm or more, gas phases may be assumed to behave ideally. Ideal gas behavior may certainly be assumed near the surface of the earth where gas phase pressures are near 1 atm.

Due to the work of Mackay and co-workers (e.g., Clark et al. 1988), interest has been growing in expressing solid and liquid phase activities in terms of their *equivalent* gas phase fugacities. This is particularly true for species like non-polar organic com-

TABLE 1.1 Summary of Nomenclature, and Model of Aquatic Systems.

Example Reactions	Thermodynamic Parameters		Names	Common Units
GAS PHASE				
	$p_i, p_j, pk, \ldots$		partial pressures	atmospheres
$H_2 + O_2 = H_2O$	$\gamma_{i,g}, \gamma_{j,g}, \gamma_{k,g}, \ldots$		activity coefficients	dimensionless
	$f_i, f_j, f_k, \ldots$		fugacities	atmospheres
AQUEOUS PHASE				
	$[i], [j], [k], \ldots$		concentrations	mol/L (M) or mol/kg (m)
$H_2CO_3 = HCO_3^- + H^+$	$\gamma_i, \gamma_j, \gamma k, \ldots$		activity coefficients†	dimensionless
	$\{i\}, \{j\}, \{k\}, \ldots$		activities	mol/L (M) or mol/kg (m)
SOLID PHASE(S) (MULTIPLE PHASES POSSIBLE)				
$Ca^{2+} + CO_3^{2-} =$	α PHASE,	β PHASE, etc.		
$CaCO_3(s)$	$X_{i,\alpha}, \zeta_{i,\alpha}$	$X_{i,\beta}, \zeta_{i,\beta}$	X = mole fraction	dimensionless
	$X_{j,\alpha}, \zeta_{j,\alpha}$	$X_{j,\beta}, \zeta_{j,\beta}$	ζ = solid phase activity coefficient	dimensionless
	$X_{k,\alpha}, \zeta_{k,\alpha}$	$X_{k,\beta}, \zeta_{k,\beta}$	ζX = activity	dimensionless
	⋮	⋮		

†The values of the aqueous solution phase activity coefficients will depend on whether units of m, M, or X are used. The most commonly used aqueous-phase concentration scales are m and M. The symbol for the activity coefficient for species i on the m scale is γ_i.

pounds (e.g., certain chlorinated pesticides) which can exist in problematic amounts in: 1) the gas phase; 2) dissolved in the aqueous phase; and 3) sorbed to soils and sediments. The reason behind this interest is the desire to compare the activities of a given species in different phases on the *same scale* so that one can understand *whether or not the species is likely to move from one of the phases to another*. For example, if there are two environmental compartments A and B, and if the fugacity of a species is larger in compartment A than in compartment B, then a portion of the species will tend to move from A to B. However, as we will discuss further, this knowledge of the thermodynamic status of the system does not tell us how fast that transfer process will tend to occur.

The equivalent gas-phase fugacity of i ($f_i \simeq p_i$) is the fugacity that species i would possess in the gas phase if there was equilibrium between the value of $[i]$ in the specific phase of interest and the gas phase. For example, in the case of an aqueous (aq) phase, this fugacity would be calculated by Henry's Gas Law:

$$f_i = \gamma_{i,\mathrm{g}} p_i = H_i \gamma_{i,\mathrm{aq}}[i] \tag{1.13}$$

where $\gamma_{i,\mathrm{g}}$ is the gas-phase activity coefficient of i, p_i is its gas-phase pressure, and H_i is the compound-, temperature-, and total pressure-dependent Henry's Gas Law constant. Common units for H_i are atm/(mol/m^3). When the gas phase is behaving ideally, Eq. (1.13) reduces to

$$f_i = p_i = H_i \gamma_{i,\mathrm{aq}}[i] \tag{1.14}$$

Equations (1.13) and (1.14) may be applied even when there is no actual gas phase present. One might want to do this in order to compare the f_i values obtained for the aqueous phase with f_i values obtained for another phase, for example, aquifer solids.

1.5 DIFFERENT APPROACHES MAY BE USED IN PREDICTING AQUEOUS CHEMISTRY SPECIATION

The main task of the aquatic chemist is to *predict the "chemical speciation"* present in aquatic systems under a wide variety of possible situations. Simply stated, the term chemical speciation refers to the set of all concentration values for the various chemical species present in the system. When more than just the aqueous phase is present, it is sometimes convenient to expand the meaning of chemical speciation to also include information on either: a) the *amounts* of each of the different phases present (e.g., the gaseous, aqueous, and solid phase(s)); or b) the *amounts of gas phase and/or solid phase(s) per liter of aqueous phase*. Different methods are available for obtaining the desired speciation information.

1.5.1 Thermodynamic Equilibrium Approach

The most widely used method for obtaining chemical speciation information is the *thermodynamic-equilibrium* approach. With this method, one assumes that the system is at equilibrium, that is, the system experiences no changes with time in chemistry, temperature, or pressure, and moreover has no *tendency* to change. The exact nature of the equilibrium state is assumed to be governed completely by the thermodynamic energy levels of all the chemical constituents involved, including all gases, dissolved species, and solids. Often, the compound adjective "thermodynamic-equilibrium" is abbreviated simply as "thermodynamic" or "equilibrium." Indeed, the nature of any *equilibrium* state is always governed by the system *thermodynamics. The thermodynamic equilibrium approach is the basis for virtually all of the discussions presented in this text.* The only exception is Section 1.5.2, which is a very brief introduction to kinetic models.

When using the thermodynamic-equilibrium approach, one assumes that all chemical reactions are governed by equilibrium constants. The Henry's Gas Law constant in Eqs. (1.13) and (1.14) is an example of an equilibrium constant. We now consider the general reaction

$$aA + bB = cC + dD. \tag{1.15}$$

At equilibrium we have

$$K \equiv \frac{\{C\}^c\{D\}^d}{\{A\}^a\{B\}^b} = \frac{\gamma_C^c[C]^c\gamma_D^d[D]^d}{\gamma_A^a[A]^a\gamma_B^b[B]^b} \tag{1.16}$$

where K is the temperature- and pressure-dependent equilibrium constant for reaction (1.15). The upper case letters refer to the specific species, and the lower case letters refer to the species' stoichiometric coefficients in the reaction. In general, for any reaction between reactants and products, at equilibrium we have

$$K \equiv \prod_j \{j\}^{\nu_j} \tag{1.17}$$

where the grand product $\prod$ in Eq. (1.17) takes place over both reactants and products. (Note that $\prod_j x_j$ equals the quantity obtained when all x_j are multiplied together.) The stoichiometric coefficients ν_j are positive for products and negative for reactants.

It is very important to realize that the results obtained from equilibrium calculations can only be as good as the equilibrium constants used in those calculations. If an error was made in the experimental determination of a given equilibrium constant, then the results obtained from using that constant will also be in error. If an incorrect equilibrium constant is tabulated in a reference book, then it will remain incorrect no matter how universally that book is used as a source of equilibrium data.

As far as activity corrections are concerned, the most often used convention for aqueous solutions is to have the activity coefficents of dissolved species equal 1.0 when the system is *infinitely dilute*. In this case, for a very dilute solution where all γ_i values are 1.0, for reaction (1.15) we have

$$K \equiv \frac{\{C\}^c\{D\}^d}{\{A\}^a\{B\}^b} = \frac{[C]^c[D]^d}{[A]^a[B]^b}. \tag{1.18}$$

1.5.1.1 Thermodynamic Equilibrium Models. It was the work of L. G. Sillén (1961) that popularized the application of equilibrium models in the solving of natural water chemistry problems. This approach continues to be a very useful way to handle a wide variety of water chemistry problems. Equilibrium models tell you what the chemistry will be at *equilibrium*, but *not* the kinetic rates according to which the system reaches the equilibrium state.

Equilibrium models will in general provide good results when the governing reactions occur quickly. Poor results will be obtained if the time to reach equilibrium is long compared to the time frame of interest. Determining the amount of time required to reach an equilibrium state is a chemical kinetics problem. The fact that equilibrium models are often very successful means that equilibrium is often obtained (or at least nearly so) for many species of interest in natural waters.

For the species ML, M, and L, consider the reaction

$$\mathrm{M} + \mathrm{L} = \mathrm{ML} \qquad K = \frac{\{\mathrm{ML}\}}{\{\mathrm{M}\}\{\mathrm{L}\}} \tag{1.19}$$

where K is the equilibrium constant for the reaction. In this simple example, we would seek to find the values of [ML], [M], and [L] that satisfy: 1) the equilibrium constant in Eq. (1.19); and 2) the mass balance conditions on the system (i.e., how much total M is in the system, and how much total L is in the system). The values of the various γ_i values (and therefore the activity values) will depend on the concentrations of the various species (including ML, M, and L) present in the water at equilibrium.

When modelling an actual natural water chemistry situation, there are likely to be *many* possible chemical reactions. At equilibrium, *all* of the corresponding K expressions and the various mass balance conditions governing that situation must be satisfied simultaneously. Such problems are usually complicated by the fact that many of the reactions are *coupled*. That is, many of the species are involved in more than one reaction. In recent years, sophisticated computer codes that compute equilibrium speciation even for very complicated situations have been developed. These models include the REDEQL model discussed by Morel et al. (1976), the MINEQL model developed by Westall et al. (1976), and the MINTEQ model of Felmy et al. (1984).

An extremely important example of a species that is involved in many different types of aqueous reactions is the proton. Fortunately, the actual transfer of protons between acids and bases tends to be very fast. Most complexation reactions are reasonably fast. When they are mediated by bacteria, many redox (reduction/oxidation) reactions may also be considered fast relative to time scales of weeks to months. However, many redox reactions can also be rather slow, and the same applies to a variety of dissolution/precipitation reactions, especially those involving very insoluble solids. The situation with solids is often further complicated by the fact that initial precipitates are often metastable, and can change into other solids. The equilibrium dissolution constants of the initial and final solids can be significantly different. For example, increasing the pH in a solution containing a significant amount of Fe^{3+} initially leads to the formation of amorphous $Fe(OH)_{3(s)}$. In time, this solid will slowly change into the less soluble mineral goethite (α-$FeOOH_{(s)}$).

If the earth were completely isolated from all energy sources, then it would not matter so much whether natural reactions were fast or slow, as the global system would be allowed to come to equilibrium. Happily for life on earth, however, our planet is not isolated, but lies in a constant stream of energy provided by the sun. *This energy stream tends to continually disrupt the state of final equilibrium that aquatic systems are always seeking to attain.* For example, the sun's energy can be stored by photosynthetic organisms through their reduction of CO_2 to form organic matter. This organic matter is unstable in the presence of atmospheric levels of oxygen. Thus, after some period of time, such organic matter will be oxidized by oxygen back to CO_2, but that move toward equilibrium will be disrupted again by another photosynthetic cycle.

1.5.1.2 The Effects of the Scales Chosen for the Concentration (Activity) of Reactants and Products on the Numerical Value of an Equilibrium Constant.

If the value for a given equilibrium constant K is to be determined, then the concentration/activity scales which are to be used for each of the reactants and products must be selected first. This is made particularly clear when it is realized that equilibrium constants are usually not unitless, but possess dimensions. When the units chosen for an equilibrium constant are changed, then the numerical value will also change. This may seem paradoxical at first since it might seem that "constants" should remain constant. This matter may be understood as follows. Consider the following reaction involving three dissolved species:

$$\mathrm{A} + \mathrm{B} = \mathrm{C} \qquad K = \frac{\{\mathrm{C}\}}{\{\mathrm{A}\}\{\mathrm{B}\}}. \tag{1.20}$$

Let us assume that the *numerical* value of the equilibrium constant is 1.0, and that the activity of the participating species is expressed in units of m. Let us also assume that the activities of all the species are equal to their concentrations. If the initial condition for the solution of interest is

$$[\mathrm{A}] = 0.0\ m \qquad [\mathrm{B}] = 0.0\ m \qquad [\mathrm{C}] = 2.0\ m$$

then the only possible equilibrium composition which would satisfy this initial condition for the value of K given above would be

$$[\mathrm{A}] = 1.0\ m \qquad [\mathrm{B}] = 1.0\ m \qquad [\mathrm{C}] = 1.0\ m.$$

(Half of the initial C dissociates to form equal amounts (1.0 m each) of A and B.) Upon substituting these values into the expression for K we note that the K has units of m^i.

What happens if we should decide to change the units in which we choose to express the concentration (activity)? Let us say we choose mols/(0.1 kg of solvent), and let us refer to a 1.0 mol/(0.1 kg solvent) solution as a 1.0 tenmolal (*tm*) solution. Intuition tells us that the composition at equilibrium *cannot* change because of this change in units. That is, the chemical species which participate in the reaction cannot know or care what units are chosen for the concentration or for the equilibrium constant. "Nature" cannot be expected to respond to changes in the choice of units (like mols or kg) that Earthlings or Venusians or anybody else should happen to devise to numerically explain nature. In less philosophical terms, by analogy, the physical height of a building 100 meters tall remains constant regardless of whether we chose to express the height in centimeters (10,000 cm tall), or feet (328.084 ft tall): the *numerical value* of the height (equilibrium constant) is dependent upon the units, but the *physical height* (distribution of species at equilibrium) remains unaffected.

Since the equilibrium composition cannot change in this example, with the new units, the equilibrium composition can be expressed

$$[A] = 0.10\ tm, \quad [B] = 0.10\ tm, \quad [C] = 0.10\ tm$$

and the expression of an equilibrium constant with these units leads to a value of 10 tm^{-1}. Clearly, the composition at equilibrium has not changed, but the numerical value and the units of the equilibrium constant have changed.

In the special case of a constant for which the concentration units *cancel* out, the K will not depend on the units. Consider for example the equilibrium constant for the reaction involving the four species in reaction (1.13). Let us assume that all of the species are dissolved. In that case, if all of the activities are expressed in the same units, for example, m, then the units cancel out when for the stoichiometric coefficients a, b, c, and d we have $(a + b) = (c + d)$.

In conclusion, after obtaining the numerical value for an equilibrium constant, one should always be sure to understand whether or not the constant has units, and if so, what those units are.

1.5.2 Kinetic Models

Kinetic models are occasionally needed to describe natural water systems, but they are often difficult to apply. Their application requires a knowledge of specific *initial conditions* as well as *all* of the pertinent kinetic rate constants (Pankow and Morgan 1981a, 1981b). Regardless of the specific nature of the system of interest, if a given kinetic model has been constructed properly, then after a sufficiently long time has elapsed for the system of interest, then the chemistry given by the kinetic model will be the same as chemistry given by the corresponding equilibrium model.

Consider the example

$$M + L \xrightarrow{k_f} ML \tag{1.21}$$

$$M + L \xleftarrow{k_b} ML. \tag{1.22}$$

Reaction (1.21) describes the bimolecular combination of M and L to form ML; the rate constant for the combination is k_f. Reaction (1.22) describes the unimolecular decomposition of ML to form M and L; the rate constant for the decomposition is k_b.

From a speciation point of view, we have three unknowns. Therefore, we need three independent equations to solve the problem. There are several possible sets of three independent equations. Each set will contain one differential equation and the two mass balance equations. Neglecting activity corrections, one possible set is:

$$d[M]/dt = k_b[ML] - k_f[M][L] \qquad \text{(differential equation)} \tag{1.23}$$

$$M_{tot} = [M]_o + [ML]_o = [M] + [ML] \qquad \text{(mass balance equation on total M)} \tag{1.24}$$

$$L_{tot} = [ML]_o + [L]_o = [ML] + [L] \qquad \text{(mass balance equation on total L)} \tag{1.25}$$

Eq. (1.23) is the differential equation for the time derivative of the species [M]. Differential equations can also be written for the time derivatives of [ML] and [L], but they are not linearly independent of the set of equations comprised by Eqs. (1.23)–(1.25).

To solve Eqs. (1.23)–(1.25), we substitute

$$[L] = L_{tot} - [ML] \tag{1.26}$$

and

$$[ML] = M_{tot} - [M] \tag{1.27}$$

into Eq. (1.23), thereby obtaining the following nonlinear differential equation

$$d[M]/dt = k_b(M_{tot} - [M]) - k_f[M](L_{tot} - M_{tot} + [M]) \tag{1.28}$$

Based on known values of k_b and k_f, and the initial conditions ($[ML]_o$, $[M]_o$, and $[L]_o$), this equation may be integrated numerically to obtain [M] as a function of time. The mass balance equations then allow the calculation of [ML] and [L] as a function of time.

At equilibrium, $d[M]/dt = 0$. This will by definition occur when [M], [L], and [ML] take on their equilibrium values. Thus, at equilibrium, Eq. (1.28) gives

$$k_b(M_{tot} - [M]) = k_f[M](L_{tot} - M_{tot} + [M]) \tag{1.29}$$

or

$$k_b[ML]_{eq} = k_f[M]_{eq}[L]_{eq} \tag{1.30}$$

where $[ML]_{eq}$, $[M]_{eq}$, and $[L]_{eq}$ represent the equilibrium values of ML, M, and L, respectively. Thus, the kinetic model will yield concentration values that satisfy

$$k_f/k_b = \frac{[ML]_{eq}}{[M]_{eq}[L]_{eq}} = K. \tag{1.31}$$

The ratio k_f/k_b may thus be identified with the equilibrium constant for reaction (1.19). We therefore see how kinetic models will predict the same species distribution at equilibrium that is predicted by the direct application of the appropriate equilibrium constant.

There are three major problems with applying kinetic models to natural water systems. Firstly, although the differential equation represented by Eq. (1.28) is not difficult to solve using numerical integration techniques, most problems of interest in natural water chemistry are much more complex than the case considered here. These situations yield sets of equations that involve more than one *coupled differential equation.* As the number of coupled species increases, the complexities of the differential equations also grow, making integration increasingly difficult. Secondly, rate constants like the k_f and k_b parameters described above are usually not known nearly as well as the corresponding equilibrium constants. For example, the presence of trace levels of species that can act as catalysts can affect rate constants greatly. Thirdly, rate expressions are often not as simple as in the case considered above. For example, there is no guarantee that the rate of a given combination reaction between two species will be simply proportional to the concentrations of the two species. Sometimes, very unexpected kinetic laws are observed.

To summarize, since: 1) equilibrium models often do a reasonably good job of describing natural water chemistry; and 2) it is often not possible to apply kinetic models to problems that are even moderately complex, this text will emphasize equilibrium models. Problems illustrating the principles of this chapter are presented by Pankow (1992).

1.6 REFERENCES

CLARK, T., K. CLARK, S. PETERSON, D. MACKAY, and R. J. NORSTROM. 1988. Wildlife Monitoring, Modeling, and Fugacity. *Env. Sci. Technol.*, **22**: 120-127.

FELMY, A. R., D. C. GIRVIN, and E. A. JENNE. 1984. *MINTEQ–A Computer Program for Calculating Aqueous Geochemical Equilibria.* US E.P.A., EPA-600/3-84-032, Washington, D.C.

MOREL, F. M. M., R. E. MCDUFF, and J. J. MORGAN. 1976. *Marine Chemistry*, **4**: 1.

PANKOW, J. F. and J. J. MORGAN. 1981. Kinetics for the Aquatic Environment, I. *Env. Sci. Technol.*, **15**: 1155–1164.

PANKOW, J. F. and J. J. MORGAN. 1981. Kinetics for the Aquatic Environment, II. *Env. Sci. Technol.*, **15**: 1306–1313.

PANKOW, J. F. 1992. *Aquatic Chemistry Problems*, Portland: Titan Press-OR, P.O. Box 91399, Portland, Oregon 97291-1399.

SILLÉN, L. G. 1961. The Physical Chemistry of Sea Water. *Oceanography*, M. Sears, ed., AAAS, **67**: 549–581.

WESTALL, J. C., J. L. ZACHARY, and F. M. M. MOREL. 1976. *MINEQL, A Computer Program for the Calculation of Chemical Equilibrium Composition of Aqueous Systems.* **Technical Note-18**, Parsons Laboratory, M.I.T., Cambridge, Mass.

2

Thermodynamic Principles

2.1 FREE ENERGY AND CHEMICAL CHANGE

2.1.1 Constant Pressure/Temperature Systems

A system that is characterized by a constant pressure (P) and a constant temperature (T) is well suited to being modeled using equilibrium thermodynamics. Fortunately, natural water systems may often be viewed in this type of context over many important time frames of interest. In the natural water environment, there are three main sources of pressure: the weight of the overlying atmosphere (= 1 atm at sea level); the weight of water overlying the water of interest (= 1 atm for each 10.7 meters of water column); and the weight of overlying rock and soil.

At sea level, the P operating on fresh surface waters such as rivers and shallow lakes is primarily due to the atmosphere, and so it is always close to 1.0 atm. The same applies to sea waters close to the interface between the atmosphere and the water. At significant depths in lakes and in the oceans, however, P can reach values that are many times greater than 1.0 atm. Nevertheless, such P values will also remain constant over time as long as the "parcel" of water of interest does not change depth. Analogous statements can be made about deep groundwaters. At elevations higher than sea level (as with mountain lakes and rivers), P can be less than 1.0 atm since the pressure of the overlying atmosphere decreases with increasing elevation. Again, however, P will remain constant as long as the water of interest remains at a constant elevation.

Although the temperatures of some surface waters are subject to fluctuations over 24-hour periods, *mean* T values are oftentimes suitable when making equilibrium calculations. These calculations will take the form of the simultaneous solution of all of the pertinent equilibrium, mass balance, and electroneutrality equations. In latitudes where the season-to-season changes in T are significant, equilibrium calculations should definitely be revised for each change in mean T of $\sim$5°C. The reason is that, as will be seen below, almost all equilibrium constants are at least somewhat T-dependent. Most equilibrium constants are also P-dependent.

As will be discussed in greater detail below, in any constant pressure/temperature system, the property that governs all thermodynamic chemical equilibria is the *Gibbs Free Energy, G*. The standard units for G are kilojoules, or simply kJ. *It is the individual free energy contents of chemicals that determines the values of equilibrium constants.* In every case, the equilibrium position that a given constant P and T system will seek to attain is the state that gives it the lowest overall possible value of G. Thus, if any given chemical process changes the total system G at constant P and T to a value that is lower than the initial G, then that process: 1) is described as being *thermodynamically possible*; and 2) will in fact *tend to occur*. The question of how rapidly the system will move towards the state giving the minimum possible G is not answered by thermodynamics. As discussed in Chapter 1, the only thing that thermodynamics tells us is that the system will, if given sufficient time, reach an equilibrium state. The rate at which a chemical reaction moves a constant P and T system towards the state where G is minimized can only be addressed by the field of chemical kinetics.

2.1.2 Constant Volume/Temperature Systems

Based on the previous discussion, one might be tempted to assume that, compared to constant P and T systems, equilibrium thermodynamic calculations based on G values are not as useful in systems in which one or more chemical reactions move the system towards an equilibrium state under conditions where the overall *volume and temperature* (V and T) are held constant. In some respects this is true. However, as we will now see, in other respects G-based calculations are just as useful as ever *once the final equilibrium state is reached.*

Although P will not necessarily remain constant during reactions in constant V and T systems, we know that in the final equilibrium state, P will reach a constant value. After all, what kind of equilibrium would we have if P was changing? An example of when V and T remain constant but P does not remain constant would be obtained when: 1) some chemicals of interest (this might include some water) are inside a confined, constant volume chamber (as might be found inside of a rigid rock mass) that is maintained at constant T; and 2) a chemical reaction is occurring for which the products, in the aggregate, tend to occupy a volume that is different than the aggregate volume of the reactants.

As an example of the second criterion, for the reaction

$$aA + bB = cC + dD \tag{2.1}$$

we would require that the volume of c mols of C plus d mols of D be either greater, or less, but not equal to the volume of a mols of A plus b mols of B. For example, c mols of C plus d mols of D might occupy more volume at some reference pressure than is occupied by a mols of A plus b mols of B. In this case, ΔV_{rxn} (i.e., the change in volume *per unit of reaction* at a constant reference pressure) is not equal to 0. Under these conditions, even if only small amounts of the reactants in reaction (2.1) are converted to products, the pressure in the system will have to change. If $\Delta V_{rxn} > 0$, then P will go up; the increase in P compresses all of the constituents present so that V can remain constant. If $\Delta V_{rxn} < 0$, then P will go down; the decrease in P allows an expansion of all of the constituents, and V remains constant.

The thermodynamic function that must be minimized for equilibrium to be attained in a constant T and V system is not G, but rather the so-called Helmholtz Free Energy, A (kJ/mol; not the same as the A as in Eq. (2.1)). The equilibrium state will be represented by whatever particular mix of reactants, products, and value of P that collectively minimizes A. This type of problem is just as inherently solvable as a constant P and T problem. It is just that if we wanted to *predict* the nature of the equilibrium state, we would now seek to determine the system characteristics that give the minimum value of A. While minimizing G will not allow us to predict the nature of the equilibrium state in a constant V and T system, as stated above, that final state will in fact be at constant P. Thus, once the constant V and T system has also become a constant P and T system, the values of the chemical equilibrium constants (which are all G-based) for that P and T may be used to compute the equilibrium speciation.

In summary, we cannot use G-based calculations to *predict the nature of the equilibrium state* in a constant V and T system, but once equilibrium is attained so that P has stabilized and can be measured, we can use G-based equilibrium constants for the final P and T to *compute the equilibrium speciation.*

2.2 A POTENTIAL ENERGY ANALOG FOR CHEMICAL ENERGY-DRIVEN CHEMICAL CHANGE

Let us assume that a certain system is characterized at time zero by values of the activities $\{A\}$, $\{B\}$, $\{C\}$ and $\{D\}$ that do not satisfy the equilibrium constant K for reaction (2.1) at a given T and P of interest. We will refer to the initial values as $\{A\}_o$, $\{B\}_o$, $\{C\}_o$, and $\{D\}_o$, respectively. Ultimately, equilibrium will be attained. The equilibrium values of those activities will be referred to as $\{A\}_e$, $\{B\}_e$, $\{C\}_e$, and $\{D\}_e$. As stated above, thermodyamics requires that the value of G as summed over the four species will be lower for the equilibrium composition than it is for the initial composition.

A good analog for the governing of chemical behavior by the free energy G is the case when mechanical behavior is governed by **potential energy** (PE). In Figure 2.1*a*, a ball is shown elevated a level with height $H + h$. The elevation H is the height of the lower level. Initially, the ball is not in its lowest possible energy state. Moreover, the initial state is not stable inasmuch as the slightest push would start the ball rolling down the slope towards the lower level. Levels below H might be possible, but we will assume that H represents the lowest energy state accessible to the ball as controlled by the nature of this particular system.

Just as thermodynamics can predict the nature of a final equilibrium chemical state, the nature of the Figure 2.1*a* system allows us to predict that the ball will no longer tend to change level once it reaches the lower elevation. However, just as thermodynamics does not tell us anything about the rate (i.e., kinetics) according to which a given chemical process will occur, our knowledge that the ball's PE governs the Figure 2.1*a* system does not tell us anything about how *quickly* the ball will roll down the slope. *Friction* is

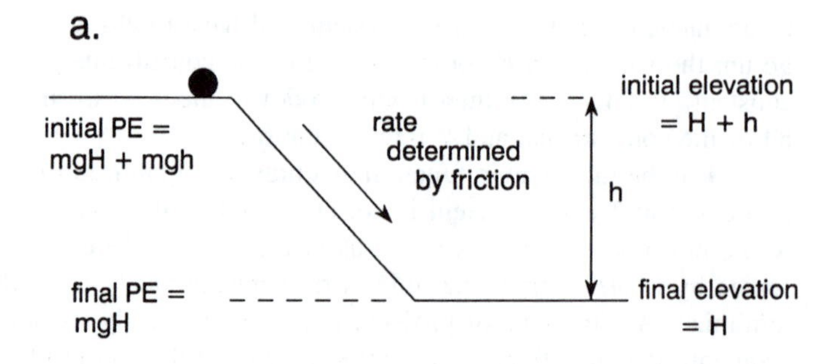

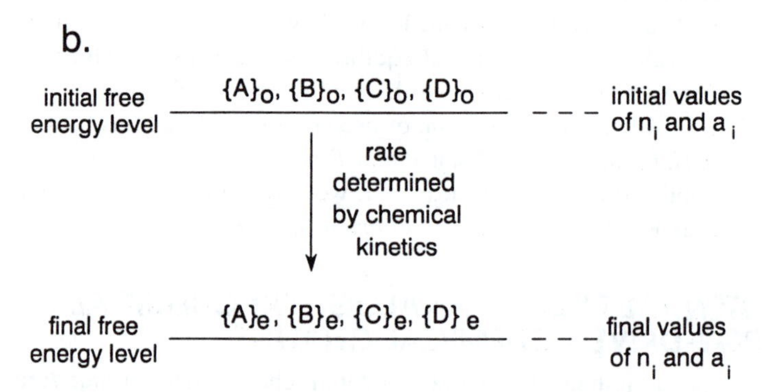

Figure 2.1 (a) A potential energy (PE) system that can move from an initial PE state to a lower PE state. The PE change is controlled by the initial and final elevations, the mass of the ball *m*, and the gravitational acceleration *g*. Friction and drag will determine the rate at which the ball will roll down the incline. (b) An initial, nonequilibrium chemical species distribution tending to move at constant *T* and *P* toward an equilibrium species distribution, i.e., toward the species distribution with the lowest total free energy possible given the initial species distribution and the *T* and *P*. Chemical kinetics will determine the rate at which the system moves toward the equilibrium state.

therefore the Figure 2.1*a* analog of the factors which resist thermodynamically-possible chemical change.

Looking at the Figure 2.1 analogy from the chemical perspective, Figure 2.1*b* shows that the initial, nonequilibrium chemical species distribution is unstable relative to the equilibrium species distribution, i.e., unstable relative to the species distribution with the lowest possible total *G*.

As may be understood based on the discussions that follow, at the *T* and *P* of the system, the height of the upper level for the total *G* will be set by the initial values for the numbers of mols and the per mol *G* values of the various species at their initial activities. Similarly, the height of the total *G* level for the equilibrium state will be set by the equilibrium values of the numbers of mols and the per mol G values of the various species at their final, equilibrium activities. As will be made more clear in subsequent chapters, the equilibrium activity values will be set by the requirement that the equilibrium speciation satisfy: 1) the equilibrium constant; 2) electroneutrality; and 3) all pertinent mass balance requirements.

2.3 THE CHEMICAL POTENTIAL AND ITS RELATIONSHIPS TO FREE ENERGY

In any given system of interest, the chemical potential of species i (μ_i, kJ/mol) is defined as the amount of free energy contained per mol of species i under the temperature, pressure, *and* chemical-composition conditions of the system. The chemical composition aspects pertain not only to species i, but also to *all* other species. For example, at $T = 298\ K$ and $P = 1$ atm, because of the salts present in sea water, the chemical potential of the hydrogen ion when $[H^+] = 10^{-7}\ M$ is significantly different in sea water than it is in pure water.

The mathematical definition of the chemical potential of species i is given by the expression

$$\mu_i = \left(\frac{\partial G}{\partial n_i}\right)_{T,\ P,\ n_j \neq n_i} . \tag{2.2}$$

Thus, μ_i is the ratio of: 1) the infinitesimal change in the total system free energy at temperature T and pressure P that results from the addition of: 2) an infinitesimal amount of species i. Since μ_i is defined as a partial derivative, we see that the composition of the system (as well as T and P) remain constant during the addition of species i. In particular, the numbers of mols of all species $j \neq i$ are held constant, and we already know that the amount of i that is being added is, by definition very small, so it will also not change the composition of the system. As implied in Eq. (2.2), μ_i values are T- and P-dependent.

Since the chemical potential represents the free energy content per mol of species under the specific conditions of the system, we can multiply μ_i by the number of mols of i (n_i) to obtain the total contribution made by species i to the total system G under those conditions. If we refer to that contribution as G_i, we have

$$G_i = n_i \mu_i . \tag{2.3}$$

Since μ_i values are T- and P-dependent, G_i values will also be T- and P-dependent.

For any given *dissolved* species (this includes water) or *gas phase* species, n_i will be given by the product of the concentration of species i ($[i]$, expressed in mol/L) and the volume V (L) of water or gas:

$$G_i = [i] V \mu_i . \tag{2.4}$$

For any given pure solid species, n_i will be given by the mass of the solid divided by the molecular or formula weight.

As a specific example of how to apply Eq. (2.4), if the chemical potential of H^+ at $10^{-5}\ M$ is y kJ/mol in a certain one-phase, aqueous system, and if the volume of that system is 5 L, then G_i for H^+ (i.e., G_{H^+}) will equal $(5 \times 10^{-5})y$ kJ. If a given system contains more than one type of phase (e.g., a gas phase and an aqueous phase), and if a given chemical species is present in all of those phases (e.g., molecular oxygen can partition across an air/water interface), then each of those phases will contribute to G_i. For the partitioning of molecular oxygen between an air + water system, we would therefore have $G_{O_2} = [O_2]_{air}V_{air}\mu_{O_2,air} + [O_2]_{water}V_{water}\mu_{O_2,water}$.

For any given system of a certain composition and at specific values of T and P, the total system G can be obtained by adding up the contributions made by *all* chemical species present, be they dissolved, solid, and/or gas phase in nature. That is, the total system G will be given by the sum of all of the individual G_i values. For a one phase system, we would have

$$G = \sum_i G_i = \sum_i n_i \mu_i. \tag{2.5}$$

2.4 PROPERTIES AND APPLICATIONS OF THE CHEMICAL POTENTIAL

A given μ_i can be positive, negative, or zero. If negative μ_i values seem a bit strange at first, it should be possible to get more comfortable with them shortly. One simple way to understand chemical potentials is to think of each species as having a certain amount in its "free energy bank account." If under certain conditions species i is in "debt" (negative μ_i), then having i in the system *under those conditions* will tend to make the total G of the system negative. Conversely, if under certain other conditions some species j has a "positive account balance" (positive μ_i), then having species j in the system *under those other conditions* will tend to make G positive. If individual μ_i values can be positive, negative, or zero, then we see by Eq. (2.5) that G can also range between negative and positive values. Thus, just as the *aggregate* financial status of a group of people: 1) will depend on the numbers of different types of people and the amounts of money those types have in their savings accounts; and 2) might be positive, negative, or zero, so too will G depend on the n_i and μ_i values in a manner that makes possible any number that is positive, negative, or zero.

The expression that gives the dependence of μ_i on activity at a given T and P is

$$\mu_i = \mu_i^o + RT \ln a_i \tag{2.6}$$

where μ_i^o is the chemical potential of species i in its *standard state* at the T and P of interest. The nature of the standard state will be discussed in more detail in the next section. For the moment we note that when $a_i = 1$, then

$$\mu_i = \mu_i^o \quad (a_i = 1). \tag{2.7}$$

Based on Eq. (2.6), we see that two things will determine the value of μ_i:

1. *The value of μ_i for some specific **standard state** value of the activity of i, i.e., the value of μ_i^o.* In most cases, μ_i^o is not zero. However, its value does always establish what a_i needs to be for μ_i to be zero and so thereby does serve as the *energy reference level* for the "free energy bank account."

2. *The activity of species i.*

2.4.1 Changes in Free Energy

As described by Eq. (2.5), G is comprised of individual contributions made by all of the species in the system. Therefore, a general equation that is valid under all circumstances is

$$\Delta G = \sum_i \Delta G_i = \sum_i [(n_i\mu_i)_{\text{final}} - (n_i\mu_i)_{\text{initial}}]. \tag{2.8}$$

For the Figure 2.1*b* process where a given mix of species undergoes a change at constant P and T that minimizes G, application of Eq. (2.8) yields

$$\Delta G = \sum_i \Delta G_i = \sum_i [(n_i\mu_i)_e - (n_i\mu_i)_o] \tag{2.9}$$

$$= \sum_i [n_{i,e}(\mu_i^o + RT\ln\{i\}_e) - n_{i,o}(\mu_i^o + RT\ln\{i\}_o) \tag{2.10}$$

Since the lower state in Figure 2.1*b* represents an *equilibrium* state, the final values of the n_i for the reactants will never be zero if the reaction is taking place in *solution*; there must be a thermodynamic balance at equilibrium, and that balance can only be achieved if some reactants remain in the system.

Oftentimes, we are interested in knowing what the ΔG is when the reactants combine to form the appropriate products, with no reactants left over. In such a case, for reaction (2.1), we would be interested in knowing what the ΔG is for the process where a mols of A combine with b mols of B to form c mols of C plus d mols of D. This ΔG would correspond to the *complete* conversion of reactants to products. In a manner similar to the definition of ΔV_{rxn} given above, we have for reaction (2.1) that

$$\Delta G_{rxn} = c\mu_C + d\mu_D - a\mu_A - b\mu_B. \tag{2.11}$$

Thus, Eq. (2.11) is a special form of Eq. (2.8) wherein the initial n_i for the products C and D are zero, and the final n_i for the reactants A and B are zero. Again, since the individual μ_i are T- and P-dependent, we conclude that ΔG_{rxn} values will be T- and P-dependent.

Usually, we do not bother including the subscript "rxn" in equations like (2.11); we rely on the context of the usage to convey when it is that we are referring to a ΔG of a reaction that goes to completion. In general, then, for the ΔG of such a reaction we have

$$\Delta G = \sum_i \upsilon_i\mu_i = \sum_i \upsilon_i(\mu_i^o + RT\ln a_i) \tag{2.12}$$

where the υ_i have sign. They are the n_i for the reaction. As discussed in Chapter 1, these *stoichiometric coefficients* are positive for products and negative for reactants. The n_i for the reactants are negative because of the minus sign that appears in Eq. (2.8) in front of the summation of the initial $n_i\mu_i$ terms.

Specific activities for the reactants and products and *specific values of T and P* are always associated with Eq. (2.12). In particular, for reaction (2.1), Eq. (2.12) gives the

ΔG that will result when a mols of A at activity a_{A} plus b mols of B at activity a_{B} are converted at T and P to c mols of C at activity a_{C} plus d mols of D at activity a_{D}. The activities are presumed to stay constant during the reaction. It does not really matter how that is done, and it is clear that it would be difficult experimentally to do it, especially if we are to start out with no C and D at the beginning of the reaction and have no remaining A and B at the end of the reaction. The important point is that Eq. (2.12) gives the difference in G that exists between one system containing c mols of C at activity a_{C} plus d mols of D at activity a_{D} and no A and B and a second system containing a mols of A at activity a_{A} plus b mols of B at activity a_{B} and no C and D; we can certainly imagine how those systems might be prepared.

Equation (2.12) can be broken down into two summations, i.e.,

$$\Delta G = \sum_i v_i \mu_i^{\mathrm{o}} + \sum_i v_i \, RT \ln a_i . \tag{2.13}$$

If we let

$$\Delta G^{\mathrm{o}} = \sum_i v_i \mu_i^{\mathrm{o}} \tag{2.14}$$

then

$$\Delta G = \Delta G^{\mathrm{o}} + RT \ln Q \tag{2.15}$$

where Q is the reaction "quotient." As mentioned in Chapter 1, it is equal to the mathematical product of the activities of the chemical products and reactants, each raised to the power of its v_i value. In the case of reaction (2.1), we would have that

$$\Delta G^{\mathrm{o}} = c\mu_{\mathrm{c}}^{\mathrm{o}} + d\mu_{\mathrm{d}}^{\mathrm{o}} - a\mu_{\mathrm{a}}^{\mathrm{o}} - b\mu_{\mathrm{b}}^{\mathrm{o}} \tag{2.16}$$

and

$$Q = \frac{a_{\mathrm{C}}^{v_{\mathrm{C}}} \, a_{\mathrm{D}}^{v_{\mathrm{D}}}}{a_{\mathrm{A}}^{v_{\mathrm{A}}} \, a_{\mathrm{B}}^{v_{\mathrm{B}}}} . \tag{2.17}$$

Since the individual μ_i^{o} are T- and P-dependent, ΔG^{o} values are T- and P-dependent.

Equation (2.15) is very important. In addition to the intrinsic statement that it makes about changes in G, it leads us to two additional important conclusions. Firstly, we note that when all of the a_i values are equal to 1.0, then

$$\Delta G = \Delta G^{\mathrm{o}} . \tag{2.18}$$

Therefore, ΔG^{o} is the change in free energy that results when the reactants at unit activity react to form products, also at unit activity. It is also the change in free energy for *all other* combinations of the a_i values that result in a Q value equal to unity; this last fact often does not receive as much attention as it should.

The second important conclusion to be drawn from Eq. (2.15) is based on the fact that when the system is at equilibrium, Q will equal the equilibrium constant K for the reaction. Thus, based on preceding discussions, we realize that G is already minimized in the equilibrium state, and so ΔG for a reaction starting from an equilibrium distribution and ending with the same equilibrium distribution must be equal to zero. We therefore have

$$0 = \Delta G^{\mathrm{o}} + RT \ln K \tag{2.19}$$

or

$$\Delta G^{o} = -RT \ln K \tag{2.20}$$

$$= -2.303\ RT \log K. \tag{2.21}$$

Since ΔG^{o} values are T- and P-dependent, we conclude that K values must also be T- and P-dependent. Combining Eqs. (2.15) and (2.21), we obtain that

$$\Delta G = 2.303\ RT \log Q/K. \tag{2.22}$$

When $Q = 1$, Eq. (2.22) reverts to Eq. (2.21). Any reaction for which $\Delta G < 0$ is favorable because the system is moving from a higher chemical energy state to a lower chemical energy state.

2.4.2 The Role and Nature of μ_i^o

As Eq. (2.8) indicates, for the case illustrated in Figure 2.1*b*, the change in G is composed of four individual changes in the product $n_i\mu_i$. A PE analogy can be constructed for μ_i, and for the portion of the change in G_i that is due to the change in μ_i. It is important to note for this analogy that μ_i represents the G content *per mol* of species i under the system conditions. The PE analog of this per mol quantity would thus be the PE *per unit mass*. As described in Figure 2.1*a*, PE is given by the quantity $mg \times$ (height). Thus, the PE per unit mass is given by the quantity $g \times$ (height). The units of $g \times$ (height) are (length)^2/(time)^2. We can reassure ourselves that these are the proper units for energy/mass as follows. Force is given as mass $\times$ acceleration ($F = ma$), and so the units of force are (mass)(length)/(time)^2. (Note that mg = gravitational force.) Work and energy are equal to force $\times$ length, that is, (mass)(length)^2/(time)^2. Therefore, units of (length)^2/(time)^2 do indeed correspond to what is needed for energy per unit mass. The units of μ_i are kJ/mol; these units are conceptually the same as energy per unit mass where mass is now measured in chemical terms, that is, in mols.

In order to illustrate the fact that both PE and the chemical potential can take on a *range* of values, Figure 2.2 depicts PE and μ_i as energy *staircases*. Although the figure shows finite heights for the steps in both of the staircases, in actuality we know that an infinitely fine gradation exists between the energy levels since both h and a_i are continuous functions.[1] Sea level is the standard reference level for PE, i.e., the level at which h is assigned a value of 0. Although it often does not *necessarily* correspond to μ_i being equal to zero, as discussed above, μ_i^o serves as the *reference level* for μ_i. Figure 2.2 illustrates a case where $\mu_i^o > 0$. Situations where $\mu_i^o < 0$ are also possible. As will be discussed in Sections 2.4.3 and 2.4.4, at $T = 298$ K and $P = 1$ atm, $\mu_i^o = 0$ for: 1) pure elements in their most stable forms at that T and P; and 2) for aqueous H^+.

Since the quantity $RT \ln a_i$ goes to $-\infty$ as a_i goes to zero, it is clear that μ_i can become negative. Morover, μ_i will be infinitely negative when species i is completely absent from the system. This is the fundamental and immutable reason why it is so difficult to prepare materials that are completely pure. The purer something gets, the more negative become the chemical potentials of *all* of the many things that are nearly absent.

[1] In advanced chemistry terms, this continuity in energy levels means that neither h nor a_i are *quantized*.

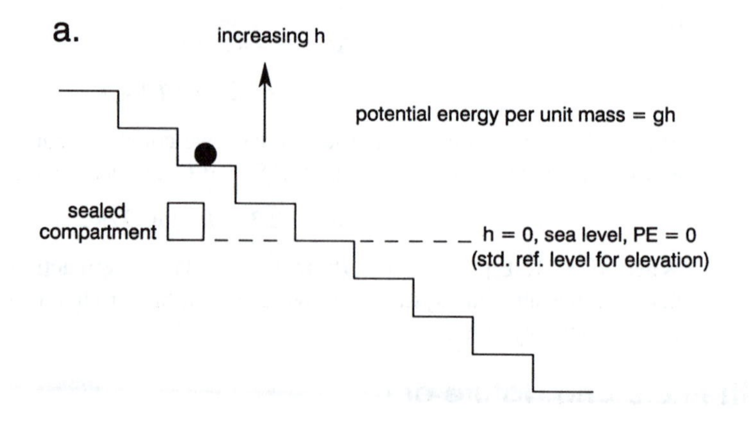

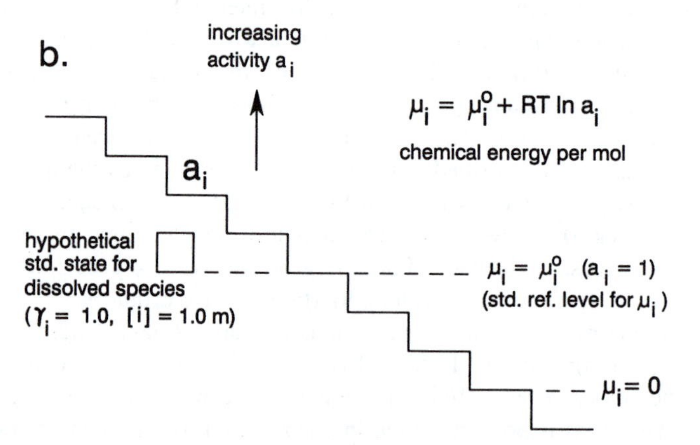

Figure 2.2 (a) Potential energy (PE) staircase. Although we cannot access the sealed compartment in the staircase, the PE level that the ball would have if it were in that compartment is perfectly well defined. (b) Chemical potential energy (μ) staircase. Although we cannot access the hypothetical standard state for which $\gamma_i = 1$ when $[i] = 1\ m$, the μ_i level that species i would have if it were in that state is perfectly well defined, and is equal to μ_i^o. Note that it is possible for $\mu_i = \mu_i^o$ without simultaneously having $\gamma_i = 1$ and $[i] = 1$. All that we need is to have the product $\gamma_i[i] = \{i\} = a_i = 1$.

Consider the example of the contamination of water by chemical A at constant T and P. Initially, A is in the surroundings ("surr"). The contamination reaction is therefore given by

$$A_{surr} = A_{water}. \tag{2.23}$$

When A is absent or nearly absent from the water, reaction (2.23) is very much favored from a thermodynamic point of view. Indeed, reaction (2.23) has a ΔG that is given by

$$\Delta G = \mu_A^{water} - \mu_A^{surr}. \tag{2.24}$$

Since μ_A^{water} is $-\infty$ in pure water, ΔG will be infinitely negative (and therefore very favorable for reaction (2.23)) for any contaminant. Pure materials are therefore seen to be like *thermodynamic magnets* for *all* other species. Achieving and maintaining *near absolute purity* is therefore difficult and expensive. (The price of a reagent chemical usually goes up rapidly as its rated purity goes up.) Moreover, it is safe to say that it will be impossible to obtain any significant quantity of any material in absolutely pure form.

2.4.3 Standard Free Energies of Formation (ΔG_f^o) for Neutral Species

Continuing the discussion of the Figure 2.1 analogy given above, we realize that the *change* in the PE of the ball in Figure 2.1*a* does not depend on the *absolute* elevation value (H) of the lower level.[2] For example, if the elevation H is assigned a value of 10 cm and if the upper level is 4 cm above the lower level, the elevation difference between the two levels (h) will obviously be 4 cm. If H is assigned a value of -11 cm, the upper level will be at -7 cm, and the elevation difference will still be 4 cm. In a similar manner, the difference in G between the lower and the upper levels in Figure 2.1*b* does not depend on the absolute free-energy value of the lower level. It is fortunate therefore that thermodynamics is only concerned with *differences* in free-energy levels since *absolute true energy levels* are not known; there is no *single absolute zero reference level* for energy in the universe. However, we *can* adopt some conventions for *assigned zero reference levels* for free energy so that we can report the results of calculations and experiments, and thereby provide the basis for calculating the ΔG values of interest.

As a simple example of how these principles might be applied, let us assume that we have measured that level A is 5 kJ above level B, and that level C is 2 units below level B. If everyone agrees that the convention should be that level B has an *assigned* zero free-energy reference level, then everyone will be able to calculate the differences in free energy between these three levels based on the data in Table 2.1. This example is summarized in Figure 2.3.

Many years ago, geographers needed to pick a reference level for measuring physical elevation. The selection was arbitrary, and was guided purely by convenience. For example, they could have picked the center of the earth, but that would have been very awkward. With that choice, all elevation numbers would have been very similar except

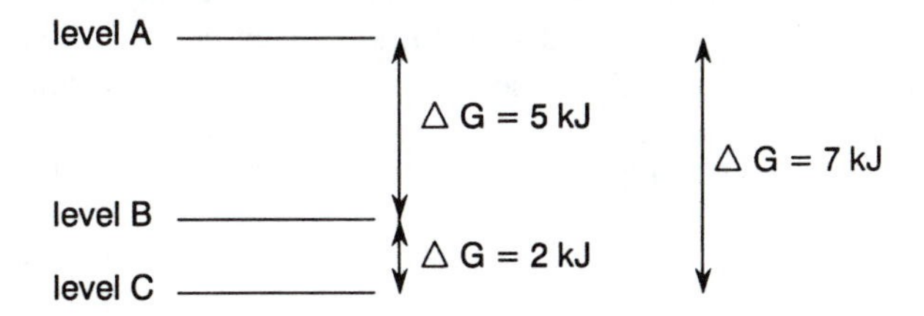

Figure 2.3 Relative free-energy levels A, B, and C.

[2]The word "absolute" here means "absolute true," and not "absolute value" as in $|-5| = 5$.

TABLE 2.1 Free Energy Levels Relative to Level B in Figure 2.3

level A	5 kJ
level B	0 kJ
level C	−2 kJ

for the very last few significant figures. Therefore, the standard reference level that has been accepted by all nations on earth is *sea level*; there is nothing magical about sea level, it is just convenient.

The convention that has been adopted by all chemists on earth for referencing free-energy measurements was also guided by convenience. This convention assigns a zero free-energy level to each element when the element is pure, and in its most stable form (i.e., lowest energy chemical form and phase type) at 25°C/1 atm. The identity of the most stable form of a given element is needed to define this special standard state. Thus, as examples, we note that for hydrogen, we choose a standard state of pure H_2 gas at 25°C/1 atm; for mercury, we choose a standard state of pure liquid Hg at 25°C/1 atm; for sulfur, we choose a standard state of pure solid sulfur in the rhombic crystal form at 25°C/1 atm; for iron, we choose a standard state of pure solid Fe at 25°C/1 atm.

Once the zero values have been assigned for the elements, then we can determine the free-energy content of any other neutral chemical species relative to this zero by measuring the change in G that occurs when one mol of the chemical species is formed from its constituent elements. For the chemical species being formed, we need to specify the standard state in which we are interested. For example, for a dissolved species like H_2S, we might choose an ideal 1.0 m solution as the standard state, or we could choose an ideal H_2S gas at 1 atm pressure. For water, we could choose an ideal H_2O gas at 1 atm pressure, or we could choose pure water, and a mole fraction of 1.0. Different standard states will have different free energy contents. All of the 25°C/1 atm free-energy values obtained in this manner (including those for the elements themselves) are referred to as "standard free energies of formation," and are given the symbol ΔG_f^o. The units are kJ/mol. In words, we define that the ΔG_f^o for a chemical equals the free energy change when one mol of the chemical is formed in its standard state at 25°C/1 atm from its constituent elements. As required, this definition yields ΔG_f^o values of zero for all the elements in their most stable forms at 25°C/1 atm.

Since the standard state for any given chemical i calls for an activity of 1.0, then according to Eq. (2.6), the chemical potential μ_i (kJ/mol) of a species in its standard state will equal the standard state chemical potential, μ_i^o. In general, μ_i^o will depend on both T and P; ΔG_f^o values refer only to 25°C/1 atm. Thus, at 25°C/1 atm, we have

$$\mu_i^o = \Delta G_f^o \qquad 25°\text{C}/1 \text{ atm only.} \tag{2.25}$$

For the elements in their most stable forms, then, the μ_i^o values are zero at 25°C/1 atm. At T and P values other than 25°C and 1 atm, the μ_i^o values for the elements in these forms will **not** be zero. Nevertheless, the difference between μ_i^o (T,P) and μ_i^o(25°C/1 atm) for a given element (or any other species) will always be measurable, in principle.

TABLE 2.2 ΔG_f^o (kJ/mol) values as determined by carrying out reactions (2.26)–(2.31), (2.39), and (2.41) at 25°C/1 atm under standard conditions (unit activity for all reactants and products). For H^+, ΔG_f^o is assigned a value of 0.0 kJ/mol.

species	standard state	ΔG_f^o (= μ_i^o(25°C/1 atm)) (kJ/mol)
$H_2S_{(aq)}$	1 m	− 22.87
$H_2S_{(g)}$	1 atm	− 33.56
$SO_{2(g)}$	1 atm	−300.2
$HgS_{(s)}$	$X = 1$	− 43.3
$H_2O_{(l)}$	$X = 1$	−237.18
$O_{2(aq)}$	1 m	16.32
$H^+_{(aq)}$	1 m	0.0
Cl^-	1 m	−131.3
Fe^{2+}	1 m	− 78.87

Let us consider now a series of reactions that involve the formation of some chemical molecules from their constituent elements:

$$H_{2(g)} + S_{(s)} = H_2S_{(aq)} \tag{2.26}$$

$$H_{2(g)} + S_{(s)} = H_2S_{(g)} \tag{2.27}$$

$$O_{2(g)} + S_{(s)} = SO_{2(g)} \tag{2.28}$$

$$Hg_{(l)} + S_{(s)} = HgS_{(s)} \tag{2.29}$$

$$H_{2(g)} + \frac{1}{2}\, O_{2(g)} = H_2O_{(l)} \tag{2.30}$$

$$O_{2(g)} = O_{2(aq)}. \tag{2.31}$$

Since the ΔG_f^o values for all of the above reactants are identically zero at 25°C/1 atm, and since each reaction only gives one product species, measurement of the ΔG^o values at 25°C/1 atm for each reaction will provide a direct determination of the ΔG_f^o for the corresponding product species.

Thus, for reaction (2.26) at 25°C/1 atm, when all of the reactants and products are in their standard states, we have

$$\Delta G_f^o = \Delta G^o = (1\ \text{mol})\mu^o_{H_2S(aq)} - 0 - 0 = (1\ \text{mol})\mu^o_{H_2S(aq)}. \tag{2.32}$$

Since the measured value of this ΔG^o is −22.87 kJ, we conclude that ΔG_f^o for $H_2S_{(aq)}$ (at 1 m) is −22.87 kJ/mol. Table 2.2 gives the results that are obtained when the ΔG^o values for reactions (2.26)–(2.31) are measured at 25°C/1 atm and the appropriate ΔG_f^o values are deduced. Stumm and Morgan (1981) and other sources such as the CRC Annual Handbook of Chemistry and Physics as well as Bard et al. (1985) provide detailed tables of standard free energies of formation.

It is important to note that the ΔG_f^o values that are measured when species are formed from their elements (i.e., utilizing reactions of the type (2.26)–(2.31)) can be used to deduce the ΔG^o value at 25°C/1 atm for any reaction of interest. For example, based on the data in Table 2.2 we can directly conclude that at 25°C/1 atm, the reaction

$$S_{(s)} + O_{2(aq)} = SO_{2(g)} \tag{2.33}$$

will have a ΔG^o that is given by

$$\begin{aligned}\Delta G^o = (1\text{ mol})(-300.2\text{ kJ/mol}) + (-1\text{ mol})(0\text{ kJ/mol}) +\\ (-1\text{ mol})(16.32\text{ kJ/mol}) = -316.52\text{ kJ.}\end{aligned} \tag{2.34}$$

As may already be clear, ΔG^o has units of kJ rather than kJ/mol because the stoichiometric coefficients have the same units as n_i, that is, "mols." Thus, at 25°C/1 atm, for the reaction

$$\frac{1}{2}\,S_{(s)} + \frac{1}{2}\,O_{2(aq)} = \frac{1}{2}\,SO_{2(g)} \tag{2.35}$$

the ΔG^o is one half that for reaction (2.33). In particular, we have that ΔG^o = (1/2 mol)(−300.2 kJ/mol) + (−1/2 mol)(0 kJ/mol) + (−1/2 mol)(16.32 kJ/mol) = −158.26 kJ = 1/2(−316.52) kJ.

Once a ΔG^o has been computed at 25°C/1 atm using tabulated ΔG_f^o values, then the value of the corresponding equilibrium constant at 25°C/1 atm can be calculated using Eq. (2.21). The value of the equilibrium constant for other combinations of T and P can then be determined using equations presented in Section 2.5. This comment applies equally well to the material discussed in the next section.

2.4.4 Standard Free Energies of Formation (ΔG_f^o) Values for Ionic Species

The species formed by reactions (2.26)–(2.31) are all neutral, that is, they are not electrically charged. In aqueous systems, however, neutral species are certainly not the only ones present. Indeed, besides water itself, ions are almost always the most abundant species. Therefore, we are clearly going to need ΔG_f^o and μ_i^o values for ions. In this section, all references to ionic species may be taken to indicate *dissolved* ions.

Of all the ions possibly present in water, it can be argued that H^+ is one of the most important ones, even when it is present at extremely low concentrations. Using the approach outlined above, we might be tempted to think that we can get at the ΔG_f^o for H^+ by measuring the ΔG^o at 25°C/1 atm for the oxidation[3] reaction

$$\frac{1}{2}\,H_{2(g)} = H^+ + e^-. \tag{2.36}$$

[3] An "oxidation" is a removal of electrons.

Since ΔG_f^o for $H_{2(g)}$ is zero, at 25°C/1 atm we would have

$$\Delta G^o = \mu^o_{H^+} + \mu^o_{e^-} - 0. \tag{2.37}$$

The problem that we encounter is that even if we could measure the ΔG^o for reaction (2.36) under standard conditions at 25°C/1 atm, there would be no way to determine how much of the ΔG^o came from $\mu^o_{H^+}$ and how much came from $\mu^o_{e^-}$. Thus, we cannot get a measure of the ΔG_f^o for H^+ by measuring the ΔG^o for its formation from elemental hydrogen at 25°C/1 atm.

Thermodynamicists have chosen to handle this particular situation as follows. Firstly, they note that in aqueous solutions, there is never any *significant* net production of electrons. The reason is that electrons are just too reactive. Therefore, before an oxidation reaction like (2.36) can produce any non-negligible concentration of electrons, some other chemical species will pick them up. If the electron donating species is $H_{2(g)}$ and the electron accepting species is $Cl_{2(g)}$, then the net reaction will be

$$\frac{1}{2}H_{2(g)} = H^+ + e^- \tag{2.36}$$

$$+ \quad \frac{1}{2}Cl_{2(g)} + e^- = Cl^- \tag{2.38}$$

$$\frac{1}{2}H_{2(g)} + \frac{1}{2}Cl_{2(g)} = H^+ + Cl^-. \tag{2.39}$$

Thus, it does not even matter what the value of ΔG^o is for reaction (2.36) under standard conditions. Since it cannot occur by itself to any significant extent in water anyway, we can just set its value at 25°C/1 atm equal to zero. We can even assign values of zero to the quantities that make up that ΔG^o. That is, we can and do assign zeros to the ΔG_f^o values for both H^+ and e^-.

If we carry out reaction (2.39) in the laboratory, we can measure

$$\Delta G^o = \mu^o_{H^+} + \mu^o_{Cl^-}. \tag{2.40}$$

If we carry out the reaction at 25°C/1 atm, then the μ values in Eq. (2.40) will be the ΔG_f^o values for H^+ and Cl^-. The fact that we have chosen to assign $\Delta G_f^o = 0$ for H^+ thus simply means that for reaction (2.39) under standard conditions and 25°C/1 atm, all of the ΔG is assigned to the formation of the chloride. That is, the ΔG_f^o for chloride ion in an ideal 1.0 m solution is assigned the value of the ΔG^o for reaction (2.39) when that reaction is carried out at 25°C/1 atm. Viewing the problem from the other direction, if we tabulate that value of ΔG_f^o for Cl^- along with $\Delta G_f^o = 0$ for H^+, then we will be able to go back at any time and correctly compute both the measured value of ΔG^o for reaction (2.39) at 25°C/1 atm, as well as the corresponding equilibrium constant at 25°C/1 atm.

Let us now consider another reaction, this time one involving Cl^- but not H^+. In particular, consider

$$Cl_{2(g)} + Fe_{(s)} = Fe^{2+} + 2Cl^- \tag{2.41}$$

with

$$\Delta G^o = \mu^o_{Fe^{2+}} + 2\mu^o_{Cl^-}. \tag{2.42}$$

As with the ΔG° of reaction (2.39), the ΔG° of reaction (2.41) is a quantity that we can physically measure. Thus, we have

$$\mu^{\circ}_{Fe^{2+}} = (\Delta G^{\circ}\ (\text{rxn } 2.41) - 2\mu^{\circ}_{Cl^-}). \tag{2.43}$$

If we apply conditions of 25°C/1 atm to Eq. (2.43), then the μ° values will be ΔG°_f values, and so

$$\Delta G^{\circ}_f \text{ for } Fe^{2+} \text{ in a 1.0 } m \text{ solution} = \Delta G^{\circ}\ (\text{rxn 2.41 at 25°C/1 atm}) - 2\Delta G^{\circ}_f \text{ for } Cl^- \text{ in a 1.0 } m \text{ solution.} \tag{2.44}$$

We can tabulate the value measured according to Eq. (2.44) along with the values for $\mu^{\circ}_{H^+}$ and $\mu^{\circ}_{Cl^-}$ at 25°C/1 atm discussed above. Clearly, we can keep on doing this until we have obtained ΔG°_f values at 25°C/1 atm for all ions of interest. Wherever possible, we would use reaction (2.36) as a direct zero reference just as we did in reaction 2.39 for the determination of ΔG°_f for Cl^-. Doing so helps to minimize propagation of errors. The resulting data may be combined with data like those in Table 2.2 to obtain a complete master list of ΔG°_f values at 25°C/1 atm for all species including ions. This is in fact the procedure that is followed.

To summarize, the adoption of $\Delta G^{\circ}_f = 0$ for H^+ at 25°C/1 atm provides the fundamental basis of the convention regarding free-energy reference levels for ions. It carries with it the important convention that $\Delta G^{\circ} = 0$ for reaction (2.36) at 25°C/1 atm.

2.4.5 Concentration Scales and Standard States

As discussed in Chapter 1, a variety of scales can be used to express solution concentration and activity. In the context of this text, the solutions of interest will include both aqueous as well as solid solutions. (For a detailed discussion of solid solutions, see Chapter 16.) For any given species i, the important scales are: mole fraction (X_i), molality (m_i), and molarity (c_i). As discussed in Chapter 1, the units of molality (mol/kg) and molarity (mol/L) are denoted with the italics m and M.

Thermodynamicists consider the mole fraction scale the most fundamental scale. They like to use it whenever they can since the degree to which a solution of dissolved species i exhibits the properties of species i is closely related to X_i. Indeed, when X_i is 1.0, the "solution" is perfectly like species i. (As the student who has some chemistry background may recall, one application of this principle is Raoult's Law which states that for an ideal solution, the vapor pressure of species i above a solution of i will be equal to the product of X_i and the vapor pressure of pure i.) As implied in Chapter 1, since water and most mineral solids are generally found as nearly pure materials (i.e., with mole fractions that are close to 1.0) in the natural water environment, it is logical that the mole fraction scale is usually chosen to express their concentrations and activities.

Despite the definite thermodynamic attractiveness of the mole fraction scale, it is usually not the scale chosen by water chemists for dealing with *dissolved species*. Often, molality and molarity are felt to be more convenient. Indeed: 1) solutions of specific molalities and molarities are easily prepared in the laboratory—one does not have to worry if one or more of the species present in the solution might be dissociating or combining with other species and thus affecting the mole fraction in an unknown way; and 2) the results of analytical measurements are most easily expressed as amount per unit volume of sample, that is, as molarities or in units that are proportional to molarity (e.g., mg/L (ppm)). *Given a choice between molality and molarity, thermodynamicists have a strong preference for the molality scale because it is not affected by the volume changes that accompany changes in T and P*. That is, a kg of water remains a kg of water essentially independent of T and P. (The qualifier "essentially" has been included since the extent of the weak dissociation of water to form H^+ and OH^- depends on both T and P.)

It is important to note that once the reference levels for G measurements have been selected for all species as described in Sections 2.4.3 and 2.4.4, then for a given system composition, T, and P, there is *one and only one* μ_i *for any given species i*. In other words, *regardless of whether mole fraction, molality, or molarity, is selected for expressing [i] and* a_i, *there is always only one* μ_i *for species i.*

The general expression for μ_i is

$$\mu_i = \mu_i^{\text{o}} + RT \ln a_i. \tag{2.6}$$

Equation (2.6) can be rewritten with a_i on each of the three different concentration/activity scales:

$$\mu_i = (\mu_i^{\text{o}})_X + RT \ln \zeta_i X_i \tag{2.45}$$

$$\mu_i = (\mu_i^{\text{o}})_m + RT \ln \gamma_i m_i \tag{2.46}$$

$$\mu_i = (\mu_i^{\text{o}})_c + RT \ln y_i c_i \tag{2.47}$$

where $(\mu_i^{\text{o}})_X$, $(\mu_i^{\text{o}})_m$, and $(\mu_i^{\text{o}})_c$ explicitly represent the standard state chemical potentials at a given T and P on the mole fraction, molality, and molarity scales, respectively.[4] ζ_i, γ_i, and y_i are the activity coefficients on the mole fraction, molality, and molarity scales, respectively. As will be discussed further in Section 2.7, these three activity coefficients have different numerical values. For solutions that are less than a few molar, the differences will be very small.

Let us consider the case of a solution that is 10^{-5} F NaCl. Ignoring the fact that some of the H_2O will dissociate to form H^+ and OH^-, the concentrations of Na^+ on the three different scales at 25°C/1 atm may be computed as follows. We first note that the density of pure water under these conditions is 0.997 g/cm^3. To a good first approximation, the contribution of the NaCl to the mass and to the volume of the solution may be ignored. Since we can also ignore the *very slight* difference between 1 mL and 1 cm^3 (0.999972 mL = 1 cm^3), one liter of the solution weighs 0.997 kg. Therefore, 1.0 L of the solution contains (0.997 g/cm^3)(1000 cm^3)/(18.01 g/mol) = 55.36 mols of H_2O. Remembering to consider both Na^+ and Cl^- when calculating the mole fraction,

[4]In order to keep the notation as simple as possible, we will normally not bother with subscripts like X, m, and c on the μ_i^{o} terms; we will rather rely on the context and identity of the species to tell us which type of μ_i^{o} is being used. The subscripts are included in this section for the sake of clarity.

we have

$$X_{Na^+} = \frac{10^{-5}}{10^{-5} + 10^{-5} + 55.36} = 1.806 \times 10^{-7}. \tag{2.48}$$

The molality may be similarly estimated as

$$m_{Na^+} = \frac{10^{-5} \text{ mol}}{0.997 \text{ kg}} = 1.003 \times 10^{-5}\ m \tag{2.49}$$

and of course

$$c_{Na^+} = 10^{-5}\ M. \tag{2.50}$$

While the above three concentration numbers are all different in magnitude, they nevertheless describe the exact same solution of Na^+ with a single specific value of μ_{Na^+}.

Usually a solute like Na^+ is considered to behave ideally (activity coefficient = 1.0) when it is totally free from interactions with other ions (i.e., when it is at infinite dilution, see Section 2.4.6). For all intents and purposes, a $10^{-5}\,F$ solution is sufficiently dilute that activity coefficients referenced to infinite dilution in this manner will be essentially equal to 1.0. Thus, for this dilute NaCl solution, not only will the activity coefficients on the three scales be very nearly the same,[5] but they will also all be nearly equal to 1.0. That is, $\zeta_{Na^+} \simeq \gamma_{Na^+} \simeq y_{Na^+} \simeq 1.0$.[6] As a result, for this solution, $[i] \simeq a_i$ for all three scales. Therefore, for this $10^{-5}\,M$ solution of Na^+, we have

$$\mu_{Na^+} \simeq (\mu^o_{Na^+})_X + RT \ln 1.806 \times 10^{-7} \tag{2.52}$$

$$\mu_{Na^+} \simeq (\mu^o_{Na^+})_m + RT \ln 1.003 \times 10^{-5} \tag{2.53}$$

$$\mu_{Na^+} \simeq (\mu^o_{Na^+})_c + RT \ln 10^{-5}. \tag{2.54}$$

We conclude then that that while μ_i is independent of the scale used to express concentration and activity, the standard state chemical potential μ_i^o must be dependent on that scale. This makes sense when taken in the context of the fact that $\mu_i = \mu_i^o$ when $a_i = 1$. That is, we would *expect* that the chemical potentials for Na^+ should all be different for solutions in which $X_{Na^+} = 1.0$, $m_{Na^+} = 1.0$, and $c_{Na^+} = 1.0$. Indeed, while we can certainly prepare solutions in which $m_{Na^+} = 1.0$ and $c_{Na^+} = 1.0$, an aqueous solution for which $X_{Na^+} = 1.0$ cannot even be prepared!

[5]As Denbigh (1981, p. 278) shows, when the molality of a species i is m_i and its mole fraction is X_i, the relationship between the activity coefficient on the molality scale (γ_i) and the activity coefficient on the mole fraction scale (ζ_i) is given by

$$\frac{\gamma_i}{\zeta_i} = \frac{X_i 1000}{m_i M_o} \tag{2.51}$$

where M_o is the molecular weight of the solvent in g/mol. For a solution comprised of just water (M_o = 18.01 g/mol) plus one species, if that species weighs 100 g/mol, then at a concentration of 1.0 m = 0.177 mole fraction, Eq. (2.51) gives the ratio of the two activity coefficients as being 0.9852. (A temperature of 298 K has been assumed.)

[6]In Chapter 16, we will consider systems involving solids in which we will adopt the convention that $\zeta_i \rightarrow 1.0$ as $X_i \rightarrow 1.0$, rather than as $X_i \rightarrow 0$, as in the case here.

An equation can be derived (Denbigh 1981, p. 278) relatively easily which gives the difference between μ_i° on the mole fraction and molality scales. It is

$$(\mu_i^{\circ})_X - (\mu_i^{\circ})_m = RT \ln \frac{1000}{M_o} \tag{2.55}$$

where M_o is the molecular weight of the solvent in g/mol. Since $R = 8.314$ J/K – mol, for water at $T = 298$ K, Eq. (2.55) gives 9.95 kJ/mol. As expected, equating Eqs. (2.52) and (2.53) and solving for the left-hand side of Eq. (2.55) also gives this result.

It is important to emphasize that if we wish to compute the ΔG° and the equilibrium constant for a given reaction at a T and P of interest, it is perfectly appropriate to use different μ_i° types in Eqs. (2.14) and (2.20) for the different reactant and product species. For example, for reaction 2.33, the μ_i° types that we would use would be: $(\mu_i^{\circ})_X$ for S_s, and $(\mu_i^{\circ})_m$ for $O_{2(aq)}$. We would use $(\mu_i^{\circ})_{atm}$ for $SO_{2(g)}$ since it is a gas. The only things that we must remember are: 1) a computed ΔG° pertains to the reaction proceeding with the reactants and products at unit activities on the various corresponding concentration/activity scales; and 2) proper corresponding units must be used for the a_i appearing in the expression for K. With regard to the second point, it is therefore apparent that the numerical value of an equilibrium constant depends on the scales selected for measuring the concentrations/activities of the reactants and products.

For species dissolved in an aqueous phase, water chemists often like to express $[i]$ in units of molarity (M). However, for reasons discussed above, thermodynamicists have usually chosen to use the molality (m) scale. As a result, most fundamental thermodynamic data for species dissolved in the aqueous phase (this includes μ_i° values and equilibrium constants) are based on the molality scale. Fortunately, *the differences between concentrations reported on the molality and molarity scales are very small except for quite concentrated solutions.* For example, a 1.0 m solution of NaCl is ~0.98 M. Since even seawater is equivalent to only ~0.6 M in total salts, we see that *concentrations expressed in units of molarity may therefore usually be used in place of the corresponding molality values in equilibrium constants without any concern for significant error.* As Stumm and Morgan (1981, p. 130) note, the difference between molarity and molality is usually small in comparison to the uncertainties involved in determining equilibrium constants and in estimating activity coefficients.

2.4.6 The Standard State and the Activity Coefficient Reference Convention

We know that μ_i is given by

$$\mu_i = \mu_i^{\circ} + RT \ln a_i \tag{2.6}$$

where μ_i° is the chemical potential of species i in a *standard state* at specific values of T and P, and a_i is the activity of species i.

We now also know that the standard state provides the energy level benchmark from which μ_i is measured. In the above discussions, we have stated that $\mu_i = \mu_i^{\circ}$ when $a_i = 1$ on the concentration/activity scale for which that μ_i° applies. However, species i is not necessarily actually *in* the standard state just because $a_i = 1$. The reason for this is that the detailed definition of the standard state is the state in which *both* the activity coefficient for species *i and* $[i]$ are equal to 1.0.

2.4.6.1 The Standard State and the Activity Coefficient Reference Convention for Dissolved Species. The concept of a standard state for a dissolved species measured on the molality scale in which both the activity coefficient (γ_i) *and* $[i]$ are *simultaneously* equal to 1.0 is one that a majority of students (as well as numerous water chemistry professionals) have significant difficulty understanding. As will now be discussed, the whole problem here is tied directly to: 1) the use of the words "and" and "simultaneously" in the definition of the standard state; and 2) the fact that units of molality (as opposed to mole fraction) are generally used for dissolved species.

In most cases, for inorganic species dissolved in water, one adopts the convention that γ_i is exactly equal to 1.0 when the solution is *infinitely dilute*. That is, a given species is said to behave ideally when it and all other species (except water of course) are, on average, so far apart that they simply do not interact with one another. The condition of infinite dilution is then said to be the *activity coefficient reference convention* for that dissolved species. (In the gas phase, the analog to this convention is the view that gases behave increasingly ideally (fugacity of $i \rightarrow$ pressure of i) as the total gas pressure tends to zero.) Now, if γ_i is defined as equalling 1.0 at infinite dilution, then it is virtually certain that γ_i will *not* equal 1.0 when $[i]$ is equal to 1.0 m. So, as implied above, the problem is a question of how can one have both γ_i *and* $[i]$ *simultaneously* equal 1.0?!

Thermodynamicists get around this difficulty by simply saying that: 1) the (γ_i = 1.0, $[i]$ = 1.0 m) standard state is a *hypothetical* state for dissolved species that does not actually exist; and 2) despite its hypothetical character, this standard state it is still meaningful as well as useful. Many people become even more confused at this point because they have difficulty understanding: 1) why it should, apparently, be disadvantageous to try to define a more realistic type of standard state; and 2) what a hypothetical standard state represents and why it is meaningful.

2.4.6.2 Disadvantages of a Real (i.e., Non-Hypothetical) a_i = 1.0 m Standard State for a Dissolved Species. It is obvious from Eq. (2.6) that if dissolved species i is in its standard state, then a_i = 1.0 m. Therefore, if we do not pick the hypothetical standard state described above, then we will have to find some *real* state in which a_i = 1.0 m for the standard state. In this case, there will be no special requirements on γ_i and $[i]$ *individually*; we will just require that their *product* be equal to 1.0 m. Now, for any given species, there will certainly be a large (actually, infinite) number of real solution compositions that will give a_i values that are equal to 1.0 m. For a system containing only NaCl and water, for the species Na^+ and Cl^-, equations that predict γ_i values for ions (see Section 2.7) may be used to calculate that one such composition is an NaCl concentration of about 1.3 F. If other species are also present, however, then the formality of the NaCl for which those a_i values will equal 1.0 m will be different, and will moreover be very dependent on the concentrations and identities of those other species.

For a solution of a salt of a 2+ cation and a 2− anion, it may be calculated using activity coefficient equations that a_i will equal ~1.0 for both the cation and the anion when the concentration of the dissolved salt is about 2 F. (As with the case of NaCl, the composition will be different when other species are present in addition to the 2+ cation and the 2− anion.) However, most $2+/2-$ salts are not even soluble enough to be able to reach 2 F! Also, for both $1+/1-$ salts and $2+/2-$ salts, the composition*s* of the systems that give $a_i = 1.0\ m$ will *never* be exactly the same for both the cation and the anion, and we can be sure that physical chemistry theory will *never* be sufficiently accurate to allow us to calculate γ values and therefore those compositions *exactly*.

We can therefore conclude that: 1) the chemical compositions of any real, $a_i = 1.0\ m$ standard states for a dissolved species would be very species dependent; 2) if we were going to use real standard states, we would have to select one of those compositions and define its exact nature for each of the many species of interest; 3) those compositions would certainly be subject to an endless series of confusing revisions as better methods are developed to calculate γ_i values; and 4) for some species (like highly charged ions), it may not even be possible to have a *real* solution in which $a_i = 1.0\ m$! The last point illustrates the fact that even when trying to establish a set of real $a_i = 1.0\ m$ standard states, for some species, we might *still* have to resort to the use of some type of hypothetical standard state. At this point, it should be clear that there are many disadvantages in trying to use real, $a_i = 1.0\ m$ standard states. The hypothetical ($\gamma_i = 1.0$, $[i] = 1.0\ m$) standard state avoids these difficulties.

2.4.6.3 Meaning of the Hypothetical ($\gamma_i = 1.0$, [i] $= 1.0\ m$) Standard State. Now that the attractiveness of the ($\gamma_i = 1.0$, $[i] = 1.0\ m$) hypothetical standard state has been established, we need to convince ourselves that this standard state represents something *meaningful*. A good way to do this is to further consider the Figure 2.2 analog. As previously discussed, this figure shows in a schematic way how the potential energy of a species varies as a function of a_i. The PE analog for the ($\gamma_i = 1.0$, $[i] = 1.0\ m$) hypothetical standard state is shown as a *sealed* compartment *inside* the Figure 2.2*a* staircase. This hypothetical standard state is *not physically accessible* to the ball. The ball is otherwise free to occupy any of the steps, provided of course that it has the proper PE. We note that although the sealed compartment is not physically accessible, its energy level is perfectly well defined. In a similar manner, although the chemical energy level of the hypothetical ($\gamma_i = 1.0$, $[i] = 1.0\ m$) standard state in Figure 2.2*b* is not physically accessible, *it is nevertheless well defined*. Finally, we note that just as the ball can occupy the same PE level as the PE hypothetical standard state, it will be possible to place a chemical species at the same energy level of the chemical standard state. As discussed in the previous section, however, doing so will require hypothetical conditions that give an a_i value of *exactly* 1.0 m.

2.4.6.4 The Standard State and the Activity Coefficient Reference Convention for Water and Solids. In the case of dissolved species, we adopted the reference convention that activity coefficient values tend to 1.0 as the solution tends to infinite dilution. The main reason for this convention was the fact that essentially all solutes of interest in aquatic chemistry are present in comparatively dilute form; even the NaCl in sea water is far from being pure NaCl.

In contrast to solutes, water and solids usually exist in nearly pure states. For example, although sea water contains significant amounts of dissolved salts, it is still mostly water with $X_{H_2O} = 0.98$. For water and solids, then, we will adopt the activity coefficient reference convention that $\zeta_i \rightarrow 1.0$ as $X_i \rightarrow 1.0$. Therefore, we also have that $a_i \rightarrow 1.0$ as $X_i \rightarrow 1.0$. Based on these considerations, the requirement that both γ_i and $[i]$ be equal to 1.0 simultaneously when solids and water are in their standard states is met by the real, $X_i = 1.0$ system composition. Thus, in such cases, there is no need for a hypothetical standard state.

2.5 EFFECTS OF TEMPERATURE AND PRESSURE ON EQUILIBRIUM CONSTANTS

Oftentimes, we are interested in the value of an equilibrium constant K at a temperature and/or pressure other than that for which it can be computed directly using Eqs. (2.14) and (2.20) and values of μ_i° that are tabulated for specific values of T and P. In such cases, we are interested in knowing the T and P dependence of K. Using basic thermodynamics (Denbigh 1981, p. 300), it is possible to show that for reactions involving condensed phase species (i.e., dissolved species and solids, but no gaseous species), then

$$R \, \mathrm{dln}\, K = \frac{\Delta H^{\circ}}{T^2} \mathrm{d}T - \frac{\Delta V^{\circ}}{T} \mathrm{d}P. \tag{2.56}$$

The need for the qualifier concerning condensed phase species originates with the term giving the pressure dependence and involving ΔV°.

The two partial derivatives that may be obtained from (2.56) are

$$\left(\frac{\partial \ln K}{\partial T}\right)_P = \frac{\Delta H^{\circ}}{RT^2} \tag{2.57}$$

and

$$\left(\frac{\partial \ln K}{\partial P}\right)_T = \frac{\Delta V^{\circ}}{RT}. \tag{2.58}$$

The parameters ΔH° and ΔV° are the net changes in enthalpy (i.e., heat content) and volume, respectively, when the reaction is occurring under standard conditions, that is, when all $a_i = 1.0$. It should be noted that both ΔH° and ΔV° are T and P dependent. In a manner that is completely analogous to Eq. (2.14) for ΔG°, we have

$$\Delta H^{\circ} = \sum_i \upsilon_i \bar{H}_i^{\circ} \tag{2.59}$$

and

$$\Delta V^{\circ} = \sum_i \upsilon_i \bar{V}_i^{\circ} \tag{2.60}$$

where $\bar{H}_i$ is the partial molar enthalpy of reactant or product i, and $\bar{V}_i$ is the partial molar volume of reactant or product i. When considering the analogies between Eq. (2.14) and Eqs. (2.59) and (2.60), it should be recalled that μ_i is the partial molar free energy of i.

As usual, the stoichiometric coefficients v_i have sign (positive for products, and negative for reactants). ΔH° is related to ΔG° through the equation

$$\Delta G^{\circ} = \Delta H^{\circ} - T\Delta S^{\circ} \tag{2.61}$$

where ΔS° is the change in entropy when the reaction is proceeding under standard conditions. Convenient units for R for Eq. (2.57) (and Eq. (2.64) on page 40) are 8.314 J/mol-K. For Eq. (2.58), convenient units for R are 0.082 L-atm/mol-K.

While Eqs. (2.56) and (2.58) are strictly valid only when all reactants and products are condensed phase species, Eq. (2.57) remains valid even when some of the reactants and/or products are gaseous. Eqs. (2.56) and (2.58) can be adapted for use when some of the reactants and/or products are gaseous by neglecting to include the partial molar volume(s) of the gaseous species when computing ΔV° (e.g., see Denbigh 1981, pp. 203–207). The fact that gases are neglected in this context is related to the fact that the μ_i value for a gaseous species is independent of pressure.

When heat is given off during the course of a reaction, ΔH° is negative. Under these conditions, the total enthalpy content of the products under standard conditions is less than the total enthalpy content of the reactants under standard conditions. When ΔH° is negative, then heat may be thought of as one of the *products* of the reaction. When heat is taken up from the surroundings during the course of a reaction, then ΔH° is positive, and heat may be thought of as one of the *reactants*. Increasing the *temperature* of a system will increase the total enthalpy of the system. Therefore, Le Chatelier's Principle[7] tells us in general terms what Eq. (2.57) tells us in specific terms, namely that increasing the temperature will tend to shift a reaction equilibrium towards the reactants (decrease K) when heat is a product (ΔH° negative), and towards the products (increase K) when heat is a reactant (ΔH° positive).

If we note that

$$d(1/T) = -dT/T^2 \tag{2.62}$$

then Eq. (2.57) can be converted into

$$\left(\frac{\partial \ln K}{\partial(1/T)}\right)_P = -\frac{\Delta H^{\circ}}{R}. \tag{2.63}$$

[7]Le Chatelier's Principle may be stated as follows:

A system at equilibrium, when subjected to a perturbation, responds in a way that tends to minimize its effect.

Many readers will be most familiar with Le Chatelier's Principle from considerations of what happens to the position of a chemical equilibrium when the level of one of the actual chemical reactants or products is changed under conditions of constant T and P. In particular, increasing the level of one or more of the reactants will shift the equilibrium towards the products, and increasing the level of one or more products will shift the equilibrium towards the reactants. For these two types of cases, it is the *constancy* of the equilibrium constant at constant T and P that yields the observed behavior. As noted in the main text, the effects on K from changing T at constant P and changing P at constant T can also be deduced based on Le Chatelier's Principle. In particular, the effects on K values from changing T and P may be deduced by considering T to be an embodiment of heat, P to be an embodiment of negative volume, and by considering enthalpy and volume as implicit reactants or products. P acts like negative volume because as P increases, volume always decreases.

Equation (2.57) or Eq. (2.63) may then be integrated if ΔH° is known as a function of T. It is often a good assumption that ΔH° will stay reasonably constant over the temperature range of interest. Under conditions of constant P, Eq. (2.63) may then be integrated to yield

$$\ln \frac{K_2}{K_1} = -\frac{\Delta H^\circ}{R}\left(\frac{1}{T_2} - \frac{1}{T_1}\right) \tag{2.64}$$

where K_1 is the *known* value of K at some temperature T_1, and K_2 is the extrapolated value of K at T_2. Eq. (2.64) can be applied with confidence as long as the extrapolation temperature range $(T_2 - T_1)$ is not too large (e.g., 15 degrees or less). As T_2 approaches T_1, Eq. (2.64) provides an increasingly good approximation of K as a function of T. More complex equations that do not make the assumption of a constant ΔH° value are available in many texts on physical chemistry.

When $\Delta V^\circ > 0$, then volume is one of the *products* of the reaction. When $\Delta V^\circ < 0$, then volume is one of the *reactants*.[8] When P is increased, by Le Chatelier's Principle, the equilibrium will be shifted towards that side of the reaction that has less total volume. When $\Delta V^\circ > 0$, the shift will be towards the reactants (K decreased), and when $\Delta V^\circ < 0$, the shift will be towards the products (K increased).

At constant temperature, when ΔV° is assumed to remain constant with pressure, Eq. (2.58) may be integrated to yield

$$\ln \frac{K_2}{K_1} = -\Delta V^\circ(P_2 - P_1)/RT \tag{2.65}$$

where K_1 is the *known* value of K at P_1, and K_2 is the extrapolated value of K at P_2. As for Eq. (2.58), convenient units for R in Eq. (2.65) are 0.082 L-atm/mol-K. In a manner that is analogous with Eq. (2.64), Eq. (2.65) will give an increasingly good estimate of K as a function of P as P_2 approaches P_1. However, even extrapolations over relatively large pressure ranges (e.g., hundreds of atmospheres) may frequently be done with confidence since ΔV° often changes little with pressure. Indeed, the compressibilities of condensed phase species are frequently very small. Thus, since the individual $\bar{V}_i$ change little with pressure, the quantity ΔV° will change little with pressure.

2.6 COMBINING EQUILIBRIUM EXPRESSIONS

Let us say that we know that at 25°C/1 atm, the equilibrium constant for the reaction

$$Fe^{3+} + OH^- = FeOH^{2+} \tag{2.66}$$

is $K_{H1} = 10^{11.8}$. Let us also say that at 25°C/1 atm, the equilibrium constant for the

[8] This is the reverse of what happens with ΔH°. The reason for this is that H is the *stored* chemical heat. When stored chemical heat is converted to thermal heat, it is lost from the chemical(s) to the environment. In contrast, when volume is released ($\Delta V^\circ > 0$), it is not "lost" from the chemicals, but remains, and in fact signifies their presence.

second complexation by hydroxide, that is,

$$FeOH^{2+} + OH^- = Fe(OH)_2^+ \tag{2.67}$$

is $K_{H2} = 10^{10.5}$.

When these two equilibrium reactions are *added* together, we obtain

	summed reactant set		summed product set		
	$Fe^{3+} + OH^-$	=	$FeOH^{2+}$	$K_{H1} = 10^{11.8}$	(2.66)
+	$FeOH^{2+} + OH^-$	=	$Fe(OH)_2^+$	$K_{H2} = 10^{10.5}$	(2.67)

$$Fe^{3+} + 2OH^- + FeOH^{2+} = FeOH^{2+} + Fe(OH)_2^+ \quad K = K_{H1}K_{H2} = 10^{22.3}. \tag{2.68}$$

Cancelling the two like $FeOH^{2+}$ terms, we have

$$Fe^{3+} + 2OH^- = Fe(OH)_2^+ \qquad K = K_{H1}K_{H2} = 10^{22.3}. \tag{2.69}$$

As has been noted, the equilibrium constant for the reaction in Eqs. (2.68) and (2.69) is given by the *product* of the equilibrium constants for the individual reactions that have been added together. This is an important general result. It may be concluded quickly that this result is correct by considering that by adding two reactions together, we sum the two sets of reaction products, and also sum the two sets of reactants. The resulting equilibrium constant will thus be given by the *product* of the activities of the overall reaction product set divided by the product of the activities of the overall reactant set. In this particular case, we have

$$\frac{\{FeOH^{2+}\}}{\{Fe^{3+}\}\{OH^-\}} \times \frac{\{Fe(OH)_2^+\}}{\{FeOH^{2+}\}\{OH^-\}} = \frac{\{Fe(OH)_2^+\}}{\{Fe^{3+}\}\{OH^-\}^2} \tag{2.70}$$

$$K_{H1} \times K_{H2} = K_{H1}K_{H2}. \tag{2.71}$$

When one equilibrium reaction is subtracted from another, the equilibrium constant for the resulting reaction is given by the *quotient* of the corresponding individual equilibrium constants. Continuing to use reactions (2.66), (2.67), and (2.69) to illustrate these principles, we have

	$Fe^{3+} + 2OH^-$	=	$Fe(OH)_2^+$	$K = K_{H1}K_{H2} = 10^{22.3}$	(2.69)
−	$Fe^{3+} + OH^-$	=	$FeOH^{2+}$	$K_{H1} = 10^{11.8}$	(2.66)

$$FeOH^{2+} + OH^- = Fe(OH)_2^+ \qquad K_{H2} = K_{H1}K_{H2}/K_{H1} = 10^{10.5}. \tag{2.67}$$

In terms of the expressions for the various equilibrium constants, we have

$$\frac{\{Fe(OH)_2^+\}}{\{Fe^{3+}\}\{OH^-\}^2} \div \frac{\{FeOH^{2+}\}}{\{Fe^{3+}\}\{OH^-\}} = \frac{\{Fe(OH)_2^+\}}{\{FeOH^{2+}\}\{OH^-\}} \tag{2.72}$$

$$K_{H1}K_{H2} \div K_{H1} = K_{H2}. \tag{2.73}$$

It is worth pointing out that subtracting a given equilibrium from a second equilibrium is equivalent to reversing the direction of the first equilibrium, then adding the result to the second. For the case at hand, we have

$$Fe^{3+} + 2OH^- = Fe(OH)_2^+ \qquad K = K_{H1}K_{H2} = 10^{22.3} \tag{2.69}$$

$$+ \quad FeOH^{2+} = Fe^{3+} + OH^- \qquad (K_{H1})^{-1} = 10^{-11.8} \tag{2.74}$$

$$FeOH^{2+} + OH^- = Fe(OH)_2^+ \qquad K_{H2} = K_{H1}K_{H2}x(K_{H1})^{-1} = 10^{10.5}. \tag{2.75}$$

2.7 INFINITE DILUTION, CONSTANT CONCENTRATION, AND "MIXED" EQUILIBRIUM CONSTANTS

2.7.1 Infinite Dilution Constants

Thermodynamically, molality is a convenient concentration scale for dissolved species. On this scale, we usually take $\gamma_i = 1$ when the solution is infinitely dilute. Thus, for dissolved species, tabulated μ_i^o values are given on the molality scale with $\gamma_i = 1$ when the solution is infinitely dilute. Thus, for equilibria involving dissolved reactants and products, the equilibrium constants that are derivable from Eqs. (2.14) and (2.20) are "infinite dilution" constants. This name emphasizes the facts that: 1) it is the *activities* of the dissolved species that appear in the quotient that defines K; and 2) we have adopted the convention that $\gamma_i \rightarrow 1.0$ as the system becomes infinitely dilute for any dissolved species i.

Consider the particular case of the acid dissociation reaction

$$HA = H^+ + A^-. \tag{2.76}$$

The expression for K will be given by

$$K = \frac{\{H^+\}\{A^-\}}{\{HA\}} = \frac{[H^+]\gamma_{H^+}[A^-]\gamma_{A^-}}{[HA]\gamma_{HA}} \tag{2.77}$$

where the $[i]$ are molalities. For dilute solutions which are at ambient T and P values, these molalities will be very close in magnitude to the corresponding molarities.

Infinite dilution constants are used more often than any other type of equilibrium constant. A big part of the reason for this is that, as noted above, this is the type of K that results from application of Eqs. (2.14) and (2.20) for reactions involving dissolved species. In addition, however, many solutions of interest are comparatively dilute, and so the γ_i values are either essentially equal to 1.0, or not very far from 1.0. Under these conditions, activity corrections will not be needed, and the infinite dilution constants may be used with concentrations substituted for activities. In non-dilute solutions of *varying* composition, it will also be convenient to work with infinite dilution constants. In such systems, the infinite dilution constants serve as a constant reference base to which varying corrections can be made.

2.7.2 Constant Ionic Medium Constants

Many important chemical equilibria in media such as seawater, biological fluids, etc. involve reactant and product species that are low in concentration relative to the other, major ionic species.[9] In these systems, the major ionic species: 1) provide the bulk of the system's ionic matrix; *and* 2) often remain very nearly *constant* in concentration. In such systems, the nature of the ionic medium *and the values of the activity coefficients that it sets* can remain very nearly constant, while the levels of species that are low in concentration (but perhaps are very important chemically) change significantly. As an example, consider sea water, which is about 0.5 m in NaCl alone. In this medium, $[H^+]$ typically varies between 10^{-9} to 5×10^{-8} m. At such levels, $[H^+]$ can vary widely without significantly altering the ionic medium.

For both ionic as well as neutral species, the property that affects activity coefficients the greatest is the ionic strength (I, m). The expression for I is

$$I = 1/2 \sum_i m_i z_i^2 \tag{2.78}$$

where m_i (mol/kg) and z_i are the concentration and charge for species i, respectively.

Let us now consider the case of a chemist who is interested in equilibrium (2.76) in a particular ionic medium that is essentially constant in composition, and in which the Eq. (2.76) species are minor. Based on a knowledge of the major species in that solution, that chemist will be able to compute estimates of the activity coefficients of H^+, A^-, and HA using equations to be presented in Section 2.8. One sequence of events experienced by the chemist might therefore occur as follows. While solving for a given system speciation, the chemist uses the known value of I to first calculate the values of γ_{H^+}, γ_{A^-}, and γ_{HA}. Then, using other known or measured variables (e.g., $A_T (= [HA]+[A^-])$, etc.), that chemist solves for the concentrations and activities of the three species using the infinite dilution constant as given in Eq. (2.77).

The next day, the chemist has another problem to solve involving the same reaction, and the same ionic medium, but for a different value of A_T. The chemist reviews his notes and finds the infinite dilution constant again. Although the system speciation is going to be somewhat different than for the previous day's problem, the system matrix is going to be the same and so the value of I is going to be essentially the same as before. The chemist therefore retrieves the activity coefficient values that were generated the previous day and solves the new problem.

When yet another similar problem comes up on the third day, the chemist becomes tired of continually keeping track of the same activity coefficient values. The chemist then reexamines Eq. (2.77), and rearranges it to read

$$K = \frac{\{H^+\}\{A^-\}}{\{HA\}} = \left(\frac{[H^+][A^-]}{[HA]}\right)\left(\frac{\gamma_{H^+}\gamma_{A^-}}{\gamma_{HA}}\right) \tag{2.79}$$

[9] Actually, the composition of sea water can change slightly, due mostly to dilution and concentration effects from upwelling, other types of mixing, evaporation at the surface, and precipitation inputs.

or equivalently,

$$K\left(\frac{\gamma_{HA}}{\gamma_{H^+}\gamma_{A^-}}\right) = \frac{[H^+][A^-]}{[HA]}. \tag{2.80}$$

We know that the infinite dilution constant K is a thermodynamic constant that is only a function of T and P. If the ionic matrix of the medium remains essentially constant, the term involving the activity coefficients on the left-hand side will also remain constant for that medium. Therefore, the entire left-hand side is a constant for that medium. This constant is referred to symbolically as ${}^cK$. Therefore, we have

$${}^cK = \frac{[H^+][A^-]}{[HA]} \tag{2.81}$$

where

$${}^cK = \frac{\gamma_{HA}}{\gamma_{H^+}\gamma_{A^-}}K. \tag{2.82}$$

The superscript c denotes that ${}^cK$ involves *concentrations* rather than *activities*, and that ${}^cK$ is a *constant* ionic medium constant.

It is important to point out that any given ${}^cK$ is valid only for a *specific* solution matrix, that is, the matrix that gives a specific value of the term $(\gamma_{HA}/\gamma_{H^+}\gamma_{A^-})$. Once a given ${}^cK$ value is calculated or measured for a given matrix, *activity corrections are completely built into* ${}^cK$ *for that equilibrium*; one can then work directly with *concentrations* in all calculations. It should be clear that anyone who is working with a constant ionic medium is likely to want to have all of the pertinent equilibrium constants expressed as ${}^cK$ values.

If one needs a ${}^cK$ but the infinite dilution K is not known, it is likely to be more attractive to simply measure ${}^cK$ rather than K. Indeed, in measuring ${}^cK$, we need not worry about what the γ_i values are; if the *concentrations* of the species at equilibrium can be measured, then one can compute ${}^cK$ directly.

2.7.3 Mixed Constants

In some circumstances, it is convenient to use equilibrium constants in which some of the reactants/products appear as concentrations, and some appear as activities. These equilibrium constants are referred to as "mixed constants." Based on the discussion in the preceding section, we can expect that these circumstances will involve constant ionic media since such media will allow us to build activity coefficients into the equilibrium constant.

The specific type of equilibrium process for which mixed constants are particularly attractive are *acid dissociations* as in Eq. (2.76). It is convenient to use [HA] and $[A^-]$ in the expression for the equilibrium constant for reasons described in the preceding section. For H^+, however, use of activity is retained. The reason for this retention is that it is easy to *measure* $\{H^+\}$ by means of the pH electrode. (Note: the pH electrode responds to H^+ *activity*, not *concentration*.) Thus, when input information on experimental acid/base equilibria in a constant ionic medium includes pH measurement data, it would make little sense to use a ${}^cK$ for that medium and then have to convert the measured $\{H^+\}$ to $[H^+]$ prior to using the constant.

Thus, for the mixed constant for reaction (2.76), we have

$$K' = \frac{\{H^+\}[A^-]}{[HA]} \tag{2.83}$$

and so

$$K' = \frac{\gamma_{HA}}{\gamma_{A^-}} K. \tag{2.84}$$

Summarizing the above results, we have

$$K = \frac{\{H^+\}\{A^-\}}{\{HA\}} = {}^{c}K \frac{\gamma_{H^+}\gamma_{A^-}}{\gamma_{HA}} = K' \frac{\gamma_{A^-}}{\gamma_{HA}}. \tag{2.85}$$

It should be clear that all of the three constants (K, ${}^{c}K$, and K') are equally correct from a fundamental point of view. They are just different ways of quantitatively expressing the same equilibrium process. Equilibrium constants found in the literature are of all three types. When a set of constants that satisfies Eq. (2.85) is used in equilibrium calculations, each of the constants will lead to the exact same result. However, if one takes a ${}^{c}K$ constant from the literature and uses it as if it were an infinite dilution constant, then errors will be incurred; one must be careful to be sure and know of which type a given constant is before using it in calculations. Also, if an infinite dilution K and a ${}^{c}K$ do not satisfy Eq. (2.85) because they are *not consistent*, they will not lead to the same result. In that case, we will be able to conclude that at least one of the equilibrium constants is fundamentally *incorrect*.

A useful discussion which illustrates how one pair of scientists dealt with the variety of acid/base equilibrium constants in the literature when compiling such information into a database may be found in the Foreword of Smith and Martell (1976) under the heading of "Log K Values." A portion of their advice to their readers in how to use the information they compiled is reproduced below:

> "Equilibria involving protons have been expressed as concentration constants [i.e., ${}^{c}K$ values] in order to be more consistent with the [manner in which] metal ion stability constants ... [are typically reported]. Concentration constants may be determined by calibrating the electrodes with solutions of known hydrogen ion concentrations or by conversion of [the directly measured] pH values using the appropriate hydrogen ion activity coefficient. When standard buffers are used, mixed constants (also known as Bronsted or practical constants) are obtained which include both activity and concentration terms. Literature values expressed as mixed constants have been converted to concentration constants by using the hydrogen ion activity coefficients determined in KCl solution before inclusion in the tables. In some cases, [the data from some] papers were ... [not included in the database] because no indication was given as to the use of concentration or mixed constants. Some papers were retained despite this lack of information when it could be ascertained which constant was used by comparing to known values or by personal communication with the authors."

2.8 ACTIVITY COEFFICIENT EQUATIONS

2.8.1 Activity Coefficient Equations for Single Ions

Different equations are available for predicting activity coefficients of individual ions. Activity coefficients are most successfully calculated when a solution is dilute. As the ionic strength increases, the difficulty of predicting activity coefficient values also increases, and increasingly empirical equations must be used. The activity coefficient equations considered in this text are presented in Table 2.3. In all of the equations, the charge on ion i is denoted as z_i. Values of the parameter a for the Extended Debye-Hückel Equation are given in Table 2.4 for various ions of interest. The version of the Davies Equation in Table 2.3 will give the same activity coefficient for all ions with charges of the same absolute value (e.g., +1 or −1, +2 or −2, etc.). A more complicated version of the Davies Equation that attempts to take the varying characteristics of particular ions into account is also available (Lloyd and Heathcote 1985).

As discussed above, for a given dissolved species, the concentration may be expressed on the mole fraction, molality, or molarity scale. That is,

$$\mu_i = (\mu_i^{\mathrm{o}})_X + RT \ln \zeta_i X_i \tag{2.45}$$

$$\mu_i = (\mu_i^{\mathrm{o}})_m + RT \ln \gamma_i m_i \tag{2.46}$$

$$\mu_i = (\mu_i^{\mathrm{o}})_c + RT \ln y_i c_i . \tag{2.47}$$

where ζ_i, γ_i, and y_i are the activity coefficients on the mole fraction, molality, and molarity scales, respectively.

In strict theoretical terms, the Debye-Hückel and the Extended Debye-Hückel Equations give activity coefficients for concentrations/activities expressed on the *mole fraction* scale, that is, they give ζ_i values. Although the Güntleberg and Davies Equations are semi-empirical in nature, they too are thought of as expressions for ζ_i. As discussed in the footnote on page 34, for dilute solutions (i.e., less than a few molar), ζ_i and γ_i values will be very nearly equal to one another. Therefore, since: 1) the concentration range over which all four equations apply is comparatively dilute; and 2) molarity and molality are essentially equal for dilute solutions, we conclude that the equations also yield good estimates of y_i values, and that each of the four equations may be used to calculate $\zeta_i \simeq \gamma_i \simeq y_i$. Since we are most interested in expressing concentration on the molality scale, we will discuss the equations in terms of their predictions of γ_i values.

Figure 2.4 presents plots of γ_i vs. $\log\sqrt{I}$ according to the: 1) Debye-Hückel Equation for $z_i = \pm 1$ and ± 2; 2) Extended Debye-Hückel Equation for H^+ (ion size parameter $a = 9$), bicarbonate (HCO_3^-) ion ($a = 4$), and carbonate (CO_3^{2-}) ion ($a = 5$); and 3) Davies Equation for $z_i = \pm 1$ and ± 2. Although the scale range for the vertical axis in Figure 2.4 extends only to 1.0, $\gamma_i > 1.0$ are physically possible. $\gamma_i \leq 0$ are not possible. As $\log\sqrt{I} \rightarrow -\infty$ (i.e., infinite dilution), all of the γ_i values tend to 1.0. At any given value of $\sqrt{I}$ in Figure 2.4, we see that γ_i for a doubly-charged ion tends to be less than that for a singly-charged ion. This effect is due to the fact that z_i is squared in the various equations.

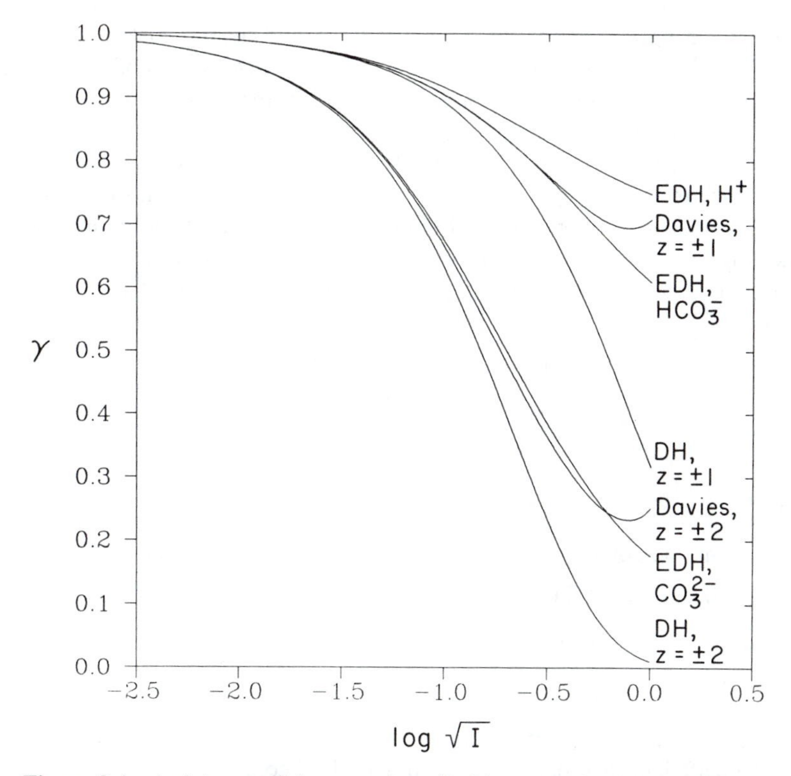

Figure 2.4 Activity coefficients vs. log $\sqrt{I}$. Lines calculated using different equations for different ions, including: 1) Debye-Hückel (DH) Equation for $z = \pm 1$; 2) DH Equation for $z = \pm 2$; 3) Extended Debye-Hückel (EDH) Equation for H^+; 4) EDH Equation for HCO_3^-; 5) EDH Equation for CO_3^{2-}; 6) Davies Equation for $z = \pm 1$; and 7) Davies Equation for $z = \pm 2$.

All of the equations try, with varying degrees of success at different $\sqrt{I}$ values, to predict the true γ_i for ion i in a given solution. It may be noted that the Debye-Hückel, Extended Debye-Hückel, and the Güntleberg Equations all lead to a monotonic decrease of γ_i with increasing I, with

$$1.0 > \gamma_i > 0. \tag{2.86}$$

The Güntleberg Equation is a generic version of Extended Debye-Hückel Equation; it assumes that an average value of the ion size parameter a is ~3.

In general, the Debye-Hückel Equation tends to underestimate γ_i values. This underestimation becomes more severe as $\sqrt{I}$ increases. The Extended Debye-Hückel Equation corrects for this underestimation up to $\sqrt{I} = 0.1$. Beyond that, theory begins to break down, and we must begin to rely on the semi-empirical Davies Equation or on other empirical equations. The field of chemical oceanography has developed some rather complex empirical equations for predicting activity coefficients in relatively concentrated, multi-ion solutions like sea water.

It is interesting to note that the term $-0.2I$ in the Davies Equation causes a minimum near $\log\sqrt{I} = 0$ in the plots of γ_i vs $\log\sqrt{I}$. This serves to reproduce the observed fact that γ_i for an ion does not simply continue to decrease to 0 as the ionic strength increases, but rather can begin to increase again for large $\sqrt{I}$. This effect is caused by the fact that *the amount of solvent available for solvation of ions decreases as* $\sqrt{I}$ *increases* (Bockris and Reddy 1970, p. 238); decreasing the amount of solvating water tends to cause ions to be more exposed to one another, and therefore more *act*ive.

As simple examples of the application of these principles, we calculate the activity coefficient for H^+ using the Debye-Hückel Equation in the following two solutions: a) 0.001 F NaCl, and b) 0.001 F $CaSO_4$:

a) Neglecting the contribution to I from the dissociation of water,

$$I = (0.5)(0.001(+1)^2 + 0.001(-1)^2) = 0.001$$

$$\log \gamma_{H^+} = -(0.5)(+1)^2(0.001)^{0.5}$$

$$\gamma_{H^+} = 0.96.$$

b) Neglecting the contribution to I from the dissociation of water,

$$I = 0.5(0.001(+2)^2 + 0.001(-2)^2) = 0.004$$

$$\log \gamma_{H^+} = -(0.5)(+1)^2(0.004)^{0.5}$$

$$\gamma_{H^+} = 0.93.$$

It may be noted that the greater I of the $CaSO_4$ solution causes more of a reduction of the activity coefficient than does the equiformal NaCl solution.

TABLE 2.3 Equations for activity coefficient γ_i where the activity $a_i = \gamma_i m_i$. For the ionic strength ranges applicable for the equations, $\zeta_i \simeq \gamma_i \simeq y_i$ where ζ_i and y_i are the activity coefficients on the mole fraction and molarity concentration/activity scales, respectively. The parameter A depends on T (K) according to the equation $A = 1.82 \times 10^6(\varepsilon T)^{-3/2}$ where ε is the temperature-dependent dielectric constant of water. $B = 50.3(\varepsilon T)^{-1/2}$. For water at 298 K (25°C), $A = 0.51$ and $B = 0.33$. Applicable ionic strength range obtained from Stumm and Morgan (1981, p.135).

Name	Equation	Applicable Ionic Strength Range
Debye-Hückel	$\log \gamma_i = -Az_i^2\sqrt{I}$	$I < 10^{-2.3}$
Extended Debye-Hückel	$\log \gamma_i = -Az_i^2[\sqrt{I}/(1 + Ba\sqrt{I})]$ a is an size parameter for ion i; it should not be confused with the activity.	$I < 10^{-1.0}$
Güntleberg	$\log \gamma_i = -Az_i^2[\sqrt{I}/(1 + \sqrt{I})]$ equivalent to Extended D-H. with an average value of $a = 3$	$I < 10^{-1.0}$
Davies	$\log \gamma_i = -Az_i^2[(\sqrt{I}/(1 + \sqrt{I})) - 0.2I]$	$I < 0.5$

TABLE 2.4 Ion size parameters a for the Extended Debye-Hückel Equation[10]

Ion	a	Ion	a	Ion	a
Ag^+	3	Fe^{2+}	6	Mn^{2+}	6
Al^{3+}	9	Fe^{3+}	9	Na^+	4
Ba^{2+}	5	H^+	9	NH_4^+	3
Be^{2+}	8	HCO_3^-	4	NO_3^-	3
Ca^{2+}	6	HPO_4^{2-}	4	OH^-	3
Ce^{3+}	9	$H_2PO_4^-$	4	Pb^{2+}	5
CH_3COO^-	4	HS^-	3	PO_4^{3-}	4
Cl^-	3	I^-	3	Sn^{2+}	6
ClO_4^-	3	K^+	3	SO_4^{2-}	4
CO_3^{2-}	5	La^{3+}	9	Sr^{2+}	5
Cu^{2+}	6	Mg^{2+}	8	Zn^{2+}	6

2.8.2 Activity Coefficient Equations for Neutral Species

The activity coefficients for neutral species in water are usually found to depend on I according to an equation of the type

$$\log \gamma_i = kI \tag{2.87}$$

where k is a constant that depends on the identity of species i, T, and P. The value of k also depends on the identity of the ions making up the ionic medium.

Usually, the value of k in Eq. (2.87) is positive, although there are cases when k is negative. When k is positive, the value of γ_i becomes increasingly greater than 1.0 as I increases. This behavior is sometimes referred to as "salting out;" the salt(s) responsible for I is (are) tending to drive species i out the solution by increasing the activity of i. In the comparatively few cases when k is negative, the value of γ_i becomes increasingly less than 1.0 as I increases. This "salting in" behavior is due to a *favorable* interaction between the species of interest and the salt(s) responsible for I. An example of salting in may be observed in the context of the effects of potassium nitrate (KNO_3) on the undissociated form of acetic acid (CH_3COOH); in this case, $k = -0.02$ (Harned and Owen 1958, p. 536).

Activity corrections for neutral species are often neglected in natural water chemistry. There are two main reasons for this: 1) the value of k in Eq. (2.87) is too dependent on the species of interest and the salt(s) responsible for I to permit the development of any universally predictive equation; and 2) k values are usually sufficiently small that such corrections may be neglected without encountering significant errors. For example, when k is positive, values usually range between 0.05 to 0.2. A typical value might be $\sim$0.15. For oxygen in solutions of NaCl, the value of k is $\sim$0.13 at 25°C (Harned and Owen 1958, p. 535). Thus, for a fresh water value of I of say 0.002, the value of γ for oxygen may be predicted to be $\sim$1.0006 at 25°C. For sea water, γ for oxygen has

[10]Data of Kielland, *J. Am. Chem. Soc.*, **59**, 1675 (1937).

been measured to be 1.30 at 25°C (Kester 1975). We will neglect activity corrections for neutral species in this text. The interested reader may wish, however, to read further into this matter. Useful discussions on this topic can be found in Harned and Owen (1958) and Pytkowicz (1983). Problems illustrating the principles of this chapter are presented by Pankow (1992).

2.9 REFERENCES

BARD, A. J., R. PARSONS, and J. JORDAN. 1985. *Standard Potentials in Aqueous Solution*. New York: Marcel Dekker.

BOCKRIS, J. O'M., and REDDY, A. K. N. 1970. *Modern Electrochemistry, Vol. 1*. New York: Plenum Press.

DENBIGH, K. 1981. *The Principles of Chemical Equilibrium*, Fourth Edition. Oxford: Cambridge University Press.

HARNED, H. S., and B. B. OWEN. 1958. *The Physical Chemistry of Electrolyte Solutions*, American Chemical Society Monograph Series. New York: Reinhold Publishing.

KESTER, D. R. 1975. *Chemical Oceanography*, Vol. 1, Second Edition, 497–556. New York: Academic Press.

KIELLAND, J. 1937. *J. Am. Chem. Soc.*, **59**: 1675.

LLOYD, J. W., and J. A. HEATHCOTE. 1985. *Natural Inorganic Hydrochemistry in Relation to Groundwater*. Oxford: Clarendon Press.

PANKOW, J. F. 1992. *Aquatic Chemistry Problems*. Portland: Titan Press-OR, P.O. Box 91399, Portland, Oregon, 97291-1399.

PYTKOWICZ, R. M. 1983. *Equilibria, Nonequilibria, and Natural Waters*, Vol. 1. New York: Wiley-Interscience.

SMITH, R. M., and MARTELL, A. E. 1976. *Critical Stability Constants, Volume 4, Inorganic Complexes*. New York: Plenum Press.

STUMM, W., and J. J. MORGAN. 1981. *Aquatic Chemistry*. New York: Wiley-Interscience.

PART II

Acid/Base Chemistry

3

$[H^+]$ As an Important and Meaningful Thermodynamic Variable in Aquatic Chemistry

3.1 GENERAL IMPORTANCE OF H^+ IN NATURAL WATERS

As discussed in Chapter 1, there are five main classes of reactions in natural water chemistry. These are:

- acid/base
- complexation
- dissolution/precipitation
- redox
- adsorption

In addition to playing its obvious role in acid/base reactions, the proton H^+ is a participant in a very large number of reactions of the other four types. The activity of H^+ therefore participates in determining the equilibrium positions of many reactions. Consequently, $\{H^+\}$ is often called a "master variable" in aquatic chemistry. Important examples of reaction involving H^+ are:

1. acid/base, e.g.,

$H_2CO_3 = H^+ + HCO_3^-$	dissociation of carbonic acid	(3.1)
$HCO_3^- = H^+ + CO_3^{2-}$	dissociation of bicarbonate ion	(3.2)
$NH_4^+ = H^+ + NH_3$	dissociation of ammonium ion	(3.3)
$RCOOH = RCOO^- + H^+$	dissociation of an organic carboxylic acid	(3.4)

$$M(H_2O)_n^{z+} = M(H_2O)_{n-1}OH^{(z-1)+} + H^+ \quad \text{loss of } H^+ \text{ by hydrated metal ion} \tag{3.5}$$

2. complexation, e.g.,

$$Fe^{3+} + H_2O = FeOH^{2+} + H^+ \quad \text{hydrolysis of } Fe^{3+} \text{ to form } FeOH^{2+} \tag{3.6}$$

$$Cu^{2+} + H_2O = CuOH^+ + H^+ \quad \text{hydrolysis of } Cu^{2+} \text{ to form } CuOH^+ \tag{3.7}$$

3. dissolution/precipitation, e.g.,

$$CaCO_{3(s)} + H^+ = HCO_3^- + Ca^{2+} \quad \text{dissolution of calcite under acidic conditions} \tag{3.8}$$

$$NaAlSi_3O_{8(s)} + H^+ + \tfrac{9}{2}\, H_2O = \tfrac{1}{2}Al_2Si_2O_5(OH)_{4(s)} + Na^+ + 2H_4SiO_4 \quad \text{interconversion of two aluminosilicates (albite to kaolinite)} \tag{3.9}$$

4. redox, e.g.,

$$O_2 + 4e^- + 4H^+ = 2H_2O \quad \text{reduction of oxygen to form water} \tag{3.10}$$

$$\tfrac{1}{4}CH_2O + H^+ + e^- = \tfrac{1}{2}CH_{4(g)} + \tfrac{1}{2}H_2O \quad \text{reduction of organic matter to form methane} \tag{3.11}$$

5. adsorption, e.g.,

$$\equiv S-OH + Pb^{2+} = \equiv S-O-Pb^+ + H^+ \quad \text{complexation of } Pb^{2+} \text{ by a surface hydroxyl group} \tag{3.12}$$

3.2 QUESTION: SHOULD WE REFER TO THE PROTON AS H^+ OR AS H_3O^+ IN AQUEOUS SOLUTIONS?

In the simple dissociation of the acid HA, we have

$$HA = H^+ + A^-. \tag{3.13}$$

In the thermodynamic equilibrium approach, the dissociation process is dealt with mathematically in terms of an acid dissociation equilibrium constant. Assuming ideality for the purposes of this discussion (i.e., $\{i\} = [i]$), the acidity constant for HA is given by

$$K = \frac{[H^+][A^-]}{[HA]}. \tag{3.14}$$

A casual interpretation of Eq. (3.14) is that the acid dissociation reaction (3.13) involves an equilibrium between HA and the *free* ions H^+ and A^-. However, species (especially ions) do not generally exist all by themselves when dissolved in a solvent such as water. For example, the fact that metal ions can become solvated and then lose a proton as in reaction (3.5) testifies to the fact that solvated ions are in fact *hydrated* in water. Actually, reactions (3.5) and (3.6) describe exactly the same type of process. In reaction (3.6), it happens that we did not bother to write in the waters of hydration, and only concerned ourselves with the net overall process.

Hydration by water occurs to some extent with all ions in solution, not just metal ions. Many chemistry texts try to recognize this explicitly in the case of the proton by writing the acid dissociation for the acid HA according to

$$HA + H_2O = H_3O^+ + A^- \tag{3.15}$$

where the creation of the "hydronium" ion H_3O^+ is assumed to occur directly upon the dissociation of HA. According to this approach, the acid dissociation equilibrium constant is written

$$K = \frac{[H_3O^+][A^-]}{[HA]} \tag{3.16}$$

where we do not bother to include $[H_2O]$ ($\simeq 1$, mole fraction scale) in the denominator. The hydration of the aqueous proton by a single water molecule may be viewed according to the equilibrium

$$H^+ + H_2O = H_3O^+ \tag{3.17}$$

where the structure of H_3O^+ is depicted schematically in Figure 3.1*a*.

For any *solution* reaction like (3.17), thermodynamics tells us that at equilibrium, there must always be some non-zero concentration of all the participating reactants and products. Therefore, *both* $[H^+]$ and $[H_3O^+]$ must be non-zero. Therefore, one is brought to the question, are *both* (3.14) and (3.16) proper conventions for the mathematical expression of the dissociation of the acid HA? Moreover, is it really possible that we can describe the acid dissociation process with two distinct equilibrium constants? How can the two approaches yield the same right answer? If both species are present, how does the sum of their concentrations add up to the total concentration of acidic protons?

3.3 "DISSOCIATED" PROTONS EXIST AS A SERIES OF SPECIES, THAT IS, H^+, H_3O^+, $H_5O_2^+$, ETC. IN AQUEOUS SOLUTION

The first step we take in answering the questions posed in the previous section is to realize that, in fact, protons in solution exist as a whole *series* of species. Some will exist as truly free, unhydrated H^+ species; some will exist as H_3O^+ species; some

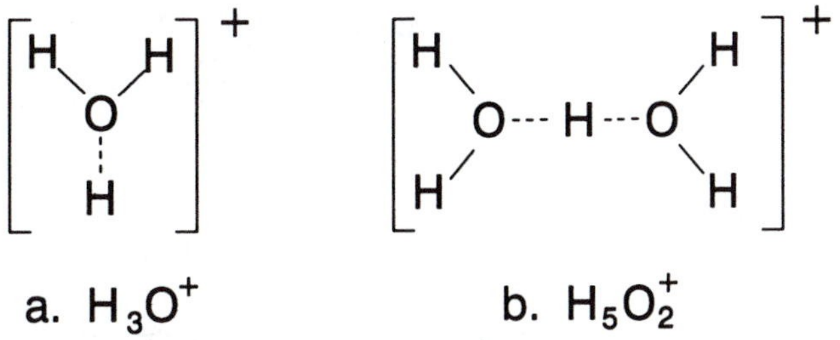

a. H_3O^+ b. $H_5O_2^+$

Figure 3.1 Schematic drawings of the species H_3O^+ and $H_5O_2^+$.

will exist in close interaction with two waters of hydration as the species $H_5O_2^+$ (Figure 3.1*b*) and so on. This range of species can be abbreviated as $H(H_2O)_n^+$ where n ranges from 0 up to about 6. Each of the solvating water molecules interacts with the positively charged H^+ through its slightly negatively-charged oxygen. For the truly free, unhydrated species H^+, n = 0. The upper limit for n is about 6 since space limitations prevent the placement of more than about 6 water molecules around a single ion. Many ions are solvated by water molecules with n = 5 or 6.

Based on the above discussion, we may therefore conclude that just referring to the "concentration" of aqueous proton is a simplification in and of itself, regardless of whether we refer to that concentration as being equal to [H^+] or [H_3O^+]. Let us then write

$$K = \frac{\text{"}[H^+]\text{"}[A^-]}{[HA]} \tag{3.18}$$

where "[H^+]" is the sum of all of the dissociated protons in solution,[1] that is

$$\text{"}[H^+]\text{"} = [H^+] + [H_3O^+] + [H_5O_2^+] + [H_7O_3^+] + \dots. \tag{3.19}$$

We now see that "[H^+]" is what chemists have traditionally just called [H^+] or [H_3O^+].

The concentration of each H^+-containing species may be assumed to be controlled by an equilibrium constant:

$$H^+ + H_2O = H_3O^+ \qquad K_1 \tag{3.20}$$

$$H_3O^+ + H_2O = H_5O_2^+ \qquad K_2 \tag{3.21}$$

$$H_5O_2^+ + H_2O = H_7O_3^+ \qquad K_3 \tag{3.22}$$

$$\vdots$$

etc.

Thus we have:

$$[H_3O^+] = K_1[H^+][H_2O] \tag{3.23}$$

$$[H_5O_2^+] = K_2[H_3O^+][H_2O] \tag{3.24}$$

$$= K_1K_2[H^+][H_2O]^2 \tag{3.25}$$

$$[H_7O_3^+] = K_3[H_5O_2^+][H_2O] \tag{3.26}$$

$$= K_1K_2K_3[H^+][H_2O]^3 \tag{3.27}$$

$$\vdots$$

etc.

[1]Strictly speaking, the proton will exist as an infinite number of different hydrated species. This is not because H^+ can be solvated with very large numbers of water molecules, but rather because the interaction distances of the each of the solvating water molecules can vary over a continuum of values.

Substituting Eqs. (3.23), (3.25), and (3.27) into Eq. (3.19), we obtain

$$\text{"[H}^+\text{]"} = [\text{H}^+] + K_1[\text{H}^+][\text{H}_2\text{O}] + K_1K_2[\text{H}^+][\text{H}_2\text{O}]^2 + K_1K_2K_3[\text{H}^+][\text{H}_2\text{O}]^3 + \ldots \tag{3.28}$$

$$= [\text{H}^+]\sum_{i=0}^{n}\left([\text{H}_2\text{O}]^i \prod_{j=0}^{i} K_j\right) \tag{3.29}$$

where K_0 is defined to be equal to 1. Thus, if

$$K = \frac{\text{"[H}^+\text{]"}[\text{A}^-]}{[\text{HA}]} \tag{3.30}$$

then

$$K = \frac{[\text{H}^+][\text{A}^-]}{[\text{HA}]}\sum_{i=0}^{n}\left([\text{H}_2\text{O}]^i \prod_{j=0}^{i} K_j\right) \tag{3.31}$$

Let

$$\sum_{i=0}^{n}\left([\text{H}_2\text{O}]^i \prod_{j=0}^{i} K_j\right) = \boldsymbol{K} \tag{3.32}$$

Then,

$$K = K^{\text{true}}\boldsymbol{K} \tag{3.33}$$

where K^{true} is the "true" equilibrium constant between the conjugate acid HA on the one side, and A^- and the unhydrated species H^+ on the other side. For dilute solutions, the $[H_2O]^i$ factors will be essentially independent of the concentrations of the dissolved species, that is, they will be constant. Since $[H_2O]$ is measured on the mole fraction scale, these factors will be very nearly equal to 1.0 for most systems of interest. Thus, Eq. (3.32) may be simplified to

$$\boldsymbol{K} \simeq \sum_{i=0}^{n}\left(\prod_{j=0}^{i} K_j\right) \tag{3.34}$$

Even though K involves "$[H^+]$" and not the true $[H^+]$ or $[H_3O^+]$, it is relatable through a temperature- and pressure-dependent constant (i.e., $\boldsymbol{K}$) to an equilibrium constant (also T- and P-dependent) that does. Therefore, we can go ahead and use K as if the situation was as simple as is implied by either

$$K = \frac{[\text{H}^+][\text{A}^-]}{[\text{HA}]} \quad (3.14) \qquad \text{or} \qquad K = \frac{[\text{H}_3\text{O}^+][\text{A}^-]}{[\text{HA}]} \quad (3.16)$$

We need to remember, though, that $[H^+]$ in Eq. (3.14) and $[H_3O^+]$ in Eq. (3.16) both really mean "$[H^+]$" as it is given by Eq. (3.19). As such, *both* Eqs. (3.14) and (3.16) are proper and consistent conventions to use in describing the acid dissociation process. Since the convention represented by Eq. (3.16) is not really any more truly "realistic" than that represented by Eq. (3.14), in this text we use the convention represented by Eq. (3.14).

At this point, we come to the realization that A^- and HA must also each exist as a series of hydrated species! However, the same type of analysis just applied to H^+ can also be applied to A^- and HA. If we did that, we would again arrive at the conclusion that Eq. (3.14) may be used directly, as long as it is realized that the three species (H^+, A^-, and HA) all actually exist as a series of hydrated species with a range of solvating water molecules. The overall result is that even though chemists use terms such as $[H^+]$, $[A^-]$, [HA] (and all other dissolved concentrations) somewhat loosely, the use of any such quantities in equilibrium constants is perfectly justified from a thermodynamic point of view; the various solvation constants (like those in Eqs. (3.20)–(3.22)) are *real quantities* with *actual values*, and so they could, in principle, always be used to get back to equilibrium expressions involving actual unhydrated species if one needed to do so; one usually does not.

TABLE 3.1 Acidity constant values for acids of interest in natural waters. Data are for 25°C/1 atm and infinite dilution.

Acid/Base Equilibrium	Common Name of Conjugate Acid	Common Name of Conjugate Base	– Log Acidity Constant for HA	– Log Basicity Constant for B^-
$HClO_4 = H^+ + ClO_4^-$	perchloric acid	perchlorate ion	~–7	~21
$HCl = H^+ + Cl^-$	hydrochloric acid	chloride ion	~–3	~17
$H_2SO_4 = H^+ + HSO_4^-$	sulfuric acid	bisulfate ion	~–3	~17
$HNO_3 = H^+ + NO_3^-$	nitric acid	nitrate ion	–1	~15
$H_2SO_3 = H^+ + HSO_3^-$	sulfurous acid	bisulfite ion	1.91	12.09
$HSO_4^- = H^+ + SO_4^{2-}$	bisulfate ion	sulfate ion	1.99	12.01
$H_3PO_4 = H^+ + H_2PO_4^-$	phosphoric acid	dihydrogen phosphate ion	2.15	11.85
$HNO_2 = H^+ + NO_2^-$	nitrous acid	nitrite ion	3.15	10.85
$CH_3COOH = H^+ + CH_3COO^-$	acetic acid	acetate ion	4.70	9.30
$H_2CO_3^* = H^+ + HCO_3^-$	carbonic acid	bicarbonate ion	6.35	7.65
$H_2S = H^+ + HS^-$	hydrogen sulfide	bisulfide ion	7.02	6.98
$HSO_3^- = H^+ + SO_3^{2-}$	bisulfite ion	sufite ion	7.18	6.82
$H_2PO_4^- = H^+ + HPO_4^{2-}$	dihydrogen phosphate ion	hydrogen phosphate ion	7.20	6.80
$HOCl = H^+ + OCl^-$	hypochlorous acid	hypochlorite ion	7.5	6.5
$HCN = H^+ + CN^-$	hydrogen cyanide	cyanide ion	9.2	4.8
$H_3BO_4 + H_2O = B(OH)_4^- + H^+$	boric acid	borate ion	9.24	4.76
$NH_4^+ = H^+ + NH_3$	ammonium ion	ammonia	9.24	4.76
$Si(OH)_4 = H^+ + SiO(OH)_3^-$	ortho-silicic acid	silicate ion	9.86	4.14
$HCO_3^- = H^+ + CO_3^{2-}$	bicarbonate ion	carbonate ion	10.33	3.67
$HPO_4^{2-} = H^+ + PO_4^{3-}$	hydrogen phosphate ion	phosphate ion	12.35	1.65
$HS^- = H^+ + S^{2-}$	bisulfide ion	sulfide ion	14 to 15	0 to –1

TABLE 3.2 Acid dissociation constant $K_w(m^2)$ for water as a function of temperature at 1 atm pressure (infinite dilution scale). From Harned and Owen (1958).

°C	K_w	log K_w
0	0.12×10^{-14}	−14.93
5	0.18×10^{-14}	−14.73
10	0.29×10^{-14}	−14.53
15	0.45×10^{-14}	−14.35
20	0.68×10^{-14}	−14.17
25	1.01×10^{-14}	−14.00
30	1.47×10^{-14}	−13.83
50	5.48×10^{-14}	−13.26

$\log K_w = -4470.99/T + 6.0875 - 0.01706T$
where T = temperature in degrees K.

The use of $[H^+]$ and other such concentration quantities directly in *mass balance equations* is also justified. For example, for H^+, we see that each of the species H^+, H_3O^+, $H_5O_2^+$, $H_7O_3^+$, etc. contains one H^+ that was liberated from some acid. These various species may thus be lumped together simply as $[H^+]$ in a mass balance equation that is designed to provide an accounting of liberated H^+. Thus, in all remaining aspects of this text, what has been referred to as "$[H^+]$" in this chapter, may and will henceforth be referred to simply as $[H^+]$; what might be referred to as "$[A^-]$" will simply be referred to as $[A^-]$; etc.

Table 3.1 summarizes some acidity constant information for a variety of acids of interest in natural waters. Table 3.2 gives the acid dissociation constant for water as a function of temperature. Problems illustrating the principles of this chapter are presented by Pankow (1992).

3.4 REFERENCES

HARNED, H. S., and OWEN, B. B. 1958. *The Physical Chemistry of Electrolyte Solutions.* New York: Reinhold.

PANKOW, J. F. 1992. *Aquatic Chemistry Problems.* Portland: Titan Press-OR, P.O. Box 91399, Portland, Oregon, 97291-1399.

4

The Proton Balance, Electroneutrality, and Mass Balance Equations

4.1 THE PROTON BALANCE EQUATION (PBE)

In many aqueous systems, it is very useful to keep track of: 1) how many protons and hydroxide ions are present; and 2) why they are present. Equations which do this can be written easily for solutions of acids (both mono- and polyprotic) and bases (both mono- and polybasic). Since these equations provide an accounting of both the protons and the hydroxide ions, in each particular case it might be most logical to call the pertinent accounting equation the "proton and hydroxide balance equation." However, this is always abbreviated to **proton balance equation** (PBE). We will now consider how the PBE is arrived at for a number of different types of solutions of acids and bases.

4.1.1 PBE for Pure Water

Water dissociates according to

$$H_2O = H^+ + OH^-. \tag{4.1}$$

For every molecule of water that dissociates, one H^+ is formed, and one OH^- is formed. We thus have

$$[H^+] = [OH^-] \qquad \text{PBE for pure water.} \tag{4.2}$$

4.1.2 PBE for a Solution of HA

There are two sources of protons in a solution of HA: the dissociation of water itself according to reaction (4.1), and the dissociation of HA according to

$$HA = H^+ + A^-. \tag{4.3}$$

As in a solution of pure water, for every molecule of water that dissociates, one H^+ is formed, and one OH^- is formed. Thus, in a solution of the acid HA, $[OH^-]$ provides a measure of proton production from the dissociation of water. For every molecule of HA that dissociates, one H^+ is formed, and one A^- is also formed. Thus, $[A^-]$ provides a measure of the production of protons from the dissociation of HA. Therefore, in a solution of HA, the total $[H^+]$ will be given by the sum of $[A^-]$ and $[OH^-]$. We conclude that

$$[H^+] = [A^-] + [OH^-] \qquad \text{PBE for solution of HA.} \tag{4.4}$$

4.1.3 PBE for a Solution of NaA

When a solution of NaA is prepared, the assumption will be that all of the NaA will first dissolve to form Na^+ and A^- ions; we will assume that there are essentially no dissolved NaA molecules in solution. There is only one source of protons in a solution of NaA, and that is from the dissociation of water. However, the PBE for this system is *not* simply Eq. (4.2). Indeed, OH^- ions are produced not only by the dissociation of water, but also by the reaction

$$A^- + H_2O = HA + OH^-. \tag{4.5}$$

Reaction (4.5) is often referred to as the "hydrolysis"[1] of A^-.

Protons, in addition to being produced by the dissociation of water, will also be formed to a very small extent by the hydrolysis of a few of the Na^+ ions by the reaction

$$Na^+ + H_2O = NaOH^o + H^+ \tag{4.6}$$

where $NaOH^o$ is a neutral, dissolved species.

The amount of OH^- that is formed by reaction (4.5) in a solution of NaA is perfectly accounted for by the amount of HA that is generated. Thus we have to *add* [HA] to $[H^+]$ if we are to provide the proper budget for the production of hydroxide ions. The amount of H^+ that is formed by reaction (4.6) is perfectly accounted for by the amount of $NaOH^o$ that is generated. Therefore, strictly speaking, for the PBE of a solution of NaA, we have

$$[H^+] + [HA] = [OH^-] + [NaOH^o]. \tag{4.7}$$

However, the amount of $NaOH^o$ that forms is so very small that we always disregard it in the PBE, and so we obtain

$$[H^+] + [HA] = [OH^-] \qquad \text{PBE for solution of NaA.} \tag{4.8}$$

The species $NaOH^o$ will also be neglected in all of the solutions containing Na^+ that are discussed in the following subsections.

[1]The term "hydrolysis" is actually a misnomer here. Strictly speaking, hydrolosis means to decompose (i.e., to undergo "lysis") through the addition of water; reaction (4.5) does not involve the decomposition of A^-. Nevertheless, the term hydrolysis is so accepted for reaction (4.5) and other similar reactions that we will use it in this text.

4.1.4 PBE for a Solution of H_2B

It should be clear by now that $[H^+]$ and $[OH^-]$ both always appear in the PBE, regardless of the type of solution under consideration. Thus, in the case of a solution of the *diprotic* acid H_2B, we need to ascertain what *other* sources of H^+ and OH^- exist besides the dissociation of water. We first note that since H_2B cannot undergo hydrolysis, the dissociation of water is the only source of OH^- in this system. However, H_2B can *dissociate* to produce HB^- and B^{2-} according to both

$$H_2B = H^+ + HB^- \tag{4.9}$$

and

$$HB^- = B^{2-} + H^+. \tag{4.10}$$

The genesis of both HB^- and B^{2-} leads to the production of protons. For each HB^- present, there is one H^+ present. For each B^{2-} present, two protons are present. Thus, we have

$$[H^+] = [OH^-] + [HB^-] + 2[B^{2-}] \qquad \text{PBE for solution of } H_2B. \tag{4.11}$$

One might think that the factor of 2 is not needed in front of $[B^{2-}]$ since the first proton liberated in its formation is counted by the presence of $[HB^-]$ in Eq. (4.11). *However, it must be noted that each* HB^- *that does go on to form* B^{2-} *is no longer in the solution.* The value of $[HB^-]$ does not therefore include a measure of how many "first protons" were released to the solution during the overall process of the formation of B^{2-}. It can only provide a measure of how many HB^- *are* in the solution and therefore how many H^+ were formed along with *those specific* HB^- species. The value of $[B^{2-}]$ on the other hand provides a direct measure of how many H_2B dissociated to form HB^- (1 proton formed) *then also dissociated again* (1 more proton formed) to form one B^{2-} plus a total of *two* protons.

One might also ask why H_2B does not appear on the left-hand side of Eq. (4.11) in view of the fact that some of the HB^- that forms could react with water to form H_2B *and* OH^- according to

$$HB^- + H_2O = H_2B + OH^-. \tag{4.12}$$

The reason is that for this to happen, some H_2B must first dissociate according to reaction (4.9) and so the net reaction (reaction (4.9) plus reaction (4.12)) is just

$$H_2O = H^+ + OH^- \tag{4.13}$$

and we know that reaction (4.13) is already accounted for in the PBE for this solution.

4.1.5 PBE for a Solution of NaHB

In a manner that is analogous to that discussed in Section 4.1.3, when a solution of NaHB is prepared, the assumption will be that all of the NaHB will first dissolve to form Na^+ and HB^- ions; we will assume there are essentially no dissolved NaHB molecules in solution.

The HB^- ions that are initially produced upon the dissolution of the salt NaHB can dissociate according to reaction (4.10). For each HB^- that dissociates, one H^+

and one B^{2-} are formed. In addition to dissociating, HB^- can hydrolyze according to reaction (4.12). For each HB^- ion that is hydrolyzed, one H_2B and one OH^- are formed. Thus, $[B^{2-}]$ serves as a tracer for the *portion* of the H^+ that is present due to HB^- dissociation, and $[H_2B]$ serves as a tracer for the *portion* of OH^- that is present due to HB^- hydrolysis. The word "portion" is used in both of the two preceding sentences since both H^+ and OH^- can also come from the dissociation of water. Therefore, we have

$$[H^+] + [H_2B] = [OH^-] + [B^{2-}] \qquad \text{PBE for solution of NaHB.} \qquad (4.14)$$

4.1.6 PBE for a Solution of Na_2B

As with the solutions of NaA and NaHB that have already been discussed, when a solution of Na_2B is prepared, it wil be assumed that all of the Na_2B will first dissolve to form Na^+ and B^{2-} ions; we will assume that there are essentially no dissolved Na_2B molecules in solution.

A solution of Na_2B is a negative analog of a solution of H_2B. In this case, there are no sources of H^+ other than the dissociation of water. Conversely, OH^- is formed not only due to the dissociation of water, but also during the hydrolysis of B^{2-} according to

$$B^{2-} + H_2O = HB^- + OH^- \qquad (4.15)$$

and during the hydrolysis of HB^- according to reaction (4.12). That is, OH^- ions are formed during the formation of both HB^- and H_2B. For each HB^- present in the solution, there is one OH^-. For each B^{2-} present, there are two OH^-. Consequently, we have

$$[H^+] + [HB^-] + 2[H_2B] = [OH^-] \qquad \text{PBE for solution of Na}_2\text{B.} \qquad (4.16)$$

4.1.7 Generalized Task Analysis for Writing the PBE for a Solution of Interest

The PBE-type equations that have been derived in Sections 4.1.1–4.1.6 represent a sampling of the types of PBEs that can apply in aqueous solutions. Although we cannot not derive every possible PBE here, we can summarize the steps used in writing them. Those steps are given in Table 4.1.

TABLE 4.1 Generalized task analysis for writing proton balance equations

Step 1.	Set up the equation $[H^+] = [OH^-]$; water is always present and so these two species will be present in *all* PBE equations.
Step 2.	Write the concentrations of all of the species whose genesis caused the production of H^+ on the $[OH^-]$ side, with the appropriate multiplying factor.
Step 3.	Write the concentrations of all of the species whose genesis caused the production of OH^- on the $[H^+]$ side, with the appropriate multiplying factor.

Table 4.2 summarizes the PBE equations for the various solutions discussed in Sections 4.1.1–4.1.6. It may be noted that the concentration of the species S used to prepare a given solution *never* appears in the PBE for that solution since its genesis was from the stock solution or reagent bottle; none of the H^+ or OH^- ions present in the solution owe their presence to the presence of S. Illustrations of this fact are summarized in Table 4.3. As a final note, as will be discussed in subsequent chapters, it may be pointed out that PBE-type equations are very useful when: 1) discussing the chemistry that is present at the equivalence points in titrations; 2) setting up the equations that can be used to solve exactly for the chemical speciation that exists at equilibrium in solutions containing acids and bases; and 3) providing a framework for both: a) the selection of the assumptions that may be used to appropriately simplify the mathematical solution for the chemical speciation that exists at equilibrium in those types of solutions; and b) the application of those assumptions using log concentration vs. pH graphs.

4.1.8 Writing a PBE for a Mixture

Consider a solution that is C_1 *F* in HA and C_2 *F* in NaA. Because some of the HA will dissociate to form H^+ and A^-, and because some of the A^- from the NaA might hydrolyze to form HA and OH^-, some might wonder whether a PBE can be written for

TABLE 4.2 Summary of PBEs, ENEs, and MBEs for examples of aqueous solutions

Solution type	PBE	ENE	MBE(s)
pure water	$[H^+] = [OH^-]$	$[H^+] = [OH^-]$	$[H^+] = [OH^-]$
HA	$[H^+] = [A^-] + [OH^-]$	$[H^+] = [A^-] + [OH^-]$	$A_T = [HA] + [A^-]$
NaA	$[H^+] + [HA] = [OH^-]$	$[Na^+] + [H^+] = [OH^-] + [A^-]$	$A_T = [HA] + [A^-]$ $[Na^+] = A_T$
H_2B	$[H^+] = [OH^-] + [HB^-] + 2[B^{2-}]$	$[H^+] = [OH^-] + [HB^-] + 2[B^{2-}]$	$B_T = [H_2B] + [HB^-] + [B^{2-}]$
NaHB	$[H^+] + [H_2B] = [OH^-] + [B^{2-}]$	$[Na^+] + [H^+] = [OH^-] + [HB^-] + 2[B^{2-}]$	$B_T = [H_2B] + [HB^-] + [B^{2-}]$ $[Na^+] = B_T$
Na_2B	$[H^+] + [HA^-] + 2[H_2A] = [OH^-]$	$[Na^+] + [H^+] = [OH^-] + [HB^-] + 2[B^{2-}]$	$B_T = [H_2B] + [HB^-] + [B^{2-}]$ $[Na^+] = 2B_T$

TABLE 4.3 Examples of how the reagent used to form a solution will not appear in the PBE for that solution

H_2O	does not appear in the PBE for water
HA	does not appear in the PBE for HA
A^-	does not appear in the PBE for NaA
H_2B	does not appear in the PBE for H_2B
HB^-	does not appear in the PBE for NaHB
B^{2-}	does not appear in the PBE for Na_2B

this solution as

$$[H^+] + [HA] = [OH^-] + [A^-] \qquad ??? \tag{4.17}$$

However, for the solution under consideration, Eq. (4.17) is not the PBE, and no PBE can in fact be written. On the LHS of Eq. (4.17), while some of the HA present in the mixture might have come from the hydrolysis of A^-, some will likely also be from the initial HA solution. Thus, the total HA does not provide a tracer for OH^- generated by A^- hydrolysis. Analogous observations can be made about the $[A^-]$ term on the RHS of Eq. (4.17).

The problem with the above situation is that we cannot differentiate between the HA that was formed from hydrolysis of the C_2 F solution of NaA, and the HA that was initially added as the C_1 F solution of HA. Similarly, we cannot differentiate between the A^- that was formed from dissociation of the C_1 F solution of HA, and the A^- that was initially added as the C_2 F solution of NaA.

If we are dealing with a solution that is C_1 F in HA and C_2 F in NaB, however, we *can* keep track of the chemistry, and write a PBE for such a mixture. In this case, then, the PBE is

$$[H^+] + [HB] = [OH^-] + [A^-]. \tag{4.18}$$

4.2 THE ELECTRONEUTRALITY EQUATION (ENE)

Under the conditions in which they are encountered in natural systems, bulk aqueous solutions do not carry any charge; they are electrically neutral. The detailed statement of this fact is known as the **electroneutrality equation** (ENE). For example, if a salt like NaCl dissolves, the amount of Na^+ that goes into the solution will be balanced by an exactly equal amount of Cl^-. Also, as discussed in the previous section, if H_2O dissociates to form H^+ and OH^-, then the amount of H^+ thereby formed will exactly balance the OH^- formed. Similarly, if HA dissociates to form H^+ and A^-, the amount of H^+ formed will exactly balance the amount of A^- formed. The overall result is that for *all* solutions,

$$\sum_i z_i c_i = 0 \qquad \text{Generalized Electroneutrality Equation (ENE)} \tag{4.19}$$

where z_i is the charge on ion i, and $c_i(M)$ is the concentration of ion i. The z_i have sign. Table 4.2 summarizes the ENE equations for the various aqueous solutions discussed in Sections 4.1.1–4.1.6. One can also write an ENE like Eq. (4.19) that is based on the molality concentration scale.

4.3 THE MASS BALANCE EQUATION (MBE)

Mass balance equations (MBEs) are statements of conservation of mass where that conservation is taking place over unit volume, or over a given kg of solvent. For example, when we prepare a solution of HA, we know that some of the HA placed in the solution will dissociate to form H^+ and A^-. We also know that both the undissociated form HA and the dissociated form A^- still have the chemical moiety A in them. Thus, if *total* A present in the solution is denoted A_T (M or m), an MBE that can be written for this solution states that for *all* conditions (i.e., whether at equilibrium or not), [HA] and $[A^-]$ must still add up to A_T. Thus, for this solution we have

$$A_T = [HA] + [A^-]. \tag{4.20}$$

Sometimes, more than one MBE may be written for a given solution. For example, for a solution of NaA, not only will Eq. (4.20) be true, but we will also have

$$[Na^+] = A_T \tag{4.21}$$

because for every unit of A that is present in solution, we will also have one unit of Na^+. The pertinent MBEs for the solutions discussed in Sections 4.1.1–4.1.6 are presented in Table 4.2. When two MBEs apply for a given solution, then both are given.

4.4 THE PBE, ENE, AND MBE(s): ARE THEY ALL LINEARLY INDEPENDENT?

As will be discussed in greater detail in later chapters, when attempting to determine the equilibrium composition of a system involving n species, n independent equations will be needed to solve the problem. Since they involve chemical variables of great interest, that is, the concentrations of the species in solution, the PBE, ENE, and MBE(s) will provide useful equations toward this end. Unfortunately, they will never all be independent. In general, the PBE and the ENE are linkable through an MBE. This means that if a given system has one pertinent MBE, then the following are possible pairs of linearly independent equations: 1) PBE and MBE; 2) ENE and MBE; and possibly but not always, 3) ENE and PBE. The reason for the qualification on the third possibility is due to the fact that sometimes, as in a solution of HA, the PBE and the ENE are identical.

Let us consider two examples:

Example 4.1. Solution of HA.

The equations are:

$$\text{PBE}: \quad [H^+] = [OH^-] + [A^-]$$

$$\text{MBE}: \quad A_T = [HA] + [A^-]$$

$$\text{ENE}: \quad [H^+] = [OH^-] + [A^-].$$

For a C F solution of NaA, $A_T = C$.

All three equations are not linearly independent. While the PBE and the MBE are linearly independent of one another, and can be used in solving for the pH of a solution of HA, they are not independent of the ENE. Alternatively, the ENE and the MBE are not independent of the PBE. As noted previously, in this particular case, the fact that all three equations are not independent of one another is due to the fact that the ENE and the PBE are identical.

Example 4.2. Solution of NaHB.

The equations are:

$$\text{PBE}: \quad [H^+] + [H_2B] = [OH^-] + [B^{2-}]$$

$$\text{MBEs}: \quad B_T = [H_2B] + [HB^-] + [B^{2-}]$$

$$[Na^+] = B_T$$

$$\text{ENE}: \quad [H^+] + [Na^+] = [OH^-] + [HB^-] + 2[B^{2-}]$$

Any three equations provide a linearly independent set. However, *all four* are not independent of one another. Indeed, 1) the PBE and the two MBEs can be combined to obtain the ENE; or equivalently, 2) the ENE and the two MBEs can be combined to obtain the PBE. In the first case we simply add the PBE as written above to the two MBEs, also as written above. In the second case, subtracting the two MBEs as written from the ENE as written will yield the PBE. The qualifier "as written" has been included in the previous two sentences because when you add or subtract an equation like $x = y$, the outcome will depend on whether the equation is written $x = y$ or $y = x$.

Other examples and problems illustrating the principles of this chapter are presented by Pankow (1992).

4.5 REFERENCES

PANKOW, J. F. 1992. *Aquatic Chemistry Problems.* Portland: Titan Press-OR, P.O. Box 91399, Portland, Oregon, 97291-1399.

5

Introduction to Quantitative Equilibrium Calculations

5.1 INTRODUCTION

As discussed in Chapter 2, the final position that a given set of chemical equilibria will take in an aqueous system will be determined by the: 1) T- and P-dependent value(s) of the pertinent equilibrium constant(s); 2) mass balance equation(s) (MBE(s)) governing the system; and the 3) manner in which the various species' activity coefficients depend on chemical composition. The specific nature of the MBE(s) will be determined by the initial chemical composition of the system.

The values of the pertinent activity coefficients will be determined by the *final* equilibrium composition of the system, and so will never be *exactly* knowable, a priori, for use in the calculations. For very dilute solutions, or when the reaction medium contains relatively large amount(s) of background dissolved salt(s), then the activity coefficients in the final equilibrium system may be predicted reasonably accurately. For very dilute solutions, the concentrations of all species will be very low, and the final activity coefficients may be predicted to be very close to unity. When background dissolved salts (e.g., NaCl, $MgSO_4$, etc.) are present in the solution, *and* when the concentrations of the species of interest (e.g., H^+) for the equilibrium will be present at concentrations that are low by comparison to the concentrations of the salts, then: 1) the ionic strength will not be affected by the equilibrating species; and 2) the values of the final activity coefficients may be predicted to be controlled by the ionic strength as it is determined by the background salts.

The cases in which the final activity coefficients will *not* be accurately predictable will involve those situations in which: 1) background salts are not present and the species involved in the equilibria lead to an equilibrium ionic strength that is $\sim 5 \times 10^{-3} M$ or higher; or 2) background salts are present and the species involved in the equilibria contribute significantly with the salts to make an equilibrium ionic strength that is $\sim 5 \times 10^{-3} M$ or higher. As will be discussed in the following sections, in such cases, a multistep solution process involving ever-improving estimates of the ionic strength is required.

This chapter will discuss how to compute the speciation of solutions of the acid HA and solutions of the base NaA. We will also consider solutions of H_2B, NaHB, and Na_2B. The formality (FW/L) of all of these different types of solutions will be represented as being equal to $C\,F$. We will discuss how to solve for the pH and speciation of solutions of these types without making any simplifying assumptions, as well as with the benefit of a variety of different simplifying assumptions. With the ready access to powerful personal computers and hand calculators that now exists, the need to educate the reader on the matter of simplifying assumptions has declined in recent years. Therefore, some readers may wish to skip some of the subsections dealing with simplifying assumptions, specifically Sections 5.2.2.2 and 5.2.2.4 and 5.3.2.2 and 5.3.2.4. Nevertheless, the ability to make good simplifying assumptions quickly can be very handy. Moreover, understanding how simplifying assumptions can be made greatly improves chemical intuition.

This chapter will consider how to carry out calculations without activity corrections, as well as calculations involving full activity corrections. For the latter, we will examine the situation in which accurate estimates of the various activity coefficients cannot be made a priori. Problems illustrating the principles of this chapter are presented in Pankow (1992).

5.2 GENERAL EQUATIONS FOR SOLVING FOR THE SPECIATION OF A SOLUTION OF THE ACID HA—NO ACTIVITY CORRECTIONS

In addition to water itself, there are a total of four species in a solution of HA. They are:

$$H^+ \quad A^- \quad OH^- \quad HA.$$

The two *chemical equilibria* governing these species are

$$H_2O = H^+ + OH^- \qquad K_w = \{H^+\}\{OH^-\} \tag{5.1}$$

$$HA = H^+ + A^- \qquad K_a = \{H^+\}\{A^-\}/\{HA\}. \tag{5.2}$$

The facts that $\{H_2O\}$ is expressed on the mole fraction scale and is essentially equal to 1.0 in most natural aqueous systems are implicit in the expression for K_w for equilibrium (5.1). The use of the subscript a for the acidity constant in equilibrium (5.2) will be limited mainly to this chapter. We are adopting this usage here to help distiguish K_a from the closely-related basicity constant K_b, which will be defined below.

5.2.1 Solving for the Speciation of a Solution of the Acid HA—No Activity Corrections, No Simplifying Assumptions

If the solution is dilute, we will not need to make activity corrections. In such a case, there will be four unknowns, $[H^+]$, $[A^-]$, $[OH^-]$, and [HA]. With four unknowns, a total of four equations is required to solve the problem at hand. They are

$$K_w = [H^+][OH] \quad \text{first chemical equilbrium equation} \tag{5.3}$$

$$K_a = [H^+][A^-]/[HA] \quad \text{second chemical equilibrium equation} \tag{5.4}$$

$$[HA] + [A^-] = C \quad \text{mass balance equation (MBE) on total A} \tag{5.5}$$

$$[H^+] = [A^-] + [OH^-] \quad \text{proton balance equation (PBE) also, simultaneously, in this case, the electroneutrality equation (ENE).} \tag{5.6}$$

The application of the MBE here takes advantage of the fact that while some of the HA originally placed in the solution will always dissociate to form H^+ and A^-, both the undissociated form HA and the dissociated form A^- still have the chemical moiety A in them. Thus, since the *total* A present in the solution is C *M*, the MBE states that at equilibrium, [HA] and $[A^-]$ must still sum to C *M*. If HA is a very strong acid (large K_a), then most of the A will tend to be present as A^-. If HA is a very weak acid (very small K_a), then most of the A will tend to be present as HA. If HA is an acid of intermediate strength (intermediate K_a), then both some A^- and some HA will tend to be present. As we will see below, in addition to K_a, the concentration C also plays a very important role in determining the pH and therefore the A^-/HA distribution.

To begin our mathematical solution, from the group Eqs. (5.3)–(5.6) we must select one starting equation for substitution of the other three equations into that one. In the starting equation, we would like to see: 1) concentration terms for as many of the different species as possible; and 2) as many first order terms as possible, that is, a minimum number of cross-products such as $[H^+][OH^-]$. PBEs and ENEs are two classes of equations that always tend to exhibit these characteristics. As noted previously, Eq. (5.6) simultaneously represents the PBE and the ENE for the case at hand. The ENE, a statement of the fact that there will be *no net charge* in a bulk aqueous solution, will always contain all of the charged species in a first order form according to

$$\sum_i z_i c_i = 0 \quad \text{Generalized Electroneutrality Equation (ENE)} \tag{5.7}$$

where the z_i are the charges on the various ions. The z_i have sign. For the species H^+, A^-, and OH^-, the values of z_i are +1, −1, and −1, respectively.

We begin with Eq. (5.6) by making the simple substitution

$$[A^-] = \frac{[A^-]}{[HA] + [A^-]}([HA] + [A^-]). \tag{5.8}$$

Equation (5.8) makes use of the fact that $[A^-]/([HA] + [A^-])$ equals the fraction of the total A in solution which is present as A^- (in Chapter 6, that fraction is called α_1). Equation (5.8) gives $[A^-]$ as the product of that fraction with the total A concentration. We now substitute for $[HA]+[A^-]$ based on the MBE that $[HA]+[A^-] = C$, and arrive at

$$[H^+] = \frac{[A^-]}{[HA] + [A^-]}C + [OH^-]. \tag{5.9}$$

Substituting using the dissociation constant for water (K_w), we obtain

$$[H^+] = \frac{1}{[HA]/[A^-]+1}C + K_w/[H^+]. \tag{5.10}$$

We now note that if activity corrections can be neglected, then all $\gamma_i = 1$ and so $[HA]/[A^-] = [H^+]/K_a$. Therefore, from Eq. (5.10) we have

$$[H^+] = \frac{K_a}{[H^+]+K_a}C + K_w/[H^+] \tag{5.11}$$

or

$$[H^+] - \frac{K_a}{[H^+]+K_a}C - K_w/[H^+] = 0. \tag{5.12}$$

By using all four equations (i.e., Eqs. (5.3)–(5.6)), we have converted the PBE into one equation in one unknown, namely $[H^+]$. The student may wish to review these steps to see where each of the four equations came into derivation of Eq. (5.12).

It is useful to note that when activity corrections are required, *and* when relatively high levels of background salts allow very good prediction of the values of the activity coefficients at equilibrium, then constant ionic medium (${}^cK$) constants can be computed. Equation (5.12) as well as all of the other equations developed in this section can then be used with the appropriate ${}^cK$ constants substituted for the infinite dilution constants. In such cases, Eq. (5.6) (the foundation of our mathematical solution) will still represent the PBE provided that the salts are neither acid nor basic (e.g., NaCl, but not $NaHSO_4$).

We now discuss how to solve Eq. (5.12) for $[H^+]$. Let

$$f([H^+]) = [H^+] - \frac{K_a}{[H^+]+K_a}C - K_w/[H^+]. \tag{5.13}$$

It is apparent that $f([H^+])$ is a function of $[H^+]$, as well as K_a, C, and K_w. For any given problem, the last three parameters will be known a priori. The *specific* positive value of $[H^+]$ that is the solution to Eq. (5.12) is the value for which $f([H^+]) = 0$. A value of $[H^+]$ for which $f([H^+]) = 0$ is called a "root" of Eq. (5.12).

There are three specific approaches that may be used to solve problems of this type, that is, to find roots of Eq. (5.12):

Option 1. Solve $f([H^+]) = 0$ directly by trial and error using: a) hand calculations (very unattractive); or b) a programmable hand calculator; or c) a personal or larger computer;

Option 2. Convert $f([H^+])$ into a polynomial in $[H^+]$ and then solve by using: a) an analytical solution; b) trial and error using a hand calculator; or c) a computer-based root solver for polynomials (i.e., a computer-based trial and error method for polynomials);

Option 3. Make assumptions to simplify the problem, then solve.

Option 1. Option 1 can be carried out simply by setting up a simple hand calculator or a computer to evaluate $f([H^+])$ with known values of the K_a, C, and K_w for a series of *guesses* for $[H^+]$. The quality of a given guess is defined by how close it makes $f([H^+])$ to zero. With this method, new and improved guesses for $[H^+]$ continue to be tried until $f([H^+])$ equals zero exactly, or *more likely*, just becomes so close to zero that only infinitesimal changes in the guessed value of $[H^+]$ bring $f([H^+])$ any closer to zero.

The operational aspects of a trial and error solution involve making two initial guesses, $[H^+]_1$ and $[H^+]_2$. Let us say that $[H^+]_1 > [H^+]_2$. Whichever of the two values of $[H^+]$ gives the smaller absolute value of $f([H^+])$ is closer to the sought-for, equilibrium-value of $[H^+]$. As an example, let us assume that $f([H^+])$ is positive for both initial guesses. If $f([H^+]_1) > f([H^+]_2)$ as shown in Figure 5.1, then $[H^+]_2$ is closer to the equilibrium value of $[H^+]$ than is $[H^+]_1$. An even better estimate of $[H^+]$ may then be obtained by making a third guess $[H^+]_3$ that is smaller than $[H^+]_2$ so that an even smaller value of $f([H^+])$ can be found. If $f([H^+]_3)$ should turn out to be negative, then we would know that $[H^+]_3$ was made too small, and a fourth guess that satisfies the criterion $[H^+]_3 < [H^+]_4 < [H^+]_2$ must be tried. The process continues until one "converges" on the solution, that is, on the $[H^+]$ value at which $f([H^+])$ essentially passes through zero.

While the term "trial and error method" does not sound very elegant, this method is usually *very effective* in finding a solution to a problem such as the one discussed here. A more sophisticated *sounding* name for "trial and error method" is "numerical method." For problems of this type, the science of numerical methods involves the development of computational schemes that pick each next trial and error guess in manner such that a computer program can converge on the solution in the fastest possible manner. Such versions of the trial and error method actually generate the series of guesses. Guesses continue to be made until: 1) the nth generated guess is essentially identical to the $(n-1)$th guess; and 2) $f((n-1)\text{th guess}) \simeq f(n\text{th guess}) = 0$. Sir Isaac Newton (1642–1727) conducted some truly pioneering work in this area. As a result, one of the best known methods for searching for roots by trial and error (i.e., numerically) is known as "Newton's Method." Butler (1964, p. 80) discusses the application of Newton's Method to problems of the type discussed in this chapter. The interested reader may also wish to refer to the related discussion in Harris (1987, pp. 183–184, 208–209). Some hand calculators are now so sophisticated that they will automatically carry out a trial and error solution for almost any mathematical expression with just one initial guess; the calculator makes the second guess itself, and goes from there. Such calculators are the Hewlett-Packard 28S and the 48S Advanced Scientific Calculators. At the time of this writing, the HP-28S or the HP-48S should be considered by the serious student of the material in this text.

One possible set of values for the parameters of interest here are $K_a = 10^{-5}m$, $C = 10^{-4}M$, and $K_w = 10^{-14}m^2$. Note that K_a has units of $m^2/m = m$. As discussed in Chapter 2, for this type of dilute solution, molality and molarity are essentially interchangeable. Thus, *equilibrium constants on the molality scale may be used in Eqs. (5.12) and (5.13) even when* $[H^+]$ *and* C *are expressed in molarity*. Let us now define

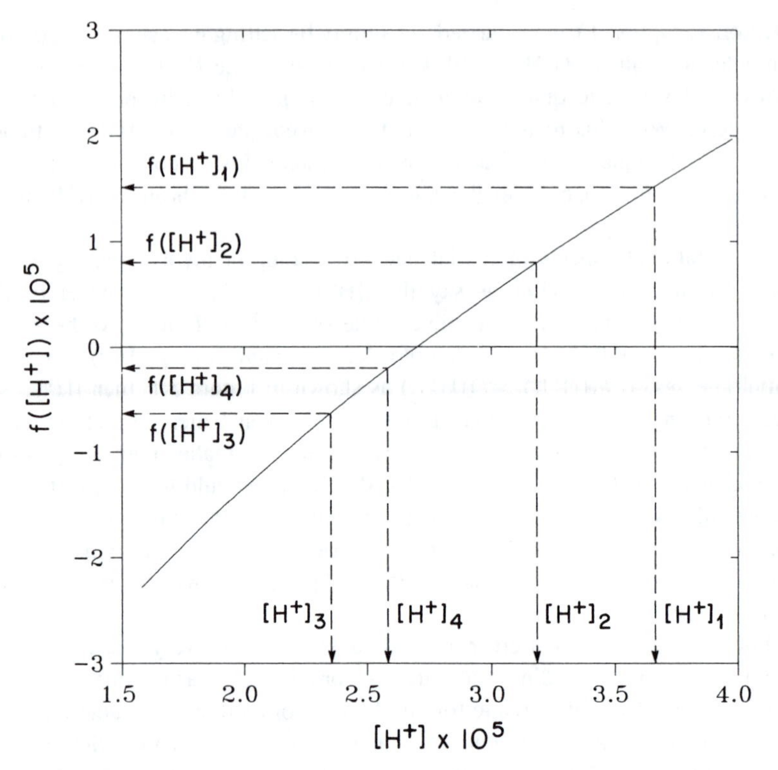

Figure 5.1 Plot of $f([H^+])$ vs. $[H^+]$ for the case when $K_a = 10^{-5}$, $C = 10^{-4}M$, and $K_w = 10^{-14}$. $[H^+]_1$ through $[H^+]_4$ represent four sequential guesses for the root of Eq. (5.12).

pH $= -\log\{H^+\}$. When activity corrections may be neglected, then pH $= -\log[H^+]$. Values of $f([H^+])$ for this case for a range of pH are given in Table 5.1. The function $f([H^+])$ is plotted vs. $[H^+]$ in Figure 5.1, and vs. pH in Figures 5.2 and 5.3. The decision to use pH on the x-axis in Figures 5.2 and 5.3 rather than $[H^+]$ as in Figure 5.1 was based on the need to clearly display the behavior of $f([H^+])$ over a wide range in pH.

It is difficult to distinguish where $f([H^+])$ passes through zero in Figure 5.2. Therefore, Figure 5.3 shows the region where $f([H^+])$ passes through zero on a much finer scale; that plot and Table 5.1 (by interpolation) illustrate that the solution for this particular problem is $[H^+] = 10^{-4.57}$ (pH = 4.57). One can see how the initial selection of any two pH values in Table 5.1 and evaluating $f([H^+])$ for those two values could initiate a trial and error solution that would converge on the desired value of $[H^+]$. Table 5.1 also illustrates that when one gets close to the desired $[H^+]$, then increasingly smaller changes in the guessed value of pH must be made to converge on the root value of $[H^+]$. In Table 5.1, pH increments of 1.0 are used far from the root value of $[H^+]$, then smaller increments near the root. Table 5.1 also illustrates that an initial guess on *either* side of the root value may be used in a trial and error solution.

TABLE 5.1 Values of $f([H^+])$ for pH between 1.0 and 13.0 for $K_a = 10^{-5}m$, $C = 10^{-4}M$, and $K_w = 10^{-14}m^2$.

pH	$f([H^+])$	pH	$f([H^+])$
1.0	0.0999999	4.9	−0.0000317
2.0	0.0099999	5.0	−0.0000400
3.0	0.0009990	5.5	−0.0000728
3.5	0.0003131	6.0	−0.0000899
4.0	0.0000909	6.5	−0.0000967
4.1	0.0000682	7.0	−0.0000990
4.2	0.0000494	7.5	−0.0001000
4.3	0.0000334	8.0	−0.0001009
4.4	0.0000197	9.0	−0.0001100
4.5	0.0000076	10.0	−0.0002000
4.6	−0.0000034	11.0	−0.0011000
4.7	−0.0000134	12.0	−0.0101000
4.8	−0.0000228	13.0	−0.1001000

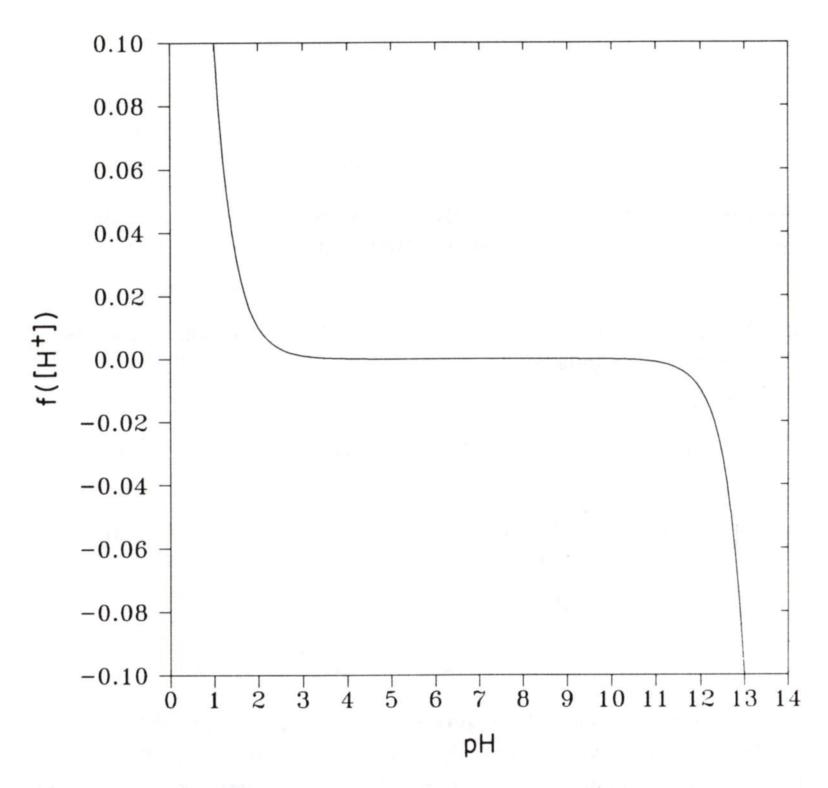

Figure 5.2 $f([H^+])$ vs. pH for $K_a = 10^{-5}$, $C = 10^{-4}M$, and $K_w = 10^{-14}$.

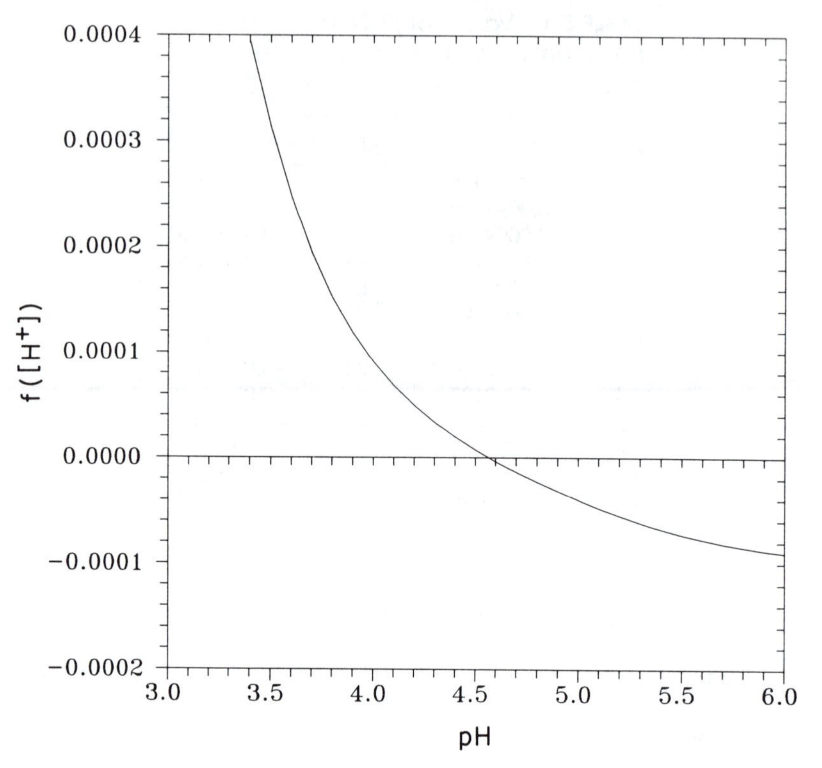

Figure 5.3 Fine-scale view of $f([H^+])$ vs. pH for $K_a = 10^{-5}$, $C = 10^{-4}M$, and $K_w = 10^{-14}$ in the region where $f([H^+]) = 0$.

With the equilibrium (root) value of $[H^+]$ in hand, the concentrations of the remaining species can be calculated for this problem. Since $K_a = 10^{-5}$, $C = 10^{-4}M$, and $K_w = 10^{-14}$, we have

$$[OH^-] = K_w/[H^+] = 10^{-14}/(2.69 \times 10^{-5}) = 3.72 \times 10^{-10}M \tag{5.14}$$

$$[A^-] = \frac{[A^-]}{[HA] + [A^-]}C = \frac{K_a}{[H^+] + K_a}C = \frac{10^{-5}}{10^{-4.57} + 10^{-5}}10^{-4}$$
$$= 2.71 \times 10^{-5}M \tag{5.15}$$

$$[HA] = C - [A^-] = 10^{-4} - 2.71 \times 10^{-5} = 7.29 \times 10^{-5}M. \tag{5.16}$$

Figure 5.4 illustrates how the equilbrium pH of a solution of HA changes with K_a and C at 25°C/1 atm. As expected, the pH decreases as either K_a or C increases. As K_a becomes very small, the pH approaches 7.00 regardless of the value of C. Similarly, as C becomes very small, the pH approaches 7.00 regardless of the value of K_a.

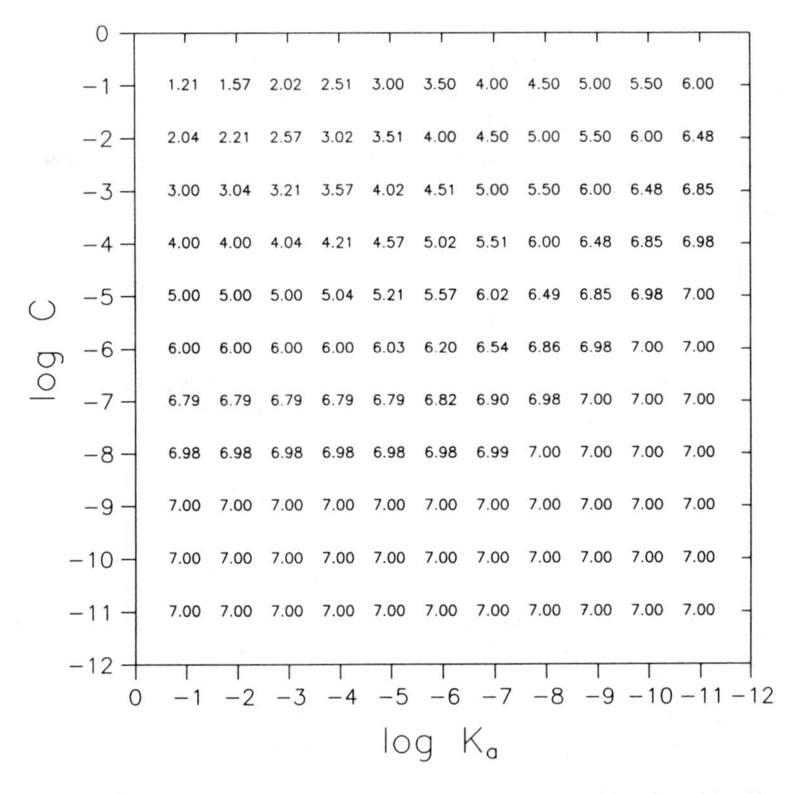

Figure 5.4 pH values for solutions of HA as a function of $\log C$ and $\log K_a$.

Option 2. The use of Option 2 in solving Eq. (5.12) will lead to a cubic equation in $[H^+]$, that is, a polynomial equation of the type

$$a[H^+]^3 + b[H^+]^2 + c[H^+] + d = 0. \tag{5.17}$$

The values of the coefficients (i.e., a, b, c, and d) will depend on the values of the parameters K_a, C, and K_w. There is an analytical solution to Eq. (5.17) that is the cubic analog of the quadratic equation[1] (Selby, 1971, p. 103). In general, there will be only *one real, positive root* for any version of Eq. (5.17) that is based in reality, that is, based on realistic, positive values of K_a, C, and K_w. Although real (i.e., non-imaginary), negative roots are also possible, negative roots clearly have no meaning in nature. Roots for Eq. (5.17) can also be imaginary. (Note: imaginary numbers involve $\sqrt{-1} = i$.)

[1]The reader may recall that the quadratic equation provides a solution to the second order equation

$$ax^2 + bx + c = 0. \tag{5.18}$$

The quadratic equation is

$$x = \frac{-b \pm \sqrt{b^2 - 4ac}}{2a}. \tag{5.19}$$

However, imaginary roots also have no meaning in nature. The real positive root of Eq. (5.17) will be the *same* value of $[H^+]$ for which $f([H^+]) = 0$.

Although the general analytical solution for a cubic equation can be used to solve (5.17), unlike the quadratic equation, the cubic solution is very complex and it will always be much easier to simply solve (5.12) by trial and error.[2] Moreover, if we are going to use a trial and error approach, it is certainly not worth the effort to convert Eq. (5.12) into an equation like Eq. (5.17) in order that a trial and error solution of the cubic polynomial can be sought; after all, Eq. (5.12) can be solved directly by trial and error. In addition to the needless effort expended in converting Eq. (5.12) into an equation like Eq. (5.17), we also run a *serious risk of making errors* in doing the algebra required to determine the values of the coefficients a, b, c, and d.

Despite the above objections, for purposes of illustrating some chemistry, we will now in fact convert $f([H^+])$ into the form given in Eq. (5.17). We first multiply Eq. (5.12) by $[H^+]([H^+] + K_a)$ to obtain

$$[H^+]^2([H^+] + K_a) = CK_a[H^+] + K_w([H^+] + K_a) \tag{5.20}$$

Therefore,

$$[H^+]^3 + K_a[H^+]^2 - (CK_a + K_w)[H^+] - K_wK_a = 0 \tag{5.21}$$

A comparison of Eqs. (5.17) and (5.21) leads to the conclusion that $a = 1$, $b = K_a$, $c = -(CK_a + K_w)$, and $d = -K_wK_a$. We can solve by trial and error for the positive real root value of $[H^+]$ that satisfies Eq. (5.21). As with Eq. (5.12), to the extent that no activity corrections are needed, the result obtained will be exactly correct since no simplifying assumptions were used to arrive at Eq. (5.21).

Let

$$g([H^+]) = [H^+]^3 + K_a[H^+]^2 - (CK_a + K_w)[H^+] - K_wK_a \tag{5.22}$$

Values of $g([H^+])$ for various values of pH and for $K_a = 10^{-5}$, $C = 10^{-4}M$, and $K_w = 10^{-14}$ are given in Table 5.2. The function $g([H^+])$ is plotted vs. pH in Figure 5.5. As expected (indeed, as required!), Figure 5.5 and the data in Table 5.2 indicate that $g([H^+]) = 0$ for the same positive, real value of $[H^+]$ for which $f([H^+]) = 0$, that is, for $[H^+] \simeq 10^{-4.57}$ (pH $\simeq 4.57$).

There are some interesting features of Figure 5.5 that are worth discussing. In particular, there is a local minimum at about pH = 4.8. This situation contrasts with that depicted in Figure 5.2 in which the $f([H^+])$ vs. pH plot shows no local minima for positive values of $[H^+]$. A result of the minimum in Figure 5.5 is that one can encounter difficulties when attempting a trial and error solution of $g([H^+])$ that would not be experienced in the trial and error solution of $f([H^+])$. For example, if one made the initial two guesses of pH to the right of the minimum for $g([H^+])$, the trial and error method would lead to continually higher values of pH with $g([H^+])$ indeed becoming smaller, but asymptotically approaching $-K_wK_a$, not zero; the method would not converge. Making

[2]An analytical solution also exists that allows one to determine the roots of a fourth order polynomial (Selby 1971, p. 106). However, as with the solution of a cubic equation, it is easier to solve a fourth order polynomial using trial and error. In the case of a fifth order polynomial, a trial and error approach will always need to be used since there is mathematical proof that no analytical solution even exists.

TABLE 5.2. Values of $g([H^+])$ for pH between 1.0 and 13.0 for $K_a = 10^{-5}$, $C = 10^{-4}M$, and $K_w = 10^{-14}$.

pH	$g([H^+])$	pH	$g([H^+])$
1.0	1.000×10^{-3}	4.9	-9.009×10^{-15}
2.0	1.001×10^{-6}	5.0	-8.000×10^{-15}
3.0	1.009×10^{-9}	5.5	-3.031×10^{-15}
3.5	3.230×10^{-11}	6.0	-9.891×10^{-16}
4.0	1.000×10^{-12}	6.5	-3.153×10^{-16}
4.1	4.848×10^{-13}	7.0	-1.000×10^{-16}
4.2	2.279×10^{-13}	7.5	-3.171×10^{-17}
4.3	1.008×10^{-13}	8.0	-1.010×10^{-17}
4.4	3.913×10^{-14}	9.0	-1.100×10^{-18}
4.5	9.999×10^{-15}	10.0	-2.000×10^{-19}
4.6	-2.961×10^{-15}	11.0	-1.100×10^{-19}
4.7	-8.029×10^{-15}	12.0	-1.010×10^{-19}
4.8	-9.356×10^{-15}	13.0	-1.001×10^{-19}

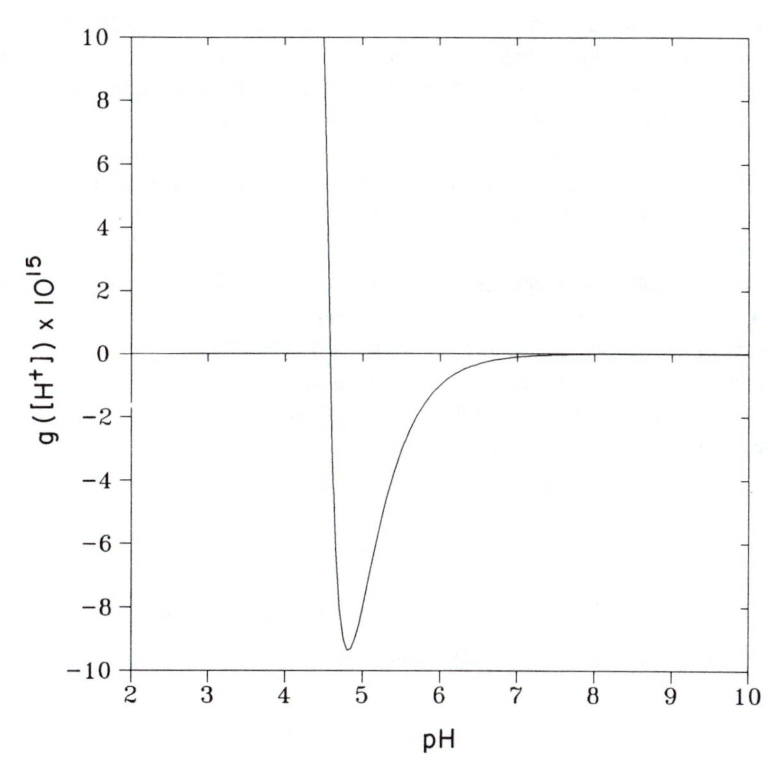

Figure 5.5 Fine-scale view of $g([H^+])$ vs. pH for $K_a = 10^{-5}$, $C = 10^{-4}M$, and $K_w = 10^{-14}$ in the region where $g([H^+]) = 0$.

the initial two guesses to the left of the minimum would allow the method to converge to the desired value of $[H^+]$. In contrast, for $f([H^+])$, the method would coverge *regardless* of where the initial two guesses were made. This is true for all reality-based values of K_a, C, and K_w. Therefore, there is in general an advantage of using Eq. (5.12) rather than (5.21) as the basis for a trial and error solution.

As one final point, we recall that any cubic equation $g(x)$ will go to $+\infty$ at one extreme of x and to $-\infty$ at the other extreme. Figure 5.4 does not look like the plot of a cubic equation. Although it does go to $+\infty$ at small pH (large $[H^+]$), it does not go to $-\infty$ for large pH (small, but *still positive* $[H^+]$). This is the result of the fact that we have plotted $g([H^+])$ vs. pH, and not vs. $[H^+]$; for large, negative $[H^+]$ (for which pH is undefined), the value of $g([H^+])$ does go to $-\infty$.

5.2.2 Solving for the Speciation of a Solution of the Acid HA—No Activity Corrections, with Simplifying Assumptions

If some significant simplifying assumptions are applied to the problem of computing the pH of a solution of HA, then we can expect that those assumptions will convert Eq. (5.21) into a polynomial with an order that is lower than a cubic. Indeed, if the simplifying assumption(s) do not even convert the cubic of Eq. (5.21) into a second order equation (which can be solved using the quadratic equation), then those assumptions have not simplified the problem. To illustrate these points, we will make a variety of assumptions and observe how they affect the complexity (i.e., order) of the resultant polynomial. As discussed below, whether or not any of these assumptions will be justifiable in any particular situation will depend upon the values of K_a *and* C.

5.2.2.1 Acidic Solution Approximation (i.e., $[OH^-]$ Negligible). For a solution of an acid HA for which K_a and C are not too small, then it is likely that $[OH^-]$ will be small and not significant relative to $[A^-]$ in the PBE. The requirement that both K_a and C should not be "too" small stems from the need to have some inherent acidity for HA, plus the need to have a non-negligible amount of the acid in the solution. Under these conditions, $[OH^-]$ can be neglected in the PBE,

$$[H^+] = [A^-] + \underbrace{[OH^-]}_{\text{assume small}} \tag{5.6}$$

that is,

$$[H^+] \simeq [A^-] \qquad \text{acidic solution approximation.} \tag{5.23}$$

With this assumption, then, the $K_w/[H^+]$ term in

$$[H^+] - \frac{K_a}{[H^+] + K_a}C - K_w/[H^+] = 0 \tag{5.12}$$

is neglected and so we obtain

$$[H^+] - \frac{K_a}{[H^+] + K_a}C \simeq 0. \tag{5.24}$$

The result is that all of the terms involving K_w drop out of the full cubic equation (Eq. (5.21)). Thus, Eq. (5.24) may be rearranged to yield

$$[H^+]^2 + K_a[H^+] - CK_a \simeq 0. \tag{5.25}$$

Application of the quadratic equation to Eq. (5.25) yields

$$[H^+] \simeq \frac{-K_a + \sqrt{K_a^2 + 4K_aC}}{2} \quad \text{assumes acidic solution, that is, neglects } [OH^-] \text{ (from dissociation of water) relative to } [A^-] \text{ in PBE} \tag{5.26}$$

where the root with the + sign in front of the square root is needed to give a positive value for $[H^+]$. For the case when $K_a = 10^{-5}$, $C = 10^{-4}M$, and $K_w = 10^{-14}$, Eq. (5.26) yields $[H^+] = 2.70 \times 10^{-5}M$ (pH = 4.57). This is the same result that was obtained by solving Eq. (5.12) by trial and error. Thus, for this particular case, making the acidic solution approximation leads to no error, at least when the pH is expressed with two figures past the decimal point. Given the limited accuracies with which the equilibrium constants (and even C) will usually be known, it will in general be meaningless to express pH to three figures past the decimal anyway. Thus, for many cases, Eq. (5.26) will provide an excellent estimate of $[H^+]$. The remaining species concentrations may be calculated as described above using Eqs. (5.14)–(5.16).

Equation (5.26) is the same second order equation that many of us first encountered in freshman chemistry when studying the solution chemistry of acids. At that time, we developed it from the following consideration of the dissociation

$$HA = H^+ + A^-. \tag{5.2}$$

First, it was said at that time that we should let the concentration of A^- that forms by Eq. (5.2) be equal to x. If the concentration of HA initially added to the solution is C, then the concentration of HA that remains after equilibrium is reached is given by

$$[HA] = C - x. \tag{5.27}$$

It was then stated that the concentration of H^+ must also equal x since every HA that dissociates forms one A^- and one H^+. So, at the time, we let

$$x = [A^-] = [H^+]. \tag{5.28}$$

Using Eqs. (5.27) and (5.28) to substitute into the equilibrium expression for Eq. (5.2), we obtained

$$K_a = \frac{x^2}{(C - x)}. \tag{5.29}$$

Or, since $x = [H^+]$,

$$K_a = \frac{[H^+]^2}{(C - [H^+])}. \tag{5.30}$$

Except for the fact that Eq. (5.30) has an = sign and not an $\simeq$ sign, Eqs. (5.24) and (5.30) are identical.

So, where did the "acidic solution approximation" come in when we were freshmen chemistry students? It came in when we let $x = [A^-] = [H^+]$, which is in fact the approximate PBE for this system. As we now know, of course, not all of the H^+ comes from the dissociation of the HA; *some of it comes the dissociation of water*. The amount of H^+ that comes from the dissociation of water is exactly equal to $[OH^-]$. As a result, $[A^-] = [H^+]$ will *never* be exactly true for a solution of HA; it can be *approximately* true when $[OH^-]$ is negligible relative to $[A^-]$.

Beginning students often find the concept of simplifying assumptions very frustrating since they have not yet developed sufficient chemical intuition to know when a given simplifying assumption might be applied, and when that assumption will not be appropriate. Under such conditions, once a given problem has been solved, the computed concentrations must be examined carefully to determine whether or not the assumption used in the solution was in fact valid.

In the case of the "acidic solution approximation," a criterion can be developed that will allow the a priori determination of whether this approximation will be valid. We first recall that the application of this approximation requires the assumption that $[OH^-]$ is small relative to $[A^-]$ in the PBE. If this assumption is valid, then by the resulting, approximate PBE, $[OH^-]$ is also small compared to $[H^+]$. The meaning of "small" is relative, so let us say that the acidic solution approximation is valid when $[OH^-]$ is 10% or less of $[H^+]$, that is, when

$$[OH^-] \leq 0.1[H^+]. \tag{5.31}$$

At 25°C/1 atm, K_w ($= [H^+][OH^-]$, neglecting activity corrections) is equal to 10^{-14}. By Eq. (5.31) then, *if* the true $[OH^-]$ is less than $0.1[H^+]$, substituting $0.1[H^+]$ for $[OH^-]$ in the expression for K_w will lead to a product that is greater than K_w. That is, we require that

$$0.1[H^+]^2 \geq 10^{-14} \tag{5.32}$$

or

$$[H^+] \geq 10^{-6.5}M. \tag{5.33}$$

Combining Eq. (5.33) with Eq. (5.30) yields the criterion that for the acidic solution approximation to apply with less than ~10% error in $[H^+]$, we must have

$$K_aC - 10^{-6.5}K_a \geq 10^{-13.0} \quad \text{criterion for applicability of Eq. (5.26).} \tag{5.34}$$

According to Eq. (5.34), as the system becomes increasingly dilute, the acid must become increasingly strong in order that dissociation from the water itself does not make a significant contribution to $[H^+]$. Thus, as $C \rightarrow 10^{-6.5}$, then $K_a \rightarrow \infty$. This means that below $C = 10^{-6.5}$, no matter how strong the acid is, $[OH^-]$ will not be negligible relative to $[A^-]$.

As an example application of Eq. (5.34), the lowest value of C that we can have for the acidic solution approximation to apply for $K_a = 10^{-5}m$ may be computed to be $10^{-6.5}M$. The acidic solution approximation (Eq. (5.26)) gives pH = 6.51 for this case. Application of the trial and error method to Eq. (5.12) yields an exact value of pH = 6.47. The error in $[H^+]$ is −9%.

When an equal sign is used, Eq. (5.34) may be plotted as a line. The result is presented in Figure 5.6*a*. The fact that $K_a \rightarrow \infty$ as $C \rightarrow 10^{-6.5}$ is evident in the plot.

The cross-hatched region, which includes all points on and above the line, satisfies Eq. (5.34). Equation (5.26) is valid in this region. The reader can test this relative validity by solving Eq. (5.26) for various values of K_a and C and comparing the results with those given in Figure 5.4 for that region; the same suggestion applies for the approximation regions considered in Figures 5.6*b*–5.6*e*. To facilitate these comparisons, a transparency of Figure 5.4 can be made and laid over each of Figures 5.6*a*–5.6*e*.

5.2.2.2 Weak Acid Approximation with [OH⁻] Not Necessarily Negligible (Optional). The previous section discussed the situation when the PBE may be simplified by neglecting $[OH^-]$. It is natural to ask what type of simplification can be obtained if we make some assumptions regarding the MBE, but leave the PBE unchanged as

$$[H^+] = [A^-] + [OH^-]. \tag{5.6}$$

Using the expressions for K_a and K_w, we can substitute into Eq. (5.6) to obtain

$$[H^+] = \frac{K_a[HA]}{[H^+]} + \frac{K_w}{[H^+]}. \tag{5.35}$$

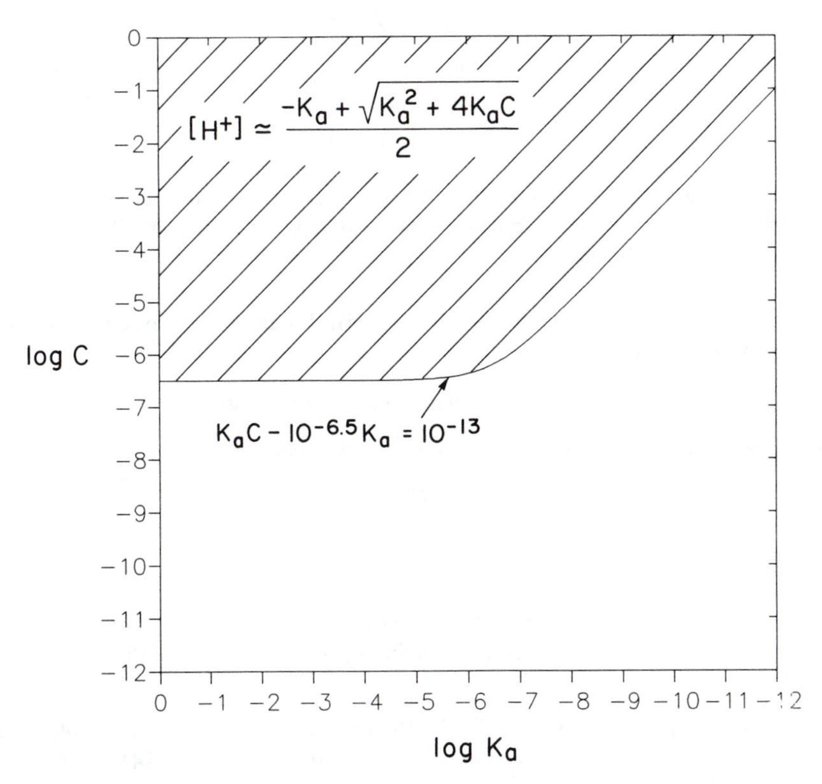

Figure 5.6*a* Region of validity for Eq. (5.26) given with cross-hatching. The equation for $[H^+]$ results from the acidic solution approximation.

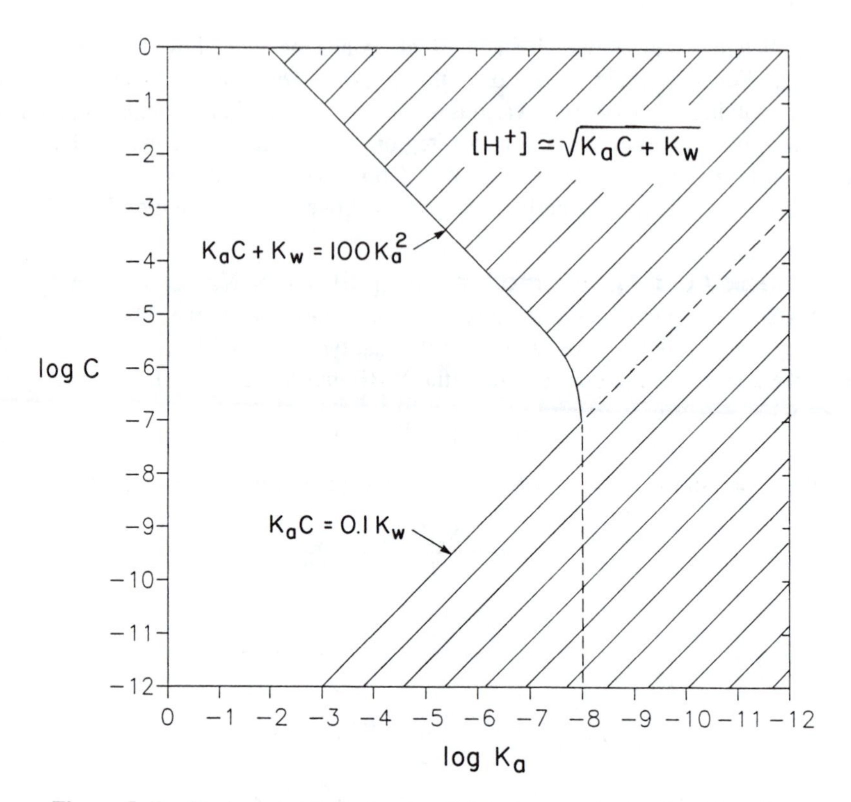

Figure 5.6*b* Region of validity for Eq. (5.38) given with cross-hatching. The equation for $[H^+]$ results from the weak acid approximation with $[OH^-]$ not necessarily negligible.

When the acid is sufficiently weak that $[A^-] \ll [HA]$, then we can assume that the MBE is given approximately by

$$[HA] \simeq C \qquad \text{weak acid approximation.} \tag{5.36}$$

Under these conditions, Eq. (5.35) becomes approximately

$$[H^+] \simeq \frac{K_aC}{[H^+]} + \frac{K_w}{[H^+]} \tag{5.37}$$

or simply

$$[H^+] \simeq \sqrt{K_aC + K_w} \qquad \text{assumes weak acid, does not neglect } [OH^-] \text{ (from dissociation of water) relative to } [A^-] \text{ in PBE.} \tag{5.38}$$

Once $[H^+]$ is computed using Eq. (5.38), then the concentrations of the other species may be calculated in a manner similar to that described by Eqs. (5.14)–(5.16).

We can develop criteria that will permit the a priori prediction of when it is permissible to use Eq. (5.38). In particular, the assumption that must now be valid is that

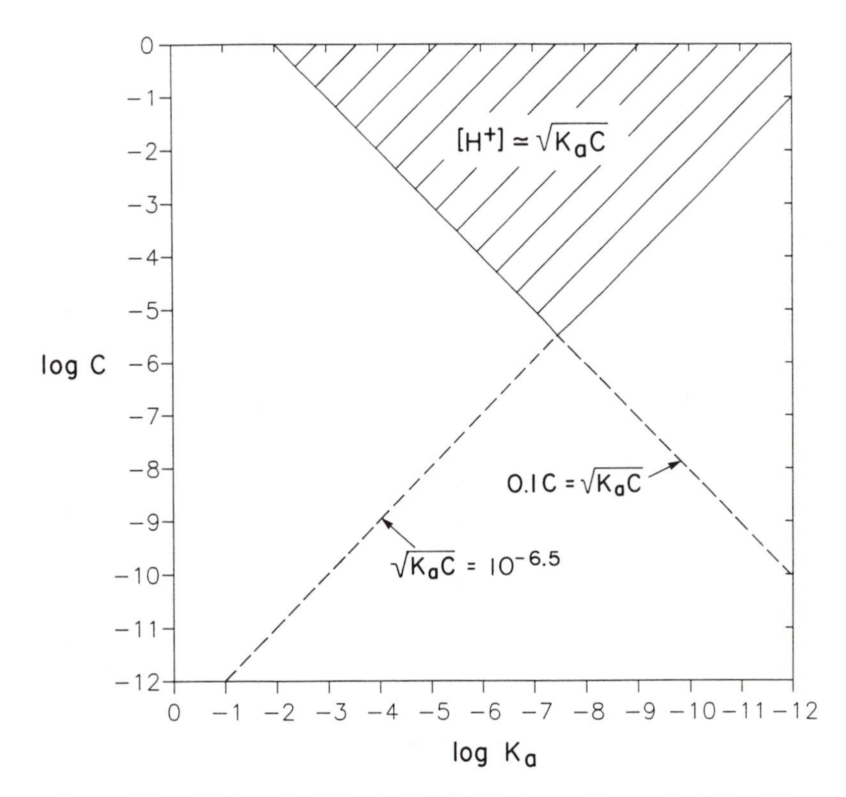

Figure 5.6*c* Region of validity for Eq. (5.45) given with cross-hatching. The equation for $[H^+]$ results from the simultaneous assumption of an acidic solution and a weak acid.

$[A^-] \ll C$. Therefore, let us take as our criterion for the validity of Eq. (5.36) that

$$[A^-] \leq 0.1C. \tag{5.39}$$

Using Eqs. (5.36), (5.38), and (5.39) together with the definition of K_a, we obtain that

$$\frac{\sqrt{K_aC + K_w}(0.1C)}{C} \geq K_a \tag{5.40}$$

or

$$K_aC + K_w \geq 100K_a^2 \qquad \text{criterion for applicability of Eq. (5.38).} \tag{5.41}$$

Equation (5.41) can also be derived by combining Eqs. (5.6), (5.38), and (5.39).

One pair of K_a and C values that satisfies Eq. (5.41) expressed with an equal sign is $K_a = 10^{-7}m$ and $C = 10^{-5}M$. Application of Eq. (5.38) yields the approximate value for pH of 6.00. Application of the trial and error method to Eq. (5.13) yields an exact pH value of 6.02. The error in $[H^+]$ is +5%.

Equation (5.38) was derived within the context of being applicable to weak acids. When C is very low it is also valid for weak *or* strong acids when the K_aC term in

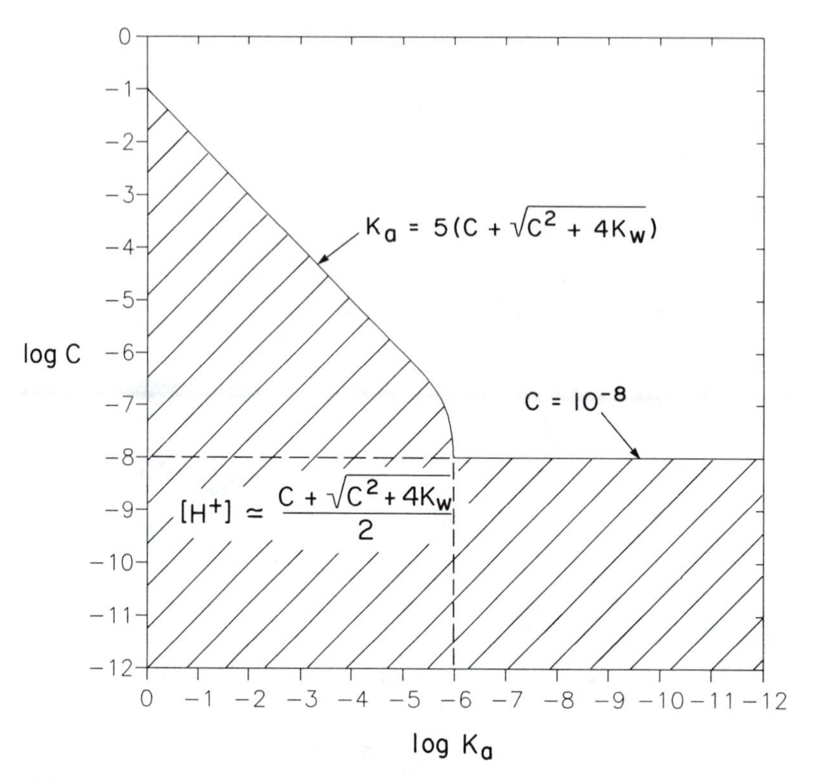

Figure 5.6*d* Region of validity for Eq. (5.52) given with cross-hatching. The equation for $[H^+]$ results from the strong acid approximation with $[OH^-]$ not necessarily neglible.

Eq. (5.38) is much less than K_w, say when

$$K_a C \leq 0.1 K_w \quad \text{alternate criterion for applicability of Eq. (5.38).} \tag{5.42}$$

When Eq. (5.42) is satisfied, Eq. (5.38) becomes simply

$$[H^+] \simeq \sqrt{K_w}. \tag{5.43}$$

Thus, Eq. (5.38) is valid when *either* Eq. (5.41) *or* Eq. (5.42) is satisfied. That $[H^+]$ is given by Eq. (5.43) whenever Eq. (5.42) is satisfied may be shown by substituting Eq. (5.42) into Eq. (5.11). The result is

$$[H^+] \leq \sqrt{K_w \frac{1.1[H^+] + K_a}{[H^+] + K_a}} \leq \sqrt{1.1 K_w} \simeq \sqrt{K_w}. \tag{5.44}$$

Since we also know that the pH of any solution of an acid HA must be less than the pH of pure water ($= 1/2\ pK_w$), Eq. (5.44) reveals that satisfaction of Eq. (5.42) will result in a pH between $1/2\ pK_w$ and $1/2\ (pK_w - \log(1.1)) = 1/2\ pK_w - 0.02$. At 25°C/1 atm, this would mean a range of 7.00 to 6.98.

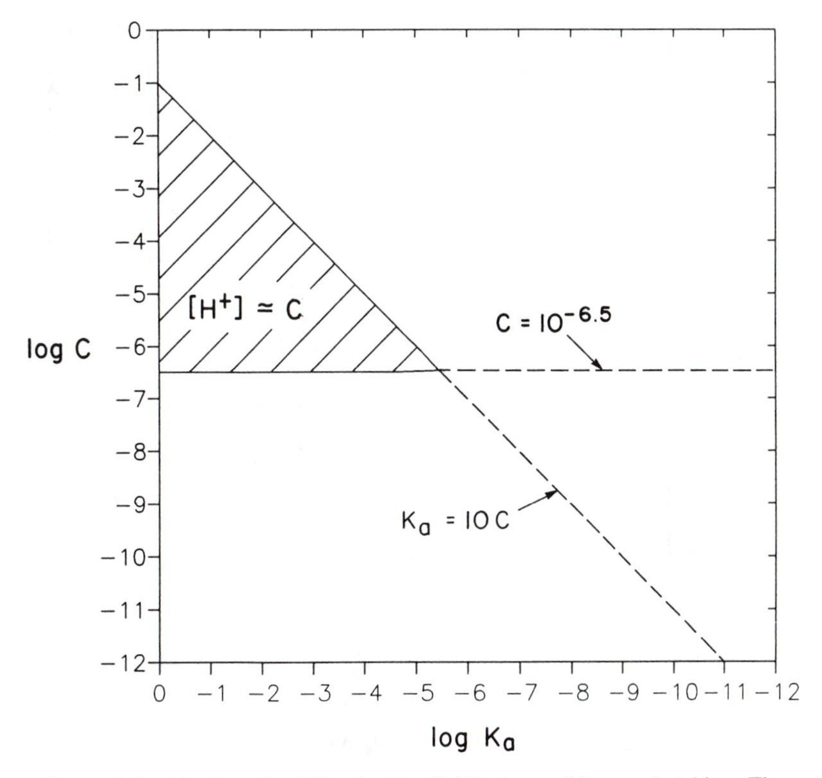

Figure 5.6*e* Region of validity for Eq. (5.49) given with cross-hatching. The equation for $[H^+]$ results from the simultaneous assumption of a strong acid and an acidic solution.

Equations (5.41) and (5.42) with equal signs are plotted in Figure 5.6*b*. The region in which either or both of these equations are satisfied includes all points on or to the right of the composite, solid line. It may be noted that the regions in Figures 5.6*a* and 5.6*b* have some area that is common to both. That common area corresponds to the set of points for which the assumptions behind Eq. (5.26) and (5.38) are *both* valid simultaneously, that is, when the weak acid approximation applies, *and* $[OH^-]$ is negligible.

5.2.2.3 Weak Acid Approximation with $[OH^-]$ Also Assumed Negligible.

If the acid is a weak acid (i.e., K_a is small), and if C is not too small, we can assume the applicability of *both* Eqs. (5.36) and (5.23). That is, we assume that $[A^-]$ is small relative to [HA] and that $[OH^-]$ is small relative to $[A^-]$. In this case, the "$-[A^-]$" correction on C in Eq. (5.29) is not needed, and Eq. (5.29) simplifies to

$$[H^+] \simeq \sqrt{K_a C} \qquad (5.45)$$

assumes weak acid, and that $[OH^-]$ (from dissociation of water) is negligible relative to $[A^-]$ in PBE.

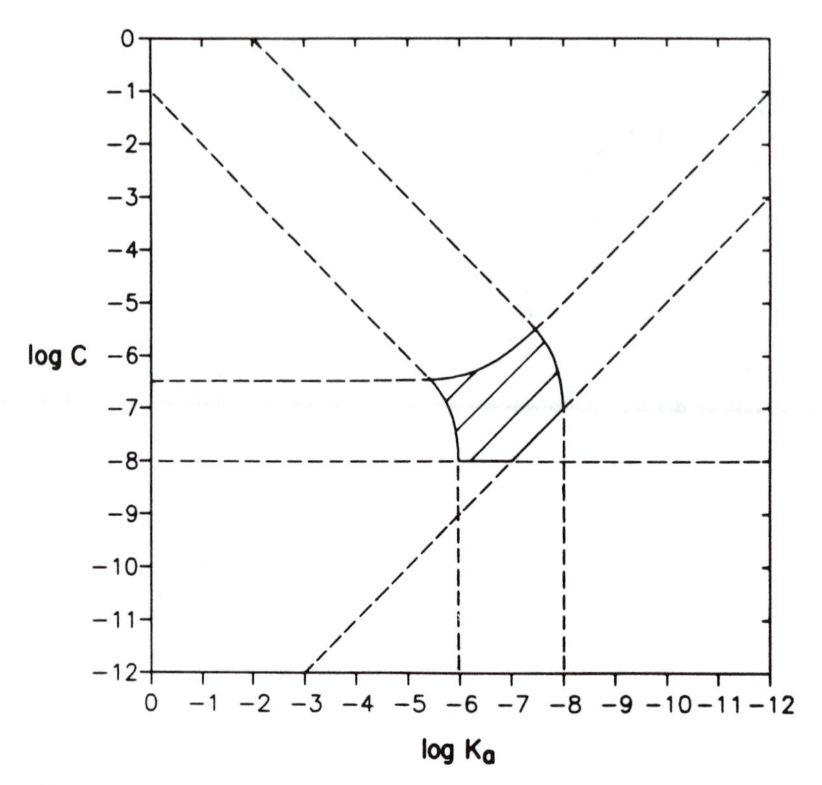

Figure 5.6*f* Region where none of the five approximation equations for $[H^+]$ will give $[H^+]$ in a solution of HA to within $\pm 10\%$ or less.

Equation (5.45) can also be obtained by neglecting $[OH^-]$ (i.e., $K_w/[H^+]$) in Eq. (5.37). As with Eq. (5.38), the original cubic equation that is (5.21) has been reduced to a first order equation.

Since: 1) Eq. (5.38) required *only* the assumption of a weak acid and *not* necessarily the validity of Eq. (5.23); and 2) is *yet* a very *simple* equation for $[H^+]$, some readers might justifiably wonder why we have even bothered to develop Eq. (5.45). The reason is that, because of its simplicity, Eq. (5.45) appears very prominently in courses in both basic and analytical chemistry. Therefore, although there is never any reason to use Eq. (5.45) rather than (5.38), Eq. (5.45) should receive some attention here.

We can develop a set of criteria that will permit the a priori prediction of when we can and cannot use Eq. (5.45). As with the simple acidic solution approximation, we again require that $[OH^-] \ll [A^-]$, and therefore

$$[H^+] \geq 10^{-6.5} M. \tag{5.33}$$

Combination of Eq. (5.45) with Eq. (5.33) yields

$$\sqrt{K_a C} \geq 10^{-6.5} M \qquad \text{criterion 1 for applicability of Eq. (5.45).} \tag{5.46}$$

The second assumption that must be valid here is that $[H^+](\simeq [A^-])$ must be small relative to C. Therefore, we have

$$0.1C \geq \sqrt{K_a C} \qquad \text{criterion 2 for applicability of Eq. (5.45).} \qquad (5.47)$$

Thus, for the weak acid approximation discussed in this section to apply with less than ~10% error in $[H^+]$, we must have

$$0.1C \geq \sqrt{K_a C} \geq 10^{-6.5}M \qquad \text{combined criteria for applicability of Eq. (5.45).} \qquad (5.48)$$

The lines given by Eqs. (5.46) and (5.47) are plotted in Figure 5.6*c*. The region satisfying Eq. (5.48) is noted as the cross-hatched region above the solid composite line. This is the region in which Eq. (5.45) is valid. As implied at the conclusion of the previous section, this region is essentially the area that is *common* to the regions in Figures 5.6*a* and 5.6*b*. It is not surprising, therefore, that Eqs. (5.46) and (5.47) are asymptotic portions of Eqs. (5.34) and (5.41).

A pair of K_a and C values which is on the boundary given by Eqs. (5.46) and (5.47) is $K_a = 10^{-8}m$, $C = 10^{-5}M$. Application of Eq. (5.45) yields the approximate value for pH of 6.50. Application of the trial and error method to Eq. (5.13) yields an exact pH value of 6.485. The error in $[H^+]$ is -3%. The reason that the error is significantly less than 10% is that the two sources of error in the weak acid approximation, that is, overestimating [HA] as C and overestimating $[A^-]$ as $[H^+]$ tend to cancel one another out when computing the *ratio* $[H^+] \simeq K_a C/[H^+]$, and so Eq. (5.48) results in providing criteria for K_a and C that give less than 10% error in $[H^+]$.[3] As seen in a comparison of Figures 5.4 and 5.6*c*, within the region of validity of Eq. (5.45), we do indeed have $\text{pH} \simeq -\frac{1}{2}(\log K_a + \log C)$.

5.2.2.4 Strong Acid Approximation with $[OH^-]$ Not Necessarily Negligible (Optional). When an acid like HA is said to be "strong," what is meant is that it will dissociate essentially completely to form H^+ and A^-. In that case, we have

$$[A^-] \simeq C \qquad \text{strong acid approximation} \qquad (5.49)$$

and so the PBE may be written as

$$[H^+] \simeq C + [OH^-]. \qquad (5.50)$$

By substituting for $[OH^-]$ and simplifying, Eq. (5.50) becomes

$$[H^+]^2 - C[H^+] - K_w \simeq 0. \qquad (5.51)$$

By the quadratic equation, and since it is the root given by the + sign in the quadratic equation that is positive, we have

$$[H^+] \simeq \frac{C + \sqrt{C^2 + 4K_w}}{2} \qquad \text{assumes strong acid, with } [OH^-] \text{ (from dissociation of water) not necessarily negligible relative to } [A^-] \text{ in PBE.} \qquad (5.52)$$

[3]It is interesting to note that the error is *very* small for the pair of K_a and C for which Eqs. (5.46) and (5.47) are simultaneously *individually* minimally satisfied, that is, the pair given by the intersection of the two lines. The pair is $K_a = 10^{-7.5}m$, and $C = 10^{-5.5}M$. Application of Eq. (5.45) yields pH $\simeq$ 6.50. Application of the trial and error method with Eq. (5.13) yields pH = 6.498. The error in $[H^+]$ is only -0.5%.

We can develop criteria that will permit the a priori prediction of when Eq. (5.52) can be used. For the applicability of Eq. (5.49), we require

$$[\mathrm{HA}] \leq 0.1\,C. \tag{5.53}$$

Substituting Eqs. (5.49), (5.52) and (5.53) into the expression for K_a we obtain

$$\frac{\dfrac{C + \sqrt{C^2 + 4K_w}}{2}\,C}{0.1C} \leq K_a. \tag{5.54}$$

The $\leq$ sign applies in Eq. (5.54) since we desire that (0.1 C) is an overestimation of [HA]. Equation (5.54) simplifies to

$$K_a \geq 5\left[C + \sqrt{C^2 + 4K_w}\right] \qquad \text{criterion for applicability of Eq. (5.52).} \tag{5.55}$$

If C is extremely low, then *no matter what the strength of the acid*, that acid will not be able to lower the pH significantly. Thus, Eq. (5.55) does not tell the whole story here since Eq. (5.52) accurately predicts that $[H^+] \simeq \sqrt{K_w}$ when C is extremely low. An alternate criterion for the applicability of Eq. (5.52) is therefore

$$\mathrm{C} \leq 10^{-8}M \qquad \text{alternate criterion for applicability of Eq. (5.52).} \tag{5.56}$$

Once C satisfies Eq. (5.56), then no acid will be able to contribute more than about 10% of the H^+ present; the remainder will come from the water, and $[H^+]$ will be close to $\sqrt{K_w}$. Equations (5.55) and (5.56) are plotted in Figure 5.6*d*. The domain of points that satisfies either of the two equations are those on and below the solid composite line.

We note that since pH $\simeq$ 7 for all HA solutions with $C \leq 10^{-8}$, for $\log K_a$ values less than about -6, Eq. (5.52) will be satisfied even though Eq. (5.49) will not. In particular, as long as we do not neglect the OH^- from the dissociation of water, whether the acid dissociates essentially completely or not becomes irrelevant for the predicted pH once C becomes very low. For example, for an acid with $\log K_a = -8$, at a concentration of $10^{-8}M$, Eq. (5.52) will correctly predict that pH $\simeq$ 7, but $[A^-]$ will only represent about 10% of C (neglecting activity corrections, $[A^-]/[HA] = 10^{-8}/10^{-pH} = 0.1$).

5.2.2.5 Strong Acid Approximation with $[OH^-]$ Assumed Negligible.

We can expect that the region that is common to both Figure 5.6*d and* Figure 5.6*a* will involve the assumptions that the acid is strong, that is,

$$[A^-] \simeq C \qquad \text{strong acid approximation} \tag{5.49}$$

and $[OH^-] \ll [A^-]$ so that

$$[H^+] \simeq [A^-]. \tag{5.23}$$

Therefore,

$$[H^+] \simeq C \qquad \begin{array}{l}\text{assumes strong acid, with } [OH^-] \\ \text{(from dissociation of water) assumed} \\ \text{negligible relative to } [A^-] \text{ in PBE.}\end{array} \tag{5.57}$$

For Eq. (5.57) to be valid to a good approximation, we require

$$[HA] \leq 0.1C \tag{5.53}$$

and

$$[H^+] \geq 10^{-6.5}M. \tag{5.33}$$

Combination of Eqs. (5.49), (5.23), and (5.33) yields

$$C \geq 10^{-6.5}M \qquad \text{criterion 1 for applicability of Eq. (5.57)} \tag{5.58}$$

which is a simplified version of Eq. (5.34) that applies when the acid is strong, and $10^{-13}/K_a$ is small relative to $10^{-6.5}$.

Combination of Eqs. (5.23), (5.49), and (5.53) with the definition of K_a yields

$$K_a \geq \frac{C^2}{0.1C} \tag{5.59}$$

which simplifies to

$$K_a \geq 10C \qquad \text{criterion 2 for applicability of Eq. (5.57).} \tag{5.60}$$

Equation (5.60) is a simplified version of Eq. (5.55) that applies when $4K_w$ is small compared to C^2, that is, when the dissociation of water (governed by K_w) to form OH^- is negligible compared to the formation of A^- (governed in part by *C*). Equations (5.58) and (5.60) are plotted in Figure 5.6*e*. The set of points that satisfies both equations are those on to the left of the composite solid line. As seen in a comparison of Figures 5.4 and 5.6*e*, within the region of validity of Eq. (5.57), we do indeed have $pH \simeq -\log C$.

The small region in which none of the approximate equations will accurately predict the pH is illustrated in Figure 5.6*f*. As for all $C \leq 10^{-5.5}M$, the pH values in that region will be between 5.5 and 7. Plots similar to Figures 5.6*a*–5.6*f* have also been discussed by Narasaki (1987).

5.3 GENERAL EQUATIONS FOR SOLVING FOR THE SPECIATION OF A SOLUTION OF THE BASE NaA—NO ACTIVITY CORRECTIONS

The material $NaA_{(s)}$ can be described as the "solid sodium salt of the acid HA." This terminology has its origins in the fact that materials of the generic formula $NaA_{(s)}$ can be made by neutralizing a portion of the acid HA with an equivalent amount of the strong base sodium hydroxide (NaOH), then allowing crystals of $NaA_{(s)}$ to form by removing (evaporating) the water. Inorganic acids like hydrochloric (HCl) as well as organic acids like acetic (CH_3COOH) can be converted into their sodium salts in this manner. The sodium salt of HCl is well known as common "table salt." The sodium salt of acetic acid is called sodium acetate (CH_3COONa). While the discussions that follow will focus on cases involving the chemistry of solutions of NaA, the conclusions that we will obtain will also be applicable to solutions of other monobasic salts, e.g. solutions of potassium (K) salts of the generic formula KA.

Since Na^+ (as well as K^+, etc.) tends to have only a rather low affinity for most singly charged negative ions of the type A^-, sodium salts are usually fairly soluble. For similar reasons, as noted in Chapter 4, Na^+ will not associate with dissolved A^- or OH^- to any significant extent. This means that the concentration of the dissolved, molecular species NaA^o and $NaOH^o$ will usually be very low compared to the concentrations of the other A-containing species, namely A^- and HA. Thus, we are concerned with a total of five species in a solution of NaA:

$$H^+ \quad A^- \quad OH^- \quad HA \quad Na^+.$$

The two chemical equilibria governing these species are

$$H_2O = H^+ + OH^- \qquad K_w = \{H^+\}\{OH^-\} \tag{5.1}$$

$$HA = H^+ + A^- \qquad K_a = \{H^+\}\{A^-\}/\{HA\}. \tag{5.2}$$

Some HA will always be present in a solution of NaA because a portion of the A^- will always react with water according to

$$A^- + H_2O = HA + OH^-. \tag{5.61}$$

Since A^- accepts protons from water in reaction (5.61), A^- acts as a base in aqueous solutions. Thus, regardless of the concentration and inherent basicity of the NaA added to water, the pH will always be at least somewhat higher than the pH of pure water. That is, in a solution of NaA, the pH will always be higher than $\frac{1}{2}pK_w$.

The equilibrium constant for reaction (5.61) can be obtained by combining reactions (5.1) and (5.2) according to

$$\begin{array}{lll} & H_2O = H^+ + OH^- & K_w = \{H^+\}\{OH^-\} \quad (5.1) \\ - & HA = H^+ + A^- & K_a = \{H^+\}\{A^-\}/\{HA\} \quad (5.2) \\ \hline & A^- + H_2O = HA + OH^- & K_b = K_w/K_a = \{HA\}\{OH^-\}/\{A^-\} \quad (5.62) \end{array}$$

Just as the subscript on K_a denotes "acid," the subscript on K_b denotes "base." Presuming that we know the value of K_w, *there is really no new information contained in the value of* K_b *that we do not already have in* K_a; the information is just packaged a little differently.

As K_b increases, A^- becomes increasingly basic. At the same time, the corresponding HA species becomes less acidic. Indeed, the more basic a given A^- species is, the more it will tend to pick up protons; the more that A^- tends to pick up protons, the less the resulting HA species may be expected to give up protons. Thus, increasing K_b means decreasing K_a. An anthropomorphic analogy of this situation may be developed as follows. Consider a group of different people (different types of A^- species) all of which do not have a dollar (proton). Those that want a dollar (proton) the most are the most greedy (basic). Assume now that all of the people can hold a maximum of one dollar, and that they are in fact each given one dollar. The most greedy ones will now be least likely to give up their dollar, that is, they will act the least acidic. The ones that were the least greedy (least basic) will tend to let go of their dollar most easily, that is, be most acidic.

As indicated in the expression for K_b given for reaction Eq. (5.62), there is a hyperbolic relationship governing K_a and K_b. In particular, we have

$$K_a K_b = K_w \tag{5.63}$$

or equivalently

$$\log K_a + \log K_b = \log K_w \tag{5.64}$$

$$pK_a + pK_b = pK_w. \tag{5.65}$$

At 25°C/1 atm, we have

$$pK_a + pK_b = 14.0. \tag{5.66}$$

The relationship between K_a and K_b that is described by Eqs. (5.63)–(5.66) is illustrated in Table 3.1 for a series of different HA and A^- species.

5.3.1 Solving for the Speciation of a Solution of the Base NaA—No Activity Corrections, No Simplifying Assumptions

If we do not need to make activity corrections, in a solution of NaA there will be five unknowns, $[H^+]$, $[A^-]$, $[OH^-]$, [HA], and $[Na^+]$. In a manner that is analogous to that discussed in Section 5.2, we will now consider how to determine the speciation in a solution of NaA when activity corrections may be neglected. The equations that will be developed may also be applied when the solution of interest is an essentially constant ionic medium, and the appropriate ${}^cK$ values for the chemical equilibria are subsitituted for the infinite dilution constants.

With five unknowns, a total of five equations are required to solve problems of this type. Neglecting activity corrections, they are

$K_w = [H^+][OH]$	first chemical equilbrium equation	(5.3)
$K_a = [H^+][A^-]/[HA]$	second chemical equilibrium equation	(5.4)
$[HA] + [A^-] = C$	mass balance equation (MBE) on total A	(5.5)
$[H^+] + [HA] = [OH^-]$	proton balance equation (PBE)	(5.67)
$[Na^+] = C$	mass balance equation (MBE) on Na^+.	(5.68)

We note that the ENE could be substituted for the PBE in the above set. Also, we could have used the expression for K_b in the above set rather than the expression for K_a. The expression for K_a was used to be consistent with Section 5.2.1. Equation (5.68) has its origins in the fact that the concentrations of the dissolved species NaA^o and $NaOH^o$ are assumed to be essentially zero. Thus, $[Na^+]$ provides an essentially perfect measure of C.

Our goal will be the derivation of a single equation in one unknown ($[H^+]$) that uses the entire set of five equations. If we started with the ENE rather than the PBE, a simple substitution would just bring us to the PBE anyway. This can be shown as follows. The ENE is

$$[H^+] + [Na^+] = [OH^-] + [A^-]. \tag{5.69}$$

Rearranging yields

$$[H^+] + [Na^+] - [A^-] = [OH^-]. \tag{5.70}$$

By Eqs. (5.5) and (5.68) we note that $([Na^+] - [A^-]) = [HA]$. With this substitution, we obtain the PBE

$$[H^+] + [HA] = [OH^-]. \tag{5.67}$$

In a manner that is analogous to Eq. (5.8), it is useful to note that because $C = [HA] + [A^-]$, we can write

$$[HA] = \frac{[HA]}{[HA] + [A^-]} C. \tag{5.71}$$

Equation (5.71) makes use of the fact that $[HA]/([HA] + [A^-])$ equals the fraction of the total A in solution which is present as HA (as will be seen in Chapter 6, that fraction is called α_0). Equation (5.71) gives [HA] as the product of that fraction with the total A concentration. Using Eq. (5.71) to substitute into Eq. (5.67) for [HA], and K_w to help substitute for $[OH^-]$, we now have

$$[H^+] + \frac{[HA]}{[HA] + [A^-]} C - K_w/[H^+] = 0. \tag{5.72}$$

We can simplify Eq. (5.72) by noting that $[A^-]/[HA] = K_a/[H^+]$. Therefore,

$$[H^+] + \frac{[H^+]}{[H^+] + K_a} C - K_w/[H^+] = 0. \tag{5.73}$$

Equation (5.73) is the desired single equation in one unknown. The reader may wish to review the above steps to see where each of the five equations in five unknowns came into its derivation.

Let

$$r([H^+]) = [H^+] + \frac{[H^+]}{[H^+] + K_a} C - K_w/[H^+]. \tag{5.74}$$

The cubic equation that can be obtained from (5.73) is

$$[H^+]^3 + (K_a + C)[H^+]^2 - K_w[H^+] - K_w K_a = 0. \tag{5.75}$$

In a manner similar to that described in Section 5.2.1, we can define a function

$$s([H^+]) = [H^+]^3 + (K_a + C)[H^+]^2 - K_w[H^+] - K_w K_a. \tag{5.76}$$

In a manner that is analogous to how we solved Eq. (5.12), we can solve Eq. (5.73) for the single positive value of $[H^+]$ that gives $r([H^+]) = 0$. The same three specific different solution options may be used as were presented for solving Eq. (5.12). For reasons similar to those discussed in Section 5.2.1, if no simplifying assumptions are made, it will be easiest to solve Eq. (5.73) directly by numerical means, that is, without bothering to convert it to Eq. (5.75) and seeking the value of $[H^+]$ that gives $s([H^+]) = 0$.

Let us consider the example case for which $K_a = 10^{-5}$, $C = 10^{-4}M$, and $K_w = 10^{-14}$ ($K_b = 10^{-9}$). Values of $r([H^+])$ for various values of pH for this case are given in Table 5.3. By interpolation, the data in Table 5.3 indicate that the equilibrium pH will be 7.52. The fact that this pH is very close to being neutral is due in part to the low value of K_b. For pH = 7.52, $[H^+] = 3.02 \times 10^{-8}M$. The concentrations of the remaining species can now be calculated. We have

$$[OH^-] = K_w/[H^+] = 10^{-14}/(3.02 \times 10^{-8}) = 3.31 \times 10^{-7}M \quad (5.77)$$

$$[A^-] = \frac{[A^-]}{[HA]+[A^-]}C = \frac{K_a}{[H^+]+K_a}C = \frac{10^{-5}}{10^{-7.52}+10^{-5}}10^{-4}$$
$$= 9.97 \times 10^{-5}M \quad (5.78)$$

$$[HA] = C - [A^-] = 10^{-4} - 9.97 \times 10^{-5} = 3.01 \times 10^{-7}M. \quad (5.79)$$

Figure 5.7 illustrates how the equilbrium pH of a solution of NaA changes with K_b and C at 25°C/1 atm. As expected, the pH increases as either K_b or C increases. As K_b becomes very small, the pH approaches 7.00 regardless of the value of C. Similarly, as C becomes very small, the pH approaches 7.00 regardless of the value of K_b.

5.3.2 Solving for the Speciation of a Solution of the Base NaA—No Activity Corrections, with Simplifying Assumptions

As with the case of an aqueous solution of the acid HA, we can expect that the application of simplifying assumptions will reduce the order of the polynomial in Eq. (5.75). We now consider the nature of the different types of simplifying assumptions that can be made for a solution of a base like NaA. Whether or not any of these assumptions will be justifiable in any particular situation will depend upon the values of K_b and C.

TABLE 5.3. Values of $r([H^+])$ for pH between 1.0 and 13.0 for $K_a = 10^{-5}$, $C = 10^{-4}M$, and $K_w = 10^{-14}$ ($K_b = 10^{-9}$).

pH	$r([H^+])$	pH	$r([H^+])$
1.0	0.10009999	7.5	0.00000003
2.0	0.01009990	7.6	−0.0000001
3.0	0.00109901	7.7	−0.0000003
4.0	0.00019091	7.8	−0.0000005
5.0	0.00006000	7.9	−0.0000007
6.0	0.00001008	8.0	−0.0000009
7.0	0.00000099	9.0	−0.0000100
7.1	0.00000074	10.0	−0.0001000
7.2	0.00000053	11.0	−0.0010000
7.3	0.00000035	12.0	−0.0100000
7.4	0.00000019	13.0	−0.1000000

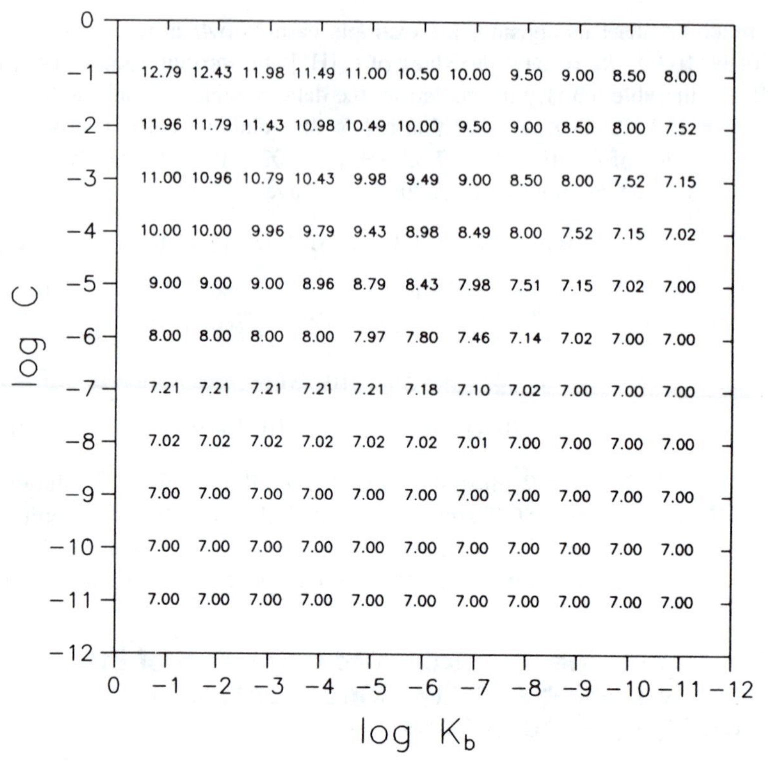

Figure 5.7 pH values for solutions of NaA as a function of $\log C$ and $\log K_b$ (note: $\log K_b = -14 - \log K_a$).

5.3.2.1 Basic Solution Approximation (i.e., [H⁺] Negligible). For a solution of a base NaA for which both C and K_b are not too small, then it is likely the solution will be *basic*. Under these conditions, $[H^+]$ will be small relative to [HA], and may be neglected in the PBE. This simplifies the problem considerably. The criterion that C should not be too small stems from the need to have a non-negligible amount of the base NaA in the solution. The value of K_b should not be too small because we need the A^- to have some inherent basicity.

When the basic solution approximation is valid, then

$$[H^+] + [HA] = [OH^-]. \quad (5.67)$$

↖ assume small

Therefore, we have

$$[HA] \simeq [OH^-] \qquad \text{basic solution approximation.} \quad (5.80)$$

Application of Eq. (5.62) together with Eq. (5.5) for $[A^-]$ then yields

$$\frac{[OH^-]^2}{C - [OH^-]} \simeq K_b. \quad (5.81)$$

There is much similarity between Eq. (5.30) for a solution of HA, and Eq. (5.81) for a solution of NaA. For the former, the variable is $[H^+]$ and the equilibrium constant is K_a. For the latter, the variable is $[OH^-]$ and the constant is K_b.

Equation (5.81) can be solved by applying the quadratic equation. The result is

$$[OH^-] \simeq \frac{-K_b + \sqrt{K_b^2 + 4K_bC}}{2} \qquad \text{assumes basic solution, i.e., neglects } [H^+] \text{ (from dissociation of water) relative to [HA] in PBE.} \tag{5.82}$$

We also note that since Eq. (5.81) can be converted to a second order polynomial in $[H^+]$, it is a simplified version of Eq. (5.75). Since we are dealing with the assumption that H^+ from the dissociation from water is negligible, the simplification involves the disappearance of all terms in Eq. (5.75) that involve K_w. Equation (5.82) is the base analog of Eq. (5.26) with K_b substituted for K_a.

Once we calculate $[OH^-]$, then we can calculate the concentrations of all other species. Indeed, if we know the value of K_w, then knowing $[OH^-]$ is obviously as good as knowing $[H^+]$ and so the remaining species can be calculated in a manner similar to that given in Eqs. (5.78) and (5.79).

Procedures analogous to those discussed in Section 5.2.2.1 can be used to develop a criterion for the a priori prediction of when Eq. (5.82) can be applied with little error. The result is

$$K_bC - 10^{-6.5}K_b \geq 10^{-13} \qquad \text{criterion for applicability of Eq. (5.82).} \tag{5.83}$$

Equation (5.83) is the base analog of Eq. (5.34). Figure 5.8*a*, which contains a plot of Eq. (5.83), is the base analog of Figure 5.6*a*. Equation (5.82) will apply with good accuracy on and above the line in Figure 5.8*a*.

The reader can test this relative validity by solving Eq. (5.82) for various values of K_b and C and comparing the results given in Figure 5.7 for that region; the same suggestion applies for the approximation regions considered in Figures 5.8*b*–5.8*e*. To facilitate in these comparisons, a transparency of Figure 5.7 can be made and laid over each of Figures 5.8*a*–5.8*e*.

5.3.2.2 Weak Base Approximation with $[H^+]$ Not Necessarily Negligible (Optional). The preceding section discussed the situation when the PBE may be simplified by neglecting $[H^+]$. We now ask what type of simplification can be obtained if we make some assumptions regarding the MBE, but leave the PBE unchanged as

$$[H^+] + [HA] = [OH^-]. \tag{5.67}$$

Substituting for $[H^+]$ in terms of K_w, we obtain

$$\frac{K_w}{[OH^-]} + \frac{K_b[A^-]}{[OH^-]} = [OH^-]. \tag{5.84}$$

We now assume that the base is sufficiently weak that most of the A^- do not pick up protons. Under these conditions, the MBE is given approximately by

$$[A^-] \simeq C \qquad \text{weak base approximation} \tag{5.85}$$

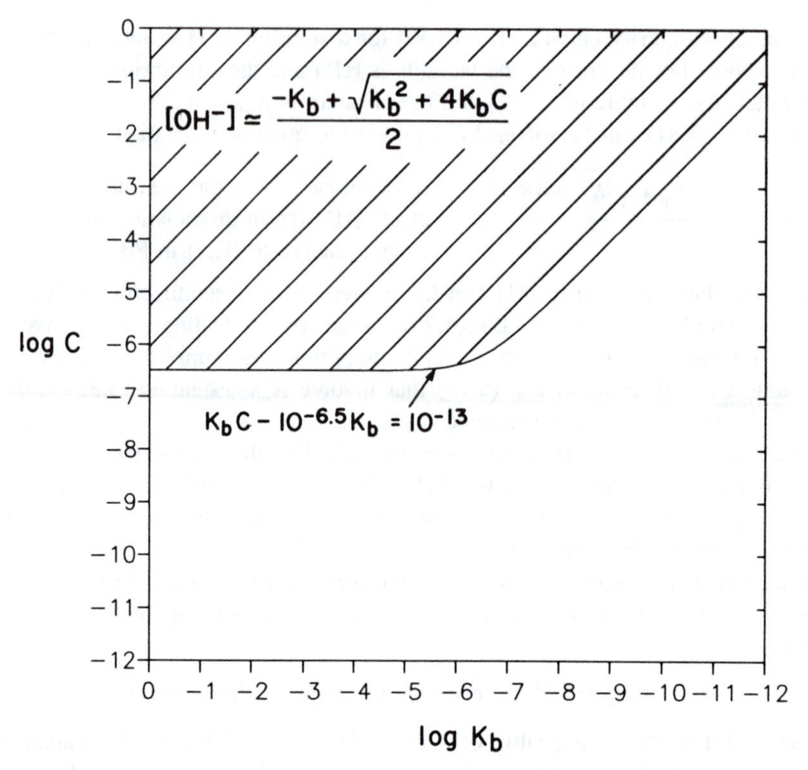

Figure 5.8*a* Region of validity for Eq. (5.82) given with cross-hatching. The equation for $[OH^-]$ results from the basic solution approximation.

Combining Eqs. (5.84) and (5.85) we obtain

$$[OH^-] \simeq \sqrt{K_b C + K_w} \qquad (5.86)$$

assumes weak base, does not neglect $[H^+]$ (from dissociation of water) relative to [HA] in the PBE

Once the value of $[OH^-]$ is known, then the remaining species can be calculated in a manner similar to that given in Eqs. (5.77) and (5.78). Equation (5.86) is the base analog for Eq. (5.38), with K_b substituted for K_a.

By analogy with Section 5.2.2.2, at least one of two criteria must be satisfied if Eq. (5.86) is to be applied with good accuracy, that is,

$$K_b C + K_w \geq 100 K_b^2 \qquad (5.87)$$

criterion for applicability of Eq. (5.86)

or

$$K_b C \leq 0.1 K_w \qquad (5.88)$$

alternate criterion for applicability of Eq. (5.86).

Equations (5.87) and (5.88) are the base analogs of Eqs. (5.41) and (5.42). Equation (5.86)

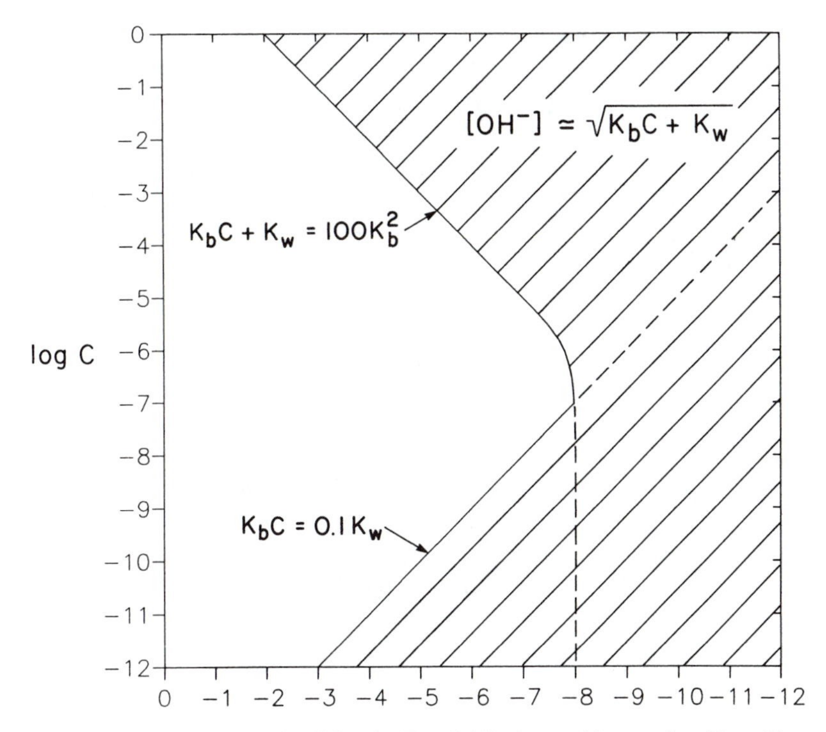

Figure 5.8*b* Region of validity for Eq. (5.86) given with cross-hatching. The equation for $[OH^-]$ results from the weak base approximation with $[H^+]$ not necessarily negligible.

provides a good approximation for all of the points in Figure 5.8*b* on and to the right of the solid, composite line.

5.3.2.3 Weak Base Approximation with $[H^+]$ Also Assumed Negligible. If the base is a weak base and K_b is small, and if C is not too small, we can assume the applicability of *both* Eqs. (5.80) and (5.85). In this case, the "$-[HA]$" correction on C in Eq. (5.81) is not needed, and Eq. (5.81) simplifies to

$$[OH^-] \simeq \sqrt{K_b C} \qquad (5.89)$$

assumes weak base, and that $[H^+]$ (from dissociation of water) is negligible relative to [HA] in PBE.

The remaining species can be calculated in a manner similar to that given in Eqs. (5.78) and (5.79). Equation (5.89) is the base analog of Eq. (5.45) with K_b substituted for K_a.

By analogy with Section 5.2.2.3, the set of criteria that will permit the a priori prediction of when we can and cannot apply Eq. (5.89) are contained in

$$0.1C \geq \sqrt{K_b C} \geq 10^{-6.5} M \qquad (5.90)$$

combined criteria for applicability of Eq. (5.89).

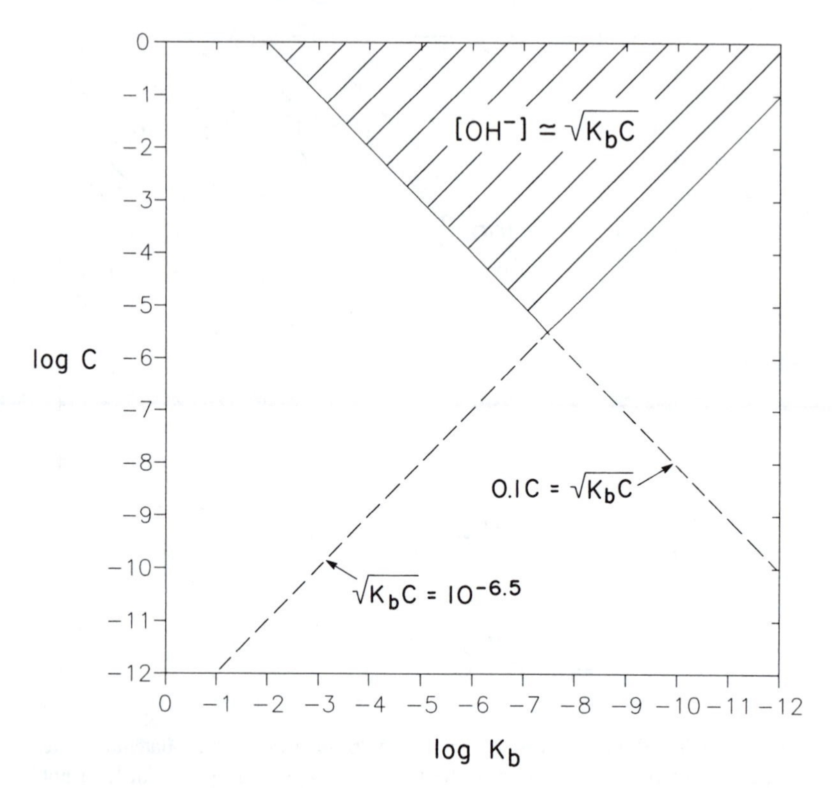

Figure 5.8*c* Region validity for Eq. (5.89) given with cross-hatching. The equation for $[OH^-]$ results from the simultaneous assumption of a basic solution and a weak base.

Equation (5.90) is the base analog of Eq. (5.48). Equation (5.89) provides a good approximation for all of the points in Figure 5.8*c* on and above the solid, composite line. As seen in a comparison of Figures 5.7 and 5.8*c*, within the region of validity of Eq. (5.89), we do indeed have $\text{pOH} \simeq -\frac{1}{2}(\log K_b + \log C)$. We note that $\text{pOH} = -\log\{OH^-\}$; when activity corrections can be neglected, $\text{pOH} = -\log[OH^-]$.

5.3.2.4 Strong Base Approximation, [H⁺] Not Necessarily Negligible (Optional). When a "strong" base NaA is placed in water, the A^- will be converted almost completely to HA according to reaction (5.62). In that case, we will have

$$[HA] \simeq C \qquad \text{strong base approximation} \tag{5.91}$$

and so Eq. (5.70) may be written as

$$[H^+] + C \simeq [OH^-]. \tag{5.92}$$

By substituting for $[H^+]$ and simplifying, Eq. (5.92) becomes

$$[OH^-]^2 - C[OH^-] - K_w \simeq 0. \tag{5.93}$$

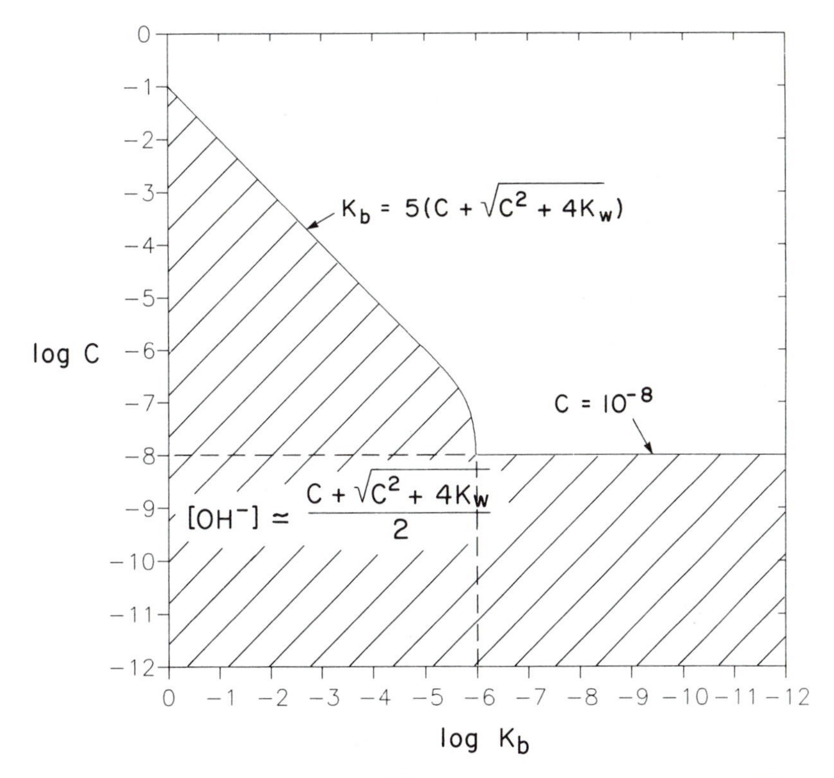

Figure 5.8*d* Region of validity for Eq. (5.94) given with cross-hatching. The equation for $[OH^-]$ results from the strong base approximation with $[H^+]$ not necessarily negligible.

Applying the quadratic equation to Eq. (5.93) gives two roots, one positive and one negative. Since we are interested in the positive root, we have

$$[OH^-] \simeq \frac{C + \sqrt{C^2 + 4K_w}}{2} \qquad \text{assumes strong base, with } [H^+] \text{ (from dissociation of water) not necessarily negligible.} \qquad (5.94)$$

The remaining species can be calculated in a manner similar to that given in Eqs. (5.78) and (5.79). Equation (5.94) is the base analog of Eq. (5.52).

By analogy with Section 5.2.2.4, the criteria for when we can use Eq. (5.94) are

$$K_b \geq 5\left[C + \sqrt{C^2 + 4K_w}\right] \qquad \text{criterion for applicability of Eq. (5.94)} \qquad (5.95)$$

or

$$C \leq 10^{-8} \qquad \text{alternate criterion for applicability of Eq. (5.94).} \qquad (5.96)$$

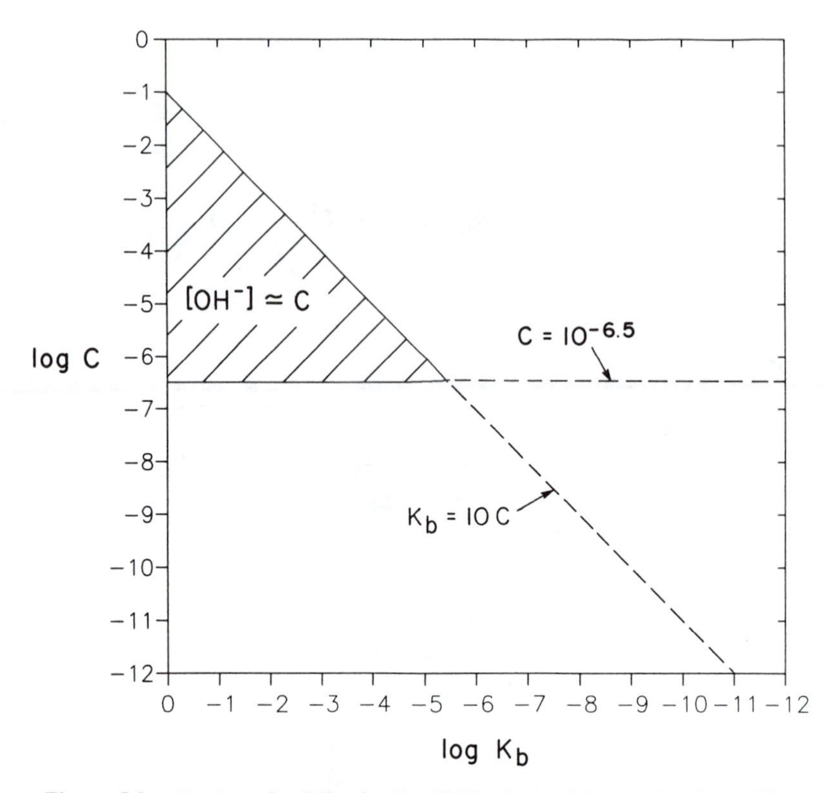

Figure 5.8e Region of validity for Eq. (5.97) given with cross-hatching. The equation for $[OH^-]$ results from the simultaneous assumption of a strong base and a basic solution.

Equation (5.95) is the base analog of Eq. (5.55) with K_b substituted for K_a. Equations (5.95) and (5.96) are plotted in Figure 5.8*d*. The domain of points that satisfies either Eq. (5.95) or (5.96) are those in Figure 5.8*d* that are on and below the solid composite line. Equation (5.94) may be applied to a good approximation in that area.

5.3.2.5 Strong Base Approximation, with $[H^+]$ Assumed Negligible. If A^- is a strong base and if $[H^+]$ is also negligible, we have that

$$[HA] \simeq C \qquad \text{strong base approximation} \tag{5.91}$$

and

$$[HA] \simeq [OH^-] \qquad \text{basic solution approximation.} \tag{5.80}$$

Therefore,

$$[OH^-] \simeq C \qquad \begin{array}{l}\text{assumes strong base, and that } [H^+] \\ \text{(from dissociation of water) is negligible} \\ \text{relative to [HA] in PBE.}\end{array} \tag{5.97}$$

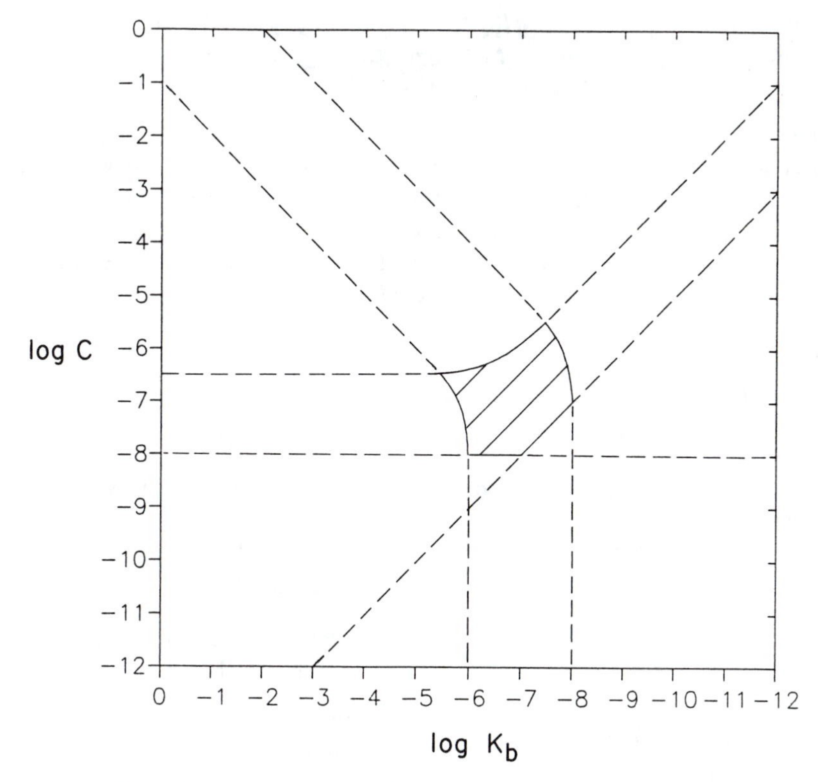

Figure 5.8*f* Region where none of the five approximation equations for $[OH^-]$ will give $[OH^-]$ in a solution of NaA to within ±10% or less.

The remaining species can be calculated in a manner similar to that given in Eqs. (5.78) and (5.79). Equation (5.97) is the base analog of Eq. (5.57).

For Eq. (5.97) to be valid to a good approximation, we require

$$C \geq 10^{-6.5} \qquad \text{criterion 1 for applicability of Eq. (5.97)} \tag{5.58}$$

and

$$K_b \geq 10C \qquad \text{criterion 2 for applicability of Eq. (5.97).} \tag{5.98}$$

Equations (5.58) and (5.98) apply to a solution of NaA in a manner similar to how Eqs. (5.58) and (5.60) apply to a solution of HA. Equations (5.58) and (5.98) are plotted in Figure 5.8*e*. The domain of points that satisfies both of these two equations are those that are on and to the left of the solid composite line. Equation (5.97) may be applied to a good approximation in that area. As seen in a comparison of Figures 5.7 and 5.8*e*, within the region of validity of Eq. (5.97), we do indeed have pOH $\simeq -\log C$. None of the approximation equations discussed in Section 5.3.2 will apply inside of the shaded domain in Figure 5.8*f*.

5.4 GENERAL EQUATIONS FOR SOLVING FOR THE SPECIATION OF SOLUTIONS OF H_2B, NaHB, AND Na_2B—NO ACTIVITY CORRECTIONS, NO SIMPLIFYING ASSUMPTIONS

For solutions of acids and bases different from HA and NaA, the reader may well suspect that solving for $[H^+]$ (and for the concentrations of all other species) will proceed in manners that are similar to those used for solutions of HA and NaA. In order to illustrate this point, we will briefly discuss how to set up problems involving solutions of chemicals with the generic formulas H_2B, NaHB, and Na_2B. In every case, we will consider solutions in which the total dissolved B is C M.

5.4.1 Solution of H_2B

The species present in a solution of H_2B are

$$H_2B \qquad HB^- \qquad B^{2-} \qquad H^+ \qquad OH^-.$$

Neglecting activity corrections, the equations needed to solve for the five unknowns in such systems are

$$K_1 = \frac{[HB^-][H^+]}{[H_2B]} \qquad \text{first acid dissociation constant of } H_2B \tag{5.99}$$

$$K_2 = \frac{[B^{2-}][H^+]}{[HB^-]} \qquad \text{second acid dissociation constant of } H_2B \tag{5.100}$$

$$K_w = [H^+][OH^-] \qquad \text{dissociation constant for water} \tag{5.3}$$

$$[H_2B] + [HB^-] + [B^{2-}] = C \qquad \text{MBE on total B} \tag{5.101}$$

$$[H^+] = [HB^-] + 2[B^{2-}] + [OH^-] \qquad \text{PBE for solution of } H_2B \text{ (simultaneously the ENE).} \tag{5.102}$$

Substituting into the PBE, using $[HB^-] = [HB^-]C/([H_2B]+[HB^-]+[B^{2-}])$ and $[B^{2-}] = [B^{2-}]\,C/([H_2B] + [HB^-] + [B^{2-}])$, we obtain

$$[H^+] = \frac{[HB^-]}{[H_2B] + [HB^-] + [B^{2-}]}C \;+\; 2\frac{[B^{2-}]}{[H_2B] + [HB^-] + [B^{2-}]}C \;+\; K_w/[H^+] \tag{5.103}$$

or by Eqs. (5.99) and (5.100),

$$[H^+] = \frac{1}{\dfrac{[H^+]}{K_1} + 1 + \dfrac{K_2}{[H^+]}}C \;+\; 2\frac{1}{\dfrac{[H^+]^2}{K_1K_2} + \dfrac{[H^+]}{K_2} + 1}C \;+\; K_w/[H^+]. \tag{5.104}$$

Equation (5.104) is one equation in one unknown, $[H^+]$. It may be solved by trial and error for any given values of K_1, K_2, K_w, and C using techniques described in Sections 5.2 and 5.3.

5.4.2 Solution of NaHB

The species present in a solution of NaHB are the same as those in a solution of H_2B, plus Na^+. Thus, there are six unknown concentrations in systems of this type. Four of the six equations needed to solve the problem are Eqs. (5.99), (5.100), (5.101), and (5.3). A fifth equation, the MBE for Na^+, is

$$[Na^+] = C \qquad \text{MBE on Na}^+ \text{ for solution of NaHB.} \tag{5.105}$$

A sixth, the PBE in this system, is

$$[H^+] + [H_2B] = [OH^-] + [B^{2-}] \qquad \text{PBE for solution of NaHB.} \tag{5.106}$$

Substituting into the PBE using $[H_2B] = [H_2B]C/([H_2B] + [HB^-] + [B^{2-}])$ and $[B^{2-}] = [B^{2-}]C/([H_2B] + [HB^-] + [B^{2-}])$ we obtain

$$[H^+] + \frac{1}{1 + \dfrac{K_1}{[H^+]} + \dfrac{K_1K_2}{[H^+]^2}}C = K_w/[H^+] + \frac{1}{\dfrac{[H^+]^2}{K_1K_2} + \dfrac{[H^+]}{K_2} + 1}C. \tag{5.107}$$

Equation (5.107) may be solved by trial and error for any given values of K_1, K_2, K_w, and C. A solution of NaHB will always have a higher pH than a solution of H_2B of the same concentration.

5.4.3 Solution of Na_2B

The species present in a solution of Na_2B are the same six as those in a solution of NaHB. Four of the six equations needed to solve the problem are Eqs. (5.99), (5.100), (5.101), and (5.3). The PBE for this solution is the fifth equation. In this case, it is

$$[H^+] + [HB^-] + 2[H_2B] = [OH^-] \qquad \text{PBE for solution of Na}_2\text{B.} \tag{5.108}$$

The sixth equation is the MBE for Na^+, and it is

$$[Na^+] = 2C \qquad \text{MBE on Na}^+ \text{ for solution of Na}_2\text{B.} \tag{5.109}$$

Substituting into the PBE using $[HB^-] = [HB^-]C/([H_2B] + [HB^-] + [B^{2-}])$ and $[H_2B] = [H_2B]C/([H_2B] + [HB^-] + [B^{2-}])$, we obtain

$$[H^+] + \frac{1}{\dfrac{[H^+]}{K_1} + 1 + \dfrac{K_2}{[H^+]}}C + 2\frac{1}{1 + \dfrac{K_1}{[H^+]} + \dfrac{K_1K_2}{[H^+]^2}}C = K_w/[H^+] \tag{5.110}$$

Equation (5.110) may be solved by trial and error for any given values of K_1, K_2, K_w, and C.

For a given value of C, a solution of Na_2B will always have a pH that is higher than the pH of a solution of NaHB, which will in turn have a pH that is higher than the pH of a solution of H_2B. Since it only has acidic properties, at 25°C/1 atm, a solution of H_2B will always have a pH that is less than 7. Since it only has basic properties, at 25°C/1 atm, a solution of Na_2B will always have a pH that is higher than 7.0. Since it has both acidic and basic properties, at 25°C/1 atm, a solution of NaHB can have a pH that is either higher or lower than 7. Whether the pH is higher or lower than 7.0 will depend on whether HB^- is more acidic ($HB^- = H^+ + B^{2-}$) than it is basic ($HB^- + H_2O = H_2B + OH^-$).

5.5 GENERAL APPROACH FOR SOLVING FOR THE SPECIATION INCLUDING ACTIVITY CORRECTIONS—IONIC STRENGTH AT EQUILIBRIUM NOT KNOWN A PRIORI

As pointed out in Section 5.1, it is the ionic strength at equilibrium that will set the values of the activity coefficients that appear in the various equilibrium constants. When the system does not contain significant amounts of background dissolved salts, and when reactions involving ions are under consideration, then the equilibrium ionic strength I will not be known accurately a priori because the concentrations of all of the ions that contribute significantly to I will not be known a priori.

The above type of situation can be handled using an iterative[4] approach. As an example, let us consider how to apply activity corrections to the aqueous solution of HA that was discussed in Section 5.2.1. For that particular system, we took $K_a = 10^{-5}\ m$, $C = 10^{-4}M$, and $K_w = 10^{-14}\ m^2$. If the only ionic species in the system are those involved in the equilibria under consideration, the first step (iteration) would be to just assume that the final equilibrium ionic strength is zero and all $\gamma_i = 1$. In other words, we would proceed just as we did in Section 5.2.1, completely neglecting activity corrections. At the conclusion of those calculations, we would end up with a second estimate of the equilibrium ionic strength computable as usual according to

$$I = \frac{1}{2}\sum_i c_i z_i^2 \tag{5.111}$$

For the system under consideration, we have

$$I = \frac{1}{2}([H^+](1)^2 + [OH^-](-1)^2 + [A^-](-1)^2] \tag{5.112}$$

Thus, the second estimate of the equilibrium I would then be given by

$$I_{\text{second estimate}} = \frac{1}{2}(2.69 \times 10^{-5} + 3.72 \times 10^{-10} + 2.71 \times 10^{-5})$$

$$= 2.70 \times 10^{-5}M \tag{5.113}$$

This is a very low number for I, so it will give γ_i values of essentially 1.0 for all ionic and neutral species. For example, according to the Debye-Hückel Law, at 298 K we would have for an ion with a ± 1 charge,

$$\log \gamma_{1,\ \text{second estimate}} = -0.51(\pm 1)^2(2.70 \times 10^{-5})^{1/2} \tag{5.114}$$

$$= 0.994 \simeq 1 \tag{5.115}$$

Thus, the implicit assumption made in Section 5.2.1 that activity coefficient corrections were not needed was justified.

However, let us assume that the ionic strength returned by Eq. (5.113) had been sufficiently high that γ_i values significantly different from 1.0 were obtained. What would we have done then? Obviously, the speciation that had been computed assuming $I = 0$ and all $\gamma_i = 1.0$ would have been in significant *error*. Nevertheless, that speciation

[4] Webster defines "iterate" as: "To utter, or do a second time, or many times; to repeat."

obviously provides better estimates of the actual concentrations than we had prior to doing any computations at all. Therefore, the key to the iterative solution approach is to use the best *first estimate* of I (zero, in this case) to allow computation of the speciation as well as the *second estimates* of I and the γ_i. The second estimates of the γ_i can then be used in a second speciation calculation. That second speciation can be used to produce better *third estimates* of I and the γ_i which may then be used in a third speciation calculation. The process is repeated (iterated) as many times as needed until the speciation and the activity coefficient values no longer change.

It is instructive to consider in specific terms how the calculations may be carried out with activity coefficients that are revised with each successive iteration. The process can be simplified by realizing that, as pointed out in Section 5.1.2, equations like Eq. (5.12), (5.73), (5.104), (5.107), and (5.110) are valid for both infinite dilution constants when activity corrections can be neglected, *and* for ${}^cK$ type constants when activity corrections are needed. Thus, when we have an *estimate* of the equilibrium ionic strength, for each iteration we can take the equilibrium constant values as being the ${}^cK$ values for the ionic strength as computed based on the *previous* iteration. The overall process is outlined in Table 5.4.

For a solution of HA, the two constants for which we need to consider activity corrections are K_a and K_w, so

$$ {}^cK_a = K_a \frac{\gamma_{HA}}{\gamma_{H^+}\gamma_{A^-}} \tag{5.116} $$

$$ {}^cK_w = K_w \gamma_{H_2O}/(\gamma_{H^+}\gamma_{OH^-}). \tag{5.117} $$

TABLE 5.4. Summary of process for applying the iterative procedure for calculating a given chemical speciation, including the effects of activity corrections.

Iteration	Steps
first	1) Assume I to be given by concentrations of background salts (if any). When no background salts are present, take initial I to be zero. Calculate all initial γ_i. When no background salts are present, then all γ_i will be 1.0. 2) Compute first estimate of speciation.
second and subsequent	1) Compute I based on speciation from previous iteration. 2) Compute second estimate γ_i values based on new estimate of I. 3) Compute second estimate of speciation using those γ_i values; and 4) Check to see if speciation has changed significantly† as compared to previous estimate. If yes, do another iteration. If no, stop and accept calculated speciation.

† "Significantly" may be taken to mean ±5 percent for any species.

Usually, we will take the activity coefficients of neutral species like HA and H_2O to be equal to 1. This will be valid under most circumstances, but not in concentrated solutions such as sea water. For the activity coefficients of charged species, it will usually be sufficient to calculate generic activity coefficients, γ_1 for charges of ±1, γ_2 for charges of ±2, etc. Under these circumstances, Eqs. (5.116) and (5.117) become

$$ {}^{c}K_a = K_a/\gamma_1^2 \tag{5.118}$$

$$ {}^{c}K_w = K_w/\gamma_1^2. \tag{5.119}$$

For each iteration, we utilize the set of γ_i values for that iteration. For a solution of HA, using ${}^{c}K$ values, Eq. (5.12) becomes

$$[H^+] - \frac{{}^{c}K_a}{[H^+] + {}^{c}K_a}C - {}^{c}K_w/[H^+] = 0 \tag{5.120}$$

Within each iteration, Eq. (5.120) is solved for $[H^+]$ by trial and error. We then calculate the speciation including γ_i corrections based on the current iteration's ${}^{c}K$ values according to the equations

$$[OH^-] = {}^{c}K_w/[H^+] \tag{5.121}$$

$$[A^-] = \frac{{}^{c}K_a}{{}^{c}K_a + [H^+]}C \tag{5.122}$$

$$[HA] = C - [A^-] \tag{5.123}$$

Since Eq. (5.123) is strictly based on MBE considerations, it does not require any explicit activity corrections. The new speciation is then used to calculate the I and γ_i values to be used in the next iteration.

We note that both $[OH^-]$ and $[A^-]$ appear as terms in Eq. (5.120). Therefore, in addition to computing $[H^+]$, the trial and error solution seeking the root value of $[H^+]$ could be easily modified to output each of the different species concentrations and therefore also I so that once the root is found, then the speciation as well as I will automatically also be available for the next iteration.

5.6 REFERENCES

Butler, J. N. 1964. *Ionic Equilibrium: A Mathematical Approach.* New York: Addison-Wesley.

Harris, D. C. 1987. *Quantitative Chemical Analysis.* New York: W. H. Freeman.

Narasaki, H. 1987. "The range of application of the approximation formulae in acid-base equilibria," *Fresenius Z. Anal. Chem.*, **328**: 633–638.

Pankow, J. F. 1992. *Aquatic Chemistry Problems.* Portland: Titan Press-OR, P.O. Box 91399, Portland, Oregon, 97291-1399.

Selby, S. N. 1971. *Standard Mathematical Tables, 19th Edition.* Cleveland: CRC Press.

6

pH as a Master Variable

6.1 INTRODUCTION

In Chapter 5, we considered how to compute the chemistry of solutions of HA and solutions of NaA. For a given value of the acidity constant K for the acid HA and for a specified background ionic composition, we learned that all such solutions will take on pH values and general speciation distributions that are both *specific* as well as *predictable*.[1] In this chapter, we will not limit our discussions to solutions of just HA or just NaA, but will also consider systems in which HA (or NaA) *plus* variable amounts of strong acid and/or base may be present. For any given value of C, this will allow us to consider systems with pH values that are *different* from the two specific pH values that are obtained for a solution of HA and a solution of NaA.

The pH of a solution of HA can be raised by adding a strong base such as sodium hydroxide (NaOH). The pH of a solution of NaA also can be raised by adding strong base. Conversely, the pH of a solution of NaA can be lowered by adding a strong acid such as hydrochloric acid (HCl). The pH of a solution of HA can also be lowered by adding strong acid. If the volume does not change significantly as a result of these changes, then the total concentration of total A (i.e., A_T) in the system will remain constant and equal to C:

$$A_T = [HA] + [A^-] = C \tag{6.1}$$

[1]In Chapter 5, we used a subscript a on the acidity constant K_a. In that chapter, the purpose of that subscript was to help distinguish K_a from the basicity constant K_b, as well as to help facilitate comparisons between the two constants. Since we can always express K_b as K_w/K_a, for this chapter and indeed for the rest of the book, we will usually not use the subscript a on the acidity constant.

When considering how the concentrations of species like HA and A^- change as a function of pH, it is very useful to deal with them in terms of fractional concentration factors. Therefore, we define the fractional factors

$$\alpha_0 = \begin{array}{l}\text{fraction that}\\ \text{has lost}\\ \text{0 protons}\end{array} = \frac{[\mathrm{HA}]}{C} = \frac{[\mathrm{HA}]}{[\mathrm{HA}]+[\mathrm{A}^-]} \tag{6.2}$$

$$\alpha_1 = \begin{array}{l}\text{fraction that}\\ \text{that has}\\ \text{lost 1 proton}\end{array} = \frac{[\mathrm{A}^-]}{C} = \frac{[\mathrm{A}^-]}{[\mathrm{HA}]+[\mathrm{A}^-]}. \tag{6.3}$$

By definition, we have

$$[\mathrm{HA}] = \alpha_0 C \tag{6.4}$$

and

$$[\mathrm{A}^-] = \alpha_1 C. \tag{6.5}$$

The two α values sum to unity, that is,

$$\alpha_0 + \alpha_1 = 1. \tag{6.6}$$

Equations (6.2) and (6.3) can be transformed into functions of $[H^+]$ as follows:

$$\alpha_0 = \frac{[\mathrm{HA}]}{[\mathrm{HA}]+[\mathrm{A}^-]} = \frac{1}{1+[\mathrm{A}^-]/[\mathrm{HA}]} \tag{6.7}$$

$$\alpha_1 = \frac{[\mathrm{A}^-]}{[\mathrm{HA}]+[\mathrm{A}^-]} = \frac{1}{[\mathrm{HA}]/[\mathrm{A}^-]+1}. \tag{6.8}$$

The $[A^-]/[HA]$ ratio is governed by the pH and by the acidity constant K. Specifically,

$$K = \frac{\{\mathrm{H}^+\}\{\mathrm{A}^-\}}{\{\mathrm{HA}\}} = \frac{[\mathrm{H}^+]\gamma_{\mathrm{H}^+}[\mathrm{A}^-]\gamma_{\mathrm{A}^-}}{[\mathrm{HA}]\gamma_{\mathrm{HA}}}. \tag{6.9}$$

Therefore,

$$[\mathrm{A}^-]/[\mathrm{HA}] = \frac{K\gamma_{\mathrm{HA}}}{[\mathrm{H}^+]\gamma_{\mathrm{H}^+}\gamma_{\mathrm{A}^-}}. \tag{6.10}$$

When activity corrections on the *neutral species* only can be neglected, we have

$$[\mathrm{A}^-]/[\mathrm{HA}] = \frac{K}{[\mathrm{H}^+]\gamma_{\mathrm{H}^+}\gamma_{\mathrm{A}^-}} \tag{6.11}$$

and so Eqs. (6.7) and (6.8) become

$$\alpha_0 = \frac{1}{1+K/[\mathrm{H}^+]\gamma_{\mathrm{H}^+}\gamma_{\mathrm{A}^-}} = \frac{[\mathrm{H}^+]\gamma_{\mathrm{H}^+}\gamma_{\mathrm{A}^-}}{[\mathrm{H}^+]\gamma_{\mathrm{H}^+}\gamma_{\mathrm{A}^-}+K} \tag{6.12}$$

$$\alpha_1 = \frac{1}{[\mathrm{H}^+]\gamma_{\mathrm{H}^+}\gamma_{\mathrm{A}^-}/K+1} = \frac{K}{[\mathrm{H}^+]\gamma_{\mathrm{H}^+}\gamma_{\mathrm{A}^-}+K}. \tag{6.13}$$

When the solution is sufficiently dilute that all activity coefficients can be neglected, then Eq. (6.9) becomes

$$K = \frac{[H^+][A^-]}{[HA]} \tag{6.14}$$

and Eqs. (6.12) and (6.13) become simply

$$\alpha_0 = \frac{1}{1 + K/[H^+]} = \frac{[H^+]}{[H^+] + K} \tag{6.15}$$

$$\alpha_1 = \frac{1}{[H^+]/K + 1} = \frac{K}{[H^+] + K}. \tag{6.16}$$

We recall that when ${}^cK$ is substituted for K, Eq. (6.14) also applies for solutions containing significant amounts of background salts. The same therefore applies for Eqs. (6.15) and (6.16).

For both Eqs. (6.15) and (6.16), the second expression in each is easiest to remember. This may be understood by reflecting on how α_0 and α_1 should behave as functions of $[H^+]$. Neglecting activity corrections, we first note that by Eq. (6.11), increasing $[H^+]$ will tend to shift the HA/A^- equilibrium in a manner that will favor the presence of relatively more HA, and relatively less A^-. Now, in the expressions under consideration, both α_0 and α_1 have the same denominator, that is, $([H^+]+K)$. Therefore, since α_0 represents the fraction of the *acid form*, it is logical that its function should have $[H^+]$ in the numerator. That way, α_0 will *increase* with increasing $[H^+]$. By Eq. (6.6), then, K must be in the numerator for α_1. If K is in the numerator for α_1, and $([H^+]+K)$ is the denominator, then α_1 will decrease with increasing $[H^+]$.

We can conclude that as strong acid is added and $[H^+]$ increases, then the value of α_0 increases and tends to 1.0, and the value of α_1 decreases and tends to 0.0. Conversely, as strong base is added, the value of α_0 decreases and tends to 0.0, and the value of α_1 increases and tends to 1.0. However, even when activity corrections are taken into consideration, neither $[H^+]\gamma_{H^+}\gamma_{A^-}$ nor K can ever be zero, and so neither α_0 nor α_1 can ever be identically equal to either 0 or 1.

6.2 LOG CONCENTRATION VS. pH PLOTS FOR MONOPROTIC SYSTEMS

It is very useful and instructive to plot the logs of the concentrations of the species in a solution containing HA and A^-. A logical independent (x-axis) variable for the plots is pH. Let us consider the four species H^+, OH^-, HA, and A^-. We recall that

$$pH = -\log\{H^+\}. \tag{6.17}$$

When we can neglect activity corrections, we have

$$pH = -\log[H^+] \equiv p_cH \tag{6.18}$$

or equivalently,

$$\log[H^+] = \log 10^{-pH} = -pH. \tag{6.19}$$

We note that

$$d\log[H^+]/dpH = -1. \tag{6.20}$$

Under the same conditions,

$$\log[OH^-] = \log(K_w/[H^+]) = \log K_w - \log[H^+] \tag{6.21}$$

$$= \log K_w + pH \tag{6.22}$$

$$d\log[OH^-]/dpH = +1. \tag{6.23}$$

Equations (6.19) and (6.22) are very simple, linear functions of pH.

For log[HA], we have

$$\log[HA] = \log\alpha_0 + \log C. \tag{6.24}$$

Neglecting activity corrections,

$$\log[HA] = \log\frac{[H^+]}{[H^+]+K} + \log C. \tag{6.25}$$

For $\log[A^-]$, we have

$$\log[A^-] = \log\alpha_1 + \log C. \tag{6.26}$$

Neglecting activity corrections,

$$\log[A^-] = \log\frac{K}{[H^+]+K} + \log C. \tag{6.27}$$

Lines for $\log\alpha_0$ and $\log\alpha_1$ based on Eqs. (6.15) and (6.16) are presented in Figure 6.1 for the case when $K = 10^{-5}$. Note that Figure 6.1 is independent of C. Lines for $\log[H^+]$, $\log[OH^-]$, log[HA], and $\log[A^-]$ as based on Eqs. (6.19), (6.22), (6.25), and (6.27) are presented in Figure 6.2 for the system $C = 10^{-3}M$ and $K = 10^{-5}$. Activity corrections for both Figures 6.1 and 6.2 have been neglected even though the plot extends to pH values as low as pH $= 1.0\,(I \simeq 0.1)$ and as high as pH $= 14\,(I \simeq 1.0)$. Except that they are shifted down by $\log C$, the curves for log[HA] and $\log[A^-]$ are identical to the curves for $\log\alpha_0$ and $\log\alpha_1$ (see Eqs. (6.24) and (6.26)). We also note that the fact that the log[HA] line is a mirror image of the $\log[A^-]$ line (as reflected around pH = pK) is due to the perfectly symmetrical nature of the functions α_0 and α_1.

In contrast to $\log[H^+]$ and $\log[OH^-]$, log[HA] and $\log[A^-]$ are not linear functions of pH. However, they both asymptotically approach linear limits in two distinct regions. The width of the region that separates the two asymptotic regions is approximately 2 pH units. Its center is determined by the value of pK. Therefore, the separating, transition region is defined by

$$pH = pK \pm 1 \quad \text{transition region separating two asymptotic regions in a log concentration vs. pH plot for an } HA/A^- \text{ system.} \tag{6.28}$$

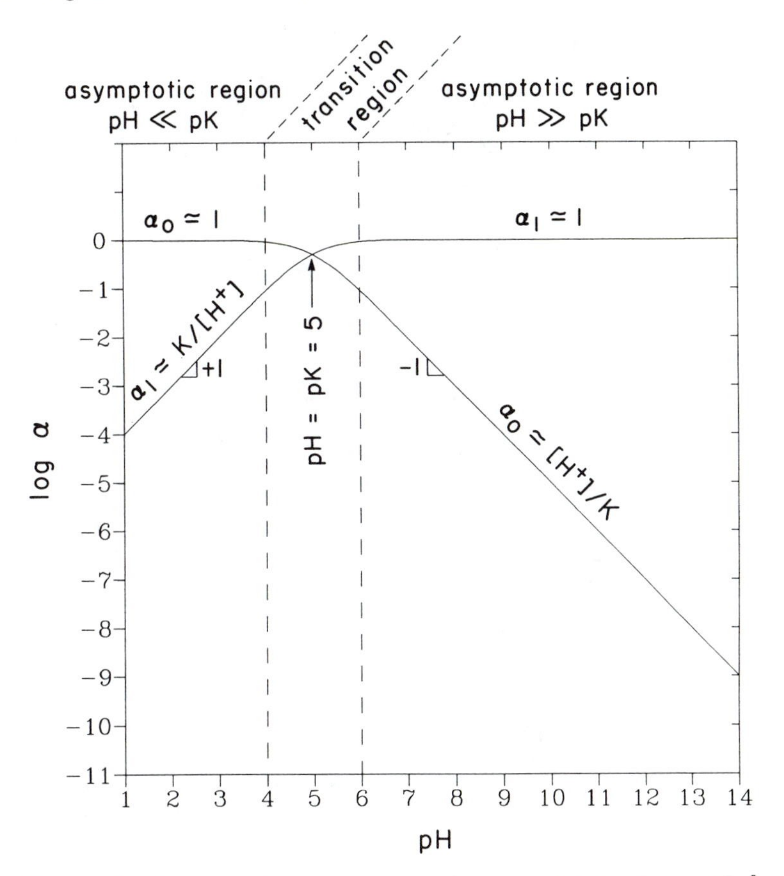

Figure 6.1. Log a_0 and log a_1 plot for monoprotic acid HA with $K = 10^{-5}$. Behaviors in asymptotic regions are given. Activity corrections neglected (not valid at very low or very high pH).

For Figure 6.2, pK = 5, and so the transition region extends from about pH 4 to 6. One asymptotic region is pH < 4; the other is pH > 6. In the two asymptotic regions, the functions for $\log \alpha_0$ and $\log \alpha_1$ become simplified. In particular, they become linear functions of pH. This is shown as follows.

6.2.1 pH ≪ p*K*

When

$$[H^+] > 10K \tag{6.29}$$

then

$$pH < pK - 1. \tag{6.30}$$

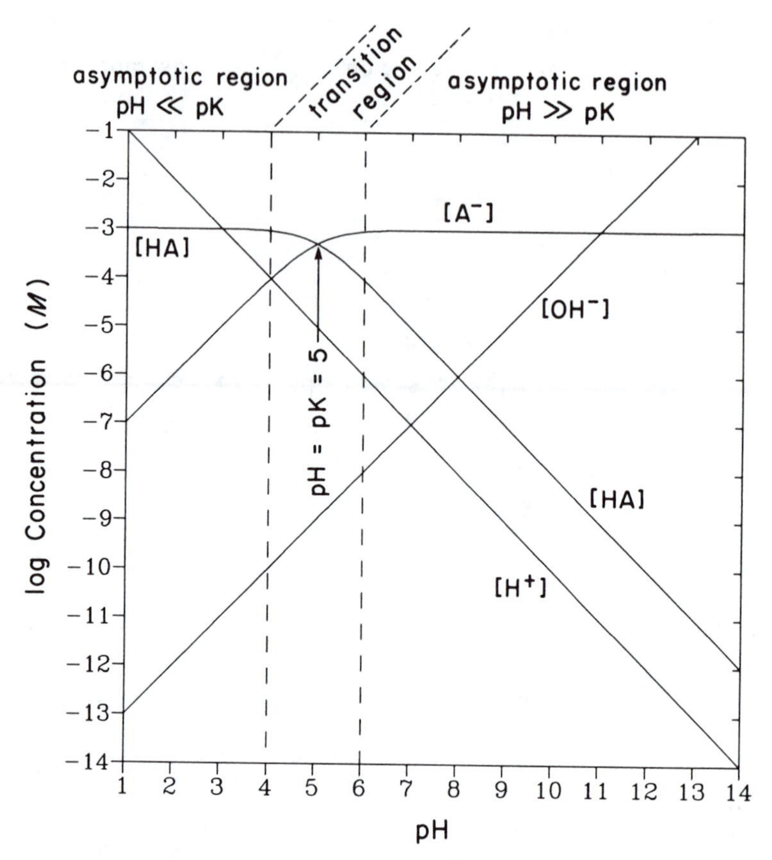

Figure 6.2. Log concentration (M) vs. pH plot for a monoprotic acid system in which $K = 10^{-5}$, $C = 10^{-3}F$, and $K_w = 10^{-14}$. Locations of asymptotic regions are given. Activity corrections neglected (not valid at very low or very high pH).

It is very common to see Eq. (6.30) abbreviated as

$$\text{pH} \ll \text{p}K. \tag{6.31}$$

Since pH values that are only about one unit less than pK will satisfy Eq. (6.30), some might feel that the $\ll$ sign in Eq. (6.31) is an overstatement of Eq. (6.30). The use of the $\ll$ sign may be traced to the fact in *chemical* terms, a pH value that is even one unit removed from the pK will create an HA/A^- balance that is very different from when pH = pK.

When Eq. (6.31) is satisfied, then by Eq. (6.25), we have

$$\log[\text{HA}] \simeq \log\frac{[\text{H}^+]}{[\text{H}^+]} + \log C \tag{6.32}$$

$$\simeq 0 + \log C = \log C \tag{6.33}$$

and by Eq. (6.15),

$$\log \alpha_0 \simeq \log \frac{[H^+]}{[H^+]} = 0. \tag{6.34}$$

We also have

$$d \log[HA]/dpH = d \log \alpha_0/dpH \simeq 0. \tag{6.35}$$

Thus, both [HA] and α_0 are essentially independent of pH when pH $\ll$ pK.

In this same region, by Eq. (6.27), we have

$$\log[A^-] \simeq \log \frac{K}{[H^+]} + \log C \tag{6.36}$$

and by Eq. (6.16),

$$\log \alpha_1 \simeq \log K/[H^+]. \tag{6.37}$$

We also have

$$d \log[A^-]/dpH = d \log \alpha_1/dpH \tag{6.38}$$

$$\simeq d \log K/dpH - d \log[H^+]/dpH \tag{6.39}$$

$$\simeq 0 + 1 = +1. \tag{6.40}$$

Thus, while [HA] and α_0 are essentially independent of pH when pH $\ll$ pK, both [A^-] and α_1 are dependent on pH in this region. These results are summarized in Tables 6.1 and 6.2.

6.2.2 pH $\gg$ pK

When

$$[H^+] < K/10 \tag{6.41}$$

then

$$pH > pK + 1. \tag{6.42}$$

It is common to see Eq. (6.42) abbreviated as

$$pH \gg pK. \tag{6.43}$$

TABLE 6.1. Asymptotic behavior of [HA] and log[HA] and [A^-] and log[A^-] for two regions, pH $\ll$ pK and pH $\gg$ pK.

pH $\ll$ pK ($[H^+] \gg K$)	pH $\gg$ pK ($[H^+] \ll K$)
$[HA] \simeq C$, independent of pH	$[A^-] \simeq C$, independent of pH
$\log[HA] \simeq \log C$, independent of pH	$\log[A^-] \simeq \log C$, independent of pH
$\log \alpha_0 \simeq 0$, $\log \alpha_1 \simeq \log(K/[H^+])$	$\log \alpha_0 \simeq \log([H^+]/K)$, $\log \alpha_1 \simeq 0$

TABLE 6.2. Asymptotic behavior of $d \log \alpha_0/\text{dpH}$, $d \log[\text{HA}]/\text{dpH}$, $d \log \alpha_1/\text{dpH}$, and $d \log[\text{A}^-]/\text{dpH}$ for $\text{pH} \ll \text{p}K$ and $\text{pH} \gg \text{p}K$.

	$\text{pH} \ll \text{p}K$ ($[\text{H}^+] \gg \text{K}$)	$\text{pH} \gg \text{p}K$ ($[\text{H}^+] \ll \text{K}$)
$d \log \alpha_0/\text{dpH}$ and $d \log[\text{HA}]/\text{dpH}$	0	−1
$d \log \alpha_1/\text{dpH}$ and $d \log[\text{A}^-]/\text{dpH}$	+1	0

When Eq. (6.43) is satisfied, by Eq. (6.25), we have

$$\log[\text{HA}] \simeq \log \frac{[\text{H}^+]}{K} + \log C \tag{6.44}$$

and by Eq. (6.15),

$$\log \alpha_0 \simeq \log \frac{[\text{H}^+]}{K}. \tag{6.45}$$

By taking the derivative of Eq. (6.44) with respect to pH, we obtain

$$d \log[\text{HA}]/\text{dpH} = d \log \alpha_0/\text{dpH} \simeq d \log[\text{H}^+]/\text{dpH} - d \log K/\text{dpH} \tag{6.46}$$

$$\simeq -1 - 0 = -1. \tag{6.47}$$

Thus, both [HA] and α_0 are dependent on pH when $\text{pH} \gg \text{p}K$.

In the same region, by Eq. (6.27),

$$\log[\text{A}^-] \simeq \log \frac{K}{K} + \log C \tag{6.48}$$

$$\simeq 0 + \log C = \log C. \tag{6.49}$$

By Eq. (6.16), in this region,

$$\log \alpha_1 \simeq \log K/K \tag{6.50}$$

$$\simeq 0. \tag{6.51}$$

Taking the derivative of Eq. (6.48) with respect to pH, we obtain

$$d \log[\text{A}^-]/\text{dpH} = d \log \alpha_1/\text{dpH} \simeq 0. \tag{6.52}$$

Thus, both $[\text{A}^-]$ and α_1 are essentially independent of pH in this region. These results are summarized in Tables 6.1 and 6.2.

6.3 HAND DRAWING THE LOG CONCENTRATION VS. pH DIAGRAM FOR THE HA/A^- SYSTEM

As discussed above, log concentration diagrams for any HA/A^- system can be prepared easily when activity corrections are neglected. Situations involving relatively large amounts of one or more dissolved salts are also straightforward inasmuch as ${}^cK$ can be used in place of K in the above equations. Once the pertinent equations are written down, and when a computer is used for the calculations, little chemical insight is needed to prepare diagrams like Figure 6.2. This is not the case when quickly preparing such a diagram by hand, as might occur in a classroom, or when examining a given HA/A^- problem on "the back of an envelope." In such cases, it is very helpful to have an understanding of the behavior of the various species HA and A^- as functions of pH, pK, and C. The steps that can be followed when hand drawing a log concentration vs. pH plot for an HA/A^- system are summarized in Table 6.3. The reader should study that information until it is well understood.

It is useful to point out that Steps 3 and 4 in Table 6.3 indicate that extrapolating the linear portion of the $\log[A^-]$ line from pH $\ll$ pK to pH = pK, and extrapolating the linear portion of the log[HA] line from pH $\gg$ pK to pH = pK, leads to an intersection of the two extrapolated lines at the point (x = pK, $y = \log C$). That this should be the case may be shown as follows. From Eqs. (6.36) and (6.44), these two linear lines are described by the equations

$$\log[A^-] = \log(K/[H^+]) + \log C \tag{6.53}$$

$$\log[HA] = \log([H^+]/K) + \log C. \tag{6.54}$$

When $[H^+] = K$, both of these equations prescribe a concentration equal to C, and so the extrapolated portions of the lines do indeed intersect at (pK, $\log C$).

6.4 USE OF LOG CONCENTRATION VS. pH DIAGRAMS IN GRAPHICAL PROBLEM SOLVING FOR SOLUTIONS OF HA AND SOLUTIONS OF NaA (NO ADDITIONAL STRONG ACID OR BASE)

Prior to the easy access to programmable hand calculators and computers that now exists, log concentration vs. pH plots like Figure 6.2 were used extensively in solving problems involving solutions of the acid HA and the base NaA (e.g., Freiser 1967, 1970; Freiser and Fernando 1963). Analogous plots were used to solve problems involving diprotic systems like H_2B, NaHB, and H_2B.

While problems involving solutions of HA, NaA, H_2B, etc. are now solved easily using calculators or computers together with trial and error-based methods (see Sections 5.2.1 and 5.3.1), the instructional utility of log concentration vs. pH plots in such problems remains undiminished. Indeed, much chemical insight can continue to be gained through their use. We will examine this matter for solutions of HA as well as for solutions of NaA.

TABLE 6.3. Steps to be followed when hand drawing a log concentration vs. pH plot for an HA/A^- system. (Activity corrections neglected.)

1. Prepare a grid with the x-axis labeled "pH" and the y-axis labeled "log concentration." For most diagrams, it will be suitable to have the x-axis extend from pH = 1 to pH = 14. The y-axis should extend over an equal number of log units, in this case from log[ ] = −1 to log[ ] = −14; the scales of the x- and y-axes are then equal, which makes drawing slopes of +1 and −1 easy.
2. Pick the value of C of interest. Then,
   a) for pH $\leq$ pK − 1, draw log[HA] = log C; and
   b) for pH $\geq$ pK + 1, draw log[A^-] = log C.
3. For pH $\leq$ pK − 1, draw log[A^-] such that d log[A^-]/dpH = +1, and such that when extrapolated to pH = pK, a dashed portion of the line intersects the horizontal line log[ ] = log C at pH = pK.
4. For pH $\geq$ pK + 1, draw log[HA] such that d log[HA]/dpH = −1, and such that when extrapolated to pH = pK, a dashed portion of the line intersects the horizontal line log[ ] = log C at pH = pK.
5. Although the log[HA] line is actually not exactly linear anywhere in the diagram, the degree of non-linearity is by far the most severe in the transition range pH = pK ± 1. At pH = pK − 1, α_0 = 0.909, and log 0.909 = −0.041. Thus, for pH = pK − 1, log[HA] = log C − 0.041 $\simeq$ log C and so is approximately given by the line drawn in Step 2a (log[HA] = log C). For pH = pK, $\alpha_0 = \alpha_1$ = 1/2. Since log(1/2) = −0.301, the actual [HA] line does not pass through the point (x = pK, y = log C), but rather through the point (x = pK, y = log C − 0.301). For pH = pK + 1, α_0 = 0.091. Note that log 0.091 = −1.041. Thus, log[HA] passes through the point (pH = pK + 1, y = log C − 1.041), which is close to the line drawn in Step 4.
6. As with the log[HA] line, the degree of non-linearity in the log[A^-] line is by far the most severe in the transition range pH = pK ± 1. At pH = pK − 1, α_1 = 0.091. Since log 0.091 = −1.041, the log[A^-] line passes through the point (pH = pK + 1, y = log C − 1.041), which is very near the line drawn in Step 3. For pH = pK, α_0 = α_1 = 1/2. Since log(1/2) = −0.301, the actual [A^-] line does not pass through the point (x = pK, y = log C), but rather through the point (x = pK, y = log C − 0.301), and intersects with the log[HA] line at that point. For pH = pK + 1, α_1 = 0.909, and log 0.909 = −0.041. Thus, for pH = pK + 1, log[A^-] = log C − 0.041 $\simeq$ log C, and so is approximately given by the line drawn in Step 2b (log[A^-] = log C). The mirror image nature of the log[HA] and log[A^-] lines is apparent.
7. Complete the plot by adding the log[H^+] and log[OH^-] lines. At 25°C/1 atm, when activity corrections are neglected, these two lines intersect at pH = 7.

6.4.1 Applying Log Concentration vs. pH Diagrams in Understanding the Chemistry of Solutions of HA

For a solution of the acid HA, when no additional strong acid or base is present, the PBE (and the ENE) is

$$[H^+] = [OH^-] + [A^-] \qquad \text{PBE (and ENE) for a solution of HA.} \tag{6.55}$$

Since there are four concentration variables in this system, we know from Chapter 5 that the equilibrium speciation will be that which simultaneously satisfies the four equations governing these variables. These equations are embodied in the two constants K and K_w, the applicable MBE on total A, and the PBE. The method of analysis that will be followed here is the graphical equivalent of the trial and error solution given in Section 5.2.1.

The beauty of diagrams like Figure 6.2 is that at *every point* along the pH axis, the relationships embodied in K, K_w, and the MBE on total A are *always simultaneously satisfied.* Therefore, neglecting activity coefficients and when $K_w = 10^{-14}$, then to determine the pH of a $10^{-3}F$ solution of HA with $K = 10^{-5}$, all we need to do is *find the specific pH value in Figure 6.2 at which the PBE (Eq. (6.55)) is also satisfied.* Values of pH that are lower than that specific pH constitute solutions that must contain some strong acid in addition to the $10^{-3}F$ HA. Values of pH that are higher than that specific pH constitute solutions that must also contain some strong base in addition to the HA. In the latter case, one of those solutions will be a simple solution of NaA with $C = 10^{-3}F$. This is because, as noted in Chapter 5, a C F solution of the salt NaA is equivalent to a C F solution of HA that also contains C F NaOH.

An examination of Figure 6.2 indicates that for these values of K and C, the PBE for a solution of HA is satisfied to a very good approximation at the intersection described by

$$\log[H^+] = \log[A^-] \qquad (6.56)$$

intersection describing one type of graphical solution to PBE for a solution of HA, assumes validity of acidic solution approximation (C and K not too low).

Indeed, at the pH of the intersection for this case, the $\log[OH^-]$ line lies very far (6 log units!) below the $\log[A^-]$ line. Thus, $[OH^-]$ is clearly negligible in Eq. (6.55). This means that the "acidic solution approximation," Eq. (5.23), is valid for this system. The equilibrium speciation for this system may therefore be accurately obtained by picking off the values of the various concentrations from a vertical line drawn at the pH of the intersection described by Eq. (6.56).

If a very detailed and accurate view of the region of the above intersection was prepared, with pH ranging from about 3 to 5 and log[ ] ranging from −5 to −3, it would be possible to graphically determine that the pH at the intersection is 4.02. Even using only Figure 6.2, it can be deduced that the equilibrium pH is very close to 4. These values may be compared to the pH value of 4.02 that can be calculated using the equation for $[H^+]$ that results from the acidic solution approximation, that is, using Eq. (5.26).

Although the acidic solution approximation is extremely well justified here, the proximity of the equilibrium pH to pH = pK (where A^- contributes significantly to C) means that the amount of HA that has dissociated is not entirely negligible relative to C. Nevertheless, the error is small in also making the weak acid assumption and computing that pH = 4.0 using Eq. (5.45). These conclusions are consistent with Figure 5.6*c*.

Indeed, $C = 10^{-3}F$ and $K = 10^{-5}$ is just on the left boundaryline of the shaded region of that figure.

As we well know from Section 5.2.2, the ability to make any given simplifying assumption(s) for solutions of HA is intimately connected to the values of both K and C. An extremely useful instructional tool for studying these matters may be constructed from: 1) a log concentration vs. pH gridded plot showing lines only for $[H^+]$ and $[OH^-]$; and 2) a generic, transparent acetate sheet with [HA] and $[A^-]$ lines. Moving the acetate sheet around on the grid allows the examination of plots for any desired pair of K and C. This tool can be prepared by photocopying Figure 6.3 onto a plain white piece of paper, and photocopying Figure 6.4 onto an acetate transparency. (Figure 5.4 is also extremely useful in these considerations.)

When the acetate sheet is laid on the grid such that the x and y axes of both figures are aligned, the resulting diagram corresponds to the case when $K = 10^{-7}$ and $C = 10^{-3}F$. For a solution of HA with these characteristics, we see that both the acidic solution *and* weak acid approximations are valid. As far as the speciation is concerned, we can immediately conclude that $[H^+] = 1\times10^{-5}$, $[A^-] = 1\times10^{-5}$, $[HA] = 1.0\times10^{-3}$, and $[OH^-] = 1.0\times10^{-9}M$. If we now move the acetate sheet to the left by 2 pH units,

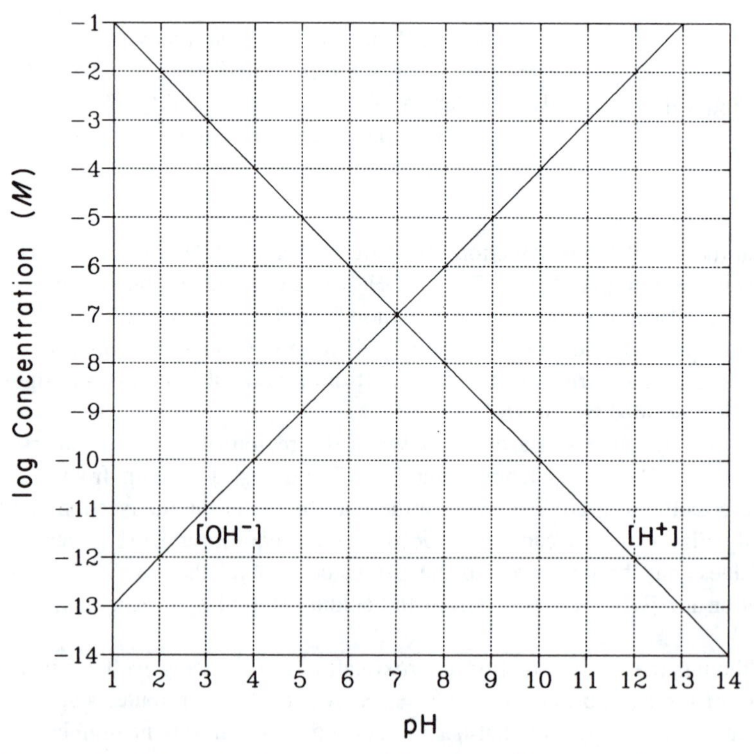

Figure 6.3. Universal grid giving $\log[H^+]$ and $\log[OH^-]$ vs. pH for $K_w = 10^{-14}$.

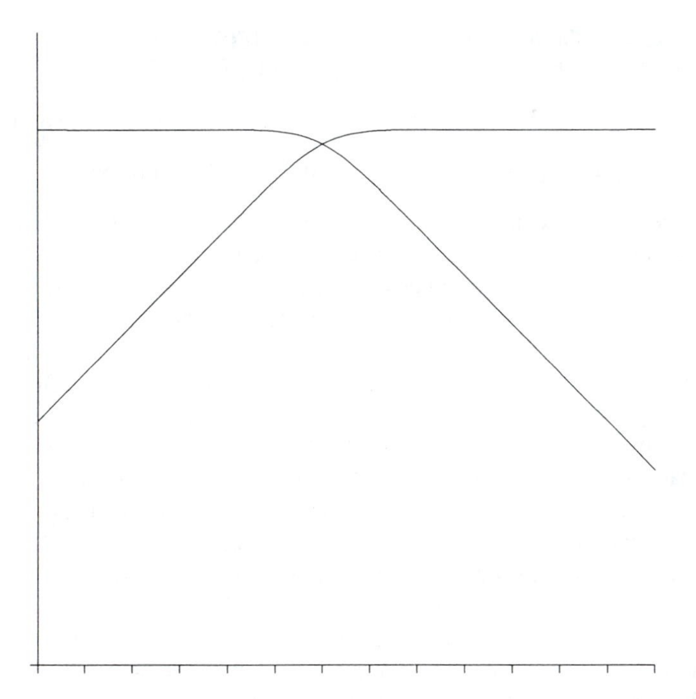

Figure 6.4. Generic plot of log[HA] and log[A^-] vs. pH. Activity corrections neglected (not valid at very low or very high pH).

we obtain Figure 6.2. Note that for all cases, the intersection of the log[HA] line with the log[A^-] line should always be aligned to occur at pH = pK.

If we now move the acetate sheet up by one unit on the y-axis, we obtain the case where $K = 10^{-5}$, and $C = 10^{-2}F$. The acidic solution approximation is clearly valid here. The weak acid approximation is also very well justified; compared to Figure 6.2, raising C for the solution of HA by a factor of 10 makes the system sufficiently acidic that the Eq. (6.56) intersection occurs at pH = 3.5; this is 1.5 pH units more acidic than pH = pK. We can conclude that the speciation for this system is $[H^+] = 10^{-3.5}$, $[A^-] = 10^{-3.5}$, $[OH^-] = 10^{-10.5}$, and $[HA] = 10^{-2}M$.

While keeping $K = 10^{-5}$, we now lower C by moving the transparency downwards. Lowering C means that the pH must increase towards $\frac{1}{2}$ pK_w, which in this case is 7.0. The pH rises because the pH at which Eq. (6.55) is satisfied increases as C decreases. The weak acid approximation is not valid below $C = 10^{-3}F$. (Since the acidic solution approximation remains valid down to $C = 10^{-6.5}M$, we can clearly see how decreasing C to that point raises the pH of the intersection given by Eq. (6.56)). Once C drops below $10^{-8}F$, the contribution that [A^-] makes in Eq. (6.55) becomes small relative to that made by [OH^-], and so the pH becomes governed essentially by the dissociation of water (pH $\simeq$ 7.0); at 25°C/1 atm, the pH cannot be raised above 7.0 simply by lowering C.

6.4.2 Applying Log Concentration vs. pH Diagrams in Understanding the Chemistry of Solutions of NaA

For a solution of NaA, when no additional strong acid or base is present, the PBE is

$$[H^+] + [HA] = [OH^-] \qquad \text{PBE for solution of NaA.} \tag{6.57}$$

Since there are five concentration variables in this system, we know from Chapter 5 that the equilibrium speciation will be that which simultaneously satisfies the five equations governing these variables. These equations are embodied in the two constants K and K_w, the MBE on total A, the MBE on Na^+, and the PBE (or the ENE).

Since the MBE on Na^+ states that $[Na^+] = C$, then once C is specified, the system is essentially reduced to a problem involving four unknowns, $[H^+]$, $[OH^-]$, [HA], and $[A^-]$. These are the same unknowns as in a solution of HA. Being able to use the MBE on Na^+ to remove Na^+ from further consideration here is why it is more direct to base our mathematical solution on the PBE, rather than on the corresponding ENE.

As in the previous section, we note that at every point along the pH axis in Figure 6.2, the relationships $K = 10^{-5}$, $K_w = 10^{-14}$, and the MBE for $10^{-3}M$ total A are always simultaneously satisfied. Therefore, neglecting activity coefficients, to determine the pH of a $10^{-3}F$ solution of NaA with $K = 10^{-5}$, all we need to do is *find the specific pH value in Figure 6.2 at which Eq. (6.57) is satisfied.* Values of pH that are higher than that specific pH constitute solutions that must contain some strong base in addition to the NaA. Values of pH that are between the two pH values at which Eqs. (6.55) and (6.57) are satisfied are either: 1) solutions of HA that also contain some strong base; or equivalently, 2) solutions of NaA that also contain some strong acid in addition to the NaA. If activity corrections may be neglected, a solution that satisfies Eq. (6.55) may be either a simple C F solution of HA, or a C F solution of NaA that also contains C F HCl. The latter is equivalent to a C F solution of HA that also contains C F NaCl. pH values that are lower than the pH value at which Eq. (6.55) is satisfied are either: 1) solutions of HA that also contain some strong acid; or 2) solutions of NaA of the same C but relatively more (C F more) strong acid than in (1).

An examination of Figure 6.2 indicates that for $K = 10^{-5}$ (that is, $K_a = 10^{-5}$, $K_b = 10^{-9}$), and $C = 10^{-3}F$, the PBE for a solution of NaA is satisfied to a very good approximation at the intersection described by

$$\log[HA] = \log[OH^-] \tag{6.58}$$

intersection describing one type of graphical solution to PBE for a solution of NaA, assumes validity of basic solution approximation (C and $K_w/K = K_b$ not too low).

Indeed, at the pH of the intersection, the $\log[H^+]$ line lies 2 log units below the log[HA] line, and so $[H^+]$ is negligible in Eq. (6.57) at that pH. This means that the "basic solution approximation," Eq. (5.80), is valid for this system. Therefore, the equilibrium speciation for this system may, to a very good approximation, be obtained by picking off the values of the various concentrations from a vertical line drawn at the pH of the intersection described by Eq. (6.58). Using Figure 6.2 it can be deduced that the equilibrium pH is very close to 8.0, and that [HA] = $[OH^-]$ = $10^{-6.0}M$, and $[A^-]$ = $10^{-3.0}M$. We obtain a pH of 8.00 using the equation for $[OH^-]$ that results from the basic solution approximation, that is, using Eq. (5.82). The fact that the equilibrium pH is 2 units removed from pH = pK (where [HA] contributes significantly to C) means that the amount of HA that forms in this system is small relative to C. Therefore, the *weak base* approximation is also valid in this case. These conclusions are consistent with Figure 5.8*c*.

The acetate sheet/grid combination described in the previous section for solutions of HA is equally applicable to solutions of NaA. (Figure 5.7 will also be very useful during this discussion.) Let us begin again with the acetate sheet laid on the grid such that the x- and y-axes of both figures are aligned. For this case, $K = 10^{-7}$ and $C = 10^{-3}M$. For a solution of NaA, we see that both the basic solution *and* weak base approximations are valid. As far as the speciation is concerned, $[OH^-] = 10^{-5.0}$, $[HA] = 10^{-5.0}$, $[A^-] = 10^{-3.0}$, and $[H^+] = 10^{-9.0}M$. If we now move the acetate sheet to the left by 2 pH units, we return again to Figure 6.2.

If we now move the acetate sheet up by one unit on the y-axis, we obtain the case where $K = 10^{-5}$, and $C = 10^{-2}F$. Both the basic solution and the weak base approximations are very well justified. The speciation for this NaA system is $[OH^-] = 10^{-5.5}$, $[HA] = 10^{-5.5}$, $[H^+] = 10^{-8.5}$, and $[A^-] = 10^{-2.0}M$.

While keeping $K = 10^{-5}$, we now lower C by moving the transparency downwards. Lowering C means that the pH must decrease towards $\frac{1}{2}\mathrm{p}K_w$, which in this case is 7.0. The pH decreases because the pH at which Eq. (6.57) is satisfied decreases as C decreases. The basic solution approximation is not valid below $C = 10^{-4}F$. However, since the pH can never fall below 7.0 at 25°C/1 atm, the weak base approximation (with $[H^+]$ not necessarily negligible) remains valid no matter how low C becomes; this is consistent with Figure 5.8*b*. Once C drops below $10^{-6}F$, the contribution that HA makes in Eq. (6.57) becomes small relative to that made by $[H^+]$, and so the pH becomes governed essentially by the dissociation of water (pH $\simeq$ 7.0); we emphasize that at 25°C/1 atm, the pH of a solution of NaA cannot be lowered below 7.0 simply by lowering C.

6.5 COMPUTING THE SPECIATION OF SOLUTIONS OF HA OR NaA PLUS SOME STRONG ACID OR STRONG BASE

6.5.1 Solution of HA Plus Some Strong Base

In Section 6.2, we discussed that log concentration vs. pH plots like Figure 6.2 apply to conditions that are not limited simply to solutions of HA and NaA. Since Section 6.4 focused on solutions of HA and NaA, it remains for us to consider what amounts of strong acid or strong base are needed to access pH values other than those that satisfy the PBEs that are Eqs. (6.55) and (6.57). We begin by considering solutions that contain a constant concentration of HA that equals C F. We will allow that these solutions also contain variable amounts of the strong base NaOH. The concentration of strong base will be represented as C_B. The ENE for these solutions is

$$[H^+] + [Na^+] = [OH^-] + [A^-]. \tag{6.59}$$

Substituting for $[Na^+]$, we have

$$[H^+] + C_B = [OH^-] + [A^-]. \tag{6.60}$$

When $C_B = 0$, we have a solution of HA, and Eq. (6.60) reduces to Eq. (6.55). When $C_B = C$, we have a solution of NaA, and Eq. (6.60) reduces to Eq. (6.57) because if $C_B = C$, then $(C_B - [A^-]) = [HA]$.

Since the meaning of the fractional value α_1 is independent of whether or not any strong acid or base is present in addition to the HA, we can rewrite Eq. (6.60) as

$$[H^+] + C_B = K_w/[H^+] + \alpha_1 C. \tag{6.61}$$

For known values of K, K_w, C, and C_B, Eq. (6.61) comprises one equation in one unknown. It can be solved readily by trial and error by finding the root value of $[H^+]$ that gives $h([H^+]) = 0$, where

$$h([H^+]) = [H^+] + C_B - K_w/[H^+] - \alpha_1 C. \tag{6.62}$$

In the specific case of a solution of NaA, then $C_B = C$, $(C_B - \alpha_1 C) = \alpha_0 C$, and so then $h([H^+])$ is equal to $r([H^+])$ as given by Eq. (5.74).

We can easily solve $h([H^+]) = 0$ using either hand calculators such as the HP-28S, the HP-48S, or a computer. Activity corrections can be incorporated as usual by inputting equilibrium constants modified to the ${}^cK$ format based on best estimates of the activity coefficients. In a manner similar to that described in Chapter 5, if a relatively large amount of background salt is not present, those estimates can be updated for each of a series of iterative solutions.

As an example, we compute the pH of a solution with the composition $10^{-3}F$ HA and $2.0 \times 10^{-4}F$ NaOH. Since the amount of NaOH is less than equivalent to the amount of HA, we know that the pH will be somewhere between the values that are obtained when Eqs. (6.55) and (6.57) are satisfied (pH = 4.02 and 8.00, respectively). Neglecting activity corrections for this comparatively dilute solution, we search for the root of $h([H^+])$ using a series of guesses given in Table 6.4. Therefore, the first guess made in Table 6.4 is 4.02, and the second guess is greater than 4.02. The final result is an equilibrium pH of 4.48. While $h([H^+])$ is not identically zero for this pH, further changes in pH will only affect the third figure past the decimal point. Given the facts that: 1) activity corrections have been neglected; and 2) any real solution will carry uncertainty in K, K_w, C, and C_B, such improvements in accuracy would be meaningless.

6.5.2 Solution of NaA Plus Some Strong Acid

We now consider solutions that contain a constant concentration of NaA that equals C F. We will allow that these solutions also contain variable amounts of the strong acid HCl. The ENE for these solutions is therefore

$$[H^+] + [Na^+] = [OH^-] + [A^-] + [Cl^-]. \tag{6.63}$$

TABLE 6.4. Trial and error solution for pH of a system in which the concentration of HA is $10^{-3}F$, the concentration of NaOH is $2.0 \times 10^{-4}F$, K for HA is 10^{-5}, and $K_w = 10^{-14}$.

pH	$h([H^+])$
4.02	2.01×10^{-4}
4.20	1.26×10^{-4}
4.40	3.91×10^{-5}
4.50	-8.63×10^{-6}
4.49	-3.72×10^{-6}
→ 4.48	1.16×10^{-6}

The concentration of strong acid will be represented as $C_A F$. Making the appropriate substitutions, we have

$$[H^+] + C = K_w/[H^+] + \alpha_1 C + C_A. \tag{6.64}$$

Because $(C - \alpha_1 C) = [HA]$, when $C_A = 0$, we have a solution of NaA, and Eq. (6.64) reduces to Eq. (6.57). When $C_A = C$, we have a solution of HA that also contains C F NaCl.

For known values of K, K_w, C, and C_A, Eq. (6.64) comprises one equation in one unknown. It can be solved readily by trial and error by finding the root value of $[H^+]$ that gives $i([H^+]) = 0$, where

$$i([H^+]) = [H^+] + \alpha_0 C - K_w/[H^+] - C_A. \tag{6.65}$$

As an example, we compute the pH of a solution with the composition $10^{-3}F$ NaA with $2.0 \times 10^{-4}F$ HCl. Since the amount of HCl is less than equivalent to the amount of NaA, we know that the pH will be somewhere between the values that are obtained when Eqs. (6.55) and (6.57) are satisfied. Neglecting activity corrections for this comparatively dilute solution, we search for the root of $i([H^+])$ using a series of guesses given in Table 6.5. Since the formality of the HCl is less than half that of the NaA, *less* than half of the NaA is neutralized back to a solution of HA, and so we know that the equilibrium pH will be greater than pH = pK = 5.00. Therefore, the first guess made in Table 6.5 is 5.00, and the second guess is greater than 5.00. The final result is an equilibrium pH of 5.61.

6.5.3 The General Case—Solution of NaA and HA Plus Strong Acid and/or Base

In the totally general case, we might have some HA in a solution, some NaA, and strong acid and/or strong base. Such solutions may be handled in a manner that is analogous to the methods outlined in the preceding two sections. If we again use $[Na^+]$ as the

TABLE 6.5. Trial and error solution for pH of a system in which the concentration of NaA is $10^{-3}F$, the concentration of HCl is $2.0 \times 10^{-4}F$, K for HA is 10^{-5}, and $K_w = 10^{-14}$.

pH	$i([H^+])$
5.00	3.10×10^{-4}
5.20	1.93×10^{-4}
5.40	8.87×10^{-5}
→ 5.60	3.27×10^{-6}
5.70	-3.17×10^{-5}
5.61	-4.49×10^{-5}

tracer for how much strong base is present (C_B), and [Cl$^-$] as the tracer for strong acid (C_A), we again obtain Eq. (6.63) as the ENE. The quantities that contribute to C (that is, to A_T) include both the HA and the NaA. The NaA that is present contributes Na^+ to the solution, and so therefore contributes to C_B. The general equation for all types of solutions containing HA, NaA, NaOH, and HCl is

$$[H^+] + C_B = K_w/[H^+] + \alpha_1 C + C_A. \tag{6.66}$$

For known values of K, K_w, C_B, and C_A, Eq. (6.66) comprises one equation in one unknown. It can be solved readily by trial and error by finding the root value of [H$^+$] that gives $j([H^+]) = 0$, where

$$j([H^+]) = [H^+] + (C_B - C_A) - K_w/[H^+] - \alpha_1 C. \tag{6.67}$$

Note that *as long as activity corrections can be neglected,* the individual values of C_B and C_A do not matter; we need only know the difference ($C_B - C_A$).

In general, if F_i denotes the formality of i, then

$$C = F_{HA} + F_{NaA} \tag{6.68}$$

$$C_B - C_A = F_{NaA} + F_{NaOH} - F_{HCl}. \tag{6.69}$$

Note that F_{NaCl} does not contribute to the net value of ($C_B - C_A$). Indeed, NaCl may be thought of as being perfectly neutralized HCl, or perfectly neutralized NaOH. Examples of how to evaluate the various input parameters for $j([H^+])$ are given in Table 6.6. Note that to the extent that activity corrections can be neglected, Case 3 is equivalent to Case 4. *If activity corrections are included, however, these two solutions will have different pH values.* Other examples and problems illustrating the principles of this chapter are presented by Pankow (1992).

TABLE 6.6. Example cases involving evaluation of input parameters for $j([H^+])$.

	Case 1	Case 2	Case 3	Case 4
Concentrations				
Formality of HA:	1.0×10^{-4}	5.0×10^{-5}	1.0×10^{-3}	5.0×10^{-4}
Formality of NaA:	5.0×10^{-5}	5.0×10^{-5}	0	5.0×10^{-4}
Formality of NaOH:	1.0×10^{-5}	0	0	1.0×10^{-4}
Formality of HCl:	0	0	0	6.0×10^{-4}
Formality of NaCl:	0	0	1.0×10^{-4}	0
Resulting Input Parameters for $j([H^+])$				
C (that is, A_T)	1.5×10^{-4}	1.0×10^{-4}	1.0×10^{-3}	1.0×10^{-3}
$C_B - C_A$	6.0×10^{-5}	5.0×10^{-5}	0	0

6.6 REFERENCES

FREISER, H., and Q. FERNANDO. 1963. *Ionic Equilibria in Analytical Chemistry*. New York: John Wiley and Sons. Reprinted by Robert E. Krieger Publishing Company, Malabar, Florida.

FREISER, H. 1967. Simplifying and Strengthening the Teaching of pH Concepts and Calculations Using Log-Chart Transparencies, *School Science and Mathematics*, March.

FREISER, H. 1970. Acid-Base Reaction Parameters, *J. Chem. Ed.*, vol. 47, 809–811.

PANKOW, J. F. 1992. *Aquatic Chemistry Problems*, Portland: Titan Press-OR, P.O. Box 91399, Portland, Oregon 97291-1399.

7

Titrations of Acids and Bases

7.1 INTRODUCTION

In Chapter 5, we discussed how to determine the specific equilibrium speciations (e.g., pH) in simple solutions of chemicals like HA, NaA, H_2B, NaHB, and Na_2B. In Chapter 6, we discussed the fact that the whole range of pH values is accessible to solutions of these chemicals, *if* we can add strong acid or strong base to them. In this chapter, then, we consider how gradual changes in strong acid or strong base cause changes in the chemistry of solutions of acids and bases like HA and NaA. That is, we discuss the mathematics of acid and base *titrations*. In this context, we discuss the fact that an *alkalimetric titration* is the process by which we: 1) add strong base to a solution of an acid (e.g., HA); and 2) record the equilibrium pH that is obtained for each increment of strong base. We also discuss the fact that an *acidimetric titration* is the process by which we: 1) add strong acid to a solution of a base (e.g., NaA); and 2) record the equilibrium pH that is obtained for each increment of strong acid. It may be noted that the word "alkalimetric" connotes the process by which base (i.e., "alkali-") is added in a measured manner to a solution. Similarly, the word "acidimetric" connotes the process by which acid is added in a measured manner to a solution.

Acidimetric titrations are usually carried out by adding *volume* increments of a relatively concentrated solution (titrant) of strong acid to the solution of interest. Similarly, alkalimetric titrations are usually carried out by adding volume increments of a relatively concentrated solution (titrant) of strong base. For both alkalimetric and acidimetric titrations, the *titration curve* is the plot of solution pH on the *y*-axis vs. amount (volume) of strong base or strong acid added on the *x*-axis.

The reason we need to develop a detailed understanding of the chemistry and the accompanying mathematics of titration curves is twofold. Firstly, since titrations can be used to measure how much acid or base is present in a given system, they will be important when one carries out quantitative analytical measurements of such chemicals in natural waters. Secondly, and perhaps more importantly, natural water chemists need to understand the chemistry and the mathematics of titrations as natural waters are continuously undergoing titration-like changes. As a very contemporary example, natural waters undergo acidimetric titration whenever acidic precipitation is added. Bases can be added to natural waters through the discharge of basic wastes, or through the dissolution of basic minerals.

7.2 TITRATION OF THE ACID HA

7.2.1 General Considerations

Consider the titration of a sample of water that contains C F of the acid HA. The titration will be carried out using a strong base, say NaOH. Before the addition of any NaOH, we know that the ENE reads

$$[H^+] = [OH^-] + [A^-]. \tag{6.55}$$

Once the titration begins, the ENE must be adjusted to take into account the fact that some Na^+ has been added, and so it becomes

$$[H^+] + [Na^+] = [OH^-] + [A^-]. \tag{6.59}$$

The mathematical nature of the alkalimetric titration curve can be readily deduced using Eq. (6.59). The reason for this is that, as discussed in Chapter 6, $[Na^+]$ provides a *quantitative tracer* for the equivalents per liter of strong base added during the course of the titration.[1] Actually, as long as the base that is added is a strong base, it does not matter whether the solution used to carry out the titration (i.e., the titrant) contains NaOH, or KOH, $Mg(OH)_2$, or any other strong base.

As in Chapter 6, C_B is the general symbol for the equivalents per liter of strong base added. If the strong base added is NaOH, then $C_B = [Na^+]$. If the base used is $Mg(OH)_2$, then $C_B = 2[Mg^{2+}]$ since two hydroxides are added for each Mg^{2+}. The general form of Eq. (6.59) is thus

$$[H^+] + C_B = [OH^-] + [A^-]. \tag{6.60}$$

If a solution of $Mg(OH)_2$ is in fact the titrant used, it may be noted that $2[Mg^{2+}]$ is exactly what we would want as C_B since the origin of Eq. (6.60) is, after all, the ENE.

Returning now to the specific example of the alkalimetric titration of HA with NaOH, we have

$$C_B = [Na^+] = [A^-] + [OH^-] - [H^+]. \tag{7.1}$$

[1]As discussed briefly in Chapter 1, concentration units of *equivalents per liter* (eq/L) are often used for acids and bases. 1.0 eq of base can produce 1.0 mol of OH^-. If the base is a strong base, then that production will tend to go to completion even when the solution pH is very high. A solution that is 1.0 F in the strong base NaOH may also be described as being 1.0 N (eq/L) in NaOH. A solution that is 1.0 F in the strong base $Mg(OH)_2$ is a 2.0 N (eq/L) in $Mg(OH)_2$. Conversely, 1.0 eq of acid can produce 1.0 mol of H^+. If the acid is a strong acid, then that production will go to tend to go to completion even when the solution pH is very low. A solution that is 1.0 F in the strong acid HCl may also be described as being a 1.0 N (eq/L) solution of HCl. A solution that is 1.0 F in the strong acid H_2SO_4 is 2.0 N (eq/L) in H_2SO_4.

Equation (7.1) is true *throughout* the titration process. This is true even before the titration begins, at which point $C_B = 0$.

7.2.2 Neglecting Dilution

As usual, we now let C equal the formality of the acid HA initially present in the solution. In the simplest case, we will assume that the volume of the solution undergoing the titration remains constant during the course of the titration. When this condition is met, then

$$[A^-] + [HA] = A_T = C \tag{7.2}$$

will remain true during the entire titration. While the relative amounts of A^- and HA will change as a function of pH during the titration, their sum remains constant. This condition can be met exactly by considering the chemistry of a variety of solutions *specifically prepared* to have varying values of C_B, but constant values of C. More practically, but not *exactly*, it can be met by taking a C F solution of HA and adding varying small volumes of a solution of base that is very concentrated relative to C.

We now let

$$f = \text{fraction of the initial acid titrated.} \tag{7.3}$$

If our assumption of a constant C is valid, then the fraction f titrated at any given point will be given by the ratio of C_B to C, that is,

$$f = C_B/C = [Na^+]/C. \tag{7.4}$$

The values of 0 and 1 hold special importance for the parameter f. It should be clear that $f = 0$ before the titration begins, that is, when $C_B = 0$. The solution at this point obviously has the chemistry of a C F solution of HA; it represents the *HA equivalence point* for the system; the system is equivalent to an HA solution that is C F.

The value of f will be 1.0 when C_B equals C. At this point, the eq/L of strong base added is exactly equivalent to the initial eq/L of acid HA. Since an equivalent amount of strong base has now been added to the HA solution, the system is said to be at the *NaA equivalence point.*

Combining Eqs. (7.1) and (7.4), we obtain

$$f = C_B/C = [A^-]/C + \frac{[OH^-] - [H^+]}{C} \tag{7.5}$$

which reduces to the very important equation

$$f = \alpha_1 + \frac{[OH^-] - [H^+]}{C}. \tag{7.6}$$

When all activity corrections may be neglected, Eq. (7.6) becomes

$$f = \frac{K}{[H^+] + K} + \frac{[OH^-] - [H^+]}{C}. \tag{7.7}$$

Since $[OH^-] = K_w/[H^+]$, Eq. (7.7) is a function of just one variable, $[H^+]$. When the titration is being conducted in a constant ionic medium, then Eq. (7.7) may still be used with ${}^cK$ substituted for K, and ${}^cK_w$ substituted for K_w.

An examination of Eqs. (7.5)–(7.7) reveals that f values *greater than* 1 are possible. This will occur when $C_B > C$, that is, when more than enough base to titrate the acid HA has been added to the system, and the system is beyond the $f = 1$ (i.e., NaA) equivalence point. As will be discussed below, f values less than 0 are also possible.

There are two methods for constructing the titration curve, that is, the curve of pH vs. f. The first method involves the selection of a range of values of f, and then solving for the corresponding set of values of $[H^+]$ that satisfy Eq. (7.6). Even assuming that activity corrections can be neglected, or alternatively that the titration is occurring in a constant ionic medium, this can be a very laborious process. Indeed, for any given value of f, Eq. (7.6) leads to a cubic equation in $[H^+]$. The calculations can obviously be facilitated by the repeated selection of a simplifying assumption that is valid for each given f value. However, as might be expected from Chapter 5, the nature of that assumption will change during the course of the titration.

One situation when the first method is relatively easy to execute is when the acid under titration is a very strong acid. Under these conditions, [HA] may be assumed to be zero for all values of pH, and so the value of α_1 remains essentially equal to 1 over the entire pH range. When activity corrections can be neglected, Eq. (7.6) then becomes

$$f \simeq 1 + \frac{[OH^-] - [H^+]}{C} \quad \text{assumes that the acid being titrated is a strong acid.} \tag{7.8}$$

The second method for constructing titration curves is much easier to use than the first method. It is just the reverse of the first method, that is, pick a series of $[H^+]$ values, and solve directly and simply for their corresponding f values. This method can be applied very easily when activity corrections can be neglected, and when the titration occurs in a constant ionic medium. Under these conditions, all of the terms on the RHS of Eq. (7.7) are very simple functions of $[H^+]$. If at the end of this process it turns out that there is a gap in a region of interest in the titration curve, then the process can be repeated for just that pH range. The use of a computer or a programmable hand calculator makes this method very easy to carry out since the equation for f as a function of $[H^+]$ can be programmed for repeated evaluation over the selected range of pH values.

7.2.3 Considering Dilution

When the titrant solution is not sufficiently concentrated to neglect the effects of dilution of the solution under titration, then C (i.e., A_T) changes during the course of the titration according to

$$C = C_oV_o/(V_o + V_t) = C_oV_o/V_T \tag{7.9}$$

where

C_o = initial value of C (F)

V_o = initial volume of solution under titration (L)

V_t = volume of titrant added (L)

V_T = total volume at any given point during titration $= V_o + V_t$ (L)

It should be clear that the same volume units must be selected for V_o, V_t, and V_T.

Equation (7.9) is a statement of the fact that the *mass amount* of total A does not change during the course of the titration. That is, during the entire titration, the number of mols of total A present must always remain equal to the *original* number of mols of A. In particular, we note that

$$C(V_o + V_t) = C_o V_o. \tag{7.10}$$

Equations (7.9) and (7.10) remain exactly true over the entire course of the titration. To obtain the equation for the titration curve when dilution must be considered, we must divide the equation for C_B (Eq. (7.1)) by the dilution-corrected expression for C, that is, by $C_o V_o/(V_o + V_t)$. The result is

$$f = \frac{C_B}{C} = \frac{[\mathrm{Na^+}]}{C_o V_o/(V_o + V_t)} = \frac{[\mathrm{A^-}]}{C_o V_o/(V_o + V_t)} + \frac{[\mathrm{OH^-}] - [\mathrm{H^+}]}{C_o V_o/(V_o + V_t)} \tag{7.11}$$

$$= \frac{[\mathrm{A^-}](V_o + V_t)}{C_o V_o} + \frac{[\mathrm{OH^-}] - [\mathrm{H^+}]}{C_o V_o/(V_o + V_t)} \tag{7.12}$$

In the first term on the RHS of Eq. (7.12), the quantity $[\mathrm{A^-}](V_o + V_t)$ represents the total number of mols of A^- present in the solution at any given point during the titration. The quantity $C_o V_o$ clearly represents the total original number of FW of acid HA. By conservation of mass, that amount still represents the total amount of A^- plus HA present at any point in the titration. Therefore, the ratio of these two quantities is α_1, and so the expression for f when dilution is taken into account simplifies to

$$f = \alpha_1 + \frac{[\mathrm{OH^-}] - [\mathrm{H^+}]}{C_o V_o/(V_o + V_t)}. \tag{7.13}$$

An examination of Eq. (7.13) reveals that when the concentration of the titrant is very large relative to C_o, then V_t is always small relative to V_o during the titration, and so C remains essentially constant and Eq. (7.13) simplifies to Eq. (7.6).

7.2.4 Relationship Between *f* and V_t

When determined experimentally, titration curves are often drawn as pH vs. V_t curves rather than pH vs. f curves. This will certainly be true when one does not begin the titration with a knowledge of C_o. When the alkalimetric titration begins at $f = 0$, the interconversion of f and V_t may be carried out according to the equation

$$f = \frac{C_t V_t}{C_o V_o} \tag{7.14}$$

where C_t is the concentration of the strong base titrant solution, in units of eq/L. Eq. (7.14) defines f as the ratio of the equivalents of base added to the FW of HA initially in the solution. This definition builds in the consideration of dilution effects, and so is exactly true at all points during the titration.

7.3 TITRATION OF THE BASE NaA

7.3.1 General Considerations

As with the titration of HA, we begin our considerations of the titration of C F of the base NaA with the pertinent ENE. Before the titration starts, the ENE is

$$[Na^+] + [H^+] = [A^-] + [OH^-]. \tag{7.15}$$

After the titration with strong acid (say HCl) begins, the ENE must be modified to read

$$[Na^+] + [H^+] = [A^-] + [OH^-] + [Cl^-]. \tag{6.63}$$

Substituting C_A as the tracer for the strong acid added, we obtain

$$C_A = [Na^+] + [H^+] - [A^-] - [OH^-]. \tag{7.16}$$

Before the titration begins, $C_A = 0$.

7.3.2 Neglecting Dilution

Neglecting any dilution caused by the addition of the strong acid titrant solution, at all points during the titration, $[Na^+] = C$, and so Eq. (7.16) becomes

$$C_A = C + [H^+] - [A^-] - [OH^-]. \tag{7.17}$$

The MBE on A_T (i.e., Eq. (7.2)) may now be invoked to substitute for C in Eq. (7.17). We obtain

$$C_A = [HA] + [A^-] + [H^+] - [A^-] - [OH^-] \tag{7.18}$$

which reduces to

$$C_A = [HA] + [H^+] - [OH^-] \tag{7.19}$$

$$= \alpha_0 C + [H^+] - [OH^-]. \tag{7.20}$$

If we let g equal the fraction of base titrated $= C_A/C$, then

$$g = \frac{C_A}{C} = \alpha_0 + \frac{[H^+] - [OH^-]}{C}. \tag{7.21}$$

When all activity corrections may be neglected, the form of Eq. (7.21) that may be used is

$$g = \frac{[H^+]}{[H^+] + K} + \frac{[H^+] - [OH^-]}{C}. \tag{7.22}$$

Since $[OH^-] = K_w/[H^+]$, Eq. (7.22) is a function of just one variable, $[H^+]$. When the titration is being conducted in a constant ionic medium, ${}^cK$ must be substituted for K, and ${}^cK_w$ must be substituted for K_w.

As was the case for the parameter f, the values of 0 and 1 hold special importance for the parameter g. It should be clear that $g = 0$ before the titration begins, that is, when $C_A = 0$. The solution at this point obviously has the chemistry of a C F solution of NaA; it represents the *NaA equivalence point* for the solution. (We recall from above that $f = 1$ for this solution.) The value of g will be 1 when C_A equals C, that is, when the eq/L of acid added is exactly equal to the initial F of the monobasic NaA. At this point, since an equivalent amount of strong acid has been added to the NaA solution, the system is at the *HA equivalence point.* (We recall from above that $f = 0$ for this solution.) An examination of the equation for g reveals the fact that g values greater than 1 are possible. This will occur when $C_A > C$, that is, more than enough acid to titrate the NaA has been added to the system, and the system is beyond the HA equivalence point. As will be discussed below, g values less than 0 are also possible.

7.3.3 Considering Dilution

If dilution must be considered, Eq. (7.22) must be modified in a manner that is closely analogous to the modifications needed for Eq. (7.6). The result is

$$g = \frac{C_A}{C} = \alpha_0 + \frac{[H^+] - [OH^-]}{C_o V_o/(V_o + V_t)}. \tag{7.23}$$

7.3.4 Relationship Between g and V_t

In a manner analogous to the case with f, when the acidimetric titration begins at $g = 0$, the interconversion of g and V_t may be carried out according to the equation

$$g = \frac{C_t V_t}{C_o V_o}. \tag{7.24}$$

In an acidimetric titration, C_t (eq/L) refers to the concentration of the strong acid in the titrant solution. Eq. (7.24) defines g as the ratio of the equivalents of acid added to the FW of NaA initially in the solution. This definition builds in all needed corrections for dilution effects, and so is exactly true throughout the entire acidimetric titration.

7.4 RELATIONSHIP BETWEEN THE PARAMETERS f AND g FOR THE HA/A^- SYSTEM

The above considerations lead us to suspect that although the various possible expressions for g look different from the expressions for f, if one knows the value of f for a given solution, then the value that g must take on for that solution must also be specifiable. Thus, there is no new information in g that is not in f; at any given point in the solution, the fraction of acid HA that has been titrated with strong base must, when viewed from the acidimetric perspective, be equal to (1 – (fraction of the base NaA that has been titrated with strong acid)).

There is in fact a great deal of symmetry in the functional relationships for f and g. Firstly, f involves α_1, and g involves α_0. Secondly, f involves ($[OH^-] - [H^+]$), and g involves the negative of that, that is, ($[H^+] - [OH^-]$). These properties lead to the conclusion that in a monoprotic system,

$$f + g = 1. \tag{7.25}$$

The reader might have begun to suspect the existence of Eq. (7.25) when it was observed towards the end of Section 7.3.2 that $g = 0$ when $f = 1$, and $g = 1$ when $f = 0$.

When the concentration of HA that is under titration is not too low, and when neither K nor K_w/K ($= K_b$) is very large, then in the range where $0 < f < 1$, the terms involving both $[H^+]$ and $[OH^-]$ can be neglected in the expressions for f and g. Under these conditions,

$$f \simeq \alpha_1, \quad g \simeq \alpha_0 \qquad \begin{array}{l} C \text{ not too low, } 0 < f < 1, \text{ neither } K \\ \text{nor } K_w/K \text{ is large} \end{array} \tag{7.26}$$

The nature of the relationship between f and g is depicted graphically using the generalized titration curve for the HA/A^- system in Figure 7.1. In this figure, sharp changes in pH occur at both the HA and NaA equivalence points. As we will see in Chapter 8, whether or not such sharp changes are present depends on the values of K and C.

Figure 7.1 has been divided into three regions, $f < 0$, $0 \leq f \leq 1$, and $f > 1$. When determining the titration curve for a solution of HA experimentally, one only sees the center and rightmost regions. Conversely, when determining the titration curve for a solution of NaA experimentally, one only sees the center and leftmost regions. When considering the whole of HA/A^- titration chemistry, we need to consider all three regions. A solution whose composition falls in the *leftmost* region may be prepared either from a solution of HA plus some strong acid, or from a solution of NaA plus more strong acid than is necessary to titrate all of the NaA. A solution whose composition falls in the *center* region may be prepared from either some HA plus some strong base, or from some NaA plus some strong acid. A solution whose composition falls in the *rightmost* region may be prepared either from a solution of NaA plus some strong base, or from a solution of HA plus more strong base than is necessary to titrate all of the HA. In general, solutions prepared by starting with an NaA solution and adding strong acid (e.g., HCl) will contain a certain amount of salt (e.g., NaCl) that will not be present in solutions of the same f (and g), but prepared from HA plus strong base.

7.5 GENERALIZED FORMS OF THE EQUATIONS FOR *f* AND *g*

It is useful to re-examine the ENE given above for the titration of a solution of NaA with a solution of HCl, that is,

$$[Na^+] + [H^+] = [A^-] + [OH^-] + [Cl^-]. \tag{6.63}$$

Rearranging, we obtain

$$[Na^+] - [Cl^-] = [A^-] + [OH^-] - [H^+]. \tag{7.27}$$

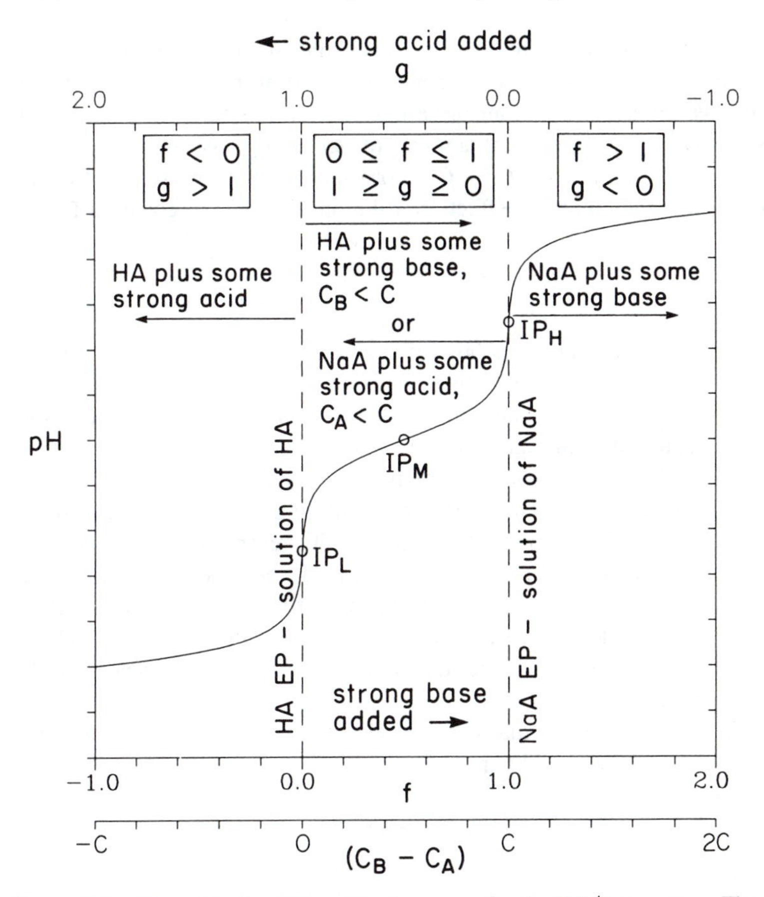

Figure 7.1 Generalized acid/base titration curve for the HA/A^- system. The values of K and C are such that: 1) sharp changes in pH occur at both the HA and the NaA equivalence points (EPs); 2) three inflection points are visible in the titration curve, one at a low pH (IP_L), one at a medium pH (IP_M), and one at a high pH (IP_H). The location of IP_L almost exactly gives the location of the HA equivalence point (HA EP). The location of IP_H almost exactly gives the location of the NaA equivalence point (NaA EP). The location of the region to the left of the HA EP can be accessed by adding some strong acid to a solution of HA. The location between the HA and NaA EPs can be accessed either by adding some strong base to a solution of HA (with $C_B < C$), or by adding some strong acid to a solution of NaA (with $C_A < C$). The location of the region to the right of the NaA EP can be accessed by adding some strong base to a solution of NaA.

In general, for any combination of strong base(s) or strong acid(s), we have that

$$C_B - C_A = [A^-] + [OH^-] - [H^+]. \tag{7.28}$$

Thus, as in Eq. (6.67), the term $(C_B - C_A)$ refers to the difference between the total eq/L of strong base(s) *minus* the total eq/L of strong acid(s). The plurals in the previous sentence indicate that C_B might have contributions from more than one strong base, and C_A might have contributions from more than one strong acid. In summary, $(C_B - C_A)$ provides a *generic* measure of the amount of *net* strong base in the system. It may be noted that a negative value of $(C_B - C_A)$ corresponds to the presence of net strong acid. We conclude that including the effects of dilution, the general equation for f is

$$f = (C_B - C_A)/C = \alpha_1 + \frac{[OH^-] - [H^+]}{C_oV_o/(V_o + V_t)}. \tag{7.29}$$

Equation (7.29) clearly illustrates how, in addition to possibly being greater than 1, the value of f can take on negative values. This can occur when $C_A > C_B$, that is, when there is net strong ("mineral") acid in the system in addition to the HA. From the point of view of a acidimetric titration of the base NaA, this will occur when the solution has been titrated beyond the HA equivalence point.

If we begin an alkalimetric titration of HA under conditions such that $C_B = C_A$, but neither are zero, then there is background salt present. If NaOH was the source of the C_B, and HCl was the source of the strong acid, then the background salt would be NaCl. As we know, since the activity coefficients of the various species will be different in the presence of background salt compared to their values with no such salt present, the exact nature of the resulting titration curve will depend on the amount of background salt initially present.

By Eq. (7.25), we know that in a monoprotic system, we always have that $f + g = 1$. Therefore, by Eq. (7.29) we obtain

$$(C_B - C_A)/C + g = 1 \tag{7.30}$$

$$g = (C_A - C_B)/C + C/C \tag{7.31}$$

$$g = \frac{(C_A - C_B) + C}{C} \tag{7.32}$$

where $(C_A - C_B)$ is now the net strong acid present in the solution, again using a solution of HA as the reference point for a zero value of $(C_A - C_B)$. Combining Eq. (7.32) with Eq. (7.23), we now obtain the general equation for g as

$$g = \frac{(C_A - C_B) + C}{C} = \alpha_0 + \frac{[H^+] - [OH^-]}{C_oV_o/(V_o + V_t)}. \tag{7.33}$$

For a simple solution of NaA, $C_A = 0$, and $C_B = C$.

In addition to possibly being greater than 1, the value of g can take on negative values when $C_B > (C + C_A)$, that is, when there is a strong base in the system in addition to the NaA. Such additional strong base is referred to as "caustic" alkalinity. From the point of view of the alkalimetric titration of the acid HA, this will occur when the solution has been titrated beyond the NaA equivalence point.

7.6 TITRATION OF POLYPROTIC SYSTEMS

Consider the alkalimetric titration of a polyprotic acid H_nB. Let the following reactions have the following K values:

$$H_nB = H_{n-1}B^- + H^+ \qquad K_1 \tag{7.34}$$

$$H_{n-1}B = H_{n-2}B^{2-} + H^+ \qquad K_2 \tag{7.35}$$

$$H_{n-2}B = H_{n-3}B^{3-} + H^+ \qquad K_3 \tag{7.36}$$

etc.

$$\alpha_0 = [H_nB]/B_T \tag{7.37}$$

$$\alpha_1 = [H_{n-1}B^-]/B_T \tag{7.38}$$

$$\alpha_2 = [H_{n-2}B^{2-}]/B_T \tag{7.39}$$

$$\alpha_3 = [H_{n-3}B^{3-}]/B_T \tag{7.40}$$

etc.

The reader is reminded that for this polyprotic system, $C = B_T$.

It can be shown based on the ENE for this system that

$$C_B - C_A = C(\alpha_1 + 2\alpha_2 + 3\alpha_3 .... + n\alpha_n) + [OH^-] - [H^+] \tag{7.41}$$

$$f = (C_B - C_A)/C = \sum_{i=1}^{n} i\alpha_i + \frac{[OH^-] - [H^+]}{C_oV_o/(V_o + V_t)}. \tag{7.42}$$

Thus, in the case of a polyprotic acid, f represents the number of acid groups titrated up to that point in the titration. To completely titrate the acid H_nB to the Na_nB equivalence point, then enough base must be added to have f reach a value of n. If:

$f = 0.25$,	enough base added to titrate 25% of first proton;
$f = 1.00$,	enough base added to titrate 100% of first proton;
$f = 1.50$,	enough base added to titrate 100% of first proton, plus 50% of second proton.

We also note that in an n-protic system,

$$f + g = n \tag{7.43}$$

It is important to note that saying, for example, that 25% of the first proton of a diprotic system is titrated does not mean that at that point, $[HB^-]$ is exactly equal to $0.25[H_2B]_o$ where $[H_2B]_o$ is the initial formality of the H_2B solution. Indeed, if K_1 is very large, then $[HB^-]$ will be larger than $0.25[H_2B]_o$. Conversely if K_1 is very small, then $[HB^-]$ will likely be closer to $0.25[H_2B]_o$. Therefore, saying that $x\%$ of a given proton is "titrated" really means that an amount of base has been added to the system that is *equivalent* to that given amount of protons. Analogous statements can be made for titrations of basic solutions.

7.7 SPECIFIC EXAMPLE OF CALCULATING A TITRATION CURVE

Task: Calculate the detailed titration curve for 0.025 L of 10^{-3} *F* HA, p*K* = 4.0, titrant 0.1 *N* NaOH. Take effects from volume changes due to addition of the strong base titrant into consideration,[2] but neglect activity corrections.
Solution:

$$f = C_t V_t / C_o V_o \tag{7.14}$$

$$f = \alpha_1 + \frac{[OH^-] - [H^+]}{C_o V_o / (V_o + V_t)}. \tag{7.13}$$

TABLE 7.1 Titration curve for 0.025 L of 10^{-3} *F* HA, p*K* = 4.0, titrant 0.1 *N* NaOH. Activity corrections neglected. Effects from volume changes due to addition of base taken into consideration, as well as neglected (numbers in parentheses).

pH	V_t (μL)		*f*		pH	V_t (μL)		*f*
3.5684	0.0	(0.0)	0.0	← HA eq. pt.	5.60	243.2	(243.2)	0.973
3.60	8.4	(8.4)	0.035		5.80	245.7	(245.7)	0.983
3.70	33.5	(33.6)	0.134		6.00	247.3	(247.3)	0.989
3.80	57.0	(57.1)	0.228		6.25	248.5	(248.5)	0.994
3.90	79.1	(79.2)	0.316		6.50	249.1	(249.1)	0.996
4.00	99.9	(100.0)	0.400		6.75	249.5	(249.5)	0.998
4.10	119.4	(119.5)	0.478		7.00	249.8	(249.8)	0.999
4.20	137.4	(137.5)	0.550	NaA	7.25	249.9	(249.9)	0.9996
4.30	153.9	(154.0)	0.616	eq. pt. →	7.50	250.0	(250.0)	1.0
4.40	168.8	(168.9)	0.675		7.75	250.1	(250.1)	1.0004
4.50	182.0	(182.0)	0.728		8.00	250.2	(250.2)	1.001
4.60	193.5	(193.5)	0.774		8.25	250.4	(250.4)	1.002
4.70	203.4	(203.4)	0.814		8.50	250.8	(250.8)	1.003
4.80	211.8	(211.8)	0.847		8.75	251.4	(251.4)	1.006
4.90	218.9	(218.9)	0.876		9.00	252.5	(252.5)	1.010
5.00	224.8	(224.8)	0.899		9.50	258.0	(257.8)	1.032
5.20	233.6	(233.6)	0.934		10.00	275.3	(275.0)	1.101
5.40	239.4	(239.4)	0.958		10.50	330.1	(329.1)	1.320

[2]When access to a programmable hand calculator or computer is available, one might as well consider dilution. It takes little added programming effort to do so. After that, we can get an exact solution with no additional effort: the calculator (computer) will be doing all of the work.

Therefore,

$$C_tV_t/C_oV_o = \alpha_1 + \frac{[OH^-] - [H^+]}{C_oV_o/(V_o + V_t)} \tag{7.44}$$

$$C_tV_t = \alpha_1 C_oV_o + ([OH^-] - [H^+])(V_o + V_t) \tag{7.45}$$

$$V_t = \frac{\alpha_1 C_oV_o + ([OH^-] - [H^+])V_o}{C_t - ([OH^-] - [H^+])} \tag{7.46}$$

$$= \frac{[K/([H^+] + K)]C_oV_o + (K_w/[H^+] - [H^+])V_o}{C_t - (K_w/[H^+] - [H^+])}. \tag{7.47}$$

We note that if the effects of dilution were to be neglected, then the term $(K_w/[H^+] - [H^+])$ would not appear in the denominator. An example of an application of Eq. (7.47) for specific input values of K, K_w, C_o, V_o, and C_t is given in Table 7.1. The numbers in parentheses in the table are the results if dilution is neglected. As one can see, the error from neglecting dilution in this case is rather small. The reason is that the concentration of the titrant (C_t) is so large relative to C_o. Specifically, it is 100 times larger. Additional problems illustrating titration principles may be found in Pankow (1992).

7.8 REFERENCES

PANKOW, J. F. 1992. *Aquatic Chemistry Problems*. Portland: Titan Press-OR, P.O. Box 91399, Portland, Oregon 97291-1399.

8

Buffer Intensity β, and the Effects of Changes in β with pH on Titration Curves

8.1 INTRODUCTION

If any amount of strong base is added to any aqueous solution, we know that the pH will go up. Conversely, if any amount of strong acid is added, we know that the pH will go down. We are often interested in knowing how large the change in pH will be for a specific incremental addition of strong base or acid. The parameter which quantifies this concept is the buffer intensity β.

As discussed in Chapters 6 and 7, it is neither C_B nor C_A alone, but rather the *net strong base* $(C_B - C_A)$ that governs the pH of any given HA/A^- solution. We therefore define that

$$\beta = \frac{d(C_B - C_A)}{d\text{pH}} \tag{8.1}$$

where $d(C_B - C_A)$ (eq/L) represents the incremental change in the eq/L of net strong base in the solution, and dpH represents the change in pH that results. For situations involving just changes in C_B, Eq. (8.1) simplifies to

$$\beta = \frac{dC_B}{d\text{pH}} \qquad \text{(just strong base added).} \tag{8.2}$$

For situations involving just changes in C_A, Eq. (8.1) simplifies to

$$\beta = -\frac{dC_A}{d\text{pH}} \qquad \text{(just strong acid added).} \tag{8.3}$$

Equations (8.2) and (8.3) illustrate how strong acid is the negative analog of strong base.

The units that one uses for β will depend on whether or not one views pH as having units. Since we always give the value of pH as a pure number with *no units* (e.g., pH = 7.0, etc.), it makes sense to take the units of β as being the same as the units for C_B and C_A, that is, "eq/L." Therefore, eq/L will be the units used for β in this text. Some texts use units of eq/(L-pH unit).

When adding strong base,

$$dC_B > 0 \text{ and } d\text{pH} > 0 \qquad \text{and so by Eq. (8.2), } \beta > 0. \tag{8.4}$$

When adding strong acid,

$$dC_A > 0 \text{ and } d\text{pH} < 0 \qquad \text{and so by Eq. (8.3), } \beta > 0. \tag{8.5}$$

Adding net strong base ($d(C_B - C_A)$ positive) causes the pH to increase (dpH positive); just adding strong base (dC_B positive) also causes the pH to increase; just adding strong acid (dC_A positive) causes the pH to decrease (dpH negative). Therefore, by Eqs. (8.1)–(8.3), the parameter β is seen to always be *positive and nonzero.*

In words, β is a measure of how well a solution is able to *resist* pH changes when *either* strong acid or strong base is added. If adding a large amount of strong acid or base causes only a small pH change, then the system has a large buffer intensity, and β is large. Conversely, if adding a small amount of strong acid or base causes a large change in the pH, then β is small. The following sections will deal with many of the specifics concerning the buffer intensity β, and also how that nature of β affects the character of titration curves. Problems illustrating the priniciples of this chapter are presented by Pankow (1992).

8.2 GENERAL CONSIDERATIONS REGARDING THE BUFFER INTENSITY β IN MONOPROTIC ACID SYSTEMS

Based on the discussions provided in Chapter 7, we know that like f, $(C_B - C_A)$ is an appropriate variable for the x-axis of an alkalimetric titration. In particular, if we focus on the alkalimetric titration curve for the monoprotic acid HA, we know that at the HA equivalence point, $(C_B - C_A) = 0$ (i.e., $f = 0$). To achieve pH values higher than those of the HA equivalence point, the value of $(C_B - C_A)$ must be increased by increasing C_B. To achieve pH values lower than those of the HA equivalence point, $(C_B - C_A)$ must be decreased by increasing C_A. These concepts are summarized in Figure 8.1. Like Figure 7.1, the titration curve in Figure 8.1 has been drawn for values of K and C that give reasonably sharp changes in pH at both the HA and NaA equivalence points.

Based on Eq. (8.1), it is clear that at any given pH, the value of β is given by the *slope of the tangent line* (i.e., by the first derivative) of the curve that gives $(C_B - C_A)$ vs. pH. This type of plot is obviously closely related to the pH vs. $(C_B - C_A)$ titration curve. In fact, for the titration curve depicted in Figure 8.1, to obtain the corresponding $(C_B - C_A)$ vs. pH curve, we just rotate Figure 8.1 by 180 degrees. The rotation is carried out with the points of the bottom left and upper right corners of the figure remaining stationary. The rotation therefore takes place on an axis located in the plane of the paper and passing through the bottom left and upper right corners. The result is Figure 8.2.

For any given pH, there is only *one* tangent line in Figure 8.2. The slope of that tangent line is the same whether one sits at the point and looks towards larger pH, or towards smaller pH. "Looking towards larger pH" corresponds to adding a small amount

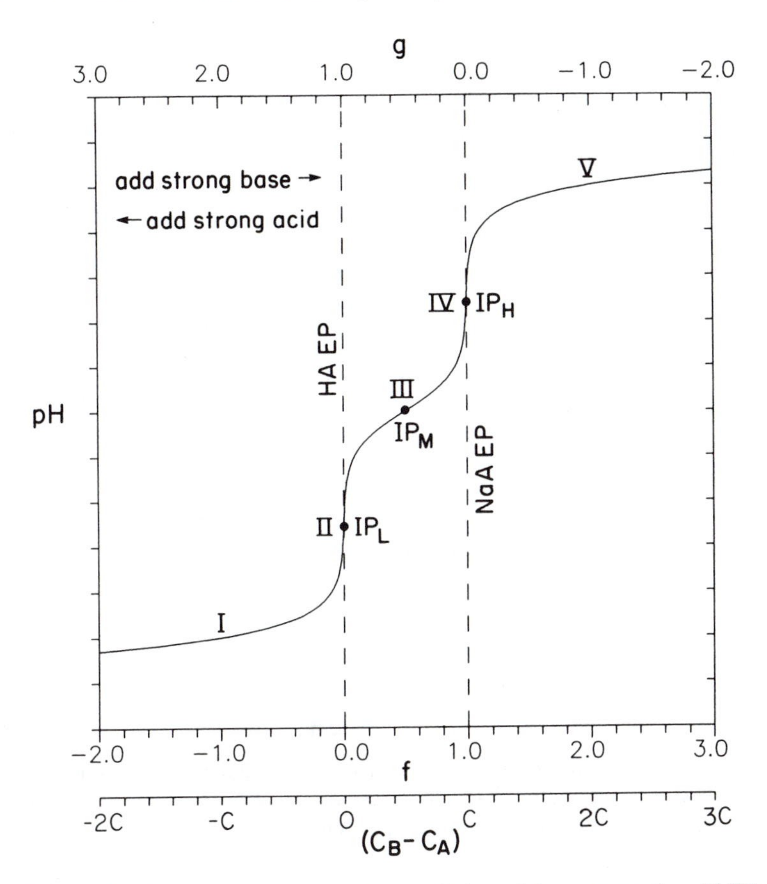

Figure 8.1 Generic titration curve for a solution of the monoprotic acid HA of concentration C F. The values of K and C are such that three inflection points are present in the titration curve. The x-axis may be labelled in terms of f, or equivalently, in terms of $(C_B - C_A)$.

of strong base. "Looking towards smaller pH" corresponds to adding a small amount of strong acid. The fact that there is only one tangent line explains why Eqs. (8.2) and (8.3) are *both* proper definitions of β.

There are five regions of the curve in Figure 8.2 that are of interest. They are labelled I–V. Since β is given by the slope of the curve of $(C_B - C_A)$ vs. pH, β will obviously be large when that slope is large, and β will be small when that slope is small. There are three sections of the Figure 8.2 curve where the slope is large, that is, sections I, III, and V. There are two sections where the slope is small, i.e., sections II and IV. Sections I–V are also labelled in Figure 8.1 for comparison.

When activity corrections can be neglected, we have

$$\mathrm{pH} = \mathrm{p}K + \log \frac{[\mathrm{A}^-]}{[\mathrm{HA}]}. \tag{8.6}$$

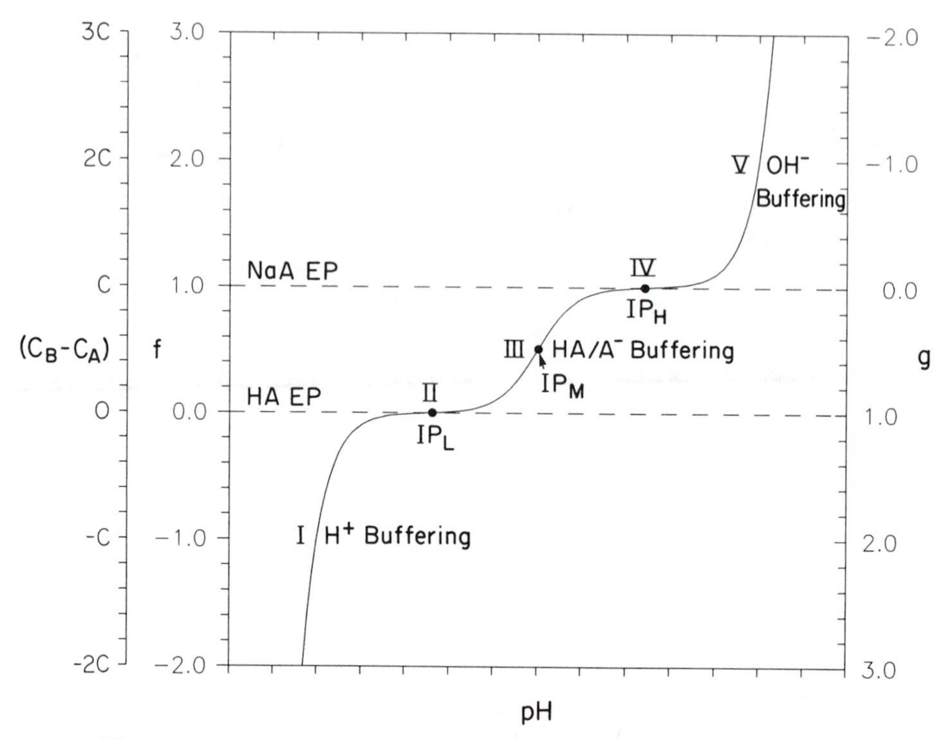

Figure 8.2 Generic rotated titration curve for a solution of the monoprotic acid HA of concentration C F. The values of K and C are such that all three inflection points are present in the titration curve.

For a constant ionic medium, Eq. (8.6) may still be used if ${}^cK$ is substituted for K. Equation (8.6) indicates that if a given addition of strong acid or base changes the $[A^-]/[HA]$ ratio significantly, then the pH will also change significantly (dpH will be large) and so β will be small. Conversely, if the $[A^-]/[HA]$ ratio does not change significantly, then the pH will not change significantly (dpH will be small) and so β will be large.

Region I. Very Low pH, $f \ll 0$. In this region, the value of β is large because there is simply so much free H^+ in the solution that adding some strong base will not decrease $[H^+]$ very much. Conversely, adding some strong acid will not increase $[H^+]$ very much. In either case, dpH is small, and so β is large. As an example, even neglecting the presence of the HA, a concentrated solution of HCl strongly resists attempts to change its pH.

Region II. Low pH, $f \sim 0$. In this region, the value of β is small because we are in the vicinity of the HA equivalence point. In particular, [HA] is relatively large here, and $[A^-]$ is relatively small. Thus, even adding a small amount of strong base causes

[A^-] to increase significantly. This leads to a significant increase in the [A^-]/[HA] ratio and so also in the pH. The value of dpH is therefore large, and so β is small. Viewing the situation from the point of view of adding strong acid, in this case since [A^-] is already small, even adding a small amount of strong acid leads to a significant decrease in [A^-]. The [A^-]/[HA] ratio and therefore the pH decrease significantly, and so β is again small.

Region III. Intermediate pH, $f \sim g \sim 0.5$. In this region, as long as C is not too low, [HA] is relatively large and can easily neutralize a small amount of strong base. Moreover, [A^-] is also large. Thus, converting some HA to A^- by adding a small amount of strong base does not change the [HA]/[A^-] ratio significantly, and so based on Eq. (8.6) we conclude that the pH will not change significantly. In a similar manner, if a small amount of strong acid is added, there is substantial A^- present to neutralize that acid. With [HA] already relatively large, we again conclude that the [HA]/[A^-] ratio will not change significantly, and so neither will the pH. Thus, since dpH is small for both the addition of a small amount of strong base or acid, β is large.

Region IV. High pH, $f \sim 1.0$. In this region, the value of β is small since we are in the vicinity of the NaA equivalence point. In particular, [A^-] is relatively large here, and [HA] is relatively small. With so little HA present to neutralize base, even adding a relatively small amount of strong base causes [HA] to decrease significantly. This leads to a significant increase in the [A^-]/[HA] ratio, and so the pH increases significantly. The value of dpH is therefore large, and so β is small. Viewing the situation from the point of view of adding strong acid, in this case even adding a relatively small amount of strong acid leads to a significant increase in [HA]. The [A^-]/[HA] ratio thus decreases significantly, and so the pH decreases significantly, and β is again small.

Region V. Very high pH, $f \gg 1$. In this region, the value of β is large because there is so much free OH^- present that adding some additional strong base does not increase [OH^-] very much. Conversely, adding some strong acid will not decrease [OH^-] very much. In either case, then, dpH is small, and so β is large. As an example, neglecting the presence of the NaA, a concentrated solution of NaOH strongly resists attempts to change its pH.

Based on the fact that the value of β in a monoprotic acid system is given by the first derivative of a curve of the type given in Figure 8.2, we see that the typical curve of $\log\beta$ vs. pH will often have the "double-minimum" shape given in Figure 8.3. At low pH, the value of β is bounded only by the physical limits on how concentrated the solution can be made in H^+. Similarly, at high pH, the value of β is bounded only by the physical limits on how concentrated the solution can be made in OH^-. As we will see below, at pH $\simeq$ pK (i.e., in region III), the maximum value that β can attain between the two minima is determined by the value of C.

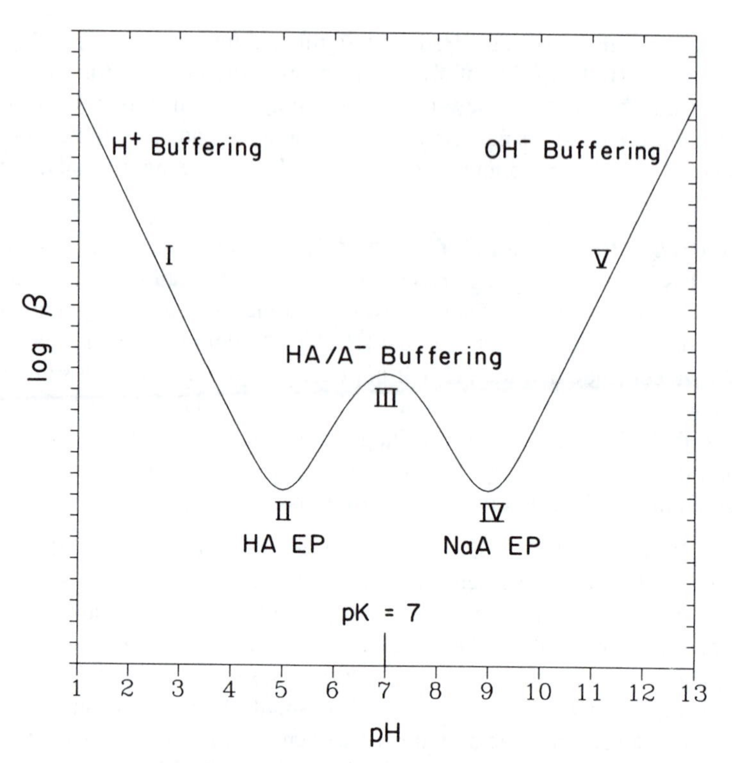

Figure 8.3 Generic plot of β vs. pH for a solution of the monoprotic acid HA. The value of C is large enough that two minima are clearly visible. (Since the curve is completely symmetrical, if the conditions are 25°C/1 atm where $K_w = 10^{-14}$, the value of K for this plot must be 10^{-7}.)

8.3 DERIVATION OF THE EXPRESSION FOR THE BUFFER INTENSITY β IN A MONOPROTIC ACID SYSTEM

The derivation of the equation for β in any given system involves nothing more than simple calculus. For the titration of a monoprotic acid, the ENE is

$$C_B - C_A = [A^-] + [OH^-] - [H^+]. \tag{8.7}$$

Taking the derivative with respect to pH, we obtain

$$\beta = \frac{d[A^-]}{dpH} + \frac{d[OH^-]}{dpH} - \frac{d[H^+]}{dpH}. \tag{8.8}$$

We now recall that

$$[A^-] = \alpha_1 C. \tag{8.9}$$

If we assume that C does not change with pH (i.e., that whatever strong base or acid is added to change the pH is sufficiently concentrated that C does not change), then combining Eqs. (8.8) and (8.9) yields

$$\beta = C\frac{d\alpha_1}{d\text{pH}} + \frac{d[\text{OH}^-]}{d\text{pH}} - \frac{d[\text{H}^+]}{d\text{pH}}. \tag{8.10}$$

The second two terms on the RHS of Eq. (8.10) can be differentiated fairly simply. Assuming that activity corrections can be neglected (or that constant ionic medium constants are used), we first note that

$$[\text{H}^+][\text{OH}^-] = K_\text{w} \tag{8.11}$$

$$\log[\text{H}^+] + \log[\text{OH}^-] = \log K_\text{w} \tag{8.12}$$

$$d(\log[\text{H}^+]) + d(\log[\text{OH}^-]) = 0 \tag{8.13}$$

$$d(\log[\text{H}^+]) = -d\text{pH} = -d(\log[\text{OH}^-]). \tag{8.14}$$

We also note that

$$\log x = (\ln x)/(\ln 10) \tag{8.15}$$

where, as usual, log x is the base 10 logarithm of x, ln x is the base e logarithm of x, and $\ln 10 = 2.303\ldots$ The value of e is $2.71828\ldots$

Based on Eqs. (8.11–8.15), we conclude that

$$d\text{pH} = d(-\log[\text{H}^+]) = -(1/2.303)d(\ln[\text{H}^+]) = \frac{-1}{2.303}\frac{d[\text{H}^+]}{[\text{H}^+]}. \tag{8.16}$$

In general then,

$$d\log x = \frac{1}{2.303}\frac{dx}{x}. \tag{8.17}$$

Therefore, by Eqs. (8.13) and (8.17), we also have

$$d\text{pH} = \frac{1}{2.303}\frac{d[\text{OH}^-]}{[\text{OH}^-]}. \tag{8.18}$$

Thus, for use in Eq. (8.10) we obtain

$$-\frac{d[\text{H}^+]}{d\text{pH}} = \frac{-d[\text{H}^+]}{-(1/2.303)d[\text{H}^+]/[\text{H}^+]} = 2.303[\text{H}^+] \tag{8.19}$$

and

$$\frac{d[\text{OH}^-]}{d\text{pH}} = \frac{d[\text{OH}^-]}{(1/2.303)d[\text{OH}^-]/[\text{OH}^-]} = 2.303[\text{OH}^-]. \tag{8.20}$$

For the derivative of α_1 in Eq. (8.10), we need to use the chain rule, that is,

$$du/dz = (du/dx)(dx/dz). \tag{8.21}$$

Applied here, we obtain

$$d\alpha_1/d\text{pH} = (d\alpha_1/d[\text{H}^+])(d[\text{H}^+]/d\text{pH}). \tag{8.22}$$

The value of the second factor in Eq. (8.22) is given by Eq. (8.19). For the first factor, we use the relation from calculus for the derivative of a quotient. That is, since

$$d(u/v) = (vdu - udv)/v^2 \tag{8.23}$$

we have for

$$\alpha_1 = K/([\mathrm{H}^+] + K) \tag{8.24}$$

that

$$d\alpha_1/d[\mathrm{H}^+] = -K/([\mathrm{H}^+] + K)^2. \tag{8.25}$$

By Eqs. (8.19), (8.22), and (8.25), we then obtain

$$C d\alpha_1/d\mathrm{pH} = C(-K/([\mathrm{H}^+] + K)^2)(-2.303[\mathrm{H}^+]). \tag{8.26}$$

Since

$$\alpha_0 = [\mathrm{H}^+]/([\mathrm{H}^+] + K) \tag{8.27}$$

by Eq. (8.24), we see that the RHS of Eq. (8.26) contains the product of α_0 and α_1, that is, we see that

$$C d\alpha_1/d\mathrm{pH} = 2.303\alpha_0\alpha_1 C. \tag{8.28}$$

Thus, combining Eqs. (8.10), (8.19), (8.20), and (8.28), for a monoprotic acid system, we obtain the important equation

$$\beta = dC_\mathrm{B}/d\mathrm{pH} = 2.303\ \ ([\mathrm{H}^+] \quad + \quad [\mathrm{OH}^-] \quad + \quad \alpha_0\alpha_1 C). \tag{8.29}$$

$[\mathrm{H}^+]$	$[\mathrm{OH}^-]$	$\alpha_0\alpha_1 C$
major term for acidic solutions	major term for basic solutions	major term when pH $\simeq$ pK, if C is not too small

Based on Eq. (8.29), we see that β can never be zero since neither $[\mathrm{H}^+]$ nor $[\mathrm{OH}^-]$ are ever zero in an aqueous solution. Thus, even when C is zero, any aqueous solution will have *some* ability to resist changes in pH. That ability can moreover be quite significant if either $[\mathrm{H}^+]$ or $[\mathrm{OH}^-]$ is large. However, under such circumstances, the pH will then be either very low or very high, respectively. Thus, though well buffered, the solution may not be buffered at a pH that is convenient either for life, or for chemical experiments of interest. Therefore, for good buffers at intermediate pH values, we rely on the presence of a significant amount of an HA/A^- pair with a pK of an appropriate value. In this manner, there is a local maximum in β in the pH region of interest. As we will now see, the local maximum occurs at pH $\simeq$ pK.

8.4 THE BUFFER INTENSITY IN A MONOPROTIC ACID SYSTEM WHEN pH $\sim$ p*K* AND *C* IS NOT TOO LOW

When the solution is neither very acidic or very basic *and* C is not too low, then both $[\mathrm{H}^+]$ and $[\mathrm{OH}^-]$ can be neglected in Eq. (8.29), and

$$\beta \simeq 2.303C\,\alpha_0\alpha_1 \tag{8.30}$$

$$\simeq 2.303C(1 - \alpha_1)\alpha_1 \tag{8.31}$$

$$\simeq 2.303C(\alpha_1 - \alpha_1^2) \tag{8.32}$$

The point at which β as given by Eqs. (8.30), (8.31), or (8.32) is maximized under these not too acidic/not too basic conditions may be determined as follows.

Since α_1 increases with increasing pH, the pH value that gives a maximum or a minimum value of β in a plot of β vs. pH will be the same pH that gives a maximum or minimum in a plot of β vs. α_1. Now, elementary calculus tells us that when the first derivative of a function equals zero, then at that point the function either has a local maximum or a local minimum. If the value of the second derivative at that point is positive, then the function is concave up, and has a local minimum there. If the value of the second derivative is negative at that point, then the function is concave down and has a local maximum. (The terms "concave up" and "concave down" are defined in Figure 8.4.) Therefore, we differentiate Eq. (8.32) with respect to α_1, and obtain

$$d\beta/d\alpha_1 \simeq 2.303C(1 - 2\alpha_1) \qquad (8.33)$$

$$d^2\beta/d\alpha_1^2 \simeq -4.6C. \qquad (8.34)$$

According to Eq. (8.33), $d\beta/d\alpha_1 = 0$ when $\alpha_1 = 1/2$. We know that $\alpha_1 = 1/2$ when $\mathrm{pH} = \mathrm{p}K$. We also note that since C is always positive, the RHS of Eq. (8.34) is always negative. Thus, when both $[H^+]$ and $[OH^-]$ may be neglected relative to $C\alpha_0\alpha_1$, we conclude from Eq. (8.34) that the equation for β (i.e., Eq. (8.29)) is concave down, and so β has a local maximum when $\mathrm{pH} = \mathrm{p}K$ and $\alpha_0 = \alpha_1 = 1/2$. Numerically, the fact that $C\alpha_0\alpha_1$ is maximized (i.e., $\alpha_0\alpha_1$ is maximized) when $\alpha_0 = \alpha_1 = 1/2$ may be verified based on the data in Table 8.1.

Table 8.1 reinforces the discussion that was given above as to why β is large in region III ($\mathrm{pH} \simeq \mathrm{p}K$) of Figure 8.2. Indeed, when C is not too low, Table 8.1 illustrates again that if some strong base is added, then there will be a significant amount of HA

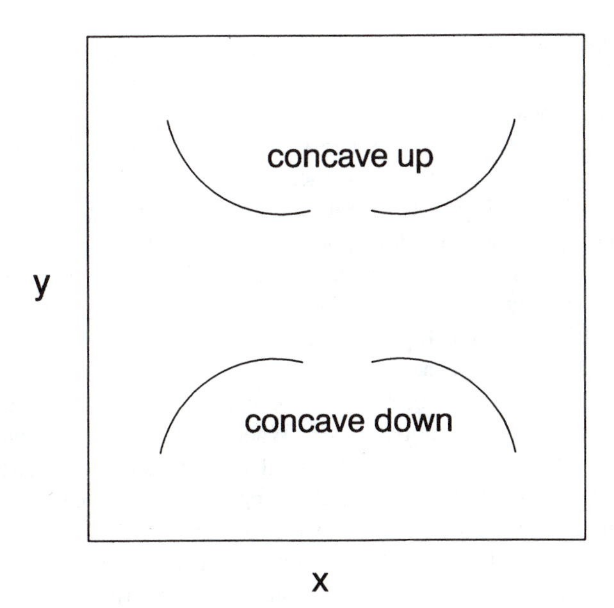

Figure 8.4 Two examples of sections of curve that are concave up, and two examples that are concave down.

TABLE 8.1. Values of α_0, α_1, $[A^-]/[HA] = \alpha_1/\alpha_0$, and $\alpha_0\alpha_1$ for arbitrary pK for α_1 between 0.05 and 0.95.

	α_0	α_1	$[A^-]/[HA]$	$\alpha_0\alpha_1$	
	0.95	0.05	0.053	0.048 - near the HA *EP*	
	0.90	0.10	0.111	0.090	
	0.75	0.25	0.333	0.188	
	0.60	0.40	0.667	0.240	approximate range for good HA/A^- buffering
	0.55	0.45	0.818	0.248	
pH = pK →	0.50	0.50	1.00	0.250	
	0.45	0.55	1.22	0.248	
	0.40	0.60	1.50	0.240	
	0.25	0.75	3.00	0.188	
	0.10	0.90	9.00	0.090	
	0.05	0.95	19.00	0.048 - near the NaA EP	

around to neutralize it. With a significant amount of A^- also present, the ratio $[A^-]/[HA]$ will not change much, and so a large pH change is thereby prevented. Conversely, if some strong acid is added, then there will be a significant amount of A^- around to neutralize it as well. With a significant amount of HA also present, the ratio $[A^-]/[HA]$ will again not change much, thereby again preventing a large pH change.

8.5 EFFECTS OF *K* AND *C* ON PLOTS OF log β VS. pH AND THEIR CORRESPONDING TITRATION CURVES

8.5.1 General

In Figure 8.1, there is a sharp change in pH at both the HA and NaA *equivalence points* (EPs). As a result of these changes, very near both of these EPs, the titration curve shifts from being concave up to concave down. Any point at which a curve shifts from concave up to concave down (or vice versa) is referred to as an *inflection point* (IP). In addition to the two IPs at the two EPs, there is a third IP in Figure 8.1. It is located very near pH = pK. The Figure 8.1 titration curve therefore represents four subsections that are sequentially concave up, down, up, and then finally down. The first concave up region is where $f < 0$. For $0 < f < 1$, the curve is first concave down and then concave up. Finally, for $f > 1$, the curve is concave down. The three sequential IPs in Figure 8.1 are identified as IP_L, IP_M, and IP_H. The subscripts L, M, and H are intended to invoke low, medium, and high pH, respectively.

Consider when the values of K and C are such that there is in fact both an IP_L and an IP_H in a given HA/A^- titration curve. When carrying out an acidimetric titration of a solution for which f is initially greater than 0, titration chemists usually equate the process of determining the amount of strong acid titrant needed to reach the HA EP with determining the amount required to reach the IP_L. (The rationale here is that an IP is an identifiable characteristic of a curve that can often be located by eye, or instrumentally.) Conversely, when carrying out an alkalimetric titration of a solution for which f is initially less than 1, locating the NaA EP is usually equated with determining the amount of strong base required to reach the IP_H.

While the IP_H and the NaA EP will usually not coincide *exactly*, often they are sufficiently close that the above approach for locating the NaA EP is valid. The same applies to IP_L and the HA EP. However, as C decreases, both IP_H and IP_L will become less and less noticeable, as well as less and less reliable as locators for the NaA EP and the HA EP. The value of K also has a role to play in this regard, but it is somewhat more complex The actual *error* that results from taking the location of an IP to be the location of a corresponding EP is discussed by Pankow (1992) in the companion problems book for this text. In general, when $K > \sqrt{K_w}$, the error in f in taking IP_H to be equivalent to the NaA EP is generally smaller (in magnitude) than –1%, and the error in g in taking IP_L to be equivalent to the HA EP is generally smaller than –26%. When $K < \sqrt{K_w} (K_b > \sqrt{K_w})$, the error in taking IP_L to be equivalent to the HA EP is generally smaller than –1%, and the error in taking IP_H to be equivalent to the NaA EP is generally smaller than –26%. The errors are all negative because, as it turns out, in a monoprotic system, all of the IPs always occur *between* the two EPs.

Given the above facts, the next section will consider the manner in which the values of K and C interact to determine: 1) whether or not either IP_L and/or IP_H will even be present in any given titration curve; 2) if present, how *noticeable* an IP will be.

8.5.2 Type of Behavior for β that Leads to the Presence of an Inflection Point (IP) in a Titration Curve for a Monoprotic System

All monoprotic system titration curves will not have three IPs. Since all titration curves will be concave up for $f < 0$ and concave down for $f > 1$,[1] if there are not three IPs, the *only* other possibility is one IP. Indeed, one IP will yield the required conversion from concave up to concave down, but two IPs will not. For mathematical reasons that need not be discussed here, a disappearance of two IPs is related to changes in K and/or C that lead to two roots of a governing fifth order polynomial going from real to imaginary (Pankow, 1992).

By elementary calculus, an IP in a pH vs. $(C_B - C_A)$ titration curve is a point where

$$\frac{d^2\text{pH}}{d(C_B - C_A)^2} = 0 \qquad \text{criterion for an IP in a pH vs. } (C_B - C_A) \text{ titration curve.} \qquad (8.35)$$

When a pH vs. $(C_B - C_A)$ titration curve is rotated to obtain the corresponding $(C_B - C_A)$ vs. pH curve, all of the IPs present in the first curve remain as IPs in the second curve. The only difference is that a concave up to concave down transition becomes a concave down to concave up transition, and vice versa. Thus, each time Eq. (8.35) is satisfied, there is an IP in the $(C_B - C_A)$ vs. pH curve, and we also have

$$\frac{d^2(C_B - C_A)}{d\text{pH}^2} = 0 \qquad \text{criterion for an IP in a } (C_B - C_A) \text{ vs. pH curve } \textit{and} \text{ in a pH vs. } (C_B - C_A) \text{ titration curve.} \qquad (8.36)$$

[1]For $f < 0$, the decrease in pH must slow as f becomes increasingly negative, and for $f > 1$, the increase in pH must slow as f increases.

Calculus also tells us that when an IP is present in the plot of a function, there is *either* a maximum or a minimum in the plot of the first derivative of that function. Thus, when there is an IP in a pH vs. $(C_B - C_A)$ titration curve, we can conclude that there is either a maximum or a minimum in the plot of β vs. pH, or equivalently, in a plot of $\log \beta$ vs. pH. The total number of IPs in a titration curve will therefore be equal to the number of minima plus maxima in a plot of $\log \beta$ vs. pH.

When C is large enough to allow the 2.303 $\alpha_0\alpha_1 C$ term to dominate in Eq. (8.29), then as in Figure 8.3, there is a local maximum near pH $\simeq$ pK in the curve of $\log \beta$ vs. pH. Near the pH values corresponding to the HA and NaA EPs, there are local minima. Their presence may be understood on the basis of Table 8.1 where we see that $\alpha_1\alpha_0$ goes to near zero near both EPs, and so beyond those points β will begin to rise again due to buffering from $[H^+]$ and $[OH^-]$ respectively. It is precisely the three local maxima and minima in Figure 8.3 that give rise to the three IPs in Figures 8.1 and 8.2. The minima give IP_L and IP_H, and the maximum gives IP_M.

Figure 8.5 illustrates for $K = 10^{-7}$ how the height of the "peak" given to β by the $2.303\alpha_0\alpha_1 C$ term depends on the value of C. As C becomes lower and lower, there

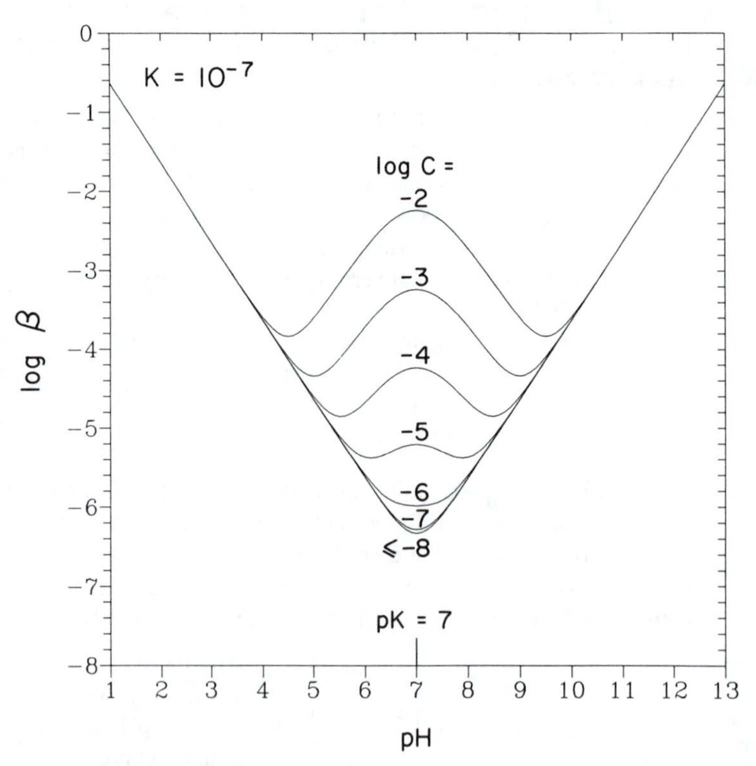

Figure 8.5 Plots of $\log \beta$ vs. pH for $K = 10^{-7}$ and varying values of C between 10^{-8} and 10^{-2} M. Activity corrections neglected (not valid at very low or very high pH).

comes a point where there is no longer any local maximum, and only one minimum remains. At this point, two of the IPs in the titration curve disappear. IP_M is always one of the IPs that disappears. (When $K > \sqrt{K_w}$, then the other IP to disappear is IP_L. When $K < \sqrt{K_w}$, then the other IP to disappear is IP_H.) As C continues to decrease, then for all pH the $\alpha_0\alpha_1 C$ term eventually becomes completely negligible relative to the buffer intensity provided by the 2.303[H⁺] term at low pH, and by the 2.303[OH⁻] term at high pH. When $C = 0$, then the β curve is given simply by $2.303([H^+]+[OH^-])$. The curve for $C = 10^{-8}$ M is essentially coincident with the curve for $C = 0$.

The contributions of the three terms $2.303[H^+]$, $2.303[OH^-]$, and $2.303\alpha_0\alpha_1 C$ are plotted individually in Figure 8.6 for the same K value (10^{-7}) and range of C values considered in Figure 8.5. Figure 8.6 illustrates why for $C = 10^{-7}M$, the contribution of the $2.303\alpha_0\alpha_1 C$ term to β is so small, even at pH = 7 where that contribution maximizes.

Many of the observations of the preceding paragraph may be further understood in terms of Figure 8.7 which contains the titration curves for most of the cases given in

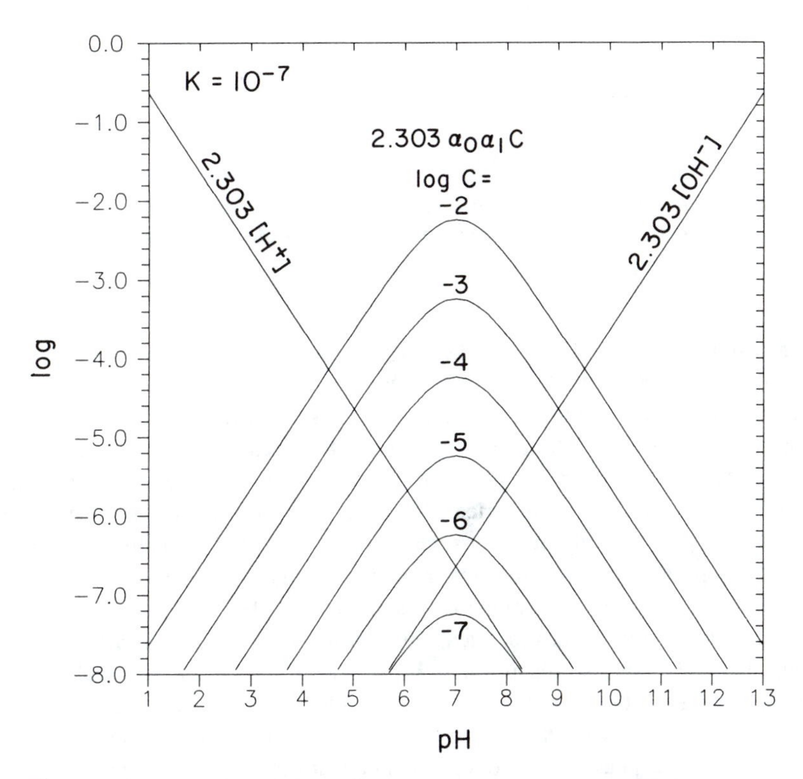

Figure 8.6 Plots of $\log(2.303[H^+])$, $\log(2.303[OH^-])$, and $\log(2.303\alpha_0\alpha_1 C)$ vs. pH for $K = 10^{-7}$ and varying values of C between 10^{-7} and $10^{-2}M$. Activity corrections neglected (not valid at very low or very high pH).

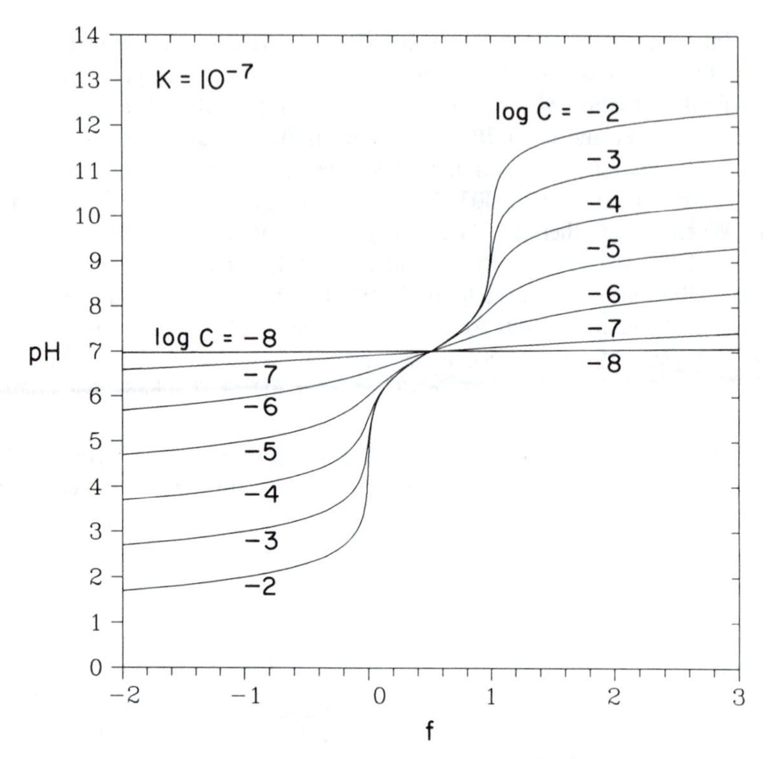

Figure 8.7 Titration curves (pH vs. f) for $K = 10^{-7}$ and varying values of C between 10^{-8} and $10^{-2}M$. Activity corrections neglected (not valid at very low or very high pH).

Figures 8.5 and 8.6. At 25°C/1 atm where $K_w = 10^{-14}$, then when $K = 10^{-7}$, both EP inflection points are present when $C = 10^{-5}M$, but not when $C = 10^{-6}M$.

Figures 8.8 and 8.9 consider a range of cases for which $K = 10^{-5}$. The shift in pK of two units to the left relative to Figure 8.5 means that the maximum in the $2.303\alpha_0\alpha_1C$ term in Figure 8.8 is also shifted two pH units to the left. Thus, the log β vs. pH curves are no longer symmetrical. While a hump at pH = pK is clearly visible at $C = 10^{-5}M$ for pK = 7 (Figure 8.5), such a hump is no longer visible at $C = 10^{-5}M$ for pK = 5 (Figure 8.8) due to the importance of the H^+ buffering term. Figures 8.10 and 8.11 consider the analogous range of cases for $K = 10^{-9}$.

Thus, as may now already be clear, in addition to having IPs disappear by virtue of C being low, IPs can also disappear if K becomes too large *or* too small for a given value of C. In order to illustrate this point, Figures 8.12 and 8.13 give log β and titration curves for $C = 10^{-3}M$, with K ranging from 10^{-4} to 10^{-10}. Whenever visible, the pH center of each $\alpha_0\alpha_1C$ peak in Figure 8.12 occurs near pH $\simeq$ pK. In Figure 8.13, when K is sufficiently large (e.g., $K = 10^{-4}$), there is no IP_L (or IP_M) and so we will obviously not be able to use the IP_L to locate the acidimetric (HA) EP in the corresponding titration curve. For large K then, by the time that an acidimetric titration starting from $f > 0$

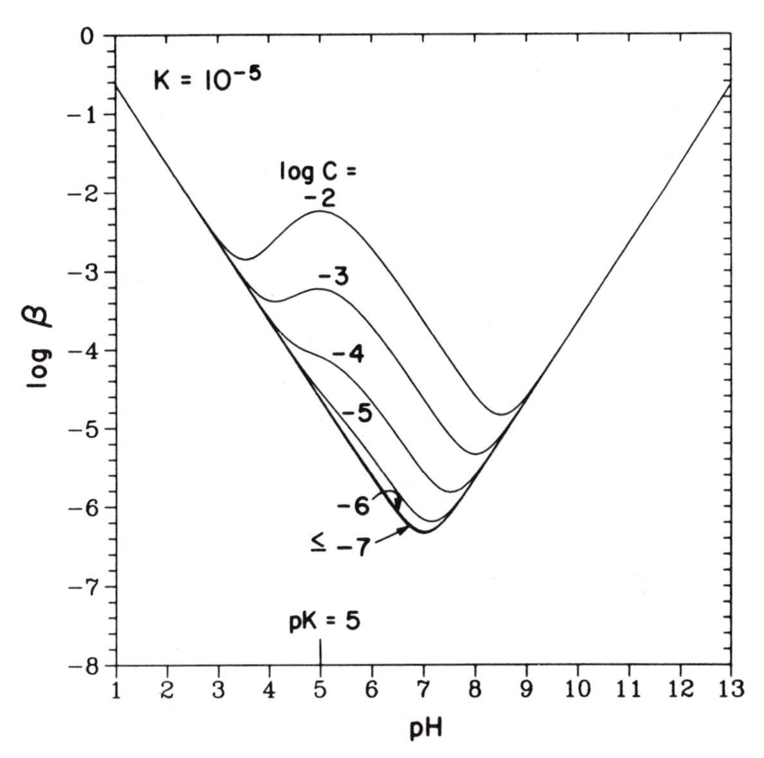

Figure 8.8 Plots of $\log \beta$ vs. pH for $K = 10^{-5}$ and varying values of C between 10^{-7} and $10^{-2}M$. Activity corrections neglected (not valid at very low or very high pH).

reaches a pH such that we can protonate significant numbers of A^- ions, the solution is already so acidic that H^+ buffering obviates the presence of IP_L (and IP_M). What is happening here is that in the corresponding plot of $\log \beta$ vs. pH (see Figure 8.12), the $2.303\alpha_0\alpha_1 C$ peak has been "pushed" too far into the H^+ buffering "shoulder"; the acid-side minimum in the $\log \beta$ vs. pH curve has disappeared. If the concentration C was increased sufficiently, however, the $2.303\alpha_0\alpha_1 C$ peak could be made to pop up again on the left side of the $\log \beta$ vs. pH curve, allowing IP_L and IP_M to reappear in the titration curve. A good every day example of a conjugate base that we cannot determine by acidimetric titration is Cl^-; the acid HCl is simply too strong ($pK \simeq -3$). In general, it can be shown (albeit laboriously) that for an IP_L to be present, we must have (Pankow 1992)

$$C/K = C/K_a > 27. \tag{8.37}$$

We also note that for a given value of C, when K is sufficiently small, there is no IP_H (or IP_M). Now, OH^- buffering dominates before β can become minimized in the region of the NaA EP; the $2.303\alpha_0\alpha_1 C$ peak has been "pushed" too far into the OH^- buffering "shoulder"; the basic-side minimum in the $\log \beta$ vs. pH curve has disappeared.

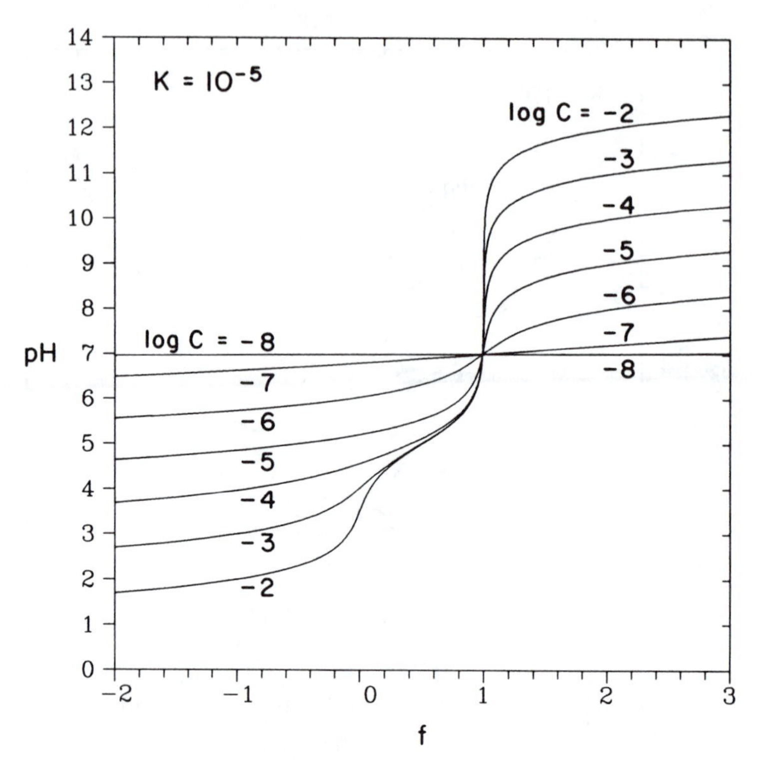

Figure 8.9 Titration curves (pH vs. f) for $K = 10^{-5}$ and varying values of C between 10^{-8} and $10^{-2}M$. Activity corrections neglected (not valid at very low or very high pH).

We conclude that just as it is not possible to titrate a solution of the conjugate base (e.g., Cl^-) of a very strong acid (e.g., HCl) to a visible acidimetric EP, it is generally not possible to titrate a solution of an extremely weak acid to a visible alkalimetric EP. For $C \simeq 10^{-3}M$ or less, you would not be able to determine the amount of $(CH_3)_3NH^+Cl^-$ (HCl salt of trimethylamine) present in a solution by alkalimetric titration; $(CH_3)_3NH^+$ is too weak of an acid (pK = 9.8 at 25°C/1 atm).[2] Indeed, before you can get to the alkalimetric EP, you run into too much buffering from OH^- to see that EP. However, if the concentration C was increased, the $2.303\alpha_0\alpha_1 C$ peak would pop up on the right side of the log β vs. pH curve, allowing IP_M and IP_H to reappear in the titration curve. In general, it can be shown that for an IP_H to be present, we must have (Pankow 1992)

$$CK/K_w = C/K_b > 27. \tag{8.38}$$

[2]Note that in this case, the conjugate acid "HA" is the charged species $(CH_3)_3NH^+$, and the conjugate base "A^-" is the neutral species $(CH_3)_3N$. Much like ammonia, $(CH_3)_3N$ reacts strongly with water to form a solution of $(CH_3)_3NH^+$ and OH^-.

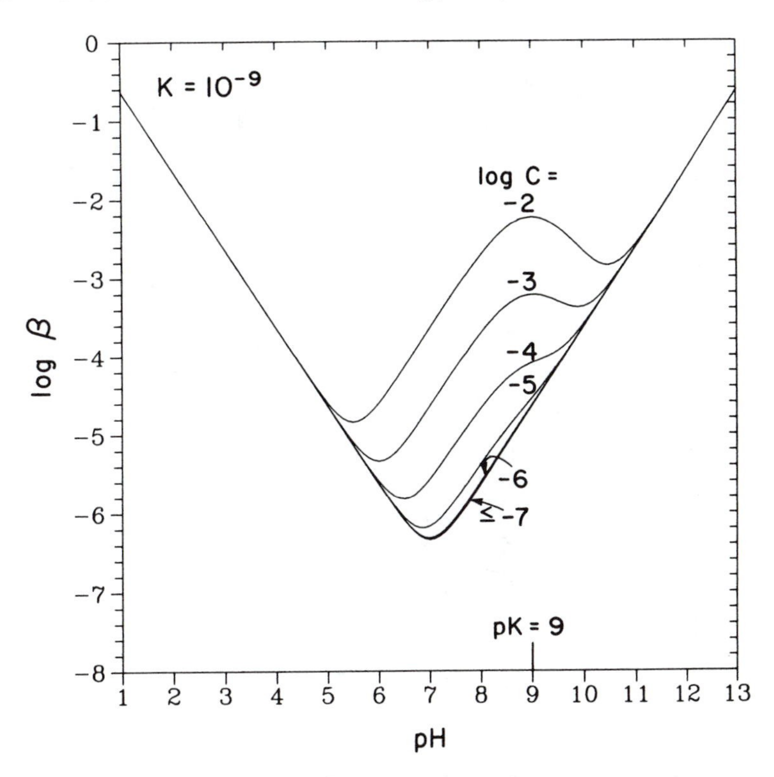

Figure 8.10 Plots of $\log \beta$ vs. pH for $K = 10^{-9}$ and varying values of C between 10^{-7} and $10^{-2}M$. Activity corrections neglected (not valid at very low or very high pH).

The one IP that remains in a solution of HA with $K = 10^{-4}$ and $C = 10^{-3}M$ will be IP_H. (Indeed, when $K > \sqrt{K_w}$, there will always be an IP_H in the titration curve. As C becomes low, however that IP may become impossibly faint to detect.) Although we cannot see an IP_L for $K = 10^{-4}$ and $C = 10^{-3}M$, we can perform an alkalimetric titration and determine how much of such an acid is present in a solution. The minimum in the Figure 8.12 plot of $\log \beta$ vs. pH for such an acid produces a clearly visible IP_H in the corresponding Figure 8.13 titration curve. That IP is essentially coincident with the NaA EP.

The one IP that will remain in a $10^{-3}M$ solution of trimethylamine ($K \simeq 10^{-10}$) will be IP_L. (When $K < \sqrt{K_w}$, there will *always* be an IP_L in the titration curve. As C becomes low, however that IP may become impossibly faint to detect.) Although we cannot see IP_H for $K = 10^{-10}$ and $C = 10^{-3}M$, we can perform an acidimetric titration and determine how much OH^- from the reaction of trimethylamine with water is present. Indeed, the minimum in the Figure 8.12 plot of $\log \beta$ vs. pH for $K = 10^{-10}$ and $C = 10^{-3}$ produces a clearly visible IP_L in the corresponding Figure 8.13 titration curve. That IP is essentially coincident with the HA EP for that case.

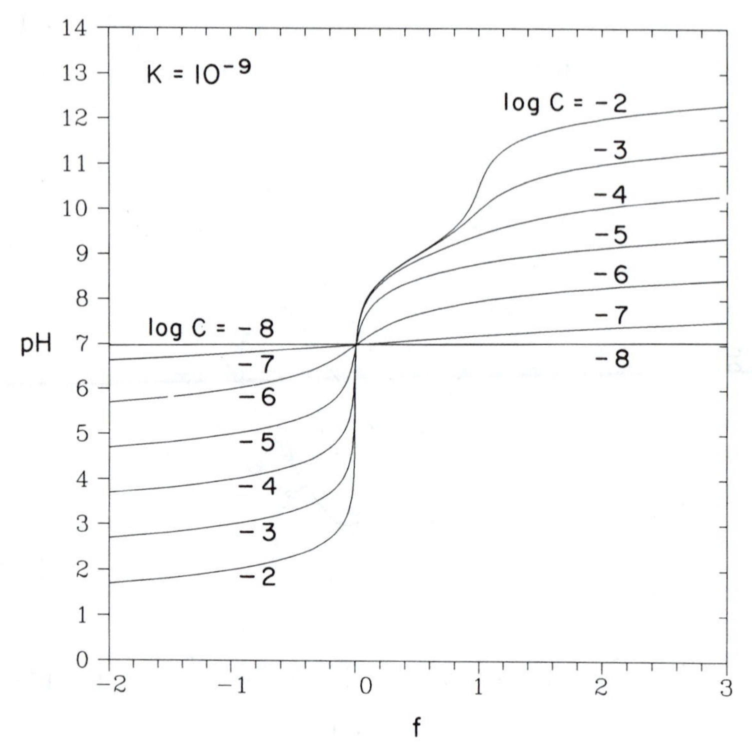

Figure 8.11 Titration curves (pH vs. f) for $K = 10^{-9}$ and varying values of C between 10^{-8} and $10^{-2}M$. Activity corrections neglected (not valid at very low or very high pH).

8.6 UNIQUENESS OF THE VALUE OF β FOR ANY GIVEN SYSTEM AT A GIVEN pH

It is interesting to ask if any given solution can be a better buffer against the addition of acid or base than it is against the addition of the other (i.e., base or acid, respectively). As discussed earlier, when answered in terms of the definition of β as the *slope* of the C_B vs. pH curve, the answer is *no*, since there is only *one slope* (i.e., tangent) at a given point, regardless of the direction toward which it is viewed. When answered in terms of a *non-infinitely small* incremental addition of either strong base or strong acid, however, the answer can be yes if C is not too small and the system is not at the special point where $\alpha_0 = \alpha_1 = 1/2$. For example, for $\alpha_0 \simeq 0.1$ and $\alpha_1 \simeq 0.9$, since there is more conjugate base A^- present than conjugate acid HA, then a non-infinitely small addition of strong acid would cause less change in pH than would a non-infinitely small addition of strong base of the same size. For this case, when $K = 10^{-5}$ and $C = 1.1 \times 10^{-3}M$, initially we have $[HA] = 10^{-4}M$, $[A^-] = 10^{-3}M$, and pH = 6.0. After the addition of 0.2×10^{-4} eq/L of strong acid, the pH changes to 5.91 (ΔpH = −0.09; note the

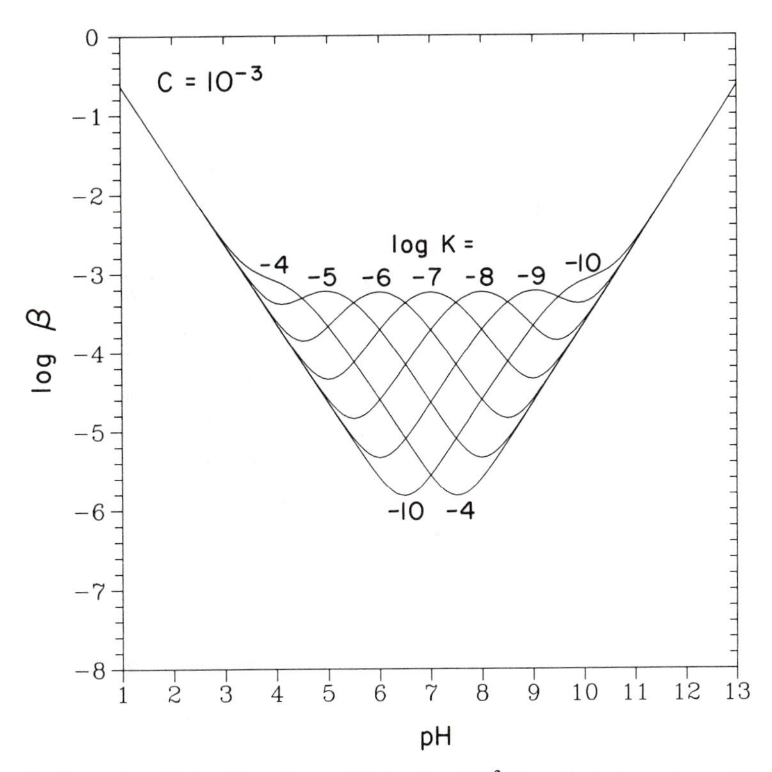

Figure 8.12 Plots of $\log \beta$ vs. pH for $C = 10^{-3}M$ and varying values of K between 10^{-10} and 10^{-4}. Activity corrections neglected (not valid at very low or very high pH).

use of ΔpH, and not dpH) and $-\Delta C_A/\Delta\text{pH} = 2.2 \times 10^{-4}$ eq/L. After the addition of 0.2×10^{-4} eq/L of strong base to the same initial solution, the pH changes to 6.11 ($\Delta\text{pH} = +0.11$) and $\Delta C_B/\Delta\text{pH} = 1.8 \times 10^{-4}$ eq/L.

8.7 BUFFER INTENSITY FOR SYSTEMS CONTAINING MORE THAN ONE ACID/BASE PAIR, OR CONTAINING A POLYPROTIC ACID

If various different acid/base pairs are present, for example, HA/A^-, HB/B^-, HC/C^-, etc., then there will be contributions to β from all of the acid/base pairs as well as from $[H^+]$ and $[OH^-]$. Thus, it may be shown that

$$\beta = 2.303([H^+] + [OH^-] + \alpha_{HA}\alpha_{A^-}C_A + \alpha_{HB}\alpha_{B^-}C_B + \alpha_{HC}\alpha_{C^-}C_C + \ldots) \qquad (8.39)$$

where C_A, C_B, C_C, etc. are the total concentrations of A, B, C, etc., and α_{HA} represents the fraction of the total A that is present as HA, α_{A^-} represents the fraction of the total A that is present as A^-, etc.

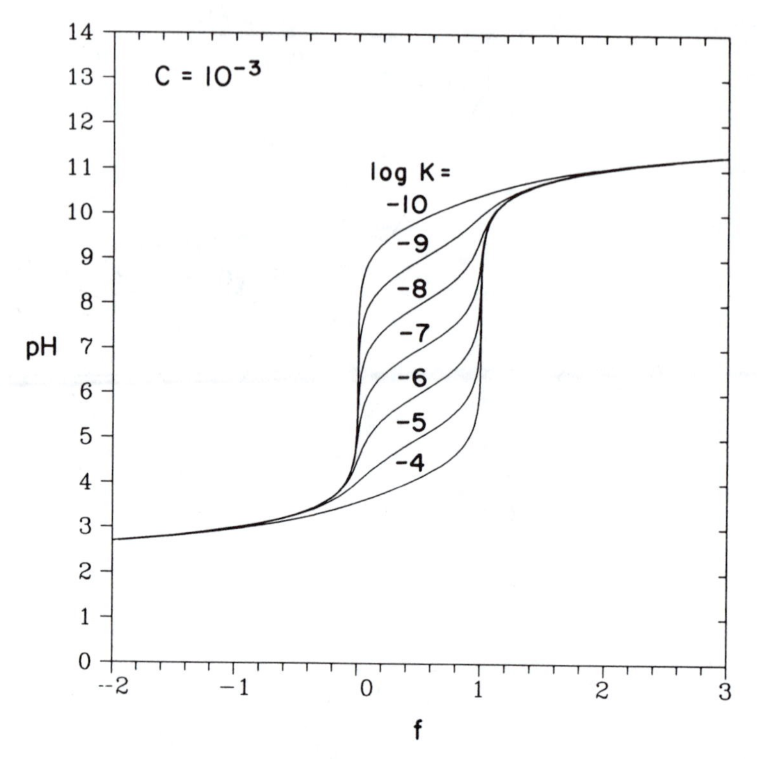

Figure 8.13 Titration curves (pH vs. f) for $C = 10^{-3}M$ and varying values of K between 10^{-10} and 10^{-4}. Activity corrections neglected (not valid at very low or very high pH).

To a first approximation a solution of a single *polyprotic* acid can be often be considered to be a solution of individual acid/base pairs all of the same total concentration. Thus, for a solution of total concentration C of the diprotic acid H_2B, we have that

$$\beta \simeq 2.303([H^+] + [OH^-] + \alpha_0\alpha_1 C + \alpha_1\alpha_2 C) \tag{8.40}$$

where α_0, α_1, and α_2 represent the fractions of the total B that have lost zero, one, and two protons, respectively. Equation (8.40) becomes increasingly accurate as the two acidity constants for H_2B become further and further apart. The exact expression for β for a system involving a polyprotic acid may be obtained by differentiating the corresponding ENE. For a diprotic system, we obtain

$$\beta = \frac{d(C_B - C_A)}{dpH} = 2.303([H^+] + [OH^-]) + C\frac{d\alpha_1}{dpH} + 2C\frac{d\alpha_2}{dpH}. \tag{8.41}$$

The completion of the two differentials on the RHS of Eq. (8.41) is left as an exercise for the reader. The result is

$$\beta = \frac{d(C_B - C_A)}{dpH} = 2.303([H^+] + [OH^-] + \alpha_0\alpha_1 C + \alpha_1\alpha_2 C + 4\alpha_0\alpha_2 C). \tag{8.42}$$

For a solution of total concentration C of the the triprotic acid H_3C, we have that

$$\beta \simeq 2.303([H^+] + [OH^-] + \alpha_0\alpha_1 C + \alpha_1\alpha_2 C + \alpha_2\alpha_3 C). \qquad (8.43)$$

Equation (8.43) becomes increasingly accurate as the three acidity constants become further and further apart.

8.8 REFERENCES

PANKOW, J. F. 1992. *Aquatic Chemistry Problems*. Portland: Titan Press-OR, P.O. Box 91399, Portland, Oregon 97291-1399.

9

Chemistry of Dissolved CO_2

9.1 INTRODUCTION

Gaseous carbon dioxide (i.e., $CO_{2(g)}$) can dissolve in water according to the reaction

$$CO_{2(g)} = CO_2 \tag{9.1}$$

where CO_2 represents *dissolved* CO_2. Since the carbon dioxide molecule has no protons associated with it, it is not an acid in and of itself. However, dissolved carbon dioxide can equilibrate with water to form the diprotic species carbonic acid (H_2CO_3) according to

$$H_2CO_3 = CO_2 + H_2O \qquad K = \frac{\{CO_2\}}{\{H_2CO_3\}}. \tag{9.2}$$

As is indicated by Eq. (9.2), both CO_2 and H_2CO_3 will be present in any solution containing carbon dioxide. Whatever CO_2 is present in a given solution has the *capability* to react with water according to Eq. (9.2) to form more H_2CO_3. Therefore, when counting how many acidic protons are present or *potentially* present in a solution of CO_2, one must count those on the H_2CO_3 molecules as well as those that can be added by Eq. (9.2) using the CO_2 molecules present. Indeed, if a strong base is added to a solution of CO_2, we can expect that the reaction in Eq. (9.2) will be driven to the left to provide protons to react with the base. In effect then, the total amount of diprotic acid in a CO_2 system will be given by the concentration of the hypothetical species $H_2CO_3^*$ where

$$[H_2CO_3^*] = [CO_2] + [H_2CO_3]. \tag{9.3}$$

The fact that $H_2CO_3^*$ is acidic is of enormous importance for all of natural water chemistry because: 1) there are significant amounts of carbon dioxide in the atmosphere and therefore also in every natural water system near the surface of the earth; 2) many subsurface geological systems that are quite isolated from the atmosphere can also contain large amounts of carbon dioxide; and 3) the prevalence of $H_2CO_3^*$ means that it will play a fundamental role in determining the pH of natural waters.

As a diprotic acid, H_2CO_3 can lose up to two protons according to the reactions

$$H_2CO_3 = H^+ + HCO_3^- \qquad K_{H_2CO_3} = \frac{\{H^+\}\{HCO_3^-\}}{\{H_2CO_3\}} \tag{9.4}$$

$$HCO_3^- = H^+ + CO_3^{2-} \qquad K_2 = \frac{\{H^+\}\{CO_3^{2-}\}}{\{HCO_3^-\}}. \tag{9.5}$$

The reason why the equilibrium constant for reaction (9.4) is not labelled K_1 is that, as will be seen below, K_1 is reserved for the first acidity constant of $H_2CO_3^*$.

On the basis of reactions (9.1), (9.2), (9.4), and (9.5), we conclude that *all* aqueous solutions containing carbon dioxide will contain the four dissolved species CO_2, H_2CO_3, HCO_3^-, and CO_3^{2-}. In the natural environment, there are a great many processes in which these carbon dioxide-related species participate. Some of the more important of these processes are listed in Table 9.1, and summarized schematically in Figure 9.1.

TABLE 9.1. Examples of important geological chemical, biological, and human-driven processes involving the dynamics of carbon dioxide species in the environment.

Process	Forward Reaction	Reverse Reaction
Processes Occurring on Land		
$6CO_{2(g)} + 6H_2O = C_6H_{12}O_6 + 6O_{2(g)}$	Plant photosynthesis	Plant respiration
$C_6H_{12}O_6 + 6O_{2(g)} = 6CO_{2(g)} + 6H_2O$	Aerobic bacterial and animal respiration	Anaerobic microbial anabolism
$CO_2(\text{geological}) = CO_{2(g)}$	Emission by volcanoes	NA*
$C_{(s)} + O_{2(g)} = CO_{2(g)}$	Combustion of coal for heat and power	NA
$C_nH_{2n+2} + \left(\frac{3}{2}n + \frac{1}{2}\right) O_{2(g)} = nCO_{2(g)} + (n+1)H_2O$	Combustion of alkane hydrocarbons (e.g., natural gas, petroleum products, etc.) for heat and power	NA
Processes Occurring in Natural Waters		
$CO_{2(g)} = CO_2$	Dissolution of atmospheric carbon dioxide	Outgassing of dissolved carbon dioxide
$6CO_2 + 6H_2O = C_6H_{12}O_6 + 6O_2$	Aquatic plant photosynthesis	Aquatic plant respiration
$Ca^{2+} + CO_3^{2-} = CaCO_{3(s)}$(water column)	Precipitation of metal carbonates in water column	Dissolution of metal carbonates in water column
$CaCO_{3(s)}$(water column) $= CaCO_{3(s)}$ (sediments)	Sedimentation of metal carbonates	Resuspension of metal carbonates
$CaCO_{3(s)}$(sediments) $= Ca^{2+} + CO_3^{2-}$	Remineralization of sedimented metal carbonates	NA

*NA = Not Applicable

Table 9.2 gives values for K_1 and K_2 as a function of temperature at 1 atm/25°C and infinite dilution. Stumm and Morgan (1981) summarize some of the ${}^cK$ data that are available for seawater. Problems illustrating the principles of this chapter are presented by Pankow (1992).

TABLE 9.2. Equilibrium constants for the carbon dioxide system as a function of temperature at 1 atm/25°C and infinite dilution.

Temp. (°C)	$\log K_1$	$\log K_2$	$\log K_H(M/\text{atm})$
0	−6.58	−10.63	−1.11
5	−6.53	−10.56	−1.19
10	−6.46	−10.49	−1.27
15	−6.42	−10.43	−1.32
20	−6.38	−10.38	−1.41
25	−6.35	−10.33	−1.47
30	−6.33	−10.29	−1.53
35	−6.31	−10.25	—
40	−6.30	−10.22	−1.64
50	−6.29	−10.17	−1.72

Data from K. Buch, *Meeresforschung*, 1.15, 1951; A. J. Ellis, *Amer. J. Sci.*, v. 257, 217, 1959; H. S. Harned and R. Davies Jr., *J. Amer. Chem. Soc.*, v. 65, 2030, 1943; and H. S. Harned and S. R. Scholes, *J. Amer. Chem. Soc.*, v. 63, 1706, 1941.

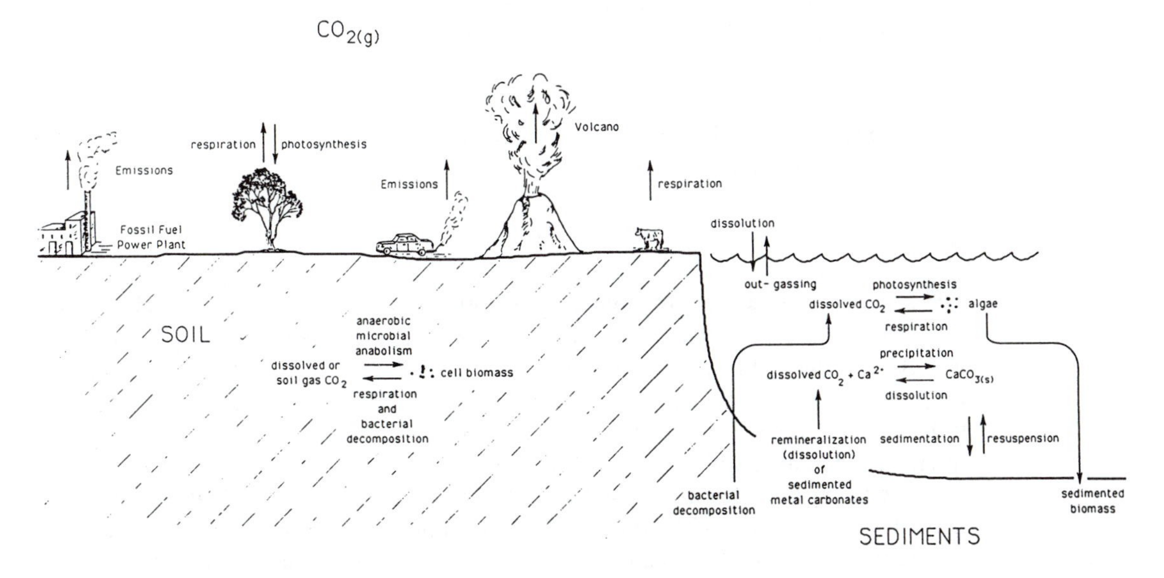

Figure 9.1 Schematic representation of the important processes governing carbon dioxide-related species in the environment.

9.2 TERMS DESCRIBING THE NET ACID/BASE (I.E., TITRATION) POSITION OF CARBON DIOXIDE SYSTEMS

9.2.1 General

Since $H_2CO_3^*$ acts like a diprotic acid, there will be three equivalence points (EPs) in the titration curve for a C_T (M or m) solution of carbon dioxide. The symbol C_T represents the total solution phase concentration of all carbon dioxide-related species. In particular,

$$C_T = [H_2CO_3^*] + [HCO_3^-] + [CO_3^{2-}]. \tag{9.6}$$

Since C_T represents the sum of individual concentrations, its units can be either m or M. Although we have not bothered much with the subscript T ("total") on C in previous chapters, it is standard notation for carbon dioxide systems.

For a system containing C_T of any acid, from the definition for f given in Chapter 7, we have

$$f = (C_B - C_A)/C_T. \tag{9.7}$$

At the EP for $f = 0$ in the CO_2 system, the solution chemistry will simply be that of a C_T F solution of $H_2CO_3^*$. At the EP for $f = 1$, the equivalents of strong base added to that point will equal the equivalents of acidity associated with *one* of the $H_2CO_3^*$ protons. Thus, if the strong base used in the titration is NaOH, the EP for $f = 1$ corresponds to a C_T F solution of $NaHCO_3$. The eq/L of NaOH that must be added to reach that point is C_T. The EP for $f = 2$ corresponds to a C_T F solution of Na_2CO_3. The eq/L of NaOH that must be added to reach that point is $2C_T$. We recall that in the titration of an acid like HA, $f = 0$ and $f = 1$ were the only EPs.

During the titration of a solution of $H_2CO_3^*$, the ENE will be

$$(C_B - C_A) = [HCO_3^-] + 2[CO_3^{2-}] + [OH^-] - [H^+]. \tag{9.8}$$

By combining Eqs. (9.7) and (9.8), and assuming that C_T remains constant during the titration (i.e., no dilution), we obtain

$$f = \alpha_1 + 2\alpha_2 + \frac{[OH^-] - [H^+]}{C_T}. \tag{9.9}$$

The parameters α_1 and α_2 represent the fractions of C_T that are present as HCO_3^- and CO_3^{2-}, respectively. (Sections 9.5 and 9.8.2 provide a detailed discussion of α values in CO_2 systems.) When dilution must be considered, C_T in Eq. (9.9) is replaced with $C_{T,o}V_o/(V_o + V_t)$ where $C_{T,o}$ and V_o represent the initial C_T and volume of the solution to be titrated, respectively, and V_t represents the volume of strong base or strong acid titrant added to achieve a particular value of f.

Equation (9.9) has been used to calculate the full pH vs. f titration curves for $C_T = 10^{-3}$, $10^{-2.5}$, and 10^{-2} M that are given in Figure 9.2. As might be expected based on Chapter 8, the fact that a *sharp* upturn in the pH does not occur in any of the curves at the Na_2CO_3 EP ($f = 2$) is due to the fact that at these concentrations, K_2 is sufficiently small that strong OH^- buffering already sets in before the $f = 2$ EP is reached. However, although an actual inflection point (IP) is not present for $C_T = 10^{-2.5}M$, a careful examination of the curve for $C_T = 10^{-2}M$ reveals a very slight IP at pH = 11.0.

As pointed out in Chapter 8, just because an IP is visible does not mean that its location will coincide *exactly* with the location of the corresponding EP as defined on the basis of a certain integral value of f. In the case of the curve for $C_T = 10^{-2}M$, while the IP occurs at $f = 1.93$ and pH = 11.00, it may be calculated using Eq. (5.110) that the $f = 2$ EP actually occurs somewhat after the IP at pH = 11.12. As measured from $f = 0$, taking the IP to be equivalent with the Na_2CO_3 EP will therefore lead to a −3.5% error in determining the value of $(C_B - C_A)$ required to reach that EP.

When discussing the chemistry of any CO_2 system *where the value of C_T is known*, stating how far that system is from any *one* of the three EPs will completely describe the position which that system has in terms of a CO_2 titration. For each EP, one could state how much more basic the system is than that EP, or one could state how much more acidic than that EP the system is; doing either would suffice. As an analogy, consider the number line given in Figure 9.3. The uniformly spaced points C, E, and G are analogous to the three EPs in a CO_2 titration curve (f = 0, 1, and 2, respectively).

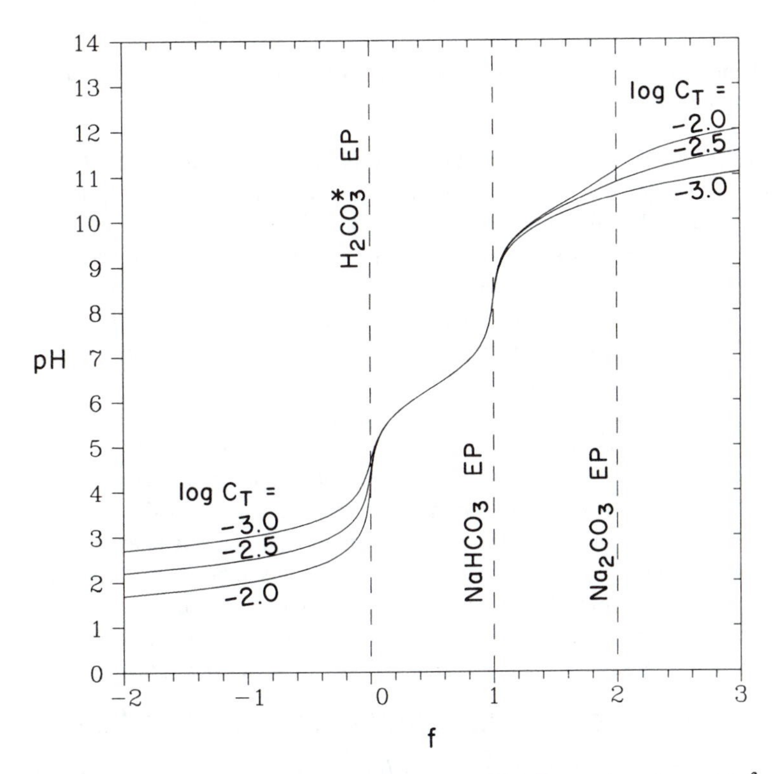

Figure 9.2 Closed system titration curves at 25°C/1 atm for $C_T = 10^{-3}$, $10^{-2.5}$, and $10^{-2}M$. Activity corrections neglected (not valid at very low or very high pH).

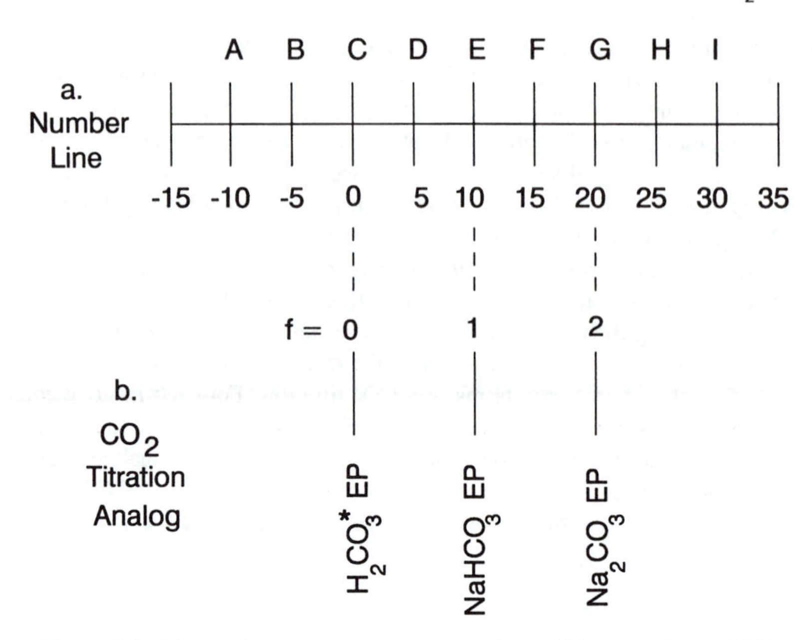

Figure 9.3 Numberline analog for terms defining acid/base position of CO_2 systems.

Consider the position of point D on the Figure 9.3 number line. If we choose the name "*Alkalinity*" (or just "*Alk*") to mean "how far to the right of 0 (point C)" a given point of interest is, then for point D, $Alk = 5$. Thus, knowing the value of *Alk* will specify exactly where on the number line you are; no other information is needed. Indeed, if we choose "H^+*-Acidity*" (or just "H^+*-Acy*") as the term for how far we are to the left of 0, then if $Alk = 5$, there is no *new* information regarding the location of D in pointing out that $H^+\text{-}Acy = -5$ for D. (Note that $Alk = -H^+\text{-}Acy$.) Similarly, there is no new information regarding D's location in pointing out that it is 5 and −5 units to the left and right, respectively, of point E, or in pointing out that it is 15 and −15 units to the left and right, respectively, of point G. However, although just one of these measures will specify where D is on the number line, we can see that it might nevertheless be convenient to have all of the different reference points available for use, with the selection of which one to use depending on where we are on the number line. For example, when driving across the United States on some specific road, one can see how the west to east progress of the trip might be recorded at different points during the trip using a series of different convenient reference points, for example, 100 km east of San Francisco, 300 km west of Chicago, 200 km east of Chicago, and 200 km west of New York City.

9.2.2 *Alkalinity* and H^+*-Acidity* — The $H_2CO_3^*$ Equivalence Point

As implied in Figure 9.3, both *Alkalinity* (*Alk*) and H^+*-Acidity* (H^+*-Acy*) are defined with respect to the $H_2CO_3^*$ EP. For a solution that has the chemistry of this EP, the PBE is

$$[H^+] = [OH^-] + [HCO_3^-] + 2[CO_3^{2-}] \qquad \text{PBE for } H_2CO_3^* \text{ EP.} \qquad (9.10)$$

If some strong base and/or strong acid has been added, then we are moved away from the $f = 0$ EP, and $(C_B - C_A)$ as given by

$$C_B - C_A = [HCO_3^-] + 2[CO_3^{2-}] + [OH^-] - [H^+] \tag{9.8}$$

is nonzero. When the right-hand side (RHS) of Eq. (9.10) is greater than its left-hand side (LHS), then $(C_B - C_A)$ is positive, and there is net alkalinity in the system relative to the $H_2CO_3^*$ EP. It is therefore logical for us to define

$$Alk = C_B - C_A = [HCO_3^-] + 2[CO_3^{2-}] + [OH^-] - [H^+]. \tag{9.11}$$

When the units of the quantity $(C_B - C_A)$ are eq/L,[1] those are also the units of *Alk*. It is important to note that at the $H_2CO_3^*$ EP, $Alk = 0$.

When the LHS of Eq. (9.10) is greater than its RHS, then $(C_B - C_A)$ is negative, and there is net strong acid in the system relative to the $H_2CO_3^*$ EP. We therefore define

$$H^+\text{-}Acy = C_A - C_B = [H^+] - [OH^-] - [HCO_3^-] - 2[CO_3^{2-}]. \tag{9.12}$$

By definition,

$$Alk = -H^+\text{-}Acy \tag{9.13}$$

The quantity H^+-*Acy* is often referred to as the *mineral acidity* since strong acids are usually inorganic acids (e.g., HCl, H_2SO_4, HNO_3, etc.). These acids are H^+ analogs of mineral materials (e.g., NaCl, $CaSO_4$, KNO_3, etc.).

Alk and H^+-*Acy* are defined in the context of a generic CO_2 system titration curve in Figure 9.4. The lengths of the various arrows in Figure 9.4 are intended to connote how much strong acid or base is required to return any given system to one of the three EPs. Like *Alk*, the units of H^+-*Acy* are usually eq/L. As with *Alk*, at the $H_2CO_3^*$ EP, H^+-Acy $= 0$. Based on Eq. (9.13), we note that H^+-*Acy* is perfectly well defined for a solution that has a positive *Alk*, it is just that H^+-*Acy* is negative for such a solution. The converse is also true.

For natural waters in which basic species *other* than HCO_3^-, CO_3^{2-}, and OH^- are present, it is customary to include certain of those bases in the expression for *Alk*. For example, the borate ion (i.e., $B(OH)_4^-$) is present in seawater at significant levels. The criterion for whether to include a given base in the expression for *Alk* is that the pK for the corresponding conjugate acid must be at least as large as pK_1 for $H_2CO_3^*$; in that manner, titration to the $H_2CO_3^*$ EP will also protonate those bases. Boric acid (i.e., $B(OH)_3$, or H_3BO_3) reacts with water according to

$$B(OH)_3 + H_2O = B(OH)_4^- + H^+ \qquad pK = 9.3. \tag{9.14}$$

Since p^cK_1 for $H_2CO_3^*$ is about 5.9 at 25°C/1 atm for seawater, for that medium we have

$$Alk = [HCO_3^-] + 2[CO_3^{2-}] + [B(OH)_4^-] + [OH^-] - [H^+]. \tag{9.15}$$

[1]Units of eq/(kg of water) are also possible.

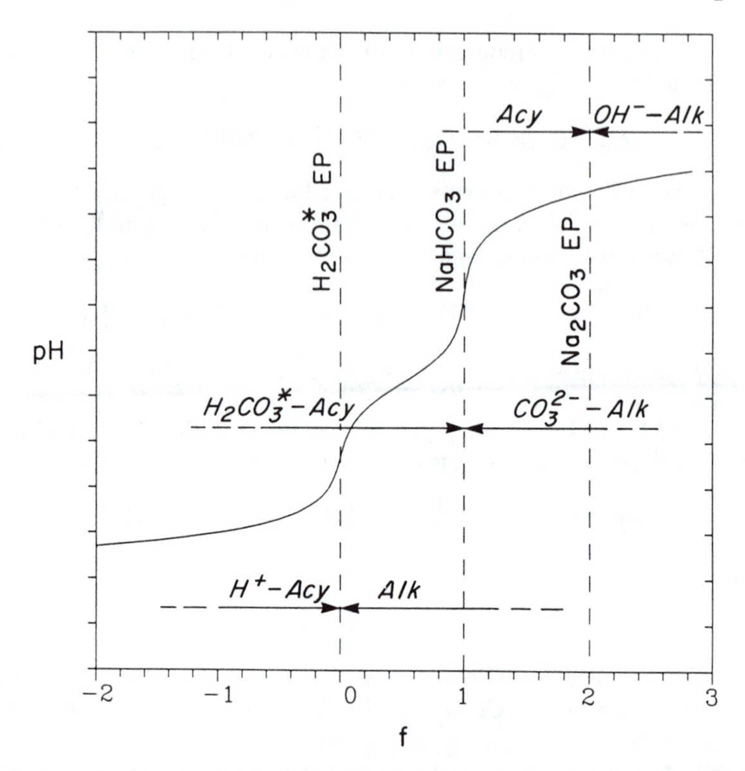

Figure 9.4 Generic titration curve for closed carbon dioxide system with C_T too low to see an inflection at the $f = 2$ equivalence point. For each point in the titration curve, the length of each arrow gives the value of the terms *Alk*, H^+-*Acy*, *Acy*, etc. Note that each of these terms can be negative. Certain of them cannot, however, be negative simultaneously (e.g., *Alk* and H^+-*Acy*).

9.2.3 CO_3^{2-}-*Alk* and $H_2CO_3^*$- *Acy* — The $NaHCO_3$ Equivalence Point

The two terms that are defined relative to the $NaHCO_3$ EP are CO_3^{2-}-*Alk* and $H_2CO_3^*$-*Acy*. For a solution of $NaHCO_3$, we know that some of the HCO_3^- will hydrolyze according to

$$HCO_3^- + H_2O = H_2CO_3 + OH^-. \tag{9.16}$$

If nothing happened to the H_2CO_3, then its concentration would provide a direct measure of the amount of OH^- formed by reaction (9.16). However, we know that *some* of the H_2CO_3 formed will dissociate according to

$$H_2CO_3 = CO_2 + H_2O. \tag{9.17}$$

Thus, to obtain a measure of the amount of OH^- formed from HCO_3^- hydrolysis, we need to add the amount of CO_2 formed to the amount of remaining H_2CO_3. This measure will be given by $[H_2CO_3^*]$. (We are starting to see how $[H_2CO_3^*]$ plays the role of the undissociated form of a diprotic acid.) Since some of the HCO_3^- can also dissociate to form H^+ and CO_3^{2-}, overall, the PBE for the $NaHCO_3$ EP is

$$[H^+] + [H_2CO_3^*] = [OH^-] + [CO_3^{2-}] \qquad \text{PBE for } NaHCO_3 \text{ EP} \tag{9.18}$$

Strong acid or base can be added to move the system away from the $NaHCO_3$ EP. If strong base is added, then the RHS of Eq. (9.18) will be greater than the LHS. The magnitude of this difference provides a measure of how much strong base was added. Therefore, we have

$$CO_3^{2-}\text{-}Alk = [OH^-] + [CO_3^{2-}] - [H^+] - [H_2CO_3^*]. \tag{9.19}$$

If strong acid is added to a solution of $NaHCO_3$, then the LHS of Eq. (9.18) will be greater than the RHS. The magnitude of this difference provides a measure of how much strong acid was added to the $NaHCO_3$ solution. Therefore, we have

$$H_2CO_3^*\text{-}Acy = [H^+] + [H_2CO_3^*] - [OH^-] - [CO_3^{2-}]. \tag{9.20}$$

By definition,

$$CO_3^{2-}\text{-}Alk = -H_2CO_3^*\text{-}Acy. \tag{9.21}$$

As with *Alk* and H^+-*Acy*, both CO_3^{2-}-*Alk* and $H_2CO_3^*$-*Acy* are usually given units of eq/L, and are defined in the context of the generic CO_2 system titration curve in Figure 9.4.

The reason that the name CO_3^{2-}-*Alk* is chosen for Eq. (9.19) is that this term represents the total eq/L for all bases in the system that are at least as basic as the first basic position on CO_3^{2-}. Similarly, the name $H_2CO_3^*$-*Acy* is chosen for Eq. (9.20) because this term represents all of the protons in the system that are at least as acidic as the first proton on $H_2CO_3^*$.

9.2.4 OH^--*Alk* and *Acy* — The Na_2CO_3 Equivalence Point

Both OH^--*Alk* and *Acy* are defined with respect to the Na_2CO_3 EP. For this EP, the PBE is

$$[H^+] + 2[H_2CO_3^*] + [HCO_3^-] = [OH^-] \qquad \text{PBE for } Na_2CO_3\text{EP}. \tag{9.22}$$

Proceeding in a manner that is analogous to the discussions given in the previous two sections, we have

$$OH^-\text{-}Alk = [OH^-] - [H^+] - 2[H_2CO_3^*] - [HCO_3^-] \tag{9.23}$$

$$Acy = [H^+] + 2[H_2CO_3^*] + [HCO_3^-] - [OH^-]. \tag{9.24}$$

As with the other alkalinity and acidity terms defined above, OH^--*Alk* and *Acy* have units of eq/L, and are defined in the context of the generic CO_2 system titration curve in Figure 9.4. By definition we have

$$OH^-\text{-}Alk = -Acy. \tag{9.25}$$

Acy refers to the total acidity up to the Na_2CO_3 EP. It gives the total eq/L (or eq/(kg of water)) of all protons in the system that are at least as acidic as HCO_3^-.

TABLE 9.3. *Alk*, *H^+-Acy*, and other CO_2 system terms

Defining Equivalence Point	Term	Meaning
$H_2CO_3^*$	*H^+-Acy* (= "*mineral Acy*")	eq/L of strong acid in excess of $H_2CO_3^*$ EP
	Alk	eq/L of strong base in excess of $H_2CO_3^*$ EP
$NaHCO_3$	$H_2CO_3^*$-*Acy*	eq/L of strong acid in excess of $NaHCO_3$ EP
	CO_3^{2-}-*Alk*	eq/L of strong base in excess of $NaHCO_3$ EP
Na_2CO_3	*Acy*	eq/L of strong acid in excess of Na_2CO_3 EP
	OH^- -*Alk* (= "*caustic Alk*")	eq/L of strong base in excess of Na_2CO_3 EP

Table 9.3 summarizes the three EPs and the six defining terms for CO_2 systems. *It is important to remember that each of the terms is defined using the PBE for one of the EPs.* For each EP, one term measures the distance that the system is from one side of the EP, and another term measures it from the other side of the EP. Of the six terms given in Table 9.3, *Alk*, *H^+-Acy*, and *OH^--Alk* are the three that are the most used.

OH^--Alk is sometimes referred to as *caustic alkalinity*. The origin of this name rests in the facts that the traditional names for two important strong bases are "*caustic* soda" for concentrated sodium hydroxide, and "*caustic* potash" for concentrated potassium hydroxide (KOH). The word "potash" has its roots in the fact that solutions of KOH were obtained in olden times by taking a *pot* and using water to extract wood *ash*es. The resulting alkaline solution was in part a solution of K_2CO_3. The actual solid KOH was obtained by heating the solution to drive off water and CO_2. Not surprisingly, *potash* is also the linguistic root of "potassium." The fact that there is a relatively large amount of potassium in wood and other plant materials is due to the fact that this element is a significant nutrient used by plants. Returning it to the soil by slash burning fields and jungles has been an important fertilization method in many primitive as well as some civilized cultures.

9.3 ACID AND BASE NEUTRALIZING CAPACITY (ANC & BNC)

The terms **Acid Neutralizing Capacity** (ANC) and **Base Neutralizing Capacity** (BNC) are often used to refer to how much strong acid or strong base a system is capable of neutralizing, respectively. Both of these terms are always defined using the PBEs for certain pertinent EPs for the system. ANC refers to how much strong acid can be added before a certain acidimetric (low f) EP is reached, and BNC refers to how much strong base can be added before a certain alkalimetric (high f) EP is reached.

For a solution of the acid/base pair HA/A^-, we know that the PBE for the HA EP is

$$[H^+] = [OH^-] + [A^-] \tag{9.26}$$

and so we define

$$ANC = [OH^-] + [A^-] - [H^+]. \tag{9.27}$$

The PBE for the NaA EP is

$$[H^+] + [HA] = [OH^-] \tag{9.28}$$

and so we define

$$BNC = [H^+] + [HA] - [OH^-]. \tag{9.29}$$

For systems containing CO_2, because *Alk* represents *all* of the alkalinity in the system, we usually choose

$$ANC = Alk \tag{9.30}$$

Similarly, because *Acy* represents all of the available acidic protons, we usually define

$$BNC = Acy \tag{9.31}$$

9.4 THE PRINCIPLE OF CONSERVATION OF *ALK*, *H⁺-ACY*, ANC, BNC, ETC.

It is clear that when the C_T of a solution of $H_2CO_3^*$ is changed by adding or removing CO_2, the result will be different values of pH as well as different carbon dioxide speciations. However, since *Alk* and H^+-*Acy* are identically equal to zero for all $H_2CO_3^*$ solutions, *Alk* and H^+-*Acy* remain identically zero in such solutions, independent of changes in C_T. Thus, while the *distribution* among the acids and bases making up *Alk* and H^+-*Acy* will change, the parameters themselves remain zero.

In a similar manner, when a solution of $H_2CO_3^*$ of a given C_T undergoes changes in T and P, the corresponding changes in K_w and the acidity constants for $H_2CO_3^*$ will cause changes in pH, changes in the distribution among the various carbon dioxide species making up that C_T, as well as changes in the degree of the dissociation of the water. Irrespective of any changes in T and P, however, the PBE for this type of solution (Eq. (9.10)) will always remain satisfied, and *Alk* and H^+-*Acy* will both remain exactly conserved at zero eq/L (and zero eq/(kg of water)).

Consider now a $10^{-3}F$ solution of $NaHCO_3$. We know that such a solution will have *Alk* and H^+-*Acy* values of $+10^{-3}$ eq/L and -10^{-3} eq/L, respectively. If some CO_2 is now added to the solution without changing the volume, C_T will increase, the pH will decrease, and the fractional amounts of the various types of carbon dioxide species will change. However, the values of *Alk* and H^+-*Acy*, will remain *conserved* at $+10^{-3}$ eq/L and -10^{-3} eq/L, respectively. The reason is that *adding (or removing)* CO_2 *(or* $H_2CO_3^*$ *for that matter) does not change the net strong base* $(C_B - C_A)$ which in this case equals $[Na^+]$. Analogous comments can be made for a solution of Na_2CO_3.

If a solution of $NaHCO_3$ or Na_2CO_3 is made to undergo changes in T and P, then as with changing the T and P of a solution of $H_2CO_3^*$, the pH and carbon dioxide speciation will change. However, if the changes in T and P are relatively small such that the density of the water does not change significantly, then the eq/L of *Alk* and H^+-*Acy* will remain essentially conserved. If we want to be very strict, we can make *Alk* and H^+-*Acy* remain *exactly* conserved during such changes by using units of eq/(kg of water) rather than eq/L.

The principle of conservation of *Alk*, *H*$^+$*-Acy*, as outlined above is not limited to solutions for which $f = 0$, 1, or 2, but indeed applies to a solution with *any* value of f. However, whenever $f \neq 0$ *and* changes in density occur, *Alk* and *H*$^+$*-Acy* will not be exactly conserved unless we employ units of eq/(kg of water).

When it applies, the principle of conservation of *Alk* provides a powerful tool for solving a wide variety of important water chemistry problems. For example, consider a sample of lake water with an initial pH equal to pH_1, a known value of *Alk*, and an initial C_T equal to $C_{T,1}$. We desire to know the pH after the lake experiences an algal bloom, that is, we desire pH_2. As indicated in Table 9.1, the photosynthetic conversion of CO_2 to organic carbon may be *approximated*[2] by

$$6CO_2 + 6H_2O = C_6H_{12}O_6 + 6O_2. \tag{9.32}$$

We recall now that as used in this chapter, C_T refers only to the sum of the various dissolved CO_2 species. When some CO_2 is simply converted to $C_6H_{12}O_6$, the value of C_T decreases, but the value of *Alk* remains unchanged (approximately).[2] Under these conditions, pH_2 may be obtained solving Eq. (9.11) with the *initial* value of *Alk* together with $C_{T,2}$. If some of the $C_6H_{12}O_6$-type biomass should later be oxidized by bacteria by the backwards version of Eq. (9.32), then C_T will go back up, but *Alk* will again remain constant (approximately).[2]

As another example of conservation of *Alk*, if water of types I and II are mixed in a proportion of x parts of I to y parts of II, then the resulting *Alk* will be given simply by the weighted mean for the mixing process, that is,

$$Alk_{mix} = \frac{xAlk_I + yAlk_{II}}{x + y} \tag{9.33}$$

$$= \frac{(x/y)Alk_I + Alk_{II}}{x/y + 1}. \tag{9.34}$$

Equations (9.33) and (9.34) embody the conservation of *Alk* because the net strong base that each of the two types of water brings to the mixture is individually conserved. They are valid for any values (i.e., positive, zero, or negative) of Alk_I and Alk_{II}. The equation for Alk_{mix} for the combination of any number of different types of water will always be a simple weighted mean expression similar to Eq. (9.33). Table 9.4 gives a few simple examples of the application of these principles for the mixing of two types of water.

[2]The fact that some nutrients like nitrogen (NO_3^- or NH_4^+) and phosphorus (e.g., HPO_4^{2-}) are required to permit the photosynthetic process to go forward makes the situation as described just a little oversimplified. Indeed, photosynthetic and respiratory activity will alter the *Alk*. However, as long as the initial *Alk* is not small and is of the order of 10^{-3} eq/L, the change in *Alk* will be small (see the detailed discussion on this topic by Stumm and Morgan (p. 195, 1981)). If NO_3^- serves as the source of the nutrient nitrogen and HPO_4^{2-} serves as the source of the phosphorus, there will be a slight increase in the *Alk*. The increase is a result of the consumption of the protons needed for the redox conversion of the NO_3^- and the HPO_4^{2-} into the lower oxidation states that they possess in organic matter. If NH_4^+ serves as the source of the nutrient nitrogen, *Alk* will decrease slightly.

TABLE 9.4 Examples of the application of Eq 9.33.

Mixture	Water Type I	Water Type II	Volume of I (x)	Volume of II (y)	x/y	$Alk_{mixture}$ (eq/L)
A	0.001 F $NaHCO_3$ $Alk_I = 0.001$ eq/L	0.002 F $NaHCO_3$ $Alk_{II} = 0.002$ eq/L	1 L	2 L	0.500	1.67×10^{-3}
B	0.001 F HCl $Alk_I = -0.001$ eq/L	0.001 F $NaHCO_3$ $Alk_{II} = 0.001$ eq/L	1 L	2 L	0.500	3.33×10^{-4}
C	0.001 F HCl $Alk_I = -0.001$ eq/L	0.002 F $NaHCO_3$ $Alk_{II} = 0.002$ eq/L	2 L	3 L	0.667	8.00×10^{-4}
D	0.001 F $NaHCO_3$ $Alk_I = 0.001$ eq/L	0.001 F Na_2CO_3 $Alk_{II} = 0.002$ eq/L	2 L	3 L	0.667	1.60×10^{-3}

Processes which can lead to a change in *Alk* include, of course, any addition of strong acid or base. Any process by which bicarbonate or carbonate salts or minerals are added to or removed from a solution will also change *Alk*, as in the dissolution of $CaCO_{3(s)}$ (*Alk* increases), or in the precipitation of $CaCO_{3(s)}$ (*Alk* decreases).

9.5 α VALUES IN CO_2 SYSTEMS

With K_1 defined according to Eq. (9.45), it is useful to define α values for the CO_2 system as follows:

$$[H_2CO_3^*] = \alpha_0 C_T \qquad [HCO_3^-] = \alpha_1 C_T \qquad [CO_3^{2-}] = \alpha_2 C_T \tag{9.35}$$

$$\alpha_0 = \frac{[H_2CO_3^*]}{[H_2CO_3^*] + [HCO_3^-] + [CO_3^{2-}]} \tag{9.36}$$

$$= \frac{1}{1 + \dfrac{[HCO_3^-]}{[H_2CO_3^*]} + \dfrac{[CO_3^{2-}]}{[H_2CO_3^*]}} \tag{9.37}$$

$$= \frac{1}{1 + K_1/[H^+] + K_1K_2/[H^+]^2} \qquad \text{requires all } \gamma_i = 1.0, \text{ or use of } {}^cK \text{ values} \tag{9.38}$$

$$\alpha_1 = \frac{[HCO_3^-]}{[H_2CO_3^*] + [HCO_3^-] + [CO_3^{2-}]} \tag{9.39}$$

$$= \frac{1}{\dfrac{[H_2CO_3^*]}{[HCO_3^-]} + 1 + \dfrac{[CO_3^{2-}]}{[HCO_3^-]}} \tag{9.40}$$

$$= \frac{1}{[H^+]/K_1 + 1 + K_2/[H^+]} \qquad \text{requires all } \gamma_i = 1.0, \text{ or use of } {}^cK \text{ values} \tag{9.41}$$

$$\alpha_2 = \frac{[CO_3^{2-}]}{[H_2CO_3^*] + [HCO_3^-] + [CO_3^{2-}]} \tag{9.42}$$

$$= \frac{1}{\frac{[H_2CO_3^*]}{[CO_3^{2-}]} + \frac{[HCO_3^-]}{[CO_3^{2-}]} + 1} \tag{9.43}$$

$$= \frac{1}{[H^+]^2/K_1K_2 + [H^+]/K_2 + 1} \quad \text{requires all } \gamma_i = 1.0\text{, or use of } {}^cK \text{ values.} \tag{9.44}$$

Equations (9.35), (9.36), (9.37), (9.39), (9.40), (9.42), and (9.43) are exact under all conditions. Equations (9.38), (9.41), and (9.44) require either the ability to neglect activity corrections, or the use of ${}^cK$ values. Figure 9.5 is a plot of $\log \alpha$ values vs. pH at 25°C/1 atm.

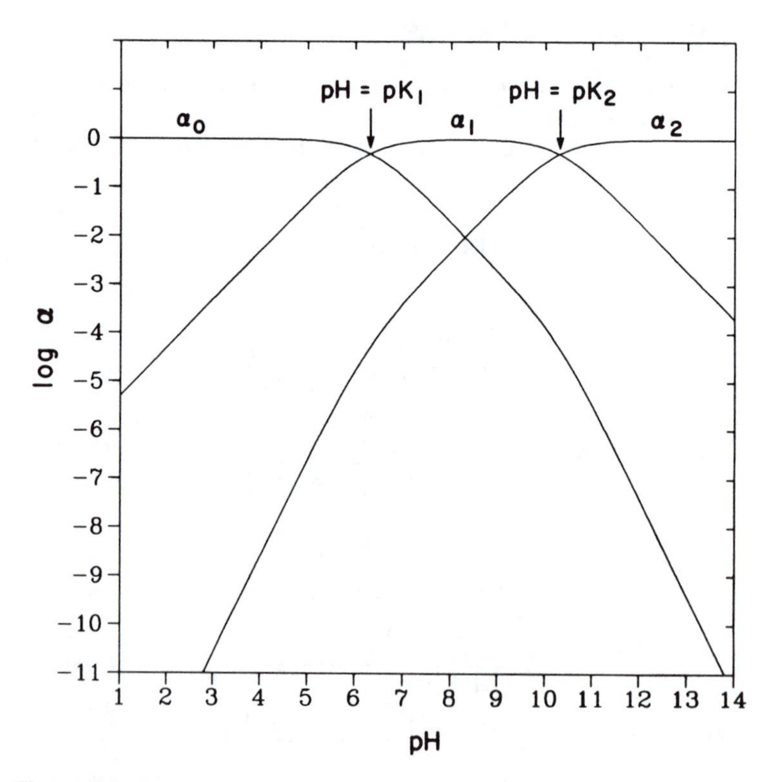

Figure 9.5 Log α_0, $\log \alpha_1$, and $\log \alpha_2$ vs. pH for 25°C/1 atm. Activity corrections neglected (not valid at very low or very high pH).

9.6 CO_2 SYSTEM FOR C_T FIXED (CLOSED SYSTEM): THE $H_2CO_3^*$ EP, I.E., NO NET STRONG ACID OR BASE

As discussed above, in addition to water itself, there are six species in a solution of $H_2CO_3^*$. These species are

$$\text{Six Unknowns} \qquad CO_2, \quad H_2CO_3, \quad HCO_3^-, \quad CO_3^{2-}, \quad H^+, \quad OH^-.$$

With the concentrations of the above six species comprising six unknowns, neglecting activity corrections, we will need six equations to solve the problem. When the system is "closed" and carbon dioxide species can neither enter or leave the system, one of these equations will be an MBE on C_T. The full set of six equations are:

1. Equilibrium constant for $H_2CO_3 = H_2O + CO_2$ (i.e., K)
2. Equilibrium constant for $H_2CO_3 = H^+ + HCO_3^-$ (i.e., $K_{H_2CO_3}$)
3. Equilibrium constant for $HCO_3^- = H^+ + CO_3^{2-}$ (i.e., K_2)
4. Equilibrium constant for $H_2O = H^+ + OH^-$ (i.e., K_w)
5. PBE *or* ENE (given the PBE, ENE, and MBE, only two out of the three will be independent)
6. MBE on C_T (i.e., $[CO_2] + [H_2CO_3] + [HCO_3^-] + [CO_3^{2-}] = C_T$).

The number of unknowns that we have to deal with in problems concerning any type of solution containing CO_2 is reduced by one if we make use of

$$K_1 = \frac{\{HCO_3^-\}\{H^+\}}{\{H_2CO_3^*\}}. \tag{9.45}$$

Neglecting activity corrections, we have

$$K_1 = \frac{[HCO_3^-][H^+]}{[CO_2] + [H_2CO_3]} \tag{9.46}$$

$$= \frac{[H^+][HCO_3^-]}{K[H_2CO_3] + [H_2CO_3]}. \tag{9.47}$$

Dividing the top and bottom of the RHS of Eq. (9.47) by $[H_2CO_3]$, we obtain

$$K_1 = \frac{K_{H_2CO_3}}{K + 1} \tag{9.48}$$

A consideration of the acid dissociation of H_2CO_3 is built into the value of K_1. Therefore, the use of K_1 to describe the first acid dissociation of $H_2CO_3^*$ allows us to replace the two unknowns $[CO_2]$ and $[H_2CO_3]$ with the *one* unknown, $[H_2CO_3^*]$. Therefore, the above lists of unknowns and equations becomes

$$\text{Five Unknowns} \qquad H_2CO_3^*, \quad HCO_3^-, \quad CO_3^{2-}, \quad H^+, \quad OH^-$$

which we can solve using:

1. Equilibrium constant for $H_2CO_3^* = H^+ + HCO_3^-$ (i.e., K_1)
2. Equilibrium constant for $HCO_3^- = H^+ + CO_3^{2-}$ (i.e., K_2)
3. Equilibrium constant for $H_2O = H^+ + OH^-$ (i.e., K_w)
4. PBE *or* ENE
5. MBE for C_T

Starting with the PBE, for a solution of $H_2CO_3^*$, we have

$$[H^+] = [HCO_3^-] + 2[CO_3^{2-}] + [OH^-] \tag{9.49}$$

$$= \alpha_1 C_T + 2\alpha_2 C_T + [OH^-]. \tag{9.50}$$

Neglecting activity corrections for $[OH^-]$ and $[H^+]$, we obtain

$$[H^+] = \alpha_1 C_T + 2\alpha_2 C_T + K_w/[H^+] \tag{9.51}$$

which is equivalent to Eq. (5.104). When activity corrections can be neglected, Eq. (9.51) can be solved easily by trial and error for the equilibrium value of $[H^+]$. That value may then be used to determine the various α values. The α values together with C_T may then be used to calculate the concentrations of the various carbon dioxide species. When activity corrections cannot be neglected, as already discussed for other problems (e.g., see Section 5.5), a series of trial and error solutions with increasingly more accurate estimations of the ionic strength may be used to determine the equilibrium speciation.

As a quick check to see where the five equations came into use in the solution process discussed above, we recall that: a) we began with the PBE; then invoked usage of the various α values which brought in b) K_1 and c) K_2 and d) the MBE requirement for C_T; finally, we used e) K_w.

9.7 THE IMPLICATIONS OF THE EQUILIBRIUM RELATIONSHIP BETWEEN CO_2 AND H_2CO_3

At 25°C/1 atm, the value of the equilibrium constant K for the dehydration conversion of H_2CO_3 to CO_2 is 650. This constant leads to a $\{CO_2\}/\{H_2CO_3\}$ ratio which is rather large, as well as *independent of pH*. This situation has two very important implications for aqueous CO_2 systems: 1) the bulk of the $H_2CO_3^*$ will always be present as CO_2, that is, $[H_2CO_3^*] \simeq [CO_2]$; and 2) $H_2CO_3^*$ acts like an acid that is weaker than H_2CO_3 by a factor of (K + 1); at 25°C, this factor equals 651. The first implication is fairly straightforward. The second implication is a direct result of Eq. (9.48).

9.8 CLOSED CO_2 SYSTEMS (FIXED C_T) WITH THE POSSIBLE PRESENCE OF NET STRONG ACID OR BASE

As we now understand, for a given fixed (i.e., closed system) value of C_T, we can change the pH by adding strong acid or base. We recall that when the value of C_T is fixed, carbon dioxide species can neither enter or leave the system. By changing the pH, we can however change the *fractions* of the different CO_2 species according the pH dependence of the various α values. Figure 9.6 is a plot of the carbon dioxide speciation

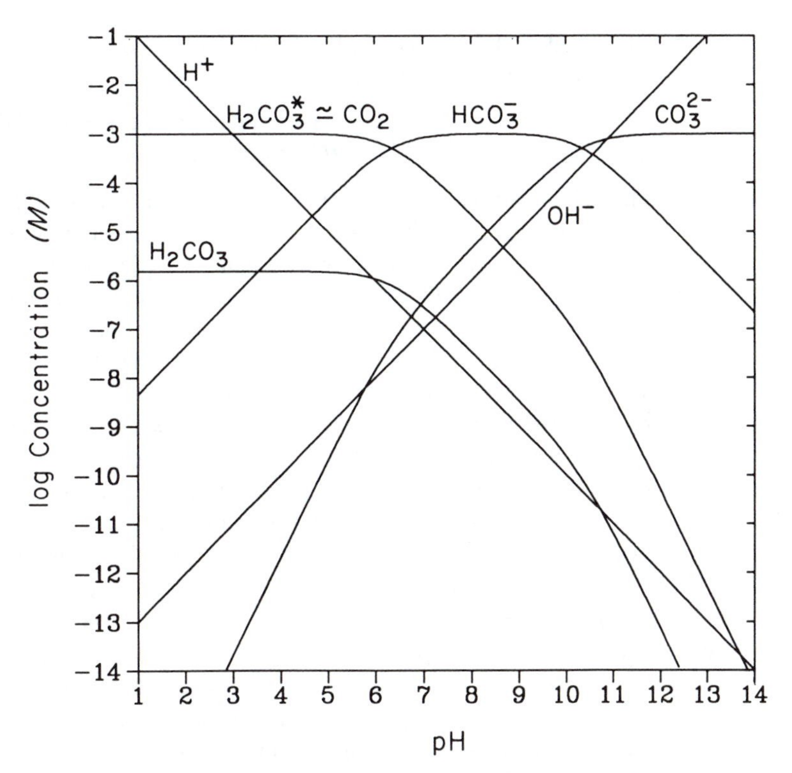

Figure 9.6 Carbon dioxide speciation as a function of pH at 25°C/1 atm for $C_T = 10^{-3}M$. Activity corrections neglected (not valid at very low or very high pH).

as a function of pH for a system in which $C_T = 10^{-3}M$. The conditions for Figure 9.6 are 25°C/1 atm.

9.8.1 Locating the Equivalence Points in a CO_2 System with the Help of a Log Concentration vs. pH Diagram

In Section 6.4, we discussed the fact that log concentration vs. pH diagrams can be used to advantage when solving speciation problems for HA/A^- solutions at the HA and NaA EPs. Similar statements can in fact be made for solutions of any type of acid/base system, and solutions containing carbon dioxide are no exception. As in Section 6.4, for a given EP, the approach that we will take is to use the log concentration vs. pH diagram to try and find where the PBE for that EP is satisfied.

9.8.1.1 The $H_2CO_3^*$ Equivalence Point. For a solution of $H_2CO_3^*$, we have

$$[H^+] = [OH^-] + [HCO_3^-] + 2[CO_3^{2-}] \qquad \text{PBE for } H_2CO_3^* \text{ EP.} \tag{9.10}$$

Therefore, for a solution of $H_2CO_3^*$, we seek the value of pH where Eq. (9.10) is satisfied. Since $H_2CO_3^*$ is in fact acidic, then when C_T is not too small, let us try and assume that both $[OH^-]$ and $[CO_3^{2-}]$ are negligible on the RHS of Eq. (9.10). Thus, for $C_T = 10^{-3}M$, we seek the point in Figure 9.6 where

$$[H^+] \simeq [HCO_3^-] \quad \text{Approximate PBE for } H_2CO_3^* \text{ EP } (C_T \text{ not too low}) \tag{9.52}$$

The pH value where Eq. (9.52) is satisfied is obviously the point where $\log[H^+] \simeq \log[HCO_3^-]$. In Figure 9.6, this pH is 4.67, and the intersection of the two lines is circled. When we use Figure 9.6 to check the assumptions made above, we see that at the pH of this intersection, the $\log[OH^-]$ and $\log[CO_3^{2-}]$ lines are indeed far below the $\log[HCO_3^-]$ line. Thus, we can conclude that the system speciation for a 10^{-3} F solution of $H_2CO_3^*$ is given to a very good approximation by the log concentration values that may be picked off of a *vertical line* drawn through the intersection of $\log[H^+]$ with $\log[HCO_3^-]$.

For the true system speciation, $[H^+]$ will be slightly greater than $[HCO_3^-]$. The reason is that both $[OH^-]$ and $2[CO_3^{2-}]$ are added to $[HCO_3^-]$ in the PBE. Therefore, the true solution speciation will in fact be given by a vertical line that is drawn slightly to the left of the intersection of $\log[H^+]$ with $\log[HCO_3^-]$.

The relative validity of the assumption that $[OH^-]$ is negligible compared to $[HCO_3^-]$ will improve as C_T increases, and degrade as C_T decreases. However, the negligibility of $[CO_3^{2-}]$ relative to $[HCO_3^-]$ remains *unchanged* as C_T changes since the curves for both of these species in a log concentration vs. pH diagram are lifted or lowered simultaneously with changing C_T. These concepts may be studied by making an acetate transparency of Figure 9.7, and raising and lowering it on a copy of the grid given in Figure 6.3. As C_T is lowered, the pH will increase as the intersection of the $\log[H^+]$ and $\log[HCO_3^-]$ lines slides down the $\log[H^+]$ line towards more neutral pH values.

Since the first proton on $H_2CO_3^*$ is weakly acidic and since the second proton is *very* weakly acidic, the pH of a not-too dilute solution of $H_2CO_3^*$ may be considered to behave as a solution of a weak *monoprotic* acid, and $[H^+]$ may be calculated using Eq. (5.45) with K_1 substituted for K_a. Thus, we have

$$[H^+] \simeq \sqrt{K_1 C_T}. \tag{9.53}$$

At 25°C/1 atm where $K_1 = 10^{-6.35}$, for $C_T = 10^{-3}$, we compute to a very good approximation that pH = 4.67.

9.8.1.2 The $NaHCO_3$ Equivalence Point. For a solution of $NaHCO_3$, we have

$$[H^+] + [H_2CO_3^*] = [OH^-] + [CO_3^{2-}] \quad \text{PBE for } NaHCO_3 \text{ EP.} \tag{9.18}$$

Therefore, for a 10^{-3} F solution of $NaHCO_3$, we seek the value of pH in Figure 9.6 where Eq. (9.18) is satisfied. Since the HCO_3^- ion can both lose and gain protons, solutions of $NaHCO_3$ are unlikely to be very acidic or very basic. For a solution of $NaHCO_3$ which is not too dilute, it therefore seems that both [OH−] and $[H^+]$ will be negligible in Eq. (9.18). Thus, we seek the point in Figure 9.6 where

$$[H_2CO_3^*] \simeq [CO_3^{2-}] \quad \text{Approximate PBE for } NaHCO_3 \text{ EP } (C_T \text{ not too low}). \tag{9.54}$$

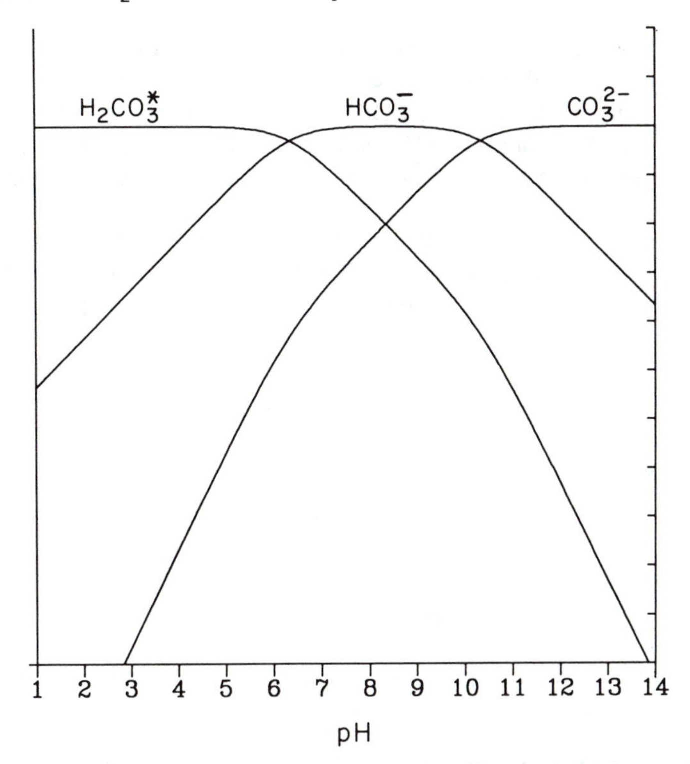

Figure 9.7 Generic log concentration (*M*) vs. pH diagram applicable at 25°C/1 atm for any value of C_T. To be used as an acetate transparency with Figure 6.3. Activity corrections neglected (not valid at very low or very high pH).

The pH value where Eq. (9.54) is satisfied is the point where $\log[H_2CO_3^*] \simeq \log[CO_3^{2-}]$. This intersection is circled in Figure 9.6; it occurs at pH = 8.34. When we check the assumptions made above, we see that at the pH of this intersection, $\log[H^+]$ is far below $\log[H_2CO_3^*]$. However, $\log[OH^-]$ is only about one log unit below $\log[CO_3^{2-}]$. We can conclude from this that the HCO_3^- is more basic than it is acidic.[3] Thus, while neglecting $[OH^-]$ in Eq. (9.18) for $C_T = 10^{-3}$ is not a bad assumption, it is also not an excellent one; the assumption would have been better if C_T were larger. We can nevertheless conclude that the system speciation for a 10^{-3} *F* solution of $NaHCO_3$ is given to a reasonable approximation by the log concentration values that may be picked off of a vertical line drawn through the intersection of $\log[H_2CO_3^*]$ with $\log[CO_3^{2-}]$. Given the greater relative importance of $[OH^-]$ vs. that of $[H^+]$ in Eq. (9.18), we can conclude that the true equilibrium pH value is slightly less than is given by the intersection discussed.

[3]The acidity of HCO_3^- is measured by K_2. The basicity of HCO_3^- is measured by its K_b value which is K_w/K_1. At 25°C/1 atm, $K_2 = 10^{-10.33}$, and $K_w/K_1 = 10^{-7.65}$. Since $K_w/K_1 \gg 10^{-10.33}$, HCO_3^- is significantly more basic that it is acidic. This is one way of understanding why a solution of $NaHCO_3$ has a pH that is greater than $(1/2)pK_w$.

As noted above, the relative validity of the assumptions that $[OH^-]$ and $[H^+]$ are negligible in Eq. (9.16) will *both* improve as C_T increases, and degrade as C_T decreases. Indeed, as C_T is increased, the intersection of $\log[H_2CO_3^*]$ with $\log[CO_3^{2-}]$ is lifted away from *both* the $\log[OH^-]$ and $\log[H^+]$ lines. As C_T is decreased, it is moved toward those lines. These may be examined by using the acetate transparency of Figure 9.7 with Figure 6.3.

It is interesting and important to note that the pH at the intersection of $\log[H_2CO_3^*]$ with $\log[CO_3^{2-}]$ is independent of pH. It is given by the simple relationship $pH = \frac{1}{2}(pK_1 + pK_2)$. Indeed, neglecting activity coefficients, multiplying the equilibrium expressions for K_1 and K_2 together yields

$$K_1 K_2 = \frac{[CO_3^{2-}][H^+]^2}{[H_2CO_3^*]}. \tag{9.55}$$

When Eq. (9.54) is valid, then

$$K_1 K_2 \simeq [H^+]^2 \tag{9.56}$$

and so

$$pH \simeq \frac{1}{2}(pK_1 + pK_2). \tag{9.57}$$

At 25°C/1 atm where $K_1 = 10^{-6.35}$ and $K_2 = 10^{-10.33}$, we obtain pH = 8.34. Solving Eq. (9.18) exactly for $C_T = 10^{-3}$, we obtain pH = 8.30.

9.8.1.3 The Na_2CO_3 Equivalence Point. For a solution of Na_2CO_3, we have

$$[H^+] + 2[H_2CO_3^*] + [HCO_3^-] = [OH^-] \qquad \text{PBE for } Na_2CO_3 \text{ EP}. \tag{9.22}$$

Therefore, for a 10^{-3} *F* solution of Na_2CO_3, we seek the value of pH in Figure 9.6 where Eq. (9.22) is satisfied. Since Na_2CO_3 is rather basic, and since 10^{-3} *F* is not low, let us try and assume that both the two most acidic species on the LHS of Eq. (9.22) (i.e., $[H^+]$ and $[H_2CO_3^*]$) are negligible relative to $[HCO_3^-]$. Thus, we seek the point in Figure 9.6 where

$$[HCO_3^-] \simeq [OH^-] \qquad \text{Approximate PBE for } Na_2CO_3 \text{ EP } (C_T \text{ not too low}). \tag{9.58}$$

The pH value where Eq. (9.58) is satisfied is obviously the point where $\log[HCO_3^-] \simeq \log[OH^-]$. In Figure 9.6, this pH is 10.57, and the intersection of these two lines is circled. When we check the assumptions made above, we see that in fact at the pH of this intersection, $\log[H^+]$ and $\log[H_2CO_3^*]$ are far below $\log[HCO_3^-]$. Thus, we can conclude that the system speciation for a 10^{-3} *F* solution of Na_2CO_3 is given to a very good approximation by the log concentration values that may be picked off of a vertical line drawn through the intersection of $\log[HCO_3^-]$ with $\log[OH^-]$. However, by Eq. (9.22) we know that the actual $[OH^-]$ value is greater than $[HCO_3^-]$. Therefore, the exact solution speciation will in fact be given by a vertical line that is drawn very slightly to the right of the intersection. (Solving Eq. (9.22) exactly for $C_T = 10^{-3}$ also gives pH = 10.57.)

The relative validity of the assumption that $[H^+]$ is negligible relative to $[HCO_3^-]$ will improve as C_T increases, and degrade as C_T decreases. In a manner that is analogous to the situation discussed in Section 9.8.1.1, however, the negligibility of $[H_2CO_3^*]$ relative to $[HCO_3^-]$ remains unchanged as C_T changes since the curves for both of these species in a log concentration vs. pH diagram are lifted or lowered simultaneously with changing C_T. As before, these points may be studied using the acetate transparency of Figure 9.7 together with Figure 6.3.

Since adding the first proton to the CO_3^{2-} ion occurs much more readily than adding the second, to a first approximation, a solution of Na_2CO_3 will behave like a solution of NaA with the monoprotic acidity constant for HA equal to K_2. Therefore, the pH of a not-too dilute solution of Na_2CO_3 may be estimated using the equation for a weak monobasic base (Eq. (5.89)). In this case, we have

$$[OH^-] \simeq \sqrt{(K_w/K_2)C_T}. \tag{9.59}$$

For $C_T = 10^{-3}$, at 25°C/1 atm, we compute pH = 10.67 which is similar to the pH of the intersection in Figure 9.6. As C_T is lowered, the pH will decrease as the intersection of the $\log[OH^-]$ and $\log[HCO_3^-]$ lines slides down the $\log[OH^-]$ line towards more neutral pH values. At the same time, the relative importance of $[H^+]$ in Eq. (9.22) will increase.

9.8.2 Slopes of the Lines in the Log Concentration vs. pH Diagram for Fixed C_T

In Figure 9.6, three asymptotic pH regions may be identified: pH $\ll$ pK_1, p$K_1 \ll$ pH $\ll$ pK_2, and pH $\gg$ pK_2. The three asymptotic regions are separated by two transition regions, namely p$K_1 \pm 1$, and p$K_2 \pm 1$. In each of the three asymptotic regions, the slopes of the log concentration lines for the three different CO_2 species take on different asymptotic values. Since C_T itself remains constant, a consideration of the relationships given in Eq. (9.35) leads to the conclusion that the reason for the slope changes must involve changes in the first derivatives of the α values with respect to pH, that is, changes in $d\alpha_0/d$pH, $d\alpha_1/d$pH, and $d\alpha_2/d$pH.

If we examine Eqs. (9.38), (9.41), and (9.44) and compare the various terms in the denominators for the three parameters, we find that in the three pH ranges identified above, the expressions for α_0, α_1, and α_2 take on asymptotic functionalities. The results are summarized in Table 9.5. When we take the first derivatives of the logarithms of the expression in Table 9.5 with respect to pH, we obtain the results that are presented in Table 9.6.

One interesting aspect of Figure 9.6 that is relevent here involves the shape of the $\log[H_2CO_3]$ line. (Note that we are referring now to the true H_2CO_3 species, not $H_2CO_3^*$.) Indeed, although we know that the acidity constant for H_2CO_3 at 25°C/1 atm is $10^{-3.54}$, there is no break at pH = 3.54 in the line for $\log[H_2CO_3]$. Rather, it appears to follow the exact same pattern exhibited by the $\log[H_2CO_3^*]$ line; its first break in slope appears at pH = 6.35, that is, at pK_1. While this result seems strange on the surface, it may be understood in terms of the facts that:

TABLE 9.5. Asymptotic functionalities for α_0, α_1, and α_2 in the three important pH regions for aqueous carbon dioxide systems.

	$pH \ll pK_1$	$pK_1 \ll pH < pK_2$	$pH \gg pK_2$
25°C/1 atm:	$pH \ll 6.35$	$6.35 \ll pH \ll 10.33$	$pH \gg 10.33$
	$\alpha_0 \simeq 1$	$\alpha_0 \simeq [H^+]/K_1$	$\alpha_0 \simeq [H^+]^2/K_1K_2$
	$\alpha_1 \simeq K_1/[H^+]$	$\alpha_1 \simeq 1$	$\alpha_1 \simeq [H^+]/K_2$
	$\alpha_2 \simeq K_1K_2/[H^+]^2$	$\alpha_2 \simeq K_2/[H^+]$	$\alpha_2 \simeq 1$

TABLE 9.6 Asymptotic values for $d\log\alpha_0/dpH$, $d\log\alpha_1/dpH$, $d\log\alpha_2/dpH$, $d\log[H_2CO_3^*]/dpH$, $d\log[HCO_3^-]/dpH$, and $d\log[CO_3^{2-}]/dpH$ in the three important pH regions for aqueous carbon dioxide systems.

pH Range Slope	$pH \ll pK_1$	$pK_1 \ll pH \ll pK_2$	$pH \gg pK_2$
$d\log\alpha_0/dpH$ and $d\log[H_2CO_3^*]/dpH$	0	−1	−2
$d\log\alpha_1/dpH$ and $d\log[HCO_3^-]/dpH$	+1	0	−1
$d\log\alpha_2/dpH$ and $d\log[CO_3^{2-}]/dpH$	+2	+1	0

1. At equilibrium, the relative activities of aqueous CO_2 species and H_2CO_3 are set by the equilibrium for reaction (9.2).
2. CO_2 greatly dominates over H_2CO_3 ($K \simeq 650$ at 25°C/1 atm), and to a large extent, controls its chemistry.

Thus, at 25°C/1 atm, the break in the $\log[H_2CO_3]$ line cannot occur until $pH \simeq pK_1 = 6.35$, even though it is "inherently" more acidic than $H_2CO_3^*$. If this result still does not appear intuitively obvious, it is also possible to show it mathematically by defining an α value for H_2CO_3. In particular, let

$$[H_2CO_3] = \alpha_{H_2CO_3} C_T \tag{9.60}$$

where

$$\alpha_{H_2CO_3} = \frac{[H_2CO_3]}{[H_2CO_3] + [CO_2] + [HCO_3^-] + [CO_3^{2-}]} \tag{9.61}$$

$$= \frac{1}{1 + \frac{[CO_2]}{[H_2CO_3]} + \frac{[HCO_3^-]}{[H_2CO_3]} + \frac{[CO_3^{2-}]}{[H_2CO_3]}} \tag{9.62}$$

$$= \frac{1}{1 + K + \frac{[HCO_3^-]}{[H_2CO_3]} + \frac{[CO_3^{2-}]}{[H_2CO_3]}}. \tag{9.63}$$

Thus, rather than just having a constant of 1, there is a constant of $(1 + K)$ in the denominator for $\alpha_{H_2CO_3}$. This causes the break in the $\log[H_2CO_3]$ line to occur when the value of $[HCO_3^-]/[H_2CO_3]$ approaches $(1 + K)$ rather than just 1. Thus, at 25°C/1 atm, the break in the $\log[H_2CO_3]$ line occurs at an $[H^+]$ value that is a factor of 651 lower than $10^{-3.54}$, that is, at a pH of ~6.35.

9.8.3 Buffer Intensity for Closed CO_2 Systems

Eq. (9.8) can be written

$$(C_B - C_A) = \alpha_1 C_T + 2\alpha_2 C_T + [OH^-] - [H^+]. \tag{9.64}$$

According to the definition of buffer intensity β that is given in Chapter 8, we know that

$$\beta = \frac{d(C_B - C_A)}{\text{dpH}} \tag{8.1}$$

Therefore, for a closed system containing CO_2 species (C_T is constant), we have

$$\beta = C_T \frac{d\alpha_1}{d\text{pH}} + 2C_T \frac{d\alpha_2}{d\text{pH}} + \frac{d[OH^-]}{d\text{pH}} - \frac{d[H^+]}{d\text{pH}}. \tag{9.65}$$

Expressions for the last two terms in Eq. (9.65) were derived in Section 8.2, and including signs, they are equal to $2.303[OH^-]$ and $2.303[H^+]$, respectively. We thus need to focus only on the derivatives of α_1 and α_2.

For α_1, starting with the chain rule, we note that

$$\frac{d\alpha_1}{d\text{pH}} = \frac{d\alpha_1}{d[H^+]} \frac{d[H^+]}{d\text{pH}}. \tag{9.66}$$

Recalling that

$$d(u/v) = (vdu - udv)/v^2 \tag{8.23}$$

for $d\alpha_1/d[H^+]$ we obtain

$$d\alpha_1/d[H^+] = \frac{-\frac{1}{K_1} + \frac{K_2}{[H^+]^2}}{\left[\frac{[H^+]}{K_1} + 1 + \frac{K_2}{[H^+]}\right]^2} = \left[-\frac{1}{K_1} + \frac{K_2}{[H^+]^2}\right]\alpha_1^2. \tag{9.67}$$

Multiplying Eq. (9.67) by $d[H^+]/dpH = -2.303[H^+]$ (Eq. (8.19)) as prescribed by Eq. (9.66), we obtain

$$d\alpha_1/dpH = 2.303 \left[\frac{[H^+]}{K_1} - \frac{K_2}{[H^+]} \right] \alpha_1^2. \tag{9.68}$$

It is relatively easy to show that $([H^+]/K_1)\alpha_1 = \alpha_0$ and that $(K_2/[H^+])\alpha_1 = \alpha_2$. Therefore, we have

$$d\alpha_1/dpH = 2.303\ (\alpha_0\alpha_1 - \alpha_2\alpha_1). \tag{9.69}$$

Proceeding in a similar manner for $d\alpha_2/dpH$ we obtain

$$d\alpha_2/dpH = 2.303 \frac{\dfrac{2[H^+]^2}{K_1K_2} - \dfrac{[H^+]}{K_2}}{\left[\dfrac{[H^+]^2}{K_1K_2} + \dfrac{[H^+]}{K_2} + 1\right]^2} = 2.303 \left[\frac{2[H^+]^2}{K_1K_2} - \frac{[H^+]}{K_2} \right] \alpha_2^2 \tag{9.70}$$

and so

$$d\alpha_2/dpH = 2.303\ (2\alpha_0\alpha_2 + \alpha_1\alpha_2) \tag{9.71}$$

When the various terms in β are added together (remembering to include the factor of 2 for the $d\alpha_2/dpH$ term), the overall result is

$$\beta = 2.303\ [C_T(\alpha_1(\alpha_0 + \alpha_2) + 4\alpha_0\alpha_2) + [OH^-] + [H^+]]. \tag{9.72}$$

Equation (9.72) applies for any diprotic acid system. In the CO_2 system, pK_1 and pK_2 are separated by four log units, and so the term $4\alpha_0\alpha_2$ is always small relative to either $\alpha_1\alpha_0$ or $\alpha_1\alpha_2$. Eq. (9.72) for CO_2 may thus be simplified to

$$\beta \simeq 2.303[C_T\alpha_1(\alpha_0 + \alpha_2) + [OH^-] + [H^+]] \tag{9.73}$$

Equation (9.73) is equivalent to Eq. (8.40).

Figure 9.8 gives $\log \beta$ as a function of pH for the three C_T values used to prepare Figure 9.2. For each curve, the minimum at low pH corresponds to the $H_2CO_3^*$ ($f = 0$) EP. The pH of this minimum depends on C_T according to the approximation $[H^+] \simeq \sqrt{(K_1C_T)}$ (Eq. (9.53)). The pH at which all three curves are subsequently maximized is essentially independent of C_T, and occurs at $pH \simeq pK_1$ because, as in a monoprotic system, that is where the product $\alpha_1\alpha_0$ is maximized. The subsequent minimum for the $NaHCO_3$ ($f = 1$) EP is also independent of C_T, and occurs at $pH \simeq 1/2\ (pK_1 + pK_2)$ (Eq. (9.57)).

At the Na_2CO_3 ($f = 2$) EP, a minimum is present for $C_T = 10^{-2}$ *M*, but not for $C_T = 10^{-2.5}$ or 10^{-3} *M*. When a minimum is present, its pH will depend on C_T, and will be given by the approximation equation $[OH^-] \simeq \sqrt{((K_w/K_2)C_T)}$ (Eq. (9.59)). The presence of the weak minimum for $C_T = 10^{-2}$ *M* allows the presence in Figure 9.2 of the weak $f = 2$ IP for $C_T = 10^{-2}$ *M*. The absence of minima for $C_T = 10^{-3}$ and $C_T = 10^{-2.5}$ *M* is due to the strength in Eq. (9.73) of the $[OH^-]$ term relative to $\alpha_1\alpha_2C_T$. It is interesting to note here that Stumm and Morgan (1981) show a strong minimum for β at the $f = 2$ EP in their Figure 4.1*f* for seawater with $C_T = 10^{-2.6}$ *M* and total borate $= 4.1 \times 10^{-4}$ *M*. The presence of this minimum is due in part to the presence of the borate, but is a result mostly of the fact that in seawater, ${}^cK_2 = 10^{-9.3}$, that is, an *order*

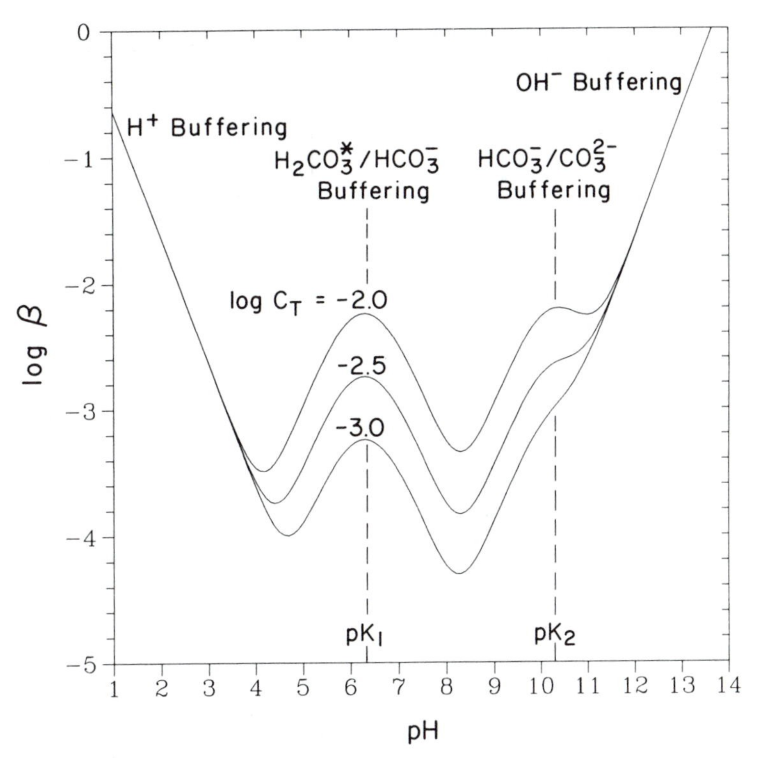

Figure 9.8 Log β vs. pH for closed systems at 25°C/1 atm for $C_T = 10^{-3}$, $10^{-2.5}$, and 10^{-2} *M*. Activity corrections neglected (not valid at very low or very high pH).

of magnitude larger than the corresponding infinite dilution constant. This acidic shift allows the $\alpha_1\alpha_2C_T$ term in Eq. (9.73) to emerge out of the OH^- buffering when C_T is as "small" as $10^{-2.6}$ *M*. Figure 9.9 collects the titration curve, the speciation plot, and the log β vs. pH plot for $C_T = 10^{-3}$ and 25°C/1 atm.

9.9 SYSTEMS OPEN TO AN ATMOSPHERE WITH CONSTANT p_{CO_2} (C_T NOT FIXED)

9.9.1 General

When equilibrium is established between an aqueous phase and a gas phase containing $CO_{2(g)}$, CO_2 will partition according to

$$CO_{2(g)} + H_2O = H_2CO_3^* \qquad K_H = \frac{\{H_2CO_3^*\}}{p_{CO_2}} \tag{9.74}$$

where K_H is the Henry's Law constant for CO_2. Strictly speaking, tabulated values of

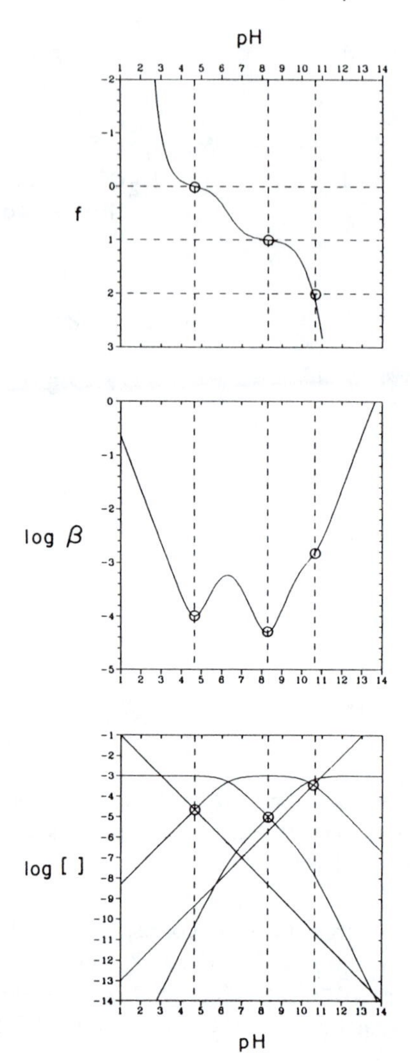

Figure 9.9 Closed system titration curve, log β vs. pH, and log concentration vs. pH for $C_T = 10^{-3}$ M and 25°C/1 atm. Activity corrections neglected (not valid at very low or very high pH).

K_H usually have units of m/atm. Our interest in dilution aqueous solutions allows us to use units of M/atm.[4] The adjective "Henry's Law" describes a relationship in which the solution phase concentration of one species (in this case CO_2) is linearly proportional to its gas phase pressure. At 25°C, the value of K_H is $10^{-1.47}$ M/atm.

[4]It may be noted that in reality, the gas/solution equilibrium is between $CO_{2(g)}$ and the dissolved species CO_2. However, just as we converted $K_{H_2CO_3}$ to K_1, we can obtain K_H using the constant for $CO_{2(g)} + H_2O = H_2CO_3$.

In contrast with closed systems, for open systems, C_T *is not fixed*. Usually, when we talk about systems that are open relative to the removal or addition of CO_2, the specific type of system that is under discussion is one in which there is a fixed volume of water in equilibrium with an atmosphere that has a constant partial pressure of $CO_{2(g)}$ (i.e., fixed p_{CO_2}). When p_{CO_2} is assumed constant no matter what we do to the aqueous solution, then the volume of the gas phase has in essence been assumed to be infinitely large relative to the aqueous solution.

The main driving force that will change C_T in an open system with constant p_{CO_2} is a change in pH. For example, if some strong base is added to a solution that is in equilibrium with an atmosphere containing $CO_{2(g)}$, C_T will increase since the acidic properties of CO_2 will tend to drive it into the solution. On the other hand, if some strong acid is added to the solution to lower the pH, C_T will decrease since some of the acidic CO_2 will be driven out of the solution.

9.9.2 Log Concentration vs. pH Diagrams for Open Systems

When they first encounter the idea of open CO_2 systems, most students think that such systems will be more complicated than closed systems. In fact, in many respects, the reverse is true. Indeed, an important simplifying fact for open systems is that the activities of all CO_2 species may be related to the value of p_{CO_2}. This may be shown as follows.

Neglecting activity corrections, from Eq. (9.74) we obtain

$$[H_2CO_3^*] = K_H p_{CO_2}. \tag{9.75}$$

In addition to Eq. (9.74), for any CO_2 system we have

$$[H_2CO_3^*] = \alpha_0 C_T. \tag{9.76}$$

Combining Eqs. (9.75) and (9.76), we have

$$C_T = K_H p_{CO_2}/\alpha_0. \tag{9.77}$$

Since α_0 is a function of pH, Eq. (9.77) illustrates how C_T will be also be a function of pH in an open system. Thus, if we know p_{CO_2} and pH for any given solution, we can calculate C_T.

For the remaining CO_2 species, we can therefore write

$$[HCO_3^-] = \alpha_1 C_T = \frac{\alpha_1}{\alpha_0} K_H p_{CO_2} = \frac{K_1}{[H^+]} K_H p_{CO_2} \tag{9.78}$$

$$[CO_3^{2-}] = \alpha_2 C_T = \frac{\alpha_2}{\alpha_0} K_H p_{CO_2} = \frac{K_1 K_2}{[H^+]^2} K_H p_{CO_2}. \tag{9.79}$$

The results of Eqs. (9.75)–(9.79) are summarized in Table 9.7.

As seen in Eqs. (9.78) and (9.79), neglecting activity corrections, *ratios of α values are simple functions of the acidity constants and* $[H^+]$. This was pointed out for closed systems in the development of Eq. (9.69), and so may be seen to be a general result that applies equally well to closed as well as open systems. This is logical because α values are not functions of either C_T or p_{CO_2}.

TABLE 9.7 Concentration relations for CO_2 species as a function of pH in open systems (activity corrections neglected).

Species		Relationship	d log concentration/dpH
$[H_2CO_3^*]$	=	$K_H p_{CO_2}$	0
$[CO_2]$	=	$\frac{K}{K+1} K_H p_{CO_2}$	0
$[H_2CO_3]$	=	$\frac{1}{K+1} K_H p_{CO_2}$	0
$[HCO_3^-]$	=	$\frac{K_1}{[H^+]} K_H p_{CO_2}$	+1
$[CO_3^{2-}]$	=	$\frac{K_1 K_2}{[H^+]^2} K_H p_{CO_2}$	+2

In an open system, when activity corrections are neglected, all log concentration vs. pH lines for *individual* CO_2 species are *straight lines*. This may be seen from a consideration of the equations in Table 9.7. At the earth's surface, the value of $p_{CO_2} = 10^{-3.5}$ atm. Using this value along with the values of K_H, K_1, and K_2 at 25°C, we obtain Figure 9.10. There are several interesting observations that can be made regarding this figure. Firstly, at constant p_{CO_2}, $[H_2CO_3]$, $[CO_2]$, and $[H_2CO_3^*]$ are constant and independent of pH. Secondly, at high pH values, $[HCO_3^-]$, $[CO_3^{2-}]$, and C_T can become quite large, so large in fact that it becomes unreasonable to make the assumption that activity coefficients can be neglected. Thirdly, the $\log[H_2CO_3^*]$ line crosses the $\log[HCO_3^-]$ line at pH $= pK_1$, and the $\log[HCO_3^-]$ line crosses the $\log[CO_3^{2-}]$ line at pH $= pK_2$.

Although the $\log C_T$ line in Figure 9.10 is not straight, within each of the three asymptotic CO_2 regions where it is dominated by one of the three CO_2 species, it is essentially straight. In particular, for pH $\ll pK_1$, the $\log C_T$ line follows $\log[H_2CO_3^*]$; for $pK_1 \ll$ pH $\ll pK_2$, it follows $\log[HCO_3^-]$; and for pH $\gg pK_2$, it follows $\log[CO_3^{2-}]$. The slope changes occur in the regions pH $\simeq pK_1$ and pH $\simeq pK_2$.

Figure 9.11 is a generic CO_2 species plot that can be used as an acetate transparency with Figure 6.3 to prepare a plot for any desired value of p_{CO_2}. When using Figures 9.10 and 6.3 in this manner, it is useful to remember that K_H for $CO_{2(g)}$ is $10^{-1.47} \simeq 10^{-1.5}$ at 25°C/1 atm. Therefore, when p_{CO_2} is 10^{-y}, the log $[H_2CO_3^*]$ line should be aligned at $\sim(-1.5 - y)$.

Like a log concentration diagram for a closed system, plots such as Figure 9.10 are useful in determining the speciation of solutions of initially pure water equilibrated with a given p_{CO_2}. In particular, as in a closed system, the ENE (and the PBE) for such a solution will be

$$[H^+] = [HCO_3^-] + 2[CO_3^{2-}] + [OH^-]. \tag{9.80}$$

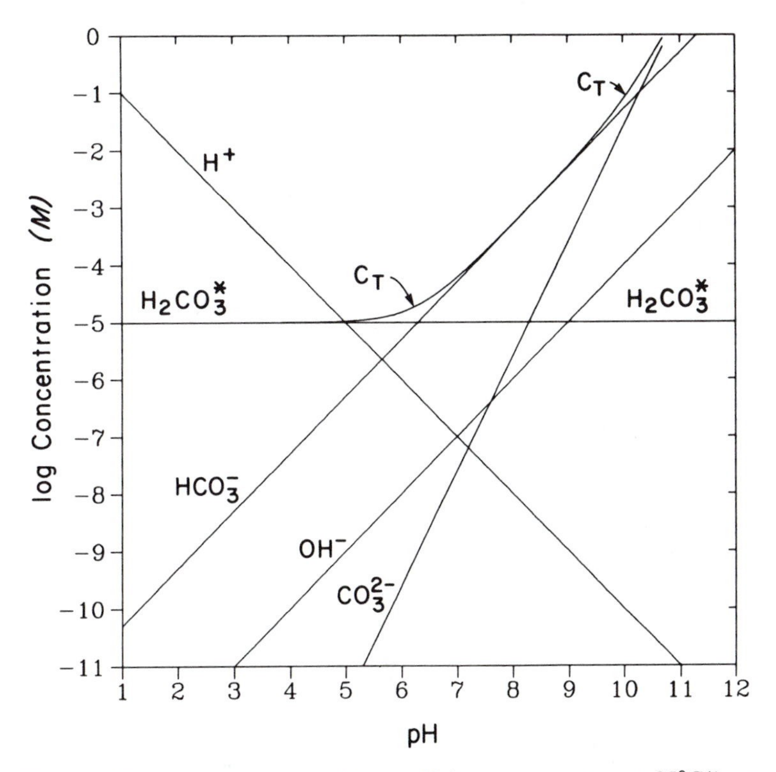

Figure 9.10 Log concentration (M) vs. pH for an open system at 25°C/1 atm with $p_{CO_2} = 10^{-3.5}$ atm. Activity corrections neglected (not valid at very low or very high pH).

Since log concentration lines for all of these species are included in Figure 9.10, to determine the pH when $p_{CO_2} = 10^{-3.5}$ atm, all we need to do is find where Eq. (9.80) is satisfied. Given the acidic nature of CO_2 solutions, we know that the pH will always be less than $\frac{1}{2}pK_w$. Therefore, as long as p_{CO_2} is not too low, we will be able to assume that both $2[CO_3^{2-}]$ and $[OH^-]$ are negligible in Eq. (9.80) relative to $[HCO_3^-]$. Thus, the solution speciation may be found to a first approximation by drawing a vertical line through the intersection of $\log[H^+]$ with $\log[HCO_3^-]$. In Figure 9.10, the pH of this intersection is 5.66, and the above assumptions are valid.

A second very useful way to look at the speciation that results when $CO_{2(g)}$ is equilibrated with initially pure water is to realize that $[H_2CO_3^*]$ is given exactly by Eq. (9.75). Assuming that $[H^+] = x \simeq [HCO_3^-]$ now gives

$$\frac{x^2}{K_H p_{CO_2}} \simeq K_1 \tag{9.81}$$

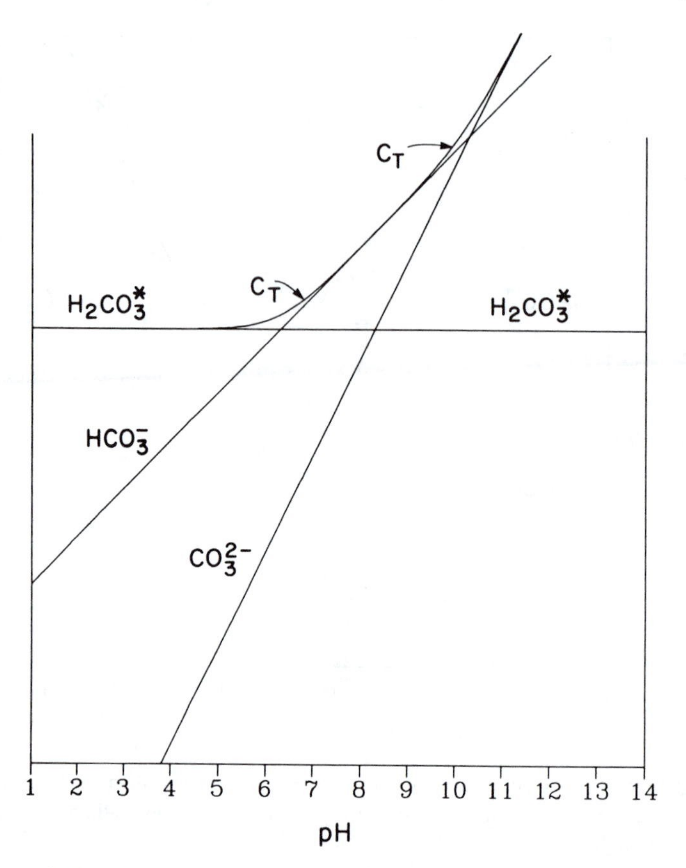

Figure 9.11 Generic log concentration (M) vs. pH for an open system at 25°C/1 atm with $p_{CO_2} = 10^{-3.5}$ atm. To be used as an acetate transparency with Figure 6.3. When $p_{CO_2} = 10^{-y}$, align $\log[H_2CO_3^*]$ on the log concentration $= (-1.47 - y)$. Activity corrections neglected (not valid at very low or very high pH).

or simply

$$[H^+] \simeq \sqrt{K_1 K_H p_{CO_2}} \qquad p_{CO_2} \text{ not too low.} \tag{9.82}$$

Note that we do not have the $-x$ correction in Eq. (9.81) that is present in the weak acid approximation (Eq. (5.29)): the gas phase replenishes whatever $[H_2CO_3^*]$ dissociates to form H^+ and HCO_3^-. When Eq. (9.82) is applied at 25°C/1 atm and $p_{CO_2} = 10^{-3.5}$ atm, we obtain pH = 5.66, that is, the same result that was obtained above by finding the pH at which the $\log[H^+]$ and $\log[HCO_3^-]$ lines cross. Obviously, the method embodied in

Eq. (9.82) and that described in the preceding paragraph involve the same fundamental approach, but different execution mechanics.[5]

Unfortunately, we cannot use Figure 9.10 for solving the speciation of a solution of $NaHCO_3$ equilibrated with $p_{CO_2} = 10^{-3.5}$, or the speciation of a solution of Na_2CO_3 equilibrated with $p_{CO_2} = 10^{-3.5}$. The reason is that, unlike a closed system, we cannot write a governing PBE for either case (the acidic properties of the $CO_{2(g)}$ that dissolves prevent this). Similarly, although we can certainly write an ENE, for these solutions we need a $\log[Na^+]$ line (or equivalently, a $\log(C_B - C_A) = \log Alk$ line), and such a line is not typically included in diagrams such as Figure 9.10. We could compute and plot $\log[Na^+]$ for diagrams like Figure 9.10 by writing the ENE as

$$(C_B - C_A) = Alk = [HCO_3^-] + 2[CO_3^{2-}] + [OH^-] - [H^+] \tag{9.84}$$

and solving it for all pH. Expressions for $[HCO_3^-]$ and $[CO_3^{2-}]$ are given in Table 9.7. Note that $Alk = 0$ for pH $= 5.66$, and $Alk < 0$ for pH < 5.66. Therefore, $\log Alk$ goes asymptotically to $-\infty$ as pH $\rightarrow 5.66$. Note also that we would have to plot $\log(-Alk)$ for pH < 5.66.

9.9.3 Alkalinity in Open CO_2 Systems

Whether or not a given system is open or closed, the expression for Alk will be given by Eq. (9.84). We can reexpress Eq. (9.84) as

$$Alk = \alpha_1 C_T + 2\alpha_2 C_T + [OH^-] - [H^+]. \tag{9.85}$$

For the specific case of an open system, substituting for C_T as given by Eq. (9.77) and neglecting activity corrections, we obtain

$$Alk = \frac{K_1 K_H p_{CO_2}}{[H^+]} + 2\frac{K_1 K_2 K_H p_{CO_2}}{[H^+]^2} + K_w/[H^+] - [H^+]. \tag{9.86}$$

Note that Alk goes up very rapidly as pH increases.

Based on Eq. (9.86), we can conclude that, a given value for Alk prescribes one and only one pH in an open system of fixed p_{CO_2}. It is exactly this point that Stumm and Morgan (1981) make in their Example 4.2. This example question asks what the pH will be for four different aqueous solutions after being opened to $10^{-3.5}$ atm CO_2:

1. 10^{-3} F KOH
2. 10^{-3} F $NaHCO_3$

[5]The fact that these two methods are exactly the same may be seen as follows. Starting with the graphical approach, when p_{CO_2} is not too low, the intersection of the $\log[H^+]$ line with $\log[HCO_3^-]$ will occur at a distance of $(\log K_H p_{CO_2} + \text{pH})$ below the $\log[H_2CO_3^*]$ $(= \log K_H p_{CO_2})$ line, and a distance of $(\text{p}K_1 - \text{pH})$ to the left of $\text{p}K_1$. (pH refers to the equilibrium pH.) The asymptotic portion of the $\log[HCO_3^-]$ line for acidic pHs intersects the $\log[H_2CO_3^*]$ line at a 45° angle. Since Tan 45° = 1,

$$\frac{\log K_H p_{CO_2} + \text{pH}}{\text{p}K_1 - \text{pH}} = 1 \tag{9.83}$$

which reduces to Eq. (9.82).

3. 5×10^{-4} *F* Na_2CO_3

4. 5×10^{-4} *F* MgO

Assuming that activity corrections can be neglected, since each of these open systems have the same *Alk*, then according to Eq. (9.86), they will all have the same pH. Using the equilibrium constants for 25°C/1 atm, that pH will be 8.32.

9.9.4 Buffer Intensity in Open CO_2 Systems

When deriving the expression for β in an open system where p_{CO_2} is a constant, we could take the derivative of Eq. (9.64) with respect to pH, but that would be inconvenient since C_T would not be constant. Alternatively, let us take the derivative of *Alk* as it is given by Eq. (9.86). In particular, using the chain rule for the first two terms, for constant p_{CO_2} we have

$$\beta = -\frac{K_1 K_H p_{CO_2}}{[H^+]^2}\frac{d[H^+]}{dpH} - 4\frac{K_1 K_2 K_H p_{CO_2}}{[H^+]^3}\frac{d[H^+]}{dpH} + 2.303([OH^-] + [H^+]). \qquad (9.87)$$

Recalling from Eq. (8.19) that $d[H^+]/dpH = -2.303[H^+]$, we obtain the remarkably simple result that

$$\beta = 2.303\left(\frac{K_1 K_H p_{CO_2}}{[H^+]} + 4\frac{K_1 K_2 K_H p_{CO_2}}{[H^+]^2} + [OH^-] + [H^+]\right) \qquad (9.88)$$

or equivalently

$$\beta = 2.303([HCO_3^-] + 4[CO_3^{2-}] + [OH^-] + [H^+]) \qquad (9.89)$$

Although Eq. (11) on page 197 of Stumm and Morgan (1981) looks much more complicated than (9.89), these two equations are in fact equivalent. However, Eq. (9.89) may be used to point out that Eq. (12) on page 197 of Stumm and Morgan (1981) is correct *only* when HCO_3^- makes the dominant contribution to C_T (i.e., when $\alpha_1 \simeq 1$). A plot of $\log\beta$ vs. pH computed for $p_{CO_2} = 10^{-3.5}$ atm, and K_H, K_1, and K_2 for 25°C/1 atm is presented in Figure 9.12.

9.10 THE CONCEPT OF THE IN-SITU p_{CO_2} FOR CLOSED CO_2 SYSTEMS

As noted above, when we refer to a system as being "closed," we usually mean that there is no gas phase present. In the aqueous systems that embody closed systems, however, it is common for significant amounts of CO_2 species to be present. The absence of an actual gas phase obviously means that there is no physical gas phase pressure for $CO_{2(g)}$ with which the closed system solution is in equilibrium. This seems somewhat unfortunate since Section 9.9 has illustrated the great convenience of having a value for p_{CO_2} that can be used in a variety of useful equations. This "deficiency" can be handled by defining an "in-situ p_{CO_2}" for closed systems. We accomplish this as follows.

Consider the beaker of water that is depicted in Figure 9.13*a*. Above the water there is an essentially infinitely large gas phase containing some $CO_{2(g)}$. Obviously, for this open system, we can use all of the open system equations involving p_{CO_2}. For example,

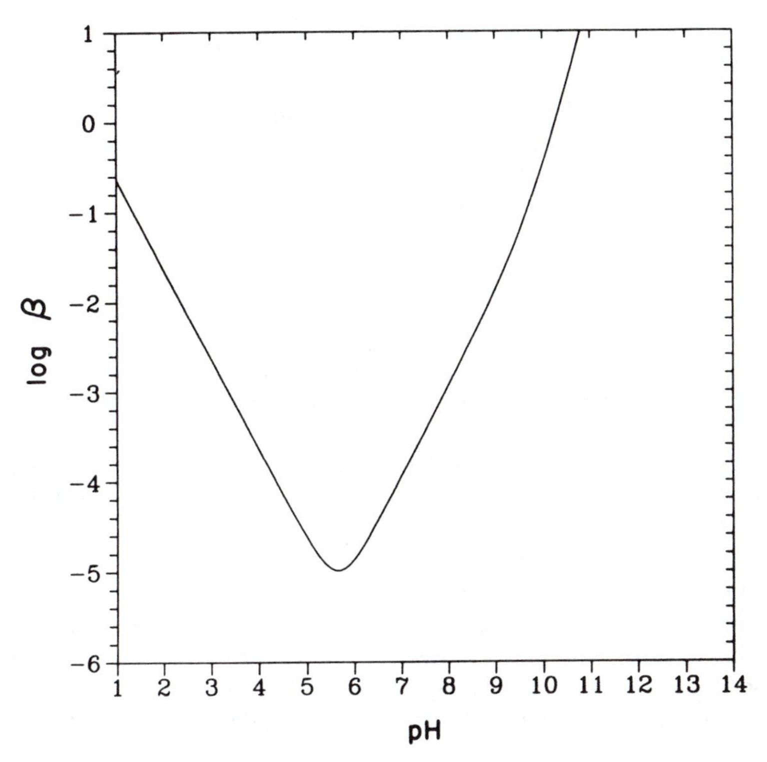

Figure 9.12 Log β vs. pH for an open system at 25°C/1 atm with $p_{CO_2} = 10^{-3.5}$ atm. Activity corrections neglected (not valid at very low or very high pH).

if we know the pH, then neglecting activity corrections, we can directly calculate *Alk*, the concentrations of a variety of individual species, and β.

We now envision taking a syringe and removing some of the water in the beaker. This produces the closed system depicted in Figure 9.13*b*. The concentrations of the various dissolved species remained constant during the transfer from the beaker to the syringe.

Because the water that was transferred to the syringe has no intelligence, it does not "know" that it is no longer in equilibrium with an actual, physical gas phase. Despite this, application of a knowledge of $[H_2CO_3^*]$ in the syringe with a knowledge of K_H will produce a value for p_{CO_2} that is *exactly* the same as the p_{CO_2} value above the beaker. Thus, while there is no actual gas phase in equilibrium with the water in the syringe, thermodynamically, the system is characterizable by a perfectly well defined p_{CO_2}. This is the "in-situ" p_{CO_2}.[5] Use of an in-situ p_{CO_2} with the pH will produce perfectly correct values for $[HCO_3^-]$, $[CO_3^{2-}]$, *Alk*, etc. It is important to note, however, that it will *not*

[5]The in-situ p_{CO_2} concept is identical to the environmental fugacity concept as employed by Clark et al. (1988).

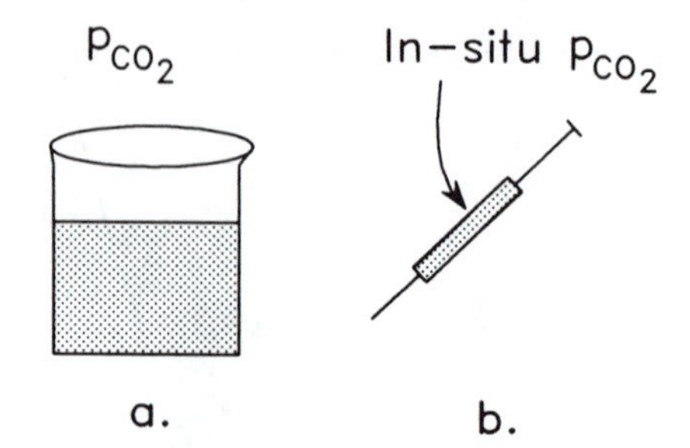

Figure 9.13 (*a*) Open system in equilibrium with an essentially infinitely large gas phase containing a constant p_{CO_2}. (*b*) Closed system (no gas phase) containing water that was removed from the beaker in Figure 9.13*a*; the chemical composition of the water in the syringe is identical to that in the beaker.

produce a correct value for β if Eq. (9.88) (which assumes that p_{CO_2} remains *constant*) is used. Indeed, once we close off the system from the gas phase, as we add strong base, the C_T will remain constant, but the in-situ p_{CO_2} will decrease. As we add strong acid, C_T will remain constant and the in-situ p_{CO_2} will build up. Thus, for a closed system like that in Figure 9.13*b*, Eq. (9.72) must be used to compute β. As discussed above, the ability of an open system to resist changes in pH is due in part to its ability to exchange CO_2 (which is acidic) with the gas phase. In the Figure 9.13*a* system, adding strong base causes C_T to go up. Adding strong acid causes C_T to go down. At any given pair of values for pH and C_T, β will be larger in an open system than in a closed system.

9.11 REFERENCES

CLARK, T, K. CLARK, S. PETERSON, D. MACKAY, and R. J. NORSTROM. 1988. Wildlife Monitoring, Modeling, and Fugacity, *Env. Sci. Technol.*, **22**, 120–127.

PANKOW, J. F., 1992. *Aquatic Chemistry Problems.* Portland: Titan Press-OR, P.O. Box 91399, Portland, Oregon 97291-1399.

STUMM, W., and J. J. MORGAN. 1981. *Aquatic Chemistry.* New York: Wiley-Interscience.

10

Gran Titrations

10.1 INTRODUCTION

When carrying out the titration of a single acid, a single base, a mixture of acids, or a mixture of bases, oftentimes *neither* a marked change in pH *nor* an IP is present in the portion of the titration curve surrounding an EP of interest. As we learned in Chapter 8, there will be no IP and therefore no directly visible EP whenever there is no minimum in the region of the EP in the plot of the buffer intensity β (or $\log \beta$) vs. pH. Even when an IP is present, it might be so weak that it is difficult to detect. As discussed in Chapter 8, one of the prime causes for these difficulties is a low concentration of the acid or base undergoing titration.

When faced with the need to locate a difficult-to-detect EP, the "Gran Titration" method can often provide a solution. This method does not provide a different way to carry out the *physical mechanics* of a titration. Rather, it is a clever way to examine titration data (i.e., pH vs. volume of titrant data) that allows one to extract information concerning both the:

1. location(s) of the EP(s); and
2. acid dissociation K value(s) for the conjugate acids involved in the titration.

When we discussed the chemistry of CO_2 systems in Chapter 9, we learned that terms like H^+-*Acy*, *Alk*, OH^--*Alk*, etc. may be defined using the PBEs of the EPs in the CO_2 system. These terms provide measures of how far away a given CO_2 system is from those EPs. Interestingly, a chemical/mathematical analysis of expressions like H^+-*Acy* reveals that data adjacent to, but not in the immediate vicinity of a given EP, can be used to locate that EP. It is analyses of this type that provide the basis for various applications of the Gran Titration method. It is thus of fundamental importance to note that in contrast to "typical" data analysis for titration curves, in a Gran Titration, one focuses on the data for pH vs. volume of titrant that are *somewhat removed* from the EP(s). Since terms like H^+-Acy are based on proton balance equations (PBEs), we can expect that the application of the Gran Titration method to any given titration problem will be based on the PBE(s) for the EP(s) of interest in that system.

10.2 APPLICATION OF THE GRAN TITRATION METHOD TO A SYSTEM CONTAINING SOME WEAK ACID, AND POSSIBLY ALSO SOME STRONG ACID

Probably the best-known application of the Gran Titration method involves the determination of the eq/L of weak acid present in a dilute solution, as well as the K value of that acid. Oftentimes, a small amount of weak acid can be present in a solution that also contains some strong acid, and so it then becomes of interest to both differentiate as well as quantitate both acids. For example, a number of investigators interested in the phenomenon of "acid rain" have sought to use the Gran Titration method to determine the importance of comparatively small amounts of weak acids in precipitation that also contains substantial strong acid. This section will therefore provide an analysis of this general problem. It may be used for an initial solution that contains: 1) a weak acid and some net strong acid (initial $f < 0$); 2) just a small amount of a weak acid (initial $f = 0$); or 3) a small amount of weak acid with some net strong base (initial $f > 0$).

Let us assume that we know that the solution to be titrated contains some weak acid, but we do not know how much. We will allow that we do not know whether or not the solution contains any net strong acid, or any net strong base. That is, as just discussed above, the value of f for the initial HA/A^- system may be less than zero, zero, or greater than zero. It will be convenient to have all of the generality just outlined, but it will *also* be convenient to have a knowledge of *roughly* where, relative to the HA EP, the actual titration is going to begin. One way to access both of these conveniences would be to add enough strong acid to the initial solution prior to beginning the actual titration so that any possible system that might be encountered for which $0 \leq f \leq 1$ would be taken back to $f < 0$.[1] An alkalimetric titration would then follow. If, unbeknownst to the analyst, the f value for the initial solution was already less than zero prior to the addition of that strong acid, then prior to the actual titration, it would just be converted to a solution for which f would be even more negative.

Let us define:

Alk_o = alkalinity of the initial solution to be analyzed before any manipulations (eq/L)

$A_{T,o}$ = concentration of total A in the initial solution (M)

[1]For solutions which are likely to have f values in excess of 1, rather than adding a great deal of strong acid and then carrying out an alkalimetric titration, it would make more sense to carry out an acidimetric titration as discussed in Section 10.3.

$(C_A - C_B)_o$ = net strong acid in the initial solution to be analyzed before any manipulations (eq/L) ($= -Alk_o$)

$c_{t,a}$ = concentration of strong acid in solution used to acidify sample prior to initiation of titration (eq/L)

$c_{t,b}$ = concentration of strong base titrant (eq/L)

V_o = volume of initial solution to be analyzed before any manipulations (mL)

V_a = volume of strong acid added to ensure that $f < 0$ prior to carrying out the actual titration (mL)

V_t = volume of basic titrant added (a variable) (mL)

V_s = volume of basic titrant needed to reach the strong acid/HA EP (mL)

V_w = volume of basic titrant needed to reach the weak acid/NaA EP (mL)

The subscripts t, s, and w imply titrant, strong acid, and weak acid, respectively.

10.2.1 F_1 and the Strong Acid/HA Equivalence Point

The PBE for a solution of HA is

$$[H^+] = [A^-] + [OH^-]. \tag{10.1}$$

When $f < 0$, there is net strong acid (mineral acidity) in the system, and so in a manner that is analogous to the CO_2 case, for this system we can define

$$H^+\text{-}Acy = [H^+] - [A^-] - [OH^-]. \tag{10.2}$$

At *any* given point in the titration, the product of H^+-Acy and the total volume of the solution must be equal to the product of $c_{t,b}$ and the volume of that basic titrant still needed to neutralize the H^+-Acy. Thus, for any point in the titration,

$$(V_o + V_a + V_t)H^+\text{-}Acy = (V_s - V_t)c_{t,b}. \tag{10.3}$$

The word "any" is emphasized in the first sentence of this paragraph because even if H^+-Acy is negative, then the volume of basic titrant still needed to neutralize that acidity (i.e., $V_s - V_t$)) will simply also be negative. Figure 10.1, which is an example of one possible titration curve, shows how H^+-Acy is defined relative to the strong acid/HA EP. In the past, we have referred to this EP as being simply the "HA EP." In this chapter, for clarity, we will refer to it as the strong acid/HA EP; when this EP is reached, the strong acid has been fully titrated, and one has a solution of HA plus some salt.

In the portion of the titration during which $f < 0$, H^+-Acy will be positive, but perhaps more importantly for this chapter, $[H^+]$ will be the dominant term on the RHS of Eq. (10.2). Under these conditions,

$$H^+\text{-}Acy \simeq [H^+]. \tag{10.4}$$

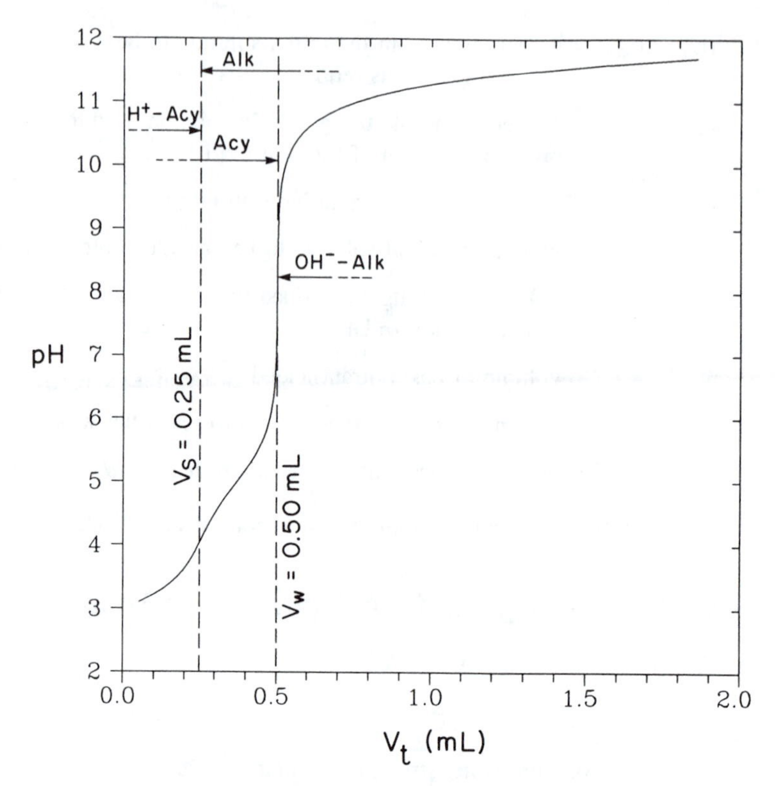

Figure 10.1 Alkalimetric titration curve for a 25 mL solution of 10^{-3} F acid HA with ${}^{c}K = 10^{-5}$. 0.25 mL of strong acid at 0.1 eq/L added prior to beginning the titration. Titration carried out using a strong base solution at 0.1 eq/L. Strong acid/HA ($f = 0$) equivalence point at $V_t = 0.25$ mL. Weak acid/NA ($f = 1$) eqivalence point at $V_t = 0.50$ mL.

Thus, by Eq. (10.3), for $f < 0$ we have

$$(V_o + V_a + V_t)[H^+] \simeq (V_s - V_t)c_{t,b}. \tag{10.5}$$

Equation (10.5) has two very interesting features. First, at any point during the titration, all the terms on the left-hand side are either known or measureable. In the particular case of $[H^+]$, a pH electrode may be used to follow the progress of the titration in terms of $\{H^+\}$ as a function of V_t. A knowledge of the ionic strength will allow an estimation of γ_{H^+}, and therefore $[H^+]$ and p_cH $(= -\log[H^+])$ as a function of V_t. The second important feature of Eq. (10.5) rests in the fact that if we set

$$F_1 = (V_o + V_a + V_t)[H^+] \qquad \simeq (V_s - V_t)c_{t,b} \tag{10.6}$$

$$= \underline{(V_o + V_a + V_t)10^{-p_cH}} \qquad \simeq (V_s - V_t)c_{t,b} \tag{10.7}$$

then, Eq. (10.7) indicates that if F_1 is evaluated as underlined, then plotted vs. V_t, the F_1 data will tend to zero linearly with a slope of $-c_{t,b}$ as V_t tends toward V_s. Eq. (10.7) also indicates that a plot of F_1 vs. V_t will pass through the V_t axis at $V_t = V_s$. The location of this x-intercept will therefore allow a determination of V_s. However, it is important to point out that the linearity of F_1 will continue only as long as the approximation embodied in Eq. (10.4) is valid. Since that approximation will tend to break down more and more as V_t approaches V_s, only the data points somewhat removed from the EP, with V_t significantly less than V_s, will plot on the straight line.

When the solution being titrated remains sufficiently dilute throughout the entire titration that activity corrections are not required, then the measured pH ($= -\log\{H^+\}$) values may be substituted directly for p_cH in Eq. (10.7). On the other hand, if activity corrections are needed, and if the ionic strength I remains essentially constant during the course of the titration, then one can either: a) estimate I, undertake a simple, constant correction for γ_{H^+}, and use F_1 as given in Eq. (10.7); or b) not worry about what I actually is, and use the directly-measured pH values to compute the points for a *modified* F_1 Gran function that equals $(V_o + V_a + V_t)10^{-pH}$. With the latter approach, the extrapolated linear intercept would still be V_s, but the slope would then be $-c_{t,b}\gamma_{H^+}$. Since $c_{t,b}$ is a parameter that would be known anyway, when I remains essentially constant, many analysts might well choose to use that modified F_1 Gran function for convenience sake. One reason to make the activity corrections and apply Eq. (10.7) as written however, would be for quality assurance purposes in order to see how closely the measured slope reproduces the known value of $c_{t,b}$.

The types of conditions that would lead to a relatively constant I during the course of a titration include a very dilute initial solution of interest, the natural presence of salts in the solution of interest, or the pre-titration addition of some salt to "buffer" the ionic strength.[2] When none of these conditions apply, or I is not known during the titration, or activity corrections are needed, then it will not be possible to apply either Eq. (10.7) or the modified F_1 Gran function exactly. Nevertheless, over any limited region within a given titration, I will not change significantly, and so the modified Gran function approach may be used to a very good approximation.

It is useful now to note that 1 milliequivalent (1 meq) of acid or base is equal, logically enough, to 10^{-3} eq. Thus, when a volume in units of mL is multiplied by a concentration expressed in units of eq/L, the resulting number has units of meq. Since the number of meq of strong acid that are titrated in the process of taking V_t from zero to V_s is equal to $V_s c_{t,b}$, and since the meq of strong acid added to the initial solution prior to carrying out the titration is equal to $V_a c_{t,a}$, we conclude that the meq of strong acid in the original initial solution is given by

$$\begin{array}{l}\text{meq of strong acid in} \\ \text{the initial solution}\end{array} = V_s c_{t,b} - V_a c_{t,a} \tag{10.8}$$

where V_s is as measured using F_1. Therefore, the concentrations in eq/L of net strong acid and Alk_o in the initial solution are given by

[2]When titrating very dilute solutions, the addition of salt to buffer I should be done very carefully, using high quality material. The reason is that low grade salt may contain acidic or basic impurities that will cause errors in the subsequent titration.

$$(C_A - C_B)_o = -Alk_o = (V_s c_{t,b} - V_a c_{t,a})/V_o. \tag{10.9}$$

As discussed above, depending on the chemistry of the original solution, the value of $(C_A - C_B)_o$ can be positive ($f < 0$), zero ($f = 0$), or negative ($f > 0$, and $(C_B - C_A)_o = Alk_o > 0$).

The Gran function F_1 is the first example of a total of four functions that will be developed in Section 10.2. As has been illustrated for F_1, each of the remaining three Gran functions to be discussed will locate an EP by examining data that is *removed from that EP*.

10.2.2 F_2 and the Weak Acid/NaA Equivalence Point

The PBE for a solution of NaA is

$$[H^+] + [HA] = [OH^-] \tag{10.10}$$

Using this EP, we can define

$$Acy = [H^+] + [HA] - [OH^-] \tag{10.11}$$

as the measure of protons in the solution that are at least as acidic as HA itself. Figure 10.1 shows how *Acy* is defined relative to the weak acid/NaA EP. In the past, we have referred to this EP as being simply the "NaA EP." In this chapter, for clarity, we will refer to it as the weak acid/NaA EP; when this EP is reached, the weak acid has been fully titrated, and one has a solution of NaA.

When $0 < f < 1$, although there is no positive net strong acid (mineral acidity) in the system, there are protons associated with the as yet untitrated HA. We realize that not all of these protons will be actually on HA molecules; some of them will be dissociated and present as H^+ in the solution. The purpose of the $[OH^-]$ in Eq. (10.11) is, of course, to make the quantity $([H^+] - [OH^-])$ that is also in that equation represent the portion of the untitrated *Acy* that is present as free H^+ ions.

At *any* given point in the titration, the product of *Acy* and the total volume of the solution must be equal to the product of $c_{t,b}$ and the volume of that basic titrant still needed to neutralize the *Acy*. Thus, for any point in the titration,

$$(V_o + V_a + V_t)Acy = (V_w - V_t)c_{t,b}. \tag{10.12}$$

When $0 < f < 1$, unless the solution is very dilute, [HA] will be the dominant term on the RHS of Eq. (10.11). Under these conditions, then,

$$Acy \simeq [HA]. \tag{10.13}$$

Thus, by Eq. (10.12), for $0 < f < 1$ we have

$$(V_o + V_a + V_t)[HA] \simeq (V_w - V_t)c_{t,b}. \tag{10.14}$$

We will use Eq. (10.14) to help develop the Gran function F_2. However, unlike F_1, consideration of just one PBE is not enough to develop F_2. We therefore turn our attention back to the HA EP for a moment. Indeed, if H^+-*Acy* is defined as given in Eq. (10.2), then for this system we have

$$Alk = [A^-] + [OH^-] - [H^+]. \tag{10.15}$$

Figure 10.1 shows how *Alk* is defined relative to the strong acid/HA EP for one possible titration curve.

At *any* given point in the titration, the product of *Alk* and the total volume of the solution must equal the product of $c_{t,b}$ and the volume of basic titrant that has been added to produce that alkalinity. Thus, for any point in the titration,

$$(V_o + V_a + V_t)Alk = (V_t - V_s)c_{t,b}. \tag{10.16}$$

When $0 < f < 1$, unless the solution is very dilute, $[A^-]$ will be the dominant term on the RHS of Eq. (10.15). Under these conditions, then,

$$Alk \simeq [A^-]. \tag{10.17}$$

Thus, by Eq. (10.16), for $0 < f < 1$ we have

$$(V_o + V_a + V_t)[A^-] \simeq (V_t - V_s)c_{t,b}. \tag{10.18}$$

We now recall that

$$[HA] = \frac{[H^+][A^-]}{{}^cK} \tag{10.19}$$

where ${}^cK$ represents the acidity constant for the acid HA at the ionic strength that applies for the solution. If the solution is very dilute, then the infinite dilution constant may be used. In a manner that is analogous to the discussion provided in the preceding section, if the ionic strength is sufficiently large that activity corrections will be required, and if it also changes during the titration due to the addition of the basic titrant, then it might be advantageous to add some very pure salt to the solution prior to beginning the titration so that the changes in ionic strength are minimal and a single value of ${}^cK$ may be used.

Combining Eqs. (10.14) and (10.19), we obtain

$$(V_o + V_a + V_t)[A^-] \simeq (V_w - V_t)c_{t,b}\,{}^cK/[H^+]. \tag{10.20}$$

The LHSs of Eqs. (10.18) and (10.20) are the same. Thus, we equate them and obtain

$$(V_t - V_s)10^{-p_cH} \simeq (V_w - V_t){}^cK. \tag{10.21}$$

If the value of V_s is evaluated using F_1, then all of the terms on the LHS of Eq. (10.21) are known, and so we define for F_2 that

$$\underline{F_2 = (V_t - V_s)10^{-p_cH}} \simeq (V_w - V_t){}^cK \tag{10.22}$$

If F_2 is evaluated as underlined above, then plotted vs. V_t in the region beyond V_s, Eq. (10.22) shows that the result will be a straight line that passes through the V_t axis at $V_t = V_w$. Thus, while F_1 can be used to evaluate V_s, F_2 can be used to evaluate V_w. In a manner that is analogous to the behavior exhibited by F_1, the linearity in F_2 will break down as V_t approaches V_w.

Equation (10.22) shows that to the extent that the assumptions embodied in that equation are valid, the slope of the F_2 line will be equal to $-{}^cK$. In a manner that is similar to that discussed above for F_1, for purposes of convenience the analyst may choose to use pH rather than p_cH in Eq. (10.22). Then, under conditions of constant ionic strength (including I $\simeq$ 0), the slope of the linear portion of a plot of the modified F_2 function, equalling $(V_t - V_s)10^{-pH}$, will be $-^cK\gamma_{H^+}$.

Now that we have mechanisms for determining both V_s and V_w, one more piece of information that may be obtained is the initial total concentration of A, that is, $A_{T,o}$ (M). Indeed, since HA is a monoprotic acid, we note that the equivalents of base needed to carry out the titration from V_s to V_w must equal the total number of mols of A. Thus, we have that

$$A_{T,o} = \frac{(V_w - V_s)c_{t,b}}{V_o}. \tag{10.23}$$

Although the product of a volume in mL and a concentration in eq/L produces meq as the units for the numerator of Eq. (10.23), dividing by mL in the denominator makes the RHS of Eq. (10.23) have units of eq/L. Since HA is a monoprotic acid, the concentration of $A_{T,o}$ in units of M on the LHS is consistent with the units of eq/L on the RHS.

10.2.3 F_3 and the Strong Acid/HA Equivalence Point

As discussed above, as long as the concentration of total A is not too low, Eq. (10.22) is valid in the region $0 < f < 1$. Using the value of V_s determined with F_1, we saw how F_2 can be used to determine V_w based on the titration datapoints for V_t that are near, but less than V_w (i.e., f near, but less than 1). However, we note that since Eq. (10.22): 1) is also valid for V_t near, but *greater* than V_s (i.e., f near, but greater than 0); and 2) involves both V_s and V_w, it may be used to develop a third Gran function, F_3, which may be used to provide a second measure of V_s.

We therefore rearrange Eq. (10.22) to read

$$\underline{F_3 = (V_w - V_t)10^{p_cH}} \simeq (V_t - V_s)/^cK. \tag{10.24}$$

If F_3 is evaluated as underlined above using the value of V_w obtained using F_2, then plotted as a function of V_t in the region beyond V_s but less than V_w, Eq. (10.24) shows that the result should be a straight line with slope $^cK^{-1}$ that passes through the V_t axis at $V_t = V_s$. In a manner that is analogous to the behavior exhibited by F_1 and F_2, the linearity will break down as V_t approaches V_s. If for purposes of convenience the analyst should choose to use pH rather than p_cH in Eq. (10.24), then under conditions of constant ionic strength the slope of the linear portion of the plot of the modified F_3 function, equalling $(V_w - V_t)10^{pH}$, will be $(^cK\gamma_{H^+})^{-1}$.

If the values of V_s and ^cK obtained using F_3 are not in reasonable agreement with the values obtained using F_1 and F_2, respectively, then there are three possible explanations: 1) the relative validities of the assumptions inherent in F_1, F_2, and F_3 are different for the problem at hand; 2) procedural errors were made during the titration; or 3) calculational errors were made while carrying out the data analysis.

Although the value of V_s obtained with F_3 using V_w from F_2 does not represent a determination that is completely independent of the value obtained using F_1 (note that the estimate of V_w obtained with F_2 depends on the value of V_s obtained with F_1), it is nevertheless based in large part on a different subset of the titration curve data, and so it may be considered quasi-independent. As we will now see, the Gran function F_4

yields an estimate of V_w that is based on titration data for $V_t > V_w$. Since that F_4-based value of V_w is fully independent of F_1, it may be used with F_3 to generate a second, fully independent estimate of V_s.

10.2.4 F_4 and the Weak Acid/NaA Equivalence Point

Using the PBE for the weak acid/NaA EP (Eq. (10.10)), we can define

$$OH^- \text{-} Alk = [OH^-] - [H^+] - [HA]. \tag{10.25}$$

Figure 10.1 shows how OH^--*Alk* is defined relative to the weak acid/NaA EP for one possible titration curve. At *any* given point in the titration, the product of OH^--*Alk* and the total volume of the solution will be equal to the product of $c_{t,b}$ and the volume of basic titrant that produced that OH^--*Alk*. Thus, for any point in the titration, we have

$$(V_o + V_a + V_t) OH^- \text{-} Alk = (V_t - V_w) c_{t,b}. \tag{10.26}$$

In a manner that is analogous to the other Gran function cases discussed above, Eq. (10.26) is valid even when $V_t < V_w$. In that case, OH^--*Alk* is negative along with $(V_t - V_w)$.

When $f \geq 0$, $[OH^-]$ will be the dominant term on the RHS of Eq. (10.25). Under these conditions,

$$OH^- \text{-} Alk \simeq [OH^-]. \tag{10.27}$$

Thus, by Eq. (10.26), for $f > 1$ we have

$$(V_o + V_a + V_t)[OH^-] \simeq (V_t - V_w) c_{t,b}. \tag{10.28}$$

We now define Gran function F_4 as

$$\underline{F_4 = (V_o + V_a + V_t) 10^{p_cH}} \simeq (V_t - V_w) c_{t,b} / {}^cK_w. \tag{10.29}$$

If F_4 is evaluated as underlined above, then plotted as a function of V_t in the region beyond V_w, Eq. (10.29) shows that the result will be a straight line with a slope of $c_{t,b}/{}^cK_w$ that passes through the V_t axis at $V_t = V_w$. In a manner that is analogous to the behavior exhibited by Gran functions F_1, F_2, and F_3, the linearity of F_4 will break down as V_t approaches V_w. If for purposes of convenience the analyst should choose to use pH rather than p_cH in Eq. (10.29), then under conditions of constant ionic strength the slope of the linear portion of a plot of the modified F_4 function, equalling $(V_o + V_a + V_t)10^{pH}$, will be $c_{t,b}/(\gamma_{H^+} {}^cK_w)$.

It may be noted that the two determinations of V_w provided by F_2 and F_4 are completely independent, and the two can be applied in any order. Actually, a clever programmer familiar with statistical algorithms would have little difficulty in developing a single computer program that would either: a) apply F_2 and F_4 together to generate a single best-fit estimate of V_w, or b) apply F_1 through F_4 together to generate single best-fit estimates for V_s, V_w, and ${}^cK$. Only the linear portions of the data for each of these functions would be used.

10.2.5 Three Example Titrations Using F_1 through F_4 for Solutions in Which Initially (i.e., Before the Addition of V_a), the Value of f is Zero

In order to illustrate the application of Gran functions F_1 through F_4, it will be useful to calculate a few representative titration curves, then carry out the Gran analysis on those curves assuming that they represent raw data. The degree to which the Gran functions can reproduce the known features of the titration curves (i.e., V_s, V_w, ${}^cK$, ${}^cK_w$, and $c_{t,b}$) may provide the reader with a better feeling for the utility of the Gran Titration method than can be extracted from simply reading this material.

Before we can proceed, however, we need to develop a governing equation for the titration of a system containing the acid HA that is somewhat more specific than Eq. (7.46). The specific case that we will consider here will be that in which the initial system contains a certain concentration $A_{T,o}$. For these calculations, the value of f for the initial solution will be assumed to be zero. In each case, we will also add some strong acid for purposes of making f less than zero. We begin with the fact that the ENE will apply throughout the titration. Therefore, we note that

$$C_B - C_A = [A^-] + [OH^-] - [H^+]. \tag{10.30}$$

Substituting for $[OH^-]$ in terms of ${}^cK_w$ and $[H^+]$, Eq. (10.30) becomes

$$C_B - C_A = [A^-] + {}^cK_w/[H^+] - [H^+]. \tag{10.31}$$

With the value of f for the initial solution assumed to be zero, the only source of base of the system is the basic titrant. Since the total volume of the system at any given point in the titration will be given by $(V_o + V_a + V_t)$, we can conclude that

$$C_B = \frac{c_{b,t}V_t}{(V_o + V_a + V_t)}. \tag{10.32}$$

Similarly, we have

$$C_A = \frac{c_{a,t}V_a}{(V_o + V_a + V_t)}. \tag{10.33}$$

Since the initial solution to which the strong acid was added was simply a solution of HA (i.e., initial $f = 0$) of concentration $A_{T,o}$, then $A_{T,o}V_o$ gives the total amount of A present at any point during the titration. The concentration of A^- at any point in the titration is therefore

$$[A^-] = \alpha_1 \frac{A_{T,o}V_o}{(V_o + V_a + V_t)}. \tag{10.34}$$

Combining Eqs. (10.32)–(10.34) with Eq. (10.31), we obtain

$$\begin{aligned} V_t c_{t,b} - V_a c_{t,a} = \alpha_1 A_{T,o}V_o &+ ({}^cK_w/[H^+] - [H^+])(V_a + V_o) \\ &+ ({}^cK_w/[H^+] - [H^+])V_t. \end{aligned} \tag{10.35}$$

Equation (10.35) may be rearranged to yield

$$V_t = \frac{\alpha_1 A_{T,o} V_o + ({}^cK_w/[H^+] - [H^+])(V_a + V_o) + V_a c_{t,a}}{c_{t,b} - ({}^cK_w/[H^+] - [H^+])} \tag{10.36}$$

Equation (10.36) gives V_t as a function of $[H^+]$ for known values of ${}^cK$, $A_{T,o}$, V_o, V_a, $c_{t,a}$, $c_{t,b}$, and ${}^cK_w$. Table 10.1 gives three sets of these parameters for which titration curves have been computed using Eq. (10.36). Figures (10.1)–(10.3) present the three corresponding titration curves. The datapoints for the curves were calculated for each incremental change in pH of 0.1. For Figure 10.1, which has already been discussed above in generic terms, $A_{T,o} = 10^{-3}$ M, $V_a = 0.25$ mL, and $c_{t,a} = c_{t,b} = 0.10$ eq/L. As a result, for Figure 10.1, V_s and V_w (the values of V_t for the two EPs) are 0.25 mL and 0.50 mL, respectively. For Figure 10.2, where $A_{T,o}$ is a factor of ten lower, $c_{t,a}$ and $c_{t,b}$ were both also made a factor ten lower, and so V_s and V_w remain at 0.25 mL and 0.50 mL, respectively. If $c_{t,b}$ were to remain at 0.10 eq/L for Figure 10.2, V_s and

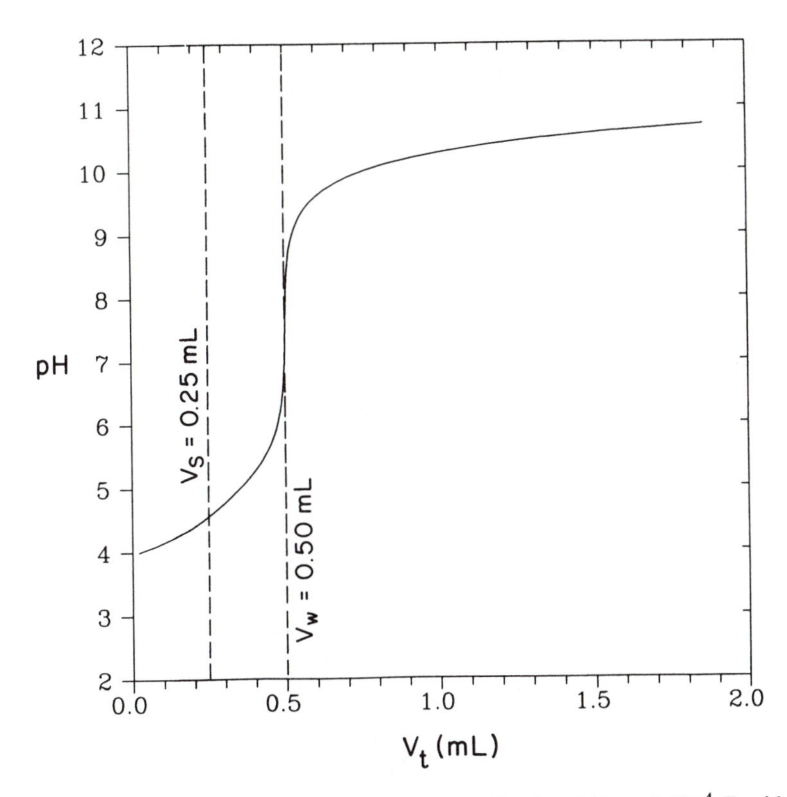

Figure 10.2 Alkametric titration curve for a 25 mL solution of 10^{-4} F acid HA with ${}^cK = 10^{-5}$. 0.25 mL of strong acid at 0.01 eq/L added prior to beginning the titration. Titration carried out using a strong base solution at 0.01 eq/L. Strong acid/HA ($f = 0$) equivalence point at $V_t = 0.25$ mL. Weak acid/NA ($f = 1$) equivalence point at $V_t = 0.50$ mL.

TABLE 10.1 Values for titration parameters for Figures 10.1–10.3. ${}^{c}K = 10^{-5}$ and ${}^{c}K_w = 10^{-14}$ for all three cases.

Figure	$A_{T,o}$	V_o	V_a	$c_{t,a}$	$c_{t,b}$	V_s	V_w
10.1	10^{-3}	25 mL	0.25 mL	0.10 eq/L	0.10 eq/L	0.25 mL	0.50 mL
10.2	10^{-4}	25 mL	0.25 mL	0.01 eq/L	0.01 eq/L	0.25 mL	0.50 mL
10.3	10^{-4}	25 mL	1.00 mL	0.01 eq/L	0.01 eq/L	1.00 mL	1.25 mL

V_w would have become 0.025 mL and 0.050 mL, respectively, and it would have been misleading to thereby imply that it is possible to easily measure out increments in V_t that are small fractions of such small volumes.

For Figure 10.1, while the $V_t = V_w$ (i.e., $f = 1$) EP is clearly visible by virtue of the sharp change in pH that occurs there, the same is not true for the $V_t = V_s$ (i.e., $f = 0$) EP. Although both EPs exhibit an IP, the exact location of the EP at $V_t = V_s$ is

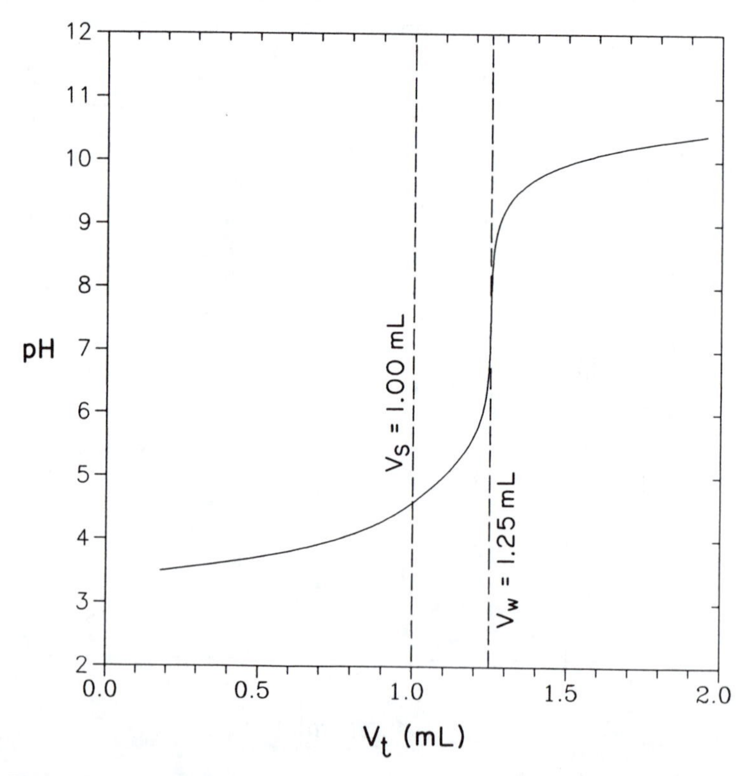

Figure 10.3 Alkalimetric titration curve for a 25 mL solution of 10^{-4} F acid HA with ${}^{c}K = 10^{-5}$. 1.00 mL of strong acid at 0.01 eq/L added prior to beginning the titration. Titration carried out using a strong base solution at 0.01 eq/L. Strong acid/HA ($f = 0$) equivalence point at $V_t = 1.00$ mL. Weak acid/NA ($f = 1$) equivalence point at $V_t = 1.25$ mL.

not nearly as apparent as is the location of the EP at $V_t = V_w$.[3] Thus, while the Gran Titration method would not be needed to locate the V_w EP, we can expect that it will be helpful in locating the V_s EP. For the Figure 10.2 titration curve, although the location of the V_w EP is still obvious, there is in this case no IP at the V_s EP. Indeed, since the concentration of strong acid is a factor of 10 lower than it is for the Figure 10.1 titration, it is not surprising that the V_s EP is even more difficult to locate by eye in Figure 10.2 than it is in Figure 10.1.

Gran functions F_1–F_4 were applied to the computed "raw data" for the Figure 10.1–10.3 titration curves, and the corresponding Gran function plots are presented in Figures 10.4–10.6. In every case, for F_2, the value of V_s used was that obtained from F_1; for F_3,

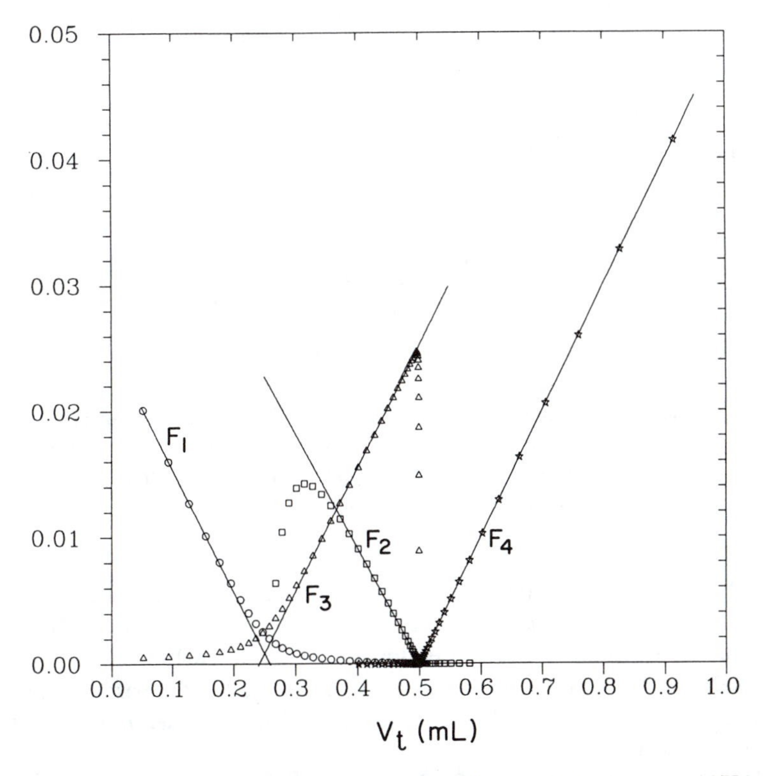

Figure 10.4 Gran plot for titration curve in Figure 10.1. Strong acid/HA ($f = 0$) equivalence point at $V_t = 0.25$ mL. Weak acid/NA ($f = 1$) equivalence point at $V_t = 0.50$ mL. For F_1 (circles), F_2 (squares), F_3 (triangles), and F_4 (stars), the scale factors are $\times 1$, $\times 10^4$, $\times 10^{-6}$, and $\times 10^{-14}$, respectively.

[3]Based on the discussions in Chapter 8, we note that the difficulty in seeing the V_s ($f = 0$) EP may be viewed as the result of either: 1) ^cK being too large (i.e., ^cK_b too small) given the concentration of HA under titration, or equivalently, 2) the concentration of HA under titration being too low given the value of ^cK.

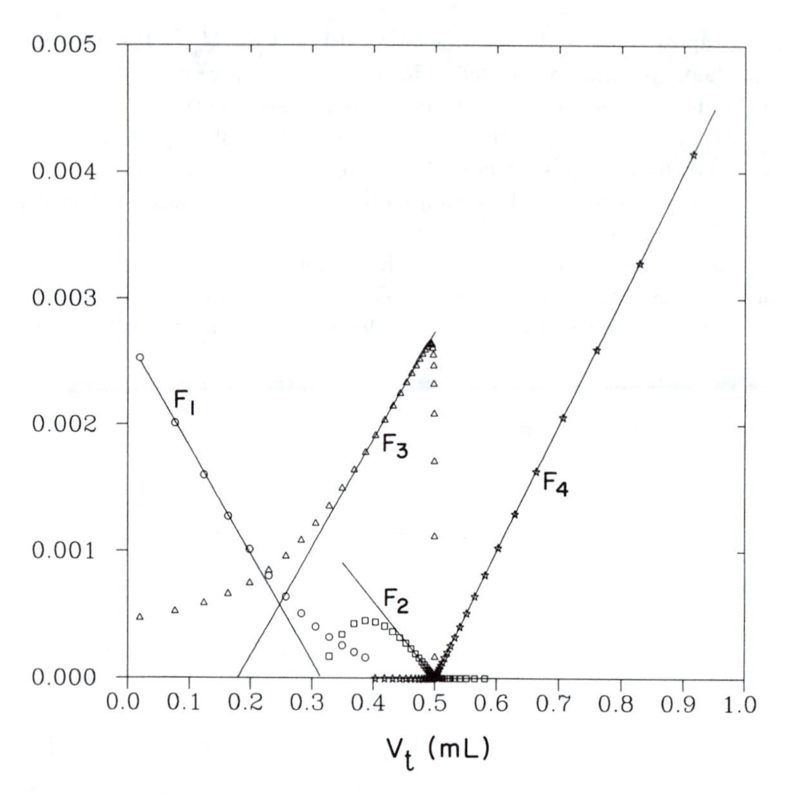

Figure 10.5 Gran plot for titration curve in Figure 10.2. Strong acid/HA ($f = 0$) equivalence point at $V_t = 0.25$ mL. Weak acid/NA ($f = 1$) equivalence point at $V_t = 0.50$ mL. Scale factors applied to F_1, F_2, F_3, and F_4 are $\times 1$, $\times 10^3$, $\times 10^{-7}$, and $\times 10^{-14}$, respectively.

the value of V_w used was that obtained using F_4. In other words, *the Gran analyses were carried out assuming that the actual values of* V_s *and* V_w *were not known.* Given the fact that the value of V_w from F_4 was chosen for use in F_3, for each case, the two estimates of V_s from F_1 and F_3 are fully independent. Table 10.2 gives the best fit parameters for the four Gran functions, as well as the ranges in V_t that were used to carry out the four corresponding linear regressions.

Using the parameters given in Table 10.2, the corresponding values of V_s, V_w, ${}^cK$, and ${}^cK_w$ have been calculated and are presented in Table 10.3. For the Figure 10.1/10.4 case, the errors in these various parameters are either very low, or essentially zero. For the Figure 10.2/10.5 case, the errors in V_w and ${}^cK_w$ remain low because the $f = 1$ titration EP is still marked by a very dramatic change in pH. The error is also low in the estimate of V_w obtained with F_2, despite the significantly-flawed estimate of V_s used in F_2 as obtained from F_1. For the $f = 0$ titration EP, however, the percent errors in V_s and ${}^cK$ are considerably larger in the Figure 10.2/10.5 case than in the Figure 10.1/10.4 case. As the reader may already suspect, the reason is that the concentration $A_{T,o}$ is a

TABLE 10.2 Best-fit slopes and intercepts for Gran functions $F_1 - F_4$, together with the ranges in V_t that were used to carry out the linear regressions ($F_i = m_i V_t + b_i$) for the Gran functions presented in Figures 10.4–10.6.

Figure	GRAN FUNCTION F_1 m_1	b_1	V_t range (mL)	F_2 m_2	b_2	V_t range (mL)
10.4	-9.71×10^{-2}	2.51×10^{-2}	0.052 – 0.155	-9.07×10^{-6}	4.55×10^{-6}	0.372 – 0.495
10.5	-8.49×10^{-3}	2.68×10^{-3}	0.020 – 0.199	-6.02×10^{-6}	3.01×10^{-6}	0.462 – 0.495
10.6	-9.75×10^{-3}	1.00×10^{-2}	0.180 – 0.755	-7.49×10^{-6}	9.37×10^{-6}	1.168 – 1.249
Figure	**F_3 m_3**	**b_3**	**V_t range (mL)**	**F_4 m_4**	**b_4**	**V_t range (mL)**
10.4	9.75×10^{4}	-2.36×10^{4}	0.357 – 0.465	1.00×10^{13}	-5.00×10^{12}	0.565 – 0.915
10.5	8.55×10^{4}	-1.54×10^{4}	0.404 – 0.484	1.00×10^{12}	-5.00×10^{11}	0.506 – 0.915
10.6	8.33×10^{4}	-7.64×10^{4}	1.098 – 1.211	9.98×10^{11}	-1.25×10^{12}	1.247 – 1.952

TABLE 10.3 Best-fit values for V_s, V_w, ${}^cK$, and ${}^cK_w$ as derived from Gran functions F_1 through F_4. Percent errors computed based on the known values used to compute the titration curves are included in parentheses.

Figure	GRAN FUNCTION F_1 V_s (mL)		F_2 V_w (mL)	${}^cK$
10.4	0.259 (+3.6%)		0.501 (+0.2%)	9.07×10^{-6} (−9.3%)
10.5	0.315 (+26%)		0.501 (+0.2%)	6.02×10^{-6} (−40%)
10.6	1.028 (+2.8%)		1.251 (0.0%)	7.49×10^{-6} (−25%)
Figure	**F_3 V_s (mL)**	**${}^cK$**	**F_4 V_w (mL)**	**${}^cK_w$**
10.4	0.242 (−3.2%)	1.03×10^{-5} (+3.0%)	0.500 (0.0%)	1.00×10^{-14} (0.0%)
10.5	0.179 (−28%)	1.17×10^{-5} (+17%)	0.500 (0.0%)	1.00×10^{-14} (0.0%)
10.6	0.917 (−8.3%)	1.20×10^{-5} (+20%)	1.249 (+0.1%)	1.00×10^{-14} (0.0%)

factor of ten lower in the Figure 10.2/10.5 case than it is in the Figure 10.1/10.4 case. Thus, the $f = 0$ titration EP is sufficiently difficult to locate in Figure 10.2 that even the Gran Titration method has some trouble in finding it. It is useful to note here that for all the data presented in Table 10.3, the estimates of V_s that can be calculated by averaging the results obtained using F_1 and F_3 are in each case better than either of the individual estimates. The same is very nearly true for the estimates of ${}^cK$ that are obtained from

F_2 and F_3. For the sake of completeness, Table 10.4 presents the best-fit values for $c_{t,b}$ as obtained from the slopes of Gran functions F_1 and F_4.

When titrating a dilute solution of a weak acid which also contains only a small amount of strong acid (as in the Figure 10.2/10.5 case), it has been suggested by Molvaersmyr and Lund (1983) that adding significant additional strong acid will improve the ability of the Gran Titration method to locate the $f = 0$ EP. The Figure 10.3/10.6 case was devised to test this concept. Its value of $A_{T,o}$ was the same as the Figure 10.2/10.5 case, but the amount of strong acid added was four times as great. As is shown in Table 10.3, the errors in the estimates of V_s as determined with both F_1 and F_3 are in fact much lower for the Figure 10.3/10.6 case than for the Figure 10.2/10.5 case. Since the other Gran functions deal with data that are in regions where all strong acidity has been neutralized, the values of the parameters that they determine are less affected by the presence of additional initial strong acid.

Although the preceding paragraph has established that F_1 and F_3 can in fact be made more capable of determining the value of the V_s when additional strong acid is

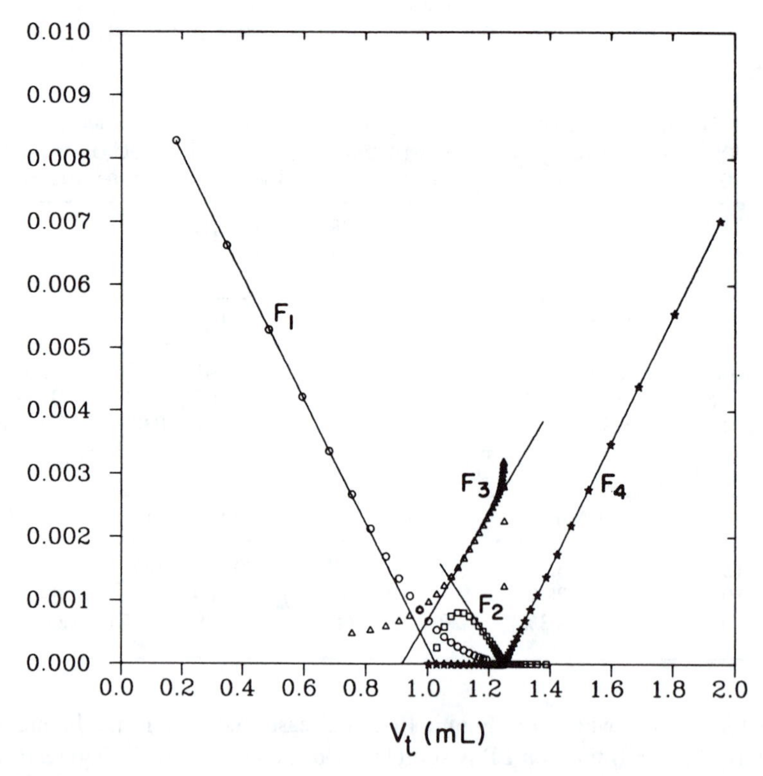

Figure 10.6 Gran plot for titration curve in Figure 10.3. Strong acid/HA ($f = 0$) equivalence point at $V_t = 1.00$ mL. Weak acid/NA ($f = 1$) equivalence point at $V_t = 1.25$ mL. Scale factors applied to F_1, F_2, F_3, and F_4 are $\times 1$, $\times 10^3$, $\times 10^{-7}$, and $\times 10^{-14}$, respectively.

TABLE 10.4 Best-fit values for $c_{t,b}$ from slopes of Gran functions F_1 and F_4. Percent errors as calculated based on the known values used to compute the titration curves are included in parentheses.

Figure	$c_{t,b}$ from F_1	$c_{t,b}$ from F_4
10.4	9.71×10^{-2} (−2.9%)	1.00×10^{-1} (0.0%)
10.5	8.49×10^{-3} (−15%)	1.00×10^{-2} (0.0%)
10.6	9.75×10^{-3} (−2.5%)	9.98×10^{-3} (0.0%)

TABLE 10.5 Values of $(C_A - C_B)_o$ as calculated using Eq. (10.9), that is, $(C_A - C_B)_o = (V_s c_{t,b} - V_a c_{t,a})/V_o$ using V_s from Gran functions F_1 and F_3.

Figure	$(C_A - C_B)_o$ from F_1	$(C_A - C_B)_o$ from F_3
10.4	$[(0.259)(0.1) - (0.1)(.250)]/25$ $= 3.60 \times 10^{-5}$ eq/L	$[(0.242)(0.1) - (0.1)(0.250)]/25$ $= -3.20 \times 10^{-5}$ eq/L
10.5	$[(0.315)(0.01) - (0.01)(.250)]/25$ $= 2.60 \times 10^{-4}$ eq/L	$[(0.179)(0.01) - (0.01)(0.250)]/25$ $= -2.84 \times 10^{-4}$ eq/L
10.6	$[(1.028)(0.01) - (0.01)(1.00)]/25$ $= 1.12 \times 10^{-5}$ eq/L	$[0.917)(0.01) - (0.01)(1.00)]/25$ $= -3.32 \times 10^{-5}$ eq/L

present in a dilute solution of HA, the real question is, however, whether adding more strong acid will allow a better determination of $(C_A - C_B)_o$. In other words, since the values V_s are different for the Figure 10.2/10.5 and 10.3/10.6 cases, we should perhaps not be comparing the ability of the Gran method to determine their different values, but rather its ability to measure the single underlying property of the initial solution that is related to the values of V_s, that is, $(C_A - C_B)_o$. The values of $(C_A - C_B)_o$ for the Figure 10.2/10.5 and 10.3/10.6 cases were therefore estimated by Eq. (10.9). The results are presented in Table 10.5. Since the results obtained with both F_1 and F_3 were closer to the true value (zero) for the Figure 10.3/10.6 case than for the Figure 10.2/10.5 case, we can conclude that adding more strong acid may indeed improve the determination of the initial amount of strong acid in a solution.

10.3 APPLICATION OF THE GRAN TITRATION METHOD TO A SYSTEM CONTAINING SOME WEAK BASE, AND POSSIBLY ALSO SOME STRONG BASE

Section 10.2 provided a detailed analysis of how to use the Gran Titration method in the determination of weak and strong acids using an alkalimetric titration. There will be situations, however, when one desires to examine solutions that contain a weak base, and possibly also some strong base. The approach that would be used under those circumstances would be exactly analogous to the approach presented in Section 10.2. In that case, however, since the titration would be an acidimetric one, the first PBE that would be used to create the first acidimetric Gran function would be the NaA EP. Also, provision would be made for adding some strong base to the initial solution so that one could be sure that the titration would be starting with an f value that is greater than 1. If we refer to the four acidimetric Gran functions as F_5 through F_8, it may be pointed out that F_5 would be applied in the same region of the titration curve as F_4, F_6 would be applied in the same region as F_3, F_7 would be applied in the same region as F_2, and F_8 would be applied in the same region as F_1.

Problems illustrating the principles of this chapter are presented by Pankow (1992).

10.4 REFERENCES

MOLVAERSMYR, K., and W. LUND. 1983. Acids and Bases in Fresh-Waters. Interpretation of Results from Gran Plots. *Water Research*, vol. 17, 303–307.

PANKOW, J. F. 1992. *Aquatic Chemistry Problems*, Portland: Titan Press-OR, P.O. Box 91399, Portland, Oregon 97291-1399.

PART III

Mineral/Solution Chemistry

11

Solubility Behaviors of Simple Mineral Salts, and Metal Oxides, Hydroxides, and Oxyhydroxides

11.1 INTRODUCTION

The dissolution of some solids can involve the production of aqueous species that have no charge, and the dissolution of other solids can lead to the production of charged (i.e., ionic) species. The dissolution of the organic compound glucose is an example of the first case:

$$C_6H_{12}O_{6(s)} = C_6H_{12}O_{6(aq)} \qquad K = \frac{\{C_6H_{12}O_{6(aq)}\}}{\{C_6H_{12}O_{6(s)}\}}. \tag{11.1}$$

The subscript "aq" on the RHS of equilibrium (11.1) is used to emphasize that although the chemical formula is the same for both the dissolved and solid species, the species on the RHS is, in fact, dissolved.

A natural inorganic mineral that predominantly forms a non-charged species when it dissolves is $SiO_{2(s)}$ (quartz), that is,

$$SiO_{2(s)} + 2H_2O = Si(OH)_4 \qquad K = \frac{\{Si(OH)_4\}}{\{SiO_{2(s)}\}}. \tag{11.2}$$

As an example of a dissolution leading to charged species, consider the dissolution of solid sodium chloride ($NaCl_{(s)}$, halite):

$$NaCl_{(s)} = Na^+ + Cl^- \qquad K_{s0} = \frac{\{Na^+\}\{Cl^-\}}{\{NaCl_{(s)}\}}. \tag{11.3}$$

The symbolism behind the subscript "s0" (s-zero) will be discussed shortly.

The dissolution of glucose and quartz both lead to a single dissolved species. Therefore, at a given T and P and assuming that the activities of the solid phases are both unity ($X = 1$), at equilibrium there will be only one value for $\{C_6H_{12}O_6\}$ and one value for $\{Si(OH)_4\}$. However, for solids like $NaCl_{(s)}$ which *dissociate* into ions when they dissolve, taking $X = 1$ again for the solid, there can be an infinite number of *combinations* of ion activities that satisfy the equilibrium solubility expression: only the *product* of the activities must equal the equilibrium constant.

Equilibria (11.1)–(11.3) explicitly consider the activity of the solid phase in the corresponding solubility equilibrium constants. Most readers of this text probably first learned how to solve simple speciation problems involving the dissolution of ionic solids in a high school or college chemistry class. At that time, the activity of the solid phase was very likely not considered. Rather, the solubility expression that was used probably just involved the product of the activities (or concentrations) of the various *dissolved* species. We can now conclude that those earlier treatments of dissolution equilibria were implicitly assuming that when a given solid phase was present, its mole fraction activity was equal to 1.0. In this chapter and also in Chapters 12–15, we will make the same assumption.[1] For convenience, we will also assume that *solution phase* activity corrections may be neglected. As usual, the equations that are derived using infinite dilution constants may also be used with constant ionic medium constants if that type of condition applies.

11.2 UNDERSATURATION, SATURATION, AND SUPERSATURATION

When the acid HA dissociates to form A^- and H^+, we know that at equilibrium, the quantity $Q = \{H^+\}\{A^-\}/\{HA\}$ will be exactly equal to the acidity constant K for HA. If $Q < K$, then some HA will tend to dissociate to form more H^+ and A^- until $Q = K$. If $Q > K$, then some H^+ will tend to combine with an equal amount of A^- to form more HA until $Q = K$.

The situation is very analogous for equilibria describing the dissolution of solids, that is, for reactions like those in Eqs. (11.1)–(11.3). For a given solid, we describe: 1) $Q < K$ as being a state of *undersaturation*; 2) $Q = K$ as a state of *saturation*; and 3) $Q > K$ as a state of *supersaturation*. When a solution is undersaturated, it can "hold" additional dissolved solid; if some solid is present, then some of it will tend to dissolve; if no solid is present, none will form. When a solution is saturated, it can hold no additional dissolved solid; if some solid is present, then none of it will dissolve; if no solid is present, none will form. When a solution is supersaturated, it is holding more than it is capable of holding at equilibrium; if some solid is present, then more of it will tend to form; if no solid is present, some will tend to form. These results are summarized in Table 11.1. If we do know that $Q = K$, we do not necessarily know that solid is present.

11.3 SOLUBILITY OF SIMPLE MINERAL SALTS

$NaCl_{(s)}$ is a classic example of a simple mineral salt. When salts like $NaCl_{(s)}$ dissolve in water, the ions in the solid are simply transferred from the solid to the dissolved phase where they are hydrated (i.e., solvated) by the water molecules. Neither the cation nor

[1] A detailed consideration of the effects of solid-phase activities that are not 1.0 is provided in Chapter 16.

TABLE 11.1 Implications concerning the relative values of Q and K for a dissolution reaction. Eq. (11.3) is used as an example. Assumptions: no solution phase activity corrections are needed; and when present, the solid is pure (mole fraction $X = 1$).

Condition	State	Implications
$[Na^+][Cl^-] < K_{s0}$	Undersaturated	If solid is present, some will tend to dissolve; if solid is not present, it will not form.
$[Na^+][Cl^-] = K_{s0}$	Saturated	If solid is present, no more will dissolve; if solid is not present, it will not form.
$[Na^+][Cl^-] > K_{s0}$	Supersaturated	If solid is present, more will tend to form; if solid is not present, some will tend to form.

the anion enter into any significant interactions with each other, or with other dissolved species. For example, for $NaCl_{(s)}$, other than activity correction effects, neither Na^+ nor Cl^- interact to any significant degree with each other, H^+, or with OH^-.

Let us consider the dissolution of two more salts, that is, $CaSO_4{\cdot}2H_2O_{(s)}$ (gypsum) and $PbCl_{2(s)}$. (The "$\cdot 2H_2O$" in the gypsum formula denotes the fact that there are two mols of water per formula weight in the gypsum crystal.) These two solids dissolve according to

$$CaSO_4{\cdot}2H_2O_{(s)} = Ca^{2+} + SO_4^{2-} + 2H_2O \qquad K_{s0} = \frac{\{Ca^{2+}\}\{SO_4^{2-}\}}{\{CaSO_4{\cdot}2H_2O_{(s)}\}} \tag{11.4}$$

$$PbCl_{2(s)} = Pb^{2+} + 2Cl^- \qquad K_{s0} = \frac{\{Pb^{2+}\}\{Cl^-\}^2}{\{PbCl_{2(s)}\}}. \tag{11.5}$$

The dissolution of $CaSO_{4(s)}$ and $PbCl_{2(s)}$ will not always be as simple as is typically observed for $NaCl_{(s)}$. For example, SO_4^{2-} will always react with some H^+ (or with some H_2O) to form some HSO_4^-, and Pb^{2+} will always react with OH^- (or H_2O) to form some $PbOH^+$. Nevertheless, there are pH conditions in which the dissolution of these two solids will essentially be "simple" inasmuch as the extents of those reactions will be limited.

Based on Eq. (11.4), when $CaSO_4{\cdot}2H_2O_{(s)}$ is present and at equilibrium with the solution phase, neglecting solution phase activity corrections, even when some HSO_4^- forms we can always write

$$\log[Ca^{2+}] = \log K_{s0} - \log[SO_4^{2-}]. \tag{11.6}$$

Based on Eq. (11.5), under similar conditions for $PbCl_{2(s)}$, even when some $PbOH^+$ forms we can always write

$$\log[Pb^{2+}] = \log K_{s0} - 2\log[Cl^-]. \tag{11.7}$$

The K_{s0} in Eq. (11.6) is of course the solubility product for $CaSO_4{\cdot}2H_2O_{(s)}$, and the K_{s0} in Eq. (11.7) is the solubility product for $PbCl_{2(s)}$. We could place a special superscript or subscript on the K_{s0} for each of these solids to identify them with their respective solids, but that should not be necessary; the reader will be able to infer from each individual context the identity of the K_{s0} that is intended.

By Eq. (11.6), we conclude that when $CaSO_4{\cdot}2H_2O_{(s)}$ is present and at equilibrium,

$$d\log[Ca^{2+}]/d\log[SO_4^{2-}] = -1. \tag{11.8}$$

By Eq. (11.7), when $PbCl_{2(s)}$ is present and at equilibrium,

$$d\log[Pb^{2+}]/d\log[Cl^-] = -2. \tag{11.9}$$

Thus, if M^{z+} is taken to refer to a general metal ion of charge z+, for $CaSO_4{\cdot}2H_2O_{(s)}$, the $\log[M^{z+}]$ vs. log[anion] plot will have a slope of –1. For $PbCl_{2(s)}$, the corresponding slope will be –2. For the general solid $M_x(\text{anion})_{y(s)}$,

$$\log[M^{z+}] = (1/x)(\log K_{s0} - y\log[\text{anion}]) \tag{11.10}$$

and

$$d(\log[M^{z+}])/d\log[\text{anion}] = -y/x. \tag{11.11}$$

For $CaSO_4{\cdot}2H_2O_{(s)}$, $-y/x = -2/2 = -1$. For $PbCl_{2(s)}$, $-y/x = -2/1 = -2$. For $Na_2SO_4{\cdot}10H_2O$ (mirabillite), $-y/x = -1/2$. A log concentration vs. log[anion] plot for a variety of solids may be found in Figure 11.1 (see also Figure 5.1 of Stumm and Morgan (1981)). The K_{s0} values for the various solids considered in Figure 11.1 are presented in Table 11.2 on page 224.

For any 1:1 solid (i.e., one cation for each anion) such as $CaSO_4{\cdot}2H_2O_{(s)}$, *when the dissolution is simple*, the speciation that results when the dissolution ocurrs into initially-pure water will correspond to equal concentrations of the cation and the anion. Indeed, each cation that enters the solution is accompanied by one anion. For the specific case of $CaSO_4{\cdot}2H_2O_{(s)}$ equilibrated with initially-pure water, neglecting the small amount of hydrolysis of SO_4^{2-} to HSO_4^-, at equilibrium we can write the MBE as

$$[Ca^{2+}] = [SO_4^{2-}] \qquad \text{MBE} \tag{11.12}$$

and so of course we also have

$$\log[Ca^{2+}] = \log[SO_4^{2-}]. \tag{11.13}$$

Thus, for the simple dissolution of a 1:1 solid, the speciation at equilibrium will be given by the intersection of the $\log[M^{z+}]$ line with the log[anion] line.

We recall from Chapter 6 that when a certain amount of the acid HA is added to pure water, the resulting equilibrium composition will be the one that satisfies the PBE for that solution. The fundamental governing role of the PBE in the HA system is analogous to the role played here by the MBE (Eq. (11.12)) in governing the simple, equilibrium dissolution of $CaSO_4{\cdot}2H_2O_{(s)}$. As a result, significant similarities underlie Figures 6.2 and 11.1. For example, a vertical line drawn in Figure 6.2 through the intersection of the $\log[H^+]$ and $\log[A^-]$ lines very nearly gives the speciation of a solution of HA. Points

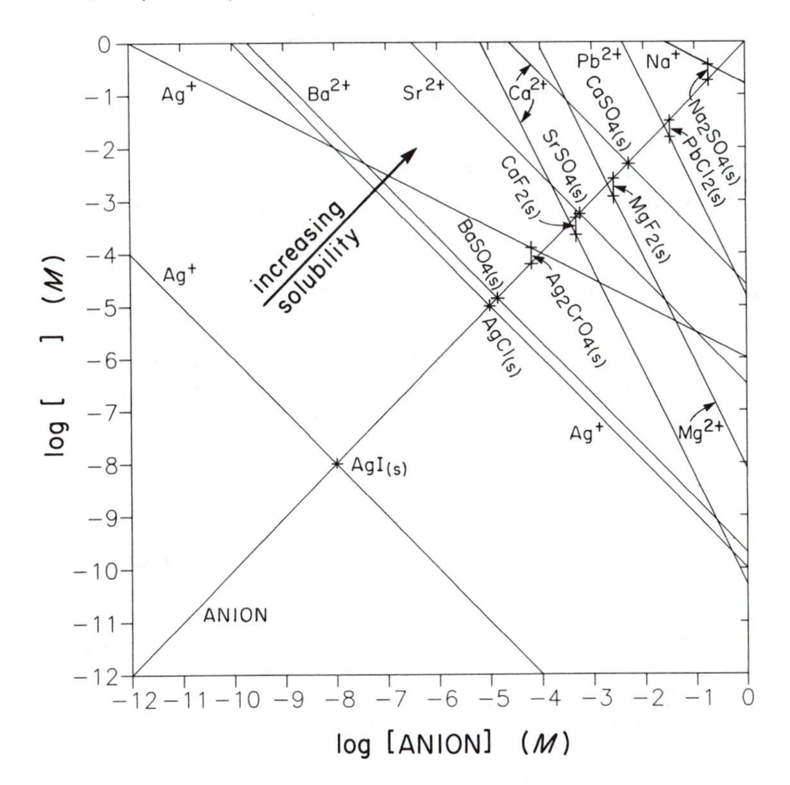

Figure 11.1 Solubility of various mineral salts as a function of the anion concentration at 25°C/1 atm. Crosses and vertical lines with crosses mark the composition of the system satisfying the mass balance equation governing the metal and the anion when the solid is equilibrated with initially-pure water. Activity corrections neglected. (Adapted from Stumm and Morgan (1981).)

to the right of that line can be accessed by adding some strong base to an HA solution, much like points to the right of the Eq. (11.13) intersection can be accessed by adding some soluble salt that contains sulfate (but not calcium) to the $CaSO_4 \cdot 2H_2O_{(s)}$/water equilibrium system. Similarly, points to the left of the vertical line giving the equilibrium speciation for the solution of HA in Figure 6.2 can be accessed by adding strong acid to an HA solution, much like points to the left of the Eq. (11.13) intersection can be accessed by adding some soluble salt that contains some calcium (but not sulfate) to the $CaSO_4 \cdot 2H_2O_{(s)}$/water equilibrium system.

For the simple dissolution of any 1:2 solid (one cation for each two anions) into initially-pure water, the equilibrium saturation speciation that results will be characterized by an anion concentration that is twice the dissolved metal ion concentration. For $PbCl_{2(s)}$, assuming that there is no significant formation of $PbOH^+$, we will have

$$[Cl^-] = 2[Pb^{2+}] \qquad \text{MBE} \tag{11.14}$$

or

$$\log[Pb^{2+}] = \log[Cl^-] - \log 2 \tag{11.15}$$

$$= \log[Cl^-] - 0.301. \tag{11.16}$$

Thus, when a 1:2 mineral salt is dissolved in pure water and the dissolution is simple, the equilibrum composition may be located on Figure 11.1 as the speciation for which the $\log[M^{z+}]$ line is 0.301 log units below the log[anion] line. When a 1:3 mineral salt is dissolved in pure water and the dissolution is simple, the equilibrium speciation will be that for which the $\log[M^{z+}]$ line is $\log 3 = 0.477$ units below the log[anion] line. For a 2:1 salt, say $Na_2SO_4 \cdot 10H_2O_{(s)}$, the equilibrium speciation will be that for which the $\log[M^{z+}]$ line is 0.301 log units *above* the log[anion] line. For all of the solids considered in Figure 11.1, the appropriate MBE is used to mark the equilibrium composition for the specific case when solid is equilibrated with initially-pure water.

TABLE 11.2 K_{s0} values at 25°C/1 atm for the solids in Figure 11.1. All constants pertain to infinite dilution.

Dissolution Equilibrium	$\log K_{s0}$
$Na_2SO_4 \cdot 10H_2O$ (mirabillite) $= 2Na^+ + SO_4^{2-} + 10H_2O$ $\log[Na^+] = 0.5(-1.6 - \log[SO_4^{2-}])$	−1.6
$PbCl_{2(s)} = Pb^{2+} + 2Cl^-$ $\log[Pb^{2+}] = -4.8 - 2\log[Cl^-]$	−4.8
$CaSO_4 \cdot 2H_2O_{(s)} = Ca^{2+} + SO_4^{2-} + 2H_2O$ $\log[Ca^{2+}] = -4.6 - \log[SO_4^{2-}]$	−4.6
$MgF_{2(s)} = Mg^{2+} + 2F^-$ $\log[Mg^{2+}] = -8.1 - 2\log[F^-]$	−8.1
$SrSO_{4(s)} = Sr^{2+} + SO_4^{2-}$ $\log[Sr^{2+}] = -6.5 - \log[SO_4^{2-}]$	−6.5
$Ag_2CrO_{4(s)} = 2Ag^+ + CrO_4^{2-}$ $\log[Ag^+] = 0.5(-12.0 - \log[CrO_4^{2-}])$	−12.0
$BaSO_{4(s)} = Ba^{2+} + SO_4^{2-}$ $\log[Ba^{2+}] = -9.7 - \log[SO_4^{2-}]$	−9.7
$AgCl_{(s)} = Ag^+ + Cl^-$ $\log[Ag^+] = -10.0 - \log[Cl^-]$	−10.0
$AgI_{(s)} = Ag^+ + I^-$ $\log[Ag^+] = -16.0 - \log[I^-]$	−16.0
$CaF_{2(s)}$(fluorite) $= Ca^{2+} + 2F^-$ $\log[Ca^{2+}] = -10.3 - 2\log[F^-]$	−10.3

11.4 NOMENCLATURE FOR SOLUBILITY EQUILIBRIUM CONSTANTS

11.4.1 K_{s0} Solubility Constants

In beginning texts, the equilibrium constants for dissolution reactions that produce dissolved ions are usually referred to as "K_{sp}" values. The subscript "sp" is an abbreviation for the words "solubility product." The word "product" refers to the product of ion activities that appears in the expression for the solubility equilibrium constant. While most of the solubility constants that we will be dealing with in this text are indeed solubility products, it will be useful to have available more than one type of symbol so that we can distinguish between the different types of dissolution constants.

One of the types of symbols used for solubility products is the "K_{s0}" encountered above. In this symbol, the s in the subscript refers again to "solubility." The zero, on the other hand, makes the statement that the constant refers to the specific solubility product for which the identities of the ions in solution are exactly the same as in the solid. For example, in Eq. (11.4), the equilibrium constant for the dissolution of $CaSO_4 \cdot 2H_2O_{(s)}$ is a K_{s0} because the dissolved ions in equilibrium with the solid are Ca^{2+} and SO_4^{2-}. Similarly, for Eq. (11.5), the equilibrium constant for the dissolution of $PbCl_{2(s)}$ is a K_{s0} because the aqueous ions are Pb^{2+} and Cl^-.

11.4.2 K_{s1} Solubility Constants

Some readers might now be wondering where Section 11.4 is leading because, when you dissolve a solid (e.g., $CaSO_4{\cdot}2H_2O_{(s)}$), what could you get initially besides the constituent ions, anyway? Moreover, if you know the value of the K_{s0} for the solid of interest as well as the values of all of the pertinent aqueous phase equilibrium constants (e.g., the acidity constant for HSO_4^-), why would you *need* any type of dissolution equilibrium constant other than the K_{s0} for that solid?

We do not, in fact, *need* any other type of constant. However, that admission still allows that other types of dissolution equilibrium constants might be *more convenient* in certain situations than a K_{s0} constant. For example, let us consider the dissolution of $Fe(OH)_{2(s)}$. The dissolution process corresponding to the K_{s0} for that solid is

$$Fe(OH)_{2(s)} = Fe^{2+} + 2OH^- \qquad K_{s0} = [Fe^{2+}][OH^-]^2. \tag{11.17}$$

Just like H^+, the Fe^{2+} ion has a certain affinity for OH^- ions. Thus, when $Fe(OH)_{2(s)}$ dissolves, we will also often need to consider the reaction of Fe^{2+} with OH^-:

$$Fe^{2+} + OH^- = FeOH^+ \qquad K_{H1} = \frac{[FeOH^+]}{[Fe^{2+}][OH^-]}. \tag{11.18}$$

Naturally, since Eq. (11.18) involves hydroxide ions, the extent to which $FeOH^+$ forms will depend on the pH.

The subscript "H1" is used for the K in Eq. (11.18) because what is described is the combination of a first hydroxide with the Fe^{2+} ion.

Let us say that we are interested in computing the total dissolved iron at a given pH under conditions where Fe^{2+} and $FeOH^+$ are the only important iron species. Using the pH, we could first make a calculation based on the K_{s0} for the concentration of Fe^{2+}. Then, using the pH again and Eq. (11.18), we could compute then add the concentration of $FeOH^+$ that must be in equilibrium with that $[Fe^{2+}]$ at that pH. That, however, is not the only approach available. Indeed, in addition to having K_{s0} for computing $[Fe^{2+}]$, it would be very convenient to have a knowledge of the constant for the equilibrium

$$Fe(OH)_{2(s)} = FeOH^+ + OH^- \qquad K_{s1} = [FeOH^+][OH^-]. \qquad (11.19a)$$

The "1" in the subscript for the constant for Eq. (11.19a) denotes that for each of the Fe^{2+} ions that dissolves, *one* of the anions of the type that is in the solid is associated with the dissolved metal ion. Therefore, when $Fe(OH)_{2(s)}$ is present and at equilibrium, K_{s0} and K_{s1} can be used to directly compute $[Fe^{2+}]$ and $[FeOH^+]$, respectively, as a function of pH.

It is important to note that when a solid such as $Fe(OH)_{2(s)}$ is present and at equilibrium with an aqueous solution, the equilibria corresponding to K_{s0}, K_{H1}, and K_{s1} must *all* be satisfied *simultaneously*. In particular, at equilibrium, the expressions for K_{s0} and K_{H1} will be satisfied. When K_{s0} and K_{H1} are both satisfied, then K_{s1} must also be satisfied since the expression for K_{s1} may be derived from the expressions for K_{s0} and K_{H1}. In other words, Eqs. (11.17) and (11.18) may be combined to obtain Eq. (11.19a). According to the principles discussed in Section 2.8, when two chemical equilibrium reactions are added or subtracted, the equilibrium constant for the net equilibrium reaction will be equal to the product or quotient, respectively, of the two individual K values. Applied here, we have

$$Fe(OH)_{2(s)} = Fe^{2+} + 2OH^- \qquad K_{s0} = [Fe^{2+}][OH^-]^2 \qquad (11.17)$$

$$+ \quad Fe^{2+} + OH^- = FeOH^+ \qquad K_{H1} = \frac{[FeOH^+]}{[Fe^{2+}][OH^-]} \qquad (11.18)$$

$$Fe(OH)_{2(s)} = FeOH^+ + OH^- \qquad K_{s1} = K_{s0}K_{H1} = [FeOH^+][OH^-]. \qquad (11.19b)$$

11.4.3 K_{s2}, K_{s3}, etc. Solubility Constants

If there is a K_{s1} value that is useful for $Fe(OH)_{2(s)}$, are there K_{s2}, K_{s3}, etc. values for this and other solids? The answer is yes. For $Fe(OH)_{2(s)}$, neglecting activity coefficients again, we have

$$Fe(OH)_{2(s)} = Fe(OH)_2^o \qquad K_{s2} = [Fe(OH)_2^o] \qquad (11.20)$$

$$Fe(OH)_{2(s)} + OH^- = Fe(OH)_3^- \qquad K_{s3} = [Fe(OH)_3^-]/[OH^-] \qquad (11.21)$$

For K_{s3} in Eq. (11.21), since there are only two OH^- ions per Fe^{2+} ion in the solid, $Fe(OH)_{2(s)}$ reacts with one more OH^- in order to get the three OH^- in the species $Fe(OH)_3^-$.

Another very important iron solid is $Fe(OH)_{3(s)}$. As will be explained in detail in the chapters on oxidation and reduction, the iron in this solid is in the III oxidation state, while the iron in $Fe(OH)_{2(s)}$ is in the II oxidation state. When oxygen is present at the types of levels found in water equilibrated with the atmosphere, the vast majority of the iron will be present as Fe(III). Fe(II) can only be found at significant levels when the level of oxygen is very low.

For $Fe(OH)_{3(s)}$, neglecting solution phase activity corrections, we can write the important series of dissolution reactions

$$Fe(OH)_{3(s)} = Fe^{3+} + 3OH^- \qquad K_{s0} = [Fe^{3+}][OH^-]^3 \tag{11.22}$$

$$Fe(OH)_{3(s)} = FeOH^{2+} + 2OH^- \qquad K_{s1} = [FeOH^{2+}][OH^-]^2 \tag{11.23}$$

$$Fe(OH)_{3(s)} = Fe(OH)_2^+ + OH^- \qquad K_{s2} = [Fe(OH)_2^+][OH^-] \tag{11.24}$$

$$Fe(OH)_{3(s)} = Fe(OH)_3^o \qquad K_{s3} = [Fe(OH)_3^o] \tag{11.25}$$

$$Fe(OH)_{3(s)} + OH^- = Fe(OH)_4^- \qquad K_{s4} = [Fe(OH)_4^-]/[OH^-]. \tag{11.26}$$

Just as we do not *need* K_{s1} if we know K_{s0} and K_{H1}, for a given hydroxide solid, we do not *need* K_{s2}, K_{s3}, etc. if we know K_{s0} and K_{H2}, K_{H3}, etc. Nevertheless, having them available for computations can be very convenient. As shown in Eq. (11.19b),

$$K_{s1} = K_{s0}K_{H1}. \tag{11.27}$$

When the numbering system for additional combinations between the metal ion M^{Z+} and OH^- is described by

$$Me(OH)_{n-1}^{(z-n+1)+} + OH^- = Me(OH)_n^{(z-n)+} \qquad K_{Hn} \tag{11.28}$$

then it is easy to show by combining the appropriate chemical equilibria that in general

$$K_{sn} = K_{s0}K_{H1}K_{H2}\cdots K_{Hn}. \tag{11.29}$$

11.4.4 $^*K_{s0}$, $^*K_{s1}$, $^*K_{s2}$, $^*K_{s3}$, etc. Solubility Constants

When one uses dissolution constants like K_{s0}, K_{s1}, K_{s2}, K_{s3}, etc. for a given solid, the concentrations of the corresponding metal ions may be expressed in terms of the associated constant and the concentration of the anion. For many solids of environmental interest, those anions are often reasonable bases. The OH^- in $Fe(OH)_{2(s)}$ and $Fe(OH)_{3(s)}$ obviously fits into this group. As a result, it is possible to also express the dissolution process from the perspective of H^+ rather than the perspective of the base. The superscript "*" is used to connote the presence of this acidity on the LHS of a dissolution equilibrium. For example, for the solid $Fe(OH)_{2(s)}$, neglecting activity corrections, we write

$$Fe(OH)_{2(s)} + 2H^+ = Fe^{2+} + 2H_2O \qquad ^*K_{s0} = [Fe^{2+}]/[H^+]^2 \tag{11.30}$$

$$Fe(OH)_{2(s)} + H^+ = FeOH^+ + H_2O \qquad ^*K_{s1} = [FeOH^+]/[H^+] \tag{11.31}$$

$$Fe(OH)_{2(s)} + H_2O = Fe(OH)_3^- + H^+ \qquad ^*K_{s3} = [Fe(OH)_3^-][H^+]. \tag{11.32}$$

A set of $^*K_s$ values can also be written for $Fe(OH)_{3(s)}$. Neglecting activity corrections, we have

$$Fe(OH)_{3(s)} + 3H^+ = Fe^{3+} + 3H_2O \qquad ^*K_{s0} = [Fe^{3+}]/[H^+]^3 \tag{11.33}$$

$$Fe(OH)_{3(s)} + 2H^+ = FeOH^{2+} + 2H_2O \qquad ^*K_{s1} = [FeOH^{2+}]/[H^+]^2 \tag{11.34}$$

$$Fe(OH)_{3(s)} + H^+ = Fe(OH)_2^+ + H_2O \qquad ^*K_{s2} = [Fe(OH)_2^+]/[H^+] \tag{11.35}$$

$$Fe(OH)_{3(s)} + H_2O = Fe(OH)_4^- + H^+ \qquad ^*K_{s4} = [Fe(OH)_4^-][H^+]. \tag{11.36}$$

For Eqs. (11.30), (11.31), and (11.33)–(11.35), the superscript "*" acidity is readily recognizable as the species H^+. For Eqs. (11.32) and (11.36), however, the "acidity" appears as H_2O. This may seem a little confusing, but this is made clear when we realize that the corresponding non-$^*K$ values, that is, Eqs. (11.21) and (11.26), involve OH^- on the LHS, and so H_2O is indeed acidic compared to that species. Since neither Eq. (11.20) nor Eq. (11.25) involve base on either side of the equilibrium, there will be no $^*K_s$ constant for either of those reactions.

It is important to emphasize that the extent of dissolution as well as the nature of the resulting speciation *will not* depend on how we choose to express the solubility constant. Thus, a knowledge of K_{s0} is equivalent to a knowledge of $^*K_{s0}$. For solids in which the anion is OH^- (i.e., hydroxide solids) or O^{2-} (i.e., oxide solids), the K_{s0} and $^*K_{s0}$ will differ by a factor involving K_w to a power involving the number of protons required to "neutralize" the OH^- or O^{2-} units. For the hydroxide $Fe(OH)_{2(s)}$, we have

$$Fe(OH)_{2(s)} = Fe^{2+} + 2OH^- \qquad K_{s0} \tag{11.37}$$

$$+ \quad 2(H^+ + OH^- = H_2O) \qquad K_w^{-2} \tag{11.38}$$

$$Fe(OH)_{2(s)} + 2H^+ = Fe^{2+} + 2H_2O \qquad ^*K_{s0} = K_{s0}/K_w^2. \tag{11.39}$$

For the hydroxide $Fe(OH)_{3(s)}$, we have

$$Fe(OH)_{3(s)} = Fe^{3+} + 3OH^- \qquad K_{s0} \tag{11.40}$$

$$+ \quad 3(H^+ + OH^- = H_2O) \qquad K_w^{-3} \tag{11.41}$$

$$Fe(OH)_{3(s)} + 3H^+ = Fe^{3+} + 3H_2O \qquad ^*K_{s0} = K_{s0}/K_w^3. \tag{11.42}$$

For the oxide $Ag_2O_{(s)}$, we have

$$\tfrac{1}{2}Ag_2O_{(s)} + \tfrac{1}{2}H_2O = Ag^+ + OH^- \qquad K_{s0} \tag{11.43}$$

$$+ \quad (H^+ + OH^- = H_2O) \qquad K_w \tag{11.44}$$

$$\tfrac{1}{2}Ag_2O_{(s)} + H^+ = Ag^+ + \tfrac{1}{2}H_2O \qquad ^*K_{s0} = K_{s0}/K_w. \tag{11.45}$$

In general, for an hydroxide or oxide solid involving a metal ion of charge z+, we have

$$^*K_{s0} = K_{s0}/K_w^z. \qquad (11.46)$$

It may be pointed out that since O^{2-} does not exist to any extent in aqueous solutions but OH^- does, the equilibrium constant for Eq. (11.43) is still considered a K_{s0} even though the dissolved anion is not O^{2-}.

11.5 SOLUBILITY OF METAL OXIDES, HYDROXIDES, AND OXYHYDROXIDES

11.5.1 Solubility of Metal Oxides, Hydroxides, and Oxyhydroxides as Measured by [M^{z+}]

Solids with metal-oxygen interactions are among the least soluble and therefore the most common natural minerals. The majority of these minerals are oxides, hydroxides, oxyhydroxides, carbonates, and sulfates. As aluminum/silicon oxides, clays and aluminosilicates like feldspar can be included in this group.

Many metal ions can form both a hydroxide solid as well as an oxide solid. For example, for Fe(II), Eq. (11.17) describes the dissolution of $Fe(OH)_{2(s)}$:

$$Fe(OH)_{2(s)} = Fe^{2+} + 2OH^- \qquad K_{s0} = [Fe^{2+}][OH^-]^2. \qquad (11.17)$$

In addition, the solid Fe(II) solid $FeO_{(s)}$ is also known, and dissolves according to

$$FeO_{(s)} + H_2O = Fe^{2+} + 2OH^- \qquad K_{s0} = [Fe^{2+}][OH^-]^2. \qquad (11.47)$$

For Fe(III), we have the possibility of a hydroxide as in

$$Fe(OH)_{3(s)} = Fe^{3+} + 3OH^- \qquad K_{s0} = [Fe^{3+}][OH^-]^3 \qquad (11.22)$$

an oxide as in

$$\frac{1}{2}Fe_2O_{3(s)} + \frac{3}{2}H_2O = Fe^{3+} + 3OH^- \qquad K_{s0} = [Fe^{3+}][OH^-]^3 \qquad (11.48)$$

and as a mixed oxyhydroxide as in

$$\alpha\text{-}FeOOH_{(s)}\ \text{(goethite)} + H_2O = Fe^{3+} + 3OH^- \qquad K_{s0} = [Fe^{3+}][OH^-]^3. \qquad (11.49)$$

Since the same number of OH^- are produced for Eqs. (11.22), (11.48), and (11.49), the slopes of the log[Fe^{3+}] lines vs. log[OH^-] will be the same for all three solids. However, since the K_{s0} values for the three solids are different for most T and P conditions, the lines themselves will not lie on top of one another.

Sometimes, there can even be more than one oxide and more than one hydroxide possible for a given metal ion. Indeed, there can be more than one crystalline form in which the metal ion and OH^- can combine, and more than one crystalline form in which the metal ion and O^{2-} can combine. For example, including the amorphous hydroxide, chemists have identified six different $Pb(OH)_{2(s)}$ solids, and two different $PbO_{(s)}$ solids (yellow and red). (When a solid is said to be "amorphous," what is meant is that it is has no regular crystal structure as might be detected by a method such as X-ray diffraction.)

The solubilities of the hydroxide, oxide, and oxyhydroxide of any given metal ion will be different. Usually, for a given metal ion, it is an oxide that has the lowest solubility. If one desires to calculate the composition which a given system would have at *ultimate* equilibrium, then *within* this type of group (i.e. excluding carbonates, etc.), the solubility constants (e.g., $^*K_{s0}$, $^*K_{s1}$, etc.) for the solid with the lowest solubility should be used. However, the time required for some of the low solubility oxides to form can be very great. As a result, for problems that involve: 1) the initial precipitation of a given metal; and 2) time frames of days to weeks, then using the solubility constants of a more soluble, less crystalline metal hydroxide is often advised.

A classic example of the scenario outlined above is provided by the chemistry of Fe(III). After initial precipitation, the $Fe(OH)_{3(s)}$ that forms is amorphous. Since amorphous solids lack crystalline structure, there are fewer cation/anion interactions than there would be in a fully crystalline structure of that stoichiometry. Interactions such as these stabilize a solid, and lower its free energy. Therefore, amorphous $Fe(OH)_{3(s)}$ (denoted (am)$Fe(OH)_{3(s)}$) is more soluble than crystalline $Fe(OH)_{3(s)}$, and even more soluble than either hematite or goethite. However, the conversion of (am)$Fe(OH)_{3(s)}$ to either goethite or hematite can take weeks to months. Therefore, if one is interested in problems involving the initial precipitation of Fe(III) and the subsequent short term equilibrium chemistry of the resulting solid, the use of the solubility constants for (am)$Fe(OH)_{3(s)}$ is preferred over the use of the constants for either hematite or goethite.

As discussed above, for the general dissolution of a metal hydroxide, neglecting activity corrections in the solution phase, we have

$$M(OH)_{z(s)} = M^{z+} + zOH^- \qquad K_{s0} = [M^{z+}][OH^-]^z. \tag{11.50}$$

Thus,

$$\log[M^{z+}] = \log K_{s0} - z\log[OH^-]. \tag{11.51}$$

Since OH^- is the anion here, Eq. (11.51) is analogous to Eq. (11.10). Under similar conditions, for the general dissolution of a metal oxide

$$\frac{1}{x}M_xO_{y(s)} + \frac{z}{2}H_2O = M^{z+} + zOH^- \qquad K_{s0} = [M^{z+}][OH^-]^z \tag{11.52}$$

which again gives Eq. (11.51), albeit with a different value of K_{s0}.

We can thus conclude that plotting $\log[M^{z+}]$ vs. $\log[OH^-]$ for various metal hydroxides, oxides, and also oxyhydroxides *will give a diagram that is similar to Figure 11.1.* However, $\log[OH^-]$ is not as common a solution parameter as is pH. Since the $^*K_{s0}$ equilibrium relationships for hydroxides, oxides, and oxyhydroxides provide the link between $\log[M^{z+}]$ and pH, we begin to see why the "*" superscript equilibrium relationships are useful. For the general hydroxide, we therefore obtain

$$M(OH)_{z(s)} + zH^+ = M^{z+} + zH_2O \qquad ^*K_{s0} = [M^{z+}]/[H^+]^z \tag{11.53}$$

and so

$$\log[M^{z+}] = \log\ ^*K_{s0} + z\ \log[H^+] \tag{11.54}$$

$$= \log\ ^*K_{s0} - z\text{pH}. \tag{11.55}$$

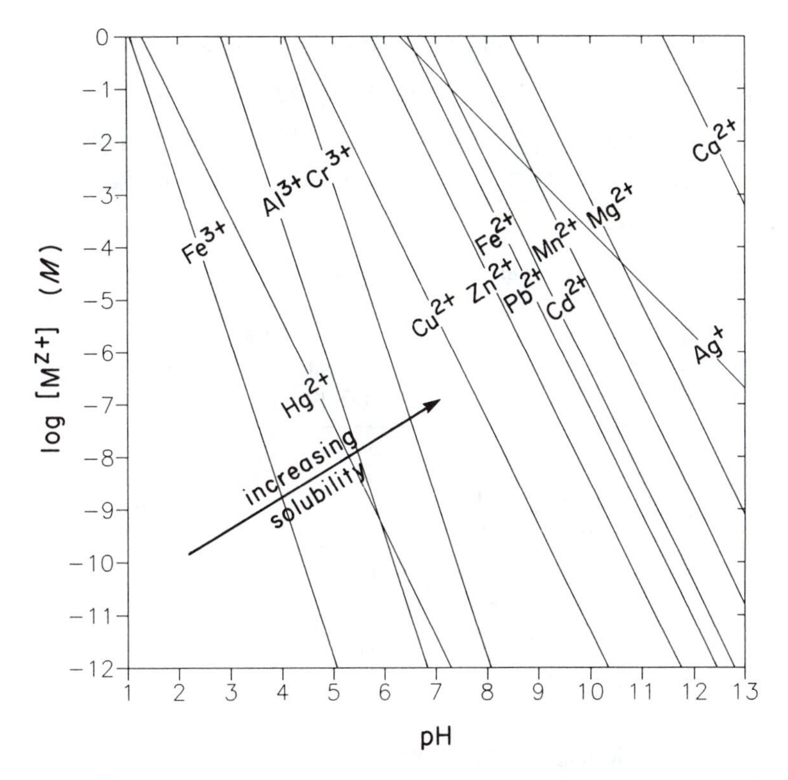

Figure 11.2 Solubility of various metal ions as controlled by their oxides or hydroxides as a function of pH at 25°C/1 atm. Activity corrections neglected (not valid at very low or very high pH). (Adapted from Stumm and Morgan (1981).)

Under similar conditions, for the dissolution of a general metal oxide

$$\frac{1}{x}\mathrm{M_xO_{y(s)}} + z\mathrm{H^+} = \mathrm{M^{z+}} + \frac{z}{2}\mathrm{H_2O} \qquad {}^*K_{s0} = [\mathrm{M^{z+}}]/[\mathrm{H^+}]^z \tag{11.56}$$

which also leads to Eq. (11.55).

Figure 11.2 is a plot of log$[M^{z+}]$ vs. pH for a number of hydroxides and oxides (see also Figure 5.3 of Stumm and Morgan (1981)). Based on Eq. (11.55), when $z = +3$, +2, and +1, the slopes in the log$[M^{z+}]$ vs. pH lines are −3, −2, and −1, respectively. Table 11.3 gives the $^*K_{s0}$ values for the solids included in Figure 11.2.

Based on Figure 11.2, it is obvious that the different solids prescribe widely varying solubilities for their respective M^{z+} ions. One cannot, however, give an order of increasing solubility *as measured by* $[M^{z+}]$ *alone* for all pH; since the log$[M^{z+}]$ lines are characterized by different slopes, the order at one pH will not necessarily be the order at another pH. For the same reason, the order of increasing solubility as measured by $[M^{z+}]$ at a particular pH will not strictly follow the order of increasing $^*K_{s0}$. For pH 7, *as measured by* $[M^{z+}]$ *alone*, the order of increasing solubility is given by:

(am)$Fe(OH)_{3(s)}$ (3.2) < α-$Al(OH)_{3(s)}$ (8.5) < $HgO_{(s)}$ (2.6) < $Cr(OH)_{3(s)}$ (12.2) < $Cu(OH)_{2(s)}$ (8.7) < $Zn(OH)_{2(s)}$ (11.5) < $Fe(OH)_{2(s)}$ (12.9) = $PbO_{(s)}$ (yellow, 12.9) < $Ag_2O_{(s)}$ (6.3) < $Cd(OH)_{2(s)}$ (13.6) < $Mn(OH)_{2(s)}$ (15.2) < $Mg(OH)_{2(s)}$ (16.9) < $Ca(OH)_{2(s)}$ (22.8).

The various $^*K_{so}$ values are given in parentheses.

TABLE 11.3 $^*K_{s0}$ values at 25°C for the solids in Figure 11.2. All constants for infinite dilution, except for $Cr(OH)_{3(s)}$ which is a $^{*c}K_{s0}$ at ionic strength = 0.1*M*.

Dissolution Equilibrium	$\log {}^*K_{s0}$
(am)$Fe(OH)_{3(s)} + 3H^+ = Fe^{3+} + 3H_2O$ $\log[Fe^{3+}] = 3.2 - 3pH$	3.2
α-$Al(OH)_{3(s)} + 3H^+ = Al^{3+} + 3H_2O$ $\log[Al^{3+}] = 3.2 - 3pH$	8.5
$Cr(OH)_{3(s)} + 3H^+ = Cr^{3+} + 3H_2O$ $\log[Cr^{3+}] = 12.2 - 3pH$	12.2
$HgO_{(s)} + 2H^+ = Hg^{2+} + H_2O$ $\log[Hg^{2+}] = 2.6 - 2pH$	2.6
$Cu(OH)_{2(s)} + 2H^+ = Cu^{2+} + 2H_2O$ $\log[Cu^{2+}] = 8.7 - 2pH$	8.7
$Zn(OH)_{2(s)} + 2H^+ = Zn^{2+} + 2H_2O$ $\log[Zn^{2+}] = 11.5 - 2pH$	11.5
$Fe(OH)_{2(s)} + 2H^+ = Fe^{2+} + 2H_2O$ $\log[Fe^{2+}] = 12.9 - 2pH$	12.9
$PbO_{(s)}$(yellow) $+ 2H^+ = Pb^{2+} + 2H_2O$ $\log[Pb^{2+}] = 12.9 - 2pH$	12.9
$Cd(OH)_{2(s)} + 2H^+ = Cd^{2+} + 2H_2O$ $\log[Cd^{2+}] = 13.6 - 2pH$	13.6
$Mn(OH)_{2(s)} + 2H^+ = Mn^{2+} + 2H_2O$ $\log[Mn^{2+}] = 15.2 - 2pH$	15.2
$Mg(OH)_{2(s)} + 2H^+ = Mg^{2+} + 2H_2O$ $\log[Mg^{2+}] = 16.9 - 2pH$	16.9
$Ca(OH)_{2(s)} + 2H^+ = Ca^{2+} + 2H_2O$ $\log[Ca^{2+}] = 22.8 - 2pH$	22.8
$\frac{1}{2}Ag_2O_{(s)} + H^+ = Ag^+ + \frac{1}{2}H_2O$ $\log[Ag^+] = 6.3 - pH$	6.3

Although there are some exceptions, in general the solubilities of the oxides and hydroxides in the above group increase as z decreases. Also, it may be noted that *within a particular valence class*, e.g., $z = +3, +2$, or $+1$, the order of increasing solubility as measured by $[M^{z+}]$ *will* follow the order of increasing $^*K_{s0}$ at *any* pH.

11.5.2 Solubility of Metal Oxides, Hydroxides, and Oxyhydroxides as Measured by the Total Dissolved Metal

When considering the solubility of metal oxides, hydroxides, and oxyhydroxides, an understanding of the type of chemistry depicted in Figure 11.2 is far from sufficient. Indeed, as we have already seen in Sections 11.3.2–11.3.4, metal ions often combine with OH^- to form dissolved metal-hydroxo complexes like $FeOH^+$. Thus, under most conditions, the *total* solubility of a given metal ion cannot be considered to be given by just $[M^{z+}]$, rather, it must include several different dissolved hydroxo complexes. When there is equilibrium between an aqueous solution and any given oxide, hydroxide, or oxyhydroxide solid, there are in fact conditions under which the *total* solubility of the metal will be many orders of magnitude larger than is just given by $[M^{z+}]$.

On the basis of this perspective, let us consider the solubility of three solids, $(am)Fe(OH)_{3(s)}$, $(am)Al(OH)_{3(s)}$, and red $PbO_{(s)}$. For all three solids, the individual ion concentrations will be given by the values of $^*K_{s0}$, $^*K_{s1}$, $^*K_{s2}$, etc. For $(am)Fe(OH)_{3(s)}$ then, the total dissolved iron (Fe_T) will be given by

$$Fe_T = [Fe^{3+}] + [FeOH^{2+}] + [Fe(OH)_2^+] + [Fe(OH)_3^o] + [Fe(OH)_4^-] \qquad (11.57)$$

$$= {}^*K_{s0}[H^+]^3 + {}^*K_{s1}[H^+]^2 + {}^*K_{s2}[H^+] + K_{s3} + {}^*K_{s4}/[H^+]. \qquad (11.58)$$

Equations (11.57) and (11.58) neglect polymeric species like $Fe_2(OH)_2^{4+}$. As is discussed in Chapter 18, such species are important only at fairly high concentrations of total dissolved metal.

Equation (11.58) will also give Al_T as a function of $[H^+]$, though at any given T and P the values of the various equilibrium constants will take on different values for these two $z = +3$ solids. For $PbO_{(s)}$, we have

$$Pb_T = [Pb^{2+}] + [PbOH^+] + [Pb(OH)_2^o] + [Pb(OH)_3^-] \qquad (11.59)$$

$$= {}^*K_{s0}[H^+]^2 + {}^*K_{s1}[H^+] + {}^*K_{s2} + K_{s3}/[H^+] \qquad (11.60)$$

Constants for the above three solids are given in Table 11.4. The form of $Al(OH)_{3(s)}$ considered in Table 11.4 is the amorphous form ($\log {}^*K_{s0} = 10.8$). The form of that solid considered in Figure 11.2 and Table 11.3 is a more crystalline form ($\log {}^*K_{s0} = 8.5$). The $^*K_{s0}$ values for these two $Al(OH)_{3(s)}$ solids provide a good example of the greater solubility of an amorphous hydroxide relative to a more crystalline hydroxide of the same stoichiometry. In a similar manner, the two $PbO_{(s)}$ solids in Tables 11.3 and 11.4 provide an example of how two crystalline solids with the same stoichiometric formula can display different solubilities ($\log {}^*K_{s0} = 12.9$ and 12.7, respectively).

Figures 11.3–11.7 were prepared using the equilibrium constants in Table 11.4. The $\log Fe_T$ and $\log Pb_T$ lines were calculated using Eqs. (11.58) and (11.60). The $\log Al_T$

TABLE 11.4 Dissolution equilibrium constant values at 25°C/1 atm for the solids in Figures 11.3–11.7. All constants for infinite dilution.

Dissolution Equilibrium	log K
$(am)Fe(OH)_{3(s)} + 3H^+ = Fe^{3+} + 3H_2O$ $\log[Fe^{3+}] = 3.2 - 3pH$	$\log\ {}^*K_{s0} = 3.2$
$(am)Fe(OH)_{3(s)} + 2H^+ = FeOH^{2+} + 2H_2O$ $\log[FeOH^{2+}] = 1.0 - 2pH$	$\log\ {}^*K_{s1} = 1.0$
$(am)Fe(OH)_{3(s)} + H^+ = Fe(OH)_2^+ + H_2O$ $\log[Fe(OH)_2^+] = -2.5 - pH$	$\log\ {}^*K_{s2} = -2.5$
$(am)Fe(OH)_{3(s)} = Fe(OH)_3^o$ $\log[Fe(OH)_3^o] < -12.0$	$\log\ K_{s3} < -12.0$
$(am)Fe(OH)_{3(s)} + H_2O = Fe(OH)_4^- + H^+$ $\log[Fe(OH)_4^-] = -18.4 + pH$	$\log\ {}^*K_{s4} = -18.4$
$(am)Al(OH)_{3(s)} + 3H^+ = Al^{3+} + 3H_2O$ $\log[Al^{3+}] = 10.8 - 3pH$	$\log\ {}^*K_{s0} = 10.8$
$(am)Al(OH)_{3(s)} + 2H^+ = AlOH^{2+} + 2H_2O$ $\log[AlOH^{2+}] = 5.8 - 2pH$	$\log\ {}^*K_{s1} = 5.8$
$(am)Al(OH)_{3(s)} + H^+ = Al(OH)_2^+ + H_2O$ $\log[Al(OH)_2^+] = 1.5 - pH$	$\log\ {}^*K_{s2} = 1.5$
$(am)Al(OH)_{3(s)} = Al(OH)_3^o$ $\log[Al(OH)_3^o] = -4.2$	$\log\ K_{s3} = -4.2$
$(am)Al(OH)_{3(s)} + H_2O = Al(OH)_4^- + H^+$ $\log[Al(OH)_4^-] = -12.2 + pH$	$\log\ {}^*K_{s4} = -12.2$
$PbO_{(s)}(red) + 2H^+ = Pb^{2+} + H_2O$ $\log[Pb^{2+}] = 12.7 - 2pH$	$\log\ {}^*K_{s0} = 12.7$
$PbO_{(s)}(red) + H^+ = PbOH^+ + H_2O$ $\log[PbOH^+] = 5.0 - pH$	$\log\ {}^*K_{s1} = 5.0$
$PbO_{(s)}(red) + H_2O = Pb(OH)_2^o$ $\log[Pb(OH)_2^o] = -4.4$	$\log\ K_{s2} = -4.4$
$PbO_{(s)}(red) + 2H_2O = Pb(OH)_3^- + H^+$ $\log[Pb(OH)_3^-] = -15.4 + pH$	$\log\ {}^*K_{s3} = -15.4$

line was calculated using an Al analog of Eq. (11.58). In Figure 11.3, for $(am)Fe(OH)_{3(s)}$, the slopes of the lines for $\log[Fe^{3+}]$, $\log[FeOH^{2+}]$, $\log[Fe(OH)_2^+]$ and $\log[Fe(OH)_4^-]$ are −3, −2, −1, and +1. These slopes are equal to the numbers of OH^- groups that the corresponding species is "away from the solid," which has +3 OH^- groups. The $\log[Fe(OH)_3^o]$ line was not drawn since the value of K_{s3} for that solid is not well known.

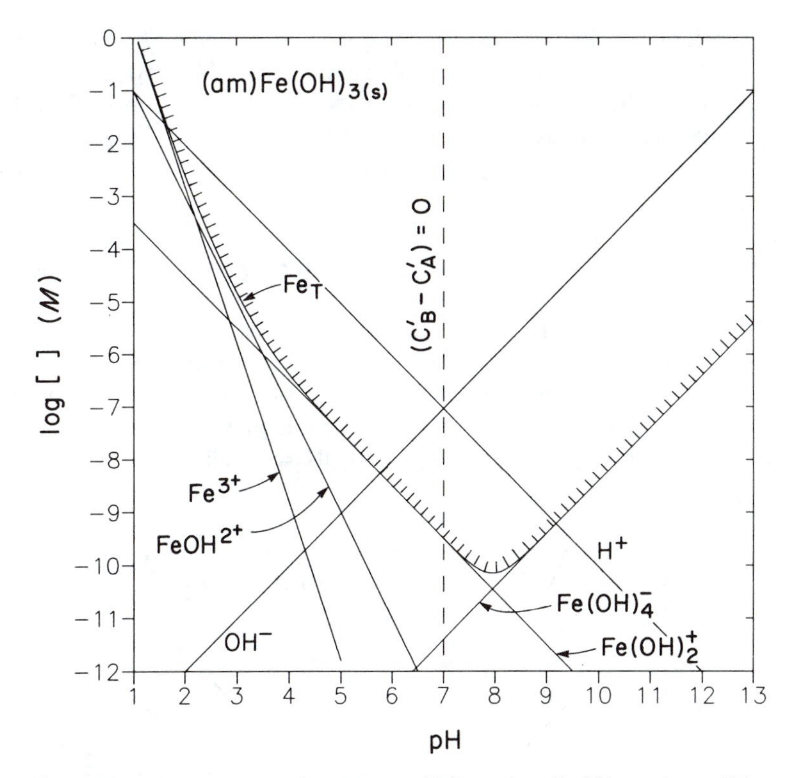

Figure 11.3 Log concentration (M) vs. pH for various Fe(III) species and Fe_T in equilibrium with (am)$Fe(OH)_{3(s)}$ at 25°C/1 atm. Dashed vertical line gives speciation for dissolution into initially-pure water (i.e., for $(C'_B - C'_A) = 0$). Polynuclear species like $Fe_2(OH)_2^{4+}$ have been neglected. Activity corrections neglected (not valid at very low or very high pH).

As is indicated in Table 11.4, however, it is known that $K_{s3} < 10^{-12}$. As a result, based on the other lines in Figure 11.3, we can conclude that: 1) there will be no pH value at which $[Fe(OH)_3^o]$ makes a significant contribution to Fe_T; and 2) one may always safely neglect $Fe(OH)_3^o$ when making speciation calculations concerning Fe(III). If the $\log[Fe(OH)_3^o]$ line could have been drawn, the slope would have been zero since $Fe(OH)_3^o$ is zero OH^- groups away from the solid. The functionality of the $\log Fe_T$ vs. pH curve indicates that (am)$Fe(OH)_{3(s)}$ becomes increasingly soluble as the pH decreases, and also as the pH increases. This type of behavior is exhibited by most metal oxides, hydroxides, and oxyhydroxides.

Figure 11.4 is the solubility diagram for the (am)$Al(OH)_{3(s)}$. In a manner that is completely analogous to the behavior observed for (am)$Fe(OH)_{3(s)}$, the lines for $\log[Al^{3+}]$, $\log[AlOH^{2+}]$, $\log[Al(OH)_2^+]$, $\log[Al(OH)_3^o]$, and $\log[Al(OH)_4^-]$ have slopes of −3, −2, −1, 0, and +1. Also, as with (am)$Fe(OH)_{3(s)}$, we see that the solubility of the solid (am)$Al(OH)_{3(s)}$ increases at both low and high pH. In contrast to the (am)$Fe(OH)_{3(s)}$

case, however, for (am)$Al(OH)_{3(s)}$ the neutral species controlled by K_{s3} is important over a fairly wide pH range. In fact, at neutral pH values, $Al_T \simeq [Al(OH)_3^o]$.

Figure 11.5 is the solubility diagram for red $PbO_{(s)}$. The slopes of the lines for $\log[Pb^{2+}]$, $\log[PbOH^+]$, $\log[Pb(OH)_2^o]$, and $\log[Pb(OH)_3^-]$ are −2, −1, 0, and +1. Since this solid is not a hydroxide, we can conclude that the doubly basic O^{2-} group in the solid "counts" for two singly basic OH^- groups, and so the *slopes* are the same as would be expected if the solid was $Pb(OH)_{2(s)}$. As in Figures 11.3 and 11.4, the solubility of $PbO_{(s)}$ increases at both low and high pH. The lines for $\log Fe_T$, $\log Al_T$, and $\log Pb_T$ are collected together in Figure 11.6.

Diagrams like Figures 11.3–11.6 are very useful because they allow one to see at a glance how the chemistry of a system containing a metal oxide, hydroxide, or oxyhydroxide solid changes with pH. Consider a system at 25°C/1 atm that contains a *total* of 10^{-6} mols of Fe(III) per liter of water. Figure 11.7 presents the $\log Fe_T$ vs. pH line for control by (am)$Fe(OH)_{3(s)}$. At pH = 1, all of the Fe(III) will be in

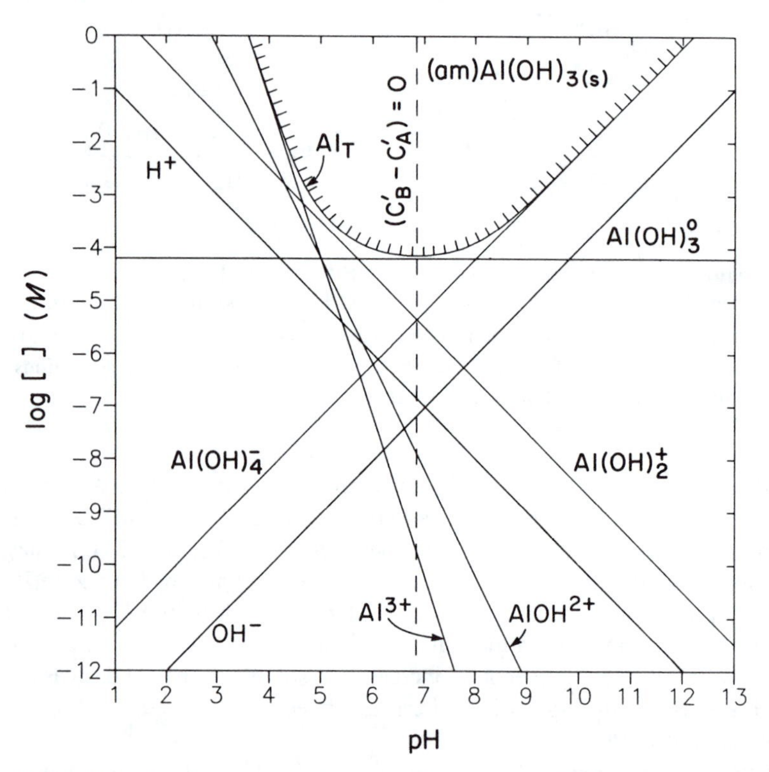

Figure 11.4 Log concentration (*M*) vs. pH for various Al(III) species and Al_T in equilibrium with (am)$Al(OH)_{3(s)}$ at 25°C/1 atm. Dashed vertical line gives speciation for dissolution into initially-pure water (i.e., for $(C_B' - C_A') = 0$). Polynuclear species like $Al_2(OH)_2^{4+}$ have been neglected. Activity corrections neglected (not valid at very low or very high pH).

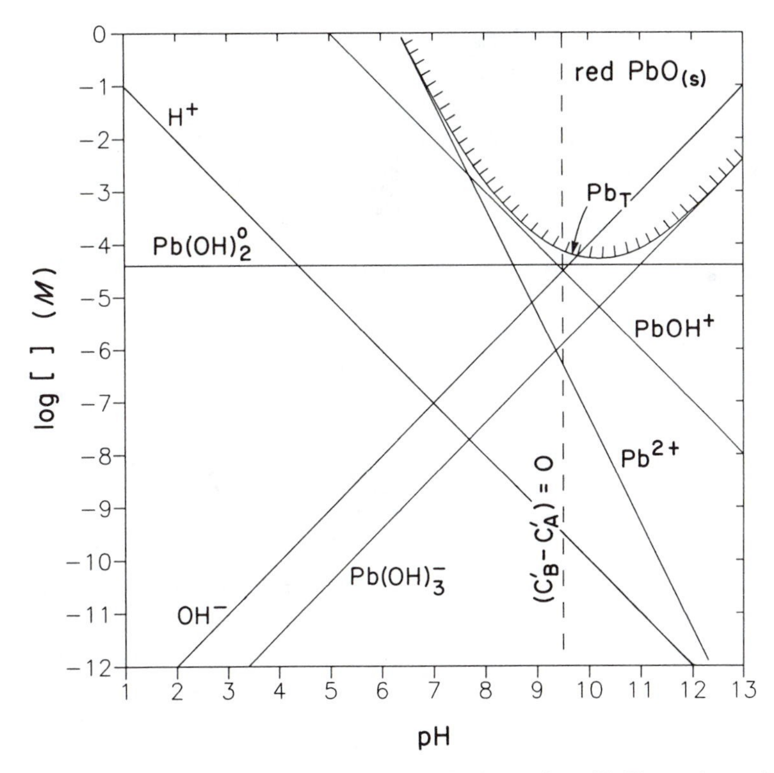

Figure 11.5 Log concentration (M) vs. pH for various Pb(II) species and Pb_T in equilibrium with red $PbO_{(s)}$ at 25°C/1 atm. Dashed vertical line gives speciation for dissolution into initially-pure water (i.e., for $(C_B' - C_A') = 0$). Activity corrections neglected (not valid at very low or very high pH).

solution. As the pH is increased, the total dissolved Fe(III) will follow the heavy solid line in Figure 11.7. As one leaves pH = 1 and approaches a pH of 3.71, the distribution between the various types of Fe(III) species will change, and there will be less and less of Fe^{3+} and more and more $Fe(OH)_2^+$. At pH = 3.71, Eq. (11.58) gives $Fe_T = 10^{-6}M$, that is, pH = 3.71 is a root of the equation

$$^*K_{s0}[H^+]^3 + {}^*K_{s1}[H^+]^2 + {}^*K_{s2}[H^+] + K_{s3} + {}^*K_{s4}/[H^+] - 10^{-6} = 0 \qquad (11.61)$$

Thus, at pH = 3.71, the solution has become *saturated* with Fe(III). As long as pH $\leq$ 3.71, no solid will precipitate, and the dissolved Fe_T will remain $10^{-6}M$. Once pH > 3.71, precipitation will begin. If the pH continues to be raised, the dissolved Fe_T will decrease and precipitation will lead to more and more solid until the minimum in the log Fe_T curve is reached at pH = 8.0. Much of the Fe(III) is now in solid form. As seen in Figure 11.3, in the region pH = 5.5 to about 7.5, $Fe(OH)_2^+$ is the dominant dissolved Fe(III) species. Beyond pH = 8.0, the dissolved Fe_T will increase again because the solubility of the solid is increasing due to the formation of more and more $Fe(OH)_4^-$. For pH $\geq\sim$ 9, $Fe(OH)_4^-$ is the dominant dissolved species. Once pH = 12.40, by

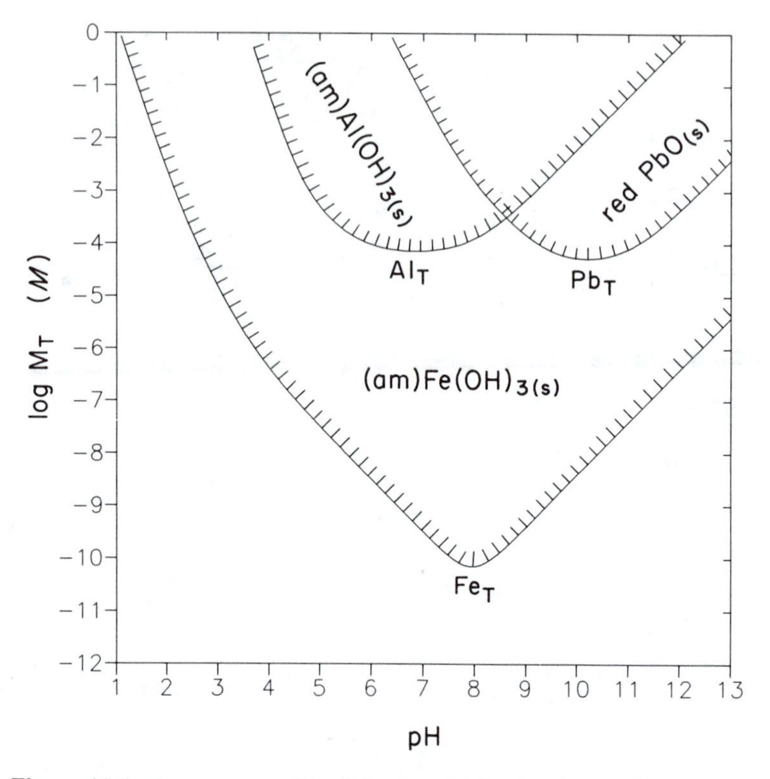

Figure 11.6 Log concentration (M) vs. pH for Fe_T in equilibrium with $(am)Fe(OH)_{3(s)}$, Al_T in equilibrium with $(am)Al(OH)_{3(s)}$, and Pb_T in equilibrium with red $PbO_{(s)}$ at 25°C/1 atm. Polynuclear species have been neglected. Activity corrections neglected (not valid at very low or very high pH).

Eq. (11.58), $(am)Fe(OH)_{3(s)}$ again prescribes that the dissolved $Fe_T = 10^{-6}M$. Thus, pH = 3.71 is a low pH root of Eq. (11.58) and pH = 12.40 is a high pH root of that equation. This means that the solid must become totally dissolved at pH $\geq$ 12.40, and the value of dissolved Fe_T returns to a constant 10^{-6} M at pH $\geq$ 12.40. To summarize, as the pH is increased, we "run into" the solid at pH $\geq$ 3.71, and the solid all redissolves at pH $\geq$ 12.40. The hashmarks on the interior of the log M_T curves in Figures 11.3–11.7 schematically represent the presence of solid material in addition to the aqueous phase.

If we consider the case in which the total dissolved Fe(III) is just equal to the value given by the minimum in Figure 11.7 ($Fe_T = 7.10 \times 10^{-11}$ M), then there is just one positive, real root (pH = 7.95) to the equation

$$^*K_{s0}[H^+]^3 + {}^*K_{s1}[H^+]^2 + {}^*K_{s2}[H^+] + K_{s3} + {}^*K_{s4}/[H^+] - 7.10 \times 10^{-11} = 0. \quad (11.62)$$

For this situation, the solution would be just saturated with $(am)Fe(OH)_{3(s)}$ at pH = 7.95, but no solid would be possible. If we consider a case wherein the total dissolved Fe(III) was less than the value given by the minimum in Figure 11.7 (e.g., 10^{-11} mols per liter), then again, $(am)Fe(OH)_{3(s)}$ cannot be formed at any pH. Mathematically, we now have

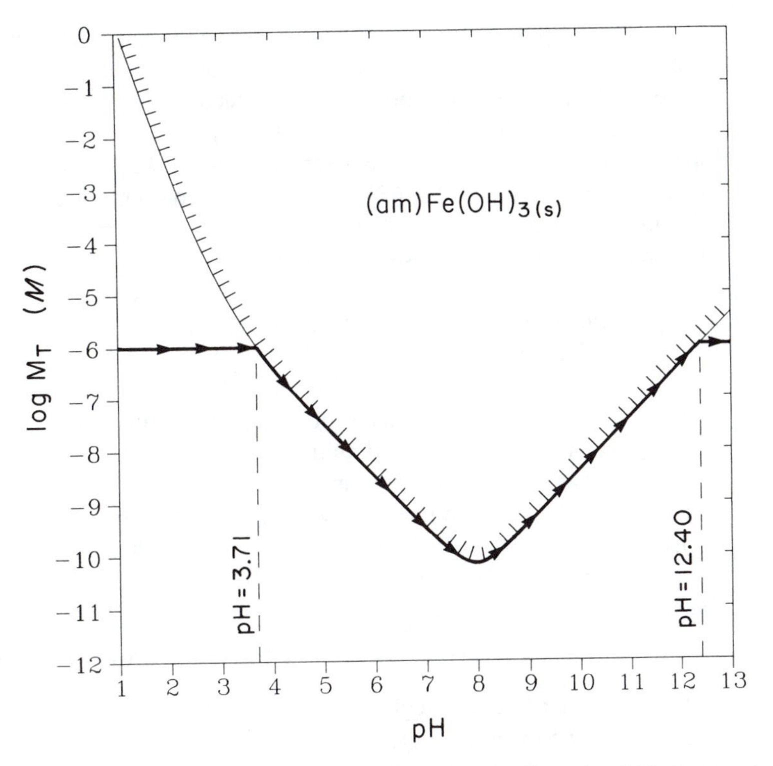

Figure 11.7 Plot of the track of log Fe_T (dissolved) as the pH is increased from 1 to 13 for a system containing a total of 10^{-6} mols of Fe(III) per liter of water at 25°C/1 atm. Polynuclear species like $Fe_2(OH)_2^{4+}$ have been neglected when computing Fe_T. Activity corrections neglected (not valid at very low or very high pH).

no positive, real roots for the equation

$$^*K_{s0}[H^+]^3 + {}^*K_{s1}[H^+]^2 + {}^*K_{s2}[H^+] + K_{s3} + {}^*K_{s4}/[H^+] - 10^{-11} = 0. \qquad (11.63)$$

The reader should never become complacent when using diagrams like Figures 11.3–11.7. The validity of any conclusions drawn using any such figure will depend on the validities of: 1) the assumptions used to draw the figure (e.g., the true identity of the solid controlling the solubility); and 2) the constants used to prepare the figures.

11.6 METAL OXIDE, HYDROXIDE, OR OXYHYDROXIDE EQUILIBRATED WITH INITIALLY-PURE WATER (I.E., FOR ($C_B' - C_A'$) = 0)—NO SIMPLIFYING ASSUMPTIONS

When we are interested in the particular speciation that results when a metal oxide, hydroxide, or oxyhydroxide is equilibrated with initially-pure water, we need to calculate the concentrations of all of the metal species (i.e., M^{z+}, $MOH^{(z-1)+}$, $M(OH)_2^{(z-2)+}$, etc.), plus $[H^+]$ and $[OH^-]$. For this set of concentration unknowns, we have a $^*K_s$ value for each of the metal species, K_w, and an appropriate ENE or PBE. We do not have an MBE since we do not know a priori how much metal will dissolve.

When considering the case of initially-pure water, we are assuming that neither any strong acid like HCl nor any strong base like NaOH is present. We will refer to this case with the statement that $(C'_B - C'_A) = 0$. We do not say that $(C_B - C_A) = 0$ because the dissolution of a metal oxide or hydroxide can carry strong basicity and alkalinity into the solution. Thus, the dissolution of a metal oxide or hydroxide will make $C_B > 0$. However, if the dissolution occurs into initially-pure water, we can still say that, at equilibrium, there are no *other* sources of basicity in the solution, and also no sources of acidity. That is, we can say that $(C'_B - C'_A) = 0$.

As noted in Section 5.2.1, when formulating the solution to a speciation problem, one should start with the equation that has: 1) the most species in it; and 2) the fewest cross-product terms (i.e., terms multiplying one another). One can then make substitutions into that equation without having things get *unnecessarily* complicated. Once the substitution process is completed, the equation is hopefully converted to a form that has only one unknown. Usually, the starting equation of choice is either the ENE or the PBE.

For dissolution of the basic solid (am)$Fe(OH)_{3(s)}$ into initially-pure water, the ENE and the PBE are the same, and we have

$$3[Fe^{3+}] + 2[FeOH^{2+}] + [Fe(OH)_2^+] + [H^+] = [OH^-] + [Fe(OH)_4^-] \tag{11.64}$$

where Eq. (11.64) again neglects polymer species like $Fe_2(OH)_2^{4+}$. Neglecting activity corrections, we must therefore solve

$$3\,^*K_{s0}[H^+]^3 + 2\,^*K_{s1}[H^+]^2 + \,^*K_{s2}[H^+] + [H^+] = K_w/[H^+] + \,^*K_{s4}/[H^+]. \tag{11.65}$$

For (am)$Al(OH)_{3(s)}$, the ENE (and PBE) is

$$3[Al^{3+}] + 2[AlOH^{2+}] + [Al(OH)_2^+] + [H^+] = [OH^-] + [Al(OH)_4^-] \tag{11.66}$$

and we must again solve Eq. (11.65), this time using the constants for (am)$Al(OH)_{3(s)}$. For $PbO_{(s)}$, the ENE (and PBE) is

$$2[Pb^{2+}] + [PbOH^+] + [H^+] = [OH^-] + [Pb(OH)_3^-] \tag{11.67}$$

and, neglecting activity corrections, we must solve

$$2\,^*K_{s0}[H^+]^2 + \,^*K_{s1}[H^+] + [H^+] = K_w/[H^+] + \,^*K_{s3}/[H^+]. \tag{11.68}$$

Using the above equations, the equilibrium pH value that is obtained at 25°C/1 atm for (am)$Fe(OH)_{3(s)}$ is 7.00; this solid is so insoluble that it has no distinguishable effect on the pH. Values of pH that are lower than 7.00 imply the presence of some strong acid like HCl so that $(C'_B - C'_A) < 0$. Values of pH that are higher than 7.00 imply the presence of some strong base like NaOH so that $(C'_B - C'_A) > 0$.

For (am)$Al(OH)_{3(s)}$ and red $PbO_{(s)}$, the pH values for initially-pure water at 25°C/1 atm are 6.85 and 9.50, respectively. For both solids, comments similar to those for (am)$Fe(OH)_{3(s)}$ can be made concerning how the pH will tend to change for $(C'_B - C'_A) < 0$ and for $(C'_B - C'_A) > 0$. A dashed vertical line is drawn in each of

Figures 11.3, 11.4, and 11.5 at the pH characterized by $(C'_B - C'_A) = 0$. For each solid, the speciation that results for dissolution into initially-pure water may be obtained by the intersection of the various log concentration lines with that vertical line.

Since the OH^- groups in $(am)Al(OH)_{3(s)}$ tend to make us think of this solid as being basic, it may have come as somewhat of a surprise that the pH for the initially-pure water in equilibrium with this solid is less than $\frac{1}{2}pK_w$. Obviously, viewing this solid as just being a simple base is incorrect. Actually, as is true for all hydroxides, oxides, and oxyhydroxides, the solids $(am)Al(OH)_{3(s)}$, $(am)Fe(OH)_{3(s)}$, and red $PbO_{(s)}$ are all *amphoteric*.[3] Like HCO_3^-, these solids can all lead to the production of OH^- *as well as* H^+. Solutions of HCO_3^- are somewhat basic because the tendency to produce OH^- by hydrolysis (as measured by K_w/K_1) is stronger than the tendency to produce H^+ (as measured by K_2). For $(am)Al(OH)_{3(s)}$, the tendency to produce H^+ by the formation of $Al(OH)_4^-$ is stronger than the tendency to produce OH^- by formation of Al^{3+}, $AlOH^{2+}$, and $Al(OH)_2^+$. For red $PbO_{(s)}$, the tendency to produce OH^- by the formation of Pb^{2+} and $PbOH^+$ is stronger than the tendency to produce H^+ by the formation of $Pb(OH)_3^-$.

11.7 METAL OXIDE, HYDROXIDE, OR OXYHYDROXIDE EQUILIBRATED WITH INITIALLY-PURE WATER PLUS STRONG ACID AND/OR STRONG BASE (I.E., FOR $(C'_B - C'_A) \neq 0$)

If we add some strong acid like HCl and/or some strong base like NaOH to a $(am)Fe(OH)_{3(s)}$/water system, then the ENE becomes

$$[Na^+] + 3[Fe^{3+}] + 2[FeOH^{2+}] + [Fe(OH)_2^+] + [H^+] = [OH^-] + [Fe(OH)_4^-] + [Cl^-]. \quad (11.69)$$

Using C'_A and C'_B as described above to denote generic strong acid like HCl and generic strong base like NaOH, respectively, Eq. (11.69) becomes

$$(C'_B - C'_A) + 3[Fe^{3+}] + 2[FeOH^{2+}] + [Fe(OH)_2^+] + [H^+] = [OH^-] + [Fe(OH)_4^-]. \quad (11.70)$$

Neglecting activity corrections, we must therefore solve

$$(C'_B - C'_A) + 3\,{}^*K_{s0}[H^+]^3 + 2\,{}^*K_{s1}[H^+]^2 + {}^*K_{s2}[H^+] + [H^+] = K_w/[H^+] + {}^*K_{s4}/[H^+]. \quad (11.71)$$

For $(am)Al(OH)_{3(s)}$, we still use Eq. (11.71), but with the constants for $(am)Al(OH)_{3(s)}$. For $PbO_{(s)}$, we use

$$(C'_B - C'_A) + 2\,{}^*K_{s0}[H^+]^2 + {}^*K_{s1}[H^+] + [H^+] = K_w/[H^+] + {}^*K_{s3}/[H^+]. \quad (11.72)$$

[3] A species is described as being amphoteric when it can lead to the production of H^+ as well as OH^-.

When activity corrections may be neglected, Eqs. (11.71) and (11.72) can be used to compute the pH (and therefore the speciation) for any value of $(C'_B - C'_A)$ for the solids in Figures 11.3, 11.4, and 11.5. For each solid, these equations indicate explicitly how to access the pH values that are off of the dashed vertical corresponding to dissolution into initially-pure water $((C'_B - C'_A) = 0)$. Analogous equations can be developed for other oxides, hydroxides, and oxyhydroxides. Problems illustrating the principles of this chapter are presented by Pankow (1992).

11.8 REFERENCES

STUMM, W., and J. J. MORGAN. 1981. *Aquatic Chemistry*. New York: Wiley-Interscience.

PANKOW, J. F. 1992. *Aquatic Chemistry Problems*. Portland: Titan Press-OR, P.O. Box 91399, Portland, Oregon 97291-1399.

12

Solubility Behavior of Metal Carbonates in Closed Systems

12.1 INTRODUCTION

In addition to the effects that CO_2 has on the simple acid/base chemistry of natural waters, the presence of this compound in the environment has a great impact on the chemistry of a variety of important metal ions through the formation of their solid carbonates. Indeed, for typical environmental levels of CO_2, the metal carbonates of many divalent (i.e., $z = 2+$) metal ions are often more stable (i.e., less soluble) than their corresponding metal hydroxides, oxides, or oxyhydroxides. For example, if CO_2 is present in a system containing Fe^{2+} or Ca^{2+}, the possibility of the formation of $FeCO_{3(s)}$ or $CaCO_{3(s)}$, respectively, must be taken into consideration. In contrast, solid carbonates of metal ions with $z = 1+$ (e.g., sodium) do not form very frequently, though they can be found in systems that are very alkaline and have very high C_T levels (e.g., alkaline playa lakes). Metal carbonates of metals with $z = 3+$ are also rare, since those metal ions are usually much more stable as oxide, hydroxide, or oxyhydroxide solids than as carbonate solids.

Examples of important metal carbonates are:

1. $CaCO_{3(s)}$ in the hexagonal crystal form, known as calcite and found in nature as ancient deposits of marble, chalk, and limestone.
2. $CaCO_{3(s)}$ in the orthorhombic crystal form, known as aragonite, and found in nature in some shells and in geologically young deposits of $CaCO_{3(s)}$ (old deposits of $CaCO_{3(s)}$ are not aragonite because aragonite is relatively less stable than calcite, and is slowly converted to calcite over time).
3. $MgCa(CO_3)_{2(s)}$, known as dolomite, and found in nature as deposits of the same name (e.g., the Dolomite mountain range in the east Alps in northeast Italy).

4. $FeCO_{3(s)}$, known as siderite, and found in nature in hydrothermal veins, and as a result of the action of iron-containing solutions on calcium carbonates.

12.2 SOLUBILITY OF $MCO_{3(s)}$ AS A FUNCTION OF pH WHEN $C_T = M_T$

Consider the dissolution of the carbonate of a typical divalent metal ion M^{2+}. Neglecting activity corrections in the aqueous phase, and also assuming that the mole fraction X for the solid equals 1, we have

$$MCO_{3(s)} = M^{2+} + CO_3^{2-} \qquad K_{s0} = [M^{2+}][CO_3^{2-}] \tag{12.1}$$

$$M^{2+} + OH^- = MOH^+ \qquad K_{H1} = \frac{[MOH^+]}{[M^{2+}][OH^-]} \tag{12.2}$$

$$MOH^+ + OH^- = M(OH)_2^o \qquad K_{H2} = \frac{[M(OH)_2^o]}{[MOH^+][OH^-]} \tag{12.3}$$

$$M(OH)_2^o + OH^- = M(OH)_3^- \qquad K_{H3} = \frac{[M(OH)_3^-]}{[M(OH)_2^o][OH^-]}. \tag{12.4}$$

Since some of the CO_3^{2-} will react to form both HCO_3^- and $H_2CO_3^*$, and since we are considering the species MOH^+, $M(OH)_2^o$, and $M(OH)_3^-$, it is clear that $MCO_{3(s)}$ does not dissolve as a simple mineral salt like $NaCl_{(s)}$. For metal ions like Ca^{2+} and Mg^{2+}, the higher hydroxo complexes $M(OH)_2^o$ and $M(OH)_3^-$ will usually not be very important; for Pb^{2+} and Cd^{2+}, they can be very important in alkaline solutions.

Neglecting activity corrections, one can fully define the position of solubility equilibrium (12.1) (i.e., undersaturated, saturated, or supersaturated) through a knowledge of any *one* of the following five sets of variables:

Set 1	Set 2	Set 3	Set 4	Set 5
$[H^+]$	$[H^+]$	p_{CO_2}	p_{CO_2}	—
$[HCO_3^-]$	*Alk*	$[H^+]$	$[HCO_3^-]$	$[CO_3^{2-}]$
$[M^{2+}]$	$[M^{2+}]$	$[M^{2+}]$	$[M^{2+}]$	$[M^{2+}]$

The first two variables in each set together specify the position of the CO_2 equilibrium and therefore $[CO_3^{2-}]$. Then, with $[M^{2+}]$ and a knowledge of K_{s0} for the T and P of interest, one can determine whether the system is undersaturated, saturated, or supersaturated. In Sets 3 and 4 above, p_{CO_2} is included as a possible variable. For an open system with a gas phase containing $CO_{2(g)}$, the meaning of p_{CO_2} is obvious. For a closed system with no gas phase, it refers to the *equivalent in-situ partial pressure* of $CO_{2(g)}$ discussed in Section 9.10.

As discussed for simple salts in Table 11.1, the types of behavior that solid metal carbonates exhibit under conditions of undersaturation, saturation, or supersaturation are dissolution, no change, and precipitation, respectively. If we somehow already know that a given system is saturated, that information reduces by one the number of variables needed to specify the chemistry of the system. For example, if a system is in fact at equilibrium with the solid, all we need is information concerning the position of the CO_2 equilibrium and a knowledge of K_{s0} to calculate $[M^{2+}]$.

For every carbon atom in $MCO_{3(s)}$ that dissolves, one metal atom dissolves. Thus, in a closed system in which $MCO_{3(s)}$ dissolves into water that initially does not contain either metal or any CO_2 species (though it might contain some strong base like NaOH and/or some strong acid like HCl), we will always have $C_T = M_T$. The carbon atoms will be distributed among $H_2CO_3^*$, HCO_3^-, and CO_3^{2-}, and the metal atoms will be distributed between M^{2+}, MOH^+, $M(OH)_2^o$, and $M(OH)_3^-$. Thus, we have

$$C_T = [M^{2+}] + [MOH^+] + [M(OH)_2^o] + [M(OH)_3^-] = M_T. \tag{12.5}$$

For equilibrium between $MCO_{3(s)}$ and water, again neglecting activity corrections,

$$[M^{2+}] = K_{s0}/[CO_3^{2-}] \tag{12.6}$$

or

$$[M^{2+}] = K_{s0}/(\alpha_2 C_T). \tag{12.7}$$

From Eqs. (12.2), (12.3), (12.4), and (12.7), we have

$$[MOH^+] = K_{H1}[OH^-]K_{s0}/(\alpha_2 C_T) \tag{12.8}$$

$$[M(OH)_2^o] = K_{H1}K_{H2}[OH^-]^2 K_{s0}/(\alpha_2 C_T) \tag{12.9}$$

$$[M(OH)_3^-] = K_{H1}K_{H2}K_{H3}[OH^-]^3 K_{s0}/(\alpha_2 C_T). \tag{12.10}$$

Combining Eqs. (12.7)–(12.10) with Eq. (12.5) yields

$$C_T = \sqrt{(K_{s0}/\alpha_2)(1 + K_{H1}K_w/[H^+] + K_{H1}K_{H2}K_w^2/[H^+]^2 + K_{H1}K_{H2}K_{H3}K_w^3/[H^+]^3)} \tag{12.11}$$

which gives C_T $(= M_T)$ as a simple function of $[H^+]$. Substituting Eq. (12.11) into Eqs. (12.7)–(12.10) allows the concentrations of the various metal species to be calculated as functions of $[H^+]$. Expressions for $[H_2CO_3^*]$, $[HCO_3^-]$, and $[CO_3^{2-}]$ may be obtained using the three α values and Eq. (12.11). All of the metal-hydroxo complexes are likely to be negligible compared to M^{2+} when MOH^+ is negligible compared to M^{2+}. Therefore, when the second term in the square root is small compared to 1, that is, when $K_{H1} \leq 0.1[H^+]/K_w$, then

$$M_T \simeq [M^{2+}] \simeq C_T \simeq (K_{s0}/\alpha_2)^{1/2}. \tag{12.12}$$

At 25°C/1 atm, as long as $K_{H1} \leq 100$, then the formation of MOH^+ may be neglected at pH values as high as 11.

Figure 12.1 contains log concentration vs. pH diagrams for four environmentally-relevant divalent metal carbonates at 25°C/1 atm. For any given plot in Figure 12.1, only one of the pH values will be the one taken on by the system for which the dissolution takes

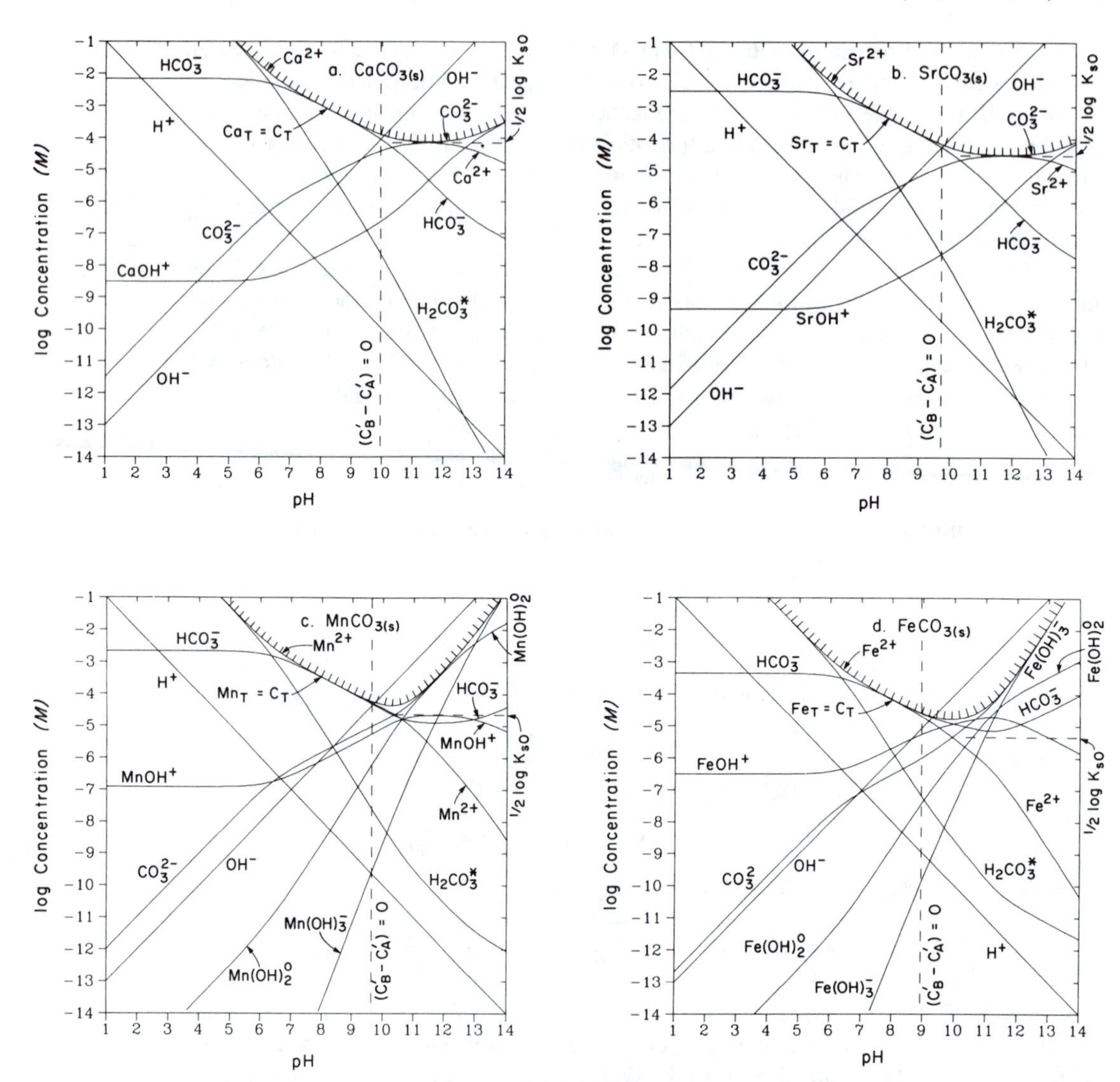

Figure 12.1 Log concentration (M) vs. pH diagrams with $C_T = M_T$ (closed system) for four divalent metal carbonates in equilibrium with water at 25°C/1 atm. Activity corrections neglected (not valid at very low or very high pH). Strong base like NaOH and/or strong acid like HCl is used to change the pH. In each case, a dashed vertical line gives the equilibrium composition obtained when metal carbonate is equilibrated with initially pure water, that is, when $(C_B' - C_A') = 0$. Note the proximity of each dashed vertical line to the pH at which $[M^{2+} \simeq C_T \simeq [OH^-]$. Note also that the solubility ($\log M_T$) increases both when the pH becomes very low, and when it becomes very high. When the metal hydroxo complexes can be neglected in a portion of the region pH $\gg$ pK_2, then $\log[M^{2+}] \simeq \frac{1}{2} \log K_{s0}$ in that portion. This type of behavior is observed for $CaCO_{3(s)}$ and $SrCO_{3(s)}$; it is not observed for either $MnCO_{3(s)}$ or $FeCO_{3(s)}$. In this type of $C_T = M_T$ system, at 25°C/1 atm, the four solids are unstable relative to their corresponding solid hydroxides at pH values above 13.8, >14, 9.8, and 8.8, respectively.

place into initially-pure water. It is the addition of a strong base like NaOH or a strong acid like HCl to such a system that allows access to the full pH range. The constants used to prepare the diagrams are contained in Table 12.1. As seen in Figure 12.1*b*, since K_{H1} for Sr^{2+} is only $10^{0.82}$, the concentration of $SrOH^+$ is negligible relative to Sr_T below pH = 12. However, $FeOH^+$ is significant far below this pH (Figure 12.1*d*) since the K_{H1} for Fe^{2+} is $10^{4.5}$. Figure 12.2 presents $\log M_T$ vs. pH for all of the metal carbonates considered in Table 12.1.

The plots in Figures 12.1 and 12.2 extend over the full pH range. However, as will be discussed in detail in Chapter 14, not all metal carbonates in closed systems of this type will be stable over this entire range. The reason is that at high pH, the in-situ p_{CO_2} can become sufficiently low that conversion to the corresponding metal hydroxide becomes possible. Thus, at a certain "transition pH," the metal carbonate is no longer stable relative to the corresponding metal hydroxide. For the generic metal carbonate, the reaction is $MCO_{3(s)} + H_2O = M(OH)_{2(s)} + CO_{2(g)}$. The reaction can become favored at low p_{CO_2}.[1]

TABLE 12.1 Metal carbonate K_{s0}, K_{H1}, K_{H2}, K_{H3}, and K_{H4} values for several divalent metals at 25°C/1 atm and infinite dilution. Also given are the pH values obtained when the various metal carbonates are equilibrated at 25°C/1 atm with initially-pure water (i.e., when $(C_B' - C_A') = 0$). The pH values are calculated *exactly* using Eq. (12.17) (i.e., with no simplifying assumptions, but neglecting activity corrections), and *approximately* using Eq. (12.26) (i.e., with several simplifying assumptions).

					pH for initially-pure water (i.e., $(C_B' - C_A') = 0$) in equilibrium @ 25°C/1 atm with a divalent metal carbonate.	
Metal Ion	$\log K_{s0}$	$\log K_{H1}$	$\log K_{H2}$	$\log K_{H3}$	exactly using Eq. (12.17)	approximately using Eq. (12.26)
Mg^{2+}	−7.46	2.58	–	–	10.19	10.29
Ca^{2+}	−8.30	1.3	–	–	9.96	10.01
Ba^{2+}	−8.30	0.64	–	–	9.96	10.01
Sr^{2+}	−9.03	0.82	–	–	9.73	9.77
Mn^{2+}	−9.30	3.4	3.4	1.0	9.63	9.68
Zn^{2+}	−10.00	5.0	6.0	2.5	9.24	9.44
Fe^{2+}	−10.68	4.5	2.9	2.6	8.93	9.22
Pb^{2+}	−13.13	6.3	4.6	3.0	8.20	8.40
Cd^{2+}	−13.74	3.9	3.8	2.6	7.88	8.20

[1]For the metal carbonates in Table 12.1, and for $M_T = C_T$ type systems like those in Figure 12.1, at 25°C/1 atm, the conversion to the metal hydroxide becomes thermodynamically favorable (i.e., the metal carbonate is no longer stable relative to the metal hydroxide) above the following pH values: $MgCO_{3(s)}$, 10.2; $CaCO_{3(s)}$, 13.8; $BaCO_{3(s)}$, 14.6; $SrCO_{3(s)}$, >14; $MnCO_{3(s)}$, 9.8; $ZnCO_{3(s)}$, 8.3; $FeCO_{3(s)}$, 8.8; $PbCO_{3(s)}$, 10.8; $CdCO_{3(s)}$, 10.6.

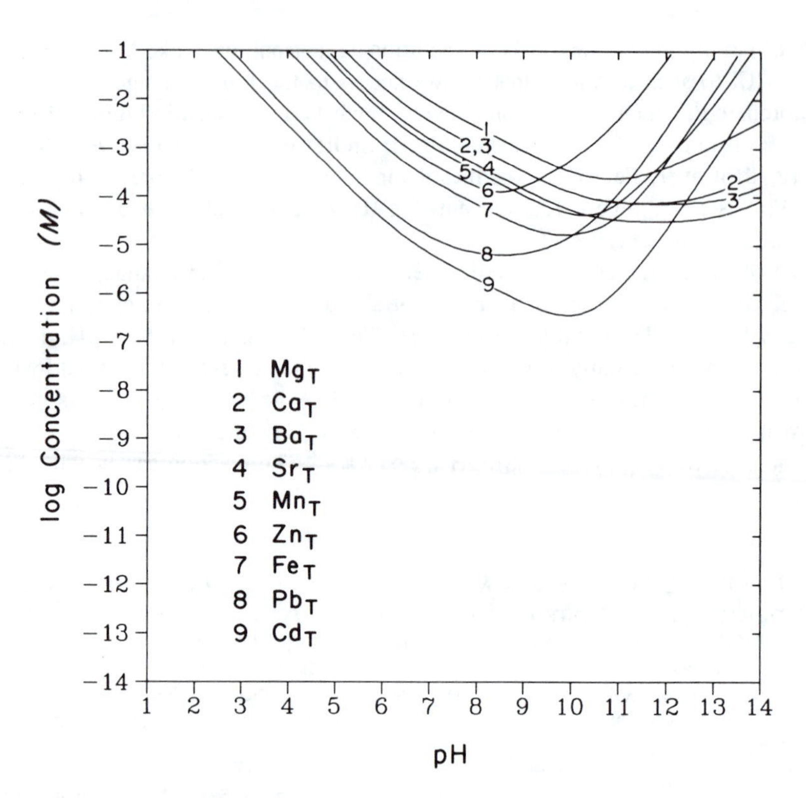

Figure 12.2 Log M_T ($= \log C_T$) vs. pH lines for nine divalent metal carbonates in closed systems in equilibrium with water at 25°C/1 atm. Strong base like NaOH and/or strong acid like HCl is used to change the pH. Activity corrections neglected (not valid at very low or very high pH). For the type of $M_T = C_T$ system considered here, the various metal carbonate solids are unstable relative to their corresponding solid hydroxides above the following pH values: $MgCO_{3(s)}$, 10.2; $CaCO_{3(s)}$, 13.8; $BaCO_{3(s)}$, 14.6; $SrCO_{3(s)}$, > 14; $MnCO_{3(s)}$, 9.8; $ZnCO_{3(s)}$, 8.3; $FeCO_{3(s)}$, 8.8; $PbCO_{3(s)}$, 10.8; $CdCO_{3(s)}$, 10.6.

When MOH^+ and the other hydroxo complexes are negligible, the behavior of the $\log[M^{2+}]$ vs. pH line will be fairly simple. In this range, by Eq. (12.12), we have

$$\log[M^{2+}] \simeq \frac{1}{2}\log(K_{s0}/\alpha_2) = \frac{1}{2}\log K_{s0} - \frac{1}{2}\log \alpha_2. \tag{12.13}$$

The slope of $\log[M^{2+}]$ as determined by Eq. (12.13) will be controlled by the behavior of α_2. In particular,

$$d\log[M^{2+}]/d\text{pH} \simeq -\frac{1}{2}d\log \alpha_2/d\text{pH} \tag{12.14}$$

We know from Chapter 9 that α_2 has three regions of different, but simple behavior. The three analogous regions that can be deduced for $\log[M^{2+}]$ vs. pH as given by Eqs. (12.13) and (12.14) are summarized in Table 12.2. Since K_{H1} for Sr^{2+} is so small, the behavior of $\log[Sr^{2+}]$ in Figure 12.1*b* follows the results expected from Eqs. (12.13) and (12.14) and Table 12.2 for all but the most basic of pH values. The same is *not true* for Fe^{2+} (Figure 12.1*d*).

As indicated in Table 12.2, when MOH^+ and the other hydroxo complexes can be neglected, then at pH $\gg pK_2$, $\log[M^{2+}] \simeq \log M_T = \log C_T \simeq \frac{1}{2}\log K_{s0}$. We note that $\frac{1}{2}\log K_{s0}$ is what would be expected for the simple dissolution of a 1:1 salt. Consider the case of $SrCO_{3(s)}$ as depicted in Figure 12.1*b*. In the region characterized by pH $\gg pK_2$, very little of the dissolved CO_2 is present as either $H_2CO_3^*$ or HCO_3^-. By the same token, for pH values less than about 12.2, $[Sr^{2+}] \gg [SrOH^+]$. Thus, when $SrCO_{3(s)}$ dissolves in the region $\sim 11.3 \leq pH \leq 12.2$, most of the carbon goes into solution as CO_3^{2-}, and most of the Sr goes into solution as Sr^{2+}: the $SrCO_{3(s)}$ is essentially dissolving as a *simple salt*, and the amount of $SrCO_{3(s)}$ that dissolves is *independent* of pH ($d\log[Sr^{2+}]/dpH \simeq d\log Sr_T/dpH = d\log C_T/dpH \simeq 0$). As the pH decreases towards and past pK_2, then the slope of the $\log[Sr^{2+}]$ ($\simeq \log Sr_T = \log C_T$) vs. pH line changes to $-1/2$ as described in Table 12.2. Finally, as the pH decreases further past pK_1, then the slope changes to -1. The high and intermediate pH limiting linear lines intersect at pH $= pK_2$. The intermediate and low pH limiting linear lines intersect at pH $= pK_1$.

Of course we know that even for a solid like $SrCO_{3(s)}$, if we make the pH high enough, then we can make the concentrations of the various hydroxo complexes larger than $[M^{2+}]$. By Eq. (12.2), $[MOH^+]$ will equal 10% of $[M^{2+}]$ when

$$[H^+] = 9K_wK_{H1} \qquad \text{criterion for when } [MOH^+] \text{ is 10\% of } [M^{2+}]. \qquad (12.15)$$

At 25°C/1 atm, this occurs when pH $= 13.05 - \log K_{H1}$. For Sr^{2+}, this pH is 12.2; for Ca^{2+}, it is 11.8; for Fe^{2+}, it is only 8.6. As is clearly seen in Figures 12.1*c* and 12.1*d*, well beyond this metal-dependent pH value, *and* when pH $\gg pK_2$, the slope of the $\log M_T = \log C_T$ line is not 0 as given in Table 12.2. Rather, the slope takes on an asymptotic value of $+n/2$ where n is the *largest number* of hydroxides that can attach to

TABLE 12.2. Behavior when $C_T = M_T$ of $d\log\alpha_2/dpH$ and $\log[M^{2+}]$ in $MCO_{3(s)}$/water systems in the three important regions for CO_2. Hydroxo complexes are assumed negligible, and strong acid and/or strong base are used to change the pH; results based on Eqs. (12.13) and (12.14).

	pH Range		
Parameter	pH $\ll pK_1$	$pK_1 \ll$ pH $\ll pK_2$	pH $\gg pK_2$
α_2	$\sim K_1K_2/[H^+]^2$	$\sim K_2/[H^+]$	~ 1
$d\log\alpha_2/dpH$	+2	+1	0
$\log[M^{2+}]$ ($\simeq \log M_T$)	$\frac{1}{2}\log K_{s0} - \frac{1}{2}\log K_1K_2 + \log[H^+]$	$\frac{1}{2}\log K_{s0} - \frac{1}{2}\log K_2 + \frac{1}{2}\log[H^+]$	$\frac{1}{2}\log K_{s0}$
$d\log[M^{2+}]/dpH$ ($\simeq d\log M_T/dpH$)	-1	$-\frac{1}{2}$	0

any significant extent to M^{2+}. (For Sr^{2+}, $n = 1$.) This may be understood by developing the form of Eq. (12.11) that is appropriate under the highest pH conditions where $\alpha_2 \simeq 1$ (i.e., $C_T \simeq \sqrt{(K_{s0}\beta_n K_w^n/[H^+]^n)}$ where β_n is the product of the individual K_H values from 1 up to n), and taking the derivative with respect to pH.

12.3 $MCO_{3(S)}$ EQUILIBRATED WITH INITIALLY-PURE WATER (I.E., FOR $(C_B'-C_A') = 0$) IN A CLOSED SYSTEM WITH $C_T = M_T$

12.3.1 No Simplifying Assumptions

When a metal carbonate $MCO_{3(s)}$ is equilibrated with initially-pure water, the value of $(C_B' - C_A')$ as defined in Chapter 11 will be zero. That is, the concentration of strong bases such as NaOH minus the concentration of strong acids such as HCl will be zero. We note that $(C_B' - C_A')$ is defined in a manner that does *not* include the contribution to a solution's alkalinity from the dissolution of the metal carbonate itself. Thus, $(C_B' - C_A')$ is different from the quantity $(C_B - C_A)$. In this system, the latter includes both: 1) the portion of the alkalinity represented by $(C_B' - C_A')$; *as well as* 2) that from the dissolved carbonate, which can be quantified as 2 M_T (eq/L).

In this system, we will need to calculate the concentrations of the species: M^{2+}, the n metal hydroxo complexes, H^+, CO_3^{2-}, HCO_3^-, $H_2CO_3^*$, and OH^-. The $6+n$ equations needed to solve this problem are: K_1, K_2, K_w, K_{s0}, an MBE (i.e., $C_T = M_T$), the ENE, and the n K_{Hi} constants. As usual, we will need the values of the various equilibrium constants at the specific T and P of interest.

As noted in Section 5.2.1, when formulating the solution to a speciation problem, we want to start with an equation that contains: 1) as many of the species of interest as possible; and 2) the fewest cross-product terms (i.e., concentrations multiplying one another). One can then make substitutions into that equation without having things get unnecessarily complicated. Once the substitution process is complete, the equation is hopefully converted to a form that has only one unknown.

According to the above criteria, in this situation, the starting equation of interest is the ENE. For the most general case, $n = 3$, and so we begin with

$$2[M^{2+}] + [MOH^+] + [H^+] = [HCO_3^-] + 2[CO_3^{2-}] + [M(OH)_3^-] + [OH^-]. \qquad (12.16)$$

We seek to express each of the terms in Eq. (12.16) as a function of $[H^+]$. For simplicity, we will neglect activity corrections. Alternatively, as usual, the following equations may be thought of as involving ${}^cK$ values rather than infinite dilution constants.

By Eqs. (12.7), (12.8), and (12.10), Eq. (12.16) becomes

$$\frac{2K_{s0}}{\alpha_2 C_T} + \frac{K_{H1}K_w K_{s0}}{\alpha_2 C_T[H^+]} + [H^+] = \alpha_1 C_T + 2\alpha_2 C_T + \frac{K_{H1}K_{H2}K_{H3}K_w^3 K_{s0}}{\alpha_2 C_T[H^+]^3} + K_w/[H^+] \qquad (12.17)$$

which is one equation in two unknowns, C_T and $[H^+]$. When Eq. (12.11) is used to substitute for C_T, Eq. (12.17) represents one equation in one unknown, and so can be solved by trial and error for any specific set of values of K_{s0}, K_1, K_2, K_w, and the K_{Hi}.

Table 12.1 gives the pH values computed at 25°C/1 atm for the metal carbonates of the ions included there. As expected, the equilibrium pH increases with increasing K_{s0}.

For the carbonates considered in Figures 12.1*a–d*, at 25°C/1 atm, the results are: $CaCO_{3(s)}$, pH = 9.96; $SrCO_{3(s)}$, pH = 9.73; $MnCO_{3(s)}$, pH = 9.63; and $FeCO_{3(s)}$, pH = 8.93. A dashed vertical line (labeled $(C'_B - C'_A) = 0$) at the appropriate pH value is drawn in each of the four diagrams; the speciation that results for equilibration of each solid with initially-pure water may be obtained from the intersections of the various log concentration lines with that vertical. pH values lower than that for each vertical imply the presence of some net strong acid *like HCl* $((C'_B - C'_A) \leq 0)$, and higher pH values imply the presence of some net strong base *like NaOH* $(C'_B - C'_A) \geq 0)$. We note that the pH values on the lines labeled $(C'_B - C'_A) = 0$ all possess positive alkalinity $((C_B - C_A) > 0)$ because metal carbonates are themselves basic.

12.3.2 Simplify by Assuming that the Metal Carbonate Is Very Insoluble

Although one can certainly always use trial and error to solve Eq. (12.17), it is instructive to consider the implications and requirements of certain simplifying assumptions. For example, it is clear that as $K_{s0} \to 0$ (i.e., as the metal carbonate becomes increasingly insoluble), then *all* of the terms involving dissolved metal and C_T must decrease in significance in the ENE. Ultimately, for very small K_{s0}, the ENE will be given approximately by

$$[H^+] \simeq K_w/[H^+] \tag{12.18}$$

Equation (12.18) describes the chemistry of essentially pure water, namely

$$[H^+] \simeq \sqrt{K_w} \tag{12.19}$$

When appropriate, $[H^+]$ as given by Eq. (12.19) may be used with Eq. (12.11) to calculate C_T. The concentrations of the various metal ion species may then be obtained with Eqs. (12.7)–(12.10). The concentrations of the various dissolved CO_2 species can be obtained using Eq. (12.11) and the various α values.

For most divalent metal ions, CO_3^{2-} is more basic than M^{2+} is acidic.[2] That is, CO_3^{2-} tends to pick up protons more readily than the hydrolysis of M^{2+} releases them. This means that a solid of the type $MCO_{3(s)}$ will usually tend to raise the pH above $\frac{1}{2}pK_w$, and moreover as K_{s0} increases upwards from zero, the pH must increase upwards from $\frac{1}{2}pK_w$. At 25°C/1 atm where $K_w = 10^{-14}$, let us set as the criterion for using Eq. (12.19) that the pH must not be greater than 7.05 (i.e., $0.9 \times 10^{-7} \leq [H^+] \leq 10^{-7}$). If we take all $K_{Hi} = 0$, we will then get the strictest possible criterion since the binding of OH^- by M^{2+} tends to limit the degree to which the pH rises. Using the value of C_T given by Eq. (12.11) with pH = 7.05, solving Eq. (12.17) with pH = 7.05 gives $K_{s0} = 10^{-18.8}$. Thus, for the general application of Eq. (12.19), we require

$$K_{s0} \leq 10^{-18.8} \tag{12.20}$$

[2]The acidity of M^{2+} is embodied in its ability to react with water to form metal-hydroxo complexes and thereby release H^+, or equivalently, by its ability to bind free OH^-.

At 25°C/1 atm, the vast majority of divalent metal carbonates are characterized by K_{s0} values larger than $10^{-18.8}$ (see Table 12.1 and Smith and Martell (1976)). We can therefore conclude that when equilibrated with water at 25°C/1 atm, most such solids will lead to pH values that are significantly higher than pH = 7.05.

12.3.3 Simplify by Assuming that $H_2CO_3^*$, CO_3^{2-}, H^+, and all Metal-hydroxo Complexes Are Negligible

We now consider a set of assumptions more generally applicable than the one that led to Eq. (12.19). Firstly, we note that many divalent metal ions tend to form metal-hydroxo complexes only weakly. If the solution of $MCO_{3(s)}$ does not lead to a very basic pH, none of the various metal-hydroxo complexes will be very significant compared to M^{2+} itself. Under these conditions, the term $2[M^{2+}]$ may be viewed as being approximately equal to $([M^{2+}]+C_T)$, and we have

$$[M^{2+}]+C_T+[H^+] \simeq \alpha_1 C_T + 2\alpha_2 C_T + K_w/[H^+]. \qquad (12.21)$$

Subtracting $(\alpha_1+\alpha_2)C_T$ from both sides of Eq. (12.21), we obtain

$$[M^{2+}]+\alpha_0 C_T+[H^+] \simeq \alpha_2 C_T + K_w/[H^+]. \qquad (12.22)$$

If a metal carbonate is at least somewhat soluble, but not *too* soluble, then the solution will be basic enough that $H_2CO_3^*$ is insignificant compared to M_T, but not so basic that CO_3^{2-} is significant. Under these conditions, $\alpha_1 \simeq 1$, and $\alpha_0 \simeq 0$ and $\alpha_2 \simeq 0$. When the solution is also sufficiently basic as well as sufficiently concentrated in dissolved metal carbonate that we can also neglect H^+, Eq. (12.22) becomes

$$[M^{2+}] \simeq C_T \simeq [HCO_3^-] \simeq K_w/[H^+] = [OH^-]. \qquad (12.23)$$

Assuming again that the metal-hydroxo complexes can be neglected compared to M^{2+}, combining Eqs. (12.12) and (12.23), we have

$$(K_{s0}/\alpha_2)^{1/2} \simeq K_w/[H^+]. \qquad (12.24)$$

If indeed $\alpha_1 \simeq 1$, then $\alpha_2 \simeq [CO_3^{2-}]/[HCO_3^-] = K_2/[H^+]$, and so

$$(K_{s0}[H^+]/K_2)^{1/2} \simeq K_w/[H^+] \qquad (12.25)$$

or simply,

$$[H^+] \simeq [K_w^2 K_2/K_{s0}]^{1/3}. \qquad (12.26)$$

Equation (12.26) may be solved directly for any value of K_{s0}. It does not depend on K_1 or any of the K_{Hi} because $H_2CO_3^*$ and metal-hydroxo complexes have been assumed to be negligible.

Table 12.1 compares the pH values computed at 25°C/1 atm using Eq. (12.26) with those obtained exactly using Eq. (12.17) and trial and error. For many of the metal carbonates, the agreement is good. For some of the metal carbonates, the agreement is in fact surprisingly good. As an example, for $CaCO_{3(s)}$, neglecting activity corrections, the exact solution pH will be 9.96, and Eq. (12.22) gives (10.01). The value of α_2 is not small at pH $\simeq$ 10. That such good agreement is still possible is a result of compensating approximation errors. In particular, for a relatively soluble metal carbonate, neglecting $\alpha_2 C_T$ on the RHS of Eq. (12.22) underestimates the RHS. However, as pH increases towards pK_2, taking $\alpha_2 \simeq [H^+]/K_2$ also underestimates the LHS of Eq. (12.24). The overall result is that Eq. (12.26) will give a good estimate of the pH for surprisingly large values of K_{s0}. The accompanying problems text (Pankow, 1992) considers these various effects and obtains the result that as long as the metal-hydroxo complexes are negligible at the equilibrium pH, Eq. (12.26) may be used provided that

$$-16.40 \leq \log K_{s0} \leq -7.34 \tag{12.27}$$

In addition to Eq. (12.27), if MOH^+ and the other metal-hydroxo complexes are truly going to be negligible relative to M^{2+}, we also require that $[H^+]$ as given by Eq. (12.26) be greater than or equal to the value given in Eq. (12.15). That is, we require

$$[K_w^2 K_2 / K_{s0}]^{1/3} \geq 9 K_w K_{H1} \tag{12.28}$$

or

$$K_{H1} \leq \frac{1}{9}\left(\frac{K_2}{K_w K_{s0}}\right)^{1/3}. \tag{12.29}$$

At 25°C/1 atm, Eq. (12.29) becomes

$$K_{H1} \leq 1.9(K_{s0})^{-1/3}. \tag{12.30}$$

For every one of the metal ions in Table 12.1 for which Eq. (12.30) is not satisfied, there is a difference of 0.1 pH unit or more between the pH values computed exactly using Eq. (12.17), and approximately using Eq. (12.26).

12.4 $MCO_{3(S)}$ EQUILIBRATED WITH INITIALLY-PURE WATER PLUS STRONG ACID AND/OR STRONG BASE (I.E., FOR $(C'_B - C'_A) \neq 0$) WHEN $C_T = M_T$

Section 12.3 discussed the determination of the specific pH and speciation that will result when $MCO_{3(s)}$ is equilibrated with initially-pure water. Section 12.2 considered the whole range of speciations that result as the pH is adjusted using strong base like NaOH and/or strong acid like HCl. We now examine the specifics of the link between a given value of $(C'_B - C'_A) \neq 0$ and the pH and speciation. That is, we examine how the full range of pH values may be accessed for the systems depicted in Figures 12.1 and 12.2 by varying $(C'_B - C'_A)$.

If we add some strong acid like HCl and/or some strong base like NaOH to a $MCO_{3(s)}$/water system, then the ENE becomes

$$[Na^+] + 2[M^{2+}] + [MOH^+] + [H^+] = [HCO_3^-] + 2[CO_3^{2-}] + [M(OH)_3^-] + [OH^-] + [Cl^-] \tag{12.31}$$

or

$$(C'_B - C'_A) + 2[M^{2+}] + [MOH^+] + [H^+] = [HCO_3^-] + 2[CO_3^{2-}] + [M(OH)_3^-] + [OH^-]. \quad (12.32)$$

Substituting as above for the various terms, Eq. (12.32) becomes

$$(C'_B - C'_A) + \frac{2K_{s0}}{\alpha_2 C_T} + \frac{K_{H1}K_w K_{s0}}{\alpha_2 C_T [H^+]} + [H^+] = \alpha_1 C_T + 2\alpha_2 C_T + \frac{K_{H1}K_{H2}K_{H3}K_w^3 K_{s0}}{\alpha_2 C_T [H^+]^3} + K_w/[H^+]. \quad (12.33)$$

Even when some NaOH or some HCl is added to a system initially containing only pure water and the solid divalent metal carbonate, the only source of both the metal ion and carbon dioxide species is the solid. Thus, Eq. (12.11) still gives C_T. When the value of $(C'_B - C'_A)$ is specified, Eq. (12.33) may then be considered to be one equation in one unknown, $[H^+]$.

Equation (12.33) [12.32] is very similar to Eq. (12.17) [12.16]. Now, however, the term $(C'_B - C'_A)$ is also present on the LHS. Using the graphical approach to solving speciation problems, the equilibrium pH when $(C'_B - C'_A) = 0$ will be the one in diagrams like Figure 12.1 that satisfies Eq. (12.16). As discussed in the previous section, often, Eq. (12.24) [12.23] is a suitable substitute for Eq. (12.17) [12.16] in this regard.

12.5 $MCO_{3(s)}$ DISSOLVED IN WATER OF FIXED C_T WITH STRONG ACID AND/OR STRONG BASE

In natural metal carbonate/water systems, C_T will surely vary in response to changes in the pH. For example, as in the $C_T = M_T$ situation discussed in Sections 12.2 and 12.4, when the pH is brought to low levels, a significant amount of a metal carbonate can dissolve and C_T will increase. Thus, systems in which C_T varies with pH are much more realistic than systems in which C_T is somehow maintained at some constant value.

The above comments notwithstanding, there is a range of C_T values ($\sim 10^{-4}$ to 10^{-2} M) that is particularly typical for natural waters. Thus, it is quite possible to find systems with a wide variety of pH values and with C_T values in this range. It is therefore worthwhile considering how metal carbonate chemistry changes with pH for specific values of C_T, for example, 10^{-3} M.

Regardless of whether or not C_T is constant, neglecting activity corrections, we have

$$[M^{2+}] = K_{s0}/(\alpha_2 C_T). \quad (12.7)$$

When C_T is constant, to determine $[M^{2+}]$, we simply need the pH to calculate α_2. Understanding how $[M^{2+}]$ changes is thus just a matter of understanding how α_2 changes as a function of pH. Taking the derivative of Eq. (12.7), we obtain

$$d\log[M^{2+}]/d\text{pH} = d\log K_{s0}/d\text{pH} - d\log\alpha_2/d\text{pH} - d\log C_T/d\text{pH}. \quad (12.34)$$

Due to the constancy of C_T for the case at hand, we have

$$d\log[M^{2+}]/d\text{pH} = -d\log\alpha_2/d\text{pH}. \quad (12.35)$$

Thus, for a system where C_T is constant, a plot of $\log[M^{2+}]$ vs. pH will be an exact mirror image of $\log \alpha_2$ vs. pH. When C_T is not a constant, but equals M_T, we saw in Section 12.2 that when the metal-hydroxo complexes can be neglected, the limiting slopes of $\log[M^{2+}]$ vs. pH will be -1, $-1/2$, and 0 (see Table 12.1). Here, for constant C_T, the limiting slopes will be -2, -1, and 0 (see Table 12.3). However, these slopes are now *independent* of any importance of species like MOH^+.

The concentrations of the various dissolved CO_2 species may be determined using the appropriate α values and C_T. The behavior of the curves for the metal hydroxo complexes can be deduced based on Eqs. (12.8)–(12.10). As in the $C_T = M_T$ case, the tendency of such complexes to increase in importance as pH increases makes M_T increase with pH at high pH.

When hand drawing the log[ ] vs. pH diagram for fixed C_T, the easiest way to approach the problem is to first evaluate $\log(K_{s0}/C_T)$. As described in Table 12.3, at high pH where $\alpha_2 \simeq 1$, this will equal $\log[M^{2+}]$, and we have $d\log[M^{2+}]/d\text{pH} \simeq 0$. As the pH decreases then passes pK_2, the slope of the $\log[M^{2+}]$ vs. pH line changes to -1. Finally, as the pH approaches and passes pK_1, the slope changes to -2. As in Figure 12.1, the high and intermediate pH limiting linear lines intersect at pH $= pK_2$; the intermediate and low pH region limiting linear lines intersect at pH $= pK_1$. Figures 12.3*a–d* illustrate these principles for four selected metal carbonates for $C_T = 10^{-3}$ *M*. Figure 12.4 plots just $\log M_T$ and $\log[CO_3^{2-}]$ vs. pH for all of the carbonates considered in Table 12.1. Not all of the metal carbonates will be stable over the entire pH range.[3]

TABLE 12.3 Behavior when C_T is constant of $d\log\alpha_2/d\text{pH}$ and $\log[M^{2+}]$ in $MCO_{3(s)}$/water systems in the three important pH regions for CO_2. It is presumed that strong acid and/or strong base are used to change the pH. The results are based on Eqs. (12.7) and (12.35). The results given do not require that the metal-hydroxo complexes be negligible.

	pH Range		
Parameter	pH $< pK_1$	$pK_1 <$ pH $< pK_2$	pH $> pK_2$
α_2	$\sim K_1K_2/[H^+]^2$	$\sim K_2/[H^+]$	~ 1
$d\log\alpha_2/d\text{pH}$	+2	+1	0
$\log[M^{2+}]$	$\log K_{s0} - \log K_1K_2$ $+2\log[H^+] - \log C_T$	$\log K_{s0} - \log K_2$ $+\log[H^+] - \log C_T$	$\log K_{s0} - \log C_T$
$d\log[M^{2+}]/d\text{pH}$	-2	-1	0

[3]For $C_T = 10^{-3}$ *M*, at 25°C/1 atm, the conversion to the metal hydroxide becomes thermodynamically favorable (i.e., the metal carbonate is no longer stable relative to the metal hydroxide) above the following pH values: $MgCO_{3(s)}$, 10.6; $CaCO_{3(s)}$, 14.1; $BaCO_{3(s)}$, >14; $SrCO_{3(s)}$, >14; $MnCO_{3(s)}$, 10.7; $ZnCO_{3(s)}$, 9.1; $FeCO_{3(s)}$, 10.1; $PbCO_{3(s)}$, 11.4; $CdCO_{3(s)}$, 12.3.

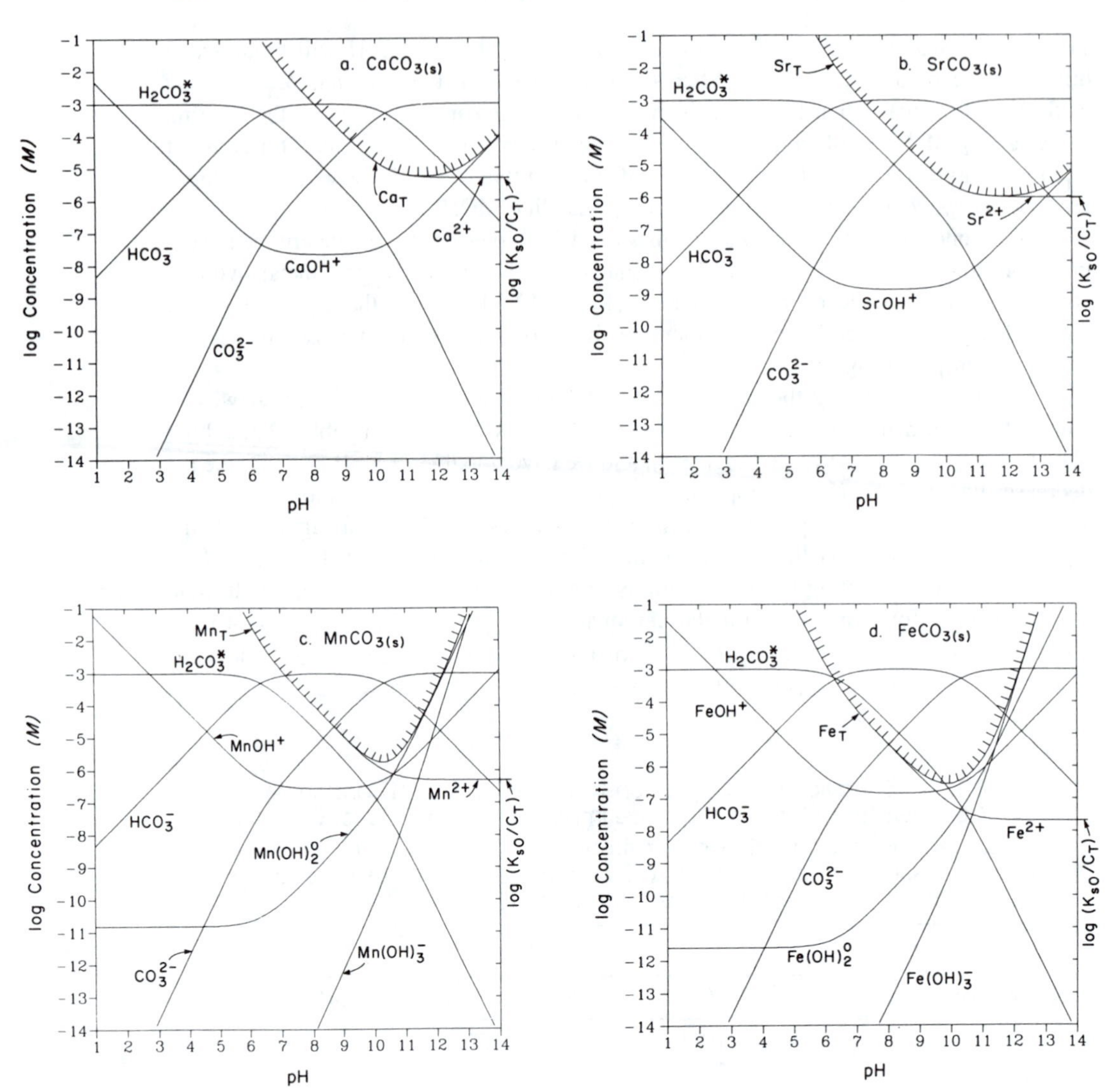

Figure 12.3 Log concentration (M) vs. pH diagrams with $C_T = 10^{-3}\ M$ (closed system) for four divalent metal carbonates in equilibrium with water at 25°C/1 atm. Activity corrections neglected (not valid at very low or very high pH). Note that the solubility ($\log M_T$) increases both when the pH becomes very low, and when it becomes very high. When the metal-hydroxo complexes can be neglected in a portion of the region pH $\gg$ pK_2, then $\log M_T \simeq \log(K_{s0}/C_T)$ in that portion. This type of behavior is observed for $CaCO_{3(s)}$ and $SrCO_{3(s)}$; it is not observed for either $MnCO_{3(s)}$ or $FeCO_{3(s)}$. In this type of $C_T = 10^{-3}\ M$ system, at 25°C/1 atm, the four solids are unstable relative to their corresponding solid hydroxides at pH values above 14.1, >14, 10.7, and 10.1, respectively.

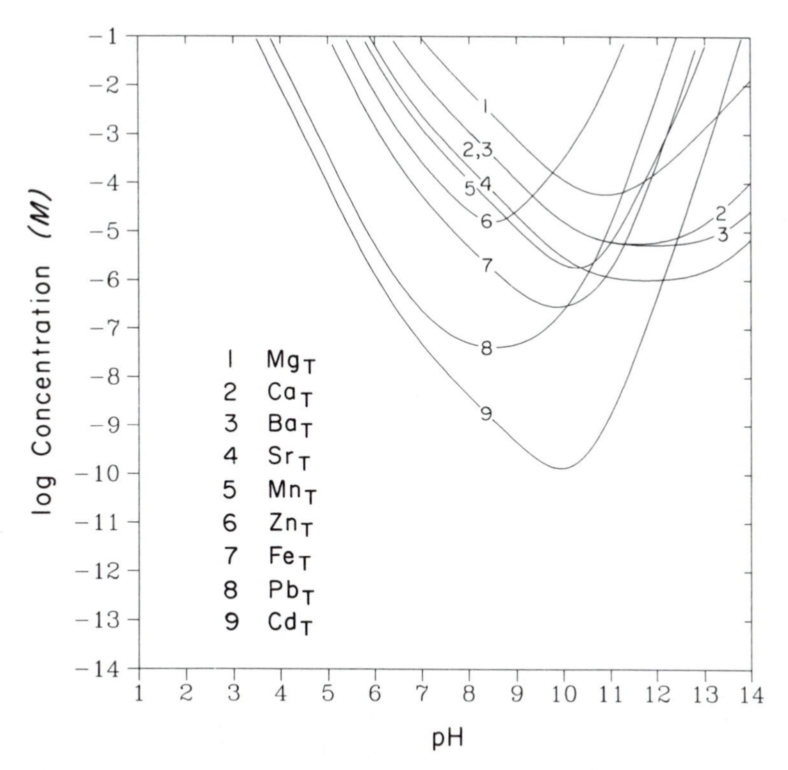

Figure 12.4 Log M_T vs. pH diagram with $C_T = 10^{-3}$ M for nine metal carbonates in equilibrium with water at 25°C/1 atm. Activity corrections neglected (not valid at very low or very high pH). In this type of system, the various metal carbonate solids are unstable relative to their corresponding solid hydroxides above the following pH values: $MgCO_{3(s)}$, 10.6; $CaCO_{3(s)}$, 14.1; $BaCO_{3(s)}$, >14; $SrCO_{3(s)}$, >14; $MnCO_{3(s)}$, 10.7; $ZnCO_{3(s)}$, 9.1; $FeCO_{3(s)}$, 10.1; $PbCO_{3(s)}$, 11.4; $CdCO_{3(s)}$, 12.3.

12.6 REFERENCES

PANKOW, J. F. (1992). *Aquatic Chemistry Problems.* Portland: Titan Press-OR, P.O. Box 91399, Portland, Oregon 97291-1399.

SMITH, R. M., and MARTELL, A. E. 1976. Critical Stability Constants, vol. 4, Inorganic Complexes. New York: Plenum Press.

13

Solubility of Metal Carbonates in Open Systems with a Gas Phase of Constant p_{CO_2}

13.1 INTRODUCTION

As discussed in Chapter 12, when a solid divalent metal carbonate is equilibrated with water in a *closed* aqueous/solid system, it will be possible to write an MBE for the aqueous phase that relates C_T to M_T. If there is neither any metal nor any CO_2-type carbon in solution prior to the equilibration, then the MBE will be

$$C_T = M_T \tag{13.1}$$

When some initial CO_2-type carbon is present in solution (say $C_{T,o}$), and/or when some initial divalent metal is present (say $M_{T,o}$), then we would have the three independent MBEs, that is,

$$C_{T,d} = M_{T,d} \tag{13.2}$$

$$C_T = C_{T,o} + M_{T,d} \tag{13.3}$$

$$M_T = M_{T,o} + C_{T,d} \tag{13.4}$$

where $C_{T,d}$ and $M_{T,d}$ represent the amounts of carbon and metal that have *dissolved* from the solid divalent metal carbonate. Since precipitation is a negative analog of dissolution, when some metal carbonate precipitates from the initial amounts of dissolved metal and carbon in solution, both $C_{T,d}$ and $M_{T,d}$ will be negative. Note that Eq. (13.1) is a simplified version of Eq. (13.3) that applies when $C_{T,o} = M_{T,o}$. Equation (13.1) and Eqs. (13.2)–(13.4) describe *closed* systems in which pH changes cause changes in C_T that are related solely to changes in how much metal carbonate has dissolved (or precipitated).

In *open* systems that contain a *gas phase*, CO_2 can *partition* across the air/water interface, and so CO_2 can be gained or lost from the solution phase by a means other than the dissolution or precipitation of metal carbonate. Indeed, when the aqueous phase is also in equilibrium with a *gas phase*, then: 1) changes in pH will cause some CO_2 to either degas and be released to the gas phase, or dissolve and be taken up from the gas phase; and 2) Eq. (13.3) will no longer apply. Nevertheless, whenever equilibrium exists between some $MCO_{3(s)}$ and the aqueous phase, then we still have that: 1) the K_{s0} equilibrium must be satisfied,

$$MCO_{3(s)} = M^{2+} + CO_3^{2-} \qquad K_{s0} = \{M^{2+}\}\{CO_3^{2-}\}; \tag{13.5}$$

and 2) any one of the five sets of parameters given in Section 12.2 will still define the position of the $MCO_{3(s)}$ solubility equilibrium, that is, whether the system is undersaturated, saturated, or supersaturated.

In an $MCO_{3(s)}$/water/gas system, if the volumes of both the gas and aqueous phases are *finite* as well as known, it will still be possible to write an MBE linking the aqueous C_T and M_T values with the p_{CO_2}. Systems of this type will occur in the subsurface where $MCO_{3(s)}$, liquid water, and gas are all simultaneously present. In contrast, if the volume of the gas phase is essentially *infinite* in comparison to the volume of the aqueous phase, then changes in the aqueous phase will not be able to cause measureable changes in the gas phase p_{CO_2}; while an MBE relating C_T, p_{CO_2}, and M_T is still certainly possible, it would be very inconvenient to actually put into use since its application would require keeping track of those *infinitesimal changes* in p_{CO_2}. As an example, if we were dealing with a system composed of just a beaker of water, some $MCO_{3(s)}$ in that beaker, and the earth's atmosphere, it would be difficult to try and use such an MBE.

When some $MCO_{3(s)}$ is equilibrated at known values of T and P with initially-pure water and a gas phase, the unknowns are exactly the same as in the analogous closed, $MCO_{3(s)}$/solid system with no gas phase, with one addition: p_{CO_2}. To help deal with the additional unknown, we are able to *add* the equilibrium governing the gas/water partitioning of CO_2 to our list of equations, that is, we now have

$$CO_{2(g)} + H_2O = H_2CO_3^* \qquad K_H = \frac{\{H_2CO_3^*\}}{p_{CO_2}}. \tag{13.6}$$

However, as discussed above, we have also *lost* the MBE on the aqueous phase C_T since the systems of interest here are in equilibrium with a gas phase. Thus, in the most general case (i.e., when the gas and aqueous volumes are not specified), we will still be short one equation for the number of unknowns with which we will be concerned. The situation will be similar when we consider the addition of strong base like NaOH and/or strong acid like HCl to the $MCO_{3(s)}$/aqueous/gas system. As in Chapters 11 and 12, we will use the symbols C_B' and C_A' to refer to the concentration of strong base like NaOH and strong acid like HCl. Thus, knowledge of the values of C_B' and C_A' carries with it two new variables, say $[Na^+]$ and $[Cl^-]$, and we are still short one equation.

In general then, we will not be able to proceed until we provide some type of additional restriction on the types of system to be considered. Based on the above discussion, we see that we have two choices. We could:

1. Allow for the input of the initial sizes of the gas and aqueous phases so that we can set up the modified MBE discussed earlier (actually, since we only need one more equation, just their *ratio* would do).
2. Simply eliminate p_{CO_2} as an unknown by assuming that the gas phase is infinitely large; we would then have p_{CO_2} = some constant, and we would limit ourselves to cases that *require* the input of the value of p_{CO_2}.

Option 1 is the only way to solve certain types of problems, for example, those involving sealed geological systems containing solid metal carbonate plus finite, known volumes of water and gas. We do not need to know how much solid metal carbonate is present since all that matters is that the aqueous system is saturated with respect to the metal carbonate. Obviously, however, Option 1 cannot be used in situations in which the volume of the gas phase is infinitely large compared to the aqueous phase. Option 2 will characterize the cases considered in this chapter. Option 1 will not be considered further; its application is fairly straightforward, though some careful mass balance algebra is needed.

Under the condition of Option 2, the value of p_{CO_2} will not change when a finite amount of $CO_{2(g)}$ dissolves, or when a finite amount of CO_2 degasses. Fortunately, shallow surface waters (both lakes and rivers), shallow ocean waters, and some near-surface groundwaters can sometimes be approximated as equilibrium systems involving water, a gas phase with a constant p_{CO_2}, and some solid metal carbonate.

In specific terms, this chapter deals with the open system analogs of the cases examined in Sections 12.2–12.4. We consider the chemistry that obtains when a solid divalent metal carbonate is equilibrated with: 1) initially-pure water ($C_{T,o} = M_{T,o} = 0$), an atmosphere of any given constant value of p_{CO_2}, plus whatever strong acid or base is needed to access the full range of pH values (compare with Section 12.2); 2) the same type of system as in (1) but with $(C_B' - C_A') = 0$ (compare with Section 12.3); and 3) the same type of system as in (1) but with variable values of $(C_B' - C_A') \neq 0$ (compare with Section 12.4). Problems illustrating the principles of this chapter are presented by Pankow (1992).

13.2 DISSOLUTION OF $MCO_{3(s)}$ AS A FUNCTION OF pH WITH EQUILIBRIUM WITH A GAS PHASE OF CONSTANT p_{CO_2}

When there is equilibrium between $MCO_{3(s)}$ and the aqueous phase, and when activity corrections can be neglected, we have that

$$[M^{2+}] = K_{s0}/[CO_3^{2-}] \qquad (13.7)$$

Substituting for $[CO_3^{2-}]$ in terms of p_{CO_2},

$$[M^{2+}] = \frac{K_{s0}}{K_1K_2K_Hp_{CO_2}/[H^+]^2}. \tag{13.8}$$

For constant p_{CO_2} at any T and P, then

$$d\log[M^{2+}]/d\text{pH} = -2. \tag{13.9}$$

Similarly, for the other M-containing species, we have

$$[MOH^+] = \frac{K_{H1}K_wK_{s0}}{K_1K_2K_Hp_{CO_2}/[H^+]} \tag{13.10}$$

$$d\log[MOH^+]/d\text{pH} = -1 \tag{13.11}$$

$$[M(OH)_2^o] = \frac{K_{H1}K_{H2}K_w^2K_{s0}}{K_1K_2K_Hp_{CO_2}} \tag{13.12}$$

$$d\log[M(OH)_2^o]/d\text{pH} = 0 \tag{13.13}$$

$$[M(OH)_3^-] = \frac{K_{H1}K_{H2}K_{H3}K_w^3K_{s0}}{K_1K_2K_Hp_{CO_2}[H^+]} \tag{13.14}$$

$$d\log[M(OH)_3^-]/d\text{pH} = +1 \tag{13.15}$$

$$M_T = \frac{K_{s0}[H^+]^2}{K_1K_2K_Hp_{CO_2}} + \frac{K_{H1}K_wK_{s0}[H^+]}{K_1K_2K_Hp_{CO_2}} + \frac{K_{H1}K_{H2}K_w^2K_{s0}}{K_1K_2K_Hp_{CO_2}} + \frac{K_{H1}K_{H2}K_{H3}K_w^3K_{s0}}{K_1K_2K_Hp_{CO_2}[H^+]}. \tag{13.16}$$

Equation (13.16) is a simple function of the equilibrium constants, p_{CO_2}, and $[H^+]$. As when M_T is controlled by an oxide or hydroxide, the solubility of a metal carbonate will increase as the pH becomes low, and as it becomes very high.

At 25°C/1 atm, Eqs. (13.8), (13.10), (13.12), and (13.14) yield

$$\log[M^{2+}] = \log K_{s0} - 2\text{pH} - \log p_{CO_2} + 18.15 \tag{13.17}$$

$$\log[MOH^+] = \log K_{s0} - \text{pH} - \log p_{CO_2} + \log K_{H1} + 4.15 \tag{13.18}$$

$$\log[M(OH)_2^o] = \log K_{s0} - \log p_{CO_2} + \log K_{H1}K_{H2} - 9.85 \tag{13.19}$$

$$\log[M(OH)_3^-] = \log K_{s0} + \text{pH} - \log p_{CO_2} + \log K_{H1}K_{H2}K_{H3} - 23.85 \tag{13.20}$$

The progression from Eq. (13.17) to Eq. (13.20) involves the repeated addition of the quantity $(\text{pH} + \log K_{Hi} + \log K_w)$. Table 13.1 contains the set of constants at 25°C/1 atm and infinite dilution that are of interest in this chapter.

Equations (13.17)–(13.20) indicate that in constrast to the cases discussed in Chapter 12, the slopes of the log concentration vs. pH lines for the various M-containing species do not undergo changes in the regions near pK_1 and pK_2. The constancy in slope is a direct result of the fact that when p_{CO_2} is constant, the slope of the $\log[CO_3^{2-}]$ vs. pH line is constant.

Figure 13.1 illustrates the manner in which the behavior of four different divalent metal carbonate systems change according to Eqs. (13.17)–(13.20) for $p_{CO_2} = 10^{-3.5}$ atm

TABLE 13.1. Metal carbonate K_{s0}, K_{H1}, K_{H2}, K_{H3}, and K_{H4} values for several divalent metals at 25°C/1 atm and infinite dilution. Also given are the pH values obtained when initially-pure water (i.e., $(C'_B - C'_A) = 0$) at 25°C/1 atm is equilibrated with the various metal carbonates at three different levels of p_{CO_2}. The pH values are calculated exactly using Eq. (13.24) (i.e., with no simplifying assumptions, but neglecting activity corrections), and also approximately using Eq. (13.34) which is based on the assumption that $2[M^{2+}] \simeq C_T \simeq [HCO_3^-]$.

					pH for initially-pure water (i.e., $(C'_B - C'_A) = 0$) in equilibrium @ 25°C/1 atm with a divalent metal carbonate and varying values of p_{CO_2}					
					exactly using Eq. (13.24)			approximately using Eq. (13.34)		
					p_{CO_2} (atm)			p_{CO_2} (atm)		
Metal Ion	$\log K_{s0}$	$\log K_{H1}$	$\log K_{H2}$	$\log K_{H3}$	$10^{-4.5}$	$10^{-3.5}$	$10^{-2.5}$	$10^{-4.5}$	$10^{-3.5}$	$10^{-2.5}$
Mg^{2+}	−7.46	2.58	–	–	9.24	8.60	7.94	9.26	8.60	7.93
Ca^{2+}	−8.30	1.3	–	–	8.98	8.32	7.66	8.98	8.32	7.65
Ba^{2+}	−8.30	0.64	–	–	8.98	8.32	7.66	8.98	8.32	7.65
Sr^{2+}	−9.03	0.82	–	–	8.74	8.07	7.41	8.74	8.07	7.41
Mn^{2+}	−9.30	3.4	3.4	1.0	8.65	7.99	7.32	8.65	7.98	7.32
Zn^{2+}	−10.00	5.0	6.0	2.5	8.44	7.76	7.09	8.42	7.75	7.08
Fe^{2+}	−10.68	4.5	2.9	2.6	8.20	7.53	6.86	8.19	7.52	6.86
Pb^{2+}	−13.13	6.3	4.6	3.0	7.41	6.72	6.05	7.37	6.71	6.04
Cd^{2+}	−13.74	3.9	3.8	2.6	7.18	6.51	5.85	7.17	6.50	5.84

at 25°C/1 atm. Figure 13.2 illustrates the case of $CaCO_{3(s)}$ for three different values of p_{CO_2}, namely $10^{-4.5}$, $10^{-3.5}$, and $10^{-2.5}$ atm.[1] Since the open system equilibria governing the CO_2-related species must be satisfied regardless of whether or not a metal carbonate is also present, neglecting activity corrections, the lines for the CO_2-related species are the same as when a metal carbonate is absent. Figure 13.3 presents the $\log M_T$ lines for all of the divalent metals considered in Table 13.1.

When *Alk* is determined solely by CO_2- and water-related species, Eq. (9.86) gives

$$Alk = \frac{K_1 K_H p_{CO_2}}{[H^+]} + \frac{2K_1 K_2 K_H p_{CO_2}}{[H^+]^2} + \frac{K_w}{[H^+]} - [H^+] \quad \text{only } CO_2 \text{ and water-related acids and bases in solution} \tag{9.86}$$

As discussed in Chapter 9, when bases other than HCO_3^-, CO_3^{2-}, and OH^- are present in the solution that are at least as basic as HCO_3^-, then the concentrations of those other species will also contribute positively to *Alk*. Similarly, when acids other than H^+ are present that are at least as acidic as $H_2CO_3^*$, then they will contribute negatively to *Alk*.

[1]These p_{CO_2} values are sufficiently high that all of the metal carbonates in Table 13.1 are thermodynamically stable relative to their corresponding hydroxides, at all pH values.

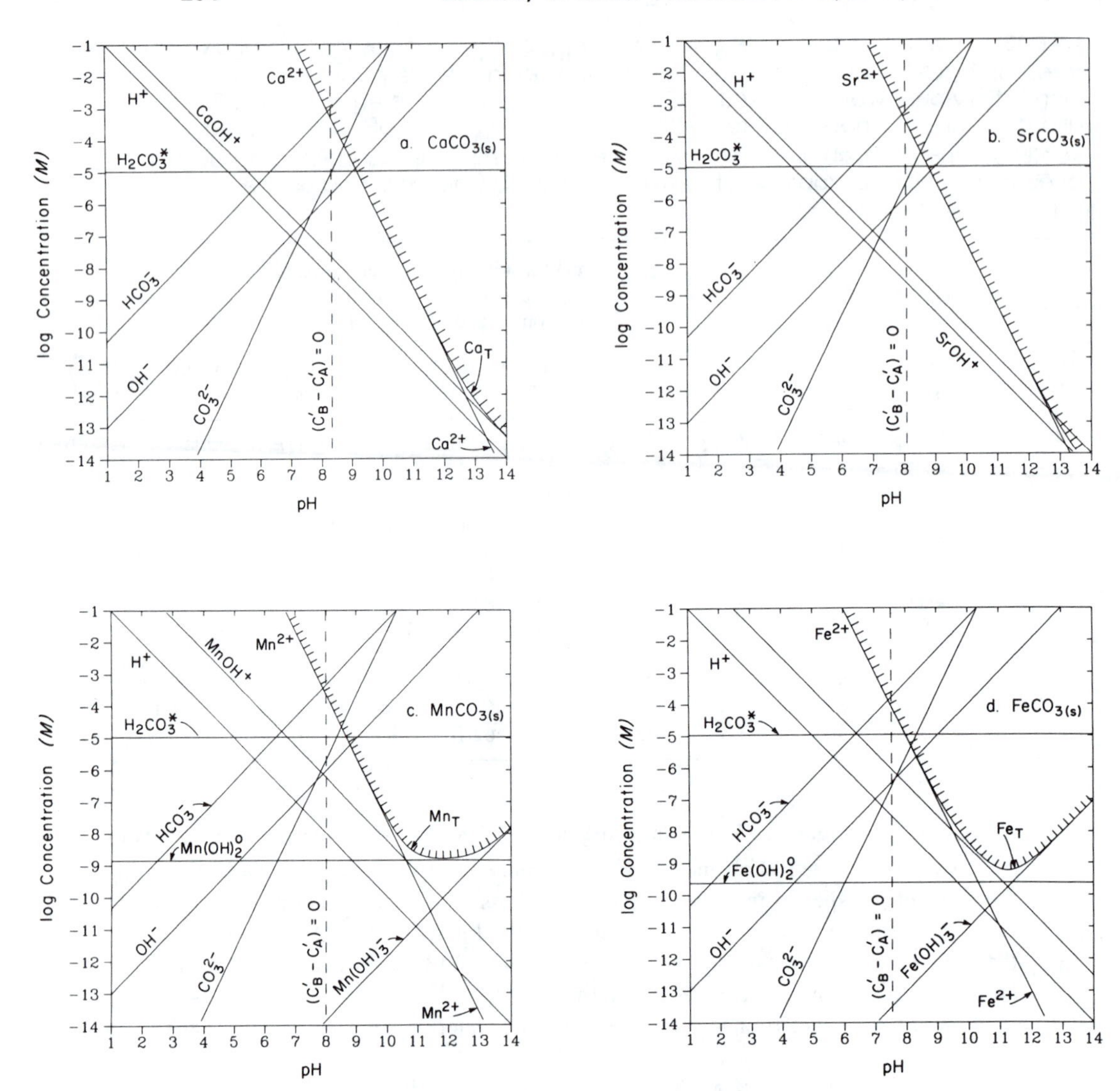

Figure 13.1 Log concentration (*M*) vs. pH diagrams with $p_{CO_2} = 10^{-3.5}$ atm (open system) for four divalent metal carbonates in equilibrium with water at 25°C/1 atm. Activity corrections neglected (not valid at very low or very high pH). Strong base (e.g., NaOH) or strong acid (HCl) is used to change the pH. In each case, a dashed vertical line gives the equilibrium composition obtained when metal carbonate is equilibrated with the gas phase and initially-pure water, i.e., when $(C_B' - C_A') = 0$. Note the proximity of each dashed vertical line to the pH at which $2[M^{2+}] \simeq [HCO_3^-]$ (approximate ENE for the $(C_B' - C_A') = 0$ case). Note also that for all of the solids, the solubility as measured by $\log M_T$ tends to increase both when the pH becomes very low, and when it becomes very high. The region where the solubility increases at high pH is not visible for $CaCO_{3(s)}$ because the values of K_{H2} and K_{H3} are so small for calcium.

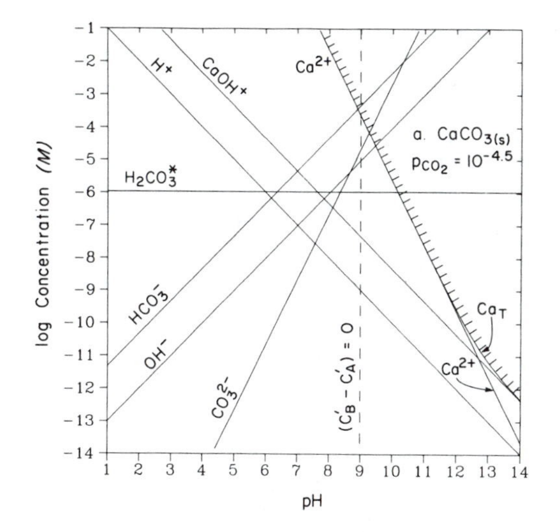

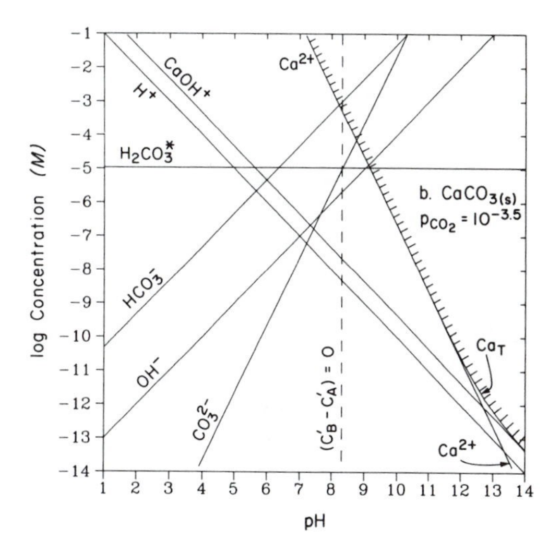

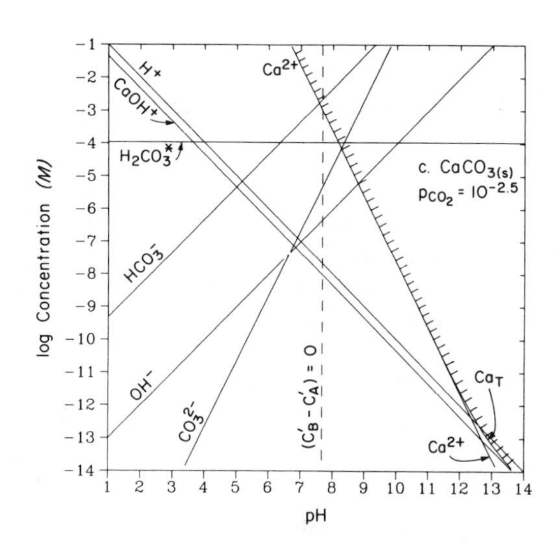

Figure 13.2 Log concentration (*M*) vs. pH diagrams for $CaCO_{3(s)}$ for $p_{CO_2} = 10^{-4.5}$, $10^{-3.5}$, and $10^{-2.5}$ atm (open system) in equilibrium with water at 25°C/1 atm. Activity corrections neglected (not valid at very low or very high pH). Strong base (e.g., NaOH) or strong acid (HCl) is used to change the pH. In each case, a dashed vertical line gives the equilibrium composition obtained when metal carbonate is equilibrated with the gas phase and initially pure water, i.e., when $(C_B' - C_A') = 0$. Note the proximity of each dashed vertical line to the pH at which $2[M^{2+}] \simeq [HCO_3^-]$ (approximate ENE for the $(C_B' - C_A') = 0$ case).

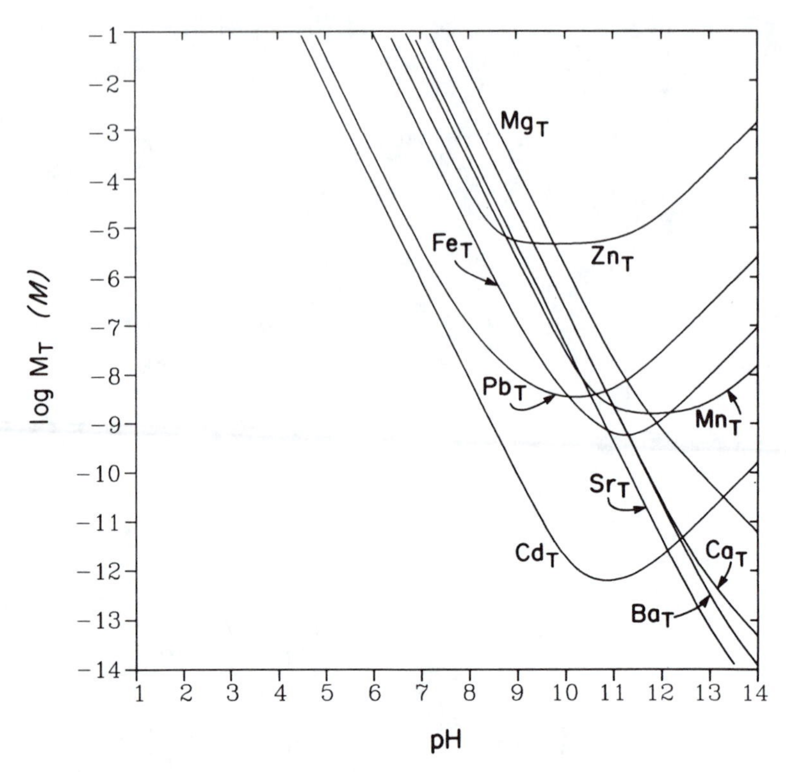

Figure 13.3 Log M_T lines for nine divalent metal carbonates for $p_{CO_2} = 10^{-3.5}$ atm in equilibrium with water at 25°C/1 atm. Strong base (e.g., NaOH) or strong acid (HCl) is used to change the pH. Activity corrections neglected (not valid at very low or very high pH).

For the case at hand, then, we have

$$\begin{aligned} Alk = &[HCO_3^-] + 2[CO_3^{2-}] + [OH^-] - [H^+] \\ &+ [MOH^+] + 2[M(OH)_2^o] + 3[M(OH)_3^-] \end{aligned} \quad \begin{matrix} CO_2\text{-, and M-, and} \\ \text{water-related acids} \\ \text{and bases in solution} \end{matrix} \tag{13.21}$$

Equation (13.21) applies for all solutions of $MCO_{3(s)}$, including: 1) solutions in which strong base like NaOH and/or strong acid like HCl is present; and 2) solutions of the closed system type as discussed in Chapter 12. From Eqs. (9.78) and (9.79), we know that in an open system, $[HCO_3^-]$ and $[CO_3^{2-}]$ are simple functions of $[H^+]$ and p_{CO_2}. With those equations, Eq. (13.21) becomes

$$\begin{aligned} Alk = &\frac{K_1 K_H p_{CO_2}}{[H^+]} + \frac{2K_1 K_2 K_H p_{CO_2}}{[H^+]^2} + K_w/[H^+] - [H^+] \\ &+ \frac{K_{H1} K_w K_{s0}}{K_1 K_2 K_H p_{CO_2}/[H^+]} + \frac{2K_{H1} K_{H2} K_w^2 K_{s0}}{K_1 K_2 K_H p_{CO_2}} + \frac{3K_{H1} K_{H2} K_{H3} K_w^3 K_{s0}}{K_1 K_2 K_H p_{CO_2}[H^+]} \end{aligned} \tag{13.22}$$

Note that the contribution made to *Alk* by $M(OH)_2^o$ is independent of pH.

13.3 $MCO_{3(s)}$ EQUILIBRATED WITH INITIALLY-PURE WATER (($C'_B - C'_A$) = 0) AND A GAS PHASE OF CONSTANT p_{CO_2}

13.3.1 No Simplifying Assumptions

The previous section has discussed how the chemistry of a metal carbonate system open to a gas phase of constant p_{CO_2} varies as a function of pH. For each such system, *one* of those pH values will be the pH that is attained when the metal carbonate and the gas phase is equilibrated with initially-pure water for which $(C'_B - C'_A) = 0$. The ten chemical variables of interest in such a system are $[M^{2+}]$, $[MOH^+]$, $[M(OH)_2^o]$, $[M(OH)_3^-]$, $[H_2CO_3^*]$, $[HCO_3^-]$, $[CO_3^{2-}]$, $[H^+]$, $[OH^-]$, and p_{CO_2}. The ten equations that we will use to solve for that specific pH are: p_{CO_2} = constant, K_{s0}, K_1, K_2, K_{H1}, K_{H2}, K_{H3}, K_w, the Henry's Law constant K_H for CO_2, and finally the ENE.[2]

As in a closed system (see Section 12.3), we begin here with the ENE. Since $(C'_B - C'_A) = 0$ for the case at hand, Eq. (12.16) applies as the ENE, that is,

$$2[M^{2+}] + [MOH^+] + [H^+] = [HCO_3^-] + 2[CO_3^{2-}] + [M(OH)_3^-] + [OH^-]. \qquad (13.23)$$

The fact that some CO_2 might move in either direction across the gas/water interface does not change the ENE relative to the form that it took in Section 12.3. We note that when $K_{s0} = 0$, then Eq. (12.16) collapses to the ENE for initially-pure water containing $H_2CO_3^*$ from the dissolution of some $CO_{2(g)}$.

For simplicity, we will neglect activity corrections. Under these conditions, Eq. (13.23) then becomes

$$\frac{2K_{s0}}{K_1K_2K_Hp_{CO_2}/[H^+]^2} + \frac{K_{H1}K_wK_{s0}}{K_1K_2K_Hp_{CO_2}/[H^+]} + [H^+] =$$

$$\frac{K_1K_Hp_{CO_2}}{[H^+]} + \frac{2K_1K_2K_Hp_{CO_2}}{[H^+]^2} + \frac{K_{H1}K_{H2}K_{H3}K_w^3K_{s0}}{K_1K_2K_Hp_{CO_2}[H^+]} + K_w/[H^+] \qquad (13.24)$$

Equation (13.24) may be solved by trial and error for the single value of $[H^+]$ that is obtained for any given values of p_{CO_2}, K_{s0}, and the other equilibrium constants. Once $[H^+]$ is determined, the values of the various individual concentrations represented in Eq. (13.24) (e.g., $[M^{2+}]$, etc.) as well as *Alk* are readily calculated. Table 13.1 gives the equilibrium pH values obtained using Eq. (13.24) for 25°C/1 atm and $p_{CO_2} = 10^{-4.5}$, $10^{-3.5}$, and $10^{-2.5}$ atm for a variety of divalent metal carbonates. The three exact pH data points for each solid are plotted in Figure 13.4. Table 13.2 contains the corresponding *Alk* values calculated according to (13.22). The three *Alk* data points for each solid are plotted in Figure 13.5.

In the presence of a gas phase of given p_{CO_2} but in the absence of any $MCO_{3(s)}$, *Alk* would be zero for initially-pure water. When some $MCO_{3(s)}$ is also present, however, *Alk* will always be positive for initially-pure water because the dissolution of each mol

[2]Note the absence of an MBE involving C_T and M_T in this group of ten equations.

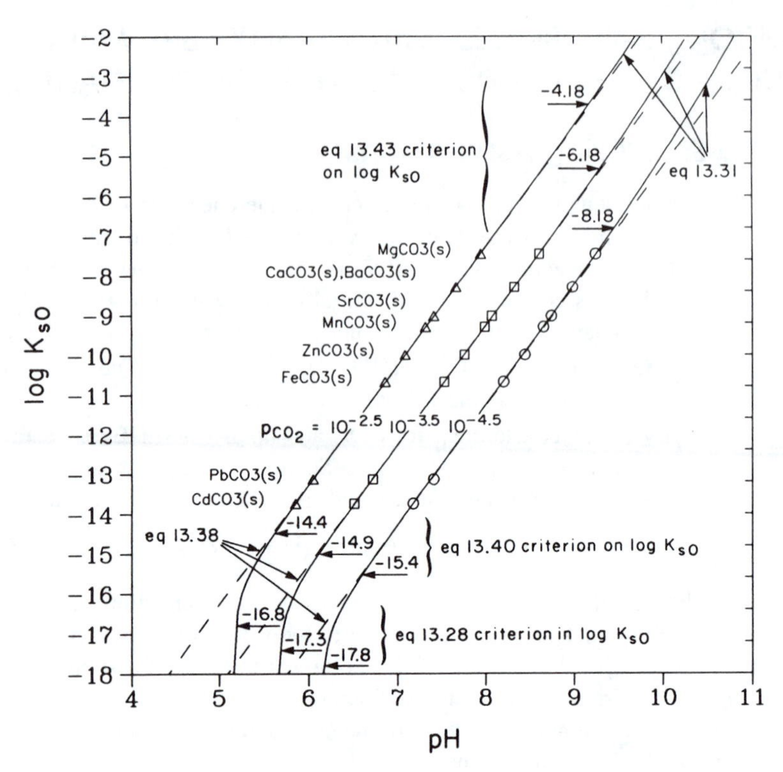

Figure 13.4 Log K_{s0} vs. pH data for nine divalent metal carbonates computed for $(C'_B - C'_A) = 0$ and 25°C/1 atm according to Eq. (13.24) for three different levels of p_{CO_2}: $10^{-2.5}$ atm (triangles); $10^{-3.5}$ atm (squares); and $10^{-4.5}$ atm (circles). The solid line for each level of p_{CO_2} has been computed according to Eq. (13.31), that is, by assuming that $[MOH^+]$ and $[M(OH)_3^-]$ are negligible in the ENE. The dashed lines have been computed according to Eq. (13.38), that is, by assuming that the ENE simplifies to $2[M^{2+}] \simeq [HCO_3^-]$. Eq. (13.38) breaks down both for very small and for very large K_{s0} values. As long as p_{CO_2} is not too small, $[H^+]$ approaches $\sqrt{(K_1 K_H p_{CO_2})}$ as K_{s0} becomes increasingly small; at 25°C/1 atm, for $p_{CO_2} = 10^{-2.5}$, $10^{-3.5}$, and $10^{-4.5}$ atm, we have pH = 5.16, 5.66, and 6.16, respectively. Activity corrections neglected (not valid at very low or very high pH).

of CO_3^{2-} brings with it two equivalents of alkalinity. The dissolution of each mol of M^{2+} does not add to the *Alk*; however, to the extent that the various metal hydroxo complexes form, it does lead to a *redistribution* of the alkalinity. We note that in the specific case when $(C'_B - C'_A) = 0$, then all of the alkalinity is traceable to the dissolution of the CO_3^{2-} in the $MCO_{3(s)}$, and so $Alk = 2M_T$.

Four observations may be made based on Tables 13.1 and 13.2 and Figures 13.4 and 13.5:

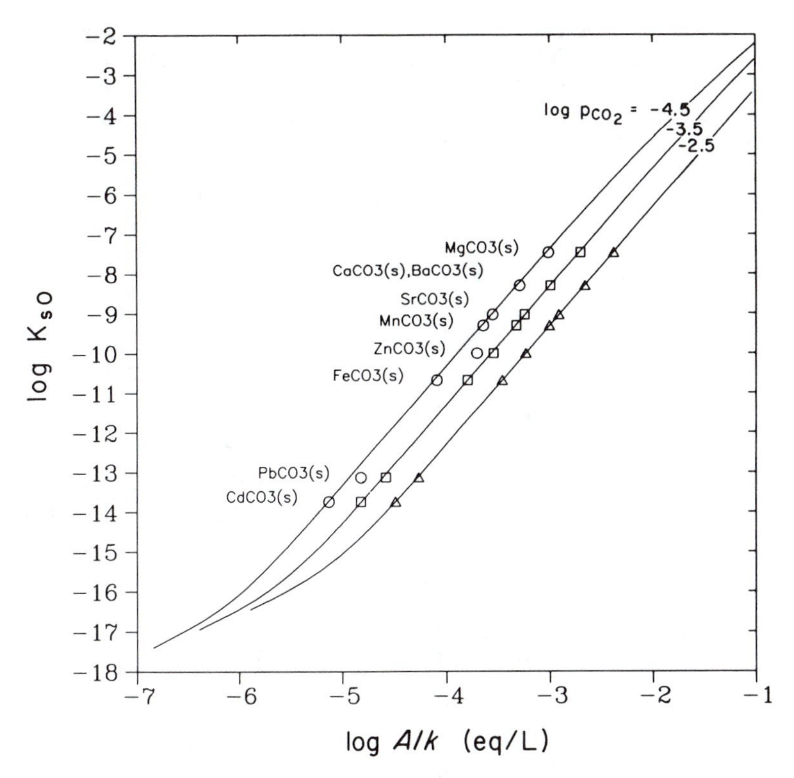

Figure 13.5 Log K_{s0} vs. log *Alk* for solutions of nine divalent metal carbonates for $(C'_B - C'_A) = 0$ and 25°C/1 atm. The data points (exact) have been computed according to Eq. (13.22) (using the equilibrium pH values from Eq. (13.24)) for three different levels of p_{CO_2}: $10^{-2.5}$ atm (triangles); $10^{-3.5}$ atm (squares); and $10^{-4.5}$ atm (circles). For constant K_{s0}, *Alk* increases as p_{CO_2} increases since more metal carbonate dissolves as p_{CO_2} increases. (In the absence of solid metal carbonate, *Alk* would not be affected by changing p_{CO_2}.) The solid line for each level of p_{CO_2} has been computed according to Eq. (9.86), using pH values obtained with Eq. (13.31). Thus, the solid lines assume that $[MOH^+]$, $[M(OH)_2^o]$ and $[M(OH)_3^-]$ do not contribute significantly to either *Alk* or the ENE. The deviation between the exact data points and the Eq. (9.86) lines is greatest when the tendency to form hydroxo complexes is the strongest (Zn^{2+} and Pb^{2+}). Activity corrections neglected (not valid at very low or very high pH).

1. For constant p_{CO_2}, the pH increases as K_{s0} increases (i.e., as the basic metal carbonate becomes more soluble).
2. For constant K_{s0}, the pH decreases as p_{CO_2} increases (i.e., as the acidic effects of the gas phase increase).
3. For constant p_{CO_2}, as K_{s0} increases and the pH increases, the *Alk* of the system increases (i.e., see Eq. (13.22)). And,

TABLE 13.2. *Alk* values for divalent metal carbonates in equilibrium at 25°C/1 atm with initially-pure water, and three different values of p_{CO_2}. *Alk* values are computed using Eq. (13.22) together with the exact pH values from Table 13.1. (The latter are reproduced here in parentheses below the *Alk* values.)

	Alk (eq/L) @ 25°C/1 atm (by Eq. (13.22)		
	p_{CO_2} (atm)		
Metal Carbonate	$10^{-4.5}$	$10^{-3.5}$	$10^{-2.5}$
$MgCO_{3(s)}$	9.88×10^{-4} (pH = 9.24)	1.98×10^{-3} (pH = 8.60)	4.20×10^{-3} (pH = 7.94)
$CaCO_{3(s)}$	5.08×10^{-4} (pH = 8.98)	1.02×10^{-3} (pH = 8.32)	2.20×10^{-3} (pH = 7.66)
$BaCO_{3(s)}$	5.08×10^{-4} (pH = 8.98)	1.02×10^{-3} (pH = 8.32)	2.20×10^{-3} (pH = 7.66)
$SrCO_{3(s)}$	2.82×10^{-4} (pH = 8.74)	5.70×10^{-4} (pH = 8.07)	1.23×10^{-3} (pH = 7.41)
$MnCO_{3(s)}$	2.28×10^{-4} (pH = 8.65)	4.74×10^{-4} (pH = 7.99)	1.00×10^{-3} (pH = 7.32)
$ZnCO_{3(s)}$	1.99×10^{-4} (pH = 8.44)	2.90×10^{-4} (pH = 7.76)	5.94×10^{-4} (pH = 7.09)
$FeCO_{3(s)}$	8.04×10^{-5} (pH = 8.20)	1.64×10^{-4} (pH = 7.53)	3.47×10^{-4} (pH = 6.86)
$PbCO_{3(s)}$	1.51×10^{-5} (pH = 7.41)	2.63×10^{-5} (pH = 6.72)	5.34×10^{-5} (pH = 6.05)
$CdCO_{3(s)}$	7.34×10^{-6} (pH = 7.18)	1.52×10^{-5} (pH = 6.51)	3.25×10^{-5} (pH = 5.85)

4. For constant K_{s0}, even though the pH will decrease as p_{CO_2} increases, *Alk* will increase since a larger amount of basic metal carbonate is brought into solution as the acidic effects of the gas phase increase.

13.3.2 Simplify by Assuming that the Metal Carbonate Is Very Insoluble

Although one can certainly always use trial and error to solve Eq. (13.24), it is instructive to consider how and when one can make simplifying assumptions. When $K_{s0} \rightarrow 0$, the pH of initially-pure water in equilibrium at constant p_{CO_2} will tend towards the value that is attained in the absence of any solid metal carbonate. When this happens, all of the metal terms in the ENE become negligible and the ENE is given approximately by

$$[H^+] \simeq [HCO_3^-] + 2[CO_3^{2-}] + [OH^-]. \qquad (13.25)$$

Equation (13.25) with an = sign is the PBE for a solution of $H_2CO_3^*$. The criterion for how small K_{s0} must become for this asymptotic behavior to be reached will depend on the value of p_{CO_2}, and also to a lesser extent on the values of K_{H1} and K_{H2}.

Since $K_1 \gg K_2$, $[CO_3^{2-}]$ may always be neglected relative to $[HCO_3^-]$ in the PBE for a solution of $H_2CO_3^*$. When p_{CO_2} is large enough so that $[OH^-]$ may also be neglected, then as shown in Section 9.9.2, the PBE for a solution of $H_2CO_3^*$ may be further simplified with the result that

$$[H^+] \simeq \sqrt{K_1 K_H p_{CO_2}} \qquad p_{CO_2} \text{ not too low.} \qquad (9.82)$$

Thus, when K_{s0} is very small *and* p_{CO_2} is not too low, then $[H^+]$ as given by Eq. (9.82) may be used with

$$C_T = K_H p_{CO_2}/\alpha_0 \qquad (13.26)$$

to obtain C_T. Then, using α_2 and K_{s0}, the value of $[M^{2+}]$ may be determined. The concentrations of the other species may be obtained in a similar manner.

The basic character of a solid metal carbonate will tend to cause the pH to be *higher* than is given by Eq. (9.82). This tendency will be mitigated to some extent by the release of protons from the formation of some MOH^+ and $M(OH)_3^-$.[3] The magnitude of that release will diminish as K_{s0} and consequently both $[M^{2+}]$ *and* the pH decrease (recall that the tendency to form MOH^+ and $M(OH)_3^-$ from M^{2+} decreases as pH decreases). Given the acidic character of CO_2, we will therefore neglect MOH^+ and $M(OH)_3^-$ in developing a criterion for the applicability of Eq. (9.82) to an $MCO_{3(s)}$/aqueous/gas system in which the water is initially-pure.

Specifically, let us require that $[M^{2+}]$ can contribute only 10% or less to the LHS of the ENE (Eq. (13.23)). When $[MOH^+]$ is neglected, we therefore have

$$\frac{2K_{s0}}{K_1 K_2 K_H p_{CO_2}/[H^+]^2} \leq 0.1\ [H^+]. \qquad (13.27)$$

[3]The protons from the formation of $M(OH)_2^o$ do not remain as net H^+ in the solution, but are converted to H_2O with the concomitant release of CO_2 according to the net reaction

$$\begin{aligned} & MCO_{3(s)} = M^{2+} + CO_3^{2-} \\ & M^{2+} + 2H_2O = M(OH)_2^o + 2H^+ \\ + \quad & 2H^+ + CO_3^{2-} = CO_{2(g)} + H_2O \\ \hline & MCO_{3(s)} + H_2O = M(OH)_2^o + CO_{2(g)} \end{aligned}$$

Viewed from another perspective, we see that the term $[M(OH)_2^o]$ does not appear in the ENE, which is the equation that we are using to solve for pH for both zero and non-zero values of $(C_B' - C_A')$.

Using Eq. (9.82) to substitute for $[H^+]$, we obtain that for Eq. (9.82) to be valid, as long as p_{CO_2} is not too low, we must have

$$K_{s0} \leq 0.05 K_2 \sqrt{K_1 K_H p_{CO_2}} \tag{13.28}$$

The values of K_{s0} that satisfy Eq. (13.28) increase as p_{CO_2} increases: the CO_2 in the gas phase can be more dominating of the solution phase chemistry as p_{CO_2} increases. For $p_{CO_2} = 10^{-4.5}$, $10^{-3.5}$, and $10^{-2.5}$ atm, at 25°C/1 atm pressure, Eq. (13.28) gives $K_{s0} \leq 10^{-17.8}$, $K_{s0} \leq 10^{-17.3}$, and $K_{s0} \leq 10^{-16.8}$, respectively. Since these K_{s0} values are smaller than the K_{s0} for the least soluble metal carbonate in Table 13.1, we can conclude that Eq. (9.82) will only be applicable for very insoluble metal carbonates and p_{CO_2} values that are much higher than typical ambient levels.

3.3.3 Simplify by Assuming that MOH^+ and $M(OH)_3^-$ Are Negligible

It may be noted that those metal carbonates in Table 13.1 that cause the pH to be relatively basic (e.g., pH $\simeq$ 8 to 9) are the ones (e.g., $CaCO_{3(s)}$ and $MgCO_{3(s)}$) for which M^{2+} has the lowest tendency to react with OH^- to form MOH^+ and $M(OH)_3^-$. At the same time, the metal carbonates for which M^{2+} is most prone to form hydroxo complexes (e.g., $PbCO_{3(s)}$ and $CdCO_{3(s)}$) are the least soluble, and so they are the least likely to cause the pH to be sufficiently basic that MOH^+ and $M(OH)_3^-$ will be important in the ENE. These results are not coincidental. Indeed, high metal ion solubility goes hand in hand with a low affinity for species like CO_3^{2-} *and* OH^-, and vice versa.

The preceding paragraph leads us to conclude that for all the metal carbonates and p_{CO_2} values considered in Table 13.1, it will be a good assumption to neglect MOH^+ and $M(OH)_3^-$ in the ENE. Under these conditions, we have

$$2[M^{2+}] + [H^+] \simeq [HCO_3^-] + 2[CO_3^{2-}] + [OH^-] \tag{13.29}$$

and so when $[MOH^+]$ and $[M(OH)_3^-]$ are neglected, Eq. (13.24) can be converted to

$$K_{s0} \simeq \frac{K_1 K_2 K_H p_{CO_2}}{2} \left(\frac{K_1 K_H p_{CO_2}}{[H^+]^3} + \frac{2 K_1 K_2 K_H p_{CO_2}}{[H^+]^4} + \frac{K_w}{[H^+]^3} - \frac{1}{[H^+]} \right) \tag{13.30}$$

Using the values of K_1, K_2, K_H and K_w at 25°C/1 atm, Eq. (13.30) becomes

$$K_{s0} \simeq p_{CO_2} \left(\frac{10^{-26.27} p_{CO_2}}{[H^+]^3} + \frac{10^{-36.30} p_{CO_2}}{[H^+]^4} + \frac{10^{-32.45}}{[H^+]^3} - \frac{10^{-18.45}}{[H^+]} \right) \tag{13.31}$$

Equation (13.31) has been plotted as solid lines in Figure 13.4 for p_{CO_2} values of $10^{-4.5}$, $10^{-3.5}$ and $10^{-2.5}$ atm. We note that each of the Table 13.1 data points plots either directly on, or very close to its corresponding line. This indicates that $[MOH^+]$ and $[M(OH)_3^-]$ are indeed negligible in the ENE for the range of K_{s0} and p_{CO_2} conditions considered in Table 13.1.

Since *Alk* depends in a simple manner on pH, the $\log K_{s0}$ vs. pH data points obtained using Eq. (13.31) can be used to obtain the dependence of *Alk* on K_{s0}. To be consistent with the assumptions inherent in Eq. (13.31) (i.e., negligible metal hydroxo complexes), the data points obtained using Eq. (13.31) were used with Eq. (9.86) to plot lines of $\log K_{s0}$ vs. $\log Alk$ in Figure 13.5. Most of the Table 13.2 data points plot very close to the lines. The two exceptions are for $PbCO_{3(s)}$ and $ZnCO_{3(s)}$ for $p_{CO_2} = 10^{-4.5}$ atm. This may be understood in terms of the facts that: 1) of all the metal ions in Table 13.1, Pb^{2+} and Cd^{2+} possess the highest K_{H1}, K_{H2}, and K_{H3} values; and 2) the pH will tend to be the highest (and so the metal hydroxo complexes will be the most important) when p_{CO_2} is the lowest.

It would now be worthwhile for the reader to re-examine the four observations made in Section 13.2.1 concerning the effects of K_{s0} and p_{CO_2} on pH and *Alk*. An additional fifth observation that we can now make is that:

5. For constant p_{CO_2}, as K_{s0} becomes very small, then: a) The pH decreases, then reaches a constant value. That is, $[H^+]$ reaches the value that it would have in the absence of any metal carbonate, and the ENE which applies is that for initially-pure water in equilibrium with the given p_{CO_2}. As described above, when p_{CO_2} is not too low, this $[H^+]$ will be given approximately by $\sqrt{(K_1 K_H p_{CO_2})}$ (Eq. (9.82)); and 2) *Alk* asymptotically approaches zero for all values of p_{CO_2}.

The Eq. (13.28) criterion on K_{s0} for the applicability of Eq. (9.82) is marked on Figure 13.4 for each of the three p_{CO_2} values. In each case, that criterion is the K_{s0} value at which the $\log K_{s0}$ vs. pH line approaches an essentially constant pH.

13.3.4 Simplify by Assuming that the Aqueous Solution Will Be Neither Very Basic Nor Very Acidic

Given that the pH values in Table 13.1 are neither very basic nor very acidic, it is tempting to further simplify the approximate ENE given by Eq. (13.29). In particular, it seems likely that $[H^+]$, $[OH^-]$, and $[CO_3^{2-}]$ will also often be negligible in the ENE, leaving only

$$2[M^{2+}] \simeq C_T \simeq [HCO_3^-] \tag{13.32}$$

Equation (13.32) is very similar to Eq. (12.23) which was developed for the analogous system that contains no gas phase. In this case, however, $[HCO_3^-]$ is not approximately equal to $[OH^-]$. *Now*, the acidic gas phase CO_2 tends to react [4] with the OH^- and CO_3^{2-} that is formed when $MCO_{3(s)}$ dissolves to form HCO_3^-. Therefore, the CO_3^{2-} carbon that dissolves is converted to almost exactly twice as much HCO_3^-, and so a factor of *two* multiplies $[M^{2+}]$ in Eq. (13.32), but not in Eq. (12.23).

[4]The conversion of OH^- to HCO_3^- occurs according to the reaction

$$CO_{2(g)} + OH^- = HCO_3^-$$

The conversion of CO_3^{2-} to HCO_3^- occurs according to

$$CO_{2(g)} + H_2O + CO_3^{2-} = 2HCO_3^-$$

Equation (13.32) can be converted to

$$\frac{2K_{s0}}{K_1K_2K_Hp_{CO_2}/[H^+]^2} \simeq \frac{K_1K_Hp_{CO_2}}{[H^+]}. \quad (13.33)$$

or simply

$$[H^+] \simeq \sqrt[3]{K_1^2K_2K_H^2p_{CO_2}^2/2K_{s0}}. \quad (13.34)$$

Equation (13.34) is easily solved for any given values of p_{CO_2} and K_{s0}. At 25°C/1 atm, we have

$$[H^+] \simeq 1.75 \times 10^{-9}(p_{CO_2}^2/K_{s0})^{1/3}. \quad (13.35)$$

The fact that p_{CO_2} appears to a 2/3 power in Eq. (13.35) explains why, for constant K_{s0}, increases in p_{CO_2} of a factor of 10 in Table 13.1 lead to decreases in pH of ~0.67.

Since the RHS of Eq. (13.33) represents $2[M^{2+}]$, using Eq. (13.34) to substitute for $[H^+]$ yields

$$[M^{2+}] \simeq 1.26K_1^{1/3}K_H^{1/3}K_2^{-1/3}p_{CO_2}^{1/3}K_{s0}^{1/3} \quad (13.36)$$

or at 25°C/1 atm,

$$[M^{2+}] = 4.33p_{CO_2}^{1/3}K_{s0}^{1/3}. \quad (13.37)$$

Thus, $[M^{2+}]$ increases as either p_{CO_2} and/or K_{s0} increases.

For the calculation of approximate $\log K_{s0}$ vs. pH lines, Eq. (13.33) may be re-expressed as

$$K_{s0} \simeq K_1^2K_2K_H^2p_{CO_2}^2/2[H^+]^3 \quad (13.38)$$

These lines will have a $d\log K_{s0}/d\text{pH}$ slope of +3. Dashed lines for $p_{CO_2} = 10^{-4.5}$, $10^{-3.5}$, and $10^{-2.5}$ atm at 25°C/1 atm are given in Figure 13.4. It is seen from Figure 13.4 that Eq. (13.38) (and therefore also Eqs. (13.32)–(13.37) will break down relative to Eq. (13.31) at both low and high values of K_{s0}. As may be understood from Section 13.3.2, the failure at low K_{s0} is due to the breakdown of the assumption that $2[M^{2+}] \gg [H^+]$. The failure at high K_{s0} is due to the breakdown of the assumption that $[HCO_3^-] \gg 2[CO_3^{2-}] + [OH^-]$. This breakdown worsens as p_{CO_2} decreases because decreasing p_{CO_2} decreases the driving force for converting CO_3^{2-} and OH^- to HCO_3^-.

Criteria for the breakdown of Eqs. (13.32)–(13.38) may be developed as follows. For *low* K_{s0} values, the actual value of $[H^+]$ will be lower than is given by Eq. (13.38). Indeed, the slope of the $\log K_{s0}$ vs. pH version of Eq. (13.38) is +3, and yet as K_{s0} continues to decrease, we cannot expect that the pH will continue to decrease along that slope. In Figure 13.4, this can be seen by noting that for each p_{CO_2}, at low K_{s0}, the $\log K_{s0}$ vs. pH line as given by Eq. (13.38) is located to the left of the corresponding more accurate line given by Eq. (13.31). As an underlying requirement for Eqs. (13.32)–

(13.38), we desire that $2[M^{2+}]$ contribute $\geq 90\%$ to the LHS of the ENE (Eq. (13.23)), that is,

$$\frac{2K_{s0}}{K_1K_2K_Hp_{CO_2}/[H^+]^2} \geq 9[H^+]. \tag{13.39}$$

Using Eq. (13.34) to substitute for $[H^+]$, we obtain

$$K_{s0} \geq 13.5K_2\sqrt{K_1K_Hp_{CO_2}} \tag{13.40}$$

Equation (13.40) involves the same group that appears in Eq. (13.28), but with a factor of 13.5 rather than 0.05. For $p_{CO_2} = 10^{-4.5}$, $10^{-3.5}$, and $10^{-2.5}$ atm, at 25°C/1 atm, Eq. (13.40) gives $K_{s0} \geq 10^{-15.4}$, $10^{-14.9}$, and $10^{-14.4}$, respectively. As expected, these are the values of K_{s0} at which the three dashed lines in Figure 13.4 given by Eq. (13.38) begin to closely approach the solid lines given by Eq. (13.31). All of the specific metal carbonates considered in Figure 13.4 plot above these K_{s0} values, though one can see that the deviation between the dashed and solid lines tends to increase as p_{CO_2} increases. Indeed, for p_{CO_2} more than a factor of ten larger than $10^{-2.5}$ atm, at 25°C/1 atm, the K_{s0} for $CdCO_{3(s)}$ would fail Eq. (13.40).

For *high* K_{s0} values, the actual value of $[H^+]$ will be greater than is given by Eq. (13.38). As the pH increases with K_{s0}, eventually $[HCO_3^-]$ will no longer bear almost all of the anionic charge in the solution, and $[CO_3^{2-}]$ will start to become significant in the ENE. (Unless p_{CO_2} is *extremely* low, we can continue to ignore $[OH^-]$ in the ENE). Since Eq. (13.38) assumes negligibility of $[CO_3^{2-}]$, as K_{s0} increases, that equation will increasingly overestimate $[HCO_3^-]$. Therefore, Eq. (13.38) will increasingly overestimate the pH required to obtain that $[HCO_3^-]$, and the $\log K_{s0}$ vs. pH lines given by Eq. (13.38) will be located increasingly to the right of the corresponding, more accurate lines given by Eq. (13.31). For high K_{s0} then, neglecting activity coefficients, we set

$$2[CO_3^{2-}] \leq 0.1[HCO_3^-] \tag{13.41}$$

or simply

$$[H^+] \geq 20K_2 \tag{13.42}$$

as the criterion for the applicability of Eq. (13.32)–(13.38) at high K_{s0}. Substituting for $[H^+]$ using Eq. (13.34), we obtain

$$K_{s0} \leq 6.25 \times 10^{-5}K_1^2K_H^2p_{CO_2}^2/K_2^2. \tag{13.43}$$

For $p_{CO_2} = 10^{-4.5}$, $10^{-3.5}$, and $10^{-2.5}$ atm, at 25°C/1 atm, Eq. (13.43) gives $K_{s0} \leq 10^{-8.18}$, $10^{-6.18}$, and $10^{-4.18}$, respectively, as criterion values. As expected, these are the values of K_{s0} below which the three dashed lines in Figure 13.4 given by Eq. (13.38) begin to approach the solid lines given by Eq. (13.31). Now, for constant K_{s0}, the deviation between the dashed and solid lines becomes greater as p_{CO_2} decreases. Most of the metal carbonates considered in Figure 13.4 plot below the criterion K_{s0} values for all three p_{CO_2} values; for $p_{CO_2} = 10^{-4.5}$ atm, the exceptions are $MgCO_{3(s)}$, $CaCO_{3(s)}$, and $BaCO_{3(s)}$.

In summary, for Eqs. (13.32)–(13.38) to provide reasonable approximations for the aqueous behavior of a solid metal carbonate in equilibrium with both initially-pure water and a gas phase containing $CO_{2(g)}$, we require that

$$13.5\ K_2\sqrt{K_1 K_H p_{CO_2}} \le K_{s0} \le 6.25 \times 10^{-5} K_1^2 K_H^2 p_{CO_2}^2 / K_2^2. \tag{13.44}$$

For $p_{CO_2} = 10^{-3.5}$ atm and 25°C/1 atm, Eq. (13.44) becomes

$$10^{-14.9} \le K_{s0} \le 10^{-6.18}. \tag{13.45}$$

Based on the pH data in Table 13.1, a dashed vertical line has been drawn in each of the plots in Figures 13.1 and 13.2 where $(C_B' - C_A') = 0$. Given that all of the metal carbonates and p_{CO_2} conditions considered in Figures 13.1 and 13.2 satisfy or come close to satisfying Eq. (13.44), these dashed vertical lines are located very near the pH values at which each $\log[M^{2+}]$ line drops to 0.301 (= log 2) log units below the $\log[HCO_3^-]$ line. Indeed, at each of those pH values, $2[M^{2+}]$ is significantly greater than the sum of $[H^+]$ and $[MOH^+]$, and $[HCO_3^-]$ is significantly greater than the sum of $2[CO_3^{2-}]$, $[M(OH)_3^-]$, and $[OH^-]$.

13.4 $MCO_{3(s)}$ EQUILIBRATED WITH INITIALLY-PURE WATER, A GAS PHASE OF CONSTANT p_{CO_2}, AND STRONG ACID AND/OR STRONG BASE

For the dissolution of $CaCO_{3(s)}$ into initially-pure water with $p_{CO_2} = 10^{-3.5}$ atm, at 25°C/1 atm, Table 13.1 indicates that the pH will be 8.32. (The approximation equation Eq. (13.34) also gives pH = 8.32). Many natural waters that are in equilibrium or near equilibrium with the atmosphere and $CaCO_{3(s)}$ will exhibit pH values that are significantly different from ~8.3; values in the 7.5–8.0 range are particularly common. At or near 25°C/1 atm, when the pH is less than ~8.3 and to within corrections due to non-unity γ values, such a system must have some net strong acid like HCl present. That is, we know that $(C_B' - C_A') < 0$. When the pH is greater than ~8.3, then $(C_B' - C_A') > 0$. Thus, the addition of strong acid or strong base to a given $MCO_{3(s)}$/aqueous/gas system allows access to the whole range of pH values, and not just the single pH that will be obtained when $(C_B' - C_A') = 0$. Of course, the specific pH attained will depend on the value of $(C_B' - C_A')$.

As in the closed systems considered in Chapter 12, computing the pH and chemical speciation in an open system when $(C_B' - C_A')$ is non-zero is no more difficult than when $(C_B' - C_A') = 0$. In particular, the governing ENE remains nearly the same, differing only by virtue of the presence of the tracers C_B' and C_A'. In particular, we now have

$$(C_B' - C_A') + \frac{2K_{s0}}{K_1 K_2 K_H p_{CO_2}/[H^+]^2} + \frac{K_{H1} K_w K_{s0}}{K_1 K_2 K_H p_{CO_2}/[H^+]} + [H^+] =$$

$$\frac{K_1 K_H p_{CO_2}}{[H^+]} + \frac{2K_1 K_2 K_H p_{CO_2}}{[H^+]^2} + \frac{K_{H1} K_{H2} K_{H3} K_w^3 K_{s0}}{K_1 K_2 K_H p_{CO_2} [H^+]} + K_w/[H^+] \tag{13.46}$$

Equation (13.46) can be solved by trial and error for a specific known value of $(C_B' - C_A')$ using exactly the same procedure employed to solve Eq. (13.24) when $(C_B' - C_A') = 0$. Equation (13.46) thus provides the link between any given pH in plots like those in Figures 13.1–13.3, and the value of $(C_B' - C_A')$ necessary to attain that pH.

For given values of the various constants and p_{CO_2}, when activity corrections are neglected, the only thing that determines the chemistry is the value of $(C_B' - C_A')$. Under these conditions, a case where both C_B' and C_A' are non-zero but are equal will give the same pH as the case when $(C_B' - C_A')$ is zero because both C_A' and C_B' are zero. In contrast, when activity corrections are taken into consideration, then the specific amounts of the background ions must be known and *different* pH values will be calculated for these two cases. Since non-ideality will always affect the chemical speciation to *some* extent, strictly speaking the concentrations of M^{2+}, $H_2CO_3^*$, HCO_3^-, etc. for these two cases will always be somewhat different regardless of how low the two ionic strengths are.

13.5 ITERATIVE ACTIVITY CORRECTIONS WHEN THE FINAL IONIC STRENGTH IS NOT KNOWN A PRIORI

As discussed in preceding chapters, when the concentrations of background salt(s) in a solution of interest is sufficiently high, then one can use ${}^cK$-type equilibrium constants to solve for the equilibrium speciation. This applies equally well for zero and non-zero values of $(C_B' - C_A')$. Under these circumstances, changes in the concentrations of minor constituents (e.g., H^+, etc.) will not alter the ionic strength to a significant degree. Thus, it will not be necessary to revise the speciation calculations using an updated value of the ionic strength and corresponding updated estimates of the ${}^cK$ values.

As discussed in Section 5.5, however, when the concentration of background salt is not sufficiently high, then a good estimate of the final equilibrium ionic strength will not be available. Thus, an iterative succession of equilibrium solutions will be needed. Each of those solutions will incorporate activity corrections corresponding to the latest estimate of the ionic strength. The iterations will continue until the ionic strength and therefore also the activity corrections and the speciation no longer change. For the problem discussed in the previous section, the solution process would begin using best first estimates of γ_1 and γ_2 (the generic activity coefficients of singly- and doubly-charged ions, respectively) and the ENE modified appropriately for activity corrections:

$$(C_B' - C_A') + \frac{2K_{s0}}{\gamma_2 K_1 K_2 K_H p_{CO_2}/(\gamma_1^2[H^+]^2)} + \frac{K_{H1}K_w K_{s0}}{K_1 K_2 K_H p_{CO_2}/[H^+]} + [H^+] =$$

$$\frac{K_1 K_H p_{CO_2}}{\gamma_1^2[H^+]} + \frac{2K_1 K_2 K_H p_{CO_2}}{\gamma_1^2\gamma_2[H^+]^2} + \frac{K_{H1}K_{H2}K_{H3}K_w^3 K_{s0}}{\gamma_1^2 K_1 K_2 K_H p_{CO_2}[H^+]} + \frac{K_w}{\gamma_1^2[H^+]}. \qquad (13.47)$$

The activity coefficient of the neutral species $H_2CO_3^*$ has been assumed to be equal to 1.0.

13.6 REFERENCES

PANKOW, J. F. 1992. *Aquatic Chemistry Problems*. Portland: Titan Press-OR, P.O. Box 91399, Portland, Oregon 97291-1399.

14

Solubility Control, Solubility Limitation, the Coexistence of Multiple Solid Phases, and Multiple-Solid Predominance Diagrams

14.1 INTRODUCTION

In Chapters 11–13, we have discussed the fact that metals can form solids that are simple salts, hydroxides, oxides, oxyhydroxides, as well as carbonates. A variety of other types of solids are also possible. How, then, does one know which solid to consider when computing the equilibrium chemistry of a given system? Using ferrous iron as an example, if it is known that there is sufficient Fe(II) in a certain system that either $Fe(OH)_{2(s)}$ or $FeCO_{3(s)}$ will definitely be present, how does one know which solid to use in the equilibrium speciation calculations? That is, how does one know which solid will actually control the speciation? Moreover, it also seems natural to ask whether it might ever be possible that both of these solids could be present and in equilibrium with the same aqueous activity of Fe^{2+}.

We first note that a given solid need not be present and on the scene for it to provide an upper bound on the solubility, that is, *limit* the solubility. Obviously, however, the solid must be present if it is going to actually *control* solubility and solution speciation. Solubility limitation and actual solubility control are thus seen as two closely-related, yet slightly-different concepts that should be kept differentiated.

As an analogy, consider a 12-year-old child who is limited by law to a maximum of 15 hours/week of paid work outside the home. As long as the child works less than 15 hours/week, there is no external control on the level of work. The level of work however, is still limited by the law. Once the child wants to work 15 hours/week or *more*, then the law steps onto the scene (perhaps in the form of a policeman), and the level of work becomes limited *and* actually controlled by the law. The various analogs in this example are: level of work ↔ metal ion activity, 15 hours/week ↔ solubility-limited metal activity, and the law ↔ the K_{s0} for the solid in conjunction with the activity of the anion for the solid.

With regard to the last analog, we note that it is possible to have more than one law on record that governs child labor. For example, there might be the one law giving the maximum number of hours as 15 per week, but there might be a second law which says that if the hours are worked in the evening, then the maximum is 10 hours per week. For evening work, then, the second law would take precedence and *limit* the amount of work. Thus, for the limitation of metal ion solubility, if more than one K_{s0}/anion pair is possible, then it is the K_{s0}/anion pair that gives the lowest saturation activity that takes precedence. For the two solids $Fe(OH)_{2(s)}$ and $FeCO_{3(s)}$, we need to compare the K_{s0} for $Fe(OH)_{2(s)}$ divided by $\{OH^-\}^2$ vs. the K_{s0} for $FeCO_{3(s)}$ divided by $\{CO_3^{2-}\}$.

Based on the above discussion, we conclude that the criterion for identifying the governing (i.e., the solubility-limiting) solid for a metal ion may be stated as:

Solubility-Limiting Criterion: The solid that will limit the solubility of a metal ion M^{z+} will be that solid which gives the lowest activity of M^{z+} at equilibrium for the conditions of interest.

A first corollary to the above criterion is:

Solubility-Limiting Corollary 1: If more than one solid is in equilibrium with a given metal ion M^{z+}, then all of those solids must specify the same activity of M^{z+} for the conditions of interest.

It may be noted that an exactly analogous situation exists for non-metal containing, anionic species like Cl^-, CO_3^{2-}, etc. That is, the solid that will limit the anion solubility will be the solid which prescribes the lowest equilibrium activity for the conditions of interest. Similarly, if more than one solid is in equilibrium with and controlling a species like sulfate, then all such solids must specify the same activity for that species.

This chapter will be consider solubility limitation and control for metals like Fe, Al, Ca, Hg, etc. rather than either major metal cations such as Na^+ or major anions such as Cl^- and CO_3^{2-}. The reasons for this perspective are that: 1) major metal cations and major anions are by nature very soluble, and do not even tend to precipitate except under the most extreme of conditions; 2) metals such as Fe, Al, Hg, etc. are of great interest for many reasons relating to nutrient, ore body, and toxicity considerations; and 3) metals such as Fe, Al, Hg, etc. are more often controlled by the chemistry that is being set by more concentrated species, rather than vice versa.

An initial, naive interpretation of the criterion given above would be the mistaken idea that the solid containing the species of interest that has the lowest K_{s0} will always be the solubility limiting solid for that species. As an example of why this thinking is incorrect, consider the case of ferrous iron (Fe(II)) at 25°C/1 atm. For the hydroxide and carbonate solids of Fe^{2+}, we have

$$Fe(OH)_{2(s)} = Fe^{2+} + 2OH^- \qquad K_{s0} = 10^{-15.1} \tag{14.1}$$

$$FeCO_{3(s)} = Fe^{2+} + CO_3^{2-} \qquad K_{s0} = 10^{-10.68}. \tag{14.2}$$

Just because the K_{s0} for $Fe(OH)_{2(s)}$ is smaller than that for $FeCO_{3(s)}$ *does not* mean that $Fe(OH)_{2(s)}$ will always precipitate to the exclusion of $FeCO_{3(s)}$. On the contrary, if $\{CO_3^{2-}\}$ gets high enough, then $FeCO_{3(s)}$ will prescribe a lower $\{Fe^{2+}\}$ than does $Fe(OH)_{2(s)}$, leading to a situation where $\{Fe^{2+}\}$ becomes limited by the K_{s0} for $FeCO_{3(s)}$.

The veracity of the fact that the solid that gives the lowest equilibrium activity of the metal ion will limit the solubility may be shown by considering what would happen if this were *not* the case. Let us assume for a given water chemistry situation that $FeCO_{3(s)}$: 1) limits solubility; 2) is at saturation equilibrium with and therefore is controlling the aqueous phase speciation; *and yet* it 3) specifies a higher $\{Fe^{2+}\}$ than does $Fe(OH)_{2(s)}$. Since $Fe(OH)_{2(s)}$ specifies a lower value of $\{Fe^{2+}\}$, $Fe(OH)_{2(s)}$ must be supersaturated in this system. Thus, the minimization of the free energy G in this system will require that some $Fe(OH)_{2(s)}$ precipitate. Since this will reduce $\{Fe^{2+}\}$, it will also in turn lead to the dissolution of some $FeCO_{3(s)}$. More $Fe(OH)_{2(s)}$ will then precipitate. Finally, either: 1) no $FeCO_{3(s)}$ will be left, and just $Fe(OH)_{2(s)}$ will be present and controlling the solubility; or 2) in accordance with Corollary 1 given above, the dissolution of the $FeCO_{3(s)}$ and the precipitation of the $Fe(OH)_{2(s)}$ will have raised $\{CO_3^{2-}\}$ and lowered $\{OH^-\}$ such that both solids finally specify the *same* $\{Fe^{2+}\}$ value. In the latter case, if both solids are present, they will *both* be in equilibrium with the solution phase.

To summarize the above results, if some generic "solid 1" should prescribe a value for $\{M^{z+}\}$ that is higher than that prescribed by some generic "solid 2" under the same conditions, then solid 1 cannot be the solubility-limiting solid; rather, solid 2 will play that role.

Actually, the use of the activity of the uncomplexed metal ion $\{M^{z+}\}$ in the wordings of the Solubility-Limiting Criterion and Corollary 1 reflects only part of the chemistry at hand. Indeed, we could have expressed both statements in terms of the activity of $MOH^{(z-1)+}$, or $M(OH)_2^{(z-2)+}$, or *any other dissolved, M-containing species*. This may be shown as follows. Let us say for the aqueous conditions of interest (i.e., T, P, $\{H^+\}$, $\{OH^-\}$, etc.) that solid 2 prescribes a value for $\{M^{z+}\}$ that is lower than that prescribed by solid 1. Now, if some M^{z+} is present, then equilibria such as

$$M^{z+} + OH^- = MOH^{(z-1)+} \qquad K_{H1} = \frac{\{MOH^{(z-1)+}\}}{\{M^{z+}\}\{OH^-\}} \tag{14.3}$$

$$MOH^{(z-1)+} + OH^- = M(OH)_2^{(z-2)+} \qquad K_{H2} = \frac{\{M(OH)_2^{(z-2)+}\}}{\{MOH^{(z-1)+}\}\{OH^-\}} \tag{14.4}$$

must be satisfied regardless of whether or not any M-containing solid(s) is (are) present.

Since the solution conditions are characterized by a *single* value of $\{OH^-\}$, if solid 2 prescribes a value for $\{M^{z+}\}$ that is lower than that prescribed by solid 1, by Eq. (14.3) it must *also* prescribe a value of $\{MOH^{(z-1)+}\}$ that is lower than that prescribed by solid 1. Eq. (14.4) may be used to make an analogous argument concerning the value of $\{M(OH)_2^{(z-2)+}\}$. The appropriate equilibria may be used to make analogous arguments concerning the activities of all other dissolved, metal-containing species (e.g., $MCl^{(z-1)+}$). These considerations bring us to

Solubility-Limiting Corollary 2: If, for the conditions of interest a given solid prescribes a value of $\{M^{z+}\}$ that is lower than that prescribed by all other possible solids, then it will also prescribe values of $\{MOH^{(z-1)+}\}$, $\{M(OH)_2^{(z-2)+}\}$, and *all* other *dissolved*, metal-containing species that are lower than are prescribed by all other possible solids for those same conditions.

Just as any given aqueous solution can be characterized by only one value of $\{OH^-\}$, only one $\{Cl^-\}$, etc., then any given species will be characterized by only one activity coefficient. Thus, the species that prescribes the lowest values of $\{M^{z+}\}$, $\{MOH^{(z-1)+}\}$, $\{M(OH)_2^{(z-2)+}\}$, $\{MCl^{(z-1)+}\}$, etc. will also prescribe the lowest values of $[M^{z+}]$, $[MOH^{(z-1)+}]$, $[M(OH)_2^{(z-2)+}]$, $[MCl^{(z-1)+}]$, etc., and *therefore*, also the lowest value of the total dissolved metal, M_T. Thus, we have

Solubility-Limiting Corollary 3: If a given solid minimizes the value of $\{M^{z+}\}$, it will also minimize the total dissolved concentration M_T; and, if two solids are coexisting in equilibrium, then they will specify the same value of M_T.

The remainder of this chapter discusses the implications of these various principles. Example problems are considered in Pankow (1992).

14.2 THE COEXISTENCE OF TWO SOLID PHASES

Let us combine the appropriate equilibria to obtain the equilibrium constant for the *interconversion* of the two solids $FeCO_{3(s)}$ and $Fe(OH)_{2(s)}$. Using the values for the various constants at 25°C/1 atm, we have

$$FeCO_{3(s)} = Fe^{2+} + CO_3^{2-} \qquad K_{s0}^{car} = 10^{-10.68} \tag{14.2}$$
$$Fe^{2+} + 2OH^- = Fe(OH)_{2(s)} \qquad (K_{s0}^{hyd})^{-1} = 1/(10^{-15.1}) \tag{14.1}$$
$$2(H_2O = H^+ + OH^-) \qquad K_w^2 = 10^{-28.0} \tag{14.5}$$
$$2H^+ + CO_3^{2-} = H_2CO_3^* \qquad 1/(K_1K_2) = 1/10^{-16.68} \tag{14.6}$$
$$+ \quad H_2CO_3^* = H_2O + CO_{2(g)} \qquad 1/K_H = 1/10^{-1.47} \tag{14.7}$$
$$\overline{FeCO_{3(s)} + H_2O = Fe(OH)_{2(s)} + CO_{2(g)} \qquad K = p_{CO_2} = 10^{-5.43}.} \tag{14.8}$$

Let us assume that when solids are present that they are pure (i.e., $X = 1$), and their activities are unity. Then, for 25°C/1 atm and infinite dilution in the aqueous phase, according to Eq. (14.6), the above two solids can coexist *only* when $p_{CO_2} = 10^{-5.43}$ atm. That is, as long as $p_{CO_2} > 10^{-5.43}$ atm, reaction (14.8) will have a positive ΔG, and it will proceed spontaneously to the *left*. Conversely, as long as $p_{CO_2} < 10^{-5.43}$ atm, then reaction (14.8) will have a negative ΔG, and it will proceed spontaneously to the *right*.[1]

Based on the thermodynamic considerations discussed above, we can therefore conclude that at 25°C/1 atm:

1. For $p_{CO_2} > 10^{-5.43}$ atm, $FeCO_{3(s)}$ is the stable solid, and it therefore specifies a lower $\{Fe^{2+}\}$ than does $Fe(OH)_{2(s)}$ for all values of pH.
2. For $p_{CO_2} = 10^{-5.43}$ atm, $Fe(OH)_{2(s)}$ and $FeCO_{3(s)}$ are both stable, and they therefore specify equal values for $\{Fe^{2+}\}$ for all values of pH.
3. For $p_{CO_2} < 10^{-5.43}$ atm, $Fe(OH)_{2(s)}$ is the stable solid, and it therefore specifies a lower $\{Fe^{2+}\}$ than does $FeCO_{3(s)}$ for all values of pH.

14.3 LOG p_{CO_2} VS. pH PREDOMINANCE DIAGRAMS FOR $FeCO_{3(s)}$, $Fe(OH)_{2(s)}$, AND DISSOLVED Fe(II) SPECIES

In Chapters 11–13, we consider how saturation equilibrium with hydroxide and carbonate solids will cause the concentrations of the metal ions and their hydroxo complexes to change in response to changes in the pH. In the preceding section, we examine how the fundamental choice between $Fe(OH)_{2(s)}$ and $FeCO_{3(s)}$ as the solubility-limiting solid will be determined by the value of p_{CO_2}. It is thus logical to consider how a $\log p_{CO_2}$ vs. pH diagram might help to improve our understanding of how different Fe(II) species (including the two solids) will dominate the Fe(II) chemistry in different regions of such a diagram.

14.3.1 Log p_{CO_2} vs. pH Predominance Diagram for 10^{-4} *M* Total Fe(II)

14.3.1.1 The $FeCO_{3(s)}/Fe(OH)_{2(s)}$ Boundary. As discussed above, for $FeCO_{3(s)}$ and $Fe(OH)_{2(s)}$ to be in equilibrium at 25°C/1 atm, we must have $p_{CO_2} = 10^{-5.43}$ atm. Since H^+ does not appear in Eq. (14.8), the coexistence of these two solids at $p_{CO_2} = 10^{-5.43}$ atm will be independent of pH. We can thus draw a horizontal, pH-independent line at $p_{CO_2} = 10^{-5.43}$ atm in the $\log p_{CO_2}$ vs. pH diagram (see Figure 14.1). *Above* that line, $FeCO_{3(s)}$ is solubility limiting and is possible at equilibrium, but $Fe(OH)_{2(s)}$

[1]For the metals discussed in Chapters 12 and 13, at 25°C/1 atm, the $\log p_{CO_2}$ values for equilibrium between the metal carbonate and a corresponding metal oxide or hydroxide are −4.31, −6.16, −6.35, −7.68, −9.39, −12.96, and −14.55 atm for Zn^{2+}, Mg^{2+}, Mn^{2+}, Pb^{2+}, Cd^{2+}, Ca^{2+}, and Ba^{2+}, respectively.

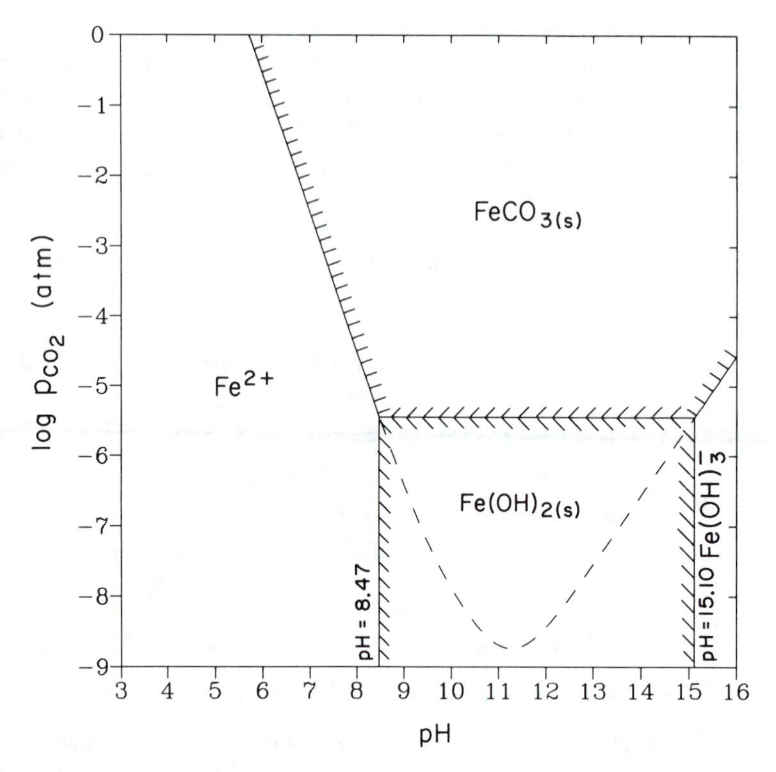

Figure 14.1 Log p_{CO_2} vs. pH predominance diagram for Fe(II) species for $Fe_T = 10^{-4}$ *M* at 25°C/1 atm. Dashed line is the portion of the boundaryline between $FeCO_{3(s)}$ and the solution that is discarded because it falls below $\log p_{CO_2} = -5.43$. Activity corrections neglected (not valid at very low or very high pH).

is not possible. Indeed, if some $Fe(OH)_{2(s)}$ was added to a system characterized by $p_{CO_2} > 10^{-5.43}$ atm, regardless of the pH, it would be unstable and eventually would: 1) dissolve completely; 2) be converted totally to $FeCO_{3(s)}$; or 3) experience some mixture of dissolution and conversion to $FeCO_{3(s)}$. Whether 1, 2, or 3 would occur would depend on the pH and on *how much* $Fe(OH)_{2(s)}$ was added.

Below the $p_{CO_2} = 10^{-5.43}$ atm line, $Fe(OH)_{2(s)}$ is solubility limiting and is possible at equilibrium, but $FeCO_{3(s)}$ is not possible; if some $FeCO_{3(s)}$ was added to a system characterized by $p_{CO_2} < 10^{-5.43}$ atm, regardless of the pH, it would be unstable and eventually would: 1) dissolve completely; 2) be converted totally to $Fe(OH)_{2(s)}$; or 3) experience some mixture of dissolution and conversion to $Fe(OH)_{2(s)}$. Whether 1, 2, or 3 would occur would depend on the pH and on how much $FeCO_{3(s)}$ was added.

As discussed in the preceding paragraph, along the $p_{CO_2} = 10^{-5.43}$ atm line itself, both solids are possible. However, just because $p_{CO_2} = 10^{-5.43}$ atm does not mean that there *will* be equilibrium between these two solids. For example, it is obvious that there can be no such equilibrium in a system characterized by $p_{CO_2} = 10^{-5.43}$ atm, but containing no iron at all. We can therefore conclude that the line $p_{CO_2} = 10^{-5.43}$ only describes the locus of points for which there *can* be equilibrium between $FeCO_{3(s)}$ and $Fe(OH)_{2(s)}$. *If* one of these two solids is present, then at equilibrium, the other one *can* also be present.

The actual presence of a solid in any given situation will be subject to mass balance requirements. If the level of Fe(II) in a system is extremely low, then *neither* $Fe(OH)_{2(s)}$ nor $FeCO_{3(s)}$ will tend to form. Also, if $p_{CO_2} = 10^{-5.43}$ atm and one of the solids is in fact present, then the other solid will only be present if the system that led to the final water + solid(s) equilibrium state had enough Fe(II) and alkalinity to also lead to the precipitation of the other solid.

14.3.1.2 The $Fe^{2+}/Fe(OH)_{2(s)}$ Boundary for 10^{-4} *M* Total Fe(II). Let us assume for the moment that we have a $Fe(OH)_{2(s)}$/aqueous system that is characterized by a specific p_{CO_2} at which $Fe(OH)_{2(s)}$ is stable. As the pH is lowered, we know from Chapter 11 that ultimately, solids like $Fe(OH)_{2(s)}$ will completely dissolve leaving a one-phase system containing only *dissolved* Fe(II) species. (We are assuming here that the supply of $Fe(OH)_{2(s)}$ is not large.) If we lower the pH while keeping p_{CO_2} constant, it is apparent that we will be moving horizontally from right to left across Figure 14.1, and that we will be moving from a two-phase, $Fe(OH)_{2(s)}$/aqueous solution system to a one-phase system in which all of the iron is present in the solution phase.

For any given value of p_{CO_2}, it will be useful to identify the pH value at which all of the $Fe(OH)_{2(s)}$ disappears and the dissolution is complete. The value of that pH will depend on the *total* amount of Fe(II) (both dissolved *plus* solid) that is present per L of water. If we wanted to be perfectly precise, we should have said "per kg of water" in the preceding sentence. Indeed, we need to know exactly how much water is present, and the density of water depends on temperature, pressure, as well as the concentrations of dissolved constituents. At the types of dilute concentrations of interest here, a liter of water is essentially all water, and since we know the density of pure water as a function of temperature and pressure, we will not in general need to bother with this detail.

The symbol for the total Fe(II) "concentration" (both dissolved plus solid) used here will be Fe_T, and it will have units of *M*. As the reader knows, the symbol Fe_T has *also* been used in this text (and in fact already in this chapter) to refer to only the total *dissolved* Fe. This duality of definitions will not be as confusing as the reader might think since the context of the application should always be clear, and in addition, our mathematical use of Fe_T in the sections that follow will always occur when all of the metal is in fact present in the solution phase.

For any given value of Fe_T, the collection of (p_{CO_2},pH) coordinates at which the dissolution is complete will form a line. Along that line, the sum of the concentrations of all dissolved Fe(II) species (as controlled by $Fe(OH)_{2(s)}$) will equal Fe_T. Each different value of Fe_T will give a different line. Let us consider the example where

$$Fe_T = 10^{-4}M. \tag{14.9}$$

Neglecting activity corrections, along the aforementioned line, we have

$$[Fe^{2+}] + [FeOH^+] + [Fe(OH)_2^o] + [Fe(OH)_3^-] = Fe_T \qquad (14.10)$$

or

$$K_{s0}^{hyd}/[OH^-]^2 + K_{s1}^{hyd}/[OH^-] + K_{s2}^{hyd} + K_{s3}^{hyd}[OH^-] = Fe_T. \qquad (14.11)$$

As in Eq. (14.2), the superscript "hyd" refers to the hydroxide solid $Fe(OH)_{2(s)}$. At 25°C/1 atm and infinite dilution, $K_{s0}^{hyd} = 10^{-15.1}$, $K_{s1}^{hyd} = 10^{-10.6}$, $K_{s2}^{hyd} = 10^{-7.7}$, $K_{s3}^{hyd} = 10^{-5.1}$. Thus, at 25°C/1 atm, for $Fe_T = 10^{-4}$ M,

$$10^{-15.1}/[OH^-]^2 + 10^{-10.6}/[OH^-] + 10^{-7.7} + 10^{-5.1}[OH^-] = 10^{-4}M. \qquad (14.12)$$

When Eq. (14.12) is solved by trial and error *using a low pH initial guess*, we obtain pOH = 5.53, and for $K_w = 10^{-14.0}$, we have pH = 8.47. It is important to note that this pH is *independent* of p_{CO_2}. We can therefore draw a vertical line in our diagram (Figure 14.1) at pH = 8.47 that begins at a p_{CO_2} value of $10^{-5.43}$ atm and extends downwards. Thus, for $p_{CO_2} < 10^{-5.43}$, for all pH $\leq$ 8.47, there is no $Fe(OH)_{2(s)}$ present. For pH values immediately to the right of 8.47, both $Fe(OH)_{2(s)}$ *and* the aqueous solution phase are present. In the latter case, since some of the Fe(II) is tied up in the solid, the total concentration of *dissolved* Fe(II) species will be *less* than 10^{-4} M.

The region to the right of the pH = 8.47 line in Figure 14.1 is labelled "$Fe(OH)_{2(s)}$." The reason is that this solid will be present for all (pH, p_{CO_2}) coordinates *inside* that domain. As noted above, aqueous solution phase will also be present inside the $Fe(OH)_{2(s)}$ domain. $Fe(OH)_{2(s)}$ *will not* be present directly on the solution phase/$Fe(OH)_{2(s)}$ boundaryline.

The type of diagram that we are drawing is called a "predominance diagram." In any solution phase region, the species used to label the region is the predominant Fe species. However, inside a region labelled with a solid, all we know from the diagram is that the solid is present and is controlling the concentrations of all of the Fe(II) species. Thus, the diagram that we are drawing might be more correctly described as a "predominance and/or solubility controlling diagram." To illustrate this matter further, consider the case of the $Fe(OH)_{2(s)}$ domain. Labelling that domain with "$Fe(OH)_{2(s)}$" does not mean that, everywhere inside that domain, there are more mols of $Fe(OH)_{2(s)}$[2] per L of water than there are of total dissolved Fe(II). In fact, inside the domain and very near the solution/solid boundaryline, we know that the concentration of some solution phase Fe species will always exceed the total FW of $Fe(OH)_{2(s)}$ present per liter of aqueous phase. When the value of Fe_T for the diagram is sufficiently large, then well inside the $Fe(OH)_{2(s)}$ domain, the FW of the solid per liter of aqueous phase can however exceed the concentration of any solution phase species. Therefore, inside a region labeled with a solid, that solid will be solubility controlling, but *may* not predominate over other forms of iron anywhere in that region.

[2]Although it might seem strange at first to talk about an aqueous concentration of a solid, it is very useful to do so whenever doing mass balance accounting for an overall solid+aqueous system.

Which dissolved species dominates for pH = 8.47? The answer to this question is the species which gives the largest concentration term in Eq. (14.12) for pH = 8.47. If we evaluate those terms at pH = 8.47, we see that at 9.12×10^{-5} *M*, that species is Fe^{2+}. As the pH is lowered beyond 8.47, $[Fe^{2+}]$ will become even larger (at the expense of the other dissolved metal species), and will asymptotically approach 10^{-4} *M*. For p_{CO_2} values less than $10^{-5.43}$ atm, we can therefore label the entire region for pH $\leq$ 8.47 with "Fe^{2+}".

14.3.1.3 The $Fe(OH)_{2(s)}/Fe(OH)_3^-$ Boundary for 10^{-4} *M* Total Fe(II). It is now useful to consider how the chemistry changes as one moves at constant p_{CO_2} from *left to right* across the $p_{CO_2} < 10^{-5.43}$ atm portion of Figure 14.1. In particular, at very low pH, we now know that almost all of the Fe(II) is present as Fe^{2+}. As the pH increases, Fe^{2+} continues to dominate the Fe(II) chemistry, though more and more of the Fe(II) hydroxo species are forming. At pH = 8.47, the solution has become saturated with respect to $Fe(OH)_{2(s)}$, but no solid is yet present. Once the pH is increased infinitesimally beyond 8.47, at equilibrium, some solid will be present. As the pH is increased further, then initially: 1) more solid will form; and 2) the total dissolved Fe(II) will decrease. The manner in which this will occur can be described using a figure for $Fe(OH)_{2(s)}$ that is analogous to Figure 11.5 for $PbO_{(s)}$. That analogous figure is Figure 14.2.[3] Thus, as the pH continues to increase, eventually the amount of solid phase will start to decrease due to the formation of relatively large amounts of $Fe(OH)_3^-$. At the same time and for the same reason, the amount of dissolved iron will start to increase. Ultimately, all of the $Fe(OH)_{2(s)}$ will redissolve. We need to identify the pH value at which that redissolution occurs for $Fe_T = 10^{-4}$ *M*.

Just as for the cases discussed in Chapter 11, when Fe_T is sufficiently high, we know that Eq. (14.11) must have *two* mathematical solutions, one at "low pH," and one at "high pH." (Note that "low" does not necessarily mean less than 7, and "high" does not necessarily mean greater than 7.) We can search by trial and error for the high-pH root for $Fe_T = 10^{-4}$ *M* by starting the trial and error solution with a very high pH guess. Doing so, we find that, like pH = 8.47, pH = 15.10 is also a root for Eq. (14.12).[4] These low and high pH roots of Eq. (14.12) are the pH values at which $Fe_T = 10^{-4}$ *M* in Figure 14.2.

The species that dominates the solution phase iron at pH = 15.10 can be identified by comparing the magnitudes of the terms in Eq. (14.12). At $\sim 10^{-4}$ *M*, the dominant species is $Fe(OH)_3^-$. Figure 14.2 can also be used to see which species dominates at a given pH; the dominant species at a given pH will be the one that has the highest concentration in that figure. At equilibrium, when there is $Fe(OH)_{2(s)}$ present, then that highest concentration will also be the concentration of the dominant species. When $Fe(OH)_{2(s)}$ is not present, then Figure 14.2 will still give the correct *relative* concentrations

[3]Figure 14.2 has been drawn out to the unrealistically high pH of 16 in order that we can see how the various curves behave in the pH region where $Fe(OH)_{2(s)}$ can give a dissolved Fe_T of 10^{-4} *M*.

[4]Actually, after the fact, we see that trial and error is not really needed here; since $Fe(OH)_3^-$ is by far the dominant species at high pH, we just need to solve an approximate form of Eq. (14.11), namely $K_{s3}^{hyd}[OH^-] \simeq 10^{-4}$.

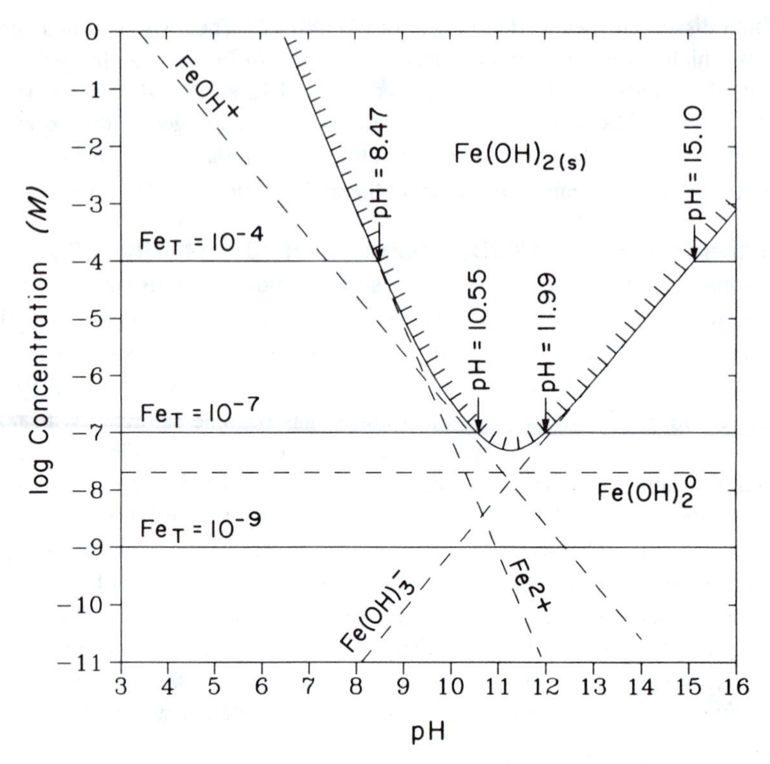

Figure 14.2 Log concentration (M) vs. pH diagram for equilibrium with $Fe(OH)_{2(s)}$. The horizontal lines for $Fe_T = 10^{-4}$ and 10^{-7} M intersect the line for log Fe_T as controlled by the solid. Thus, $Fe(OH)_{2(s)}$ can form in systems containing those levels of Fe_T when $\log p_{CO_2} \leq -5.43$. The horizontal line for $Fe_T = 10^{-9} M$ does not intersect the line for log Fe_T as controlled by $Fe(OH)_{2(s)}$. Thus, for $Fe_T = 10^{-9} M$, $Fe(OH)_{2(s)}$ will not be stable at equilibrium for any value of $\log p_{CO_2}$. Activity corrections neglected (not valid at very low or very high pH).

of the various Fe-containing species; as with the α values for other acid/base species, those relative concentrations depend only on the pH.

Given the dominance of $Fe(OH)_3^-$ at pH = 15.10, the $Fe(OH)_{2(s)}$/solution phase boundaryline is also the $Fe(OH)_{2(s)}/Fe(OH)_3^-$ boundaryline. We should realize, however, that we have not really obtained an accurate location for the position of this line. The reason is that activity corrections cannot be neglected at such pH values; the amount of strong base present makes for a very high ionic strength. Figure 14.1 has been extended to pH = 16 in order to illustrate the location of this line, approximate though it may be.

14.3.1.4 The $Fe^{2+}/FeCO_{3(s)}$ Boundary for 10^{-4} *M* Total Fe(II). Let us now assume that we have a $FeCO_{3(s)}$/aqueous system that is characterized by a specific p_{CO_2} at which $FeCO_{3(s)}$ is stable. As the pH is lowered, all of the $FeCO_{3(s)}$ will ultimately dissolve leaving only the aqueous phase, and therefore only dissolved Fe(II) species. If the lowering of the pH takes place in a manner that maintains p_{CO_2} constant, we will be moving horizontally from right to left across the diagram, and we will move from a two-phase, $FeCO_{3(s)}$/aqueous solution system to a one-phase system in which all of the Fe(II) is present in the aqueous phase.

As with $Fe(OH)_{2(s)}$, for a given value of Fe_T, the "low-pH" collection of (pH, p_{CO_2}) coordinates at which the dissolution is complete will form a line. Along that line, the sum of the concentrations of all dissolved Fe(II) species (as controlled by $FeCO_{3(s)}$) will equal Fe_T as given by Eq. (14.10). Each different value of Fe_T will give a different line.

Since CO_2 chemistry obviously depends on p_{CO_2}, the value of the pH at which all of the $FeCO_{3(s)}$ disappears and the dissolution is complete will depend on the value of p_{CO_2}. (In contrast, since $Fe(OH)_{2(s)}$ does not contain any CO_2 species, the two $Fe(OH)_{2(s)}$/solution boundarylines for $Fe_T = 10^{-4}$ M were found to be independent of p_{CO_2}.) Instead of Eq. (14.11), we now use

$$\frac{K_{s0}^{car}}{[CO_3^{2-}]} + \frac{K_{s0}^{car}}{[CO_3^{2-}]}K_{H1}[OH^-] + \frac{K_{s0}^{car}}{[CO_3^{2-}]}K_{H1}K_{H2}[OH^-]^2 + \frac{K_{s0}^{car}}{[CO_3^{2-}]}K_{H1}K_{H2}K_{H3}[OH^-]^3 = Fe_T \tag{14.13}$$

where activity corrections are again neglected, and the superscript "car" refers to the carbonate solid $FeCO_{3(s)}$. Substituting $[CO_3^{2-}] = K_1K_2K_Hp_{CO_2}/[H^+]^2$, we obtain

$$\frac{K_{s0}^{car}}{K_1K_2K_Hp_{CO_2}}([H^+]^2 + K_{H1}K_w[H^+] + K_{H1}K_{H2}K_w^2 + K_{H1}K_{H2}K_{H3}K_w^3/[H^+]) = Fe_T. \tag{14.14}$$

From Table 13.1, at 25°C/1 atm and infinite dilution, we have $K_{s0}^{car} = 10^{-10.68}$, $K_{H1} = 10^{4.5}$, $K_{H2} = 10^{2.9}$, $K_{H3} = 10^{2.6}$. Under the same conditions, $K_1K_2 = 10^{-16.68}$, $K_H = 10^{-1.47}$, and $K_w = 10^{-14.00}$. Using these equilibrium constants, and substituting 10^{-4} M for Fe_T, Eq. (14.14) becomes

$$\frac{10^{7.47}}{p_{CO_2}}([H^+]^2 + 10^{-9.50}[H^+] + 10^{-20.60} + 10^{-32.00}/[H^+]) - 10^{-4} = 0. \tag{14.15}$$

For $p_{CO_2} = 10^{-5.43}$ atm, solving Eq. (14.15) by trial and error gives pH = 8.47. This is the *same* pH value that was obtained for the low pH root of Eq. (14.12). Thus, at the p_{CO_2} at which there can be equilibrium between $Fe(OH)_{2(s)}$ and $FeCO_{3(s)}$, the boundaryline between the solution phase and $FeCO_{3(s)}$ intersects *exactly* with the intersection of the $Fe(OH)_{2(s)}/FeCO_{3(s)}$ boundaryline and the solution phase/$Fe(OH)_{2(s)}$ boundaryline. (This is a general result: for adjacent predominance regions A, B, and C, the intersection of the A/B boundaryline with the B/C boundaryline will always be the same point as the intersection of B/C and C/A boundarylines.)

While we could continue to solve Eq. (14.15) by trial and error for other values of $p_{CO_2} > 10^{-5.43}$ atm to obtain other points on the solution phase/$FeCO_{3(s)}$ boundaryline, it makes much more sense to pick a range of pH values less than 8.47 and solve directly for the corresponding p_{CO_2} values. Plotted in Figure 14.1, the resulting line indicates that for every unit increase in $\log p_{CO_2}$, the pH decreases by ~ 0.5. Thus, the slope $(dp_{CO_2}/d\text{pH}) \simeq -2$. The reason for this is that for pH ≤ 8.47, Fe^{2+} is the dominant dissolved Fe(II) species. Thus, under these conditions $[Fe^{2+}]$ dominates over the other dissolved Fe(II) species, and so Eq. (14.15) may be approximated as

$$[Fe^{2+}] = \frac{10^{7.47}}{p_{CO_2}}[H^+]^2 \simeq 10^{-4} = Fe_T \tag{14.16}$$

which gives the straight line

$$\log p_{CO_2} \simeq 11.47 - 2\text{pH}. \tag{14.17}$$

As may be seen in Figure 14.1, as the pH decreases and Fe^{2+} becomes increasingly more dominant in the solution phase, Eq. (14.17) is obeyed more and more closely.

On and to the left of the solution phase/$FeCO_{3(s)}$ boundaryline, there is no $FeCO_{3(s)}$ present. Since: 1) Fe^{2+} dominates over all other dissolved Fe(II) species at pH ≤ 8.47; and 2) the pH values in the region to the left of the solution phase/$FeCO_{3(s)}$ boundaryline are all ≤ 8.47, that entire region can be labelled with the predominating species, Fe^{2+}. This region combines with the Fe^{2+} region for $p_{CO_2} < 10^{-5.43}$ atm to form a single, contiguous Fe^{2+} predominance region.

To the right of the solution phase/$FeCO_{3(s)}$ boundaryline, both $FeCO_{3(s)}$ and an aqueous solution phase are present. As for the $Fe(OH)_{2(s)}$ domain, inside the $FeCO_{3(s)}$ domain the total concentration of dissolved Fe(II) species is *less* than 10^{-4} *M* since some of the Fe(II) is tied up as solid. Near the solution phase/$FeCO_{3(s)}$ boundaryline, there will be fewer mols of $FeCO_{3(s)}$ per liter of water than of Fe^{2+}.

14.3.1.5 The $FeCO_{3(s)}/Fe(OH)_3^-$ Boundary for 10^{-4} *M* Total Fe(II). Inside the $FeCO_{3(s)}$ region, as one moves at constant p_{CO_2} from *left to right* across Figure 14.1, the pH will increase, and ultimately, all of the $FeCO_{3(s)}$ will dissolve due to the increasing importance of $Fe(OH)_3^-$. We now need to determine the location of the *right-hand* portion of the boundary between $FeCO_{3(s)}$ and the solution for $Fe_T = 10^{-4}$ *M*.

Just as Eq. (14.12) was found to have two mathematical solutions, for any given value of p_{CO_2}, so too will Eq. (14.15). At any given value of $p_{CO_2} \geq 10^{-5.43}$ atm, we can find the high pH root of Eq. (14.15) by trial and error. At $p_{CO_2} = 10^{-5.43}$, the root is pH = 15.10. As required for this special value of p_{CO_2}, this is the same value as the high pH root of Eq. (14.12).

For values of p_{CO_2} that are greater than $10^{-5.43}$ atm, we can substitute a range of pH values greater than 15.10 into Eq. (14.15), solve for the corresponding p_{CO_2} values, and obtain this right-hand portion of the boundary between the $FeCO_{3(s)}$ domain and the solution domain. The reader may now recall that a similar procedure was suggested for obtaining the solution/$FeCO_{3(s)}$ (left-hand side) boundary for pH ≤ 8.47. Thus, it is very important to realize that we can obtain *all* of the boundaryline between the solution and $FeCO_{3(s)}$ by substituting the *full* range of pH values into Eq. (14.15), solving for p_{CO_2}, then plotting the results. Whatever portion of the line falls below $p_{CO_2} = 10^{-5.43}$ atm is discarded since in that region $Fe(OH)_{2(s)}$ is the stable phase. The discarded portion of the line is drawn as a dashed line in Figure 14.1.

For the pH values lying along the right-hand side of the boundary between $FeCO_{3(s)}$ and the solution, $Fe(OH)_3^-$ is, by far, the dominant Fe species. Thus, this right-hand side of the boundary appears to be very nearly a straight line because under these conditions, Eq. (14.15) takes on the approximate form $(10^{7.47}/p_{CO_2})(10^{-32.00}/[H^+]) \simeq 10^{-4}$, or pH $\simeq 20.53 + \log p_{CO_2}$. As p_{CO_2} increases, the pH must increase since it becomes increasingly difficult to form $Fe(OH)_3^-$ from the CO_2-containing solid $FeCO_{3(s)}$. Of course, at such high pH values, this right-hand side of the boundaryline is only approximate since activity corrections cannot be neglected in this region, and in fact we cannot even physically access a pH of 16 in water.

14.3.2 Dependence of p_{CO_2} vs. pH Predominance Diagram on Assumed Total Fe(II) Value

Since Eq. (14.8) does not depend on Fe_T, the location of the $FeCO_{3(s)}/Fe(OH)_{2(s)}$ boundaryline does not depend on Fe_T. However, both Eqs. (14.11) and (14.14) depend on Fe_T, and so the locations of the boundarylines between the solution and the solids will depend on Fe_T. Viewing the phases separated by the line $\log p_{CO_2} = -5.43$, at $Fe_T = 10^{-4}$ *M*, both solids have a solution-phase/solid-phase boundaryline, and a solid-phase/solution-phase boundaryline. For the solution-phase/$FeCO_{3(s)}$ line and the solution-phase/$Fe(OH)_{2(s)}$ line, lowering the value of Fe_T will shift both lines to the right; lowering the amount of iron in the system makes it increasingly difficult to precipitate an Fe(II) solid. ("Increasingly difficult" means "increasingly higher pH" for the basic solids $FeCO_{3(s)}$ and $Fe(OH)_{2(s)}$.) For the $FeCO_{3(s)}$/solution-phase and $Fe(OH)_{2(s)}$/solution-phase lines, lowering Fe_T shifts both lines to the left; lowering the amount of iron in the system makes it increasingly easier to redissolve these Fe(II) solids. ("Increasingly easier" means "increasingly lower pH.")

For 25°C/1 atm, $p_{CO_2} > 10^{-5.43}$ atm, and $Fe_T = x$, the width of the $FeCO_{3(s)}$ region will equal the width of the hatched region across log concentration $= \log x$ in a figure like Figure 13.1d, prepared for the p_{CO_2} value of interest. At $p_{CO_2} < 10^{-5.43}$ atm, the width of the $Fe(OH)_{2(s)}$ region will equal to the width of the hatched region across log concentration $= \log x$ in Figure 14.2.

Since lines that depend on the total metal concentration do not overlap one another, it is common to see predominance diagrams drawn for a range of M_T values. In this manner, one can use such a single diagram to communicate a great deal of information.

14.3.3 Log p_{CO_2} vs. pH Predominance Diagram for 10^{-7} *M* Total Fe(II)

14.3.3.1 The $FeCO_{3(s)}/Fe(OH)_{2(s)}$ Boundary. As noted above, the location of the $FeCO_{3(s)}/Fe(OH)_{2(s)}$ boundaryline does not depend on Fe_T. Therefore, its location for $Fe_T = 10^{-7}$ *M* will be identical to that for $Fe_T = 10^{-4}$ *M*, that is, $p_{CO_2} = 10^{-5.43}$ atm.

However, since the solution-phase/solid-phase boundarylines will be shifted, its *length* will be different (shorter) than in Figure 14.1.

14.3.3.2 The $Fe^{2+}/FeOH^+$ and $FeOH^+/Fe(OH)_{2(s)}$ Boundaries for 10^{-7} *M* Total Fe(II). We now solve Eq. (14.11) for the "low pH" root for $Fe_T = 10^{-7}$ *M*. Neglecting activity corrections again, at 25°C/1 atm, we obtain pH = 10.55. As with the pH obtained for $Fe_T = 10^{-4}$ *M*, this pH value is independent of p_{CO_2}. We therefore draw a vertical line in Figure 14.2 at pH = 10.55 that extends downwards from a starting p_{CO_2} value of $10^{-5.43}$ atm. The region to the right of the pH = 10.55 line is labelled "$Fe(OH)_{2(s)}$" since that solid will be present for all pH and p_{CO_2} values *inside* that domain, but not directly on the solution-phase/$Fe(OH)_{2(s)}$ boundaryline.

Which dissolved species dominates on and to the left of the pH = 10.55 line? The answer is the species that gives the largest concentration term in Eq. (14.11) at pH = 10.55. Evaluating the various terms at pH = 10.55 gives $[Fe^{2+}] = 6.33 \times 10^{-9}$, $[FeOH^+] = 7.09 \times 10^{-8}$, $[Fe(OH)_2^o] = 2.00 \times 10^{-8}$, and $[Fe(OH)_3^-] = 2.81 \times 10^{-9}$ *M*. The fact that $FeOH^+$ dominates at pH = 10.55 can also be seen from Figure 14.2. Thus, in the region on and to the left of the solution-phase/$Fe(OH)_{2(s)}$ boundaryline, the dominant species is now $FeOH^+$. As the pH decreases, however, we know that Fe^{2+} will ultimately become dominant. Since $K_{H1} = 10^{4.5}$ at 25°C/1 atm, with $K_w = 10^{-14.00}$, the pH value at which this dominance will change is 9.50. Figure 14.3 therefore includes a second vertical line at pH = 9.50 which defines the $Fe^{2+}/FeOH^+$ dominance boundaryline.

14.3.3.3 The $Fe(OH)_{2(s)}/Fe(OH)_3^-$ Boundary for 10^{-7} *M* Total Fe(II). We now solve Eq. (14.11) for the "high pH" root for $Fe_T = 10^{-7}$ *M*. Neglecting activity corrections again, at 25°C and 1 atm total pressure, we obtain pH = 11.99. As with the value obtained for $Fe_T = 10^{-4}$ *M*, this pH value is independent of p_{CO_2}. (It is also still a bit high to be neglecting activity coefficients.) We therefore draw a vertical line in Figure 14.2 at pH = 11.99 that extends downwards from a starting p_{CO_2} value of $10^{-5.43}$ atm. Given the values of K_{H1}, K_{H2}, and K_{H3}, the dominant species at pH = 11.99 is $Fe(OH)_3^-$. Thus, the $Fe(OH)_{2(s)}$/solution-phase boundaryline is also the $Fe(OH)_{2(s)}/Fe(OH)_3^-$ boundaryline. A comparison of Figures 14.1 and 14.3 illustrates how lowering Fe_T from 10^{-4} to 10^{-7} *M* has expanded the Fe^{2+} region, created an $FeOH^+$ region, greatly narrowed the $Fe(OH)_{2(s)}$ region, as well as created a region for $Fe(OH)_3^-$.

14.3.3.4 The $Fe^{2+}/FeCO_{3(s)}$ and $FeOH^+/FeCO_{3(s)}$ Boundaries for 10^{-7} *M* Total Fe(II). We now focus on Eq. (14.14) for $Fe_T = 10^{-7}$ *M*. Thus, neglecting activity corrections again, 25°C/1 atm, instead of solving Eq. (14.15), we use

$$\frac{10^{7.47}}{p_{CO_2}}([H^+]^2 + 10^{-9.50}[H^+] + 10^{-20.60} + 10^{-32.00}/[H^+]) - 10^{-7} = 0 \qquad (14.18)$$

As expected, for $p_{CO_2} = 10^{-5.43}$ atm, Eq. (14.18) gives a low pH root of 10.55, and a high pH root of 11.99.

As with Figure 14.1, we do not need to go through the trouble of picking a series of p_{CO_2} values and solving by trial and error for the corresponding low and high roots of

Eq. (14.18). Rather, we do the inverse, and solve Eq. (14.18) for the p_{CO_2} values given by the full range of pH values. We thereby obtain the curved line given in Figure 14.3. The dashed portion of the line below $p_{CO_2} = 10^{-5.43}$ atm is discarded. At pH $\ll$ 9.50, where $[Fe^{2+}]$ dominates over the other dissolved Fe(II) terms in Eq. (14.18), again the slope $(dp_{CO_2}/d\text{pH}) \simeq 2$. In particular, at pH $\ll$ 9.50, Eq. (14.18) simplifies to

$$[Fe^{2+}] = \frac{10^{7.47}}{p_{CO_2}}[H^+]^2 \simeq 10^{-7} = Fe_T \tag{14.19}$$

which gives

$$\log p_{CO_2} \simeq 14.47 - 2\text{pH}. \tag{14.20}$$

At pH values near to and greater than 9.50, however, there is a very noticeable departure from linearity. As the reader may well suspect, that nonlinearity is caused by the importance of $FeOH^+$ and $Fe(OH)_2^o$ in that pH range.

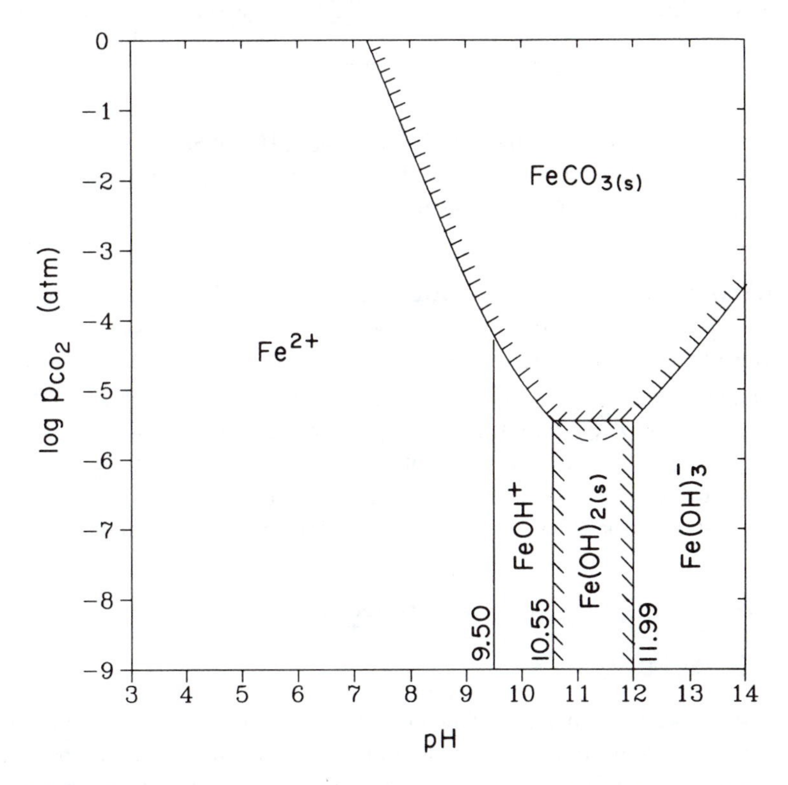

Figure 14.3 Log p_{CO_2} vs. pH predominance diagram for Fe(II) species for $Fe_T = 10^{-7}$ *M* at 25°C/1 atm. Dashed line is the portion of the boundaryline between $FeCO_{3(s)}$ and the solution that is discarded because it falls below $\log p_{CO_2} = -5.43$. Activity corrections neglected (not valid at very low or very high pH).

TABLE 14.1 Expressions used to calculate Fe(II) species concentrations for a hybrid $FeCO_{3(s)}/Fe(OH)_{2(s)}$ log concentration vs. pH diagram (activity corrections neglected).

	pH < pH_{trans}		pH > pH_{trans}	
Species	General Expression	Expression at 25°C/1 atm	General Expression	Expression at 25°C/1 atm
Fe^{2+}	$\frac{K_{s0}^{car}}{\alpha_2 C_T}$	$\frac{10^{-10.68}}{\alpha_2 C_T}$	$\frac{K_{s0}^{hyd}}{[OH^-]^2}$	$\frac{10^{-15.1}}{[OH^-]^2}$
$FeOH^+$	$\frac{K_{s0}^{car} K_{H1}[OH^-]}{\alpha_2 C_T}$	$\frac{10^{-6.18}[OH^-]}{\alpha_2 C_T}$	$\frac{K_{s1}^{hyd}}{[OH^-]}$	$\frac{10^{-10.6}}{[OH^-]}$
$Fe(OH)_2^o$	$\frac{K_{s0}^{car} K_{H1} K_{H2}[OH^-]^2}{\alpha_2 C_T}$	$\frac{10^{-3.28}[OH^-]^2}{\alpha_2 C_T}$	K_{s2}^{hyd}	$10^{-7.7}$
$Fe(OH)_3^-$	$\frac{K_{s0}^{car} K_{H1} K_{H2} K_{H3}[OH^-]^3}{\alpha_2 C_T}$	$\frac{10^{-0.68}[OH^-]^3}{\alpha_2 C_T}$	$K_{s3}^{hyd}[OH^-]$	$10^{-5.1}[OH^-]$

The reader will note that the $Fe^{2+}/FeOH^+$ line (pH = 9.50) terminates at the point at which it touches the solution/$FeCO_{3(s)}$ line. Nevertheless, even inside the $FeCO_{3(s)}$ region, if pH = 9.50, then $[Fe^{2+}] = [FeOH^+]$.

14.3.3.5 The $FeCO_{3(s)}/Fe(OH)_3^-$ Boundary for 10^{-7} *M* Total Fe(II). Since $Fe(OH)_3^-$ is by far the dominant species at the types of pH values found along the right-hand portion of the boundaryline between $FeCO_{3(s)}$ and the solution, along that portion we can use an approximate form of Eq. (14.18), namely $(10^{7.47}/p_{CO_2})10^{-32.00}/[H^+]) \simeq 10^{-7}$, or pH $\simeq 17.53 + \log p_{CO_2}$. Again, as the p_{CO_2} increases, the pH increases since it becomes increasingly difficult to form $Fe(OH)_3^-$ from the CO_2-containing $FeCO_{3(s)}$. We also note that while we have chosen to neglect activity corrections in Figure 14.3, doing so introduces some error along the $FeCO_{3(s)}/Fe(OH)_3^-$ boundary at high pH.

14.3.4 Log p_{CO_2} vs. pH Predominance Diagram for 10^{-9} *M* Total Fe(II)

Figure 14.4 is the predominance diagram for $Fe_T = 10^{-9}$ *M*. As is evident from Figure 14.2, for this value of Fe_T, we cannot saturate the system with respect to $Fe(OH)_{2(s)}$. Thus, there is no $Fe(OH)_{2(s)}$ region in Figure 14.4. Now, the curve describing the solution/$FeCO_{3(s)}$ boundaryline is located totally above the line $p_{CO_2} = 10^{-5.43}$ atm, and $FeCO_{3(s)}$ is the only solid in the diagram. The boundarylines for transition in dominance between the various dissolved Fe(II) species are located at pH = 9.50 ($[Fe^{2+}] = [FeOH^+]$), 11.10 ($[FeOH^+] = [Fe(OH)_2^o]$), and 11.40 ($[Fe(OH)_2^o] = [Fe(OH)_3^-]$).

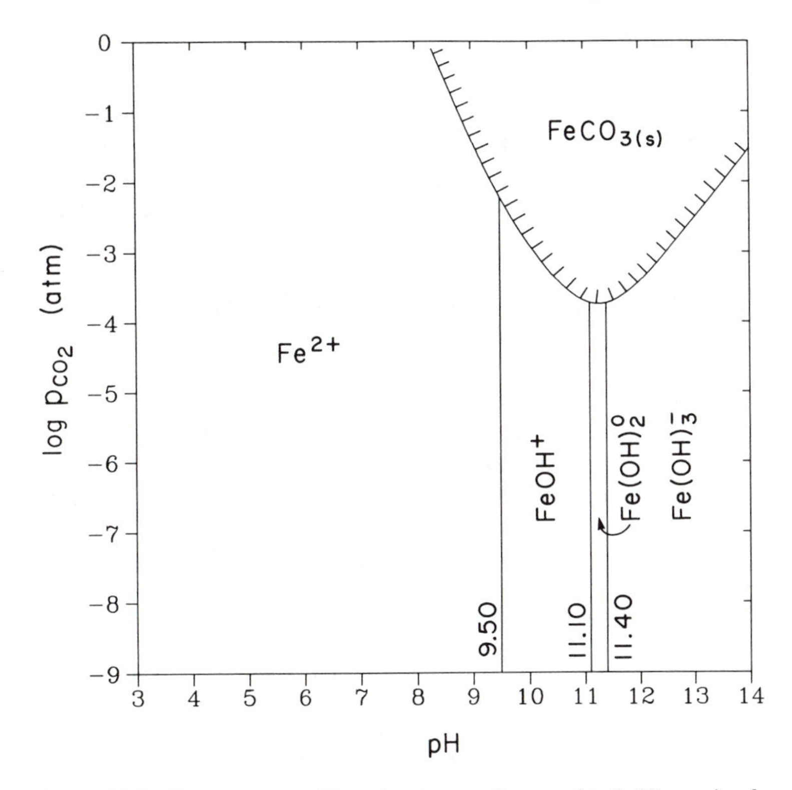

Figure 14.4 Log p_{CO_2} vs. pH predominance diagram for Fe(II) species for $Fe_T = 10^{-9}$ M at 25°C/1 atm. All of the boundaryline between $FeCO_{3(s)}$ and the solution is pertinent because none of it falls below $\log p_{CO_2} = -5.43$. $Fe(OH)_{2(s)}$ is not stable anywhere in the diagram. Activity corrections neglected (not valid at very low or very high pH).

14.4 LOG CONCENTRATION VS. pH DIAGRAM FOR A $FeCO_{3(s)}/Fe(OH)_{2(s)}$ SYSTEM FOR A CLOSED SYSTEM WITH CONSTANT C_T

14.4.1 Determination of pH_{trans}

In previous chapters, we discuss how to draw a log concentration vs. pH diagram for a metal hydroxide such as $Fe(OH)_{2(s)}$, and how to draw a similar diagram for a metal carbonate such as $FeCO_{3(s)}$ at a given constant C_T value. We now know that the solid that gives the lowest $\{Fe^{2+}\}$ under a given set of conditions will be the solubility controlling solid. This means that for a given C_T value, the $FeCO_{3(s)}$ diagram will apply wherever it specifies the lowest $\{Fe^{2+}\}$, and the $Fe(OH)_{2(s)}$ diagram will apply wherever it specifies the lowest $\{Fe^{2+}\}$. If C_T is very small, then it is possible for the $Fe(OH)_{2(s)}$ diagram to apply over the whole natural water pH range. The reason for this is that $K_{s0}^{hyd}/\{OH^-\}^2$ can then remain smaller than $K_{s0}^{car}/\alpha_2 C_T$ over the entire pH range.[5]

If both of the two solids have regions of pertinence, then the overall log concentration vs. pH diagram will be a hybrid diagram incorporating portions of the *two* individual diagrams. The pH at which the transition in control takes place (i.e., pH_{trans}) will be the pH at which both solids specify equal values for each dissolved Fe(II) species, and also for Fe_T. Thus, for each Fe(II) species (including Fe_T), the two lines for that species as specified by the two solids will cross at pH = pH_{trans}. Let us imagine that this intersection as an X (albeit possibly a tilted X) where the location of one slash (e.g., /) of the X is determined by one solid,and the location of the other slash (\) is determined by the other solid. The hybrid line that applies for that species will then be composed of the two *lowest* arms of the X, that is, the ∧ portion of the X.

We now consider how a diagram of this type will appear for $C_T = 10^{-3}$ *M*. We can determine the value of pH_{trans} by setting the expressions for $\{Fe^{2+}\}$ as determined by the two solids equal to one another. Neglecting activity corrections, we have

$$K_{s0}^{car}/(\alpha_2 C_T) = K_{s0}^{hyd}/[OH^-]^2. \tag{14.21}$$

At 25°C/1 atm and $C_T = 10^{-3}$ *M*, we have

$$10^{-10.68}/(\alpha_2 10^{-3}) = 10^{-15.1}[H^+]^2/10^{-28.00}. \tag{14.22}$$

When Eq. (14.22) is solved by trial and error, we obtain $pH_{trans} = 10.06$. (For use below, pH_{trans} to four decimal places is 10.0624.)

For pH values less than 10.06, the LHS of Eq. (14.22) is less than the RHS, and so the diagram for $FeCO_{3(s)}$ (i.e., Figure 12.3*d*) applies. For pH values greater than 10.06, the RHS of Eq. (14.22) is less than the LHS, and so the diagram for $Fe(OH)_{2(s)}$ (i.e., Figure 14.2) applies. Increasing the value of C_T will increase the value of pH_{trans} since increasing C_T favors $FeCO_{3(s)}$, and extends its dominance region. (For 25°C/1 atm, increasing C_T increases the value of the pH required to reduce the in-situ p_{CO_2} value to a value less than $10^{-5.43}$ atm.) The resultant hybrid log concentration vs. pH diagram is given in Figures 14.5 and 14.6. (See also Figure 5.11a of Stumm and Morgan (1981).) In Figure 14.5, a dashed line is used wherever a log concentration line for a species as prescribed by one solid lies above the line for the same species as prescribed by the other solid. Thus, at equilibrium, only the solid lines have any meaning. For clarity, Figure 14.6 includes only the pertinent portions of the various lines. Figure 14.6 can be obtained by taking a scissors and cutting both Figures 12.3*d* and 14.2 along the line pH = 10.06, then combining the left-hand portion of Figure 12.3*d* with the right-hand portion of Figure 14.2. The expressions used to prepare Figures 14.5 and 14.6 are summarized in Table 14.1.

[5]We choose C_T = constant for this discussion so that the potential exists for a transition in stability between $Fe(OH)_{2(s)}$ and $FeCO_{3(s)}$. As discussed above, choosing p_{CO_2} equal to some constant value would likely lead to the stability of just one of these solids.

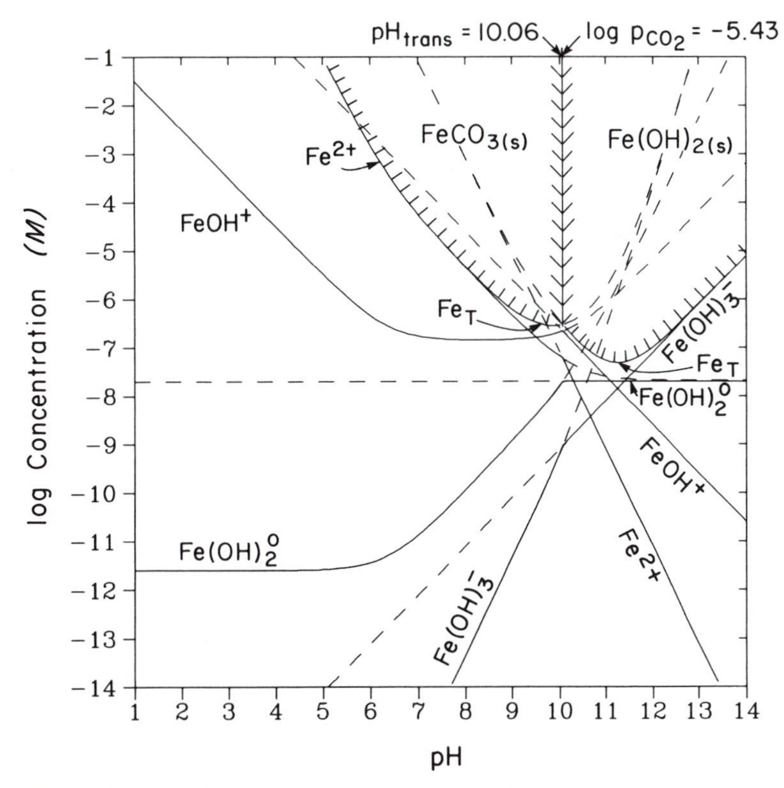

Figure 14.5 Hybrid log concentration vs. pH diagram for Fe(II) for $C_T = 10^{-3}$ *M* at 25°C/1 atm. At pH_{trans}, $\log p_{CO_2} = -5.43$. The dashed lines give lines for the various species for pH > pH_{trans} for $Fe(OH)_{2(s)}$, and for pH < pH_{trans} for $Fe(CO)_{3(s)}$. At equilibrium, only the solid lines are relevant. Note that for *each species* (including Fe_T), pH = pH_{trans}, the line for $FeCO_{3(s)}$ intersects with the line for $Fe(OH)_{2(s)}$. Activity corrections neglected (not valid at very low or very high pH).

14.4.2 Consistency of pH_{trans} with the Concept of a Specific p_{CO_2} for Equilibrium between $FeCO_{3(s)}$ and $Fe(OH)_{2(s)}$

In Section 14.2, we discussed how at 25°C/1 atm the two solids $FeCO_{3(s)}$ and $Fe(OH)_{2(s)}$ can coexist only when $p_{CO_2} = 10^{-5.43}$ atm. We also recall that in Section 9.10 we discussed that a solution that is not in equilibrium with any gas phase can still nevertheless be thought of as being characterized by an in-situ p_{CO_2} value. Thus, in order for Figures 14.5 and 14.6 to be consistent with the p_{CO_2} vs. pH predominance diagrams that we have been considering in this chapter, we conclude that the value of p_{CO_2} at pH_{trans} in Figures 14.5 and 14.6 *must* be $10^{-5.43}$ atm. To check this, we test to see if we obtain the same $[CO_3^{2-}]$ for both a pH of 10.0624 and a $C_T = 10^{-3}$ *M*, and for a pH of 10.0624

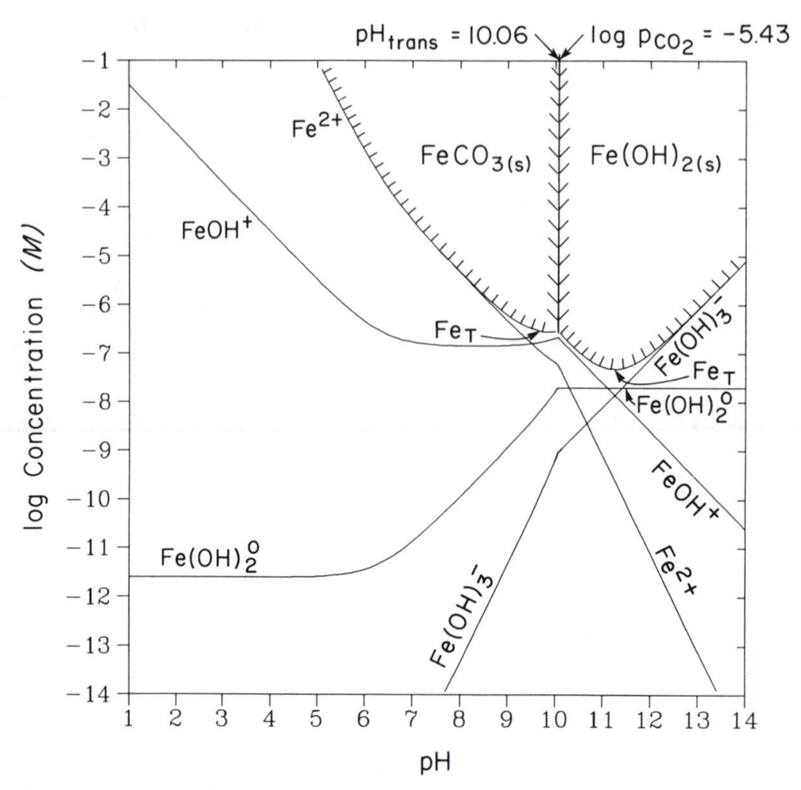

Figure 14.6 Hybrid log concentration vs. pH diagram for Fe(II) for $C_T = 10^{-3}$ *M* at 25°C/1 atm. Only the lines that are relevant at equilibrium are shown. Activity corrections neglected (not valid at very low or very high pH).

and an in-situ p_{CO_2} of $10^{-5.43}$ atm. That is, since

$$[CO_3^{2-}] = \alpha_2 C_T = K_1 K_2 K_H p_{CO_2}/[H^+]^2 \tag{14.23}$$

at 25°C/1 atm, we test if

$$\alpha_{2,10.0624} 10^{-3} =?= 10^{-6.35} 10^{-10.33} 10^{-1.47} 10^{-5.43}/(10^{-10.0624})^2 \tag{14.24}$$

where $\alpha_{2,10.0624}$ represents α_2 evaluated at pH = 10.0624. Since $\alpha_{2,10.0624} = 0.3506$ under the conditions of interest, Eq. (14.24) becomes

$$3.506 \times 10^{-4} = 3.506 \times 10^{-4}. \tag{14.25}$$

Thus, we conclude for pH = pH_{trans} in Figure 14.3, that indeed $p_{CO_2} = 10^{-5.43}$ atm. In addition to illustrating the consistency between the concept of a pH_{trans} and the concept of a specific p_{CO_2} for equilibrium between $FeCO_{3(s)}$ and $Fe(OH)_{2(s)}$, this result illustrates the usefulness of the in-situ partial pressure concept. Indeed, pH_{trans} could have been determined by solving Eq. (14.23) with $p_{CO_2} = 10^{-5.43}$ atm.

14.5 KINETICS VS. EQUILIBRIUM IN DETERMINING THE SOLUBILITY-CONTROLLING SOLID

It is useful to remind ourselves that all of the above comments presume that it is the attainment of equilibrium that determines the properties of a given system. In other words, we have assumed that as specific system properties are changed (e.g., p_{CO_2}, pH, etc.), slow kinetics that will never hinder: 1) the conversion of one solid to another; 2) the precipitation of a solid; or 3) the dissolution of a solid. Because some solids are associated with very slow kinetics, however, it will sometimes be of interest to prepare predominance diagrams drawn ignoring those solids, and considering certain less stable (i.e., more soluble) solids. For example, if this chapter had focussed on Fe(III) chemistry (i.e., an Fe(III) oxide or hydroxide vs. an Fe(III) carbonate), we may very well have chosen to carry out the discussion considering (am)$Fe(OH)_{3(s)}$ rather than hematite ($Fe_2O_{3(s)}$); while hematite does not form on short time scales, (am)$Fe(OH)_{3(s)}$ does.

14.6 REFERENCES

PANKOW, J. F. 1992. *Aquatic Chemistry Problems*. Portland: Titan Press-OR, P.O. Box 91399, Portland, Oregon 97291-1399.

STUMM, W., and J. J. MORGAN. 1981. *Aquatic Chemistry*, New York: Wiley-Interscience.

15

Solubility as a Function of Particle Size

15.1 INTRODUCTION

The solubility of particles of any given material in a liquid such as water is not independent of the sizes of the particles, but increases with decreasing particle size. This chapter will focus on the solubility behavior of solids in water. However, it is important to realize that the very same theory can be used to describe how: 1) the solubility of an organic liquid (e.g., trichloroethylene (TCE), benzene, PCBs, etc.) in water will be affected by the organic droplet size; and 2) the vapor pressure of a droplet of any liquid (e.g., water) suspended in a gas (e.g., the atmosphere) is affected by the droplet size.

Although unaware at the time, some readers may have already encountered particle-size dependent solubility in laboratory courses in organic or analytical chemistry. Indeed, such courses usually include a few experiments which involve the crystallization of a precipitate from a liquid medium. Upon the initial precipitation of the solid, the reader may have witnessed the rapid formation of a great many very small particles. Although such particles are relatively more soluble than larger crystals of the same solid, they can nevertheless form since the liquid medium in which the precipitation occurred was initially *grossly supersaturated*.[1] After allowing the liquid/solid slurry to sit undisturbed for a few weeks, the reader may then have discovered that the great many small particles of solid had grown to larger (but fewer) crystals as illustrated in Figure 15.1.

The process described in Figure 15.1 is often referred to in analytical chemistry texts as "Ostwald ripening." Smaller crystals are converted to larger crystals because: 1) the

[1] Since we know from discussions in previous chapters that an amorphous solid is always more soluble than a more crystalline solid of the same type, it is useful to point out that highly supersaturated solutions often lead to the initial formation of particles that are not only small, but also *amorphous*.

smaller crystals are more soluble than the larger crystals; and 2) we know from Chapter 14 that a solid that is present but does not specify the lowest possible solution phase activity for a given species will be converted spontaneously to the solid that does specify that lowest activity. A review of the role played by Ostwald ripening in the formation of crystals in sediments is provided by Morse and Casey (1988).

The crystal growth process illustrated in Figure 15.1 involves the dissolution of the smallest crystals and the reprecipitation of that dissolved material onto the largest crystals initially present. As illustrated in Figure 15.2, the larger solubility possessed

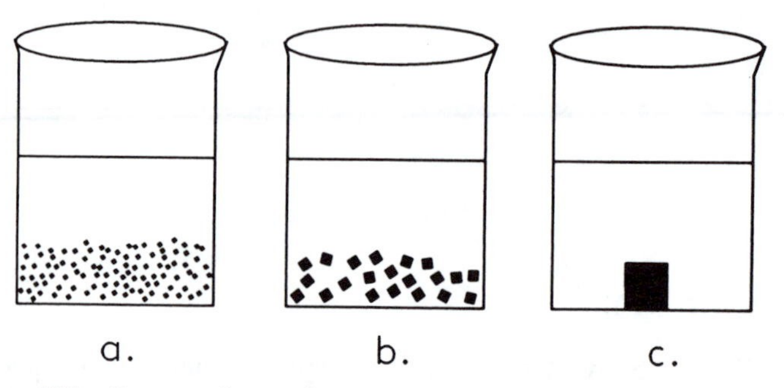

Figure 15.1 Process of crystal growth ("Ostwald ripening") in a beaker of liquid from: (*a*) a large number of very small, possibly amorphous particles; to (*b*) a small number of larger crystals; to, finally, (*c*) one relatively large crystal.

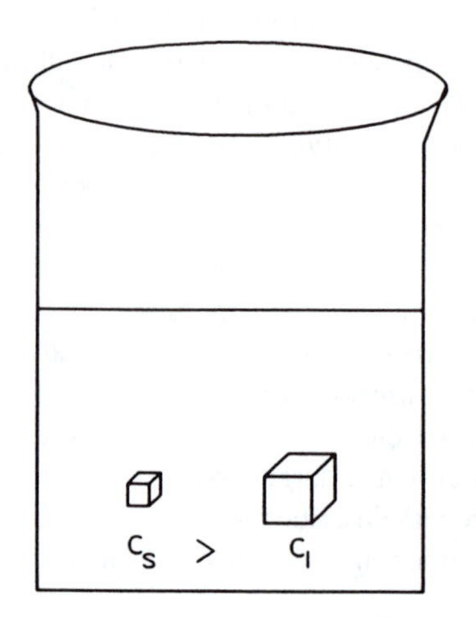

Figure 15.2 The concentration c_s of dissolved substance in the medium surrounding a small crystal or particle of solid is greater than the concentration c_l in the liquid surrounding a large particle. The resulting concentration gradient drives the diffusion of material away from the region surrounding the small particle to the region surrounding the large particle, that is, it drives the overall conversion of many small crystals to one large crystal.

by the very small crystals creates concentration gradients between them and the larger crystals. These concentration gradients drive the dissolution/reprecipitation process.

For a given small crystal/large crystal pair, as the small crystal dissolves and gets smaller, the intially-large crystal accretes material and gets larger. As a result of the changes in particle size, the magnitude of the concentration difference between the regions surrounding the two then gets larger and larger until the small crystal is completely gone and the large crystal prescribes a single saturation concentration in the solution in that region. However, if the large crystal is still smaller than some other crystals elsewhere in the system, then it will in turn be subject to dissolution for the sake of the growth of those larger crystals. If this process were allowed to proceed indefinitely, as illustrated schematically in Figure 15.1, the end result would be one large crystal in equilibrium with the liquid medium. Morse and Casey (1988) present kinetic equations which indicate that, in sediments, the number of crystals subject to this type of precipitative growth will tend to decrease according to 1/time, and that the mean crystal radius will increase according to $(\text{time})^{1/3}$.

For any given solid, the dependence of the solubility constant on the crystal radius r (cm) will follow a plot similar to that presented in Figure 15.3. The solubility of an infinitely small crystal is infinitely large. For large r, on the other hand, the solubility will approach an asymptotic limit. We also note that the slope of the $\log K_{s0}$ vs. r line is negative for all values of r. However, once r gets larger than 10 μm or so, the magnitude of that slope also becomes very small and the driving force for the conversion of many 10 μm-sized crystals to a few 100 μm-sized crystals is very small, albeit still favorable. Likewise, the driving force for the conversion of some 10 mm-sized crystals to even fewer 100 mm-sized crystals is even smaller, but is again, still favorable. As

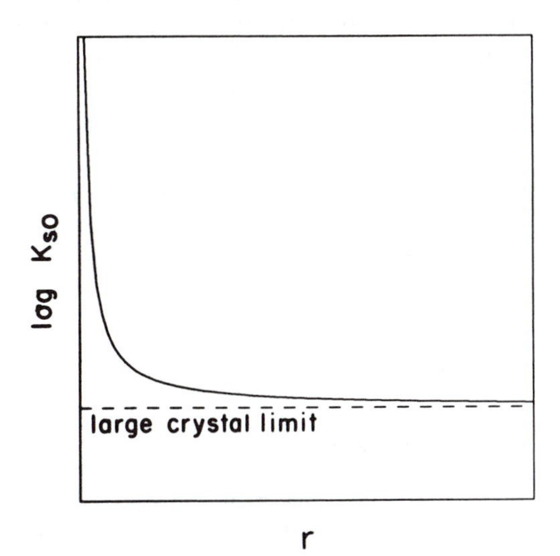

Figure 15.3 Qualitative dependence of a solubility constant on the crystal radius, r. K_{s0} approaches an asymptotic limit for large r. This limit is the value that we calculate using tabulated free energy values.

can be witnessed in many museums, over geologic time, very large, beautiful crystals with dimensions of tens of centimeters and more can be formed.

Two factors will limit the size of the single crystal that would form after infinite time in Figure 15.1: the mass of solid that is already present and available for growth, and the volume of the solution phase. The latter is important since it will determine how much mass initially remaining in solution will eventually accrete to the solid phase as the mean crystal size increases and the solubility decreases. Since the aqueous phase concentration is highest in Figure 15.1*a* and lowest in Figure 15.1*c*, we can conclude that the most solid phase is present in Figure 15.1*c* and the least is present in Figure 15.1*a*.

For a given ionic solid, the value of K_{s0} that is found in a given literature or reference text will often have been determined using crystals or particles of large enough r so that it is essentially the asymptotically-attained value of K_{s0}. However, as the solubility decreases, then the likelihood increases that a tabulated K_{s0} in a text might reflect some particle-size effects. The reason for this is that as the inherent solubility decreases, it becomes increasingly difficult to prepare large crystals of the solid with which to carry out solubility determinations. Indeed, as the solubility decreases, then: 1) the initial precipitation of the solid is increasingly likely to occur under the types of gross supersaturation that can lead to very small initial particles; and 2) the steepness of the concentration gradient driving the subsequent crystal growth decreases. (While c_s in Figure 15.2 will still be much larger than c_l, it will be small in *absolute terms*, and so the absolute magnitude of the concentration gradient between particles of different sizes will still be small.)

The fundamental principles governing the solubility behavior of small crystals are of interest in the field of water chemistry because a variety of natural and human-related processes can lead to the formation of some very small particles. Certainly water treatment is an area where the effects of solid particle size can become important. Finely-dispersed organic liquids in water will also be affected.

If we desire to model their interaction with the solution phase, we must understand how the solubility of small crystals relates to the solubility data obtained using comparatively large crystals. Thus, for very small crystals, we will need to take a K_{s0} value obtained in the laboratory using large crystals and correct it in a manner that gives the correct K_{s0} for the particle size of interest.

15.2 THE INTERFACIAL TENSION σ

15.2.1 The Origins of Interfacial Tension

As discussed in Section 15.1, the fact that small particles of a given solid are more soluble than larger particles of the same solid is exactly analogous to what happens to droplets of a liquid compound suspended in another liquid such as water, or in a gas such as air. As the droplet size decreases, the water solubility of the compound will increase, and the vapor pressure in the air will increase.

The effects described above are due to the presence of an *interfacial tension* (σ) between the solid and the water in the case of the solubility of solids in water, between the two liquids in the case of the solubility of one liquid in another, and between the liquid and the gas phase in the case of the vaporization of a liquid. For gas/liquid interfaces, σ is commonly referred to, albeit loosely, as the "surface tension" because one tends to forget about the presence of the gas. We will use the more precise and general "interfacial tension" throughout our discussions here. As already indicated, the symbol used for this parameter will be σ. Other texts and references often use γ, but that symbol has been reserved in this text for activity coefficients on the molal scale.

The value of σ will depend on the identities of the two adjacent phases, be they solid/liquid, liquid/liquid, or liquid/gas. Thus, the value of σ for the $Fe(OH)_{3(s)}$/water interface will be different from that for the $Al(OH)_{3(s)}$/water interface. In fact, even for the same type of solid, changing the nature of the water will change the value of σ. Thus, the value of σ for the $Fe(OH)_{3(s)}$/water interface will be somewhat different for waters of different composition.

The cgs units of σ are ergs/cm^2. These units are equivalent to work per unit area. *Thus, the interfacial tension is a measure of how many ergs are required to increase the interfacial area by one cm^2*. Energy per unit area is equivalent to force per unit length, and so σ also has units of dyne/cm. We note that 1 erg = 1 dyne-cm. A handy way to remember this equivalence is the mnemonic that "erg" is what might be said by a dying centipede.

In order to better understand the origins of interfacial tension, let us consider Figure 15.4. There, we model a given parcel of material as being composed of a number of cubical molecules. We will assume that any two adjacent cube faces can interact with one another by means of some collection of attractive forces such as hydrogen bonding, van der Waals forces, etc. In Figure 15.4*a*, we have a total of 64 cubical molecules arranged as a single large cube. If each adjacent-face interaction counts as one I_{af}, we have 144 I_{af}. The total surface area of the overall cube is 6×16 faces (F_s) = $96F_s$. If each interfacial interaction counts as one I_{if}, we have $96I_{if}$. If the same number of cubic molecules are rearranged to form the configuration shown in Figure 15.4*b*, we obtain $136I_{af}$, $112F_s$, and $112I_{if}$. Reducing the number of I_{af} takes energy (work) since it involves breaking interactions. Increasing the number of I_{if} releases energy since it forms interactions.

According to the above analysis, the amount of work required to achieve the transition from the system in Figure 15.4*a* to that in Figure 15.4*b* is $(8I_{af} - 16I_{if})$. In most cases of interest in this chapter, interactions between the face of the two adjacent cubes will be stronger than the interactions between the face of a cube and the surrounding medium.[2] Under such conditions, $8I_{af}$ will be larger than $16I_{if}$, and positive work will be required for the transition from Figure 15.4*a* to Figure 15.4*b*. That is, $(8I_{af} - 16I_{if})$

[2]E.g., consider a crystal of glucose suspended in some glucose-saturated water; the interaction between two adjacent glucose molecules in the crystal is stronger than the interaction between a glucose molecule on the surface and the surrounding aqueous medium.

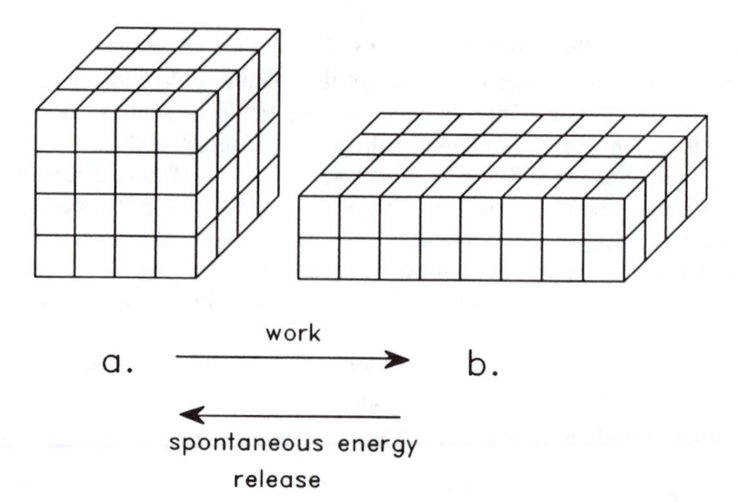

Figure 15.4 (*a*) A cubic parcel of material composed of 64 smaller cubic units: 144 adjacent-face interactions, and a surface area of 96 faces. (*b*) Rearranged parcel of 64 units: 136 adjacent-face interactions, and a surface area of 112 faces.

will be positive.[3] A finite difference estimate of the work required per unit increase in area (i.e., the interfacial tension) will then be given by

$$\sigma \simeq \frac{(8I_{\mathrm{af}} - 16I_{\mathrm{if}})}{(112 - 96)F_{\mathrm{s}}}. \tag{15.1}$$

More precisely, we define σ as the *differential* amount of work required to increase the surface area, that is,

$$\sigma \equiv \frac{\mathrm{d}W}{\mathrm{d}A}. \tag{15.2}$$

Since work per unit area is equivalent to force per unit length, we can consider that along a given length x, the interface exerts a force equal to $x\sigma$. That is,

$$\text{interfacial force exerted over distance } x = x\sigma. \tag{15.3}$$

The tension in the interface is therefore analogous both with the tension in the skin on a drum, and with the tension in the skin of a ballon. The "water strider" insect can walk on water because its legs are hydrophobic, and the legs do not get wet: the surface tension keeps the water from increasing its area and wetting the legs.

The interfacial tension causes a given volume of material to *pull in on itself* to *minimize the surface area.* We now know, for example, that the conversion of the Figure 15.4*b* system to the Figure 15.4*a* system will tend to occur spontaneously. Therefore, unless acted upon by other forces (e.g., gravity, which can force a liquid to conform to the shape of the vessel containing it), a given volume of liquid will tend to occupy the

[3]Situations in which σ is negative will lead to complete miscibility of the two adjacent phases (e.g., acetone and water), and therefore no possibility for an interface.

shape that has the minimum surface area for that volume, that is, a *perfect sphere*. (Note that a cube measuring x on a side will have a volume of x^3 and a surface area of $6x^2$; a sphere containing the same volume will have less surface area, i.e., $4.84x^2$.)

Those who have seen videotapes of astronauts in space may recall seeing them playing with undulating blobs of water. The undulations are due to the interplay of the initial internal momentum distribution in the blobs with interfacial forces. As the initial momentum distorts a blob away from a spherical shape and towards a larger surface area, then interfacial forces increase and act to return the blob to sphericity. Eventually, viscosity will damp out all of the undulations, and such a blob will take on the shape of a perfect sphere.

15.2.2 The Effects of Surfactants on Interfacial Tension

The reduction of the interfacial tension of a surface can be accomplished through the use of chemicals known as "surfactants." The name surfactant is a hybrid of *surface active*. Soaps and detergents are surfactants. Since a surfactant reduces the water/air interfacial tension, the surfactant-modified water/air interface has a lower free energy per unit area compared to the clean water/air interface.

Once the interfacial tension of a water/air surface has been reduced using a surfactant, comparatively little work needs to be done to take a small amount of the liquid and *greatly increase* its surface area, as might be done when forcing air into a film of soapy water when blowing bubbles. Conversely, since only a little energy is released when a large bubble collapses back to a liquid drop, large soap bubbles can drift through the air for long periods before ultimately being burst by the thinning of the upper portion of the bubble that occurs due to gravity and desiccation.

Another example of how surfactants allow the surface area of materials to be increased easily is provided by the action of dishwashing detergents on food grease. In that case, it is the σ of the oil/water interface that is reduced. Thus, in the presence of a good dishwashing detergent, a single globule of grease can be more easily broken up through agitation into many finer droplets. As a collection of many fine droplets, the grease is "emulsified," and can be washed away.[4] With respect to the water strider insect discussed above, it may be noted that if some surfactant was added to the water, the decrease in the surface tension would allow the water to increase its surface area and wet the insect's legs, causing the insect to sink.

15.3 THE PRESSURE INCREASE ACROSS THE INTERFACE OF A SPHERE OF MATERIAL DUE TO INTERFACIAL TENSION

We now know that interfacial tension will cause a volume of material to pull itself into a spherical shape. This pulling creates a *pressure* inside the sphere that is greater than that which is present outside of the sphere, or equivalently, than that which would be present if the surface were flat and not spherical. The magnitude of the pressure increase can be determined by carrying out a force balance on a spherical drop. We first divide

[4]Although the single globule of grease can be broken up easily in the presence of sufficient surfactant, it is still more stable as a single globule.

the drop into two hemispheres as shown in Figure 15.5. The force generated by the upper hemisphere is exactlycounterbalanced by the force generated by the lower hemisphere. Along its circular edge, the upper hemisphere in Figure 15.5 may be thought of as pulling up on the lower hemisphere, and the lower hemisphere may be thought of as pulling down on the upper hemisphere. The force generated by the upper hemisphere is equal to $2\pi r\sigma$ since the surface tension σ is operating on the circumference $2\pi r$. The balancing force from the lower hemisphere is also $2\pi r\sigma$. The net result is that the two balanced hemispheres compress the material contained within them.

In descriptive terms, a pressure is a force applied over some area, that is, pressure = force/area. The magnitude of the force creating the pressure increase inside the droplet is $2\pi r\sigma$. We do not *add* the forces from both hemispheres to obtain $4\pi r\sigma$ since counterbalancing forces do not increase the pressure, they just make for a static system. As an analogy, if we are swimming under water in a quiescent pool, the pressure we can initially feel in our ears (i.e., before the pressure equalizes in our eustachian tubes) is being generated by the height of the water column above us. The fact that the water below us is also pressing on us with a force equal to that experienced from above does not increase the local pressure, it just balances it, and prevents the upper water from pushing us deeper. Thus, we neither move upwards or downwards except for whatever motion we experience due to our swimming, or our inherent positive or negative bouyancy.

Returning to the consideration of our drop, the area over which the force $2\pi r\sigma$ is acting is the circular area on the bottom of the hemisphere, that is, πr^2. Thus,

Pressure differential across spherical interface due to surface tension

$$= \Delta P = \text{Force}/\text{Area} \tag{15.5}$$

$$= 2\pi r\sigma/(\pi r^2) \tag{15.6}$$

$$= 2\sigma/r. \tag{15.7}$$

There are several interesting comments that can be made about Eq. (15.7). Firstly, as $r \to \infty$, then $\Delta P \to 0$. We recognize $r = \infty$ as corresponding to a perfectly flat surface. Thus, there is no pressure differential across an infinitely flat interface. However, as $r \to 0$, then $\Delta P \to \infty$ and the pressure differential increases rapidly as r becomes

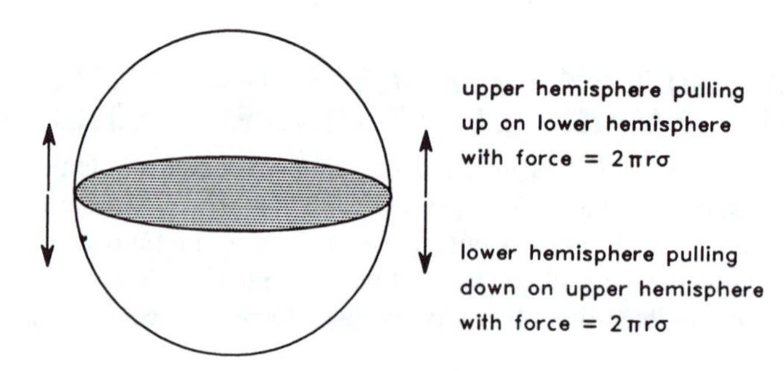

Figure 15.5 Surface-tension balance across two hemispheres comprising a sphere of material surrounded by another phase.

small. The same results would be obtained for a sphere of solid material suspended in water; in that case, σ would refer to the solid/water interfacial tension.

Let us consider an example that illustrates the magnitude of this effect. At 20°C, the value of σ for a clean water/air interface is 72.75 erg/cm^2. Therefore, for $r = 5 \times 10^{-6}$ cm (= 500 Å), we have by Eq. (15.7) that

$$\Delta P = \frac{2(72.75 \text{ erg/cm}^2)(9.8695 \times 10^{-10} \text{ L–atm/erg})(1000 \text{ cm}^3/\text{L})}{5 \times 10^{-6} \text{ cm}} = 28.7 \text{ atm.} \tag{15.8}$$

Since hydrostatic pressure increases at the rate of 1 atm for every 10.4 m of water depth, to achieve this pressure by going under water one would have to be about 300 m deep.

15.4 EFFECTS OF PRESSURES DUE TO CURVED SURFACES ON THE CHEMICAL PROPERTIES OF THE ENCLOSED MATERIAL

15.4.1 General

The dependence of the chemical potential of a pure chemical substance on pressure and temperature is given by the fundamental thermodynamic equation

$$d\mu = \overline{V}\,dP - \overline{S}\,dT \tag{15.9}$$

where $\overline{V}$ and $\overline{S}$ are the volume per mole (i.e, "molar volume") and entropy per mole (i.e., "molar entropy") of the substance of interest at the temperature and pressure of interest. At constant temperature ($dT = 0$), Eq. (15.9) reduces to

$$d\mu = \overline{V}\,dP. \tag{15.10}$$

Since molar volumes are always positive, Eq. (15.10) states that at constant T, increasing the pressure on *any* material will always increase its chemical potential. As an analogy, all physical materials act like springs when you try to compress them; pushing in on them stores energy, and μ goes up. As the material is compressed, the increased μ is stored in repulsive forces between the molecules (or ions). Thus, the water inside a droplet and therefore under interfacial tension-generated pressure will have a higher chemical potential than water that is in a pan and has a flat surface. This stored energy then makes the escape of molecules (or ions) into the surrounding gas phase (or solution) easier than it would be if the confining interface was not curved. Thus, the *vapor pressure* for a liquid inside of a droplet is greater than for that beneath a flat surface; the *solubility* of a tiny droplet of TCE in water is larger than that of a large droplet; and, the *solubility* of a small crystal of solid is greater than that for a large crystal.

The actual magnitude of the above effect can be determined by further examination of Eq. (15.10). Upon integration, we obtain

$$\int d\mu = \int \overline{V}\,dP. \tag{15.11}$$

Assuming that $\overline{V}$ is essentially independent of pressure (a good assumption for both solids and liquids for the pressure range over which we are interested), we obtain

$$\Delta\mu = \overline{V}\Delta P = \overline{V}2\sigma/r. \tag{15.12}$$

$\Delta\mu$ in Eq. (15.12) represents the difference in partial molar free energy for the dissolving or vaporizing substance between: 1) when it is inside of a curved surface of radius r; and 2) when it is beneath a flat surface. Equation (15.12) describes the specific manner in which the free energy increases as r decreases. We now consider how this increase is mathematically related to changes in the equilibrium constant for mass exchange across the interface.

15.4.2 Equilibrium Across a Flat Interface

Consider a flat surface of phase β in equilibrium with phase α. The β phase may be either liquid or a solid, and the α phase may be either aqueous or gaseous. Since the main target of discussion in this chapter is dissolution equilibria, for such cases the α may be used to mentally invoke "aqueous."

If we are dealing with a non-dissociating material A, we are interested in the chemical equilibrium

$$\mathrm{A}_\beta = \mathrm{A}_\alpha. \tag{15.13}$$

Schematically, this process is as described in Figure 15.6. At equilibrium, we require that the chemical potential in the two different phases are equal. That is,

$$\underbrace{\mu^\beta(T,P)}_{\mu \text{ in } \beta \text{ phase}} = \underbrace{\mu^{\mathrm{o},\alpha}(T,P) + RT \ln a_\mathrm{A}^\alpha}_{\mu \text{ in } \alpha \text{ phase}}. \tag{15.14}$$

Since the β (dissolving or vaporizing) phase will be considered to be a pure material, the concentration scale selected for that phase will be mole fraction and so the value of the equilibrium constant for equilibrium (15.13) will be given by $K = a_\mathrm{A}^\alpha/1 = a_\mathrm{A}^\alpha$. By Eq. (15.14) we then have

$$\mu^\beta(T,P) = \mu^{\mathrm{o},\alpha}(T,P) + RT \ln K. \tag{15.15}$$

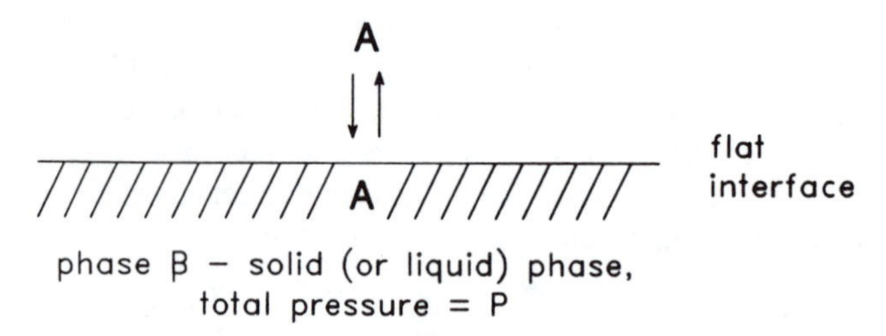

Figure 15.6 Equilibrium exchange across a flat interface for the reaction $\mathrm{A}_\beta = \mathrm{A}_\alpha$.

If the material can dissociate after it escapes the interface (as in the dissolution of a mineral), we might have

$$(\mathrm{AB})_\beta = \mathrm{A}_\alpha + \mathrm{B}_\alpha \tag{15.16}$$

and schematically we are interested in the case illustrated in Figure 15.7. At equilibrium, we have

$$\mu_{\mathrm{AB}}^{\beta}(T,P) = \mu_{\mathrm{A}}^{\mathrm{o},\alpha}(T,P) + RT \ln\ a_{\mathrm{A}}^{\alpha} + \mu_{\mathrm{B}}^{\alpha,\mathrm{o}}(T,P) + RT \ln\ a_{\mathrm{B}}^{\alpha} \tag{15.17}$$

The first two terms on the RHS of Eq. (15.17) represent the chemical potential of A in the α phase. The second two terms represent the chemical potential of B in the α phase. Now, if the β phase is a pure (AB), its mole fraction will be one, and so the value of the equilibrium constant for equilibrium (15.16) will be given by $K = a_{\mathrm{A}}^{\alpha} a_{\mathrm{B}}^{\alpha}/1 = a_{\mathrm{A}}^{\alpha} a_{\mathrm{B}}^{\alpha}$.

In general then, if $\mu^\beta(T,P)$ represents the chemical potential of the pure β phase, for equilibrium across a flat interface we have

$$\mu^\beta(T,P) = \sum_i \nu_i \mu_{\mathrm{i}}^{\mathrm{o},\alpha}(T,P) + RT \ln\ K(T,P). \tag{15.18}$$

15.4.3 Equilibrium Across a Spherical Interface of Radius *r*

For equilibrium of (AB) between the α phase and the β phase when the latter is spherical in shape, the situation of interest is as described schematically in Figure 15.8. For this case, we have in general that

$$\mu^\beta(T,P+\Delta P) = \sum_i \nu_i \mu_{\mathrm{i}}^{\mathrm{o},\alpha}(T,P) + RT \ln\ K(T,P+\Delta P). \tag{15.19}$$

Since the standard state chemical potentials at T and P in the α phase are independent of whether or not the surface of the β is curved, the summation in Eq. (15.19) is identical to

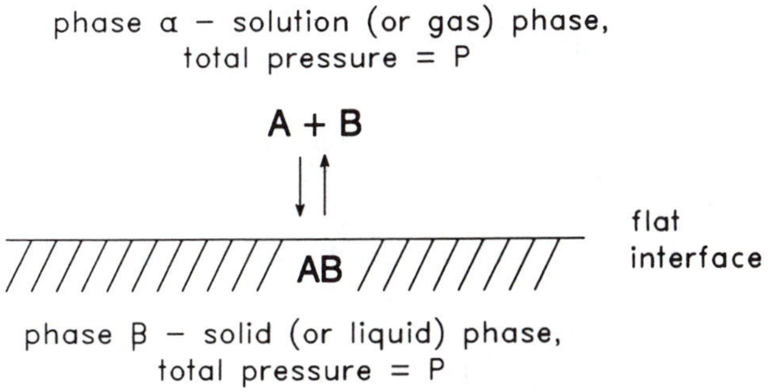

Figure 15.7 Equilibrium exchange across a flat interface for the reaction $(\mathrm{AB})_\beta = \mathrm{A}_\alpha + \mathrm{B}_\alpha$.

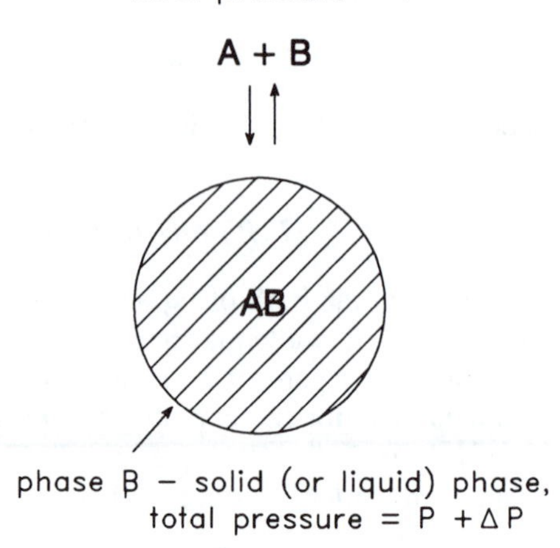

Figure 15.8 Equilibrium exchange across a spherical interface for the reaction $(AB)_\beta = A_\alpha + B_\alpha$.

that appearing in Eq. (15.18). The term $\mu^\beta(T,P+\Delta P)$ represents the chemical potential of the material *inside* the pressurized sphere, and $K(T,P+\Delta P)$ represents the equilibrium constant for dissolution (or vaporization) when the solid (or liquid) is at pressure $P+\Delta P$ and the solution (or gas) is at total pressure P.

To obtain a measure of the effect of the interface-induced pressure on μ^β, that is, to obtain $\Delta\mu^\beta$, we subtract Eq. (15.18) from Eq. (15.19), yielding

$$\Delta\mu^\beta = \mu^\beta(T,P+\Delta P) - \mu^\beta(T,P) = RT \ln \frac{K(T,P+\Delta P)}{K(T,P)}. \tag{15.20}$$

Using Eq. (15.12) to substitute for $\Delta\mu^\beta$, we obtain

$$\ln \frac{K(T,P+\Delta P)}{K(T,P)} = 2\sigma\overline{V}/(rRT) \qquad \text{The Kelvin Equation.} \tag{15.21}$$

Thus, the equilibrium constant K for dissolution (or vaporization) increases as r decreases. We reiterate that for dissolution of a mineral into water, K is the solubility product. For the dissolution of a liquid like TCE into water, K is the TCE saturation concentration in water. For liquid droplets suspended in a gas phase, K is the equilibrium saturation vapor pressure. In all cases, the quantity $K(T,P+\Delta P)$ represents the equilibrium constant for exchange across a curved interface of radius r, and $K(T,P)$ is the equilibrium constant for exchange across a flat interface, that is, for $r = \infty$.

If σ is expressed in ergs/cm^2, $\overline{V}$ is expressed in cm^3/mol, and r is expressed in cm, Eq. (15.21) yields

$$\log K(T, P + \Delta P) =$$

$$\log K(T, P) + \frac{2}{2.303} \frac{(\sigma \text{ ergs/cm}^2)(\overline{V} \text{ cm}^3/\text{mol})}{(r \text{ cm})(8.31441 \text{ J/K-mol})(T\ K)(10^7 \text{ ergs/J})} \tag{15.22}$$

or realizing that a pressure of $P + \Delta P$ corresponds to a radius of r, and that the ambient pressure P pertains to $r = \infty$, we have

$$\log K(T, r) = \log K(T, r = \infty) + 1.04 \times 10^{-8} \frac{\sigma \overline{V}}{rT} \tag{15.23}$$

If σ, r, and $\overline{V}$ are all known, the application of Eq. (15.23) is straightforward. Of the three parameters, the most easily evaluated is certainly $\overline{V}$; if it is not tabulated in reference texts for a material of interest, it can be determined simply by measuring the volume occupied by one mol of the material in the pressure range of interest.

For particles of solid, the value of r might be measured by electron microscopy. (Since particle-size effects really come into play only for sizes of the order of a micron and smaller (see Section 15.4.6), optical microscopy may not be of much use except for verifying that particle-size effects may be neglected.) However, it is very unlikely that a *single* characteristic particle radius will be found. That is, although the particle size distribution of natural solids are on rare occasions close to being *monodisperse* (i.e., essentially characterized by a single size), usually two orders of magnitude in r are required to bracket 95% of the particles (on a number basis). Thus, some average r will usually need to be used in Eq. (15.23) to determine the order of magnitude of particle size effects on the solubility constant.

An alternate way to obtain an average measure of the particle size is to determine the molar surface area $\overline{\mathbf{S}}$. (The bold face $\overline{\mathbf{S}}$ is used here to distinguish molar surface area from molar entropy.) Since: 1) $\overline{V}/r$ can be expressed in terms of $\overline{\mathbf{S}}$; and 2) $\overline{\mathbf{S}}$ values can be measured for solid powders by the BET isotherm method, it will be useful to recast some of the above equations in terms of $\overline{\mathbf{S}}$. We begin by continuing to assume that all of the particles are spherical. (Later we will show that the same result obtains for cubic particles.) We also assume that the particles are the same size (i.e., monodisperse). Since the volume and surface area of a single spherical particle are given by $4\pi r^3/3$ and $4\pi r^2$, respectively, according to these assumptions, we can therefore write

$$\overline{V} = (4/3)\pi r^3 \overline{N} \tag{15.24}$$

$$\overline{\mathbf{S}} = 4\pi r^2 \overline{N} \times 10^{-4} \tag{15.25}$$

where $\overline{N}$ is the number of particles required for 1 mol of material, and the factor of $10^{-4}(\text{m}^2/\text{cm}^2)$ in Eq. (15.25) accounts for the fact that $\overline{\mathbf{S}}$ is expressed in units of m^2/mol. Note also that $\overline{N}$ is *not* Avogadro's Number. If each particle contained 10^{-4} mol of material, then $\overline{N}$ would equal 10^4/mol. A combination of Eqs. (15.24) and (15.25) yields

$$\overline{\mathbf{S}} \times 10^4/3 = \overline{V}/r. \tag{15.26}$$

Equation (15.26) states that $\overline{\mathbf{S}}$ increases rapidly as r decreases. This is easy to understand inasmuch as the exposed surface area of 1 mol of material will increase rapidly as it is chopped up into smaller and smaller pieces.

Combining Eq. (15.26) with Eq. (15.22), with $\overline{S}$ expressed in units of m^2/mol and $\overline{\sigma}$ in units of ergs/cm^2, we have

$$\log K(T, \overline{S}) = \log K(T, \overline{S} = 0) + \frac{2}{6.908} \frac{(\sigma \text{ erg/cm}^2)(\overline{S} \text{ m}^2/\text{mol})(10^4 \text{ cm}^2/\text{m}^2)}{(8.31441 \text{ J/K-mol})(T \text{ K})(10^7 \text{ erg/J})} \quad (15.27)$$

or

$$\log K(T, \overline{S}) = \log K(T, \overline{S} = 0) + 3.482 \times 10^{-5} \sigma \overline{S}/T \quad (15.28)$$

where $K(T, \overline{S} = 0)$ pertains to the equilibrium constant for the material of interest at total pressure P and infinitely large particles.

Before proceeding on, it is probably useful to summarize that we have referred to the equilibrium constant for the particles of size r in three different, but totally equivalent ways. Namely, we have

$$\log K(T, P + \Delta P) = \log K(T, r) = \log K(T, \overline{S}). \quad (15.29)$$

Similarly, we have referred to the equilibrium constant for the particles of size $r = \infty$ (flat interface) in three different, but totally equivalent ways. Namely, we have

$$\log K(T, P) = \log K(T, r = \infty) = \log K(T, \overline{S} = 0)) \quad (15.30)$$

15.4.4 Equilibrium Between the Aqueous Phase and a Cubic Particle of Dimension 2*r*

While small *spherical* solid particles do occasionally form in the environment (e.g., see Gschwend and Reynolds, 1987), there is some ambiguity as to how Eqs. (15.23) and (15.28) might be applied to *non-spherical* particles of mineral solid. Consider then a small cubic crystal of mineral whose dimension x is set equal to $2r$. If we proceed in a manner that is similar to that used in Figure 15.5 and slice the cube in half along an imaginary square-shaped plane, the result is the system in Figure 15.9.

The different types of faces on a given type of crystal may be expected to have different characteristic arrangements of the constituent species making up the solid (e.g., the ions). It may therefore be expected that these different faces will have somewhat different interfacial tension with water. Let us denote the average interfacial tension of the different faces as $\overline{\sigma}$. Since the perimeter over which that surface tension will be acting is $4x$, then according to the same type of reasoning used above for spherical geometry, we obtain that the interfacial tension-generated force acting on the contents of the cubic crystal is given by $4x\overline{\sigma}$. Since the surface area over which that force will be acting is x^2, we obtain that the pressure differential between the inside of the cubic crystal and the solution is given by

$$\Delta P = 4x\overline{\sigma}/x^2 = 4\overline{\sigma}/x = 4\overline{\sigma}/2r = 2\overline{\sigma}/r \quad (15.31)$$

which is analogous to the result obtained for the case of the sphere.

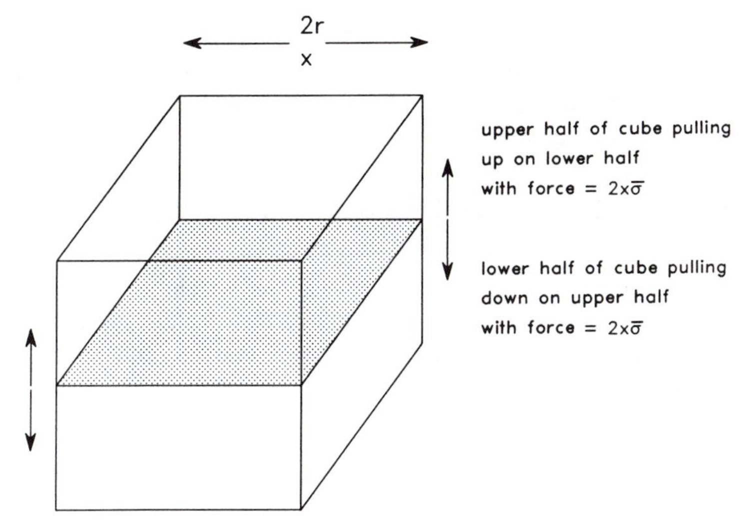

Figure 15.9 Surface-tension balance across two halves of a cube of solid in equilibrium with a solution phase. The average interfacial tension for the different crystal faces is $\overline{\sigma}$.

Now, for a monodisperse collection of cubes, we have

$$\overline{V} = x^3\overline{N} \tag{15.32}$$

$$\overline{\boldsymbol{S}} = 6x^2\overline{N} \times 10^{-4}. \tag{15.33}$$

Therefore,

$$\overline{V}/r = 2\overline{V}/x = 2x^3\overline{N}/x = 2x^2\overline{N} \tag{15.35}$$

$$\overline{V}/r = \overline{\boldsymbol{S}} \times 10^4/3 \tag{15.36}$$

as before.

Thus, under the assumptions cited above, the equations derived earlier for monodisperse spherical particles may also be used for monodisperse cubic crystals simply by substituting $\overline{\sigma}$ for σ. Crystals of other geometries will give equations similar to Eqs. (15.23) and (15.28), but with slightly different coeffcents on the terms involving $\overline{\sigma}$.

Equation (15.28) is of the same form found by Schindler et al. (1965) for $Cu(OH)_{2(s)}$, $CuO_{(s)}$, and $ZnO_{(s)}$. In particular, at 25°C/1 atm total pressure, those workers found that the solubilities of these solids varied according to

$Cu(OH)_{2(s)}$:
$$\log {}^*K_{s0} = 8.92 + 4.8 \times 10^{-5}\overline{\boldsymbol{S}} \tag{15.37}$$
$$\rightarrow \overline{\sigma} = 410 \pm 130 \text{ ergs/cm}^2$$

$CuO_{(s)}$:
$$\log {}^*K_{s0} = 7.89 + 8.0 \times 10^{-5}\overline{\boldsymbol{S}} \tag{15.38}$$
$$\rightarrow \overline{\sigma} = 690 \pm 150 \text{ ergs/cm}^2$$

$ZnO_{(s)}$:
$$\log {}^*K_{s0} = 16.8 + 9.0 \times 10^{-5}\overline{\boldsymbol{S}} \tag{15.39}$$
$$\rightarrow \overline{\sigma} = 770 \text{ ergs/cm}^2.$$

The above data suggests that $\overline{\sigma}$ might typically be higher for oxides than it is for hydroxides. The "error bar" information on $\overline{\sigma}$ in Eqs. (15.37) and (15.38) was reported by Schindler et al. (1965). Equations (15.37) and (15.38) are plotted in Figure 15.10 as a function of $\overline{S}$ (see also Figure 5.23 in Stumm and Morgan (1981)).

15.4.5 Transition of Solubility Control Due to Particle Size Effects

Equations (15.37) and (15.38) describe the manner in which particle size affects the solubility of two Cu(II)-containing solids. Both of these solids are dibasic, and they differ only by virtue of one water of hydration; one is an oxide, and one is a hydroxide. For large crystals, at 25°C/1 atm, for $Cu(OH)_{2(s)}$ we have $\log{}^* K_{s0} = 8.92$, and for $CuO_{(s)}$ we have $\log{}^* K_{s0} = 7.89$. Thus, at equilibrium, large particles of $Cu(OH)_{2(s)}$ can never

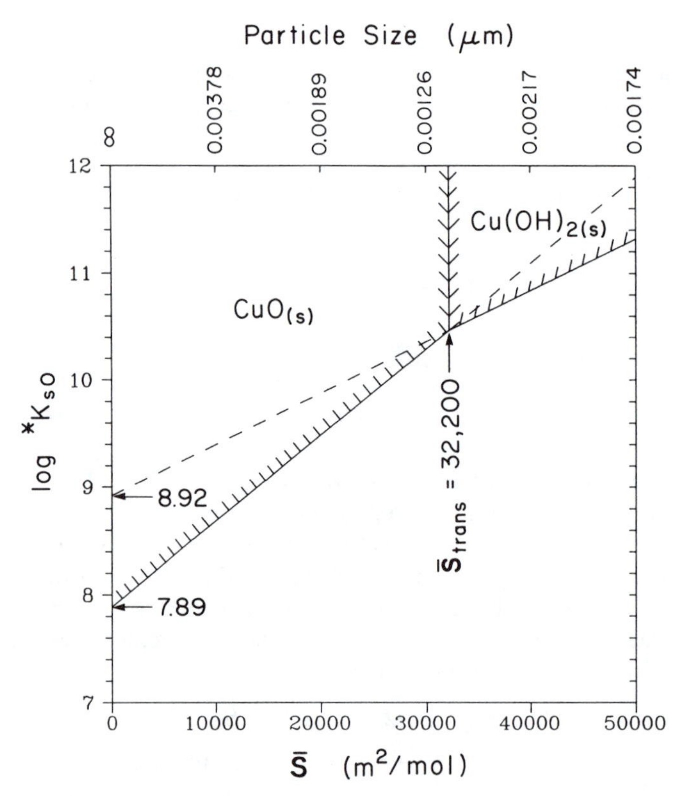

Figure 15.10 Dependence of $\log{}^*K_{s0}$ for $CuO_{(s)}$ and $Cu(OH)_{2(s)}$ on $\overline{S}$. Neglecting the possibility of crystal growth, the two solids can coexist when $\overline{S} = 32{,}200\ m^2/mol = \overline{S}_{trans}$. On the upper x axis and to the right of $\overline{S}_{trans}$, the equivalent values of r are given for $Cu(OH)_{2(s)}$. To the left of $\overline{S}_{trans}$, the equivalent values of r are given for $CuO_{(s)}$. The large crystal $\log{}^*K_{s0}$ values are 7.89 and 8.92 for $CuO_{(s)}$ and $Cu(OH)_{2(s)}$, respectively.

be solubility controlling, and $CuO_{(s)}$ may be solubility controlling for some conditions of interest, providing that no other solid (e.g., $CuCO_{3(s)}$) specifies a lower $\{Cu^{2+}\}$ for those conditions.

As the particle size begins to decrease and $\overline{S}$ increases, since the slopes of the two lines given by Eqs. (15.37) and (15.38) are different, the two lines will eventually cross, and equilibrium solubility control will be transferred to $Cu(OH)_{2(s)}$. The value at which this transfer takes place may be termed $\overline{S}_{trans}$. Therefore, by Eqs. (15.37) and (15.38), at 25°C/1 atm, total pressure we have

$$\overline{S}_{trans} = \frac{8.92 - 7.89}{(8.0 - 4.8) \times 10^{-5}} \tag{15.40}$$

$$= 32{,}200 \text{ m}^2/\text{mol}. \tag{15.41}$$

Neglecting the possibility of crystal growth, the two solids can coexist when $\overline{S}$ = 32,200 m^2/mol. This possible coexistence at $\overline{S}_{trans}$ is marked in Figure 15.10 as a vertical line much as we denoted the coexistence of two phases at a pH_{trans} in Figure 14.6. On the upper x axis and to the right of $\overline{S}_{trans}$, the equivalent values of r are given for $Cu(OH)_{2(s)}$ taking $\overline{V}$ = 28.95 cm^3/mol (FW = 97.56 and d = 3.37 g/cm^3). To the left of $\overline{S}_{trans}$, the equivalent values of r are given for $CuO_{(s)}$ taking $\overline{V}$ = 12.59 cm^3/mol (FW = 79.54 and d = 6.32 g/cm^3). Thus, near $\overline{S}_{trans}$, the sizes of these particles are predicted to be very small indeed, of the order of only a few tens of angstroms. Whether or not Eqs. (15.37) and (15.38) truly provide accurate predictors of log $^*K_{s0}$ behavior at such small r values is doubtful. Indeed, σ is a property that requires some minimum number of units to be meaningful. Thus, just as one cannot speak of the σ for a single ion in water, it difficult to speak of σ for a very small collection of molecules or ions. Nevertheless, although we may not know exactly what it is, there is an $\overline{S}_{trans}$ for these two solids.

As is always the case, a consideration of thermodynamic principles will not tell us anything directly about the kinetics of interconversion between various species of interest. For example, if a system contains some $CuO_{(s)}$ with $\overline{S} > \overline{S}_{trans}$, then while we do know that the $CuO_{(s)}$ is only metastable relative to $Cu(OH)_{2(s)}$, we do not know how quickly the $CuO_{(s)}$ will be converted to $Cu(OH)_{2(s)}$. For another example from nature, although coarse-grained hematite ($Fe_2O_{3(s)}$) is metastable relative to coarse grained goethite (γ-$FeOOH_{(s)}$), coarse grained hematite can persist in sediments for millions of years (Murray 1979).

15.4.6 Typical Dimension at Which Particle Size Effects Can Become Important for the Dissolution of Solids

At 20°C, and assuming a typical $\overline{V}$ value of ~33 cm^3/mol (FW = ~100 and d ~3 g/cm^3), the average $\overline{\sigma}$ value of ~625 ergs/cm^2 obtained by Schindler et al. (1965) for the two Cu(II) solids may be substituted in Eq. (15.23) to obtain

$$\log K(T, r) = \log K(T, r = \infty) + 7 \times 10^{-7} r^{-1}. \tag{15.42}$$

Equation (15.42) may be employed to develop a useful criterion for determining when we need to concern ourselves with the effects of particle-size on the solubility of solids. If an effect on K_{s0} of 10% is taken for this criterion, we then have

$$\log[K(T, r)/K(T, r = \infty)] = \log 1.1 = 0.041. \tag{15.43}$$

Thus, we should begin to be concerned about particle-size effects as r decreases and begins to approach

$$r \simeq 2 \times 10^{-5} \text{ cm } (= 0.2\mu\text{m}). \tag{15.44}$$

Solid-to-solid variability in the factor $\overline{\sigma V}$ probably means that size effects can begin to be important for particles as large as $\sim 1\mu$m. Problems illustrating the principles of this chapter are presented by Pankow (1992).

15.5 REFERENCES

GSCHWEND, P. M. and M. D. REYNOLDS. 1987. Monodisperse ferrous phosphate colloids in an anoxic groundwater plume. *J. Contam. Hydrology*, **1**: 309–327.

MORSE, J. W., and W. H. CASEY. 1988. Ostwald processes and mineral paragenesis in sediments. *Am. J. Science*, **288**: 537–560.

MURRAY, J. W. 1979. Iron Oxides, in Ribbe, P. H., *Marine Minerals*, Mineralog. Soc. America, Washington, D.C., 47–98.

PANKOW, J. F., 1992. *Aquatic Chemistry Problems*, Portland: Titan Press-OR, P.O. Box 91399, Portland, Oregon 97291-1399.

SCHINDLER, P., H. ALTHAUS, F. HOFER, and W. MINDER. 1965. *Helv. Chimica Acta*, **48**: 1204–1215.

STUMM, W., and J. J. MORGAN. 1981. *Aquatic Chemistry*, New York: Wiley-Interscience.

16

Solid/Solid and Liquid/Liquid Solution Mixtures

16.1 INTRODUCTION

Up to this point in this text, whenever solids have been participants in equilibrium reactions, the only possibility that we have considered is that the mole fraction of any solid present is 1.0. Since the standard state for all solids has moreover been chosen to involve pure solid, this has meant that we have up to now only considered cases when the solid activities are 1.0. Given the many elements present in natural systems, however, *completely* pure solids will never be found there. We therefore need to understand why non-pure solids can form, and also how they behave.

The theoretical basis for understanding solid solutions applies equally well to explaining how one liquid compound (e.g., trichloroethylene (TCE)) dissolves in another liquid (e.g., water). By symmetry, the theory also applies to the dissolution of water into TCE. Much of the theory in this chapter will therefore be presented in a general context that applies to solid solutions as well as to the dissolution of liquids like TCE and water in one another.

While the idea of TCE dissolved in water and vice versa is readily understood, some may find the idea of solid solutions a little less intuitive. When a mass containing solid A is not pure A, then there are obviously other materials in that mass. This might happen in either of two ways:

1. The regions where solid A is present are pure, but there are cavities within or coatings on the surface of the solid mass that contain other solid(s).
2. Elements other than those comprising solid A are physically dispersed within it, i.e., actually *dissolved* within it.

From a *mining* point of view, the first case might be considered to involve impure solid A. However, from a thermodynamic point of view, solid A in that case is still pure. It is the second case that represents a true solid solution.

This chapter will analyze what happens when true *dissolution* takes place between two solids, and between two liquids. Since we will be limiting ourselves to cases involving equilibrium, we will only consider cases involving *homogeneous* solutions; inhomogeneous solutions are not at equilibrium because the activities (and therefore the chemical potentials) of some constituents are higher in some regions of the solution than in others.

16.2 THERMODYNAMIC EQUATIONS GOVERNING THE FORMATION OF SOLID AND LIQUID SOLUTION MIXTURES

16.2.1 General Equations

For any chemical process, we have

$$\Delta G = \Delta H - T\Delta S. \tag{16.1}$$

Thus, for the **mixing** (mxg) of two gases, two liquids, or two solids, we have

$$\Delta G_{\text{mxg}} = \Delta H_{\text{mxg}} - T\Delta S_{\text{mxg}}. \tag{16.2}$$

Mixing will occur when ΔG_{mxg} is less than zero. The generic reasons why solid solutions can form are the same as the reasons why two liquids or two gases can mix.

Let us say that we are mixing some A with some B. *If the A–B interactions are of the same strength as the average of the A–A and B–B interactions, then the mixing of A and B will neither take up or release any heat.* That is, $\Delta H_{\text{mxg}} = 0$. As an analogy, let us say that we were mixing some single adults of sex A from northern New York with some single adults of sex A from southern New York. Being from the same state, and of the same sex, the two groups have similar characteristics. Thus, the northern New Yorkers interact with each other to the same degree as they do with the southern New Yorkers, and vice versa. Thus, $\Delta H_{\text{mxg}} = 0$ is zero in this case.

Whenever $\Delta H_{\text{mxg}} = 0$, two pure gases, two pure liquids, or two pure solids will mix because mixing decreases the free energy G of the overall system by increasing the entropy. Because the entropy S is a measure of the disorder in a system, the overall entropy *always* increases when two pure (i.e., highly ordered) materials are combined to form one larger, impure (i.e., less ordered) phase. Under these conditions, $\Delta S_{\text{mxg}} > 0$, $\Delta G_{\text{mxg}} < 0$, and the mixing proceeds.

Although $\Delta H_{\text{mxg}} = 0$ is a possibility, it is much more likely that either $\Delta H_{\text{mxg}} > 0$ ("endothermic," heat taken up during the mixing), or $\Delta H_{\text{mxg}} < 0$ ("exothermic," heat released during the mixing). Continuing the analogy started above, a case where we could have $\Delta H_{\text{mxg}} > 0$ would be for the mixing of single adults of sex A from two different, traditionally competitive locales, for example, New York City and Boston. In that case, strong opinions about things like who had the best basketball team might mean that the two groups would prefer not to mix, and that for the mixing to proceed, some source of kick-in-the-pants energy for the endothermic mixing process would need to be provided. Thus, $\Delta H_{mxg} > 0$ inhibits mixing. If $\Delta H_{mxg} \ggg 0$, the two groups will be only partially miscible, that is, only partially soluble in one another. That means that if we had just one New Yorker and a lot of Bostonians, then the Bostonians could perhaps absorb the one New Yorker. However, if there were numerous New Yorkers and numerous Bostonians, then "phase separation" would occur, and we would have two groups, one that was *mostly* New Yorkers, and another that was *mostly* Bostonians.

To observe $\Delta H_{mxg} < 0$, where heat would be released during the mixing, we might mix single adults of sex A from northern New York with some single adults of sex B from southern New York. Indeed, the interactions between the opposite sexes will be, on average, much stronger than with either the sex A adults among themselves, or the sex B adults among themselves. Thus, $\Delta H_{mxg} < 0$ promotes mixing.

Sulfuric acid (H_2SO_4) and H_2O provide an example of two liquids that give $\Delta H_{mxg} \ll 0$. In the early 1970s, the author of this book learned first hand just how exothermic this process is when he mixed about four liters of each of these liquids in a cylindrical *Pyrex* pipette vessel. The mixture became hot very rapidly, and the bottom of the vessel cracked and broke out in a nice clean circle. The freshly combined H_2SO_4/water mixture then proceeded to pour onto the author, the lab bench, and the floor, in that order. It is the very strong interaction between the protons in the H_2SO_4 and the H_2O molecules in the water that make this mixing process very exothermic.

To summarize, for every case involving two initially-pure materials, the portion of ΔG_{mxg} that is due to the entropy change on mixing is always negative: entropy effects always favor mixing. Depending on the nature of the chemical (enthalpic) interactions between the two mixing components, the mixing can *either* be promoted, or inhibited; $\Delta H_{mxg} < 0$ favors mixing, and $\Delta H_{mxg} > 0$ inhibits mixing.

Gases always mix because the intermolecular distances are large enough that the ΔS_{mxg} term always overwhelms any possible unfavorable interactions. On the other hand, liquids and solids, being more condensed, are much more subject to molecule-molecule interactions. Therefore, there are many pairs of liquids and many pairs of solids that are not completely miscible with one another, that is, not miscible over all proportions. Nevertheless, even in the event of immiscibility for some given proportion of materials, *some* interdissolution (i.e., dissolution of two liquids in each other, or dissolution of two solids in one another) must *always* occur. For example, no matter how insoluble a given organic liquid might be in water, the solubility can *never* be zero; the degree of dissolution (i.e., mixing) which does result will represent the minimization of the system G over the sum of the opposing entropy (favorable) and enthalpic (repulsive) interactions.

Alloys between metals like copper and zinc (e.g., brass) and tin and lead (e.g., lead-based solder) provide examples of solid pairs that are miscible over all proportions at ambient T and P. Although most *ionic* solids are usually not very miscible, based on the above, we do know that some dissolution between any two ionic solids will always occur. We consider now the thermodynamics of the mixing of two chemicals, A and B. These two chemicals might be two solids, two liquids, or even two gases.

Let n_A mols of pure A be mixed with n_B mols of pure B. Before the A and B are mixed, the free energy of the system is

$$G = n_A \mu_A^o + n_B \mu_B^o. \tag{16.3}$$

The mixing process is

$$n_A A + n_B B = A_{n_A} B_{n_B}. \tag{16.4}$$

We recall that when we are dealing with pure phases, it is convenient to use the mole fraction scale for the activity. Thus, with ζ_i representing the activity coefficient of species i on the mole fraction scale, in general we have

$$\mu_i = \mu_i^o + RT \ln \zeta_i X_i \tag{16.5}$$

where ζ_i is the activity coefficient of species i and X_i is its mole fraction. For pure i, $X_i = 1$, $\zeta_i = 1$, and so $\mu_i = \mu_i^o$.

After mixing, we have

$$X_A = n_A/(n_A + n_B) \qquad X_B = n_B/(n_A + n_B) \qquad X_A + X_B = 1. \tag{16.6}$$

The overall G of the system, that is, the G of the mixture will be given by:

$$G_{mxt} = n_A\mu_A^o + n_A RT \ln X_A + n_A RT \ln \zeta_A + n_B\mu_B^o + n_B RT \ln X_B + n_B RT \ln \zeta_B. \tag{16.7}$$

The first three terms on the RHS of Eq. (16.7) represent the G of the n_A mols of A in the solution and the second three terms represent the G of the n_B mols of B in the solution. For the mixing process described by Eq. (16.4), the change in free energy of mixing will be given by the difference between the G values given by Eqs. (16.7) and (16.3), that is, by

$$\Delta G_{mxg} = n_A\mu_A^o + n_A RT \ln X_A + n_A RT \ln \zeta_A + n_B\mu_B^o + n_B RT \ln X_B + n_B RT \ln \zeta_B - n_A\mu_A^o - n_B\mu_B^o. \tag{16.8}$$

Cancelling like terms in Eq. (16.8), we obtain

$$\Delta G_{mxg} = n_A RT \ln X_A + n_A RT \ln \zeta_A + n_B RT \ln X_B + n_B RT \ln \zeta_B. \tag{16.9}$$

If ΔG_{mxg} is divided by the total number of mols, that is, by $(n_A + n_B)$, then the ΔG per mole of mixture, that is, the *molar* ΔG of mixing ($\Delta\overline{G}_{mxg}$) is

$$\Delta\overline{G}_{mxg} = X_A RT \ln X_A + X_A RT \ln \zeta_A + X_B RT \ln X_B + X_B RT \ln \zeta_B. \tag{16.10}$$

The sum of the terms involving just the mole fractions represents the entropic contribution to $\Delta\overline{G}_{mxg}$. We know this because for A, in the absence of interactions that are different than those which A feels in pure A, then $\zeta_A = 1.0$, and only entropic effects exist; a similar statement can be made for B. The sum of the two terms in Eq. (16.10) involving activity coefficients represents the enthalpic contribution to $\Delta\overline{G}_{mxg}$. To summarize,

$$-T\Delta\overline{S}_{mxg} = X_A RT \ln X_A + X_B RT \ln X_B. \tag{16.11}$$

$$\Delta\overline{H}_{mxg} = X_A RT \ln \zeta_A + X_B RT \ln \zeta_B \tag{16.12}$$

16.2.2 Ideal Solutions

In the special case in which all ζ values equal 1 over all possible compositions, the solution is ideal (idl): chemical A feels equally at home in the mixture as it does in pure A, and similarly for B, and $\zeta_A = \zeta_B = 1$. Under these conditions, Eq. (16.10) reduces to the entropy terms alone, that is,

$$\Delta\overline{G}_{mxg} = \Delta\overline{G}_{mxg,idl} = X_A RT \ln X_A + X_B RT \ln X_B. \quad (16.13)$$

With $R = 8.314$ J/mol-K, $\Delta\overline{G}_{mxg,idl}$ may be calculated easily using Eq. (16.13) for $0 < X_A < 1$. (Note the use of "$<$" and not "$\leq$" in the previous sentence; the "$=$" case will be discussed shortly.) The results of such calculations are presented in Table 16.1 and are plotted in Figure 16.1. Note that for this purely entropy-driven case, $\Delta\overline{G} \leq 0$ for all X_A. Note also the symmetry in $\Delta\overline{G}_{mxg,idl}$ inasmuch as the same values are obtained for $X_A = 0.1$ and 0.9, for $X_A = 0.2$ and 0.8, etc. This is as expected since we have not stated that there were any differences in the natures of A and B. Indeed, we have just given them different names ("A" and "B"), and as Juliet Capulet once noted on her balcony, names alone are not enough to cause differences in nature.

The quantity $\ln X_A$ is mathematically undefined when $X_A = 0.0$ ($X_B = 1.0$), and $\ln X_B$ is undefined when $X_B = 0.0$ ($X_A = 1$). This does not cause a problem when evaluating Eq. (16.13) since there simply is no mixture when either X_A or X_B equals zero; when there is no mixture, $\Delta\overline{G}_{mxg}$ must by definition be zero. Moreover, even from a strict mathematical point of view there is no problem since the products $X_A \ln X_A$ and $X_B \ln X_B$ approach zero as X_A and X_B approach zero, respectively. Taking the former as an example, this is the case because X_A becomes small faster than $\ln X_A$ approaches $-\infty$.

TABLE 16.1 $\Delta\overline{G}_{mxg,idl}$ (J/mol) as a function of composition at 298 K.

X_A	X_B	$\Delta\overline{G}_{mxg,idl}$ (J/mol)
0.00	1.00	0.0
0.01	0.99	−138.7
0.05	0.95	−491.9
0.10	0.90	−805.5
0.20	0.80	−1239.8
0.30	0.70	−1513.5
0.40	0.60	−1667.5
0.50	0.50	−1717.4
0.60	0.40	−1667.5
0.70	0.30	−1513.5
0.80	0.20	−1239.8
0.90	0.10	−805.5
0.95	0.05	−491.9
0.99	0.01	−138.7
1.00	0.00	0.0

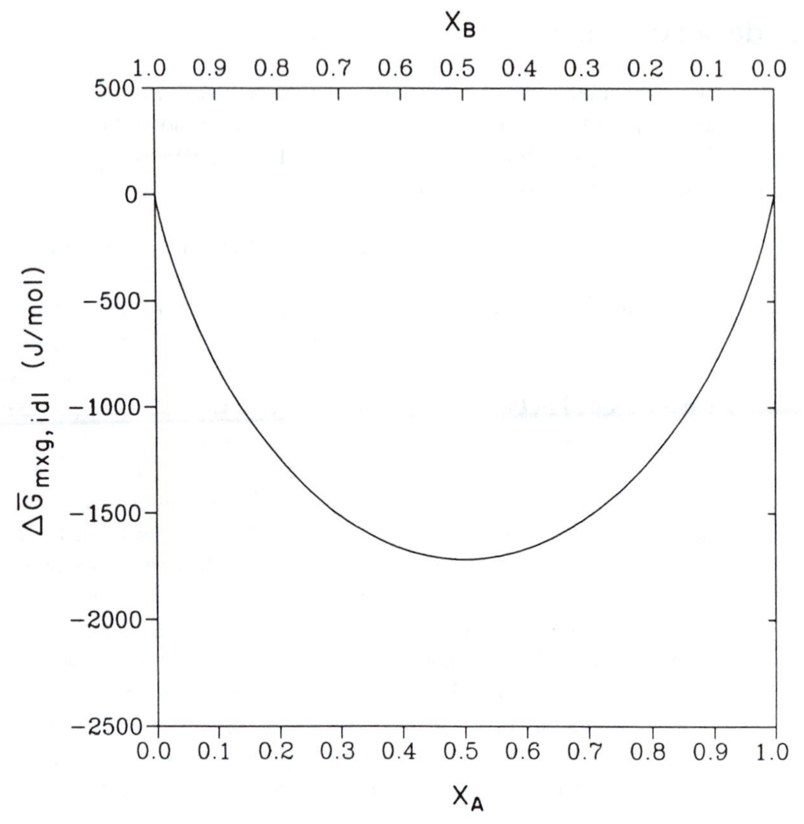

Figure 16.1 $\Delta\overline{G}_{mxg,idl}$ (J/mol) as a function of X_A for an ideal solid solution at 298 K.

Indeed, for $X_A = 0.1$, 0.01, and 0.001, etc., we have $X_A \ln X_A = -0.23, -0.046, -0.0069$, etc. When $X_A = 0$, a computer will, nevertheless, return an error for $X_A \ln X_A$.

16.2.3 Non-Ideal Solutions

Although they favor mixing, the values given for $\Delta\overline{G}_{mxg,idl}$ in Table 16.1 are not large in an absolute sense. Indeed, the maximum value given in that table is only about -1.7 kJ/mol. In comparison, covalent bonds range in strength from about -200 to -400 kJ/mol, and an ionic "bond" such as that between Na^+ and Cl^- in $NaCl_{(s)}$ at 25°C/1 atm is about -600 kJ/mol. What this tells us is that if there are any signficant interactions at all in a solution formed from two solids or from two liquids, then the formation of the solution will be either: a) much more strongly favored than indicated in Table 16.1; or b) strongly disfavored. In either case, the solution phase activity coefficients will not be 1.0.

In the general case where the possibility of non-ideality is allowed, activity coefficients cannot be assumed to be 1.0 under all conditions. $\Delta\overline{G}_{mxg}$ must then be evaluated using Eq. (16.10). Combining Eqs. (16.10) and (16.13), we have

$$\Delta\overline{G}_{\mathrm{mxg}} = \Delta\overline{G}_{\mathrm{mxg,idl}} + X_{\mathrm{A}}RT\ln\zeta_{\mathrm{A}} + X_{\mathrm{B}}RT\ln\zeta_{\mathrm{B}}. \tag{16.14}$$

Although we know how to calculate $\Delta\overline{G}_{\mathrm{mxg,idl}}$, we now need functional relations for ζ_{A} and ζ_{B} in terms of the composition (i.e., in terms of either X_{A} or X_{B}).

Since we are considering the non-ideal case, we are interested in situations in which ζ_{A} and ζ_{B} do not equal to 1.0. We realize, though, that the requirement that $\zeta_{\mathrm{A}} = 1.0$ when $X_{\mathrm{A}} = 1.0$ (reference state) (and similarly for B) means that ζ_{A} cannot equal a *non-unity* constant over the whole range $0 \leq X_{\mathrm{A}} \leq 1$. Rather, there will have to be some dependence of the ζ values on composition, with ζ_{A} approaching 1.0 as X_{A} approaches 1.0, and conversely with ζ_{B} approaching 1.0 as X_{B} approaches 1.0. One possible set of functions which demonstrate this type of behavior is:

$$\ln\zeta_{\mathrm{A}} = (\boldsymbol{A}/RT)X_{\mathrm{B}}^2 \tag{16.15}$$

$$\ln\zeta_{\mathrm{B}} = (\boldsymbol{A}/RT)X_{\mathrm{A}}^2 \tag{16.16}$$

where the boldface $\boldsymbol{A}$ is a constant which has units of J/mol. Since R has units of J/mol-K, it is useful to note that the quantity $\boldsymbol{A}/RT$ is *dimensionless*.

Equations (16.15) and (16.16) are called the "two-suffix Margules equations" (Prausnitz 1969, p 193). Although they are somewhat empirical in nature, they are nevertheless often used with success in describing the thermodynamics of binary liquid mixtures. The adjective "two-suffix" originates in the facts that: 1) a series expansion may be used to express the non-ideal portion of the RHS of Eq. (16.14); and 2) when the *two* coefficients B and C of that expansion satisfy $\boldsymbol{B} = -\boldsymbol{C} = \boldsymbol{A}$, then Eqs. (16.15) and (16.16) are obtained (Nordstrom and Munoz 1986).

In addition to being applied successfully with liquid mixtures, two-suffix (as well as higher-order) Margules equations have also been used with solid solutions of a wide variety of types. These include alkali feldspars, (Thompson and Waldbaum 1969), pyroxenes (Lindsley et al, 1981), and Fe-Ti-oxides (Andersen and Lindsley 1981). One very important reason why Margules equations represent reasonable possible functionalities for ζ_{A} and ζ_{B} is that they satisfy the very important Gibbs-Duhem equation (Prausnitz 1969, p 212).

For binary solutions (i.e., mixtures of components like A and B) that follow Eqs. (16.15) and (16.16), the value of $\boldsymbol{A}$ may be expected to vary as the identities of A and B are changed. When $\boldsymbol{A}$ is zero, the solution is ideal because then both ζ_{A} and ζ_{B} are equal to 1.0 for all X_{A}. When $\boldsymbol{A}$ is positive, the solution is characterized by $\Delta H_{\mathrm{mxg}} > 0$, and mixing is inhibited. When $\boldsymbol{A}$ is negative, the solution is characterized by $\Delta H_{\mathrm{mxg}} < 0$, and mixing is favored. Substituting Eqs. (16.15) and (16.16) into Eq. (16.14) yields

$$\Delta\overline{G}_{\mathrm{mxg}} = \Delta\overline{G}_{\mathrm{mxg,idl}} + \mathrm{X}_{\mathrm{A}}\boldsymbol{A}X_{\mathrm{B}}^2 + X_{\mathrm{B}}\boldsymbol{A}X_{\mathrm{A}}^2. \tag{16.17}$$

Some thermodynamicists like to define the quantity $\Delta\overline{G}_{\mathrm{mxg,exs}}$ as the "**excess**" (exs) free energy in the system due to non-ideality. For the case described by Eq. (16.17), we have

$$\Delta\overline{G}_{\mathrm{mxg,exs}} = X_{\mathrm{A}}\boldsymbol{A}X_{\mathrm{B}}^2 + X_{\mathrm{B}}\boldsymbol{A}X_{\mathrm{A}}^2. \tag{16.18}$$

Thus, we have

$$\Delta\overline{G}_{\text{mxg}} = \Delta\overline{G}_{\text{mxg,idl}} + \Delta\overline{G}_{\text{mxg,exs}}. \tag{16.19}$$

For an ideal mixture, $\Delta G_{\text{mxg,exs}} = 0$. The RHS of Eq. (16.18) can be simplified by factoring $X_A X_B$ out of both terms. Thus, for the two-suffix Margules case, we have

$$\Delta\overline{G}_{\text{mxg,exs}} = \boldsymbol{A} X_A X_B (X_A + X_B). \tag{16.20}$$

Since the sum $(X_A + X_B)$ is always 1, Eq. (16.20) simplifies to

$$\Delta\overline{G}_{\text{mxg,exs}} = \boldsymbol{A} X_A X_B. \tag{16.21}$$

Just as the term $\alpha_0\alpha_1$ in the buffer intensity of a monoprotic acid maximizes when $\alpha_0 = \alpha_1 = 0.5$, so too will the *magnitude* of $\Delta\overline{G}_{\text{mxg,exs}}$ maximize when $X_A = X_B = 0.5$.

Combining Eqs. (16.19) and (16.21), for the two-suffix Margules case we now have

$$\Delta\overline{G}_{\text{mxg}} = X_A RT \ln X_A + X_B RT \ln X_B + \boldsymbol{A} X_A X_B. \tag{16.22}$$

Since $X_B = 1 - X_A$, Eq. (16.22) is easily evaluated as a function of X_A for any given $\boldsymbol{A}$ value. We note now that the expressions in Eqs. (16.15) and (16.16) giving the ζ values as a function of X_A are symmetrical. Since $\Delta\overline{G}_{\text{mxg,idl}}$ is also symmetrical, then when the two-suffix Margules equations describe the activity coefficients, the $\Delta\overline{G}_{\text{mxg}}$ curve as given by Eq. (16.22) will be symmetrical as in the ideal case. In many real situations, the $\Delta\overline{G}_{\text{mxg}}$ curve will not be symmetrical, and more complicated, higher order Margules equations will be needed to model the behavior.

Figure 16.2 presents curves for $\boldsymbol{A}/RT$ values of -5, -3, -1, 0 (ideal solution), 1, 3, 5, and 10 (we recall that $\boldsymbol{A}/RT$ is dimensionless). As $\boldsymbol{A}/RT$ becomes more and more negative, the solution process becomes more and more favored. However, for sufficiently large, positive values of this parameter, a maximum in the $\Delta\overline{G}_{\text{mxg}}$ curve appears, and there is a region in which $\Delta\overline{G}_{\text{mxg}}$ is concave down. As will be discussed in the next section, whenever the $\Delta\overline{G}_{\text{mxg}}$ curve has a portion that is concave down, then A and B will not be miscible over all proportions, and *separation into two phases* will be possible whenever significant amounts of A and B are present simultaneously.

Before proceeding, it is very important to note that no matter how repulsively non-ideal a given system is, very near $X_A = 0$ ($X_B = 1.0$), $\Delta\overline{G}_{\text{mxg}}$ will always be negative due to dominance by the entropy portion of $\Delta\overline{G}_{\text{mxg}}$. Thus, sufficiently close to $X_A = 0$, the slope $d\Delta\overline{G}_{\text{mxg}}/dX_A$ will always be negative. What this means is that some interdissolution will *always* take place regardless of how repulsively non-ideal a given pair of chemicals are. In the specific case of when the two-suffix Margules equations apply, we can show that no matter how large $\boldsymbol{A}/RT$ becomes, the slope $d\Delta\overline{G}_{\text{mxg}}/dX_A$ will always be negative near $X_A = 0$. Taking the derivative of Eq. (16.22) and using the standard calculus relationship that $d\ln x = dx/x$, we obtain

$$d\Delta\overline{G}_{\text{mxg}}/dX_A = RT \ln X_A - RT \ln X_B + \boldsymbol{A} - 2\boldsymbol{A} X_A. \tag{16.23}$$

As $X_A \to 0$, the RHS of Eq. (16.23) is not only negative, but tends to $-\infty$. As $X_A \to 1$ ($X_B \to 0$), the slope goes to $+\infty$, or equivalently $d\Delta\overline{G}_{\text{mxg}}/dX_B$ goes to $-\infty$.

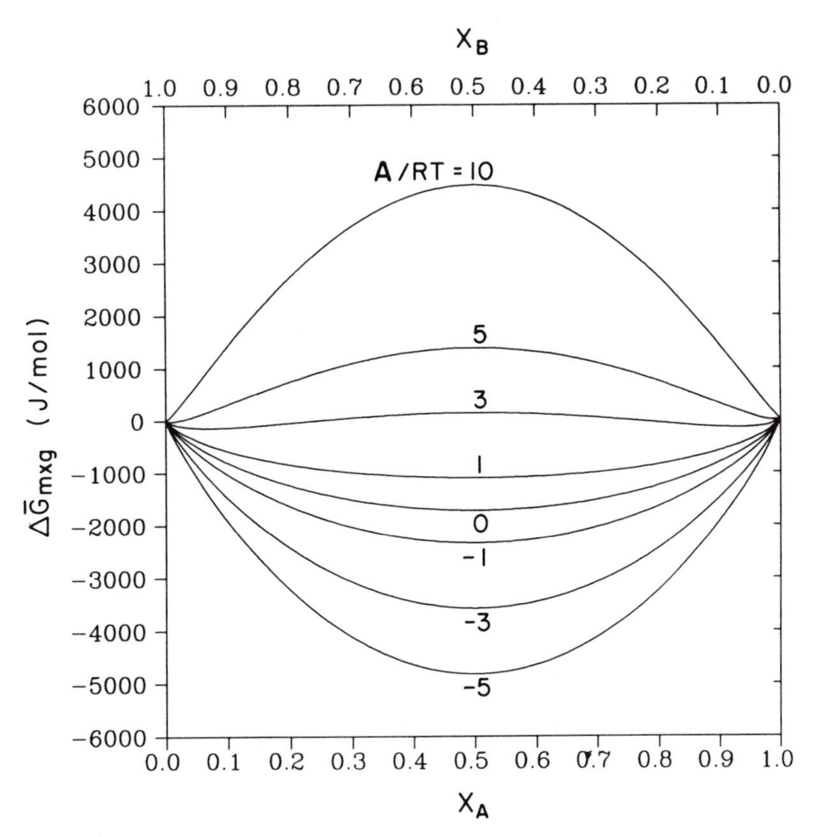

Figure 16.2 $\Delta\overline{G}_{mxg}$ (J/mol) at 298 K as a function of X_A for various values of the dimensionless, two-suffix Margules parameter A/RT. For an ideal solid solution, $A/RT = 0$.

Thus, both *pure* A and *pure* B are seen to be infinitely strong chemical magnets for B and A, respectively, regardless of any finite amount of mutual repulsiveness between A and B. Thus, while the amount of interdissolution that takes place may be small at equilibrium in a very repulsively non-ideal case, there will *always* be some interdissolution. The negative nature of the slope near $X_A = 0$ is illustrated for the $A/RT = 5$ case in the expanded view given in Figure 16.3. However, neither that figure nor Figures 16.1 and 16.2 examine data points sufficiently close to $X_A = 0$ to illustrate the infinitely steep slopes that obtain there.

16.2.4 Differences in Behavior When Chemical A Is Pure and When It Is Present in a Solution

When chemical A (liquid or solid) is pure, it will exert the full thermodynamic activity that corresponds to a mole fraction of 1.0. When it is dissolved with some B, then at equilibrium, the thermodynamic activity will be lower than for the pure A. It is clear that this is true when attractive non-ideality ($\Delta\overline{H}_{mxg} < 0$) is active. In that case, then

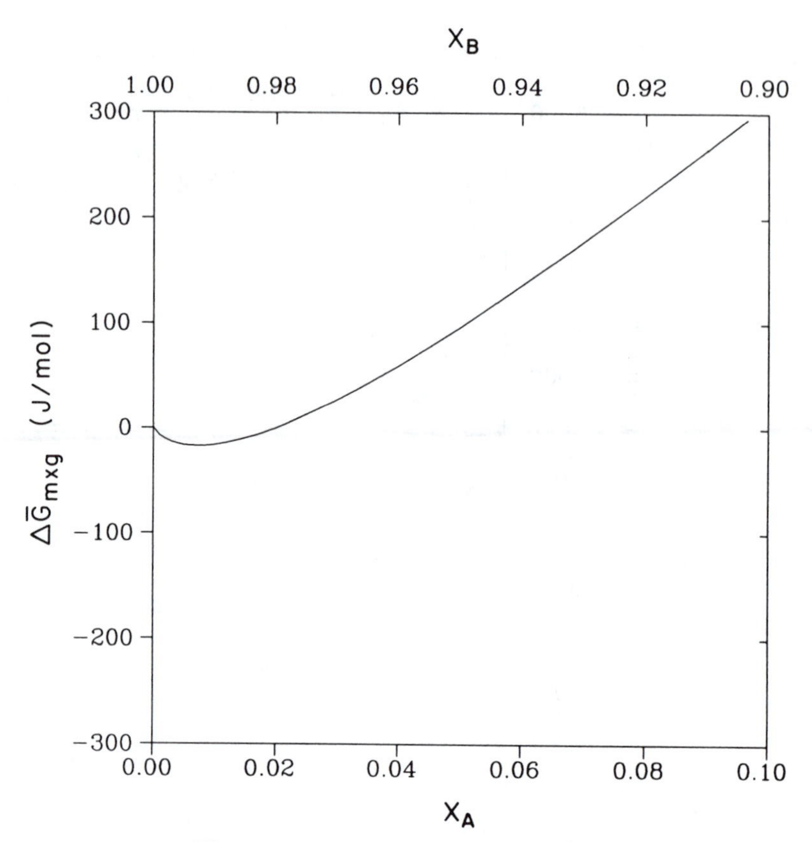

Figure 16.3 $\Delta\overline{G}_{mxg}$ (J/mol) near $X_A = 0$ for the two-suffix Margules case with $A/RT = 5$. Note that for this value of A/RT and for all other values of A/RT, $\Delta\overline{G}_{mxg}$ (J/mol) is negative for X_A values very close to zero.

not only will the mole fraction X_A be less than 1.0, but so too will ζ_A, and therefore we know $X_A\zeta_A < 1$.

When repulsive non-ideality ($\Delta\overline{H}_{mxg} > 0$) is active, then while the mole fraction X_A must be less than 1.0 in the mixture, we know that $\zeta_A > 1$. Even in this case, however, we know that the product $X_A\zeta_A$ must be less than 1 because the system has been assumed to be at equilibrium, and the formation of the mixture from some pure A can only occur when the activity (or equivalently, the chemical potential) of A is lower in the solution than in the pure phase.

Taking $SrCO_{3(s)}$ as an example for chemical A, for dissolution into water at 25°C/1 atm, we have

$$\frac{\{Sr^{2+}\}\{CO_3^{2-}\}}{\{SrCO_{3(s)}\}} = 10^{-9.03}. \tag{16.24}$$

In general, at 25°C/1 atm, dissolution into water will occur according to

$$\{Sr^{2+}\}\{CO_3^{2-}\} = 10^{-9.03}\{SrCO_{3(s)}\}. \tag{16.25}$$

When the $SrCO_{3(s)}$ is pure, it will dissolve into water according to

$$\{Sr^{2+}\}\{CO_3^{2-}\} = 10^{-9.03}1 = 10^{-9.03}. \tag{16.26}$$

However, if we had a *solid solution* of some $SrCO_{3(s)}$ in *another carbonate phase*, maybe $CaCO_{3(s)}$, then if the $SrCO_{3(s)}$ mole fraction was 0.01 and the activity coefficient was 0.1 ($\Delta\overline{H}_{mxg} < 0$), then $\{SrCO_{3(s)}\} = 0.001$, and dissolution into water will occur according to

$$\{Sr^{2+}\}\{CO_3^{2-}\} = 10^{-9.03}0.001 = 10^{-12.03}. \tag{16.27}$$

If we had a solid solution of some $SrCO_{3(s)}$ in some $CaCO_{3(s)}$ with a $SrCO_{3(s)}$ mole fraction was 0.01 and the activity coefficient was 10, ($\Delta\overline{H}_{mxg} > 0$), then $\{SrCO_{3(s)}\} = 0.1$, and dissolution into water will occur according to

$$\{Sr^{2+}\}\{CO_3^{2-}\} = 10^{-9.03}0.1 = 10^{-10.03}. \tag{16.28}$$

While activity coefficients of mineral A in a solid solution containing another mineral will in general be greater than 1 rather than less than 1, we can conclude that at equilibrium the formation of the solid solution will cause mineral A to have an activity that is lower than for pure A. Thus, for dissolution into water, we will always have

$$IAP \leq K_{s0} \tag{16.29}$$

where *IAP* represents the ion activity product in the water (for the case discussed above, $IAP = \{Sr^{2+}\}\{CO_3^{2-}\}$). When solid A is pure, the equal sign in Eq. (16.29) will apply. When the solid is not pure A, then the $<$ in Eq. (16.29) will apply.

The above discussion can be extended to the case when chemical A is an organic liquid (e.g., chlorobenzene) dissolved in a liquid solution involving another organic (e.g., some polychlorinated biphenyls (PCBs)). Now, rather than Eq. (16.29), when the organic solution is exposed to water and some A dissolves into water, we have

$$C_A \leq C_{sat,A} \tag{16.30}$$

where C_A is the equilibrium concentration of A in the water, and $C_{sat,A}$ is the saturation concentration of pure A in water.

16.3 PHASE SEPARATION IN NON-IDEAL SOLUTIONS

16.3.1 The $\Delta\overline{G}_{mxg}$ Curve and Phase Separation

We naturally expect that as a binary system becomes more and more repulsively non-ideal (i.e., as *A/RT* becomes larger and larger for the two-suffix Margules case), then ultimately, the system will not be miscible over all proportions. That is, although one could always succeed in dissolving some small amount of A in B and vice versa, it would not be possible, for example, to obtain a *stable* mixture composed of equal amounts of A and B. That is, if equal amounts of A and B were successfully mixed by some unspecified means, then after allowing them to stand for a sufficiently long period of time they would separate into *two phases*. One of the two phases would be predominantly A in composition, and the other phase would be predominantly B. The amount of B in the predominately A phase would be the "saturation" amount for the T and P of the system, and vice versa. Thus, when we equilibrate a liter of trichloroethylene (TCE) with a liter of water, one of the phases will be mostly water with a little TCE dissolved in it, and the other phase will be mostly TCE with a little water dissolved in it.

As alluded to in the previous section, phase separation of the type just discussed will only occur for A/B pairs that are characterized by a $\Delta\overline{G}_{mxg}$ curve that is concave down in some region of X_A. This may be understood in terms of Figure 16.4 which depicts a general *non-symmetrical* mixing case that is concave down over certain range of X_A values. The two minima in the Figure 16.4 curve occur at X_A^* and X_A^{**}. For a mixed system initially in the region of $X_{A,a}$, phase separation into two phases will occur. The compositions of the resultant two phases will be those which minimize the overall G of the system. These compositions are denoted X_A^α and X_A^β where the α denotes one phase, and the β denotes the second phase. In the α phase, we have $X_A^\alpha + X_B^\alpha = 1$, and in the β phase we have $X_A^\beta + X_B^\beta = 1$. In the non-symmetrical case, the points X_A^* and X_A^{**} are *not* the same as X_A^α and X_A^β. This will be explained momentarily. The manner in which the separation occurs may be studied by determining how phase separation can help a system reach a minimum G value.

16.3.2 Minimization of the Free Energy *G* by Phase Separation

For given, constant total amounts of A and B, that is, for given values of

$$n_A^\alpha + n_A^\beta = n_A \text{ (constant)} \tag{16.31}$$

and

$$n_B^\alpha + n_B^\beta = n_B \text{ (constant)}, \tag{16.32}$$

the *two* variables with respect to which G must be minimized are n_A^α and n_B^α. These two quantities represent the number of mols of A and B in the α phase. (By Eqs. (16.31) and (16.32), a knowledge of n_A^α and n_B^α will give n_A^β and n_B^β.) After the phase separation of the initially-mixed $X_{A,a}$ system, the G of the two-phase system will be given by

$$G = (n_A^\alpha + n_B^\alpha)\overline{G}_{mxt}^\alpha + (n_A^\beta + n_B^\beta)\overline{G}_{mxt}^\beta \tag{16.33}$$

where $\overline{G}_{mxt}^\alpha$ and $\overline{G}_{mxt}^\beta$ represent the molar free energies for mixtures of the compositions present in the α and β phases, respectively. (We recall that the molar G for a mixture is given by the total G for the mixture divided by the number of mols in that mixture.) Thus, with the help of Eq. (16.7), these two quantities are seen to be given by

$$\begin{aligned}\overline{G}_{mxt}^\alpha &= X_A^\alpha(\mu_A^o + RT\ln X_A^\alpha + RT\ln\zeta_A^\alpha)\\ &\quad + X_B^\alpha(\mu_B^o + RT\ln X_B^\alpha + RT\ln\zeta_B^\alpha)\end{aligned} \tag{16.34}$$

and

$$\begin{aligned}\overline{G}_{mxt}^\beta &= X_A^\beta(\mu_A^o + RT\ln X_A^\beta + RT\ln\zeta_A^\beta)\\ &\quad + X_B^\beta(\mu_B^o + RT\ln X_B^\beta + RT\ln\zeta_B^\beta).\end{aligned} \tag{16.35}$$

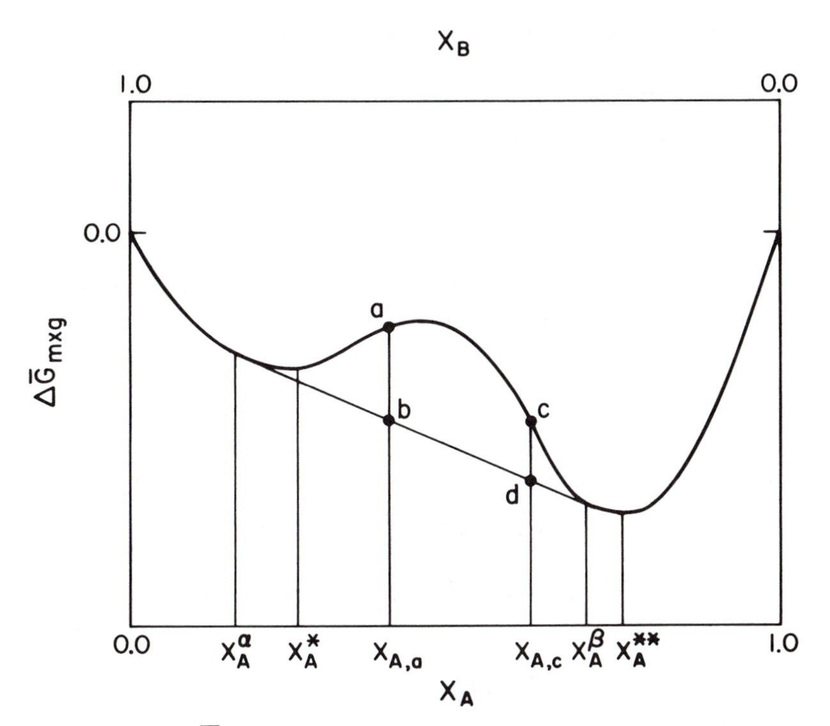

Figure 16.4 $\Delta\overline{G}_{\text{mxg}}$ (J/mol) as a function of X_{A} for a general, asymmetric, repulsively non-ideal mixing case. X_{A}^{α} and X_{A}^{β} are the compositions that obtain when the maximum possible amount of B is dissolved in A, and vice versa. X_{A}^{α} and X_{A}^{β} are therefore the compositions that obtain after phase separation occurs for any mixture having an initial composition that is intermediate between these two values, and is therefore unstable. X_{A}^{*} and X_{A}^{**} are the compositions at the two minima. For an asymmetric, repulsively non-ideal case such as this, $X_{\text{A}}^{\alpha} \neq X_{\text{A}}^{*}$ and $X_{\text{A}}^{\beta} \neq X_{\text{A}}^{**}$. However, for a symmetric case (e.g., a two-suffix Margules case), $X_{\text{A}}^{\alpha} = X_{\text{A}}^{*}$ and $X_{\text{A}}^{\beta} = X_{\text{A}}^{**}$.

If we subtract the total initial free energy of the unmixed components from the expression given by Eq. (16.33), the result will be the overall ΔG_{mxg}. That is,

$$\Delta G_{\text{mxg}} = (n_{\text{A}}^{\alpha} + n_{\text{B}}^{\alpha})\overline{G}_{\text{mxt}}^{\alpha} + (n_{\text{A}}^{\beta} + n_{\text{B}}^{\beta})\overline{G}_{\text{mxt}}^{\beta} - n_{\text{A}}\mu_{\text{A}}^{\text{o}} - n_{\text{B}}\mu_{\text{B}}^{\text{o}}. \tag{16.36}$$

Thus, since G differs from ΔG_{mxg} only by the quantity $(n_{\text{A}}\mu_{\text{A}}^{\text{o}} + n_{\text{B}}\mu_{\text{B}}^{\text{o}})$, which will be constant for any given case, the conditions that minimize ΔG_{mxg} for the formation of the two phases from the initially-pure solids will also minimize G. We can therefore investigate when the overall ΔG_{mxg} is minimized.

When phase separation does occur, the overall ΔG_{mxg} can be viewed as the result of the formation of two separate mixtures from the original n_{A} mols of A and the n_{B}

mols of B. Thus, in addition to Eq. (16.36), we can write

$$\Delta G_{\text{mxg}} = (n_{\text{A}}^{\alpha} + n_{\text{B}}^{\alpha})\Delta\overline{G}_{\text{mxg}}^{\alpha} + (n_{\text{A}}^{\beta} + n_{\text{B}}^{\beta})\Delta\overline{G}_{\text{mxg}}^{\beta}. \tag{16.37}$$

By Eqs. (16.31) and (16.32), we have

$$\partial n_{\text{A}}^{\alpha} = -\partial n_{\text{A}}^{\beta} \tag{16.38}$$

$$\partial n_{\text{B}}^{\alpha} = -\partial n_{\text{B}}^{\beta}. \tag{16.39}$$

Using standard calculus for finding a minimum in a function, we differentiate ΔG_{mxg} (as it is given by Eq. (16.37)) with respect to n_{A}^{α} and set the result equal to zero:

$$\partial\Delta G_{\text{mxg}}/\partial n_{\text{A}}^{\alpha} = (n_{\text{A}}^{\alpha} + n_{\text{B}}^{\alpha})\partial\Delta\overline{G}_{\text{mxg}}^{\alpha}/\partial n_{\text{A}}^{\alpha} + \Delta\overline{G}_{\text{mxg}}^{\alpha} - (n_{\text{A}}^{\beta} + n_{\text{B}}^{\beta})\partial\Delta\overline{G}_{\text{mxg}}^{\beta}/\partial n_{\text{A}}^{\beta} - \Delta\overline{G}_{\text{mxg}}^{\beta} = 0. \tag{16.40}$$

Equation (16.40) can be expressed in terms of mole fractions by noting that

$$\partial\Delta\overline{G}_{\text{mxg}}^{\alpha}/\partial n_{\text{A}}^{\alpha} = (\partial\Delta\overline{G}_{\text{mxg}}^{\alpha}/\partial X_{\text{A}}^{\alpha})(\partial X_{\text{A}}^{\alpha}/\partial n_{\text{A}}^{\alpha}) \tag{16.41}$$

$$\partial\Delta\overline{G}_{\text{mxg}}^{\beta}/\partial n_{\text{A}}^{\beta} = (\partial\Delta\overline{G}_{\text{mxg}}^{\beta}/\partial X_{\text{A}}^{\beta})(\partial X_{\text{A}}^{\beta}/\partial n_{\text{A}}^{\beta}) \tag{16.42}$$

and that

$$(\partial X_{\text{A}}^{\alpha}/\partial n_{\text{A}}^{\alpha}) = n_{\text{B}}^{\alpha}/(n_{\text{A}}^{\alpha} + n_{\text{B}}^{\alpha})^2 \tag{16.43}$$

$$(\partial X_{\text{A}}^{\beta}/\partial n_{\text{A}}^{\beta}) = n_{\text{B}}^{\beta}/(n_{\text{A}}^{\beta} + n_{\text{B}}^{\beta})^2. \tag{16.44}$$

Substituting Eqs. (16.41)–(16.44) into Eq. (16.40), we obtain

$$\partial\Delta G_{\text{mxg}}/\partial n_{\text{A}}^{\alpha} = X_{\text{B}}^{\alpha}\partial\Delta\overline{G}_{\text{mxg}}^{\alpha}/\partial X_{\text{A}}^{\alpha} + \Delta\overline{G}_{\text{mxg}}^{\alpha} - X_{\text{B}}^{\beta}\partial\Delta\overline{G}_{\text{mxg}}^{\beta}/\partial X_{\text{A}}^{\beta} - \Delta\overline{G}_{\text{mxg}}^{\beta} = 0. \tag{16.45}$$

Differentiation of ΔG_{mxg} with respect to n_{B}^{α} and similar substitutions gives

$$\partial\Delta G_{\text{mxg}}/\partial n_{\text{B}}^{\alpha} = X_{\text{A}}^{\alpha}\partial\Delta\overline{G}_{\text{mxg}}^{\alpha}/\partial X_{\text{B}}^{\alpha} + \Delta\overline{G}_{\text{mxg}}^{\alpha} - X_{\text{A}}^{\beta}\partial\Delta\overline{G}_{\text{mxg}}^{\beta}/\partial X_{\text{B}}^{\beta} - \Delta\overline{G}_{\text{mxg}}^{\beta} = 0. \tag{16.46}$$

Subtracting Eq. (16.46) from (16.45), then noting that

$$\partial X_{\text{B}}^{\alpha} = -\partial X_{\text{A}}^{\alpha} \quad \text{and} \quad \partial X_{\text{B}}^{\beta} = -\partial X_{\text{A}}^{\beta} \tag{16.47}$$

and noting further that $X_{\text{A}}^{\alpha} + X_{\text{B}}^{\alpha} = 1$ and that $X_{\text{A}}^{\beta} + X_{\text{B}}^{\beta} = 1$, we obtain the very interesting and surprisingly simple result that the G of the two phase system is minimized when

$$\partial\Delta\overline{G}_{\text{mxg}}^{\alpha}/\partial X_{\text{A}}^{\alpha} = \partial\Delta\overline{G}_{\text{mxg}}^{\beta}/\partial X_{\text{A}}^{\beta}. \tag{16.48}$$

Equation (16.48) states that at equilibrium, the slope of the $\Delta\overline{G}_{\text{mxg}}$ vs. X_{A} curve at the value of X_{A} present at equilibrium in the α phase is equal to the slope of the same curve at the value of X_{A} present at equilibrium in the β phase.

The application of Eq. (16.47) again in Eq. (16.46) along with Eq. (16.48) leads to

$$\partial\Delta\overline{G}^{\alpha}_{\text{mxg}}/\partial X^{\alpha}_{\text{A}} = \partial\Delta\overline{G}^{\beta}_{\text{mxg}}/\partial X^{\beta}_{\text{A}} = (\Delta\overline{G}^{\beta}_{\text{mxg}} - \Delta\overline{G}^{\alpha}_{\text{mxg}})/(X^{\beta}_{\text{A}} - X^{\alpha}_{\text{A}}). \tag{16.49}$$

This remarkable result states that not only are the slopes equal, but *also* that the slope is given by the line drawn between the points $(X^{\alpha}_{\text{A}}, \Delta\overline{G}^{\alpha}_{\text{mxg}})$ and $(X^{\beta}_{\text{A}}, \Delta\overline{G}^{\beta}_{\text{mxg}})$. That line is given in Figure 16.4 for that particular non-symmetrical case.

16.3.3 The Linear Mixing Curve for $\Delta\overline{G}_{\text{mxg}}$ vs. X_{A} in the Region Where Phase Separation Can Occur

In addition to its interesting slope properties, the line discussed above represents the "linear mixing curve" giving the *actual* $\Delta\overline{G}_{\text{mxg}}$ of any system whose *overall* composition falls between X^{α}_{A} and X^{β}_{B}. The word "actual" means "even if there are two phases," and the words "overall composition" mean "the composition if there was only one, totally mixed phase." This is important since the actual ΔG_{mxg} for the overall system is directly related to the actual $\Delta\overline{G}_{\text{mxg}}$ according to

$$\Delta G_{\text{mxg}} = (n_{\text{A}} + n_{\text{B}})\Delta\overline{G}_{\text{mxg}}. \tag{16.50}$$

To examine the above linear mixing property, we consider an unstable, one phase mixture with composition $X_{\text{A,a}}$ as given in Figure 16.4. After the mixture separates into the α and β phases having the compositions X^{α}_{A} and X^{β}_{A}, then the overall ΔG_{mxg} of the system will be lowered since the material is now distributed into two phases each of which has a lower $\Delta\overline{G}_{\text{mxg}}$. The overall $\Delta\overline{G}_{\text{mxg}}$ of the system will be that at point b. If the initial composition of the one phase mixture is that corresponding to $X_{\text{A,c}}$, then the mixture will again separate into two phases. The compositions of the two phases will be the same X^{α}_{A} and the other X^{β}_{A} as before. The overall $\Delta\overline{G}_{\text{mxg}}$ will be that at point d. In this case, the majority of the material ends up in the β phase since $X_{\text{A,c}}$ is closer to X^{β}_{A} than it is to X^{α}_{A}. Thus, even though some of the material has to rise to a $\Delta\overline{G}_{\text{mxg}}$ that is higher than $\Delta\overline{G}_{\text{mix,c}}$, the fact that most of the material can lower its $\Delta\overline{G}_{\text{mxg}}$ is what drives the phase separation.

That it would even be possible to separate *any* one phase system with known overall composition X_{A} into two phases that have the specific compositions X^{α}_{A} and X^{β}_{A} may be understood as follows. There are *four* unknowns in the system, n^{α}_{A}, n^{α}_{B}, n^{β}_{A}, and n^{β}_{B}. Equations (16.31) and (16.32) are two equations that govern any system of this type. The thermodynamics of the system provide the other two equations, i.e., that for every case, X^{α}_{A} and X^{β}_{A} are the two equilibrium compositions of the two phases.

16.3.4 When Phase Separation Occurs, the Same Two Values of X^{α}_{A} and X^{β}_{A} Are Obtained Regardless of Values of n_{A} and n_{B}

That the *same* two phases would form whenever a phase separation occurs has been shown above by demonstrating that a specific pair of compositions will minimize the overall G of the system. In addition, the following argument may be advanced. At equilibrium between the two phases, the chemical potentials of the two components must be equal in both phases:

$$\mu_A^\alpha = \mu_A^o + RT\ln\zeta_A^\alpha X_A^\alpha = \mu_A^\beta = \mu_A^o + RT\ln\zeta_A^\beta X_A^\beta \tag{16.51}$$

$$\mu_B^\alpha = \mu_B^o + RT\ln\,\zeta_B^\alpha X_B^\alpha = \mu_B^\beta = \mu_B^o + RT\ln\,\zeta_B^\beta X_B^\beta . \tag{16.52}$$

Since the same standard state μ values are used for both the α and β phases, Eqs. (16.51) and (16.52) reduce to an equating of the two activities of A and B in the two phases, that is,

$$\zeta_A^\alpha X_A^\alpha = \zeta_A^\beta X_A^\beta \tag{16.53}$$

$$\zeta_B^\alpha X_B^\alpha = \zeta_B^\beta X_B^\beta . \tag{16.54}$$

Equations (16.53) and (16.54) may be used to solve any phase separation problem for a given pair of components. They contain the four unknown mole fractions. (The four activity coefficients are implicit functions of the mole fractions and do not therefore constitute additional unknowns for the problem.) The two additional equations needed are provided by the statements that the mole fractions sum to one for each phase. Thus, when phase separation occurs, the four equations used to solve the problem will always be exactly the same. Therefore, the two phases which form will always have the same composition no matter what the absolute amounts n_A and n_B are used to create the system. The *absolute* amounts of the two phases, will, of course, depend on n_A and n_B. Thus if we equilibrate two liters of TCE and one liter of water, we will get the same saturation concentrations of TCE in water and of water in TCE that we would get if we equilibrated one liter of TCE with one liter of water.

It may be noted that for a non-symmetrical $\Delta\overline{G}_{mxg}$ curve, the compositions X_A^α and X_A^β are not equal to the compositions which occur at the two minima at X_A^* and X_A^{**}. This may be shown as follows. Consider the expression for the $\Delta\overline{G}_{mxg}$ for fixed values of n_A and n_B:

$$\Delta G_{mxg} = (n_A^\alpha + n_B^\alpha)\Delta\overline{G}_{mxg}^\alpha + (n_A^\beta + n_B^\beta)\Delta\overline{G}_{mxg}^\beta . \tag{16.37}$$

Thus, at equilibrium,

$$\partial\Delta G_{mxg}/\partial n_A^\alpha = (n_A^\alpha + n_B^\alpha)\partial\Delta\overline{G}_{mxg}^\alpha/\partial n_A^\alpha + \Delta\overline{G}_{mxg}^\alpha - (n_A^\beta + n_B^\beta)\partial\Delta\overline{G}_{mxg}^\beta/\partial n_A^\beta - \Delta\overline{G}_{mxg}^\beta = 0. \tag{16.40}$$

Substitution of Eqs. (16.41) and (16.42) yields

$$\partial\Delta\overline{G}_{mxg}/\partial n_A^\alpha = (n_A^\alpha + n_B^\alpha)(\partial\Delta\overline{G}_{mxg}^\alpha/\partial X_A^\alpha)(\partial X_A^\alpha/\partial n_A^\alpha) + \Delta\overline{G}_{mxg}^\alpha - (n_A^\beta + n_B^\beta)(\partial\Delta\overline{G}_{mxg}^\beta/\partial X_A^\beta)(\partial X_A^\beta/\partial n_A^\beta) - \Delta\overline{G}_{mxg}^\beta = 0. \tag{16.55}$$

The quantities $(\partial\Delta\overline{G}_{mxg}^\alpha/\partial X_A^\alpha)$ and $(\partial\Delta\overline{G}_{mxg}^\beta/\partial X_A^\beta)$ are identically zero at X_A^* and X_A^{**}, respectively. Since $(\partial X_A^\alpha / \partial n_A^\alpha)$ and $(\partial X_A^\beta / \partial n_A^\beta)$ are finite at the same points, respectively, then if the minima X_A^* and X_A^{**} represented the composition at equilibrium, then

$$\partial \Delta \overline{G}_{mxg} / \partial n_A^\alpha = \Delta \overline{G}_{mxg}^\alpha - \Delta \overline{G}_{mxg}^\beta = 0. \tag{16.56}$$

Since the quantities $\Delta \overline{G}_{mxg}^\alpha$ and $\Delta \overline{G}_{mxg}^\beta$ will not *in general* be equal, $\partial \Delta G / \partial n_A^\alpha$ will not be zero, and the X_A values for the minima in the $\Delta \overline{G}_{mxg}$ curve cannot represent the equilibrium compositions of the two phase system. A similar conclusion could be reached based on the examination of the derivative $\partial \overline{G}_{mxg} / \partial n_B^\alpha$.

Considering the discussion just given above, it may be noted that in the special case of *symmetrical* non-ideality (e.g., that described by the two suffix Margules equations), the minima in the $\Delta \overline{G}_{mxg}$ curve *will* be the equilibrium compositions. That is the case since at those points: 1) the slopes are equal (see above); and 2) we satisfy the requirement of zero slopes that $\Delta \overline{G}_{mxg}^\alpha = \Delta \overline{G}_{mxg}^\beta$.

16.3.5 Tendency for Correlation Between X_A^α and X_B^β

In a system in which phase separation can occur, the more repulsively non-ideal that a given A/B pair becomes, the more the value of X_A^α will tend to move towards $X_A = 0$, and the more the value of X_A^β will tend to move towards $X_A = 1$. In other words, when phase separation can occur, as the system becomes more repulsively non-ideal, then the solubility of A in B and the solubility of B in A will both tend to go down.

One can envision how the above effects are manifested with the help of Figures 16.2 and 16.5. As discussed in Section 16.2.3, near $X_A = 0$ and 1, the $\Delta \overline{G}_{mxg}$ curve is always negative, no matter how repulsively non-ideal the system. However, as a system becomes more and more repulsively non-ideal (higher and higher ***A/RT*** values), the $\Delta \overline{G}_{mxg}$ curve is pulled upwards, the height of the maximum increases, and the positions of the solubility mole fractions X_A^α and X_A^β are pulled closer to 0 and 1, respectively. Equivalently, we can say that as X_A^α tends to decrease towards 0, then X_B^β also tends towards 0. Figure 16.5 illustrates how X_A^α tends to 0 for a range of increasing ***A/RT*** values for systems following two-suffix Margules behavior. (We recall that in symmetrical systems of this type, $X_A^\alpha = X_A^*$.)

The above comments lead to the conclusion that X_A^α and X_B^β values ought to be correlated. Indeed, as chemical A becomes more unlike and therefore inhospitable to B, it is only logical to expect that chemical B is becoming more unlike and inhospitable to A. While there is not much data for solid solutions that can be examined in search of this correlation, there are data for liquid systems. Table 16.2 presents some data for a variety of organic compounds and water that illustrate this correlation tendency, and Figure 16.6 plots the data in a log X_B^β vs. log X_A^α plot. For these data, A is taken as referring to the organic compound, and the identity of the B is held constant as water. The α phase is the primarily water phase, and the β phase is the primarily organic phase. Thus, X_B^β represents the concentration of water in the primarily organic phase, and X_A^α represents the concentration of organic in the primarily water phase.

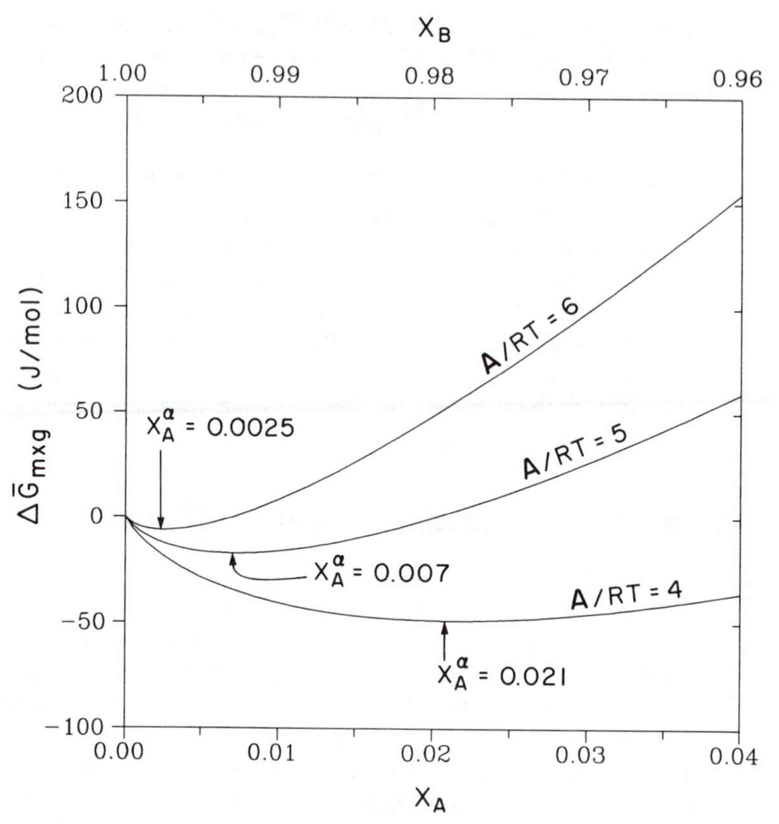

Figure 16.5 $\Delta\overline{G}_{mxg}$ (J/mol) near $X_A = 0$ for the two-suffix Margules case with $A/RT = 4$, 5, and 6. As A/RT increases, the minimum in the curve moves closer to $X_A = 0$. Because the two-suffix Margules equations yield a symmetrical curve of $\Delta\overline{G}_{mxg}$ vs. X_A, the value of X_A for the minimum (X_A^*) equals X_A^α.

While the correlation is not perfect in Figure 16.6, there is a clear trend relating the two sets of mole fractions. Thus, the solubility of water in methylethylketone (MEK) is high, as is the solubility of MEK in water. And, since the solubility of water in heptane is low, it is not surprising that the solubility of heptane in water is also low. Since water and MEK are alike in that they are both polar compounds, their mutual solubilities are relatively high; since water is polar but heptane is not, their mutual solubilities are very low. This is the origin of the chemical saying "*Like dissolves like.*"

16.4 CONCLUSIONS

The physical chemistry discussed in this chapter is considerably more complex than in many of the other chapters in this book. Unfortunately, we will usually not be able to actually apply this level of complexity to problems of interest. For example, although

TABLE 16.2. Solubilities of various organic liquids in water, and solubilities of water in various organic liquids at 20–25°C (Burdick and Jackson, 1984).

Organic Compound	Molec. Wt.	Concentration of organic in the primarily water phase		Concentration of water in the primarily organic phase	
		wt %	$\log X_A^\alpha$	wt %	$\log X_B^\beta$
Trimethylpentane	114.2	0.0002	−6.50	0.006	−3.42
Heptane	100.2	0.0003	−6.27	0.01	−3.26
Hexane	86.2	0.001	−5.68	0.01	−3.32
1,2,4-Trichlorobenzene	181.4	0.0025	−5.60	0.02	−2.70
Pentane	72.1	0.004	−5.00	0.009	−3.44
Cyclohexane	84.2	0.01	−4.67	0.01	−3.33
Cyclopentane	70.1	0.01	−4.59	0.01	−3.41
Decahydronaphthalene	140.2	0.02	−4.59	0.0063	−3.31
Chlorobenzene	112.5	0.05	−4.10	0.04	−2.60
Toluene	92.1	0.074	−3.84	0.03	−2.82
Carbon tetrachloride	153.8	0.08	−4.03	0.008	−3.17
n-Butyl Chloride	92.6	0.08	−3.81	0.08	−2.39
Trichlorethylene (TCE)	131.3	0.11	−3.82	0.32	−1.64
Benzene	78.1	0.18	−3.38	0.063	−2.57
n-Butylacetate	118.1	0.43	−3.18	1.86	−0.96
Methyl Isoamylketone	114.1	0.54	−3.07	1.3	−1.11
Ethylenedichloride	96.9	0.81	−2.82	0.15	−2.10
Chloroform	119.3	0.815	−2.91	0.056	−2.43
Methylene Chloride	84.9	1.6	−2.46	0.24	−1.95
Methylisobutylketone (MIBK)	100.1	1.7	−2.51	1.9	−1.01
Methyl-t-butylether	88.1	4.8	−1.99	1.5	−1.16
Methylpropylketone	86.1	5.95	−1.88	3.3	−0.85
Ethylether	90.1	6.89	−1.84	1.26	−1.22
1-Butanol	74.1	7.81	−1.69	20.07	−0.29
Isobutylalcohol	74.1	8.5	−1.66	16.4	−0.35
Ethylacetate	88.1	8.7	−1.72	3.3	−0.85
2-Butanol	74.1	12.5	−1.47	44.1	−0.12
Methlyethylketone (MEK)	72.1	24	−1.14	10	−0.51

a fair amount of information is available for liquid solutions, one often does not know whether a given binary solid solution system can be described in a two-suffix Margules manner, much less what the value of A/RT would be for the system of interest. The goal for this chapter is therefore that the reader understand intuitively how solid and liquid solutions behave, that is, understand:

1. Figures 16.1–16.6 and Table 16.2.
2. Why solid and liquid solutions form.
3. Why some degree of interdissolution will always occur (i.e., there is no such thing as a zero solubility of A in B.
4. Why non-ideality with $\Delta\overline{H}_{mxg} < 0$ will favor the formation of solutions.

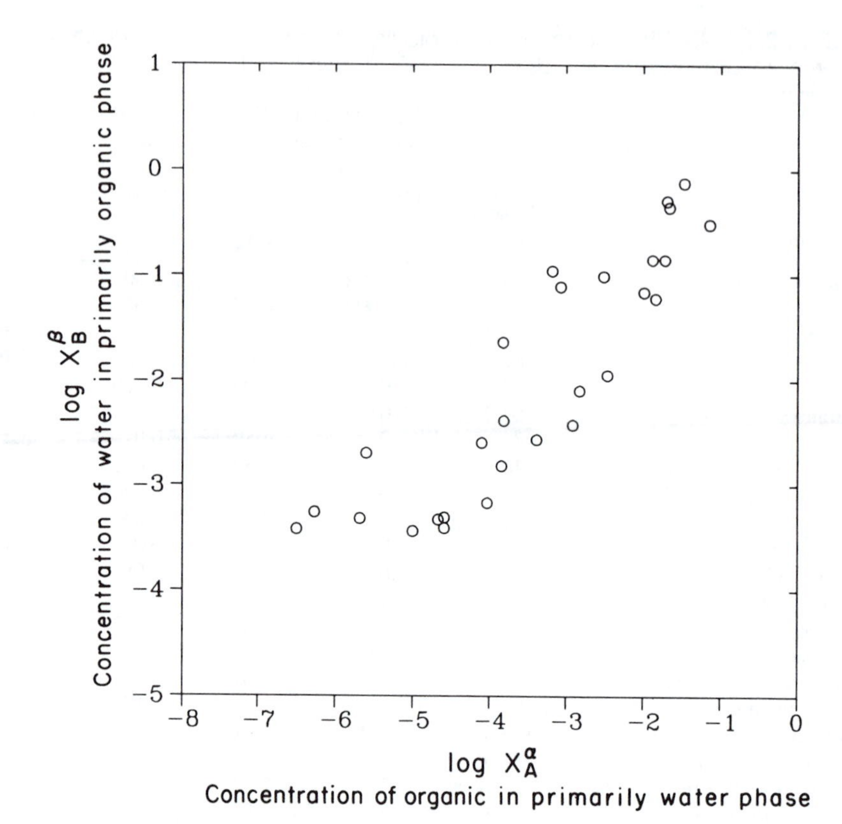

Figure 16.6 Log X_B^β vs. log X_A^α where A represents a range of organic compounds, B represents water. The data pertain to the compositions of the two phases that result when water is equilibrated with more organic compound than can be dissolved in the amount of water used. The compositions of the two phases that result are the same as would be obtained if the organic and the water were successfully mixed, then underwent phase separation. X_B^β represents the concentration of water in the primarily organic phase, and X_A^α represents the concentration of organic in the primarily water phase. Data from Burdick and Jackson (1984).

5. Why non-ideality with $\Delta\overline{H}_{mxg} > 0$ will inhibit the formation of solutions.
6. Why chemical A in a solution will *always* exert a thermodynamic activity at equilibrium that is lower than it would if it were pure.
7. Why non-ideality with $\Delta\overline{H}_{mxg} > 0$ can cause two chemicals A and B to be only partially miscible in one another, that is, cause limited solubility in one another, that is, cause phase separation over the composition range between X_A^α and X_A^β.

Problems illustrating the principles of this chapter are presented by Pankow (1992).

16.5 REFERENCES

ANDERSEN, D. J., and LINDSLEY, D. H. 1981. A Valid Margules Formulation for an Asymmetric Ternary Solution: Revision of the Olivine-Ilmenite Thermometer, with Applications. *Geochim. Cosmochim. Acta*, **45**: 847–853.

BURDICK and JACKSON. 1984. *Solvent Guide*. Burdick and Jackson Laboratories, Muskegon, Michigan.

LINDSLEY, D. H., GROVER, J. E., DAVIDSON, P. M. 1981. The Thermodynamics of $Mg_2Si_2O_6$ – $CaMgSi_2O_6$: A Review and Improved Model. *Thermodynamics of Minerals and Melts*, R. C. Newton, A. Navrotsky, and B. J. Wood, Eds., Springer-Verlag, 149–197.

NORDSTROM, D. K., and J. L. MUNOZ. 1986. *Geochemical Thermodynamics*, Palo Alto, California: Blackwell Scientific Publications, 158–163.

PANKOW, J. F. 1992. *Aquatic Chemistry Problems*. Portland: Titan Press-OR, P.O. Box 91399, Portland, Oregon 97291-1399.

PRAUSNITZ, J. M. 1969. *Molecular Thermodynamics of Fluid Phase Equilibria*. Englewood Cliffs: Prentice Hall.

THOMPSON, J. B. Jr., and WALDBAUM, D. R. 1969. Mixing Properties of Sanidine Crystalline Solutions. III. Calculations Based on Two-Phase Data. *Am. Mineral.*, **54**: 811–838.

17

The Gibbs Phase Rule

17.1 INTRODUCTION

When studying natural waters, the chemistry of the systems under consideration can exhibit varying degrees of chemical flexibility. That is, the degree to which chemical (e.g., concentration) information and physical information (T and/or P) is needed to tie down the chemistry can vary. For pure liquid water, we know that the chemistry will be specified once we specify T and P. However, as a system becomes more complicated, it becomes more difficult to quickly determine how much freedom we have to adjust variables of interest. If $\boldsymbol{P}$ represents the total number of all phases in the system, including possibly an aqueous phase, a gas phase, and various different solid phases, then it is the **Gibbs Phase Rule** (GPR) which relates the number of freedoms (F) which we have to vary chemical parameters to the number of chemical components (C) present. While it is important to remember that the GPR does not provide any new information on how to actually solve a problem, it does provide a framework for telling us how many parameters we will need to specify before a problem can be made solvable. For the case of pure liquid water, we need to specify two parameters (i.e., T and P) so that K_w is specified and the pH can be determined.

17.2 DEVELOPING AN INTUITION FOR THE GIBBS PHASE RULE

Before we get started on the actual mathematical derivation of the GPR, let us consider a few more cases involving the single component water. Three cases are illustrated schematically in Figures 17.1*a–c*. For the liquid water illustrated in Figure 17.1*a*, $C = 1$, $\boldsymbol{P} = 1$, and we know intuitively that we can vary the T and the P of such a system. However, as noted above, once values for T and P are specified, then K_w is specified and we can compute the chemistry of the solution. Thus, in Figure 17.1*a*, $F = 2$.

In Figure 17.1*b*, both liquid and gaseous water are present. Our familiarity with basic chemistry and/or the behavior of the natural environment tells us that the vapor pressure (p^o, Torr) of water (= total pressure P in the system in this case) is linked to T through the **Clausius-Clapeyron** (C-C) equation for vapor/liquid equilibrium. In particular, when the enthalpy of vaporization can be assumed to be constant, the integrated form of this equation is

$$\log p^o = -\frac{0.2185\,\overline{H}_v}{T} + B \qquad (17.1)$$

where T is temperature in degrees K, $\overline{H}_v$ is the molar heat of vaporization for the compound of interest, and B is another compound-dependent constant. Thus, we can vary either T or P (= vapor pressure p^o of the water), but not both simultaneously: now, $F = 1$. Once either T or P is specified, then both T and P are specified.

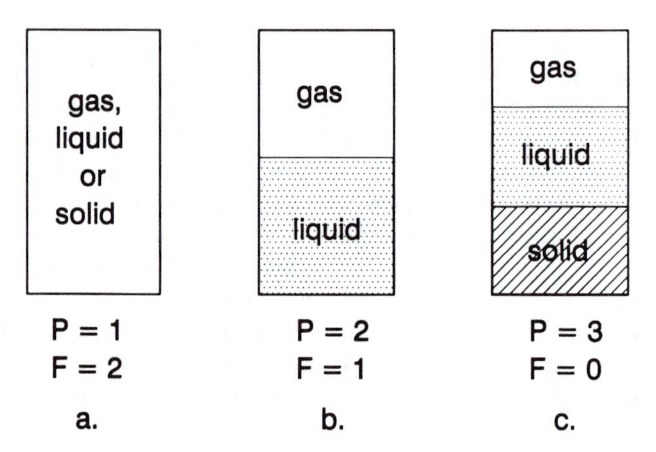

Figure 17.1 Three different examples of a $C = 1$ system. (*a*) When $\boldsymbol{P} = 1$, we can vary T and P simultaneously, so we have two "freedoms." (*b*) When $\boldsymbol{P} = 2$ and both liquid and gas are present, the system is governed by a Clausius-Clapeyron (C-C) equation relating T and P, so we can vary either T or P (= vapor pressure of the liquid), but not both simultaneously: there is only one "freedom" (i.e., $F = 1$). (*c*) When $\boldsymbol{P} = 3$ and solid, liquid, and gas are all present: we are at a triple point. We cannot vary either T or P. The values of T and P correspond the intersection for two C-C equations, one for liquid/gas equilibrium and one for solid/gas equilibrium (see point A in Figure 17.2 for water). There are no "freedoms" (i.e., $F = 0$).

In Figure 17.1*c*, solid,[1] liquid, and gaseous water are all present. Now, in addition to vapor/liquid equilibrium, we also have vapor/solid equilibrium, and so two C-C type equations must be satisfied simultaneously. It may be noted that if there is simultaneous vapor/liquid equilibrium and vapor/solid equilibrium, there is automatically liquid/solid equilibrium. Both C-C equations are of the form given in Eq. (17.1), though for vapor/solid equilibrium we need to replace $\overline{H}_v$ with the enthalpy of sublimation $\overline{H}_s$. We also note that the value of B for vapor/solid equilibrium for water is different from that for vapor/liquid equilibrium. Simultaneous vapor/liquid and vapor/solid equilibrium can only occur at the vapor/liquid/solid "triple point" for water. The T and P of this triple point are given by the intersection of the two C-C equations discussed. As indicated in Figure 17.2, for water, at this triple point we have $T = 273.16°\text{K}$ and $P = 0.006$ atm. Now, we cannot vary either T or P; now $F = 0$.[1]

As we progress from Figure 17.1*a* to 17.1*c*, what is happening is that more and more chemical equilibria must be satisfied. First, no multiphase equilibria must be satisfied, then one such equilibrium, then two such equilibria. As each equilibrium requirement is added, the number of freedoms decreases by one. The situation is somewhat similar to what happens with open and closed CO_2 systems. Indeed, open system CO_2 problems with constant p_{CO_2} are more easily solved than closed system problems in which the in-situ p_{CO_2} can vary; with the former, a freedom has been removed because setting the p_{CO_2} which at a given T and P sets $\{H_2CO_3^*\}$.

17.3 A LINEAR ALGEBRA ANALOG OF THE GIBBS PHASE RULE

Let V represent the number variables for a system. In any given situation, the number of freedoms that we have will be given by the difference between the number of variables V and the number of independent equations E involving those variables, that is

$$F = V - E. \tag{17.2}$$

Since all of the E equations must be linearly independent, we know that $E \leq V$. Thus, we cannot have more linearly independent equations than we do variables.

If we consider a case involving the two variables x and y, then $V = 2$. If $E = 0$ and we have no equations relating these two variables, then we can independently vary both x and y, and we have two freedoms ($F = 2$). Now, let us say that the algebraic equation

$$x + 2y = 5 \tag{17.3}$$

governs x and y. With one equation (i.e., $E = 1$), then $F = V - E = 2 - 1 = 1$. We can still arbitrarily select values for one of the variables. Once we do that, however,

[1]At pressures less than ~1700 atm, the form of ice that is stable is referred to as "Ice I." At higher pressures, other forms of ice (e.g., Ice II and Ice III) become possible. Thus, triple points at points other than point A are present in Figure 17.2. For example, there is a triple point at B for Ice I/Ice II/Ice III, and at point C for Ice I/Ice III/liquid. As long as three different phases of water are in equilibrium, $F = 0$. Note that each triple point occurs at a different pair of T and P values. Thus, the specific values of T and P for a given triple point are set by the actual identities of the three phases involved, and not simply by the fact that three different phases are in equilibrium.

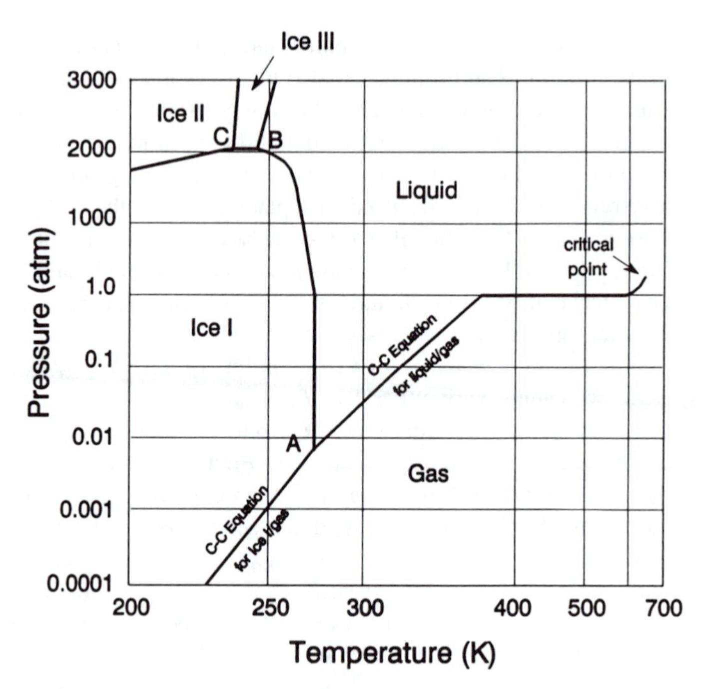

Figure 17.2 Phase diagram for pure water showing gas, liquid, and several different ice phases. Ice I, Ice II, and Ice III have different crystal structures. Three triple points are present in the diagram at points A, B, and C. The triple point for gas, liquid, and Ice I occurs at $T = 273.16$ K and $P = 0.00603$ atm (point A). Point A occurs at the intersection for two C-C equations, one for liquid/gas equilibrium and one for Ice I/gas equilibrium. The apparent discontinuity of slopes that occurs for two of the lines at $P = 1$ atm is due to a change in scales in the y-axis at $P = 1$ atm. At temperatures above the critical point, the water becomes a "supercritical fluid," and there is no longer any distinction between liquid and gas. Adapted with permission from McGraw-Hill (1983).

then we can use Eq. (17.3) to solve for the second variable. For example, if as just one possibility we set $x = 1$, then by Eq. (17.3) we have $y = 2$.

If instead of just Eq. (17.3) as a governing equation we have two linearly independent equations, say

$$x + 2y = 5 \tag{17.3}$$

and

$$x + 3y = 8 \tag{17.4}$$

then we still have $V = 2$, but now $E = 2$, and so now $F = 2 - 2 = 0$. We can no longer arbitrarily specify one of the variables and solve for the other since Eqs. (17.3)–(17.4) are completely self-contained. As at the "triple point" for water, there are no longer any freedoms, and the *only* set of x and y that satisfies Eqs. (17.1) and (17.2) is $x = -1$, $y = 3$. The values $x = -1$, $y = 3$ represent the intersection of the two lines. Thus, Eqs. (17.3) and (17.4) act like the two C-C equations in Figure 17.2.

We conclude that every time we add a linearly independent equation, we reduce the number of freedoms by one. This is a perfect analog of what is going on in Figure 17.1; every time we add a phase, we are adding an equilibrium equation that must be satisfied, and so the number of freedoms goes down by one.

17.4 DERIVATION OF THE GIBBS PHASE RULE FOR NONREACTIVE COMPONENTS

In the derivation and application of the GPR, the situation is exactly the same as that discussed in the previous section. If we analyze a system and find that there are no freedoms, the *chemical and physical properties* of the system are fully specified and we do not have the freedom to say anything about how we would like the system to be. On the other hand, if we find that F is greater than zero, then we can specify values for F different system parameters. Examples of system parameters would include T, P, pH, p_cH, pOH, p_cOH, $\{H_2CO_3^*\}$, $[H_2CO_3^*]$, $\{HCO_3^-\}$, $[HCO_3^-]$, $\{CO_3^{2-}\}$, $[CO_3^{2-}]$, etc. The following derivation is adapted from the one given by Denbigh (1981).

Consider a system of C components. Each component is a distinct chemical species. We will assume for the moment that there are no reactions relating the different components. All components are assumed to be present in all phases, though some concentrations may be exceedingly low. Thus, for each phase we must concern ourselves with the composition of that phase, as well as its T and P. Thus for each phase, we have $C+2$ variables, that is, C composition variables plus two.

$$\text{Number of variables for each phase} = C + 2. \tag{17.5}$$

Since we have $\boldsymbol{P}$ phases, the total number of independent variables is given by

$$V = \boldsymbol{P}(C+2). \tag{17.6}$$

If the system is in equilibrium, the temperature and pressure in each of the different phases must be the same, that is, we have $2(\boldsymbol{P}-1)$ equations of the type

$$\begin{array}{c|ll} 2(\boldsymbol{P}-1) & T_\alpha = T_\beta = T_\gamma \dots. & \boldsymbol{P}-1 \text{ equations for temperature} \quad (17.7) \\ \text{Equations} & P_\alpha = P_\beta = P_\gamma \dots. & \boldsymbol{P}-1 \text{ equations for pressure.} \quad (17.8) \end{array}$$

There are $\boldsymbol{P}-1$ equations for each phase for temperature (and also for pressure) since the $\boldsymbol{P}$th equation is derivable from the 1st and the $(\boldsymbol{P}-1)$th equation, and so is not linearly independent.

For chemical equilibrium between the various phases, the chemical potential of each of the species must be the same in each of the phases.

$$\begin{array}{c|ll} & \mu_1^\alpha = \mu_1^\beta = \mu_1^\gamma \dots & \boldsymbol{P}-1 \text{ equations for } \mu \text{ for species 1} \quad (17.9) \\ C(\boldsymbol{P}-1) & \mu_2^\alpha = \mu_2^\beta = \mu_2^\gamma \dots & \boldsymbol{P}-1 \text{ equations for } \mu \text{ for species 2} \quad (17.10) \\ \text{Equations} & \vdots & \\ & \multicolumn{2}{l}{\text{etc. for all } C \text{ components (recall that the } \mu \text{ values are functions of the composition).}} \end{array}$$

Finally, for each phase we know that we have an equation that states that the sum of the mole fractions equals 1.0, that is, we have $\boldsymbol{P}$ equations of the type

$$\left.\begin{array}{l} \\ P \\ \text{Equations} \\ \\ \end{array}\right| \begin{array}{ll} \sum_i X_i^\alpha = 1 & (17.11) \\ \sum_i X_i^\beta = 1 & (17.12) \\ \vdots & \\ \text{etc. for all } \boldsymbol{P} \text{ phases.} & \end{array}$$

Thus, the total number of independent equations E among the variables is

$$E = 2(\boldsymbol{P} - 1) + C(\boldsymbol{P} - 1) + \boldsymbol{P} \tag{17.13}$$

$$= (C + 2)(\boldsymbol{P} - 1) + \boldsymbol{P}. \tag{17.14}$$

As noted above, the equations in μ for any given phase indeed provide an interrelation of the T, P, and X values since in each phase, the chemical potential of each species i is a function of T, P, and the the composition, that is,

$$\mu_i = \mu(T, P, X_1 \ldots X_C). \tag{17.15}$$

By Eqs. (17.2), (17.6), and (17.14), we then have

$$F = \boldsymbol{P}(C + 2) - [(C + 2)(\boldsymbol{P} - 1) + \boldsymbol{P}] \tag{17.16}$$

or simply

$$\boldsymbol{P} + F = C + 2 \qquad \text{Gibbs Phase Rule.} \tag{17.17}$$

When it is read out using the names of the various terms, Eq. (17.17) has a ring to it that assists the memory.

Equation (17.17) states that the number of freedoms in any nonreacting system equals the number of chemical components plus 2 minus the number of phases present. Thus, as seen in Figure 17.1, for every unit increase in $\boldsymbol{P}$, the value of F goes down by one. Indeed, for $C = 1$, application of Eq. (17.17) for $\boldsymbol{P} = 1$, 2, and 3 gives 2, 1, and 0, respectively.

17.5 GIBBS PHASE RULE FOR NONREACTIVE COMPONENTS WITH EXCLUSION OF A COMPONENT FROM A PHASE

In the previous section, it was assumed every component is in every phase. What if we *exclude* some component from some phase? As an example, we might want to assume that the salt NaCl (call it component 1) is excluded from existing in the gas phase (call that phase α). In this case, the number of composition variables for the α phase will be reduced by one (i.e., we no longer have to worry about X_1^α since it is by definition zero). At the same time, the number of *equations* between the variables will be reduced by one since Eq. (17.9) becomes

$$\mu_1^\alpha > \mu_1^\beta = \mu_1^\gamma \text{ etc.} \tag{17.18}$$

Thus, Eq. (17.16) becomes

$$F = \boldsymbol{P}(C + 2) - 1 - [(C + 2)(\boldsymbol{P} - 1) + \boldsymbol{P} - 1] \tag{17.19}$$

which again reduces to the GPR as it is given in Eq. (17.17).

Actually, it must be pointed out that strictly speaking, it is *impossible* to exclude any given component from any given phase. As has been discussed in Chapter 2 and again in Chapter 16, whenever a given component is completely absent from some phase α, the chemical potential of that component in α will be $-\infty$, and there will be an infinitely strong driving force for the entry of the component into α. Nevertheless, when a given component is present at an exceedingly low concentration in phase α, then just as we are very likely to ignore the presence of that component in phase α in any mass balance calculations, the application of the GPR can be made consistent with that type of assumption. As examples, we could apply the GPR to: 1) a system composed of an aqueous solution of NaCl that is in equilibrium with water vapor; and 2) a system composed of an aqueous solution of NaCl that is in equilibrium with both water vapor and $NaCl_{(s)}$. In the first example, NaCl is assumed to be virtually excluded from the gas phase, and for all intents and purposes, $F = 2$. In the second example, H_2O is assumed to be virtually excluded from the $NaCl_{(s)}$, and for all intents and purposes, $F = 1$.

17.6 GIBBS PHASE RULE FOR REACTIVE COMPONENTS

17.6.1 General

Let us now say that we have N chemical species, some inert, some reactive. Let R be the number of independent chemical reactions involving some or all of those species. Let $\boldsymbol{P}$ again be the number of phases. In a manner that is analogous to Eq. (17.6), we now have

$$V = \boldsymbol{P}(N+2). \tag{17.20}$$

Instead of Eq. (17.14), we now have

$$E = (N+2)(\boldsymbol{P}-1) + \boldsymbol{P} + R \tag{17.21}$$

since there are now R chemical equilibrium constant equations governing the system. Application of Eq. (17.2) now yields

$$F = \boldsymbol{P}(N+2) - (N+2)(\boldsymbol{P}-1) - \boldsymbol{P} - R \tag{17.22}$$

or

$$\boldsymbol{P} + F = N - R + 2 \qquad \text{Gibbs Phase Rule for reactive components.} \tag{17.23}$$

The quantity $(N - R)$ in the reactive case thus plays the role of C as it appears in Eq. (17.17). This is no coincidence: $(N - R)$ in fact represents the minimum number of components required to manufacture N species. Indeed, even if we start out with only $(N - R)$ species, since our R reactions can produce R more species, we will end up with a total of N species. It is also important to note that there are only R conditions of chemical equilibrium, *not* $\boldsymbol{RP}$ (i.e., not R conditions for each phase). Indeed, if a reaction is at equilibrium in one phase, it must also be at equilibrium in all of the phases by virtue of Eqs. (17.9) and (17.10).

17.6.2 Additional Special Restrictions

17.6.2.1 ENE, PBE, and MBE(s) as Special Restrictions for Aqueous Systems. In aqueous systems, when individual ionic species are considered explicitly, and therefore counted when determining N, then the ENE must be considered as an additional restrictive equation. For example, consider pure, liquid water. Since $C = 1$ and $\boldsymbol{P} = 1$, then $F = 2$ (e.g, T and P). If we examine the situation more carefully, allowing that water can dissociate into H^+ and OH^-, then we have $N = 3$, and $R = 2$ (one for K_w and one for the ENE), and so F again equals 2. In this case, since both the PBE and MBE describing the dissociation (i.e, $[H^+] = [OH^-]$) are the same as the ENE, neither comprises an additional restriction beyond that of the ENE.

For a solution of NaCl in liquid water, $\boldsymbol{P} = 1$. Neglecting the dissociation of the salt and the water, we have $C = 2$ and so $F = 3$ (e.g., T, P, and [NaCl]). Explicitly considering the dissociation of both water and the NaCl, we have $N = 5$ and $R = 3$ (K_w, ENE, and PBE; or, K_w, ENE, and MBE for NaCl dissolution (i.e., $[Na^+] = [Cl^-]$)), and so F again equals 3. (The MBE, ENE, and PBE are not *all* restrictions since only two at a time will provide a linearly independent set.)

An interesting illustration of the application of the ENE as a restricting equation may be drawn from Example 5-16 of Stumm and Morgan (1981). That example considers a general system comprised of the following species: H_2O, Mg^{2+}, K^+, Na^+, H^+, Cl^-, OH^-, SO_4^{2-}. If we assume that no gas phase is present, the example asks the following question: If T and P are still allowed to remain as freedoms, and if there is no net strong acid or strong base in the system (i.e., the system just contains salts dissolved in water), what is the maximum number of phases possible at equilibrium in this system?

We first solve the problem without using the phase rule. Before proceeding, however, it is useful to ask what "maximum number of phases" means for the above conditions. The answer is that it means that the system is completely determined for any arbitrary T and P. Once T and P are specified, then we can solve for the concentrations of all of the species using the appropriate equations, including some involving K_{s0} values. The equilibrium composition will, of course, be a function of the K_{s0} values which are, in turn, functions of T and P. According to the above list, there are 8 concentration unknowns in the system. (Water is included as an unknown since its mole fraction will not be exactly 1.0.) Therefore, we need eight equations. Three equations that we can write down immediately for the aqueous phase are:

$$\sum_{i=1}^{8} X_i = 1 \tag{17.24}$$

$$2[Mg^{2+}] + [K^+] + [Na^+] + [H^+] = [Cl^-] + 2[SO_4^{2-}] + [OH^-] \qquad \text{(ENE)} \tag{17.25}$$

and

$$[H^+] = [OH^-] \qquad \text{(PBE).} \tag{17.26}$$

A fourth equation is provided by K_w. Equation (17.26) applies as the PBE for this system since: 1) it has been explicitly assumed that no net strong acid or base is present; and 2) the list of the unknowns given has implicitly assumed that none of the metal ions and none of the salt anions participate in acid/base reactions (e.g, HSO_4^- is not included as an unknown).[2] We still need 4 more equations. They must come from solubility equilibria, so for 4 independent solubility equilibria, we need 4 solids. Thus,

$$\boldsymbol{P} = 4 \text{ solids } + 1 \text{ aqueous phase } = 5 \text{ phases.} \tag{17.27}$$

The actual identities of the solids present would be determined by T and P plus the nature of the system (i.e., how much of each of the species was used to form the system).

We now use the GPR to solve the above problem. First, we note that $N = 8$. With $R = 3$ (ENE, PBE, and K_w), we have that $N - R = 5$. If we want to retain only T and P as freedoms, then $\boldsymbol{P} = (N - R) + 2 - F = 5 + 2 - 2 = 5$, or four solids plus water as shown above. If we were willing to completely remove all freedoms from the system, then we could have a total of 7 phases, with 6 solids present, and both T and P would be set. In a manner that is analogous to Figure 17.2, wherein there is more than one triple point, we note that for the problem at hand, different sets of solids will give different T, P points where seven phases can exist at equilibrium.

Some might ask why we do not need to include any of the solid salts when we compute N for the above problem. The reason is that the salts are just phases wherein the various components may reside. That is, if $NaCl_{(s)}$ is one of the solids, then Na^+ and Cl^- may reside in the $NaCl_{(s)}$ phase just as well as they do in the aqueous phase. As an analogy, for the system described by Figure 17.1c, H_2O can reside in three different phases, that is, $H_2O_{(g)}$, $H_2O_{(l)}$ and $H_2O_{(s)}$. It is interesting and reassuring to note that since we do not know how many solid salts are present a priori, we could not count them into N even if we wanted to.

As a final note for Example 5-16 of Stumm and Morgan (1981), it is useful to emphasize that we have not explicitly considered the possibility of the K_{s0} values being altered by the presence of solid solutions. For example, if $NaCl_{(s)}$ is in fact present at equilibrium, some $KCl_{(s)}$ could be dissolved in the $NaCl_{(s)}$. The GPR handles this type of situation just fine, and without any further modifications.

17.6.2.2 Additional Restriction When the Level of a Species in Some Phase Is Constant.

Any condition that somehow restricts the species distribution that a chemical system can take on will play the same type of role in reducing freedoms that an equilibrium reaction does. Thus, such restrictions must be included when evaluating R. To illustrate this concept, consider Figure 17.3 where liquid water containing dissolved O_2 is in equilibrium with a gas phase also containing water and O_2. Let us assume that the O_2 partial pressure is maintained at some constant value, for example, 0.21 atm. Perhaps we are requiring that the system stay in equilibrium with the level of O_2 found in the atmosphere.

There are a total of 4 species, H_2O, H^+, OH^-, and O_2 in the Figure 17.3 system. Thus, $N = 4$. Two of the restrictions governing this system are provided by K_w and the

[2]Choosing to include a species like HSO_4^- as another unknown would not alter the conclusions because that would just bring in another equation, i.e., the K_2 dissociation constant for H_2SO_4. This would allow the extent of formation of HSO_4^- to be computed.

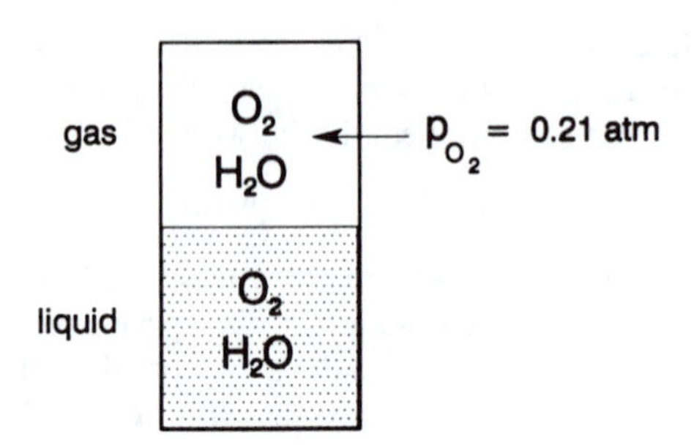

Figure 17.3 Two phase system composed of water and O_2. The fact that we are restricting our considerations to cases involving $p_{O_2} = 0.21$ atm is an additional restriction on the chemistry.

PBE (or the ENE). We recall that the various *interphase* equilibrium reactions $i_{(l)} = i_{(g)}$ are already considered in the derivation of the GPR, and so they do not need to be counted in R.) We have, however, stated that the O_2 level in the gas phase is somehow being restricted to a partial pressure of 0.21 atm. That constitutes a special restriction on the system since it provides an equation governing the composition of the gas phase. Thus, $N = 4$, $\boldsymbol{P} = 2$, $R = 3$, and so $F = 1$. With only one freedom, specification of any one of the several variables will fix the total system chemistry. For example, we could specify T, p_{H_2O}, $[H^+]$, or $[OH^-]$. Note that $\{H^+\}$ and $\{OH^-\}$ are other possible variables. Indeed, once either $\{H^+\}$ or $\{OH^-\}$ will be set, the speciation will be set, and so the ionic strength I will be set. This means that the activity coefficients for all species will be set, and so knowing either $\{H^+\}$ or $\{OH^-\}$ translates into knowing $[H^+]$ and $[OH^-]$ (as well as T, P, etc.).

17.6.2.3 Additional Restriction When the Concentration Ratio Between Two Species in Some Phase Is Constant. Let us now go on to assume that the gaseous compound in equilibrium with the water is not made up of one compound, but in fact contains two species, B1 and B2 at some fixed, specified ratio, say 2:1. For this case, then, $N = 5$. Let us assume that both B1 and B2 are insoluble in liquid water. If the total pressure of B1 and B2 is still maintained constant at 0.21 atm, then one restrictive equation for the system is that $p_{B1} + p_{B2} = 0.21$ atm, and the second is $p_{B1}/p_{B2} = 2$. Thus, with $R = 4$ and $\boldsymbol{P} = 2$, then $F = 1$. Again, the chemistry of the whole system becomes specified once the total pressure $P(= p_{B1} + p_{B2} + p_{H_2O} = 0.14 + 0.07 + p_{H_2O})$, or p_{H_2O}, or T is set.

17.6.2.4 No Additional Restriction for a Constant Ratio $n_{B1,tot}/n_{B2,tot}$ When B1 and B2 Can Be Present in Both Phases. We now consider the case where: 1) B1 and B2 are soluble in liquid water; and 2) we no longer know the ratio of B1 and B2 in the gas phase, but we do know the ratio $n_{B1,tot}/n_{B2,tot}$ where $n_{B1,tot}$ and $n_{B2,tot}$ represent the total numbers of mols of B1 and B2 in the *overall* gas/liquid system, respectively. For this system, the only restrictions are provided by K_w and the PBE (or ENE). That is, R = 2. Knowing the ratio $n_{B1,tot}/n_{B2,tot}$ does not comprise a restriction because we have no way of turning that value into an equation that can be used when computing the system speciation in any *general* case. That is, we do not, in general, know the volume of either the gas or the liquid phase. By the same token, however, the specification of either volume together with a knowledge of $n_{B1,tot}/n_{B2,tot}$ would constitute a restriction.

17.7 EXAMPLES

Table 5.6 of Stumm and Morgan (1981) contains a number of examples of how the GPR may be applied to chemical systems of interest in natural water systems. Table 17.1 expands several of those examples and adds a few additional cases. For each case considered, the table provides a detailed explanation so that the reader may derive the most possible instruction. For every case, we identify: 1) a *subset of the variables* governing the system; and 2) the *subset of equations* inter-relating those variables that one would actually use in solving the problem. Thus, we do not give the full list of V variables and E equations. We only present subsets because when we actually solve problems: 1) we always just assume that equations like Eqs. (17.7)–(17.12) are satisfied; and 2) we do not bother to write down the same temperature $\boldsymbol{P}$ times and the same pressure $\boldsymbol{P}$ times for the $\boldsymbol{P}$ different phases.

For each case in Table 17.1, the difference between the number of equations in the equations subset and the number of variables in the variables subset is equal to the number of freedoms. For simple examples involving the aqueous phase, the concentrations $[H^+]$ and $[OH^-]$ are not included as variables; at the same time, in those cases, K_w and the PBE are not included as equations. Thus, if the reader should choose to take that type of more detailed, reactive approach for the simpler examples, he/she will find that as each successively greater level of detail introduces new variables, a corresponding number of additional equations can be identified that govern those variables. The more detailed, reactive approach is taken in the cases considered in Examples 7, 8, and 9 of Table 17.1. Problems illustrating the principles of this chapter are presented by Pankow (1992).

17.8 REFERENCES

DENBIGH, K. 1981. *The Principles of Chemical Equilibrium.* Cambridge University Press, Cambridge, U.K.

PARKER, S. P., ed. 1983. *McGraw-Hill Encyclopedia of Chemistry.* New York: McGraw-Hill.

PANKOW, J. F. 1992. *Aquatic Chemistry Problems.* Portland: Titan Press-OR, P.O. Box 91399, Portland, Oregon 97291-1399.

STUMM, W., and J. J. MORGAN. 1981. *Aquatic Chemistry.* New York: Wiley-Interscience.

TABLE 17.1 Gibbs Phase Rule examples (adapted from Table 5.6 of Stumm and Morgan (1981)).

Example	Components		Phases Present		Freedoms	Variable Subset	Equation Subset
	Number	Type	Number	Type			
1	1	H_2O	1	L	2	T, P, p_{H_2O}	$P = p_{H_2O}$
			2	L,G	1	T, P, p_{H_2O}	$P = p_{H_2O}$ $p_{H_2O,L/G} = f_1(T)$ (i.e., C-C Equation for L/G equilibrium)
			3	S,L,G	0	T, P, p_{H_2O}	$P = p_{H_2O}$ $p_{H_2O,L/G} = f_1(T)$ (i.e., C-C Equation for L/G equilibrium) $p_{H_2O,S/G} = f_2(T)$ (i.e., C-C Equation for S/G equilibrium)
2	2	$SiO_{2(s)}$, H_2O	1	L	3	$T, P, [H_4SiO_4], X_{H_2O}$	$\sum_i X_i = 1$ (aqueous phase)
			2	quartz, and L (aqueous)	2	$T, P, [H_4SiO_4], X_{H_2O}$	$\sum_i X_i = 1$ (aqueous phase) $[H_4SiO_4] = K_{s0}^{quartz}(P,T)$
			3	quartz, cristobalite, and L (aqueous)	1	$T, P, [H_4SiO_4], X_{H_2O}$	$\sum_i X_i = 1$ (aqueous phase) $[H_4SiO_4] = K_{s0}^{quartz}(P,T)$ $[H_4SiO_4] = K_{s0}^{crist}(P,T)$
3	4	H_2O, CO_2, N_2, O_2	2	G (all species), L (aqueous, all species)	4	$T, P, p_{H_2O}, p_{CO_2}, p_{O_2}, p_{N_2}, X_{H_2O}, [O_2], [N_2], [CO_2]$	$P = p_{H_2O} + p_{CO_2} + p_{O_2} + p_{N_2}$ (gas phase) $\sum_i X_i = 1$ (aqueous phase) $p_{H_2O,L/G} = f_1$ (T,P, aq. comp.) $p_{CO_2,L/G} = f_2$ (T,P, aq. comp.) $p_{O_2,L/G} = f_3$ (T,P, aq. comp.) $p_{N_2,L/G} = f_4$ (T,P, aq. comp.)
			3	S (ice), L (aqueous, all species), G (all species)	3	ditto (10)	ditto (6), plus $p_{H_2O,S/G} = f_5$ (T,P, aq. comp.)

TABLE 17.1 (continued)

Example	Components		Phases Present		Freedoms	Variable Subset	Equation Subset
	Number	Type	Number	Type			
4	2	$CuSO_4$, H_2O	2	$CuSO_{4(s)}$, $G(H_2O_{(g)})$	2	T, P, p_{H_2O}	$P = p_{H_2O}$
			2	$CuSO_4 \cdot H_2O_{(s)}$, $G(H_2O_{(g)})$	2	T, P, p_{H_2O}	$P = p_{H_2O}$
			3	$CuSO_4 \cdot H_2O_{(s)}$, $CuSO_{4(s)}$, $G(H_2O_{(g)})$	1	T, P, p_{H_2O}	$P = p_{H_2O}$; $p_{H_2O} = f(T)$ (i.e., p_{H_2O} is a function of T for equilibrium between the two copper solids—note: analogous to p_{CO_2} being controlled by $CaO_{(s)}/CaCO_{3(s)}$ equilibrium, or by $Fe(OH)_{2(s)}/FeCO_{3(s)}$ equilibrium.)
5	2	$CaO_{(s)}$ and CO_2; or $CaCO_{3(s)}$ and CO_2; or $CaCO_{3(s)}$, and $CaO_{(s)}$.	3	$CaCO_{3(s)}$, CaO(s), and $G(CO_{2(g)})$.	1	T, P, p_{CO_2}	$P = p_{CO_2}$; $p_{CO_2} = f(T)$ (i.e., p_{CO_2} is a function of T for equilibrium between the two calcium solids)
6*	2	$Fe_3O_{4(s)}$ and O_2; or $Fe_2O_{3(s)}$ and O_2; or $Fe_3O_{4(s)}$ and $Fe_2O_{3(s)}$	2	S (solid solution of $Fe_2O_{3(s)}$ and $Fe_3O_{4(s)}$ and G $(O_{2(g)})$	2	$T, P, p_{O_2}, X_{Fe_2O_3}, X_{Fe_3O_4}$	$P = p_{O_2}$; $X_{Fe_2O_3} + X_{Fe_3O_4} = 1$; $K_{hem/mag}(T,P)$
			3	$Fe_2O_{3(s)}$, $Fe_3O_{4(s)}$ and G $(O_{2(g)})$	1	T, P, p_{O_2}	$P = p_{O_2}$; $K_{hem/mag}(T,P)$

*Note: For Example 6, the solid $Fe_3O_{4(s)}$ (magnetite) may be thought of as being a mixed valence-state iron solid. Indeed, we can write out its formula as $Fe_2O_3 \cdot FeO_{(s)}$. Thus, since oxygen in this compound (and also in $Fe_2O_{3(s)}$, hematite) is in the (-II) oxidation state, we see that each mol of $Fe_3O_{4(s)}$ contains the equivalent of 2 mols of Fe(III), and one mol of Fe(II). The interconversion reaction between hematite and magnetite is:

$$3\ Fe_2O_{3(s)} = 2\ Fe_3O_{4(s)} + \frac{1}{2}O_{2(g)} \qquad K_{hem/mag}$$

TABLE 17.1 (continued)

Example	Components		Phases Present		Freedoms	Variable Subset	Equation Subset
	Number	Type	Number	Type			
7	3	$CaO_{(s)}$, CO_2, and H_2O; or $Ca(OH)_{2(s)}$, $CaCO_{3(s)}$ and H_2O; or $CaCO_{3(s)}$, $CaO_{(s)}$, and H_2O†	1	L (aqueous)	4	T, P, $[Ca^{2+}]$, $[CaOH^+]$, $[CO_3^{2-}]$, $[HCO_3^-]$, $[H_2CO_3^*]$, $[H^+]$, $[OH^-]$, X_{H_2O}	$\sum_i X_i = 1$ (aqueous phase), K_{H1}, K_1, K_2, K_W, ENE
			2	calcite and L (aqueous)	3	ditto (10)	ditto (6), plus K_{s0} for calcite
			3	calcite, L (aqueous), and G ($CO_{2(g)}$ and $H_2O_{(g)}$	2	ditto (10), plus p_{CO_2}, p_{H_2O}	ditto (7), plus $P = p_{CO_2} + p_{H_2O}$, $p_{CO_2,L/G} = f_1$ (T, P, aq. comp.) $p_{H_2O,L/G} = f_2$ (T, P, aq. comp.)
			4	calcite, $Ca(OH)_{2(s)}$, L (aqueous), and G ($CO_{2(g)}$ and $H_2O_{(g)}$)	1	ditto (12)	ditto (10), plus K_{s0} for $Ca(OH)_{2(s)}$
			5	calcite, $Ca(OH)_{2(s)}$, L (aqueous), ice, and G ($CO_{2(g)}$ and $H_2O_{(g)}$)	0	ditto (12)	ditto (11), plus $p_{H_2O,S/G} = f(T$, aq. comp.)
8	4	$CaCO_{3(s)}$, H_2O, C_A = some set value C_B = some set value	1	L (aqueous)	3	T, P, $[Ca^{2+}]$, $[CaOH^+]$, $[CO_3^{2-}]$, $[HCO_3^-]$, $[H_2CO_3^*]$, $[H^+]$, $[OH^-]$, X_{H_2O}, $[Na^+]$, $[Cl^-]$	$\sum_i X_i = 1$ (aqueous phase), K_{H1}, K_1, K_2, K_W, ENE, $[Na^+] = C_B$ = some set value $[Cl^-] = C_A$ = some set value $C_T = Ca_T$
			2	calcite and L (aqueous)	2	ditto (12)	ditto (9), plus K_{s0} for calcite

† All three of these combinations of components are equivalent. For example, any system initially containing $CaCO_{3(s)}$ and $CaO_{(s)}$ can be generated from a corresponding system initially containing $CaO_{(s)}$ and CO_2.

TABLE 17.1 (continued)

Example	Components		Phases Present		Freedoms	Variable Subset	Equation Subset
	Number	Type	Number	Type			
8 (cont'd)			3	calcite, L (aqueous), and G ($CO_{2(g)}$ and $H_2O_{(g)}$)	2	ditto (12), plus p_{CO_2}, p_{H_2O}	ditto (10), *minus* $C_T = Ca_T$ plus $P = p_{CO_2} + p_{H_2O}$ $p_{CO_2,L/G} = f_1$ (T, P, aq. comp.) $p_{H_2O,L/G} = f_2$ (T, P, aq. comp.)
			4	calcite, $Ca(OH)_{2(s)}$, L (aqueous), and G ($CO_{2(g)}$ and $H_2O_{(g)}$)	1	ditto (14)	ditto (12), plus K_{s0} for $Ca(OH)_{2(s)}$
			5	calcite, $Ca(OH)_{2(s)}$, L (aqueous), ice, and G ($CO_{2(g)}$ and $H_2O_{(g)}$)	0	ditto (14)	ditto (13), plus $p_{H_2O,S/G} = f_3$ (T, P, aq. comp.)
9	5	$CaCO_{3(s)}$, CO_2, H_2O, C_A = some set value C_B = some set value	1	L (aqueous)	4	T, P, $[Ca^{2+}]$, $[CaOH^+]$ $[CO_3^{2-}]$, $[HCO_3^-]$, $[H_2CO_3^*]$, $[H^+]$, $[OH^-]$, X_{H_2O}, $[Cl^-]$, $[Na^+]$	$\sum_i X_i = 1$ (aqueous phase), K_{H1}, K_1, K_2, K_W, ENE, C_A = some set value C_B = some set value
			2	calcite and L (aqueous)	3	ditto (12)	ditto (8), plus K_{s0} for calcite
			3	calcite, L (aqueous), and G ($CO_{2(g)}$ and $H_2O_{(g)}$)	2	ditto (12), plus p_{CO_2}, p_{H_2O}	ditto (9), plus $P = p_{CO_2} + p_{H_2O}$, $p_{CO_2,L/G} = f_1$ (T, P, aq. comp.) $p_{H_2O,L/G} = f_2$ (T, P, aq. comp.)
			4	calcite, $Ca(OH)_{2(s)}$, L (aqueous), and G ($CO_{2(g)}$ $H_2O_{(g)}$)	1	ditto (14)	ditto (12), plus K_{s0} for $Ca(OH)_{2(s)}$
			5	calcite, $Ca(OH)_{2(s)}$, L (aqueous), ice, and G ($CO_{2(g)}$ and $H_2O_{(g)}$)	0	ditto (14)	ditto (13), plus $p_{H_2O,S/G} = f_3$ (P,T, aq. comp.)

PART IV

Metal/Ligand Chemistry

18

Complexation of Metal Ions by Ligands

18.1 INTRODUCTION

The subdiscipline in chemistry concerned with reactions between metal ions and species like OH^-, Cl^-, SO_4^{2-}, NH_3, etc. is known as "coordination chemistry." In this context, any species like OH^- etc. is referred to as a "ligand." Like the word ligament, the word ligand derives from the latin verb *ligare* which means "to bind." Ligands are often negatively charged, but this is not always the case. Indeed, as implied above, NH_3 can act as a ligand. The combined metal/ligand species is referred to as a "complex." As an example, for the reaction between Fe^{2+} and Cl^-, we have

$$\begin{aligned} Fe^{2+} + Cl^- &= FeCl^+ \\ \text{metal ion} + \text{ligand} &= \text{complex.} \end{aligned} \tag{18.1}$$

Based on this, we recognize that reactions like

$$Fe^{3+} + OH^- = FeOH^{2+} \tag{18.2}$$

$$FeOH^{2+} + OH^- = Fe(OH)_2^+ \tag{18.3}$$

$$\text{etc.}$$

that have been discussed in previous chapters involve the formation of coordinative bonds between metals and OH^-; it should be clear that OH^- is a very important ligand in aqueous solutions.

A chemical bond is formed when a pair of electrons is shared between two atoms. With carbon-carbon bonds, carbon-oxygen bonds, and many other types of bonds, both atoms participating in the bond bring one electron to the bond. For the type of coordination bond that forms between a metal ion and a ligand, however, *both* of the electrons in the pair originate with the ligand. For such an electron pair to be available for interaction with a metal, it must obviously not be involved in any bonding within the ligand itself. Once the metal/ligand complex is formed, the ligand and the metal ion share the electron pair in a manner that is virtually always unequal in character, with the ligand retaining the majority of the "control" on the electron pair.

Acid/base chemistry is fundamentally very similar to coordination chemistry. In particular, in a conjugate acid, the electron pair that makes up the bond between H^+ and the conjugate base may be thought of as originating with the conjugate base. (Indeed, H^+ does not have any electrons at all.) Thus, the proton plays the role of a metal ion, and the conjugate base plays the role of a ligand. Therefore, when the conjugate base is negatively charged, we have

$$\underset{\substack{\text{proton} \\ \text{(``metal ion'')}}}{H^+} \quad + \quad \underset{\substack{\text{conjugate base} \\ \text{(ligand)}}}{A^-} \quad = \quad \underset{\substack{\text{conjugate acid} \\ \text{(complex).}}}{HA} \qquad (18.4)$$

Since all conjugate bases have an electron pair available for sharing, it is very important to note that all conjugate bases will also act as ligands if there are any true metal ions in solution. Thus, when a metal ion is present in a solution, a competition will be established with H^+ for any ligands (i.e., conjugate bases) present. Two parameters that will help determine whether H^+ or the metal ion will *dominate* a given competition for A^- (neither will ever completely bind up A^- to the exclusion of the other) are: 1) the acid dissociation constant for HA; and 2) the constant for the formation of the metal/A^- complex.

The "metal ion"-like properties of the proton are sometimes emphasized in compilations of equilibrium constant data by listing H^+ as one of the metal ions (e.g, see Smith and Martell 1976). Since H^+ is so small in size, its charge-to-size ratio (and therefore the inherent attraction felt by an electron pair) is considerably larger than that for most actual metal ions. Thus, the sharing of an electron pair between H^+ and most conjugate bases is much more equal in character than is the case for most metal/ligand bonds. Nevertheless, the sharing is still not totally equal inasmuch as HA molecules always have significant polarity.

From discussions in previous chapters, we know that acids can be diprotic, triprotic, and so on. Thus, we conclude that the conjugate bases of such acids have more than one electron pair with which to bind to protons, and therefore also with which to bind to metal ions. Since H^+ is so small in size, even when multiple electron pairs are located rather closely on a conjugate base, there will usually be enough space to accommodate one proton at each electron pair site. Metal ions are, however, much bigger than protons, and so the formation of more than one coordination bond between a multi-electron pair ligand and a single metal will require that: 1) there is adequate (but not too much) distance between the electron pairs; and 2) each different electron pair is located such that it can approach the metal ion from a different, appropriate angle. As texts in inorganic chemistry describe in detail, the latter criterion is necessary since each electron pair on the ligand must interact with a different, vacant electron orbital on the metal ion. Thus, just because a given acid is multiprotic does not mean that all of the basic sites on the corresponding fully-dissociated conjugate base will be able to geometrically orient themselves so as to be able to simultaneously interact with a given metal ion. For example, while CO_3^{2-} possesses two locations where two different H^+ can attach themselves, CO_3^{2-} will not usually be able to use both of those positions to attach to one metal ion.

When a ligand has one electron pair available for binding to a metal, it is said to be "monodentate" or "unidentate" (i.e., "single-toothed"). Ligands with two, three, four, etc. available electron pairs that are located geometrically on the ligand such that they are capable of binding simultaneously with a metal ion are referred to as being "bidentate," "tridentate," "tetradentate," etc., respectively. Generically, the terms "multidentate" and "polydentate" are used to refer to any ligand with two or more binding sites.

A focal point of the discussion provided in Chapter 3 was the fact that protons in aqueous solution are surrounded by waters of hydration. Metal ions are also, of course, hydrated in aqueous solution. For most types of metal ions, the predominant hydrated species will be the one with six waters of hydration. The fact that water molecules tend to hydrate protons and metal ions is a direct result of the fact that each oxygen on a water molecule has two unshared pairs of electrons (see Figure 18.1). Thus, H_2O *itself* can always act as a ligand towards any aqueous metal ion, and may also of course be thought of as the conjugate base for the species H_3O^+. H_2O can only act as a monodentate ligand towards a particular metal ion however because its two unshared electron pairs on the oxygen atom are located very close to one another as well as oriented in different directions.

Given the ligand role played by water, strictly speaking, a more accurate representation of Eq. (18.1) would be to write it as a reaction in which the Cl^- ion exchanges for one of the six hydrating water ligands surrounding the Fe^{2+} ion:

$$Fe(H_2O)_6^{2+} + Cl^- = Fe(H_2O)_5Cl^- + H_2O. \tag{18.5}$$

Thus, in a solvent like water, eqs. (18.1) and (18.5) do not simply involve the formation of a complex from a bare metal ion, but rather the exchange of one type of complex for another. However, as in reaction (18.1), when discussing complexation reactions, the waters of hydration are usually not written out. Indeed, once it has been pointed out that reactions written as in (18.1) are really a shorthand for the type of representation given in reaction (18.5), the format used in (18.1) is usually felt to suffice. One should just remember that water molecules will tend to occupy all coordination sites on a metal ion that are not filled by other ligands.

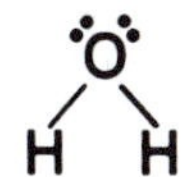

Figure 18.1 Structure of water molecule showing two unshared pairs of electrons on the oxygen.

18.2 FORMATION CONSTANTS VS. DISSOCIATION CONSTANTS (I.E., STABILITY CONSTANTS VS. INSTABILITY CONSTANTS)

When two or more species combine to form a third, composite species, one can write the equilibrium either as a *formation* reaction, i.e., with an associated *stability* constant,

$$\mathrm{A} + \mathrm{B} = \mathrm{C} \qquad K = \frac{\{\mathrm{C}\}}{\{\mathrm{A}\}\{\mathrm{B}\}} \tag{18.6}$$

or as a *dissociation* reaction with an associated *instability* constant

$$\mathrm{C} = \mathrm{A} + \mathrm{B} \qquad 1/K = \frac{\{\mathrm{A}\}\{\mathrm{B}\}}{\{\mathrm{C}\}}. \tag{18.7}$$

As noted, for metal/ligand complexation reactions, the accepted approach is to write the reactions in a "stability" format. Thus, the larger the equilibrium constant, the more stable the complex. The reason for this convention is that when the coordination chemists first began studying metal/ligand complexes, their minds were already geared towards thinking about the problem from the formation point of view. In contrast, the early chemists who were interested in acid/base reactions were concerned with measuring how strong different acids were, and so it is logical that they should have viewed their problem from the dissociation point of view; that perspective causes the equilibrium constant to increase as the acid increases in strength. It may also be pointed out that it is logical that the chemists who first studied how ionic solids dissolve should have chosen the instability format for their reactions.

After having considered the chemistry of acids with a range of dissociation constants in previous chapters, the reader will have begun to understand that acids with dissociation constants of $>10^{-2}$, $\sim 10^{-5}$, and $<10^{-8}$ may be characterized as strong, weak, and very weak, respectively. We now discuss how to translate this level of intuition and make it useful in understanding what kind of metal/ligand formation (i.e., stability) constant values characterize weak, moderately strong, and very strong complexes. This translation requires nothing more than a consideration of the value that a metal/ligand equilibrium constant would have if it was expressed in the same way as an acidity constant, that is, in the dissociation (i.e., instability) format.

Consider the formation of the first chloro complex of Cd(II):

$$\mathrm{Cd}^{+2} + \mathrm{Cl}^{-} = \mathrm{CdCl}^{+} \qquad K = 21. \tag{18.8}$$

As noted, at 25°C/1 atm, the value for the formation (stability) constant is 21. Consider now the value for the corresponding dissociation (instability) constant, that is, the equilibrium constant for the reaction

$$\mathrm{CdCl}^{+} = \mathrm{Cl}^{-} + \mathrm{Cd}^{2+} \qquad 1/K = 1/21 = 5 \times 10^{-2}. \tag{18.9}$$

An acid with a dissociation constant of $5 \times 10^{-2} = 10^{-1.30}$ would certainly be recognized as being moderately strong, that is, moderately strongly unstable. Thus, a formation constant of 21 is seen to correspond to a rather weak complex. Overall, then, we can conclude that a complexation constant may be characterized as being weak, moderately strong, and strong for $1/K$ values of $>10^{-2}$, $\sim 10^{-5}$, and $<10^{-8}$ respectively, that is, for formation constant values of $<10^{2}$, $\sim 10^{5}$, and $>10^{8}$, respectively. These considerations

TABLE 18.1 Consideration of the values of metal/ligand equilibrium constants as both formation and dissociation constants for purposes of classifying metal/ligand stability constants as being weak, moderately strong, or very strong.

	Strength of Complex		
Type of Constant	Weak	Moderately Strong	Strong
Metal/ligand *formation* constant	$<10^2$	$\sim 10^5$	$>10^8$
Metal/ligand *dissociation* constant	$>10^{-2}$	$\sim 10^{-5}$	$<10^{-8}$

are summarized in Table 18.1. It should be noted that this analysis applies to formation constants involving the combination of a metal ion with a single ligand, where the latter is in the unprotonated form. That is, it does not apply to constants of the $^*K$ type, or to constants of the β type, as introduced below.

Examples can be found of complexes with formation constants that span a very wide range of values. In general, complexes between monodentate ligands and metal ions from the first column of the periodic table (e.g., Na^+, K^+, etc.) are rather weak. For example, the complexes $NaOH^o$ and $NaCl^o$ are extremely weak. An interesting example of an extremely strong complex is $Co(NH_3)_6^{3+}$. This species is so stable that the core Co^{3+} ion can hold onto its NH_3 ligands in a hot solution of HCl; most metal ions would not be able to hold onto the basic NH_3 ligands in such an acidic medium.

18.3 NOMENCLATURE FOR COMPLEXATION CONSTANTS

There are several types of equilibrium constants that can be used to formulate the process of complexation of metal ions. In a manner that is consistent with how the complexation of metal ions by OH^- has been considered in previous chapters, when a free (i.e., unprotonated) ligand species combines with a metal ion, a simple K is used, with subscript indexing to keep track of how many total ligands have been added. Thus, without bothering to consider charges on either the metal ion or the ligand, we have:

$$M + L = ML \qquad K_1 = \frac{\{ML\}}{\{M\}\{L\}} \tag{18.10}$$

$$ML + L = ML_2 \qquad K_2 = \frac{\{ML_2\}}{\{ML\}\{L\}} \tag{18.11}$$

$$ML_2 + L = ML_3 \qquad K_3 = \frac{\{ML_3\}}{\{ML_2\}\{L\}} \tag{18.12}$$

etc.

When several of the reactions given in Eqs. (18.10)–(18.12) are added together so that an *overall* addition of *multiple* ligands can be considered, then a β is often used rather than a K to symbolize the equilibrium constant. This leads to

$$\mathrm{M + L = ML} \qquad \beta_1 = K_1 = \frac{\{\mathrm{ML}\}}{\{\mathrm{M}\}\{\mathrm{L}\}} \tag{18.13}$$

$$\mathrm{M + 2L = ML_2} \qquad \beta_2 = K_1 K_2 = \frac{\{\mathrm{ML_2}\}}{\{\mathrm{M}\}\{\mathrm{L}\}^2} \tag{18.14}$$

$$\mathrm{M + 3L = ML_3} \qquad \beta_3 = K_1 K_2 K_3 = \frac{\{\mathrm{ML_3}\}}{\{\mathrm{M}\}\{\mathrm{L}\}^3} \tag{18.15}$$

etc.

It is often handy to have the β notation available. The reason is that just as the products of the two acid dissociation constants for $H_2CO_3^*$ arise when expressing the α values for that diprotic acid system, the products of the various complexation K values arise when discussing chemistry the chemistry of the species ML, ML_2, ML_3, etc.

When the reacting ligand is considered to be in its protonated form, an asterisk (*) is used on the K values to denote the effective presence of an acidic reactant. Thus, we have

$$\mathrm{M + HL = ML + H^+} \qquad {}^*K_1 = \frac{\{\mathrm{ML}\}\{\mathrm{H^+}\}}{\{\mathrm{M}\}\{\mathrm{HL}\}} \tag{18.16}$$

$$\mathrm{ML + HL = ML_2 + H^+} \qquad {}^*K_2 = \frac{\{\mathrm{ML_2}\}\{\mathrm{H^+}\}}{\{\mathrm{ML}\}\{\mathrm{HL}\}} \tag{18.17}$$

$$\mathrm{ML_2 + HL = ML_3 + H^+} \qquad {}^*K_3 = \frac{\{\mathrm{ML_3}\}\{\mathrm{H^+}\}}{\{\mathrm{ML_2}\}\{\mathrm{HL}\}} \tag{18.18}$$

etc.

By analogy with Eqs. (18.13)–(18.15), we also have

$$\mathrm{M + HL = ML + H^+} \qquad {}^*\beta_1 = {}^*K_1 = \frac{\{\mathrm{ML}\}\{\mathrm{H^+}\}}{\{\mathrm{M}\}\{\mathrm{HL}\}} \tag{18.19}$$

$$\mathrm{M + 2HL = ML_2 + 2H^+} \qquad {}^*\beta_2 = {}^*K_1^*K_2 = \frac{\{\mathrm{ML_2}\}\{\mathrm{H^+}\}^2}{\{\mathrm{M}\}\{\mathrm{HL}\}^2} \tag{18.20}$$

$$\mathrm{M + 3HL = ML_3 + 3H^+} \qquad {}^*\beta_3 = {}^*K_1^*K_2^*K_3 = \frac{\{\mathrm{ML_3}\}\{\mathrm{H^+}\}^3}{\{\mathrm{M}\}\{\mathrm{HL}\}^3} \tag{18.21}$$

etc.

The choice of using constants of the type given in Eqs. (18.10)–(18.12), or Eqs. (18.13)–(18.15), or Eqs. (18.16)–(18.18), or Eqs. (18.19)–(18.21) is just a matter of selecting which group is most convenient for any given circumstance.

For the reaction of a metal ion with OH^-, since hydroxide is not protonated, the equilibrium constant would be denoted with a simple K. For the reaction of say Mg^{2+} with OH^-, with the subscript H denoting hydroxide, we therefore have

$$Mg^{2+} + OH^- = MgOH^+ \qquad K_{H1} = \frac{\{MgOH^+\}}{\{Mg^{2+}\}\{OH^-\}}. \tag{18.22}$$

When the reaction is written with the reacting species being H_2O rather than OH^-, since H_2O is the HL version of the ligand OH^-, we have

$$Mg^{2+} + H_2O = MgOH^+ + H^+ \qquad {}^*K_{H1} = \frac{\{MgOH^+\}\{H^+\}}{\{Mg^{2+}\}}. \tag{18.23}$$

We realize that unlike Eq. (18.16), for the special case where water is the protonated form of the ligand, $\{H_2O\}$ is not written in explicitly in the equilibrium constant expression for Eq. (18.19). Indeed, we use mole fraction to express $\{H_2O\}$, and the assumption has been made implicitly here that we are interested in comparatively dilute solutions where $X_{H_2O} \simeq 1$.

It is straightforward to show that K_{H1} and ${}^*K_{H1}$ are related according to the relation ${}^*K_{H1} = K_{H1}K_w$ where K_w is of course the acid dissociation constant for water. The general relationship between a ${}^*K_n$ and the corresponding K_n for the addition of the nth ligand, we have

$${}^*K_n = K_n K_a = \frac{\{ML_n\}\{H^+\}}{\{ML_{n-1}\}\{HL\}}. \tag{18.24}$$

Also,

$${}^*\beta_n = K_a^n \prod_{i=1}^{n} K_i = \frac{\{ML\}\{H^+\}^n}{\{M\}\{HL\}^n} \tag{18.25}$$

where K_a is the acid dissociation constant for HL. We will not bother with the special subscript H for β complexes involving hydroxide.

18.4 HYDROLYSIS OF METAL IONS

In the field of natural water chemistry, the term hydrolysis is defined as the process by which water reacts with metal ions to form hydroxo complexes. Actually, this use of the term hydrolysis is a bit of a misnomer. When one disects the word hydrolysis, it is seen to refer to a *lysis* (i.e., breaking apart) of something through the action of water. In particular, Webster defines "hydrolysis" as "A chemical process of decomposition involving addition of the elements of water."[1] In contrast, in the aquatic chemistry

[1]Examples of usages of "hydrolysis" that are faithful to a strict interpretation of its meaning include the "hydrolysis of an protein (amide) linkage," for example, R-(CO)-(NH)-R′ + H_2O = RCOOH + (NH_2)-R′, and the "hydrolysis of an ester linkage," for example, R-(CO)-O-R′ + H_2O = RCOOH + HOR′. In both of these usages, water is added across a bond, and the bond is broken. For hydrolysis of an amide, a carboxylic acid and an amine are obtained. For hydrolysis of an ester, an acid and an alcohol are obtained.

definition for "hydrolysis of metal ions," there is a lysis of *water itself* by the metal ion, not vice versa (see reaction (18.23)).

18.4.1 Effects of pH on the Extent of Hydrolysis

18.4.1.1 Effects of Dilution on a Solution of a Metal Salt. Chemists familiar with aqueous metal ion chemistry sometimes refer to the fact that when solutions of metal ions are diluted, the extent of hydrolysis tends to increase. We define the term "extent of hydrolysis" here as referring to the degree to which complexes containing hydroxide ions are important. For example, for Fe(III), a solution in which $Fe(OH)_2^+$ is the dominant dissolved Fe(III) species may be said to be more extensively hydrolyzed than a solution in which $FeOH^{2+}$ is the dominant Fe(III) species. A mathematical measure of the extent of hydrolysis would be the average number of OH^- units per Fe(III) in solution.

The fact that the extent of hydrolysis changes with dilution is a result of the fact that diluting a solution of a metal salt raises the pH of that solution. However, because simple dilution can never increase the pH of water above $\frac{1}{2}pK_w$, the extent of hydrolysis will no longer change significantly once the solution becomes very dilute. Access to pH values higher than $\frac{1}{2}pK_w$ and more highly hydrolyzed states is only possible through the addition of strong base.

As an example of the above principles, consider a solution of $HgCl_2$. As is indicated by Eq. (18.23), a solution of a metal salt like $HgCl_2$ is capable of exerting acidity. In this case, we have

$$Hg^{2+} + H_2O = HgOH^+ + H^+ \qquad {}^*K_{H1} = 10^{-3.7} \tag{18.26}$$

$$HgOH^+ + H_2O = Hg(OH)_2^o + H^+ \qquad {}^*K_{H2} = 10^{-2.6}. \tag{18.27}$$

The values of the constants given in Eqs. (18.26) and (18.27) are those for 25°C/1 atm. Although reactions (18.26) and (18.27) are like acidity constants in that each of them produces an H^+, they are still complex formation (stability) constants since they react a metal-containing ion with a ligand-containing species.

It is interesting to note that for Hg(II), ${}^*K_{H2}$ is larger than ${}^*K_{H1}$. In other words, the second "acidity" constant for Hg(II) is larger than the first. Other metals that behave in this manner include Ni(II), Co(II), Zn(II), and Cd(II). This type of behavior does not occur for conventional acids simply because the dissociation constants for such acids are numbered in order of decreasing acidity.

Computing the pH of a solution of an Hg(II) salt is equivalent to a simple, diprotic acid problem where ${}^*K_{H1}$ and ${}^*K_{H2}$ play the role of the first and second acidity constants, respectively. The PBE that may be written for the solution is

$$[H^+] = [HgOH^+] + 2[Hg(OH)_2^o] + [OH^-]. \tag{18.28}$$

If we employ α values to represent the fractions of the Hg species, then neglecting activity corrections, Eq. (18.28) becomes

$$[H^+] = \alpha_1 Hg_T + 2\alpha_2 Hg_T + \frac{K_w}{[H^+]} \tag{18.29}$$

where

$$\alpha_0 = \frac{[Hg^{2+}]}{Hg_T} = \frac{[Hg^{2+}]}{[Hg^{2+}] + [HgOH^+] + [Hg(OH)_2^o]} \tag{18.30}$$

$$= \frac{1}{1 + \frac{*\beta_1}{[H^+]} + \frac{*\beta_2}{[H^+]^2}} \tag{18.31}$$

$$\alpha_1 = \frac{[HgOH^+]}{Hg_T} = \alpha_0 \frac{[HgOH^+]}{[Hg^{2+}]} = \alpha_0 \frac{*\beta_1}{[H^+]} \tag{18.32}$$

$$\alpha_2 = \frac{[Hg(OH)_2^o]}{Hg_T} = \alpha_0 \frac{[Hg(OH)_2^o]}{[Hg^{2+}]} = \alpha_0 \frac{*\beta_2}{[H^+]^2}. \tag{18.33}$$

We note that the subscript on each of the α terms refers to the number of OH^- on that species. Complexation by Cl^- is assumed to be negligible.

The counting system represented by Eqs. (18.30), (18.32), and (18.33) contrasts somewhat with the one used in earlier chapters for acids. As we know, for acids, the subscript on an α refers to the number of protons *lost* relative to the most protonated form of the acid. For complexes, we count the number of ligands *gained* relative to the bare ion. This convention for complexes makes sense for two reasons: 1) the bare ion represents a well-defined starting point for counting ligands; and 2) as we continue to increase the concentration of the ligand, we can continue to push more ligands onto a metal ion. The second reason means that we do not have a single species that can provide a well-defined starting point for counting the number of ligand species that have been lost. As an example, in one solution of Fe(III), $Fe(OH)_4^-$ might be the logical point for starting to count the loss of OH^-; in another, more basic solution, we might have a non-negligible amount of $Fe(OH)_5^{2-}$, and that might be viewed as the logical starting point for counting. Thus, when considering complexation, it is much more straightforward to use α values that have subscripts that count the *addition* of successive ligands.

We continue now our consideration of the chemistry of solutions of $HgCl_2$. Since we are interested in $[H^+]$ for a range of Hg_T values and not just one Hg_T, for calculation purposes a more convenient form of Eq. (18.29) is

$$Hg_T = \frac{[H^+] - K_w/[H^+]}{\alpha_1 + 2\alpha_2} \tag{18.34}$$

which provides Hg_T as a function of pH. Values of pH that are lower than 3.28 give a value for Hg_T that is supersaturated with respect to the solid $Hg(OH)_{2(s)}$ ($K_{s0} = 10^{-25.32}$). Thus, at pH < 3.28, the product $\alpha_0 Hg_T[OH^-]^2$ exceeds $10^{-25.32}$.

As Hg_T decreases, that is, as the solution becomes more and more dilute, both of the Hg(II) terms in Eq. (18.29) decrease in importance, and the pH increases toward $\frac{1}{2}pK_w$. This process is depicted in Figure 18.2 where pH is plotted as a function of $\log Hg_T$. To further emphasize the perfect analogy that exists between this system and a simple diprotic acid, we recall that as a solution of H_2B is diluted, the pH will increase asymptotically toward $\frac{1}{2}pK_w$, and the relative importance of the less protonated species will necessarily increase vis à vis the more protonated species.

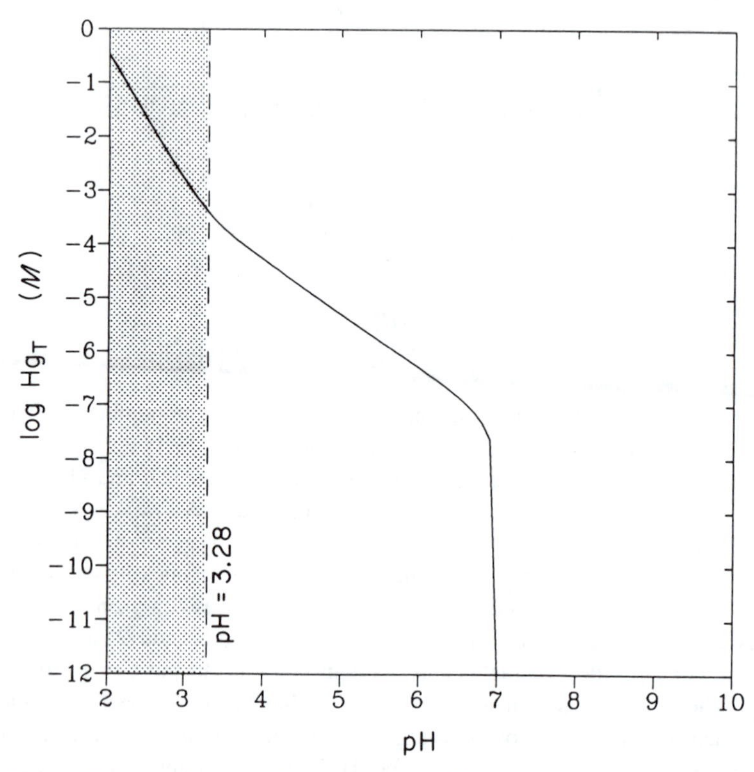

Figure 18.2 Log Hg_T vs. pH for solutions of $HgCl_2$. The hydrolysis of Hg^{2+} to form $HgOH^+$ and $Hg(OH)_2^o$ leads to the liberation of protons and therefore pH values that are lower than $\frac{1}{2}pK_w$ for all nonzero values of Hg_T. Conditions are 25°C/1 atm. Activity corrections are neglected. The solution is unstable relative to precipitation of $Hg(OH)_{2(s)}$ at pH < 3.28.

In this example, as Hg_T decreases and pH increases, the importance of the more hydrolyzed species increases relative to less hydrolyzed species. Figures 18.3*a* and 18.3.*b* illustrate how the three α values vary for a range of pH values, including those for pH $> \frac{1}{2}pK_w$. Figure 18.3*a* gives the curves for the functions $\log\alpha_0$, $\log\alpha_1$, and $\log\alpha_2$ vs. pH. Figure 18.3*b* gives the same curves, but in a linear format. Finally, Figure 18.3*c* presents curves for α_0, $(\alpha_0 + \alpha_1)$, and $(\alpha_0 + \alpha_1 + \alpha_2)$ vs. pH. At any given pH value, the distance between the *x*-axis and the first curve equals α_0. Since the second curve plots $(\alpha_0 + \alpha_1)$ vs. pH, the vertical distance between the first and second curves equals α_1. Finally, along the top margin of the plot, the vertical value of $1.0 = (\alpha_0 + \alpha_1 + \alpha_2)$ is obtained, and so the distance between the second curve and the top margin is α_2. The Figure 18.3*c* plot may therefore be divided into regions labelled with the various different Hg(II) species. Texts dealing with aqueous solution chemistry often choose to present plots of α values for metal ions in the format used in Figure 18.3*c*.

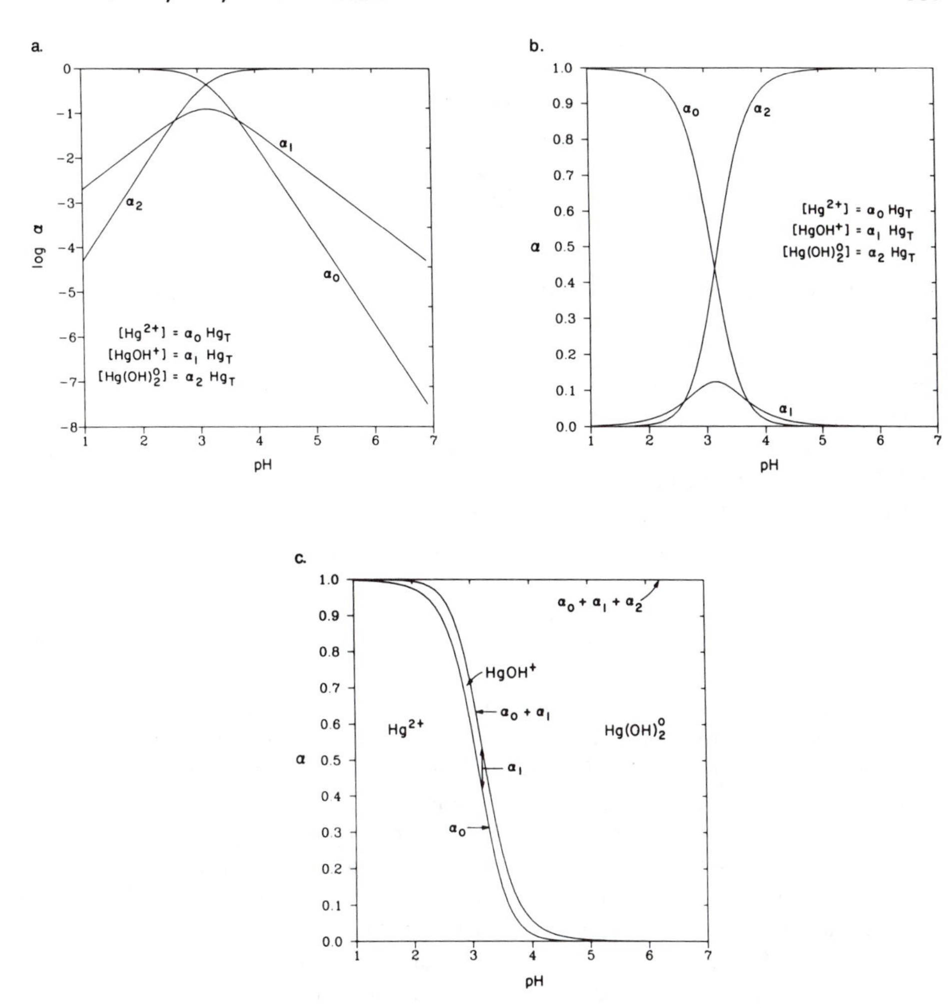

Figure 18.3 (*a*) Log α_0, log α_1, and log α_2 vs. pH for the hydrolysis of Hg(II). (*b*) α_0, α_1, and α_2 vs. pH for the hydrolysis of Hg(II). (*c*) α_0, $(\alpha_0 + \alpha_1)$, and $(\alpha_0 + \alpha_1 + \alpha_2) = 1$ vs. pH for the hydrolysis of Hg(II). For plots *a*, *b*, and *c*, the conditions are 25°C/1 atm. Activity corrections are neglected.

18.4.1.2 Polynuclear Hydroxo Complexes. A polynuclear complex is a complex containing more than one metal center. For us, probably the most important polynuclear complexes are those involving the ligand OH^- together with high-oxidation state metals like Fe(III), Cr(III), and Al(III). For Fe(III), the most important polynuclear Fe(III) species is $Fe_2(OH)_2^{4+}$. Like most other M(III) polynuclear species, $Fe_2(OH)_2^{4+}$ is characterized by OH bridges between the metal centers. When H_2O ligand molecules surrounding the Fe(III) centers are neglected, the $Fe_2(OH)_2^{4+}$ species may be represented as given in Figure 18.4. Polynuclear complexes of this type are only important in relatively concentrated solutions of metal (e.g., high Fe_T), and so they were neglected in Chapter 11.

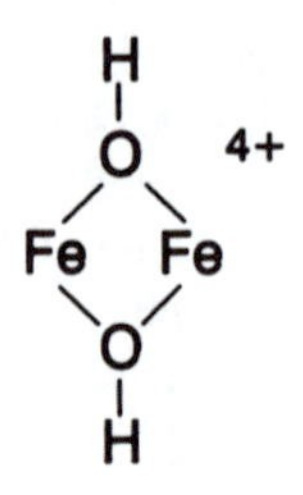

Figure 18.4 Structure of the $Fe_2(OH)_2^{4+}$ species. Note the presence of OH bridges. Waters of hydration are neglected.

We consider them now because a chapter on complexation chemistry would not be complete without a discussion of these types of complexes.

The nomenclature for the equilibrium constants governing polynuclear complexes must denote not only the number of ligand units, but also the number of metal ions for which a given constant is applicable. Therefore, for $Fe_2(OH)_2^{4+}$, we have

$$2Fe^{3+} + 2H_2O = Fe_2(OH)_2^{4+} + 2H^+ \qquad {}^*\beta_{22} = 10^{-2.9} \tag{18.35}$$

and so

$$\frac{\{Fe_2(OH)_2^{4+}\}\{H^+\}^2}{\{Fe^{3+}\}^2} = {}^*\beta_{22}. \tag{18.36}$$

A β is used for reaction (18.35) rather than a K because more than one ligand is being added. The first subscript in the ${}^*\beta$ refers to the number of ligands. The second subscript refers to the number of metals. The specific value given for the constant for reaction Eq. (18.35) is the one that applies at 25°C/1 atm.

It is the fact that the formation of polynuclear complexes requires more than one metal ion that makes their importance relative to mononuclear species *concentration dependent*. For example, since the driving force for the formation of $Fe_2(OH)_2^{4+}$ depends on the square of the Fe^{3+} activity, lowering that activity at constant pH will affect on the concentration of $Fe_2(OH)_2^{4+}$. Thus, while the α values for a metal M that does not form polynuclear complexes will be completely independent of the value of M_T, all of the α values for a metal that does form polynuclear complexes will depend on M_T.

Proceeding with the Fe(III) system, we write

$$\begin{aligned} Fe_T = {} & [Fe^{3+}] + [FeOH^{2+}] + [Fe(OH)_2^+] + [Fe(OH)_3^o] \\ & + [Fe(OH)_4^-] + 2[Fe_2(OH)_2^{4+}]. \end{aligned} \tag{18.37}$$

While Eq. (18.37) is more exact than is Eq. (11.57) which did not include $[Fe_2(OH)_2^{4+}]$, the difference between the two will only matter at relatively high Fe_T and at the types of pH values where $Fe_2(OH)_2^{4+}$ can be important.

For equilibrium with the solid (am)$Fe(OH)_{3(s)}$, modification of Eq. (11.58) by means of Eq. (18.36) yields

$$\begin{aligned} Fe_T = {} & {}^*K_{s0}[H^+]^3 + {}^*K_{s1}[H^+]^2 + {}^*K_{s2}[H^+] + K_{s3} \\ & + \frac{{}^*K_{s4}}{[H^+]} + 2\,{}^*K_{s0}^2\;{}^*\beta_{22}[H^+]^4 \end{aligned} \tag{18.38}$$

where activity corrections are neglected. Figure 11.3 neglected $Fe_2(OH)_2^{4+}$. Figure 18.5 includes the line for $\log[Fe_2(OH)_2^{4+}]$ and also considers the contribution which that species makes to $\log Fe_T$. The differences between the $\log Fe_T$ lines in Figures 11.3 and 18.5 are indistinguishable. Even at pH = 1, Fe^{3+} is the dominant contributor to Fe_T.

The dependence of the various α values on Fe_T may be understood as follows. In general, regardless of whether we have equilibrium with (am)$Fe(OH)_{3(s)}$ or not, we have

$$\alpha_0 = \frac{[Fe^{3+}]}{Fe_T} \tag{18.39}$$

$$\alpha_1 = \frac{[FeOH^{2+}]}{Fe_T} \tag{18.40}$$

$$\alpha_2 = \frac{[Fe(OH)_2^+]}{Fe_T} \tag{18.41}$$

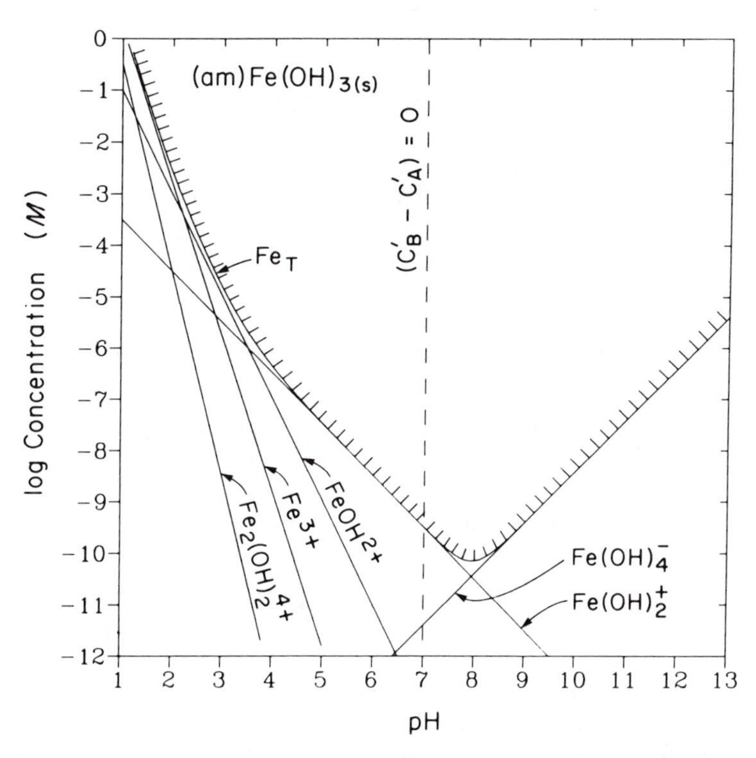

Figure 18.5 Log concentration vs. pH for various dissolved Fe(III) species (including both $Fe_2(OH)_2^{4+}$ and Fe_T) at 25°C/1 atm. Activity corrections are neglected.

$$\alpha_3 = \frac{[\mathrm{Fe(OH)_3^o}]}{\mathrm{Fe_T}} \tag{18.42}$$

$$\alpha_4 = \frac{[\mathrm{Fe(OH)_4^-}]}{\mathrm{Fe_T}} \tag{18.43}$$

$$\alpha_{22} = \frac{2[\mathrm{Fe_2(OH)_2^{4+}}]}{\mathrm{Fe_T}}. \tag{18.44}$$

A factor of 2 is included in the expression for α_{22} since there are two Fe(III) in each $Fe_2(OH)_2^{4+}$ species. Assuming of course that we have identified all of the important Fe(III) species, the various α values given above will sum to 1.0 for all pH values.

We can rewrite Eq. (18.37) as

$$\mathrm{Fe_T} = [\mathrm{Fe^{3+}}]\left(1 + \frac{[\mathrm{FeOH^{2+}}]}{[\mathrm{Fe^{3+}}]} + \frac{[\mathrm{Fe(OH)_2^+}]}{[\mathrm{Fe^{3+}}]} + \frac{[\mathrm{Fe(OH)_3^o}]}{[\mathrm{Fe^{3+}}]} + \frac{[\mathrm{Fe(OH)_4^-}]}{[\mathrm{Fe^{3+}}]} + \frac{2[\mathrm{Fe_2(OH)_2^{4+}}]}{[\mathrm{Fe^{3+}}]}\right). \tag{18.45}$$

Neglecting activity corrections (or, with constant ionic medium equilibrium constants substituted for the infinite dilution constants), Eq. (18.45) becomes

$$\mathrm{Fe_T} = [\mathrm{Fe^{3+}}]\left(1 + \frac{{}^*K_{\mathrm{H1}}}{[\mathrm{H^+}]} + \frac{{}^*\beta_2}{[\mathrm{H^+}]^2} + \frac{{}^*\beta_3}{[\mathrm{H^+}]^3} + \frac{{}^*\beta_4}{[\mathrm{H^+}]^4} + \frac{2[\mathrm{Fe^{3+}}]\,{}^*\beta_{22}}{[\mathrm{H^+}]^2}\right). \tag{18.46}$$

Since $[Fe^{3+}] = \alpha_0 Fe_T$, neglecting the term for $Fe(OH)_3^o$, we have

$$\alpha_0 = \left(1 + \frac{{}^*K_{\mathrm{H1}}}{[\mathrm{H^+}]} + \frac{{}^*\beta_2}{[\mathrm{H^+}]^2} + \frac{{}^*\beta_4}{[\mathrm{H^+}]^4} + \frac{2\alpha_0 \mathrm{Fe_T}\,{}^*\beta_{22}}{[\mathrm{H^+}]^2}\right)^{-1}. \tag{18.47}$$

Equation (18.47) may be rearranged to

$$\frac{2\mathrm{Fe_T}\,{}^*\beta_{22}}{[\mathrm{H^+}]^2}\alpha_0^2 + \left(1 + \frac{{}^*K_{\mathrm{H1}}}{[\mathrm{H^+}]} + \frac{{}^*\beta_2}{[\mathrm{H^+}]^2} + \frac{{}^*\beta_4}{[\mathrm{H^+}]^4}\right)\alpha_0 - 1 = 0 \tag{18.48}$$

which is second order in α_0, that is, it is of the type $a\alpha_0^2 + b\alpha_0 + c = 0$. Given specific values of dissolved Fe_T and $[H^+]$, the quadratic equation $x = (-b \pm \sqrt{(b^2 - 4ac)})/2a$ may be used to solve for α_0. Since the coefficient b is positive, the root $x = (-b + \sqrt{(b^2 - 4ac)})/2a$ will always be the one of interest. Interestingly, the quadratic equation approach to solving for α_0 works fine as long as the pH of interest is not too high. However, for pH values beyond about 8.2, the $Fe(OH)_4^-$ term in b (i.e., ${}^*\beta_4/[H^+]^4$) becomes so large that even precise computer calculations cannot compute an accurate value for the quantity $(-b + \sqrt{(b^2 - 4ac)})$. For these pH values, α_0 may be computed by neglecting the dimer, which removes the need to solve the above quadratic equation. In any case, once α_0 is obtained, the remaining α values may be computed by using the exact but simple relationships that exist between ratios of α values. Specifically, realizing that $\alpha_1/\alpha_0 = [FeOH^{2+}]/[Fe^{3+}]$, $\alpha_2/\alpha_0 = [Fe(OH)_2^+]/[Fe^{3+}]$, $\alpha_4/\alpha_0 = [Fe(OH)_4^-]/[Fe^{3+}]$, and $\alpha_{22}/\alpha_0 = 2[Fe_2(OH)_2^{4+}]/[Fe^{3+}]$, a comparison of corresponding terms in Eqs. (18.45) and (18.46) yields

$$\alpha_1 = \frac{\alpha_0 \, {}^*K_1}{[H^+]} \tag{18.49}$$

$$\alpha_2 = \frac{\alpha_0 \, {}^*\beta_2}{[H^+]^2} \tag{18.50}$$

$$\alpha_4 = \frac{\alpha_0 \, {}^*\beta_4}{[H^+]^4} \tag{18.51}$$

$$\alpha_{22} = \frac{\alpha_0^2 \, 2 \, Fe_T \, {}^*\beta_{22}}{[H^+]^2}. \tag{18.52}$$

At 25°C/1 atm, the values of the constants in these equations are ${}^*K_{H1} = 10^{-2.2}$, ${}^*\beta_2 = 10^{-5.7}$, ${}^*\beta_4 = 10^{-21.6}$, and ${}^*\beta_{22} = 10^{-2.9}$.

Since α_0 depends on Fe_T (as well as pH), all of the other α values also depend on Fe_T (and pH). This contrasts with the Hg(II) case discussed above for which, as for conventional acids, the α values are *independent* of the total amount of material in solution. The above equations indicate that as Fe_T decreases, α_{22} tends to zero rapidly, and so the situation tends toward one in which the other α values become essentially independent of Fe_T. Figure 6.5 in Stumm and Morgan (1981) illustrates these principles for $Fe_T = 10^{-2}$ *M* and 10^{-4} *M*.

18.5 STEPWISE ATTACHMENT OF SEVERAL MONODENTATE LIGANDS

18.5.1 Complexation of Cd(II) by Chloride

The stepwise formation of complexes involves the repeated addition of the same ligand species. It is a very important type of process. It bears many similarities, both chemically and mathematically, to the stepwise ionization of a polyprotic acid. The formation of $HgOH^+$ and $Hg(OH)_2^0$ from Hg^{2+} and OH^- is one example of stepwise complexation. The analogous action of OH^- on Fe^{3+} is another example.

Butler (1964) has examined the case of the stepwise complexation of Cd^{2+} with Cl^- at $I = 4.5$ *M*. We consider the case when $I = 1.0$ *M*. The following constants are available at 25°C/1 atm and $I = 1.0$ *M*:

$$Cd^{2+} + Cl^- = CdCl^+ \qquad {}^cK_1 = 34.7 \qquad {}^c\beta_1 = {}^cK_1 = 34.7 \tag{18.53}$$

$$CdCl^+ + Cl^- = CdCl_2^0 \qquad {}^cK_2 = 4.57 \qquad {}^c\beta_2 = {}^cK_1 \, {}^cK_2 = 158 \tag{18.54}$$

$$CdCl_2^o + Cl^- = CdCl_3^- \quad {}^cK_3 = 1.26 \quad {}^c\beta_3 = {}^cK_1\,{}^cK_2\,{}^cK_3 = 200 \tag{18.55}$$

$$CdCl_3^- + Cl^- = CdCl_4^- \quad {}^cK_4 = 0.20 \quad {}^c\beta_4 = {}^cK_1\,{}^cK_2\,{}^cK_3\,{}^cK_4 = 40.0. \tag{18.56}$$

If we restrict our consideration to cases where the pH is low enough that hydroxo complexes like $CdOH^+$ are not important,[2] we have

$$Cd_T = [Cd^{2+}] + [CdCl^+] + [CdCl_2^o] + [CdCl_3^-] + [CdCl_4^{2-}] \tag{18.57}$$

and so

$$\alpha_0 = \frac{[Cd^{2+}]}{Cd_T} = \frac{1}{1 + {}^c\beta_1[Cl^-] + {}^c\beta_2[Cl^-]^2 + {}^c\beta_3[Cl^-]^3 + {}^c\beta_4[Cl^-]^4} \tag{18.58}$$

$$\alpha_1 = \frac{[CdCl^+]}{Cd_T} = {}^c\beta_1[Cl^-]\alpha_0 \tag{18.59}$$

$$\alpha_2 = \frac{[CdCl_2^o]}{Cd_T} = {}^c\beta_2[Cl^-]^2\alpha_0 \tag{18.60}$$

$$\alpha_3 = \frac{[CdCl_3^-]}{Cd_T} = {}^c\beta_3[Cl^-]^3\alpha_0 \tag{18.61}$$

$$\alpha_4 = \frac{[CdCl_4^{2-}]}{Cd_T} = {}^c\beta_4[Cl^-]^4\alpha_0. \tag{18.62}$$

As with the addition of OH^- to Hg^{2+} and Fe^{3+}, we note that we have defined the α values with respect to the *addition* of Cl^- units to the starting unit Cd^{2+}, rather than with respect to the stepwise loss of Cl^- from some highly complexed species. We also note again that like α values for acid/base species, ratios of complexation α values are simple functions.

Figures 18.6*a* and 18.6*b* present the α functions for complexation of Cd^{2+} by Cl^- in a log format and in a linear format, respectively. As pCl changes, the identity of the dominant Cd-containing species changes accordingly. At very low pCl, $CdCl_4^{2-}$ dominates, and at high pCl, Cd^{2+} dominates. For intermediate values of pCl, the other cadmium chloro complexes can dominate. Figure 18.6*c* plots α_0, $(\alpha_0+\alpha_1)$, $(\alpha_0+\alpha_1+\alpha_2)$, $(\alpha_0 + \alpha_1 + \alpha_2 + \alpha_3)$ and $(\alpha_0 + \alpha_1 + \alpha_2 + \alpha_3 + \alpha_4)$ vs. pCl.

It is illustrative to determine the concentrations of Cl^- required to achieve a 1:1 ratio of the various successive complex pairs. The $[Cl^-]$ values required may be obtained directly from the ${}^cK$ values; they are presented in Table 18.2. In the context of Figures 18.6*a* and 18.6*b*, these values correspond to the pCl values at which the α_0 curve crosses the α_1 curve ($\log\alpha_0$ crosses $\log\alpha_1$), the α_1 curve crosses the α_2 curve ($\log\alpha_1$ crosses $\log\alpha_2$) etc.

[2] The value of K_1 for the formation of the complex $CdOH^+$ is $10^{4.16}$ at 25°C/1 atm. Thus, hydroxocomplexes can be neglected at pH values less than about 9.

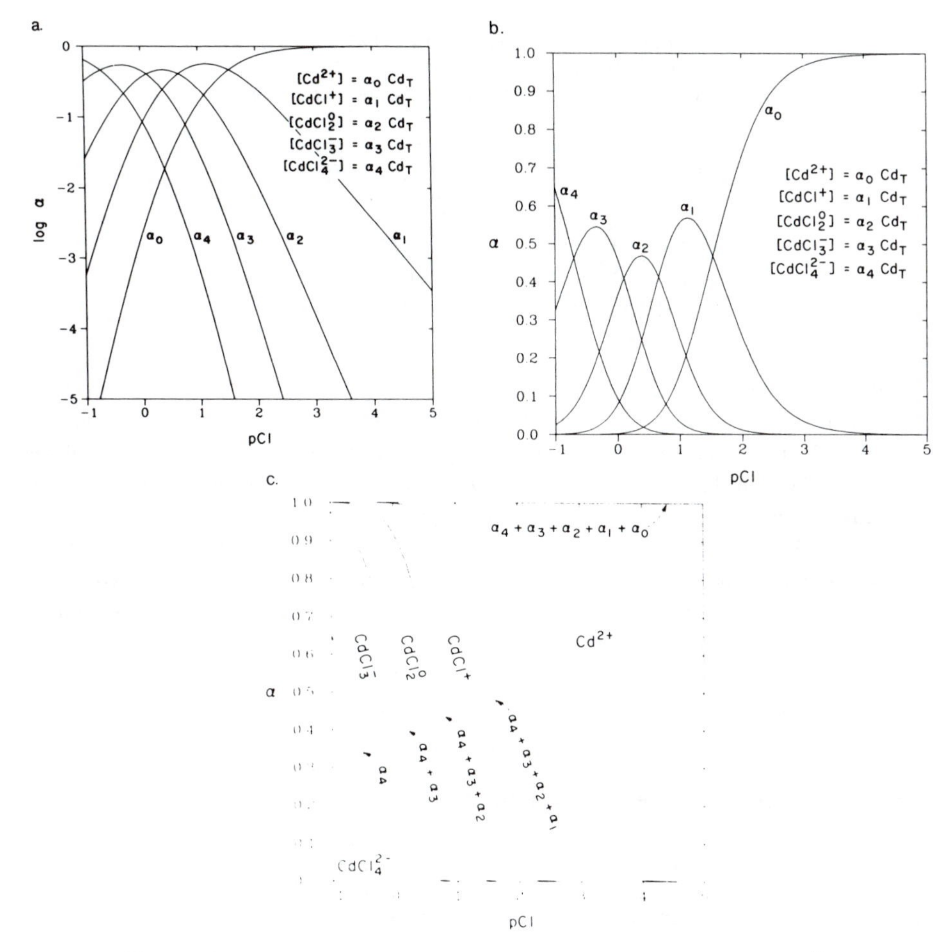

Figure 18.6 (*a*) Log α vs. pCl plot for stepwise complexation of Cd^{2+} by Cl^-. (*b*) α vs. pCl plot for stepwise complexation of Cd^{2+} by Cl^-. (*c*) α_4, $(\alpha_4 + \alpha_3)$, $(\alpha_4 + \alpha_3 + \alpha_2)$, $(\alpha_4 + \alpha_3 + \alpha_2 + \alpha_1)$, and $(\alpha_4 + \alpha_3 + \alpha_2 + \alpha_1 + \alpha_0) = 1$ vs. pCl for stepwise complexation of Cd^{2+} by Cl^-. For plots *a*, *b*, and *c*, the conditions are 25°C/1 atm and an ionic strength $I = 1.0\ M$. The pH is assumed to be sufficiently low that complexation of Cd^{2+} by OH^- can be neglected. Activity corrections are taken into account through the use of ^cK constants valid at $I = 1.0\ M$.

The "ligand number" $\bar{n}$ is defined as the *average* number of ligands per metal ion. In this case, it is given by the relation

$$\bar{n} = \frac{[CdCl^+] + 2[CdCl_2^0] + 3[CdCl_3^-] + 4[CdCl_4^{2-}]}{Cd_T} \tag{18.63}$$

$$= \alpha_1 + 2\alpha_2 + 3\alpha_3 + 4\alpha_4. \tag{18.64}$$

TABLE 18.2 $[Cl^-]$ values required in a constant ionic medium with ionic strength = 1.0 M to obtain 1 : 1 ratios of the successive cadmium chloro complex pairs.

1:1 Ratio	$[Cl^-]$ required	pCl required
$[Cd^{2+}]/[CdCl^+]$	$1/34.7 = 0.029\ M$	1.54
$[CdCl^+]/[CdCl_2^o]$	$1/4.57 = 0.22\ M$	0.66
$[CdCl_2^o]/[CdCl_3^-]$	$1/1.26 = 0.81\ M$	0.09
$[CdCl_3^-]/[CdCl_4^{2-}]$	$1/0.20 = 5.0\ M$	−0.70

Presuming that the maximum number of chlorides that can add to Cd^{2+} is 4, then $\bar{n}$ can never reach a value of exactly 4.0. The reason is that it is *impossible* to ever have so much chloride in the system that there is not still *some* of the lower chlorinated complexes still present. For similar reasons, a value of zero can never be achieved as long as there is *any* amount of Cl^- in the system: $\bar{n} = 0$ can only be achieved when $[Cl^-] = 0$, that is, $pCl = \infty$. Figure 18.7 gives $\bar{n}$ as a function of pCl. In the case of complexation of a metal ion by OH^-, the $\bar{n}$ for OH^- would provide a measure of the extent of hydrolysis.

The functionality of $\bar{n}$ for Cl^- complexation of Cd^{2+} in terms of pCl bears much, though not total similarity to the function g in the acidimetric titration of a polyprotic acid system. For example, if we define an $\bar{n}_H$ for the phosphoric acid system as being the average number of protons per phosphate unit, we would have

$$\bar{n}_H = 3\alpha_0 + 2\alpha_1 + \alpha_2 + 0\alpha_3. \tag{18.65}$$

We recall that g for this system is given by

$$g = 3\alpha_0 + 2\alpha_1 + \alpha_2 + \frac{[H^+] - [OH^-]}{C} \tag{18.66}$$

where C is the total concentration of phosphate in the system. Thus,

$$\bar{n}_H = g - \frac{[H^+] - [OH^-]}{C}. \tag{18.67}$$

Figure 18.8 gives both $\bar{n}_H$ and g as a function of pH for a phosphoric acid system in which total phosphate equals $10^{-3}\ M$. In a manner that is analogous to $\bar{n}$ in the cadmium-chloro complexation case, the value of $\bar{n}_H$ can never reach a value of 3, no matter how low the pH, and it can never reach 0, no matter how high the pH. The function g can, however, reach these values since it takes into consideration the possible presence in the solution of H^+ from the dissociation of the acid, and OH^- from the hydrolysis of the conjugate base.

18.5.2 Examples

We now consider some examples of complexation of Cd(II) by Cl^-. All of the solutions will be assumed to be sufficiently acidic so that hydroxo complexes of Cd(II) can be neglected. Also, all of the solutions will be assumed to be at 25°C/1 atm and ionic strength $\simeq 1.0\ M$ so that the values of the constants given above may be used. We will

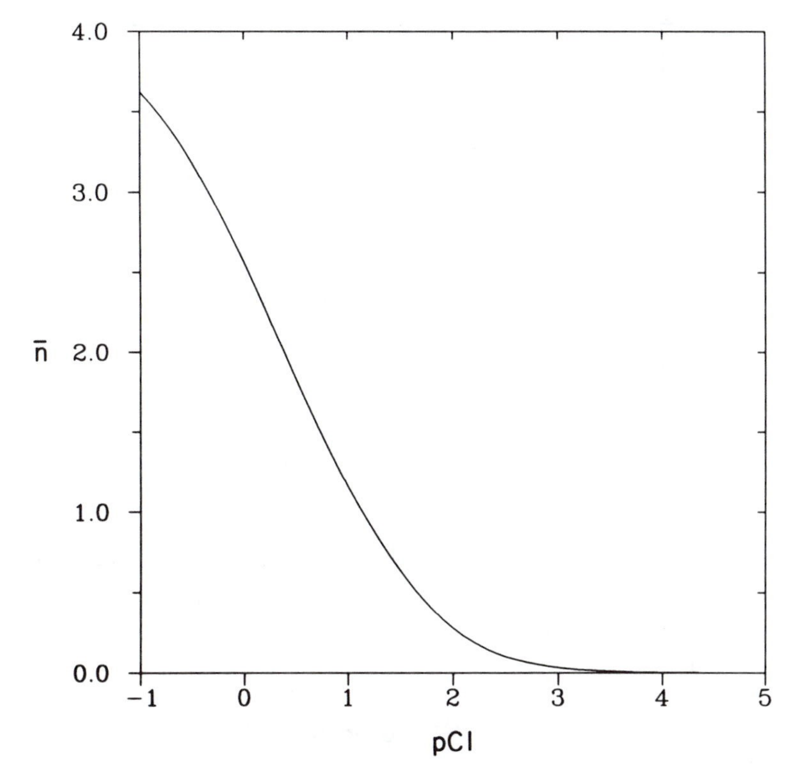

Figure 18.7 Plot of the ligand number $\bar{n}$ vs. pCl for the complexation of Cd^{2+} by Cl^-. The conditions are 25°C/1 atm and an ionic strength $I = 1.0\ M$. The pH is assumed to be sufficiently low that complexation of Cd^{2+} by OH^- can be neglected. Activity corrections are taken into account through the use of ${}^cK$ constants valid at $I = 1.0\ M$.

examine how to approach each of the problems with the help of assumptions, and then a second time without assumptions. Several of the examples considered are taken from Problem 1 of Butler (1964, p. 272).

18.5.2.1 1.0 *F* NaCl, 5.0 × 10^{-3} *F* $Cd(NO_3)_2$.

Solution with Assumptions. Since the value of $[Cl^-]$ is so large relative to Cd_T, we can safely assume that the formation of cadmium chloro complexes will not significantly deplete the concentration of Cl^-, and so

$$[Cl^-] \simeq 1.0\ M. \tag{18.68}$$

By Eq. (18.58), we therefore have

$$\alpha_0 = \frac{1}{1 + {}^c\beta_1 + {}^c\beta_2 + {}^c\beta_3 + {}^c\beta_4} = \frac{1}{432.7} = 2.31 \times 10^{-3} \tag{18.69}$$

$$[Cd^{2+}] \simeq (2.31 \times 10^{-3})(5.0 \times 10^{-3}) = 1.16 \times 10^{-5} M. \tag{18.70}$$

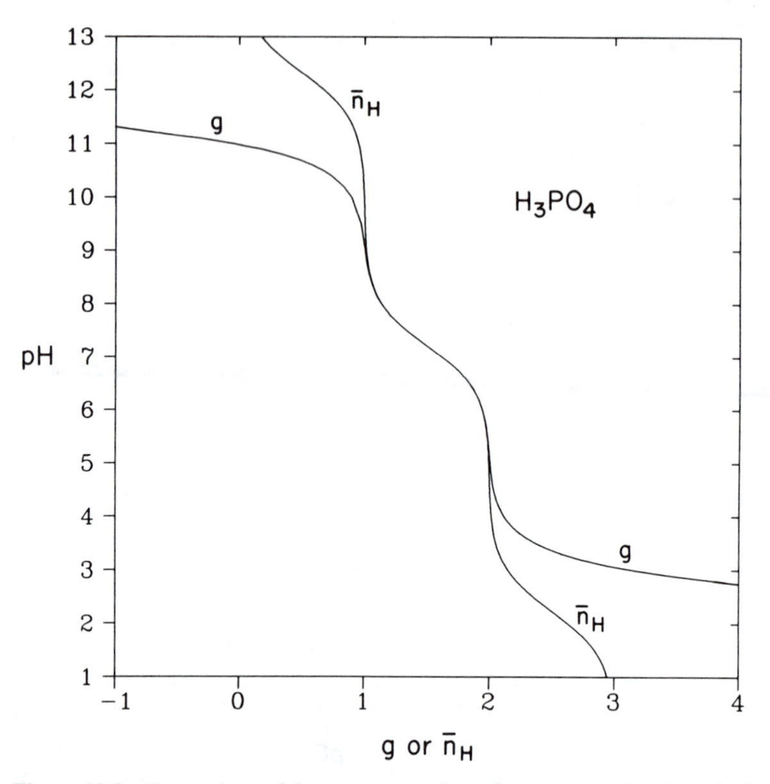

Figure 18.8 Comparison of the average number of protons per phosphate ($\bar{n}_H$) and the titration function g for the phosphoric acid (H_3PO_4) system. Conditions are 25°C/1 atm. Activity corrections are neglected.

By Eqs. (18.53)–(18.56) we therefore have

$$[CdCl^+] = {}^cK_1[Cd^{2+}][Cl^-] \simeq (34.7)(1.16 \times 10^{-5})(1.0)$$
$$= 4.01 \times 10^{-4} M \qquad (18.71)$$

$$[CdCl_2^o] = {}^cK_2[CdCl^+][Cl^-] \simeq (4.57)(4.01 \times 10^{-4})(1.0)$$
$$= 1.83 \times 10^{-3} M \qquad (18.72)$$

$$[CdCl_3^-] = {}^cK_3[CdCl_2^o][Cl^-] \simeq (1.26)(1.83 \times 10^{-3})(1.0)$$
$$= 2.31 \times 10^{-3} M \qquad (18.73)$$

$$[CdCl_4^{2-}] = {}^cK_4[CdCl_3^-][Cl^-] \simeq (0.20)(2.31 \times 10^{-3})(1.0)$$
$$= 4.62 \times 10^{-4} M. \qquad (18.74)$$

Let us check to see if the species distribution calculated using the the assumptions described above reasonably satisfies the two MBEs governing this system. We first check the MBE on Cd_T and see if the total of all Cd(II) species comes close to 5.0×10^{-3} M.

We obtain

$$\sum_{i=0}^{4}[\mathrm{Cd(Cl)}_i^{(2-i)+}] = 1.16 \times 10^{-5} + 4.01 \times 10^{-4} + 1.83 \times 10^{-3} + 2.31 \times 10^{-3} + 4.62 \times 10^{-4} = 5.01 \times 10^{-3} M \tag{18.75}$$

which is quite close to 5.0×10^{-3} M.

We now check to be sure that the concentration $[\mathrm{Cl}^-]$ is indeed close to 1.0 M, that is, we check to see whether the total amount of chloride bound up with the Cd(II) is small compared to 1.0 M. We obtain

$$\sum_{i=0}^{4} i[\mathrm{Cd(Cl)}_i^{(2-i)+}] = 4.01 \times 10^{-4} + 2(1.83 \times 10^{-3}) + 3(2.31 \times 10^{-3}) + 4(4.62 \times 10^{-4}) = 1.28 \times 10^{-2} M. \tag{18.76}$$

This is indeed small compared to 1.0 M. Nevertheless, we can now say that rather than 1.0 M, a better estimate of $[\mathrm{Cl}^-]$ would be $(1.0 - 1.28 \times 10^{-2}) = 0.987$ M. This suggests that an iteration with this value for $[\mathrm{Cl}^-]$ would lead to very good second estimates for the various concentrations.

For the ligand number, we could use Eq. (18.64), but Eq. (18.63) is more convenient as we already know the concentrations. Indeed, we have already summed the concentrations that we need for Eq. (18.63), and so we conclude that

$$\bar{n} = \frac{1.28 \times 10^{-2}}{5.0 \times 10^{-3}} = 2.56. \tag{18.77}$$

Solution without Assumptions. The MBE for total chloride can be written

$$\alpha_1 \mathrm{Cd_T} + 2\alpha_2 \mathrm{Cd_T} + 3\alpha_3 \mathrm{Cd_T} + 4\alpha_4 \mathrm{Cd_T} + [\mathrm{Cl}^-] = \mathrm{Cl_T}. \tag{18.78}$$

Dividing both sides by α_0, we obtain

$$\frac{\alpha_1}{\alpha_0}\mathrm{Cd_T} + 2\frac{\alpha_2}{\alpha_0}\mathrm{Cd_T} + 3\frac{\alpha_3}{\alpha_0}\mathrm{Cd_T} + 4\frac{\alpha_4}{\alpha_0}\mathrm{Cd_T} + \frac{[\mathrm{Cl}^-]}{\alpha_0} - \frac{\mathrm{Cl_T}}{\alpha_0} = 0 \tag{18.79}$$

or,

$$\mathrm{Cd_T}({}^c\beta_1[\mathrm{Cl}^-] + 2\,{}^c\beta_2[\mathrm{Cl}^-]^2 + 3\,{}^c\beta_3[\mathrm{Cl}^-]^3 + 4\,{}^c\beta_4[\mathrm{Cl}^-]^4) + \frac{[\mathrm{Cl}^-]}{\alpha_0} - \frac{\mathrm{Cl_T}}{\alpha_0} = 0. \tag{18.80}$$

Eq. (18.80) is only a function of $[\mathrm{Cl}^-]$, and so can be solved for any given values of $\mathrm{Cd_T}$ and $\mathrm{Cl_T}$. For the case at hand, using the above values of the various constants, we obtain the solution that

$$[\mathrm{Cl}^-] = 0.987\ M. \tag{18.81}$$

Interestingly, this is exactly the *corrected* (i.e., second iteration) estimate of $[\mathrm{Cl}^-]$ that we obtained when we solved the problem using assumptions. For $[\mathrm{Cl}^-] = 0.987$, we have $\bar{n} = 2.55$, and $\alpha_0 = 2.38 \times 10^{-3}$. Thus,

$$[Cd^{2+}] = (2.38 \times 10^{-3})(5.0 \times 10^{-3}) = 1.19 \times 10^{-5} M. \tag{18.82}$$

As might be expected, this is close to the first iteration value obtained when the problem was solved with the help of assumptions. The concentrations of the remaining species may be calculated using Eqs. (18.53)–(18.56).

18.5.2.2 0.1 *F* $CdCl_2$.

Solution with Assumptions. Since some of the chloride will be bound up in the various Cd(II) complexes, and since $[Cl_T] = 0.2\ M$, we know that the equilibrium value of $[Cl^-]$ must be less than 0.2 *M*. Since the constants for Eqs. (18.53)–(18.56) are not very large, chloride may be seen to be a fairly weak complexing agent for Cd(II), and so the equilibrium value of $[Cl^-]$ is not likely to be even one order of magnitude below 0.2 *M*. Thus, at equilibrium, we can conclude that in this case, $0.7 \leq pCl \leq 1.7$. An examination of Figure 18.6 indicates that while we can neglect $CdCl_4^{2-}$ in this region, all of the other species will be at least somewhat important. Thus, being able to neglect just $CdCl_4^{2-}$ does not help much; if we do try to solve the problem based on that assumption alone, we are still going to end up with having to carry out a trial and error solution. Therefore, in this case, we should simply use Eq. (18.80) directly, that is, solve the problem without assumptions.

Solution without Assumptions. Application of Eq. (18.80) with $Cd_T = 0.1\ M$ and $Cl_T = 0.2\ M$ yields $[Cl^-] = 8.97 \times 10^{-2}\ M$, and therefore $\alpha_0 = 1.81 \times 10^{-1}$. Thus, by Eqs. (18.53)–(18.56), we have

$$[Cd^{2+}] = (1.81 \times 10^{-1})(1.0 \times 10^{-1}) = 1.81 \times 10^{-2} M \tag{18.83}$$

$$[CdCl^+] = {}^cK_1[Cd^{2+}][Cl^-] = (34.7)(1.81 \times 10^{-2})(8.97 \times 10^{-2}) = 5.63 \times 10^{-2} \tag{18.84}$$

$$[CdCl_2^o] = {}^cK_2[CdCl^+][Cl^-] = (4.57)(5.63 \times 10^{-2})(8.97 \times 10^{-2}) = 2.31 \times 10^{-2} \tag{18.85}$$

$$[CdCl_3^-] = {}^cK_3[CdCl_2^o][Cl^-] = (1.26)(2.31 \times 10^{-2})(8.97 \times 10^{-2}) = 2.61 \times 10^{-3} \tag{18.86}$$

$$[CdCl_4^{2-}] = {}^cK_4[CdCl_3^-][Cl^-] = (0.20)(2.61 \times 10^{-3})(8.97 \times 10^{-2}) = 4.68 \times 10^{-5} \tag{18.87}$$

and

$$\bar{n} = 1.10. \tag{18.88}$$

As we suspected, at equilibrium, several of the Cd(II) species are important in this solution.

18.5.2.3 5.0×10^{-3} *F* $CdCl_2$.

Solution with Assumptions. If all the chloride was present as Cl^-, then $[Cl^-]$ would be 10^{-2} *M*. For reasons that are analogous to those presented in the previous section, however, $[Cl^-]$ must be less than 10^{-2} *M*. As in the previous case, however, we can probably conclude that $[Cl^-]$ is not an order of magnitude below 10^{-2} *M*, and so it is likely that the equilibrium system will be characterized by a system such that $2 \leq pCl \leq 3$. A quick glance at Figure 18.6 shows that the major species in this region will be Cd^{2+} and $CdCl^+$. Therefore, a good approximation would be to assume that

$$[Cd^{2+}] + [CdCl^+] \simeq Cd_T = 5.0 \times 10^{-3} M \tag{18.89}$$

and

$$[CdCl^+] + [Cl^-] \simeq Cl_T = 10^{-2} M. \tag{18.90}$$

We also know that

$$\frac{[CdCl^+]}{[Cd^{2+}][Cl^-]} = 34.7. \tag{18.91}$$

Equations (18.89)–(18.91) comprise three equations in three unknowns. Substituting into Eqs. (18.89) and (18.91) for $[CdCl^+]$ using Eq. (18.90), we obtain

$$[Cd^{2+}] + 10^{-2} - [Cl^-] = 5.0 \times 10^{-3} M \tag{18.92}$$

and

$$[Cd^{2+}] = \frac{10^{-2} - [Cl^-]}{34.7[Cl^-]}. \tag{18.93}$$

Now, we substitute Eq. (18.93) into Eq. (18.92) yielding

$$34.7[Cl^-]^2 + 0.826[Cl^-] - 10^{-2} = 0 \tag{18.94}$$

which by the quadratic equation yields $[Cl^-] = 8.83 \times 10^{-3}$ *M*, or $pCl = 2.05$. This value of pCl falls in the range that we expected. Thus, by Figure 18.6, Cd^{2+} and $CdCl^+$ are indeed the major Cd species. The equilibrium value of $[Cl^-]$ can be combined with (18.53)–(18.56) to compute the concentrations of all of the species. The results that can be obtained satisfy both the MBE for Cd_T and the MBE for Cl_T to a good approximation.

Solution without Assumptions. When Eq. (18.80) is used to solve this $Cd_T = 0.005$ *M* and $Cl_T = 0.01$ *M* problem, the exact solution that is obtained yields $[Cl^-] = 8.70 \times 10^{-3} M$ ($pCl = 2.06$). This is very close to the result obtained above using approximations. The value of $\bar{n}$ is 0.248.

18.5.2.4 1.0×10^{-5} *F* $CdCl_2$.

Solution with Assumptions. At equilibrium in this solution, we know that $[Cl^-]$ will be less 2×10^{-5} *M*, that is, pCl will be greater than 4.7. Consideration of

Figure 18.6 for pCl > 4.7 indicates that Cd^{2+} is by far the most important species in this region. That is, we conclude that

$$[Cd^{2+}] \simeq Cd_T = 1.0 \times 10^{-5} M \quad (18.95)$$

$$[Cl^-] \simeq Cl_T = 2.0 \times 10^{-5} M \quad (18.96)$$

and therefore

$$[CdCl^+] = (34.7)(1.0 \times 10^{-5})(2.0 \times 10^{-5}) = 6.94 \times 10^{-9} M \quad (18.97)$$

$$[CdCl_2^o] = (4.57)(6.94 \times 10^{-9})(2.0 \times 10^{-5}) = 6.34 \times 10^{-13} M \quad (18.98)$$

$$[CdCl_3^-] = (1.26)(6.34 \times 10^{-13})(2.0 \times 10^{-5}) = 1.60 \times 10^{-17} M \quad (18.99)$$

$$[CdCl_4^{2-}] = (0.20)(1.60 \times 10^{-17})(2.0 \times 10^{-5}) = 6.39 \times 10^{-23} M \quad (18.100)$$

$$\bar{n} \simeq \alpha_1 = 6.94 \times 10^{-4}. \quad (18.101)$$

The concentration of $[CdCl_4^{2-}]$ corresponds to

$$6.39 \times 10^{-23} M \times \frac{6.02 \times 10^{23} \text{ ions}}{\text{mol}} = \sim 34 \text{ ions/liter}. \quad (18.102)$$

This is obviously very low.

Solution without Assumptions. When Eq. (18.80) is used to solve this problem exactly, the result that is obtained is that $[Cl^-] = 2.0 \times 10^{-5}$ M, and $[Cd^{2+}] = 1.0 \times 10^{-5}$ M, that is, exactly the same result that was obtained above. Other complexation problems are presented in Pankow (1992).

18.6 CHELATES

18.6.1 Types of Chelates

Multidentate ligands are often called chelates. The word "chelate" originates from the Greek word "chele" ($\chi\epsilon\lambda\epsilon$) which means "crab's claw." This etymology is based on the fact that a crab's claw may be thought of as being bidentate. Two very well known, man-made chelates are **nitrilotriacetic acid** (NTA) and **ethylenediaminetetraacetic acid** (EDTA). The structures of the protonated and fully deprotonated forms of these two compounds are given in Figures 18.9*a,b* and 18.10*a,b*, respectively. Although both oxygens on the protonated carboxylic acid groups (i.e., on the –COOH groups) of both NTA and EDTA have unshared pairs of electrons that can be used for binding to a metal ion, *carboxylate* groups (i.e., $-COO^-$ groups) are significantly more effective as ligands. NTA has a total of four possible ligand sites: three are formed when the three carboxylic acid groups are deprotonated, and one is on the nitrogen. The fully deprotonated NTA species is denoted NTA^{3-}. EDTA has a total of six possible ligand sites: four are formed when the four carboxylic acid groups are deprotonated, and two are on the nitrogens. The fully deprotonated EDTA species is denoted $EDTA^{4-}$.

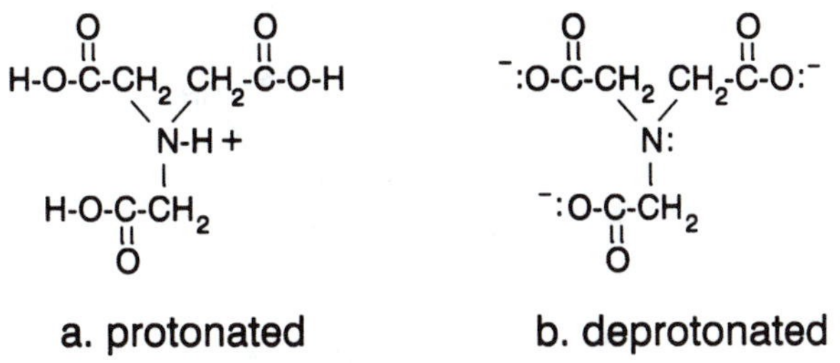

```
     O            O
     ||           ||
H-O-C-CH2   CH2-C-O-H
        \   /
         N-H+
         |
H-O-C-CH2
     ||
     O
```

```
        O            O
        ||           ||
 -:O-C-CH2   CH2-C-O:-
           \   /
             N:
             |
 -:O-C-CH2
        ||
        O
```

a. protonated   b. deprotonated

Figure 18.9 NTA in the fully protonated (*a*) and deprotonated (*b*) forms. Each of the electron pairs used for binding to a metal ion is denoted. The deprotonated form has four binding sites.

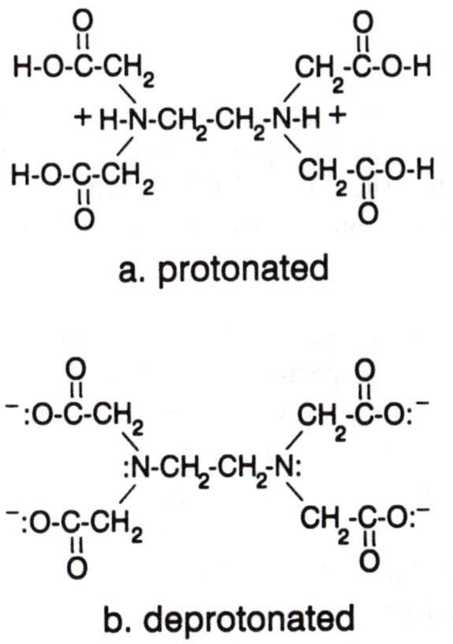

```
      O                    O
      ||                   ||
H-O-C-CH2              CH2-C-O-H
          \           /
       + H-N-CH2-CH2-N-H +
          /           \
H-O-C-CH2              CH2-C-O-H
      ||                   ||
      O                    O
```

a. protonated

```
      O                    O
      ||                   ||
-:O-C-CH2              CH2-C-O:-
          \           /
          :N-CH2-CH2-N:
          /           \
-:O-C-CH2              CH2-C-O:-
      ||                   ||
      O                    O
```

b. deprotonated

Figure 18.10 EDTA in the fully protonated (*a*) and deprotonated (*b*) forms. Each of the electron pairs used for binding to a metal ion is denoted. The deprotonated form has six binding sites.

$EDTA^{4-}$ is a particularly powerful complexing species because its ligand sites can occupy all *six* of the coordination locations of a typical metal ion. In addition, EDTA's four identical "arms" and the $-CH_2-CH_2-$ segment between the two nitrogens are the precise length to allow the ligand sites to orient themselves and thereby interact in an optimum manner with the coordination sites of most metal ions. As a result, K values with $EDTA^{4-}$ are usually quite large. For Mg^{2+} and Ca^{2+}, at 20°C and 1 atm pressure, we have $K = 10^{8.8}$ and $10^{10.7}$, respectively. (K_n values for $EDTA^{4-}$ with n greater than 1 are not relevant since there is only room for one $EDTA^{4-}$ on any given metal ion.) The three-dimensional orientation which the $EDTA^{4-}$ species takes around a metal center M is illustrated in Figure 18.11.

With four binding sites, NTA^{3-} is also characterized by large formation constants; these constants are nevertheless generally noticeably smaller than those for $EDTA^{4-}$. For

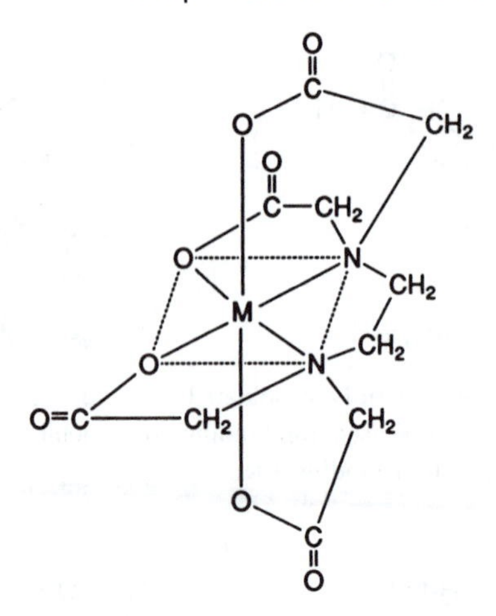

Figure 18.11 Geometric orientation of the $EDTA^{4-}$ chelate around a metal center. The dashed lines do not represent bond interactions, but rather indicate that four of the binding sites of the $EDTA^{4-}$ species lie in a plane the corners of which yield a square. (Adapted with permission from Butler, 1964).

example, at 20°C/1 atm pressure, binding of NTA^{4-} with Mg^{2+} and Ca^{2+} is characterized by K values of $10^{6.5}$ and $10^{7.6}$, respectively (Smith and Martell, 1989).

Although neither NTA or EDTA are formed naturally in the environment, both are of interest within the context of this text. Specifically, NTA has been placed in a variety of soap products as a replacement for the eutrophication-causing phosphates. NTA, like phosphates, binds naturally-occurring Ca^{2+} and Mg^{2+}, the two species that are responsible for making water "hard." In the absence of a chemicals like NTA and phosphates, the anionic end groups on soaps and surfactants will tend to coordinate to the metal ions, and then precipitate as soap scum. When that occurs, the ability of such soaps and surfactants to form micelles and emulsify grease and oils will obviously be greatly reduced. In some countries, NTA-containing detergents have been used quite extensively, and so NTA may now be readily found in the natural waters that received the corresponding muncipal wastewater effluents.

EDTA is of interest to many natural water chemists because its strong complexing powers allow it to be used in experiments when it is desired to control metal ion activities at very low levels. This has proven especially useful to researchers interested in investigating the biological effects of metal ions that are toxic at extremely low levels (see also Section 18.7).

A wide variety of naturally-found chemical species also act as bidentate and tridentate chelates. Some simple examples include salicylic acid and oxalic acid. More complex examples are provided by the polymeric "humic" and "fulvic" acids which contain multiple ligand groups that are oriented sufficiently closely that they can simultaneously bind to a single metal ion. The structures of the neutral and deprotonated forms of salicylic and oxalic acid are given in Figure 18.12; humic and fulvic acid have no single molecular structure.

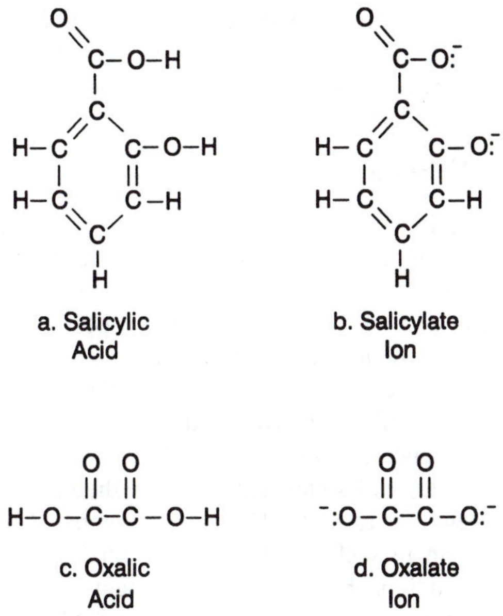

a. Salicylic Acid

b. Salicylate Ion

c. Oxalic Acid

d. Oxalate Ion

Figure 18.12 Salicylic acid and oxalic acid in their neutral and fully deprotonated forms. Each of the electron pairs used for binding to a metal ion is denoted. In their deprotonated forms, both compounds are bidentate.

18.6.2 The Special Stability of Chelates

A complex between a single, multidentate ligand and a metal ion has a special degree of stability as compared to a complex between the corresponding number of unidentate ligands and the same metal ion. Consider for example the two complexation reactions

$$\mathrm{L-L+M=M}\begin{array}{l}/\mathrm{L}\\ \backslash \mathrm{L}\end{array} \tag{18.103}$$

$$\mathrm{2L+M=ML_2}. \tag{18.104}$$

In Eq. (18.103), the bidentate chelate L–L has two L groups with which to bind the metal ion. In Eq. (18.104), two of the same type of ligand group (though this time not bound together) react to form the species ML_2.

There are two ways to understand why the chelate formed by Eq. (18.103) is relatively more stable than the complex ML_2. First, consider that in Eq. (18.103), *two* species combine to form one species, and in Eq. (18.104) *three* species combine to form one species. Thus, while the ΔS^o values for both reactions will be negative, the ΔS^o for reaction (18.103) will be less negative than that for reaction (18.104). Thus, all other factors being equal (e.g., the strength of each L-metal bond), the former will be more stable.

The thermodynamic argument outlined above may be augmented by looking at the situation from a kinetic point of view. Specifically, we note that if one ligand "arm" of a chelate complex breaks loose, the other attachment point(s) will serve to hold the majority of the chelate molecule in the immediate vicinity of the metal ion, thereby facilitating reattachment of the dissociated binding site. For ML_2 on the other hand, if one of the L units dissociates, there will be a significant chance that it will diffuse into solution thereby leading to a reduction in the concentration of ML_2.

18.7 METAL ION BUFFERS

In the preceding sections, we discussed some of the ways in which the complexation of metal ions by ligands is similar to the attachment of a conjugate base species to a proton. Another very important similarity between metal/ligand chemistry and acid/base chemistry is manifested in the fact that it is possible to prepare solutions that are *buffered* with respect to the metal ion concentration.

For a conventional HA/A^- buffer that is well buffered somewhere around the pK of the acid HA, both [HA] and $[A^-]$ must be much larger than $[H^+]$. When this condition is satisfied, neither the addition nor removal of amounts of H^+ as large or even larger than $[H^+]$ will significantly change the pH. We therefore expect that a *metal ion buffer* system that regulates [M] may be prepared by including both M and the ligand L in a system in such a manner that both [ML] and [L] greatly exceed [M]. Thus, neither the addition nor removal of amounts of M as large or even much larger than [M] will significantly change pM ($= -\log\{M\}$). Neglecting activity corrections, we have

$$pM = \log K_1 + \log \frac{[L]}{[ML]}. \tag{18.105}$$

Changing pM can thus only be accomplished if the ratio [L]/[ML] can be changed. If both [L] and [ML] are large relative to [M], then even if an amount of M many times greater than [M] was added or removed from the solution, pM would not change significantly. In this section, we will allow L to represent a complexing agent with any number of ligand sites.

As an illustration of the above principles, we consider the case where a system is initially 1.0×10^{-2} *F* in Mn^{2+}, and 2.0×10^{-2} *F* in total EDTA ($EDTA_T$). Let us take p_cH as being set at 8.0. At 25°C/1 atm, and an ionic strength of 0.1 *M*, we have

$$Mn^{2+} + EDTA^{4-} = MnEDTA^{2-} \qquad {}^cK = 10^{13.9} \tag{18.106}$$

$$H_6EDTA^{2+} = H^+ + H_5EDTA^+ \qquad {}^cK_1 = 10^{0.0} \tag{18.107}$$

$$H_5EDTA^+ = H^+ + H_4EDTA \qquad {}^cK_2 = 10^{-1.5} \tag{18.108}$$

$$H_4EDTA = H^+ + H_3EDTA^- \qquad {}^cK_3 = 10^{-2.0} \tag{18.109}$$

$$H_3EDTA^- = H^+ + H_2EDTA^{2-} \qquad {}^cK_4 = 10^{-2.66} \tag{18.110}$$

$$H_2EDTA^{2-} = H^+ + HEDTA^{3-} \qquad {}^cK_5 = 10^{-6.16} \tag{18.111}$$

$$HEDTA^{3-} = H^+ + EDTA^{4-} \qquad {}^cK_6 = 10^{-10.24} \tag{18.112}$$

The first four acid dissociations refer to the four carboxylic acids. The last two are for the two ammonium protons.

Following principles outlined in previous chapters, it is possible to derive that the fraction of $EDTA^{4-}$ will be given by

$$\alpha_6 = \frac{{}^cK_1\,{}^cK_2\,{}^cK_3\,{}^cK_4\,{}^cK_5\,{}^cK_6}{[H^+]^6 + [H^+]^5\,{}^cK_1 + [H^+]^4\,{}^cK_1\,{}^cK_2 + [H^+]^3\,{}^cK_1\,{}^cK_2\,{}^cK_3 + [H^+]^2\,{}^cK_1\,{}^cK_2\,{}^cK_3\,{}^cK_4 + [H^+]\,{}^cK_1\,{}^cK_2\,{}^cK_3\,{}^cK_4\,{}^cK_5 + {}^cK_1\,{}^cK_2\,{}^cK_3\,{}^cK_4\,{}^cK_5\,{}^cK_6}. \quad (18.113)$$

At p_cH 8 and $I = 0.1\ M$, we have $\alpha_6 = 5.6 \times 10^{-3}$.

Since the formation constant for $MnEDTA^{2-}$ is so large, and since there is more EDTA than metal ion, a reasonable assumption would be that nearly all of the Mn is bound in the $MnEDTA^{2-}$ complex, that is,

$$[MnEDTA^{2-}] \simeq 10^{-2}M. \quad (18.114)$$

The amount of *uncomplexed* $EDTA_T$ would thus be $(2.0 \times 10^{-2}) - (1.0 \times 10^{-2}) = 1.0 \times 10^{-2}\ M$. Thus, since

$$\frac{[MnEDTA^{2-}]}{[Mn^{2+}][EDTA^{4-}]} = 10^{13.9} \quad (18.115)$$

we have

$$\frac{[MnEDTA^{2-}]}{[Mn^{2+}]\alpha_6(1.0 \times 10^{-2})} = 10^{13.9} \quad (18.116)$$

or

$$\frac{[MnEDTA^{2-}]}{[Mn^{2+}]} = 10^{9.62}. \quad (18.117)$$

Thus, the complex formation is indeed nearly complete, and

$$[Mn^{2+}] = 2.38 \times 10^{-12}M. \quad (18.118)$$

If one wanted to regulate the level of Mn^{2+} at this type of level, then even if a source of Mn^{2+} equivalent to many times $10^{-12}\ M$ was acting on the system, the relatively large amount of residual unbound EDTA would be capable of binding that Mn^{2+}, thereby maintaining: 1) $[MnEDTA^{2-}]$ near $10^{-2}\ M$; 2) the $[MnEDTA^{2-}]/[EDTA^{4-}]$ ratio essentially unchanged; and *therefore* 3) $[Mn^{2+}]$ very near $2.4 \times 10^{-12}\ M$. Conversely, if some sink for Mn^{2+} equivalent to many times $10^{-12}\ M$ was acting on the system, the relatively large amount of $MnEDTA^{2-}$ present would release an amount of Mn^{2+} very nearly equal to the amount of Mn^{2+} lost to the sink. Again, however, this would occur without significantly changing: 1) $[MnEDTA^{2-}]$; 2) $[MnEDTA^{2-}]/[EDTA^{4-}]$; or 3) $[Mn^{2+}]$. Thus, we conclude that this system is very well buffered with respect to maintaining $[Mn^{2+}]$ constant at the level given above. We state again that this is made possible by virtue of the fact that both the complexed form of the metal ion and the residual uncomplexed EDTA are present at a concentrations that are much higher than that of the free metal ion.

With an acid/base buffer, one can select the value of pH at which the buffer regulates the system by selecting the $[HA]/[A^-]$ ratio and the identity of the acid HA (i.e., the value of pK) to be used. With a metal ion buffer, the situation is very analogous, with both the [complex]/[*free* ligand or chelate] ratio and the formation constant for the complex determining the value of pM at which the metal ion buffer will function. In the case of complexation by a compound like EDTA, the ability to adjust the two concentrations [complex] and [uncomplexed chelate] as well as the value of α_6 gives considerable latitude in adjusting the activity at which the metal ion is buffered.

A buffer intensity function for a metal ligand system can be defined and a corresponding expression derived. The logical definition for the buffer intensity of a system for a given metal ion is

$$\beta_M = \left[\frac{dM_T}{dpM}\right]_{pH,L_T} \tag{18.119}$$

Equation (18.119) provides a quantitative definition for how well a solution (at a given pH and a given total concentration of complexing species, L_T) is capable of resisting changes in pM when the total amount of metal in solution is changed. Note that L_T here may refer to any type of complexing species, including a simple monodentate ligand as well as a multidentate chelate.

18.8 EFFECT OF A COMPLEXING LIGAND ON THE DISSOLUTION OF A SOLID

By virtue of the fact that it can affect the speciation of a metal ion in solution, the presence of a complexing ligand can alter how much total metal dissolves when there is equilibrium with a solid phase. We have already considered some of the implications of this issue inasmuch as the proper determination of the solubility of metal hydroxides and oxides in Chapter 11 required a consideration of the large role played by hydroxide *complexes*.

In general, the consideration of the solubility of a metal solid $MB_{z(s)}$ in the presence of a complexing ligand C^- might be formulated as:

$$MB_{z(s)} = M^{z+} + zB^- \tag{18.120}$$

$$M^{z+} + B^- = MB^{(z-1)+} \tag{18.121}$$

$$M^{z+} + C^- = MC^{(z-1)+} \tag{18.122}$$

$$HB = H^+ + B^- \tag{18.123}$$

$$HC = H^+ + C^- \tag{18.124}$$

where we allow that the dissolved metal ion has charge z^+.

In this system, for simplicity, only two complexes are assumed to be important, one which is formed with the anion of the solid, $MB^{(z-1)+}$, and $MC^{(z-1)+}$, which is formed with C^-, a supplementary complexing ligand. The equations and unknowns in this system are:

9 Unknowns	*9 Equations*	
$[M^{z+}]$	K_{s0}	
$[B^-]$	K_a for HB	
$[MB^{(z-1)+}]$	K_a for HC	
$[MC^{(z-1)+}]$	K_w	
$[C^-]$	ENE	Only 3 of these 4 equations are independent: take your pick.
[HB]	MBE on M	
[HC]	MBE on B	
$[OH^-]$	MBE on C	
$[H^+]$	K for $MB^{(z-1)+}$	
	K for $MC^{(z-1)+}$	

The method used to solve this problem is similar to that used in acid/base and solubility problems considered earlier: you seek to express the problem as one equation in one unknown, then solve. A trial and error solution will almost certainly be required in cases like this one even if some simplifying assumptions can be made.

If higher complexes with B and C are possible, then the consideration of the concentrations of these additional species would be accompanied by additional complexation K values for them, and the problem could still be solved, although it might be considerably more complicated. In such cases, and in other complicated problems, even when a trial and error solution is sought, it is sometimes even difficult to set up one equation in one unknown. What one can do then is set up two complicated equations in two unkwnowns (say x and y), and solve them simultaneously for the value of the pair that satisfies both equations. Mechanistically, this is done by assuming a value for x, then solving one of the equations (say, equation 1) by trial and error for y. That value of y together with the initial guess for x will probably not satisfy equation 2. Therefore, equation 2 is solved by trial and error for a new value of x using the most recent value of y. The process is iterated until the solution converges on a constant pair of x and y values. In the case of very complicated problems, then one might wish to turn to a large-scale computer program such as MINEQL (Westall et al. 1976).

18.9 REFERENCES

BUTLER, J. N. 1964. *Ionic Equilibrium, A Mathematical Approach.* Reading: Addison-Wesley.

PANKOW, J. F. 1992. *Aquatic Chemistry Problems.* Portland: Titan Press-OR, P.O. Box 91399, Portland, Oregon 97291-1399.

SMITH, R. M., and A. E. MARTELL. 1989. *Critical Stability Constants*, Volume 6: Second Supplement. New York: Plenum Press.

STUMM, W., and J. J. MORGAN. 1981. *Aquatic Chemistry.* New York: Wiley-Interscience.

WESTALL, J. C., J. L. ZACHARY, and F. M. M. MOREL. 1976. *MINEQL, A Computer Program for the Calculation of Chemical Equilibrium Composition of Aqueous Systems*, Technical Note - 18, Cambridge: Parsons Laboratory, M.I.T. Press.

PART V

Redox Chemistry

19

Redox Reactions, pe, and E_H

19.1 INTRODUCTION

19.1.1 Half-Cell Reactions

A "redox" reaction is a reaction involving electrons. As may already be obvious, the term redox is a shortened form of "reduction/oxidation." A generalized redox "half-reaction" may be represented as

$$\underset{\xleftarrow[]{\text{oxidation}}}{\overset{\xrightarrow[\text{reduction}]{}}{OX_1 + ne^- = RED_1}} \qquad \text{half-reaction 1} \tag{19.1}$$

where OX_1 represents some oxidized species, RED_1 represents some reduced species, and e^- represents an electron. We can see then that the reduction of a species involves the acceptance of one or more electrons by that species, and the oxidation of a species involves the loss of one or more electrons.

The origin of the word "reduction" can be traced to the simple fact that when chemical species accept electrons, which are of course negatively charged, the fundamental charge on some atomic unit in the species is reduced. The origin of the word "oxidation" can be traced to the fact that in the early days of chemistry, all known oxidation reactions involved *oxygen* as the oxidant. Examples of such reactions include the conversion of elemental iron to rust, elemental sulfur to sulfate, and elemental hydrogen to water. Half reactions can be written either as reduction half reactions as in Eq. (19.1), or as oxidation half reactions as in Eq. (19.2).

Equation (19.1) is a *half*-reaction since it can only tell half of the story for a full redox reaction. That is, in an aqueous solution, since there is never any significant concentration of free electrons available for consumption by a reaction such as (19.1), such a reaction can never occur alone to any significant extent; it must be accompanied by the simultaneous occurrence of another half-reaction proceeding in a manner that produces electrons, for example,

$$RED_2 = OX_2 + ne^- \qquad \text{half-reaction 2} \tag{19.2}$$

When we sum Eqs. (19.1) and (19.2), the result is

$$OX_1 + RED_2 = RED_1 + OX_2 \tag{19.3}$$

We do not see the explicit presence of electrons in the reaction given in Eq. (19.3). There are nevertheless, electrons being exchanged between OX_1 and RED_2 to produce RED_1 and OX_2, respectively; the electrons balance out. In the above reaction, OX_1 is called the oxidant because it is oxidizing RED_2 to OX_2, and RED_2 is called the reductant because it is reducing OX_1 to RED_1. Of course when the reaction occurs in the backwards direction, then OX_2 oxidizes RED_1 to OX_1, and RED_1 reduces OX_2 to RED_2.

The fact that two redox half-reactions are needed to obtain a complete redox reaction in water is analogous to the fact that two acid/base reactions are needed to obtain overall acid/base equilibrium in water. For example, we recall from Chapter 3 that when the *acid* HA dissociates to form a truly free, unhydrated H^+ ion and A^-, the solvating properties of water make water itself a conjugate *base* that then hydrates the proton to form species of the formula $H(H_2O)_n^+$. The net conclusion is that both truly-free protons and truly-free electrons will always exhibit very low concentrations in water. However, unlike hydrated protons which can reach relatively large concentrations in water (e.g., 10^{-2} *M* at $p_cH = 2$), even the total concentration of hydrated electrons will always be very low in water.

A specific example of a redox half-reaction of interest in many natural water systems is the reduction of Fe^{3+} to Fe^{2+}:

$$Fe^{3+} + e^- = Fe^{2+} \qquad K = \frac{\{Fe^{2+}\}}{\{Fe^{3+}\}\{e^-\}}. \tag{19.4}$$

The common convention that is used in tabulating redox *K* values is to write the half reactions as *reductions* as in Eqs. (19.1) and (19.4). It is apparent that this approach follows the *stability*, that is, *formation* constant convention as was chosen for expressing complexation reactions. The redox *K* value for Eq. (19.4) may therefore be thought of as the stability constant for the complexation of Fe^{3+} by one electron.

19.1.2 The Concept of pe

Although the *K* values for acid/base reactions are tabulated as instability constants, there is otherwise much similarity between redox half reactions and the generalized acid/base reaction

$$HA = H^+ + A^- \qquad K = \{H^+\}\{A^-\}/\{HA\}. \tag{19.5}$$

For example, when we discussed acid/base equilibria in aqueous solutions, we learned that the quantity

$$pH \equiv -\log\{H^+\} \tag{19.6}$$

is very useful for describing the equilibrium position for all acid/base pairs in a system. In this manner, pH may be thought of as a "master variable" for the system. For redox equilibria, then, it should come as no surprise that pe is also very useful as a master variable, where

$$pe \equiv -\log\{e^-\}. \tag{19.7}$$

The pe will be seen to describe the equilibrium position for all redox pairs (i.e., all OX/RED pairs) in a given system.

Besides just allowing us to define pe, treating the electron like any other reactant or product species has another advantage. Indeed, it allows redox reactions like Eq. (19.4) and their corresponding equilibrium constants to be combined *directly* with other reactions. In this manner, we can determine the equilibrium constants for the combined reactions of interest. While the E_H approach (to be described below) also allows such combinations, the process is much less direct.

19.1.3 Thermodynamic Soundness of the pe Concept

A few geochemists and natural water chemists object to the use of pe since they insist that electrons do not exist in aqueous solution at the types of *concentrations* that are represented by pe values that can be calculated by equations like Eq. (19.4). It is important to state clearly and forcefully that this objection is *unfounded.* Firstly, one *cannot* calculate a concentration for e^- using Eq. (19.4), *only* an activity. In particular, by Eq. (19.7),

$$pe = -\log[e^-]\gamma_{e^-} \tag{19.8}$$

where $[e^-]$ is the sum of the concentrations for all hydrated electrons, and γ_{e^-} is the electron activity coefficient. Thus, we do not even need to concern ourselves with what $[e^-]$ really is; all we need to know is what the *activity* of the electron is, namely the value of the product $[e^-]\gamma_{e^-}$. For Eq. (19.4), then, we see that the $\{Fe^{2+}\}/\{Fe^{3+}\}$ ratio together with the value of K that is adopted by convention for Eq. (19.4) will set the value of the product $[e^-]\gamma_{e^-}$. Thus, whether or not $[e^-]$ can ever actually approach 10^{-pe} is not even an appropriate question to ask: it is the responsibility of the product $[e^-]\gamma_{e^-}$ to equal 10^{-pe}, not the responsibility of $[e^-]$ alone.[1]

We recall from Chapter 3 that while pH is used to refer to the activity of *protons* in solution, it actually quantifies a composite activity involving all hydrated protons. As noted above, the concentration of the true, totally unhydrated proton will in fact always

[1] It may be pointed out that γ_{e^-} will not equal 1.0 in infinitely dilute aqueous solutions. The state for which γ_{e^-} equals 1.0 is, in fact, a hypothetical activity coefficient reference state. Chapter 2 discussed the fact that hypothetical standard states for chemical potentials are completely sound from a thermodynamic point of view. The same applies for hypothetical reference states for activity coefficients, but it is beyond the scope of this text to discuss this interesting topic any further.

be quite small. Despite this, we showed in Chapter 3 that $[H^+]$, $\{H^+\}$, and pH as we use them are all: 1) relatable to both the concentration and the activity of the true, unhydrated H^+ species; and 2) very useful and understandable quantities. The situation with pe is similar to that with pH. It is not necessary for there to be a significant number of electrons in solution at low pe for pe to be a meaningful thermodynamic quantity; pe is meaningful even if the value of $[e^-]$ *never* approaches 10^{-pe}. Although a given electron may not have a long lifetime in aqueous solution, and it might not, on average, get to stray very far from the volume surrounding a given reductant/oxidant pair, there will *always* be some electrons in all aqueous solutions. Indeed, electrons are always being exchanged between redox pairs just as protons are always being exchanged between acid/base pairs. In the case of pure water at equilibrium, the participating redox species are H_2O, O_2, and H_2. (Although we have not yet mentioned this fact, even in pure water there will be non-zero concentrations of O_2 and H_2 formed by the dissociation of H_2O.)

In dealing with redox reactions in aqueous solutions, since the total concentration of hydrated electrons (i.e., $[e^-]$) will always be negligible relative to the other charge-bearing species in solution, we never need to write *electron* balance equations for redox systems like we write *proton* balance equations for acid/base systems. For the same reason, we do not need to include $[e^-]$ in the ENE. Thus, as pointed out above, we never need to know what $[e^-]$ actually is, and a knowledge of the product $[e^-]\gamma_{e^-}$ is adequate.[2]

19.2 RELATIONSHIP BETWEEN pe AND E_H

19.2.1 The E_H Approach

Consider the overall electrochemical cell depicted in Figure 19.1. Let us assume for the moment that when mixed together, the chemicals in the two half cells react in a forward direction according to Eq. (19.3). Under these conditions, the electrons would be passed directly from RED_2 to OX_1. In the electrochemical cell, however, the two half cells are separated. This means that direct reaction cannot occur. Now, if there is to be an exchange of electrons, then that exchange will have to take place *through the wire* that connects the two half cells.

The meter in Figure 19.1 gives a reading of E in units of V (volts). One half reaction occurs at each of the electrodes. Arrows indicate the direction of the flow of electrons. The meter that is used to measure the potential difference between the two electrodes is designed to indicate a positive voltage (i.e., $E > 0$) when electrons flow

[2]We note that while the total concentration of solvated electrons will be low in water, the same is not the always case for other solvents. For example, in liquid ammonia (NH_3), alkali metals such as sodium will actually dissolve somewhat according to

$$Na_{(s)} = Na^+ + e^- \tag{19.9}$$

Both Na^+ and e^- are both strongly solvated by ammonia molecules.

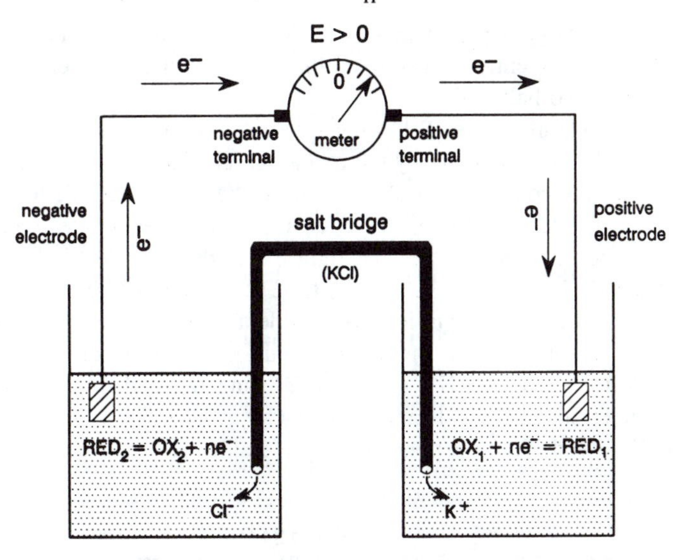

Figure 19.1. General electrochemical cell. By convention, the negative terminal of the meter is on the left, and the positive terminal is on the right. In this case, oxidation is taking place at the left electrode, and reduction is taking place at the right electrode. Arrows indicate the direction of the flow of electrons. The volt meter measures a positive voltage. The symbols OX_1, RED_1, OX_2, and RED_2 may each represent just one species, or more than one species. For a reaction in the right cell of $Ag^+ + e^- = Ag_{(s)}$, OX_1 would be Ag^+ and $\{OX_1\} = \{Ag^+\}$. For a reaction in the right cell of $O_{2(g)} + 4H^+ + 4e^- = 2H_2O$, OX_1 would be $O_{2(g)} + 4H^+$, and $\{OX_1\} = p_{O_2}\{H^+\}^4$.

into the negative (black) terminal of the meter (on the left), and out of the positive (red) terminal (on the right). Thus, when in fact the reaction at the right-hand electrode is a reduction and the reaction at the left-hand electrode is an oxidation, the meter will read a positive value. When the electrons move in a direction opposite to that indicated in Figure 19.1, then the reaction at the left electrode will be a reduction, and the reaction at the right electrode will be an oxidation. Any cell that gives a non-zero E value can be used as a battery.

The salt bridge tube between the two cells contains a physically stabilized (e.g., with agar), concentrated solution of a salt such as KCl. The salt bridge allows ionic species to diffuse into and out of the two half cells, and permits each cell to remain electrically neutral. Without the salt bridge, because of charge balance requirements, the flow of electrons out of the left half cell would lead to a net buildup of positive ions in that cell, and the introduction of electrons into the right half cell would lead to a net buildup of negative ions in that cell. For example, if RED_2 is neutral, then OX_2 will be positively charged, and if OX_1 is neutral, then RED_1 will be negatively charged. In this case, the buildup of positive charge in the left half cell is prevented by diffusion of OX_2 species into the salt bridge and by diffusion of Cl^- ions out of the salt bridge. The buildup of negative charge in the right half cell is prevented by diffusion of RED_1

species into the salt bridge and diffusion of K^+ ions out of the salt bridge. The solution in the salt bridge is sufficiently concentrated that there is little or no exchange or redox species between the two half cells.

The left electrode in Figure 19.1 is negative relative to the right electrode because electrons are being left behind there by the oxidation reaction proceeding in the left cell. By the same token, the consumption of electrons at the right electrode tends to make it positive relative to the left electrode. As an example of a specific electrochemical cell, Figure 19.2 describes a system where the right electrode is made of elemental copper and the left electrode is made of platinum (Pt). The Pt does not participate directly in the reaction at the left electrode, but it does provide a conducting metal surface. $H_{2(g)}$ is being bubbled in the solution near the left electrode. As a result, dissolved H_2 is being oxidized at the left electrode to H^+, and Cu^{2+} is being reduced to elemental Cu at the right (copper), electrode thereby forming more $Cu_{(s)}$. The meter reads +0.34 V. The activity of the dissolved H_2 corresponds to $p_{H_2} = 1$ atm.

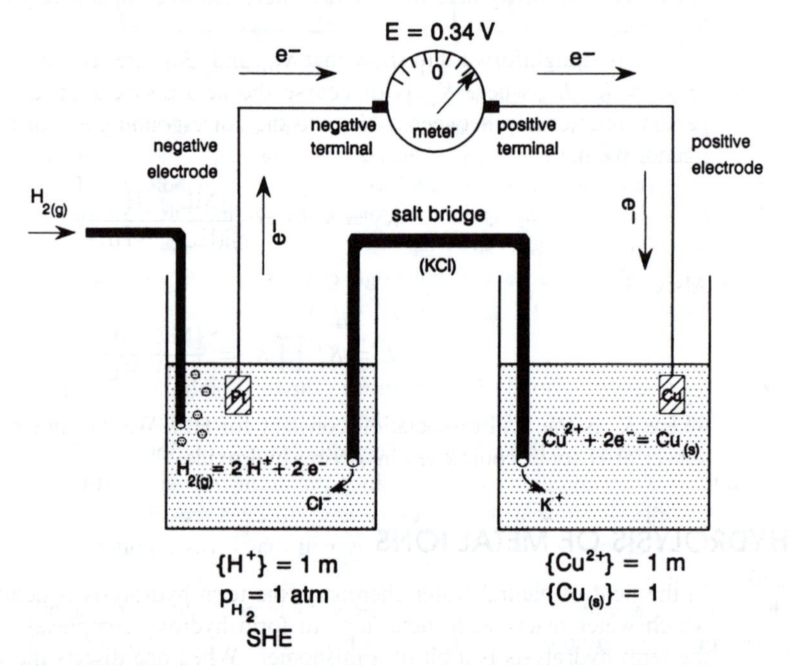

Figure 19.2. Electrochemical cell with a standard hydrogen electrode (SHE) as the left electrode. By convention, the negative terminal of the meter is on the left, and the positive terminal is on the right. Hydrogen gas at 1 atm pressure is being bubbled through the solution near the left electrode (Pt) to keep the solution phase activity of H_2 equivalent to 1 atm. $\{H^+\}$ in the left cell is 1 *m*. The right electrode is made of $Cu_{(s)}$. $\{Cu^{2+}\}$ in the right cell is 1 *m*. At the left electrode, H_2 is being oxidized to 2 H^+. At the right electrode, Cu^{2+} is being reduced to $Cu_{(s)}$. Arrows indicate the direction of the flow of electrons. The volt meter measures a positive voltage of 0.34 volts.

The fundamental thermodynamic equivalence between the ΔG for an *overall* electrochemical reaction and the electrical work done by that reaction is

$$\Delta G = -nFE \qquad (19.10)$$

where n is the number of mols of electrons exchanged in the reaction, F is the "Faraday" constant = 96,444.6 C/mol (C = coulombs of electrical charge), and E is the measured difference in electrical potential between the two electrodes. F is a measure of the magnitude of electrical charge held on one mol (= N_a = Avogadro's Number = 6.023×10^{23}) of elementary charges e_o. The charge on a proton is $+e_o$ and the charge on an electron is $-e_o$. The value of e_o is (96,444,6 C/mol)/N_a = 1.60×10^{-19} C.

Since ΔG is usually expressed in kJ/mol, we can make use of the exact conversion equivalence

$$1\text{C-V} = 1 \times 10^{-3} \text{ kJ}. \qquad (19.11)$$

Substituting 1 C = 1×10^{-3} kJ/V into the definition for F yields

$$F = \frac{96{,}444.6 \text{ C}}{\text{mol}} = \frac{96.4446 \text{ kJ}}{\text{V-mol}} \qquad (19.12)$$

which makes the manner in which the units of Eq. (19.10) work out more apparent. Indeed, by Eq. (19.12), we see that multiplying charge (nF) times volts (E) does end up with units of energy, in this case kJ. Thus, one C-V is a unit of energy just like one electron-volt ("eV") is a unit of energy. In particular, one C-V is the energy released when one coulomb worth of positive charge falls through an electrical potential drop of 1 volt; one eV is the energy released when a single elementary charge e_o falls through an electrical potential drop of 1 volt. Figure 19.3 describes the analogy that exists between gravitational potential energy and electrical potential energy.

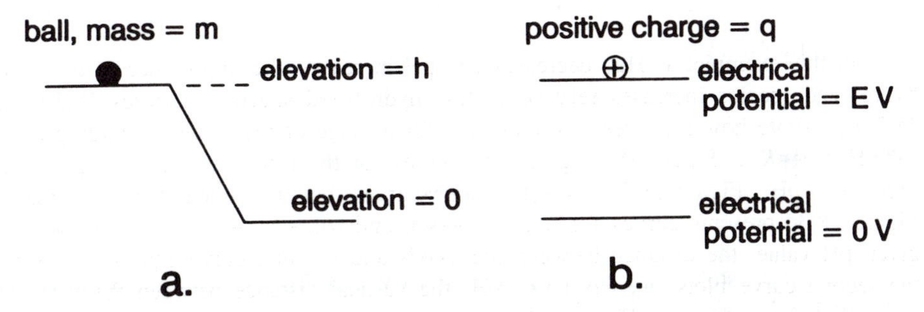

Figure 19.3. (*a*) When a ball of mass m drops through distance h, the gravitational potential energy released is given by the product $m \times gh$ where g is the gravitational acceleration constant. The quantity gh represents the potential energy (PE) per unit mass. (*b*) When a positive electrical charge of magnitude q drops through the electrical potential difference of E V, the electrical potential energy released is given by the product $q \times E$. The quantity E represents the PE per unit charge.

Let the reaction occurring at the left-hand electrode in the cell in Figure 19.1 be

$$aA = cC + ne^- \tag{19.13}$$

and let the reaction at the right-hand electrode be

$$bB + ne^- = dD. \tag{19.14}$$

The overall reaction is then

$$aA + bB = cC + dD. \tag{19.15}$$

The reason for augmenting the OX_1, RED_1, etc. notation in this manner is to emphasize that OX_1, RED_1, etc. may represent any possible type of stoichiometry among the reactants and products (i.e., *a, b, c,* and *d* not necessarily equal 1). In this case, then, we have

$$RED_2 = aA \qquad \{RED_2\} = \{A\}^a \tag{19.16}$$

$$OX_2 = cC \qquad \{OX_2\} = \{C\}^c \tag{19.17}$$

$$RED_1 = dD \qquad \{RED_1\} = \{D\}^d \tag{19.18}$$

$$OX_1 = bB \qquad \{OX_1\} = \{B\}^b. \tag{19.19}$$

As discussed in Chapter 2, when all of the reactant and product species are at unit activity, then

$$\Delta G = \Delta G^o. \tag{19.20}$$

Under the same conditions, for an electrochemical cell, by analogy with Eq. (19.20) we have

$$E = E^o \tag{19.21}$$

Therefore, by Eq. (19.10),

$$\Delta G^o = -nFE^o \tag{19.22}$$

where, as usual,

$$\Delta G^o = \sum_i \nu_i \mu_i^o. \tag{19.23}$$

The values of the stoichiometric coefficients ν_i have sign. They are positive for products, and negative for reactants.

For this reaction, we have

$$\Delta G = c\mu_C + d\mu_D - a\mu_A - b\mu_B \tag{19.24}$$

$$= c\mu_C^o + c\,RT \ln a_C + d\mu_D^o + d\,RT \ln a_D - a\mu_A^o - a\,RT \ln a_A - b\mu_B^o - bRT \ln a_B. \tag{19.25}$$

If we let

$$Q = \frac{\{C\}^c\{D\}^d}{\{A\}^a\{B\}^b} \tag{19.26}$$

then

$$\Delta G = \Delta G^o + RT \ln Q \tag{19.27}$$

In terms of the OX/RED notation, for the cell in Figure 19.1,

$$Q = \frac{\{OX_2\}\{RED_1\}}{\{RED_2\}\{OX_1\}}. \tag{19.28}$$

At equilibrium, $\Delta G = 0$, and at that point,

$$Q = K. \tag{19.29}$$

Therefore, as noted in Chapter 2, by Eq. (19.27),

$$\Delta G^o = -RT \ln K. \tag{19.30}$$

Substitution of Eq. (19.30) into Eq. (19.27) yields

$$\Delta G = -RT \ln \frac{K}{Q}. \tag{19.31}$$

As required, when $Q = 1$, then $\Delta G = \Delta G^o$.

By combining Eqs. (19.10), (19.22), and (19.27), we obtain

$$-nFE = -nFE^o + RT \ln Q \tag{19.32}$$

Equation (19.32) may be rearranged to give

$$E = E^o - \frac{RT}{nF} \ln Q \qquad \text{Nernst Equation} \tag{19.33}$$

which is the famous Nernst Equation. For log base 10, we can write

$$E = E^o - \frac{2.303\ RT}{nF} \log Q \qquad \text{Nernst Equation.} \tag{19.34}$$

At $T = 298$ K (25°C),

$$\frac{2.303\ RT}{nF} = \frac{0.05916\ \text{V}}{n}. \tag{19.35}$$

We recall now that Q represents the activity quotient for *products* over *reactants*. Thus, as Q for the cell in Figure 19.1 becomes larger, E becomes smaller. When E becomes smaller, then the driving force for the reaction decreases. If Q is made large enough, then E can in fact become negative, with the result that the flow of electrons will reverse direction, and begin to flow from right to left across the meter. Conversely, as Q is made smaller, then the driving force for the formation of products increases, and so the electrical potential registered by the meter increases. When $Q = 1$, as would occur when all of the reactants and products are at unit activity, then $E = E^o$. When Q equals the equilibrium constant for reaction (19.3), then the cell is at equilibrium, and electrons flow neither to the right nor to the left: the battery is dead.

Since the reaction leading to the E in Eq. (19.33) occurs in the context of two half reactions, we can break up this E value into *two half cell E values*. The contribution to the overall E from the right electrode may be denoted E_r, and that from the left electrode E_l. Thus, like a couple of mini-batteries acting in series, we have

$$E = E_r + E_l. \tag{19.36}$$

When considered individually, each of the two half cells has a certain tendency to proceed forward as a reduction or as an oxidation. For each half cell, the nature of that tendency will depend on the activities of the OX/RED species in that half cell.

The values of E, E_r, and E_l can be positive or negative. If E_r is positive, then we know that the right half cell in fact "desires" to proceed as a reduction and it is helping to pull electrons from left to right. However, if E_r is negative, the right half cell "desires" to proceed as an oxidation and it is tending to push electrons back into the right electrode. If E_l is positive, then we know that the left half cell "desires" to proceed as an oxidation and it is in fact helping to push electrons from left to right. If E_l is negative, the left half cell "desires" to proceed as an reduction and it is tending to pull electrons out of the left electrode. As an example, if E is positive, E_r is positive, but E_l is negative, then we know that $E_r > |E_l|$, and the "desire" of the reaction in the right half cell to proceed as a reduction is stronger than the "desire" of the left half cell to proceed as a reduction. Thus, the reaction in the right half cell pulls the electrons from left to right.

The words "desires" and "desire" have been put in quotes throughout the previous paragraph because we need to remember that any desire to proceed as a reduction or as an oxidation will be *relative*, that is, relative to some *standard*. As an analogy, if we had ten thieves, each of their desires to steal (take electrons) would depend on which standard person was being robbed (oxidized). If a baby was the standard, then all of the thieves might exhibit large positive desires to rob. If the standard person being robbed was an adult with a stick, then maybe some of the thieves might still have a positive desire to rob, but some might be so intimidated that they would rather *give* that adult some money rather than risk being clobbered (*negative* desire to rob). Note that the inherent personal desires of the thieves to steal do not change as we change our standard. However, our quantitative measures of those desires do change.[3] We will see shortly that our standard for measuring the chemical desire of half reactions to occur as reductions is provided by the **standard hydrogen electrode** (SHE).

For any given overall electrochemical cell, once it has been established which direction the electrons will flow, then as in Figure 19.1 and 19.2, the convention in electrochemistry is that for discussion purposes, the schematic diagram of the cell should be drawn in a manner such that the reduction reaction occurs at the right electrode, and the oxidation reaction occurs at the left electrode. As noted, when the reactions occur in this manner, the meter will give a positive reading for E since electrons will be flowing from left to right across the meter.

[3] As an analogy, note that the physical height of the Empire State Building in New York City does not change when we use different standard reference levels for elevation = 0, e.g., sea level = 0, *or* the sidewalk in front of the building = 0. However, for two different record books using the two different standards, the tabulated measures of the height would be different.

When the chemical species at their respective activities in the right half cell react as $OX_1 \rightarrow RED_1$, that is, occurring in a manner that pulls electrons from left to right across the meter, the contribution of the right half cell to the overall E will be denoted E_R. Given the discussion above, we conclude that $E_r = E_R$. For the species and activities in the left half cell, we let E_L represent the tendency of that half cell to occur as a reduction, that is, according to $OX_2 \rightarrow RED_2$. By analogy, $-E_L$ is a measure of the tendency of $RED_2 \rightarrow OX_2$ to occur. Given the discussion above, we can conclude that $E_l = -E_L$. Thus, by Eq. (19.36) we have

$$E = E_R - E_L \tag{19.37}$$

The reason for switching from Eq. (19.36) to Eq. (19.37) is that prediction of the value of E for many different overall cells will be facilitated when we can calculate the tendency of each of the different half cells to occur *as a reduction*, and relative to the *same standard half cell*, that is, relative to the SHE.

Consider now the case where $RED_2 = H_{2(g)}$, $OX_2 = 2H^+$, and $n = 2$. Then, by Eq. (19.3),

$$OX_1 + H_{2(g)} = RED_1 + 2H^+. \tag{19.38}$$

$H_{2(g)}$ is being oxidized to $2H^+$ (e.g., see the electrochemical cell depicted in Figure 19.2). Making the appropriate substitutions into Eq. (19.28) for Q, Eq. (19.34) becomes

$$E = E^\circ - \frac{2.303\,RT}{2F}\log\frac{\{H^+\}^2\{RED_1\}}{p_{H_2}\{OX_1\}}. \tag{19.39}$$

We now note that just as the *overall* E can be broken up into a standard E° component and a second component that involves the activities of the participating species (see Eq. (19.33)), both E_R and E_L may also be so divided. When the reaction at the left electrode is the oxidation of $H_{2(g)}$ to $2H^+$, we obtain

$$E = \overbrace{\left[E^\circ(OX_1 \rightarrow RED_1) - \frac{2.303\,RT}{2F}\log\frac{\{RED_1\}}{\{OX_1\}}\right]}^{E_R} - \underbrace{\left[E^\circ(2H^+ \rightarrow H_{2(g)}) - \frac{2.303\,RT}{2F}\log\frac{p_{H_2}}{\{H^+\}^2}\right]}_{E_L} \tag{19.40}$$

$E^\circ(OX_1 \rightarrow RED_1)$ is the *standard reduction half cell E value* for the half reaction given by Eq. (19.1). The word "standard" should be taken to mean that all half cell reactants and products are at unit activity. $E^\circ(2H^+ \rightarrow H_{2(g)})$ is the standard reduction half cell E value for the reaction

$$2H^+ + 2e^- = H_{2(g)} \tag{19.41}$$

For $E^\circ(2H^+ \rightarrow H_{2(g)})$, H^+ is at 1 *m*, and $H_{2(g)}$ at 1 atm pressure is bubbling through the solution.

We now see that the overall E° value contains contributions from the two standard reduction half cell values E°_R and E°_L. In particular, we have

$$E^\circ = E^\circ_R - E^\circ_L \tag{19.42}$$

When we are referring to *half-cell reduction* E° values as opposed to an *overall* cell E° value, that reference will be made clear through the use of the adjective "half-cell," or in a moment, through the use of the subscript H. We will not in general bother to include the adjective "reduction" for half-cell E° values; its applicability will now be assumed.

Conceptually, any given half-cell E° value is the contribution to the overall E that an electrode would make if the half reaction was proceeding as a reduction at the right electrode, and with all the reactants and products at unit activity. As discussed above, it is impossible to have any given half reaction occurring alone. Thus, any given half-cell E° value can only be *measured* by *comparing* it to our standard reference half reaction which is also characterized by specific activities for its reactants and products.

When a given half cell E° value is measured, the corresponding chemicals (at unit activity) and electrode are placed into the right half cell, and the standard reference half cell is used as the left half cell. The overall E that is measured is then set equal to the quantity [(half-cell E° of interest) – (half-cell E value of the standard reference half cell)].

The convention is to set $E^\circ = 0$ at 25°C/1 atm for the standard reference half cell comprised of a Pt electrode in a solution containing: a) $\{H^+\}$ at 1.0 *m* (OX); and b) bubbles of $H_{2(g)}$ at 1.0 atm (RED). This is the system that is referred to as the SHE. Just as we are free to set the elevation of sea level equal to zero, we can in fact set the E° value for the SHE equal to zero. However, as with physical height measurements, we will need to remain consistent in what we take as our redox zero.

Half-cell E° values measured relative to the SHE are designated E°_H values. Of course, half-cell E values with activities of the reactants and products that are not at unity may also be measured relative to the SHE. Thus, any half-cell E value measured relative to the SHE is designated as an E_H value, with E°_H referring to the special case where $\{OX\}$ and $\{RED\}$ equal 1.0 (or at least, $\{OX\}/\{RED\} = 1.0$).

For the overall E of a cell, by analogy with Eq. (19.39), we can now write

$$E = E_{H,R} - E_{H,L}. \tag{19.43}$$

Since E_H for the SHE is 0.0, we conclude that when the potential of the reaction given in Eq. (19.1) is measured relative to the SHE, $E_L = E_{H,L} = 0$ and so by Eq. (19.40) we have

$$E = E_H = E^\circ_H(OX_1 \rightarrow RED_1) + \frac{2.303\,RT}{2F}\log\frac{\{OX_1\}}{\{RED_1\}} - 0 \tag{19.44}$$

$$= E^\circ_H(OX_1 \rightarrow RED_1) + \frac{2.303\,RT}{F}\log\frac{\{OX_1\}^{1/2}}{\{RED_1\}^{1/2}}. \tag{19.45}$$

In general, n can equal any rational number (i.e., any number that can be expressed as a ratio of two other numbers). We have used $n = 2$ up to now in these examples simply because it is convenient to do so with the SHE, and so having the interconversion of OX_1 and RED_1 also involve the exchange of two electrons makes the above discussion a bit simpler. This can always be arranged for *any* redox half reaction. For example, to accomplish this for the reaction $NO_3^- + 6H^+ + 5e^- = \frac{1}{2}N_{2(g)} + 3H_2O$, we would take $OX_1 = \frac{2}{5}NO_3^-$, and $RED_1 = \frac{1}{5}N_{2(g)}$. Application of Eq. (19.45) would then yield $\{OX_1\}^{1/2} = \{NO_3^-\}^{1/5}$ and $\{RED_1\}^{1/2} = (p_{N_2})^{1/10}$ which is exactly the same as would have been obtained if we had just used the original reaction with five electrons and taken OX and RED for that reaction to the 1/5 power. Thus, we do not need to go through any intermediate step such as just described for the reduction of NO_3^- to $N_{2(g)}$, and can simply use the value of n for the reaction as written when determining E_H. That is, for any OX_1/RED_1 pair involving the exchange of n electrons where n can be different from 2, we obtain

$$E_H = E_H^o(OX_1 \rightarrow RED_1) + \frac{2.303\ RT}{F} \log \frac{\{OX_1\}^{1/n}}{\{RED_1\}^{1/n}}. \tag{19.46}$$

Equation (19.46) illustrates that the electrochemical potential of a solution (as in a cell) goes up as the $\{OX_1\}/\{RED_1\}$ ratio increases. As this ratio increases, the system becomes increasingly oxidizing (i.e., inclined to consume electrons). This is the case because as $\{OX_1\}/\{RED_1\}$ increases, either the activity of the species that tends to accept electron increases, and/or the activity of the species that tends to release electrons decreases. *Increasing E therefore equates with increasingly oxidizing conditions. By the same token, then, decreasing E equates with increasingly reducing conditions.*

In practical terms, the applicability of Eq. (19.46) may be understood as follows. Let us say that we are interested in knowing about the redox chemistry of a given sample of water. The way which that chemistry could be investigated would be to first obtain a sample of that water in a manner that does not allow *any* redox-active chemicals to either enter or leave the sample. Because of its high concentration in the atmosphere, O_2 is an example of an oxidant species which can be difficult to keep out of samples that are very reducing. In samples that are very reducing, H_2S is an example of a volatile reductant species that might escape from the sample.

Once the sample of interest is obtained, a Pt electrode can be placed in the water to serve as a conducting medium for the exchange of electrons with some OX/RED pair in the sample (e.g., Fe^{3+}/Fe^{2+}). Then, the Pt electrode would be connected to the right (positive) side of a volt meter of the type used in Figures 19.1 and 19.2. The left (negative) side of the meter would be connected to an SHE, a salt bridge would be installed to connect the two cells, and the value of E ($= E_H$) would be measured on the meter. That is, we would have

$$E = E_H = E_{H,sample} - E_{H,SHE} \tag{19.47}$$

$$= E_{H,sample} - 0 \tag{19.48}$$

$$= E_{H,sample}. \tag{19.49}$$

E_H is then seen to be identically equal to $E_{H,sample}$. The value of E_H can be positive or negative, and will depend on the chemical characteristics of the sample.

One problem with the approach outlined in the previous paragraph involves the fact that the Pt electrode *cannot* always be relied upon to provide an accurate measure of the E_H in a sample. This is a complicated and controversial area.

Secondly, even if the Pt electrode could accurately measure the true redox potential in the water, many redox reactions are slow, and so there may not even be full redox equilibrium in the sample of interest. That is to say, while there will be one true value of the electron activity and therefore one true value of E_H in the sample, application of the RHS of Eq. (19.46) using the various $\{OX\}/\{RED\}$ ratios for the sample may all give different results, and none of those results may equal the true E_H. This type of problem is not generally encountered with pH and different acid/base pairs; proton exchange reactions are sufficiently fast that overall pH equilibrium is reached fairly quickly. In the redox case, since one must take special care not to put too much confidence into measured E_H values, it is typical to resort to actually measuring the concentrations of different redox pairs (e.g., O_2/H_2O, SO_4^{2-}/H_2S, etc.) so as to obtain an estimate of the oxidizing or reducing properties of the system. *Equilibrium* redox predictions of E_H and pe for real systems should not, therefore, be overinterpreted, and should be viewed only as providing general guidance for: 1) where any given system is headed; and 2) what types of concentrations are feasible in the system.

A practical problem with the above approach for measuring the E_H of a sample is that the SHE, while convenient from a thermodynamic point of view, is very inconvenient in practical terms. There are several reasons for this, not the least of which is the highly explosive nature of $H_{2(g)}$. As an alternative to the SHE, the **saturated calomel electrode** (SCE) has been developed. This electrode, which is depicted in Figure 19.4*a*, involves the redox half reaction

$$Hg_2Cl_{2(s)} + 2\ e^- = 2\ Hg_{(l)} + 2\ Cl^- \tag{19.50}$$

Equation (19.50) describes the reduction of the solid $Hg_2Cl_{2(s)}$, known as "calomel," to elemental, liquid mercury. At 25°C/1 atm, the half-cell E_H^o value for the reaction in Eq. (19.50) is +0.268 V. The equation for the E_H for this reaction is therefore

$$E_H = 0.268 + \frac{2.303\ RT}{2\ F} \log \frac{\{Hg_2Cl_{2(s)}\}}{\{Hg_{(l)}\}^2\{Cl^-\}^2} \tag{19.51}$$

In the SCE, a Pt wire is surrounded by a paste made up of saturated aqueous solution of KCl, calomel, and liquid mercury. When the activities (mole fraction scale) of the $Hg_{(l)}$ and the calomel are taken to be 1.0, then at 25°C/1 atm,

$$E_{H,SCE} = 0.268 - \frac{0.05916}{2} \log\{Cl^-\}^2. \tag{19.52}$$

At 25°C/1 atm, the activity of Cl^- in a saturated solution of KCl is 2.86 *m*. Therefore

$$E_{H,SCE} = 0.268 - \frac{0.05916}{2} \log(2.86)^2 = 0.241\ \text{V}. \tag{19.53}$$

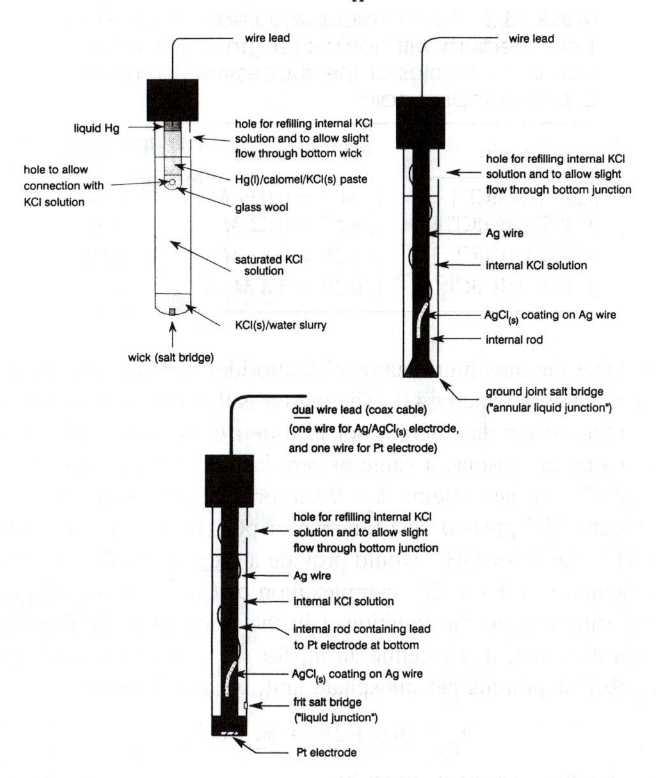

Figure 19.4 (*a*) A saturated calomel electrode (SCE). (*b*) A silver/silver chloride electrode (SSCE). (*c*) Redox combination electrode.

The use of an SCE is very straightforward since it is self-contained; all one has to do is place it in a sample as depicted in Figure 19.5. The built-in KCl salt bridge in the SCE provides the needed electrical connection between the SCE and the sample.

Another very popular reference electrode is based on the redox couple $AgCl_{(s)}$ + $e^- = Ag_{(s)} + Cl^-$. As in the SCE, the activity of Cl^- alone sets the potential of the so-called **silver/silver chloride electrode** (SSCE) (see Figure 19.4*b*).

When the electrical potential of a sample is measured relative to the SCE with the latter hooked to the negative terminal of the volt meter, the potential measured will be

$$E_{SCE} = E_H - E_{H,SCE}. \tag{19.54}$$

Thus, although the E_H of a sample would probably not actually be measured vs. the SHE, it could be obtained indirectly using the SCE and the equation

$$E_H = E_{SCE} + E_{H,SCE}. \tag{19.55}$$

At 25°C/1 atm, Eq. (19.55) becomes

$$E_H = E_{SCE} + 0.241. \tag{19.56}$$

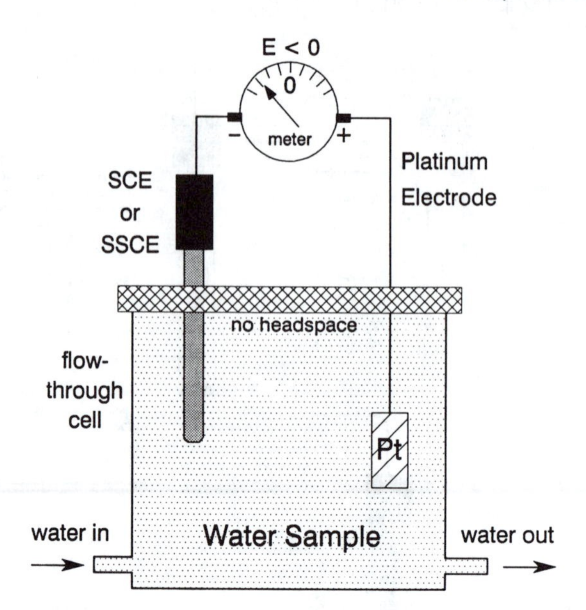

Figure 19.5 Testing the E_H or pe of a water sample using a platinum electrode and an SCE in a flow-through cell. The integrity of the sample must be protected so that contamination by and/or loss of redox active species (e.g., O_2, H_2S, etc.) does not occur. The continual pumping of fresh sample (e.g., groundwater) helps ensure that little or no loss of integrity can occur during the relatively short residence time of the sample in the cell. For the particular case depicted here, the sample is giving $E_{SCE} < 0$, and so by Eq. (19.56), at 25°C/1 atm, $E_H < 0.241$ V. By Eq. (19.69), pe < 4.07.

The steps used to measure the E_H of a water sample when using an SCE are summarized in Table 19.1. When using an SSCE as the reference electrode, the steps would be the same except that an SSCE analog of Eq. (19.55) employing the value of $\{Cl^-\}$ used in the cell would be needed to convert the measured E value back to an E_H.

Redox measurements are often carried out using modern pH meters since such meters are really just volt meters (also called "potentiometers"). For such instruments, the positive side of the meter is designated for use with the probe of interest, and the negative side is designated for the reference electrode. With this type of equipment, it becomes difficult to impossible to connect the leads to the meter in the wrong way. An important feature of a potentiometer is that the tendency of current to flow through the electrodes is balanced electronically by a second "bucking" potential generated by the instrument. In this manner, the potential across a system can be measured without any current actually flowing. This is an attractive feature because allowing current to flow would change the concentrations in the two half cells, and therefore the potential that one is trying to measure.

Instrument manufacturers have developed redox probes that contain both the sensing Pt electrode and the reference electrode *all in one unit* (see Figure 19.4*c*). The sensing Pt electrode is on the tip of the probe, and a salt-bridge provides connection between

TABLE 19.1 Steps used to determine the E_H of a water sample using an SCE.

Step	Task	Comments
1	Obtain water sample	Preserve integrity of sample by preventing either the loss or gain of any redox active species (e.g., O_2)
2	Place a Pt electrode in sample.	Recognize that an electrochemical potential measured with a Pt electrode may not reflect the exact E_H conditions in the water sample, and that the water sample may not be redox equilibrium.
3	Connect volt meter to Pt electrode.	Use positive terminal of meter.
4	Connect volt meter to SCE.	Use negative terminal of meter.
5	Place SCE in sample and measure E_{SCE}.	Note that SCE has a built in salt bridge.
6	Calculate E_H of the sample.	In general, use Eq. (19.55). At 25°C/1 atm, use Eq. (19.56).

the solution and the reference electrode (usually of the SSCE type) which is housed internally in the probe. Connecting the leads to the potentiometer is straightforward, as the cable from the probe is usually of the "BNC" type, and connection is only possible in one way. One then simply places this single probe in the solution of interest, and reads the potential from the meter.

In order to prevent any oxygen from getting into the sample, and in order to prevent the loss of any significant amounts of redox species from the sample, flow-through cells are often used in redox measurements (see Figure 19.5). For example, one might pump reducing (anoxic) groundwater through such a cell. A short travel time from the aquifer to the cell and a short residence time in the cell helps to ensure that the water in the cell is as similar as possible to that in the aquifer.

19.2.2 The pe Approach

The equilibrium constant for

$$OX_1 + n\ e^- = RED_1 \tag{19.1}$$

will be

$$\frac{\{RED_1\}}{\{OX_1\}\{e^-\}^n} = K \tag{19.57}$$

where K is the stability constant for the "complexation" of OX_1 by n electrons.

Taking the logarithm of both sides of Eq. (19.57) yields

$$-n\log\{e^-\} = \log K + \log\frac{\{OX_1\}}{\{RED_1\}} \tag{19.58}$$

or

$$-\log\{e^-\} = \frac{1}{n}\log K + \log\frac{\{OX_1\}^{1/n}}{\{RED_1\}^{1/n}} \tag{19.59}$$

$$pe = \frac{1}{n}\log K + \log\frac{\{OX_1\}^{1/n}}{\{RED_1\}^{1/n}}. \tag{19.60}$$

By Eq. (19.46), under the special conditions when $\log[\{OX_1\}^{1/n}/\{RED_1\}^{1/n}] = 0$, then E_H is equal to the E_H^o for the OX_1/RED_1 couple. It is logical therefore that we should use a special symbol for the value of pe that obtains when $\log[\{OX_1\}^{1/n}/\{RED_1\}^{1/n}] = 0$. We will refer to this value as pe° for the OX_1/RED_1 couple. Thus, in general, we have

$$pe^o = \frac{1}{n}\log K \tag{19.61}$$

and so by Eq. (19.60),

$$pe = pe^o + \log\frac{\{OX_1\}^{1/n}}{\{RED_1\}^{1/n}}. \tag{19.62}$$

Like E_H^o, the parameter pe° only has a specific value when we have a specific redox couple in mind.

If we multiply both sides of Eq. (19.62) by the quantity (2.303 RT/F), we obtain

$$\frac{2.303RT}{F}pe = \frac{2.303RT}{F}pe^o + \frac{2.303RT}{F}\log\frac{\{OX_1\}^{1/n}}{\{RED_1\}^{1/n}}. \tag{19.63}$$

There is much similarity between Eq. (19.63) and Eq. (19.46)

$$E_H = E_H^o + \frac{2.303RT}{F}\frac{\{OX_1\}^{1/n}}{\{RED_1\}^{1/n}}. \tag{19.46}$$

By comparing terms, we realize that the following must be true:

$$\frac{2.303RT}{F}pe - \frac{2.303RT}{F}pe^o = E_H - E_H^o. \tag{19.64}$$

By Eq. (19.64), we see that the pe and E_H scales are *linearly related.* However, based on what we know so far, we do not know that a plot of pe vs. E_H would pass through the point (0,0). That is, we do not know that pe = 0 when $E_H = 0$. While there is no a priori requirement that the pe scale should be set up in this manner, it would in fact be very convenient if this were the case. The convenience would stem from the fact that under these conditions, for any given system, we would have

$$\frac{2.303RT}{F}pe = E_H \tag{19.65}$$

and for any particular OX/RED couple,

$$\frac{2.303RT}{F}\text{pe}^{\circ} = E_{\text{H}}^{\circ}. \tag{19.66}$$

The manner in which the pe and E_H scales are set to the same chemical zero may be understood as follows.

In a solution where $\{H^+\}^{1/2}/(p_{H_2})^{1/2} = 1$, then $E_H = 0$ at 25°C/1 atm. We can arrange that the pe of that solution will also be zero by selecting an appropriate value for the K for the reaction in Eq. (19.41) (i.e., $K_{19.41}$)

$$2H^+ + 2e^- = H_{2(g)} \qquad K_{19.41} = \frac{p_{H_2}}{\{H^+\}^2\{e^-\}^2}. \tag{19.41}$$

By Eq. (19.60), this can be done by requiring that in that solution,

$$\text{pe} = (1/2)\log K_{19.41} + \log(1) = 0. \tag{19.67}$$

The value of pe in this solution will therefore equal 0, and Eq. (19.67) will also be satisfied if $K_{19.41}$ is set identically equal to 1 at 25°C/1 atm. As is shown in Table 19.2, the convention is indeed to set the K for the reaction in Eq. (19.41) equal to 1 at 25°C/1 atm. As a result, both Eqs. (19.65) and (19.66) are valid. The reduction half reactions in Table 19.2 are arranged in order of increasing pe° (which is the same order as increasing E_{H}°).

The fact that we should even be free to arbitrarily set the value of $K_{19.41}$ is exactly analogous to the fact that we can arbitrarily set the value of the half-cell E_{H}° value for the SHE to 0.0. We were able to do that because all we can measure is the *overall* potential difference between two half-cells; adopting a convention that assigns a value of zero to one of the half cells (under specific activity conditions in that half cell) will just lead to the adoption of a convention that assigns the overall cell potential to the other half-cell. Similarly, just as no reaction in a single half-cell can proceed by itself, reactions like those in Eqs. (19.1) and (19.41) cannot proceed individually. They can, however, proceed simultaneously, though in opposite directions:

$$OX_1 + ne^- = RED_1 \qquad K_{19.1} = \frac{\{RED_1\}}{\{OX_1\}\{e^-\}^n} \tag{19.1}$$

$$-\quad (n/2)[2H^+ + 2e^- = H_{2(g)}] \qquad (K_{19.41})^{n/2} = \frac{(p_{H_2(g)})^{n/2}}{\{H^+\}^n\{e^-\}^n} \tag{19.41}$$

$$OX_1 + \tfrac{n}{2}H_{2(g)} = RED_1 + nH^+ \qquad K_{\text{overall}} = (K_{19.1})/(K_{19.41})^{n/2}. \tag{19.68}$$

Just as the E for the overall cell characterized by Eq. (19.68) can be measured, so too can the equilibrium constant K_{overall} for the reaction in Eq. (19.68). Since the individual half-cell quantities $K_{19.1}$ and $K_{19.41}$ are not measurable, we are free to arbitrarily assign, by convention, that $K_{19.41} = 1.0$ at 25°C/1 atm. At the same time, however, we must also say that at 25°C/1 atm, the K for any given half reaction as in Eq. (19.1) will be set equal to the measured value of the K_{overall} for the reaction in Eq. (19.68). That is, $K_{19.1} = K_{\text{overall}}$.

TABLE 19.2. Data for selected redox reactions at 25°C/1 atm in order of increasing pe° and E_H^o. (Data from Bard et al. (1985) and Stumm and Morgan (1981).) The reducing strength of the RED species tends to increase towards the top of the table, and the oxidizing strength of the OX species tends to increase towards the bottom of the table.

Reduction Half Reaction	log K	pe°	pe°(W)	E_H^o
$OX + ne^- = RED$	$\frac{\{RED\}}{\{OX\}\{e^-\}^n}$	$\frac{1}{n}\log K$	$pe^o - \frac{n_H}{n_e}7$	0.05916 pe°
$Na^+ + e^- = Na_{(s)}$	−46.0	−46.0	−46.0	−2.71
$Zn^{2+} + 2e^- = Zn_{(s)}$	−26.0	−13.0	−13.0	−0.76
$FeCO_{3(s)} + 2e^- = Fe_{(s)} + CO_3^{2-}$	−25.58	−12.79	−12.79	−0.76
$Fe^{2+} + 2e^- = Fe_{(s)}$	−14.9	−7.45	−7.45	−0.44
$CO_{2(g)} + H^+ + 2e^- = HCOO^-$	−9.66	−4.83	−8.33	−0.29
$CO_{2(g)} + 4H^+ + 4e^- = CH_2O + H_2O$	−4.8	−1.2	−8.2	−0.071
$CO_{2(g)} + 4H^+ + 4e^- = \frac{1}{6}C_6H_{12}O_6(\text{glucose}) + H_2O$	−0.8	−0.2	−7.2	−0.012
$2H^+ + 2e^- = H_{2(g)}$	0.0	0.0	−7.0	0.00
$N_{2(g)} + 6H^+ + 6e^- = 2NH_3$	9.5	1.58	−5.42	0.093
$S_{(s)} + 2H^+ + 2e^- = H_2S$	4.8	2.4	−4.6	0.14
$Cu^{2+} + e^- = Cu^+$	2.7	2.7	2.7	0.16
$HCOO^- + 3H^+ + 2e^- = CH_2O + H_2O$	5.64	2.82	−7.68	0.17
$CO_{2(g)} + 8H^+ + 8e^- = CH_{4(g)} + 2H_2O$	23.0	2.87	−4.13	0.17
$AgCl_{(s)} + e^- = Ag_{(s)} + Cl^-$	3.7	3.7	3.7	0.22
$CH_2O + 2H^+ + 2e^- = CH_3OH$	8.0	4.0	−3.0	0.24
$SO_4^{2-} + 9H^+ + 8e^- = HS^- + 4H_2O$	34.0	4.25	−3.63	0.25
$Hg_2Cl_{2(s)} + 2e^- = 2Hg_{(l)} + 2Cl^-$	9.06	4.53	4.53	0.268
$N_{2(g)} + 8H^+ + 6e^- = 2NH_4^+$	28.1	4.68	−4.65	0.28
$SO_4^{2-} + 10H^+ + 8e^- = H_2S + 4H_2O$	41.0	5.13	−3.62	0.30
$Cu^{2+} + 2e^- = Cu_{(s)}$	11.4	5.7	5.7	0.34
$HSO_4^- + 7H^+ + 6e^- = S_{(s)} + 4H_2O$	34.2	5.7	−2.47	0.34
$SO_4^{2-} + 8H^+ + 6e^- = S_{(s)} + 4H_2O$	36.2	6.03	−3.3	0.36

TABLE 19.2. (continued)

Reduction Half Reaction	log K	pe°	pe°(W)	E_H^o
$OX + ne^- = RED$	$\frac{\{RED\}}{\{OX\}\{e^-\}^n}$	$\frac{1}{n}\log K$	$pe^o - \frac{n_H}{n_e}7$	0.05916 pe°
$CH_2O + 4H^+ + 4e^- = CH_{4(g)} + H_2O$	27.8	6.94	−0.06	0.41
$Cu^+ + e^- = Cu_{(s)}$	8.8	8.8	8.8	0.52
$CH_3OH + 2H^+ + 2e^- = CH_{4(g)} + H_2O$	19.8	9.88	2.88	0.58
$Fe^{3+} + e^- = Fe^{2+}$	13.0	13.0	13.0	0.77
$Ag^+ + e^- = Ag_{(s)}$	13.5	13.5	13.5	0.80
$NO_2^- + 7H^+ + 6e^- = NH_3 + 2H_2O$	81.5	13.58	5.41	0.80
$NO_3^- + 2H^+ + 2e^- = NO_2^- + H_2O$	28.3	14.15	7.15	0.84
$NO_3^- + 10H^+ + 8e^- = NH_4^+ + 3H_2O$	119.2	14.9	6.15	0.88
$NO_2^- + 8H^+ + 6e^- = NH_4^+ + 2H_2O$	90.8	15.14	5.82	0.90
$MnO_{2(s)} + HCO_3^- + 3H^+ + 2e^- = MnCO_{3(s)} + 2H_2O$	25.8	15.9	5.4	0.94
$(\alpha)FeOOH_{(s)} + HCO_3^- + 2H^+ + e^- = FeCO_{3(s)} + 2H_2O$	13.15	13.15	−0.85	0.78
$(\alpha)FeOOH_{(s)} + 3H^+ + e^- = Fe^{2+} + 2H_2O$	13.5	13.5	−7.5	0.80
$(am)Fe(OH)_{3(s)} + 3H^+ + e^- = Fe^{2+} + 3H_2O$	16.2	16.2	−4.8	0.96
$O_{2(g)} + 4H^+ + 4e^- = 2H_2O$	83.1	20.78	13.78	1.23
$NO_3^- + 6H^+ + 5e^- = \frac{1}{2}N_{2(g)} + 3H_2O$	105.3	21.05	12.65	1.25
$MnO_{2(s)} + 4H^+ + 2e^- = Mn^{2+} + 2H_2O$	43.6	21.8	7.8	1.29
$Fe^{3+} + CO_3^{2-} + e^- = FeCO_{3(s)}$	23.68	23.68	23.68	1.40
$Cl_2 + 2e^- = 2Cl^-$	47.2	23.6	23.6	1.40
$HOCl + H^+ + e^- = \frac{1}{2}Cl_2 + H_2O$	26.9	26.9	19.9	1.59
$ClO^- + 2H^+ + 2e^- = Cl^- + H_2O$	57.8	28.9	21.8	1.71
$H_2O_2 + 2H^+ + 2e^- = 2H_2O$	59.6	29.80	22.80	1.76

The fact that $K_{19.41} = 1.0$ at 25°C/1 atm means, by Eq. (19.61), that $pe^o = 0$ for the conditions in the SHE. This conclusion is consistent with Eq. (19.66). As a final note, we conclude that at 25°C (298 K), Eqs. (19.65) and (19.66) become

$$pe = (1/0.05916)\,E_H = 16.90\,E_H \qquad (T = 298\text{ K}) \qquad (19.69)$$

$$pe^o = (1/0.05916)\,E_H^o = 16.90\,E_H^o \qquad (T = 298\text{ K}). \qquad (19.70)$$

19.2.3 pe° Values and Relative Oxidative Capabilities

Consider Eqs. (19.1) and (19.2). At 25°C/1 atm, if redox half reaction 1 has a pe^o (E_H^o) value higher than the pe^o (E_H^o) of half reaction 2, then if the activities for all of the species in both half cells are either unity, or at least give $\{RED_1\}\{OX_2\}/\{OX_1\}\{RED_2\} = 1$, half reaction 1 can lead to some net oxidation of RED_2 to OX_2. Under these conditions, the activity of the electron is lower in half cell 1 that it is in half cell 2 (pe^o for half cell 1 is greater than pe^o for half cell 2), and so electrons tend to flow from half-cell 2 to half-cell 1. Thus, pe^o values provide a measure of the relative oxidative capabilities, or strengths, of different redox couples.

As an example, we ask whether the conversion of Cu^{2+} to $Cu_{(s)}$ (OX_1 and RED_1) can oxidize $H_{2(g)}$ to H^+ (RED_2 and OX_2) under conditions such that all reactants and products are at unit activity, or at least give $\{RED_1\}\{OX_2\}/\{OX_1\}\{RED_2\} = 1$. The data from Table 19.2 that we use are:

	log K	pe^o	E_H^o	
$Cu^{2+} + 2e^- = Cu_{(s)}$	11.4	5.7	0.34	(19.71)
$2H^+ + 2e^- = H_{2(g)}$	0	0	0	(19.72)

We see that the hydrogen redox half reaction has a pe^o value that is lower than that for copper, and so by the rule stated above, we know that the process

$$H_{2(g)} + Cu^{2+} = 2H^+ + Cu_{(s)} \qquad \log K = 11.4 \qquad (19.73)$$

will proceed under standard conditions. The value of the equilibrium constant for 25°C/1 atm may be determined using the data in Eqs. (19.71) and (19.72). Since the log of the K is positive, the ΔG for the reaction under standard conditions is negative (recall from Eq. (2.21) that $\Delta G^o = -2.303\,RT\log K$). Thus, when $\{Cu_{(s)}\}\{H^+\}^2/\{Cu^{2+}\}p_{H_2} = 1$, the species Cu^{2+} can oxidize $H_{2(g)}$ to H^+. In Figure 19.2, then, $E = E_{H,L}^o - E_{H,R}^o$ is positive (0.34 V), and electrons move from left to right across the cell.

19.3 MATHEMATICAL SIMILARITIES BETWEEN REDOX AND ACID/BASE EQUILIBRIA

19.3.1 Basic Considerations

Consider the case of the reaction

$$Fe^{3+} + e^- = Fe^{2+} \qquad K = \frac{\{Fe^{2+}\}}{\{Fe^{3+}\}\{e^-\}}. \qquad (19.4)$$

The Fe^{3+} species is an Fe(III) species, and Fe^{2+} is an Fe(II) species. Since we will only be examining one redox half reaction K value at a time in this section, we will not bother with equation number subscripts on the K values. By Eq. (19.60), then,

$$\text{pe} = \log K + \log\{Fe^{3+}\}/\{Fe^{2+}\} \tag{19.74}$$

$$= \text{pe}^{\circ} + \log\{Fe^{3+}\}/\{Fe^{2+}\}. \tag{19.75}$$

At 25°C/1 atm, pe° for the Fe^{3+}/Fe^{2+} couple is 13.0.

Equations (19.74) and (19.75) are very similar to the equation that can be developed for pH in the presence of the acid/base pair HA/A^-, namely

$$\text{pH} = -\log K_{HA} + \log\{A^-\}/\{HA\} \tag{19.76}$$

where K_{HA} is the acid dissociation constant for the acid HA. The facts that the $\log K_{HA}$ term is preceded by a minus sign in Eq. (19.76), but the $\log K$ term in Eq. (19.74) has no minus sign are direct results of the facts that K is a stability constant, and K_{HA} is an instability constant.

In addition to the comparabilities of Eqs. (19.74) and (19.75) with (19.76), α values may be defined in redox cases just as they can in acid/base cases. For the acid HA, we have that

$$\alpha_0 = \frac{[HA]}{[HA]+[A^-]} = \frac{[H^+]}{[H^+]+K_{HA}} \tag{19.77}$$

$$\alpha_1 = \frac{[A^-]}{[HA]+[A^-]} = \frac{K_{HA}}{[H^+]+K_{HA}} \tag{19.78}$$

where the second portions of Eqs. (19.77) and (19.78) neglect activity corrections. The next section introduces the mathematics of redox α values.

.3.2 Fe^{3+}/Fe^{2+} Equilibrium

For the case of Fe^{3+}/Fe^{2+} equilibrium, assuming activity coefficients of 1.0 for the dissolved iron species, the K for Eq. (19.4) yields

$$K = \frac{[Fe^{2+}]}{[Fe^{3+}]\{e^-\}}. \tag{19.79}$$

If the solution of interest is sufficiently acidic that all Fe(II) and all Fe(III) hydroxo complexes are negligible, and also assuming negligibility for all other iron/ligand complexes, we have

$$Fe_T = [Fe^{2+}] + [Fe^{3+}]. \tag{19.80}$$

In the case of this redox reaction, α_{III} will be defined as the fraction that of the total amount of dissolved iron that is in the III oxidation state, and α_{II} will be defined as the fraction that is in the II oxidation state. Thus, *under acidic conditions*, we have

$$\alpha_{III} = \frac{[Fe^{3+}]}{Fe_T} \tag{19.81}$$

$$\alpha_{II} = \frac{[Fe^{2+}]}{Fe_T} \tag{19.82}$$

TABLE 19.3 Asymptotic behavior in an acidic Fe^{3+}/Fe^{2+} system for $pe \ll pe^o$, and for $pe \gg pe^o$.

$pe \ll pe^o$	$pe \gg pe^o$
$\{e^-\} \gg K^{-1}$	$\{e^-\} \ll K^{-1}$
$\alpha_{II} \simeq 1$	$\alpha_{III} \simeq 1$
$d \log \alpha_{II}/d\text{pe} \simeq 0$	$d \log \alpha_{III}/d\text{pe} \simeq 0$
$\log[Fe^{2+}] \simeq \log Fe_T$	$\log[Fe^{3+}] \simeq \log Fe_T$
$d \log[Fe^{2+}]/d\text{pe} \simeq 0$	$d \log[Fe^{3+}]/d\text{pe} \simeq 0$
$\log \alpha_{III} \simeq -\log K - \log\{e^-\}$	$\log \alpha_{II} \simeq \log K + \log\{e^-\}$
$\simeq -pe^o + pe$	$\simeq pe^o - pe$
$d \log \alpha_{III}/d\text{pe} \simeq +1$	$d \log \alpha_{II}/d\text{pe} \simeq -1$
$\log[Fe^{3+}] \simeq \log Fe_T - pe^o + pe$	$\log[Fe^{2+}] \simeq \log Fe_T + pe^o - pe$
$d \log[Fe^{3+}]/d\text{pe} \simeq +1$	$d \log[Fe^{2+}]/d\text{pe} \simeq -1$

$$\alpha_{III} = \frac{[Fe^{3+}]}{[Fe^{2+}] + [Fe^{3+}]} = \frac{1}{\{e^-\}/K^{-1} + 1} = \frac{K^{-1}}{\{e^-\} + K^{-1}} \tag{19.83}$$

$$\alpha_{II} = \frac{[Fe^{2+}]}{[Fe^{2+}] + [Fe^{3+}]} = \frac{1}{1 + K^{-1}/\{e^-\}} = \frac{\{e^-\}}{\{e^-\} + K^{-1}}. \tag{19.84}$$

Taking the log of both sides of Eqs. (19.81) and (19.82) yields

$$\log[Fe^{3+}] = \log \alpha_{III} + \log Fe_T \tag{19.85}$$

$$\log[Fe^{2+}] = \log \alpha_{II} + \log Fe_T. \tag{19.86}$$

As seen in Eqs. (19.83) and (19.84), both α_{III} and α_{II} are simple functions of $\{e^-\}$ and K^{-1}. Plots of $\log[Fe^{3+}]$ vs. pe and $\log[Fe^{2+}]$ may be prepared easily based on Eqs. (19.85) and (19.86) and a given value of Fe_T. The mathematical properties of Eqs. (19.85) and (19.86) are very similar to the properties of the functions $\log[HA] = \log \alpha_0 + \log A_T$ and $\log[A^-] = \log \alpha_1 + \log A_T$ for the acid/base pair HA/A^-.

When the system is relatively reducing and $\{e^-\} \gg K^{-1}$ (i.e., when $pe \ll pe^o$), then $\alpha_{II} \simeq 1$, $\alpha_{III} \simeq K^{-1}/\{e^-\}$, $d \log \alpha_{II}/d\text{pe} \simeq 0$, and $d \log \alpha_{III}/d\text{pe} \simeq +1$. When the system is relatively oxidizing, and $\{e^-\} \ll K^{-1}$ (i.e., when $pe \gg pe^o$), then $\alpha_{II} \simeq \{e^-\}/K^{-1}$, $\alpha_{III} \simeq 1$, $d \log \alpha_{II}/d\text{pe} \simeq -1$, and $d \log \alpha_{III}/d\text{pe} \simeq 0$. These results are summarized in Table 19.3. For those parameters that are characterized by slopes of +1 or −1 in the two different regions, the fact that the slope is ±1 is a result of the fact that *one* electron is being exchanged between the two iron species. A plot of $\log[Fe^{3+}]$ and $\log[Fe^{2+}]$ vs. pe is presented in Figure 19.6 for $Fe_T = 10^{-5}$ *M*.

19.3.3 Hypochlorite/Chloride Equilibrium

The hypochlorite ion has the formula ClO^-. The charge on the oxygen in this ion is considered to be −2. Thus, the chlorine in hypochlorite is considered to have a +1 charge, and be in the I oxidation state. The chlorine in chloride (Cl^-) has a −1 charge

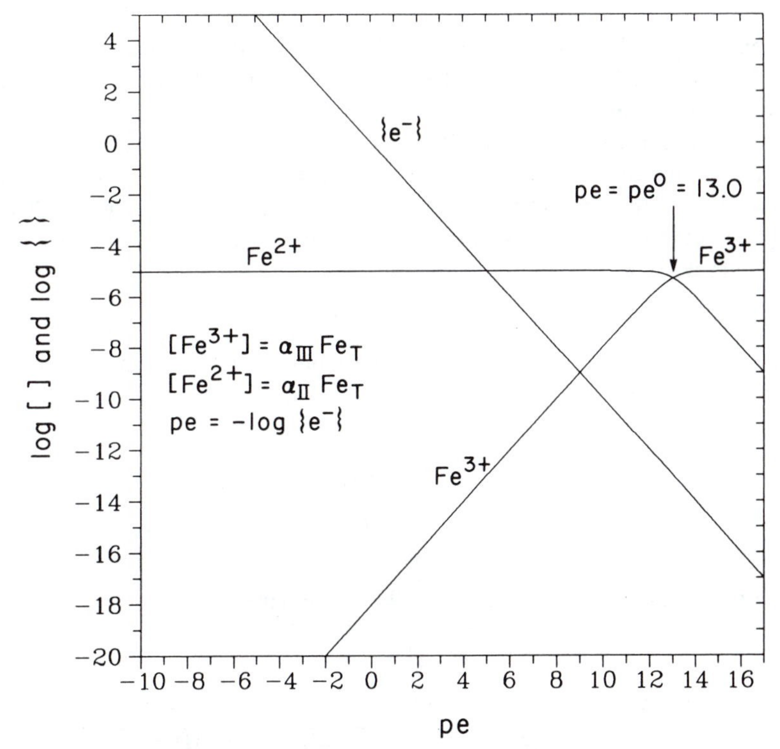

Figure 19.6 Log [ ] and log$\{e^-\}$ vs. pe in the Fe^{3+}/Fe^{2+} system. The pH is assumed to be low enough that Fe^{3+} is by far the dominant Fe(III) species, and that Fe^{2+} is by far the dominant Fe(II) species. Note that the line for e^- represents the electron *activity*, not the electron concentration.

and is in the −I oxidation state. The reduction of hypochlorite to chloride therefore involves the addition of two electrons. The reaction proceeds according to

$$ClO^- + 2H^+ + 2e^- = Cl^- + H_2O. \tag{19.87}$$

Assuming unit activity coefficients for the dissolved chlorine species, the K for the reaction given in Eq. (19.87) becomes

$$K = \frac{[Cl^-]}{[ClO^-]\{H^+\}^2\{e^-\}^2} \tag{19.88}$$

From Table 19.2, at 25°C/1 atm, we obtain that $K = 10^{57.8}$. Thus, by Eq. (19.61), $pe° = (1/2)57.8 = 28.9$. When $\{H^+\} = 1.0$ (i.e., pH = 0) the ratio of $[ClO^-]$ to $[Cl^-]$ will be 1.0 when pe = pe° = 28.9. However, a pH value of 0 is not very environmentally relevant. At pH = 7.0, Eq. (19.88) may be used to calculate that the $[ClO^-]/[Cl^-]$ ratio will be 1.0 when pe = pe(7) = 21.9. As will be discussed in Chapter 23, pe° values adjusted to pH 7 are referred to as pe°(W) values. The W is intended to invoke the idea of "water," although that is a bit misleading since aqueous solutions can obviously take on a whole range of pH values. In general, a pe° value adjusted to any given pH will be denoted pe°(pH) where a specific value can be substituted for "pH." As noted, pe°(7) ≡ pe°(W). Half reactions like $Fe^{3+} + e^- = Fe^{2+}$ that do not involve protons do not show any direct dependence on pH.

At 25°C/1 atm, the pK for HClO is 7.3. Thus, for pH values significantly higher than 7.3, the portion of the I oxidation state chlorine that is present as HClO will be very small in relation to the amount that is present as ClO^-. As we will see in Chapter 20, at such pH values, $[Cl_2]$ will also be very low for typical levels of Cl_T. Thus, to a good approximation we will have

$$Cl_T = [ClO^-] + [Cl^-]. \tag{19.89}$$

Under these conditions, following the α nomenclature system discussed in the previous section, for this system we have

$$\alpha_I = \frac{[ClO^-]}{Cl_T} \tag{19.90}$$

$$\alpha_{-I} = \frac{[Cl^-]}{Cl_T}. \tag{19.91}$$

Under the assumptions pertaining to Eq. (19.89), we have

$$\alpha_I + \alpha_{-I} = 1 \tag{19.92}$$

$$\alpha_I = \frac{[ClO^-]}{[Cl^-] + [ClO^-]} = \frac{1}{\{e^-\}^2\{H^+\}^2/K^{-1} + 1} = \frac{K^{-1}}{\{e^-\}^2\{H^+\}^2 + K^{-1}} \tag{19.93}$$

$$\alpha_{-I} = \frac{[Cl^-]}{[Cl^-] + [ClO^-]} = \frac{1}{1 + K^{-1}/\{e^-\}^2\{H^+\}^2} = \frac{\{e^-\}^2\{H^+\}^2}{\{e^-\}^2\{H^+\}^2 + K^{-1}}. \tag{19.94}$$

For both Eqs. (19.90) and (19.91), taking the log of both sides yields

$$\log[ClO^-] = \log \alpha_I + \log Cl_T \tag{19.95}$$

$$\log[Cl^-] = \log \alpha_{-I} + \log Cl_T. \tag{19.96}$$

Plots of $\log[ClO^-]$ vs. pe and $\log[Cl^-]$ vs. pe may be prepared easily based on Eqs. (19.93)–(19.96) for any given values of Cl_T *and* pH. Figure 19.7*a* has been prepared for $Cl_T = 10^{-3}$ *M* and pH = 9, and Figure 19.7*b* for $Cl_T = 10^{-3}$ *M* and pH = 11. (pH values significantly greater than 7.3 have been selected for these plots to ensure that $[ClO^-] \gg [HClO]$.)

Although similar to Figure 19.6, Figure 19.7 differs in some respects. Two major differences are: 1) the steeper slopes of portions of the log concentration vs. pH lines; and 2) the pH dependence. The steeper slopes are the result of the fact that *two* electrons separate ClO^- and Cl^-, and so $\{e^-\}$ appears as a square in the two α expressions. The pH dependence is a result of the presence of $\{H^+\}$ in the expressions for α_I and α_{-I}. In particular, the pe at which these two α values are equal depends on the pH according to the expression $\{e^-\}^2\{H^+\}^2 = K^{-1}$. Thus, the two α values are equal when

$$\{e^-\} = \frac{K^{-1/2}}{\{H^+\}} \tag{19.97}$$

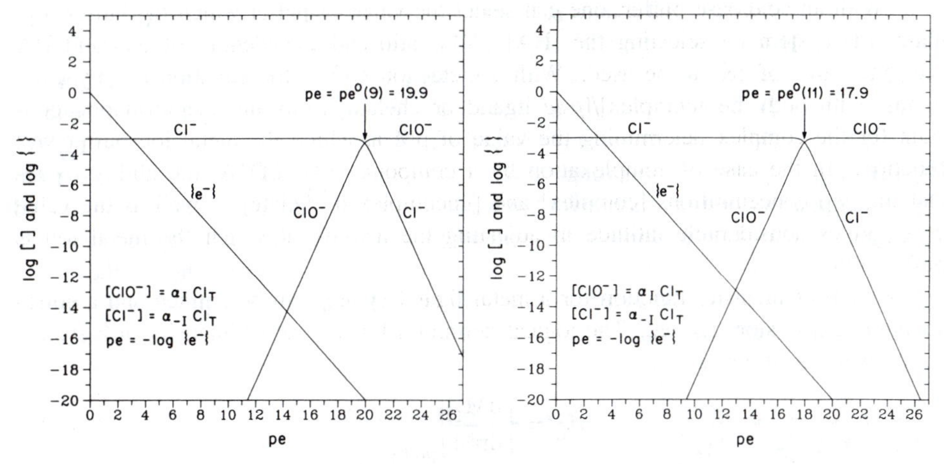

Figure 19.7 Log [ ] and log $\{e^-\}$ vs. pe in the ClO^-/Cl^- system. The pH is assumed to be high enough (e.g., greater than 8.3) so that ClO^- and Cl^- are the only important chlorine-containing species (see also Figure 20.3). (*a*) pH = 9; (*b*) pH = 11. Note that the lines for e^- represent the electron *activity*, not the electron concentration.

that is, when

$$pe = pe^\circ(pH) = pe^\circ - pH. \tag{19.98}$$

As noted, at 25°C/1 atm, $pe^\circ(7) = 21.9$. Under the same T and P conditions, $pe^\circ(9) = 19.9$ and $pe^\circ(11) = 17.9$. Increasing the pH therefore favors ClO^- over Cl^-, a result which may also be deduced directly from the expression for K given in Eq. (19.88). Table 19.4 provides a summary of these results.

TABLE 19.4 Asymptotic behavior in the ClO^-/Cl^- system for $pe \ll pe^\circ(pH)$ and $pe \gg pe^\circ(pH)$

$pe \ll pe^\circ(pH)$	$pe \gg pe^\circ(pH)$
$\{e^-\} \gg K^{-1/2}/\{H^+\}$	$\{e^-\} \ll K^{-1/2}/\{H^+\}$
$\alpha_{-I} \simeq 1$	$\alpha_I \simeq 1$
$d \log \alpha_{-I}/d pe \simeq 0$	$d \log \alpha_I/d pe \simeq 0$
$\log[Cl^-] \simeq \log Cl_T$	$\log[ClO^-] \simeq \log Cl_T$
$d \log[Cl^-]/d pe \simeq 0$	$d \log[ClO^-]/d pe \simeq 0$
$\log \alpha_I \simeq -\log K - \log\{e^-\}^2\{H^+\}^2$	$\log \alpha_{-I} \simeq \log K + \log\{e^-\}^2\{H^+\}^2$
$\simeq -2pe^\circ + 2pe + 2pH$	$\simeq 2pe^\circ - 2pe - 2pH$
$d \log \alpha_I/d pe \simeq +2$	$d \log \alpha_{-I}/d pe \simeq -2$
$\log[ClO^-] \simeq \log Cl_T - 2pe^\circ + 2pe + 2pH$	$\log[Cl^-] \simeq \log Cl_T + 2pe^\circ - 2pe - 2pH$
$d \log[ClO^-]/d pe \simeq +2$	$d \log[Cl^-]/d pe \simeq -2$

Figures 19.6 and 19.7.a may be used to reveal the relative oxidizing strengths of Fe^{3+} at low pH vs. those of ClO^- at pH 9. We note that ClO^- can dominate chlorine chemistry only when the pe is very high. In contrast, Fe^{3+} can dominate iron chemistry at much lower pe values. Adding a strong reducing agent to any system will tend to lower the pe. (This is exactly analogous to the fact that adding a strong acid will lower the pH of any system.) The strongest oxidants present will pick up the electrons from the reducing agent preferentially over the weaker oxidants. (Analogously, protons added to a system containing multiple bases will be picked up preferentially by the strongest bases.) Since ClO^- at pH 9 tends to be converted to Cl^- at pe values higher than those at which Fe^{3+} is converted to Fe^{2+}, we can conclude that ClO^- at pH 9 is a stronger oxidant than is Fe^{3+}.

19.3.4 α Values for Redox Species Often Either Near 1.0, or Exceedingly Small

The previous section has described the fact that there are many similarities between the mathematical properties of the α values for pe, metal complexation, and pH problems. However, α values for redox species are more often near 1.0 or exceedingly small than is the case for the α values for acid/base pairs and metal complexes. This is due to the fact that *the range of values that pe can take in natural waters is fairly large*, from about -10 to 17 (see Section 19.4). This is much wider than the range that bounds typical pH values. As a result, for any given OX/RED pair, it is in general not very likely that the pe for a given system will be near the specific value that gives non-negligible values for *both* the OX and the RED species; very often, one of the α values will be essentially 1.0, and the other value(s) will be exceedingly small. Under these circumstances, MBEs, ENEs, and other equations where redox α values might otherwise be useful will often be readily simplified to forms that involve *only one* redox species.

For acid/base problems in natural waters, the pH values most commonly encountered are typically found in a range of about 6 units, from about 4 to about 10. Thus, for an acid with a pK of ~6, the probability of being in the pH region where both α_0 and α_1 are important (about 5 to 7) is comparatively high (~2/6), and so we will frequently need to consider both α_0 and α_1. Consider now an OX/RED pair that exchanges one electron and exhibits a $pe^\circ(W)$ value of 10. Both redox species may be presumed to be important over an approximate pe range of 9 to 11. If all pe values from -10 to 17 were equally likely (a somewhat gross, but nevertheless useful assumption here), the probability of being in the 9 to 11 range is only 2/27. On either side of the range $9 < pe < 11$, one of the redox α values will be close to one, and the other will be small.

19.3.5 At Equilibrium, an Aqueous Solution Has One pe

At equilibrium, any given aqueous solution has one and only one pH. Thus, if more than one conjugate acid/conjugate base pair is present, at equilibrium, each one of those pairs must specify the same pH value. If K_{HA}, K_{HB}, K_{HC}, etc. are the acid dissociation constants of the acids HA, HB, HC, etc, then at equilibrium we have

$$\text{pH} = \text{p}K_{\text{HA}} + \log\frac{\{\text{A}^-\}}{\{\text{HA}\}} = \text{p}K_{\text{HB}} + \log\frac{\{\text{B}^-\}}{\{\text{HB}\}} = \text{pK}_{\text{HC}} + \log\frac{\{\text{C}^-\}}{\{\text{HC}\}} = \text{etc.} \tag{19.99}$$

A knowledge of the pH as determined based on a knowledge of the one conjugate acid/conjugate base ratio allows the computation of all other conjugate acid/conjugate base ratios.

The situation with pe is exactly analogous to that described by Eq. (19.99). Namely, since there can be only one *equilibrium* pe in any given solution at any given time, at full equilibrium, all redox pairs must prescribe the same pe. That is, if K_1, K_2, K_3, etc. are the half-cell reduction constants for redox pairs OX_1/RED_1, OX_2/RED_2, OX_3/RED_3, etc., then we have

$$\begin{aligned}\text{pe} &= \frac{1}{n_1}\left(\log K_1 + \log\frac{\{\text{OX}_1\}}{\{\text{RED}_1\}}\right)\\ &= \frac{1}{n_2}\left(\log K_2 + \log\frac{\{\text{OX}_2\}}{\{\text{RED}_2\}}\right)\\ &= \frac{1}{n_3}\left(\log K_3 + \log\frac{\{\text{OX}_3\}}{\{\text{RED}_3\}}\right) = \text{etc.}\end{aligned} \tag{19.100}$$

where n_1, n_2 and n_3 are the numbers of electrons involved in reducing OX_1 to RED_1, OX_2 to RED_2, OX_3 to RED_3, respectively. Thus, at equilibrium: 1) once we know the pe for one redox couple, we know it for all others; 2) a knowledge of pe for one redox couple will allow us to calculate the OX/RED ratio for any other redox couple, provided of course that we know the value of the K for that couple.

The reader may have noted the emphasis given the fact that full equilibrium must be present for all acid/base equilibria to prescribe the same pH, and for all redox half cells to prescribe the same pe. The fact that acid/base proton exchange is a relatively fast process usually makes the assumption of full pH equilibrium a good one. However, as mentioned above, many electron exchange processes are not fast, and as a result *real-world* water samples are often *not* at full redox equilibrium. For example, a sample of water that contains a few mg/L of dissolved organic carbon (DOC) as well as a few mg/L of dissolved oxygen is definitely not at equilibrium: the DOC is unstable relative to oxidation by the O_2 to CO_2.[4]

In general, if one were to obtain a real-world water sample and were to measure $\{OX_1\}$ and $\{RED_1\}$, $\{OX_2\}$ and $\{RED_2\}$, $\{OX_3\}$ and $\{RED_3\}$, etc. for the various redox pairs present, then calculate the pe indicated by each pair and the assumption of equilibrium, more than one pe value might very well be obtained. In some cases, these differences will be small, but in other cases, the differences will be quite large (as in the above example involving the redox pairs CO_2/DOC and O_2/H_2O).

When full redox equilibrium does not exist, one net set of redox half reactions (e.g., OX_1/RED_1 and OX_2/RED_2) might be very nearly in equilibrium, and acting together to

[4]By the same token, we can all be grateful that oxidation of organic matter by O_2 is a relatively slow process; the organic matter in our bodies is unstable in the presence of the O_2 environment in which we live. When the forces of life cease, in the continued presence of O_2, conversion to CO_2 will begin.

set the pe (E_H) of the water. Other redox half reactions could be out of equilibrium with that pe. For example, one might come across some oxygenated water at pH 8 in which the pe was being set by redox equilibrium between 1/2 $O_{2(g)}$ (OX_1) and H_2O (RED_1), and 2 H^+ (OX_2) and $H_{2(g)}$ (RED_2). For oxygenated water with an in-situ p_{O_2} of ~0.2 atm, the resulting pe value would be fairly high (~13). Oxygenated waters of this type can be found in which some elemental sulfur (RED_3, due to bacterial activity) and some sulfate (OX_3) are also present. In such a case, the system would not be at full redox equilibrium, and the sulfate/sulfur couple would be undergoing some net conversion to sulfate, with a concomitant lowering of the pe. As the pe was being lowered, it is possible that the $O_{2(g)}/H_2O$ and 2 $H^+/H_{2(g)}$ reaction could have sufficiently fast kinetics to allow it to "keep up" and stay at equilibrium with the true, changing, nonequilibrium pe.

19.4 OXIDATION AND REDUCTION OF WATER

19.4.1 Basic Considerations

In water, one very important redox process is the reduction of the water, that is,

$$2H_2O + 2e^- = H_{2(g)} + 2OH^- \qquad \log K = -28.0. \tag{19.101}$$

Equation (19.101) may be converted to

$$2H^+ + 2e^- = H_{2(g)} \qquad \log K = 0.0. \tag{19.102}$$

Equation (19.101) describes the reduction of the H(I) in water to H(0) in $H_{2(g)}$, and Eq. (19.102) describes the reduction of protons in solution, also H(I), to H(0) in $H_{2(g)}$.

The values of the constants given in Eqs. (19.101) and (19.102) and in the remainder of this section are those for 25°C/1 atm. The difference between the K values given in Eqs. (19.101) and (19.102) is the constant for the dissociation 2 $H_2O = 2H^+ + 2OH^-$, that is, $K_w^2 = 10^{-28.0}$. Either Eq. (19.101) or (19.102) may be used to describe the acceptance of electrons by H(I) in aqueous solutions. Since Eq. (19.102) involves protons rather than hydroxide, is often more convenient to use Eq. (19.102) rather than Eq. (19.101). At 25°C/1 atm, the equilibrium constant for Eq. (19.102) gives

$$\log p_{H_2} = 0.0 - 2pH - 2pe \tag{19.103}$$

$$pe = -\frac{1}{2}\log p_{H_2} - pH \tag{19.104}$$

$$\left(\frac{\partial \log p_{H_2}}{\partial pe}\right)_{pH} = -2. \tag{19.105}$$

As noted, the partial derivative in Eq. (19.105) is carried out at constant pH.

Water may be oxidized according to

$$O_{2(g)} + 4H^+ + 4e^- = 2H_2O \qquad \log K = 83.1 \tag{19.106}$$

where the oxidation takes place when viewing the reaction *backwards*, that is, from right to left. By taking into consideration the dissociation of water, Eq. (19.106) may be converted to

$$O_{2(g)} + 2H_2O + 4e^- = 4OH^- \qquad \log K = 27.1. \qquad (19.107)$$

Equation (19.106) describes the oxidation of O(-II) in water to O(0) in O_2, and Eq. (19.107) describes the oxidation of O(-II) in hydroxide to O(0) in O_2. The difference between the K values given in Eqs. (19.106) and (19.107) is the constant for the combination of 4 H^+ and 4 OH^- to form 4 H_2O, that is, K_w^{-4}. At 25°C/1 atm, the equilibrium constant in Eq. (19.103) gives

$$\log p_{O_2} = -83.1 + 4\text{pH} + 4\text{pe} \qquad (19.108)$$

$$\text{pe} = 20.78 - \text{pH} + \frac{1}{4}\log p_{O_2} \qquad (19.109)$$

$$\left(\frac{\partial \log p_{O_2}}{\partial \text{pe}}\right)_{\text{pH}} = 4. \qquad (19.110)$$

As noted, the partial derivative in Eqs. (19.110) is carried out at constant pH. The manner in which log p_{H_2} and log p_{O_2} vary as a function of pe for different pH values can be calculated using Eqs. (19.103) and (19.108), respectively. Results for pH = 4, 7, and 10 are presented for 25°C/1 atm in Figure 19.8.

For both the reduction and oxidation of water, we note that we were able to obtain related reactions ((19.102) and (19.107)) simply by subtracting or adding in a certain number of water dissociation reactions. As noted above, being able to simply add/subtract non-redox equations to/from redox reactions is an extremely useful advantage that the pe approach has over the E_H approach.

In general, for a given value of either p_{H_2} or p_{O_2}, it may be concluded that the more acidic the water is, the higher the pe will be. Indeed, for the reactions in both (19.102) and (19.106), the protons are located on the same side of the reaction as the electrons. As a result, by Le Chatelier's principle, increasing $\{H^+\}$ will tend to lead to a consumption of electrons and therefore a more oxidizing state for the system.

19.4.2 p_{H_2} and p_{O_2} Limited to 1.0 atm at Sea Level: pe Ranges from about −10 to 17 in Natural Waters

At sea level at the surface of the earth, the pressure produced by the overlying atmosphere is 1.0 atm. Unless a system is contained within a closed vessel that can be pressurized, total gas pressures significantly larger than 1.0 atm cannot be obtained. For example, it is well known that the vapor pressure of water increases with temperature. If the temperature is increased to slightly above 100°C, the vapor pressure of some water in a pot on a stove will then be slightly greater than the overlying pressure of 1.0 atm. At this point, bubbles of pure $H_2O_{(g)}$ can form and expand against the overlying pressure of 1.0 atm: the water will boil[5] (see Figure 19.9). In this open system, adding more heat

[5]Actually, at 1 atm pressure, bubbles can start to form in water at a temperature below 100°C due to the fact that increasing the temperature tends to drive dissolved gases like O_2 out of solution. Once those gases have been stripped out, the water will boil at exactly 100°C. Note that the vapor pressure of H_2O may be thought of as the equivalent in-situ pressure of dissolved H_2O.

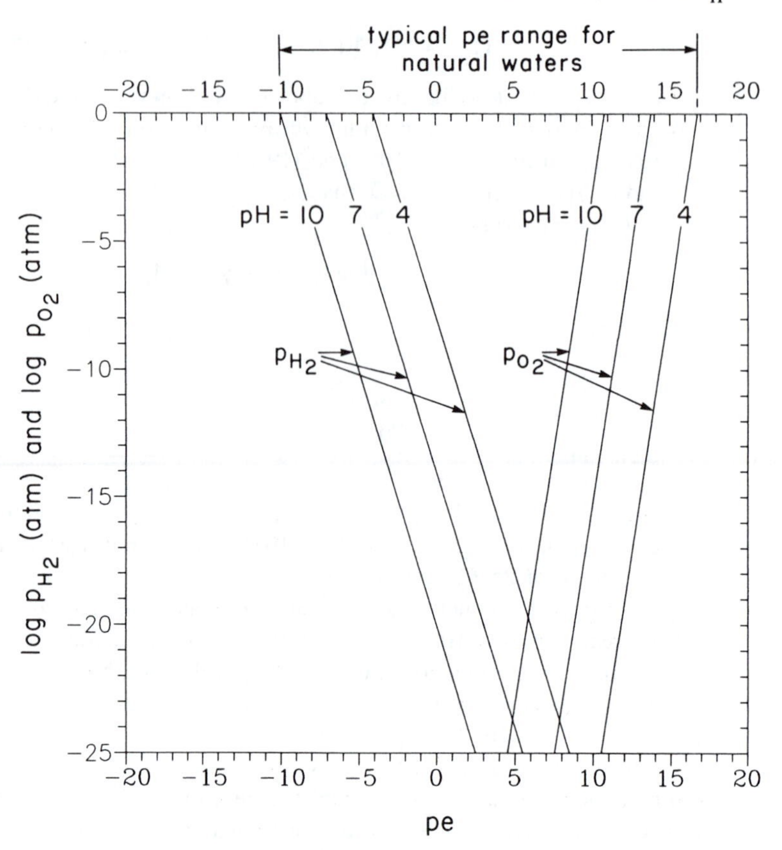

Figure 19.8 Log p_{H_2} (atm) and log p_{O_2} (atm) vs. pe at 25°C/1 atm at pH = 4, 7 and 10.

to the water *will not increase the vapor pressure further*. In this open system, adding more heat just permits the boiling process to continue. The continued vaporization absorbs the added heat. At the water/atmosphere interface, the pure $H_2O_{(g)}$ will be dissipated into the atmosphere. The boiling of water on a stove may be viewed as *thermal boiling*.

We see then that if some process serves to increase the in-situ pressure of a dissolved gas, then at the surface of the earth, that process can continue only to the point that an in-situ pressure of 1.0 atm is reached. Beyond that, the process will lead to the production of bubbles of the gas that are at slightly larger than 1.0 atm, with those bubbles then forming, rising, and being dissipated into the atmosphere.

The situation with redox processes is exactly analogous to what happens when we heat water. At sea level, if we try to reduce the pe of some water at a given constant pH, we can only lower the pe and reduce water to the point that the in-situ p_{H_2} becomes as large as 1.0 atm. Beyond that, *chemical boiling* of the water will occur, and bubbles of $H_{2(g)}$ at 1.0 atm will start to form (see Figure 19.10*a*). At 25°C/1 atm, for pH = 4, 7,

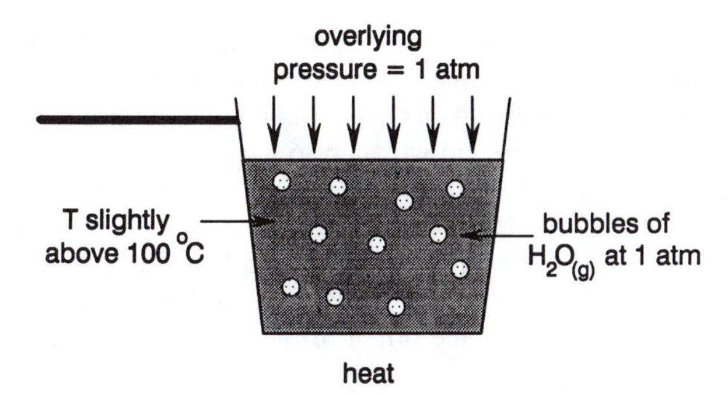

Figure 19.9 Water can be boiled in a pot on a stove. The vapor pressure of the water reaches 1 atm when the temperature reaches 100°C, and exceeds 1 atm at all temperatures above 100°. Thus, if the overlying pressure is 1 atm, the water can boil as soon as the temperature exceeds 100°C by the slightest amount. This process may be thought of as *thermal boiling*.

and 10, Eq. (19.104) indicates that this will occur at pe = −4, −7, and −10, respectively. These values may be recognized as the top x-axis intercepts for the $\log p_{H_2}$ vs. pe lines in Figure 19.8. Since the vast bulk of natural waters have pH values less than 10, a pe value of −10 may be considered as an approximate lower limit on natural water pe. This is the origin of the lower limit of −10 on the pe range that was discussed in Section 19.3.4.

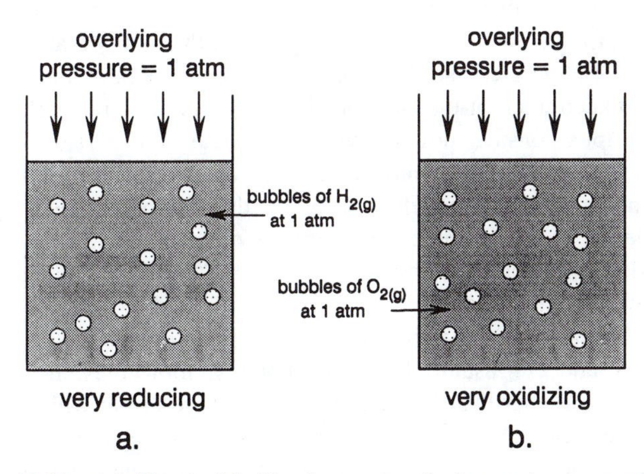

Figure 19.10 (*a*) *Chemical boiling* in a very reducing environment when the overlying pressure equals 1 atm. Bubbles of $H_{2(g)}$ can form when the in-situ p_{H_2} exceeds 1.0 atm by the slightest amount. When pH = 4, 7, or 10, this can occur when pe drops below −4, −7, or −10, respectively (see Figure 19.8). (*b*) *Chemical boiling* in a very oxidizing environment when the overlying pressure equals 1 atm. Bubbles of $O_{2(g)}$ can form when the in-situ p_{O_2} exceeds 1.0 atm by the slightest amount. When pH = 4, 7, or 10, this can occur when pe rises above 16.78, 13.78, and 10.78, respectively (see Figure 19.8).

By the same token, if we try to raise the pe of some water at a given constant pH, at sea level, we can only oxidize water and raise the pe to the point that the in-situ p_{O_2} becomes as large as 1.0 atm. Beyond that, chemical boiling of the water will again occur. This time, bubbles of $O_{2(g)}$ at 1.0 atm will start to form. (See Figure 19.10*b*.) At 25°C, for pH = 4, 7, and 10, Eq. (19.109) indicates that this will occur at pe = 16.78, 13.78, and 10.78, respectively. These values may be recognized as the top *x*-axis intercepts for the $\log p_{O_2}$ vs. pe lines in Figures 19.8. Since the vast bulk of natural waters have pH values greater than 4, a pe value of 16.78 may be considered as an approximate upper limit on natural water pe. This is the origin of the upper limit of 17 on the pe range discussed in Section 19.3.4.

19.5 COMPUTATION OF ASSIGNED OXIDATION STATES

In general, a covalent bond is formed between two atoms when each of the atoms brings one electron to the bond. When the two atoms are identical, then the two electrons in the bond are shared equally between the two atoms. Thus, the oxidation state does not change when two atoms of oxygen combine to form an O_2 molecule; both the atoms and the O_2 contain oxygen in the zero oxidation state. Similarly, Cl_2 contains chlorine in the zero oxidation state. By extension, all pure elements whether in atomic or molecular form are in the zero oxidation state.

When a bond is formed between two dissimilar atoms, then for purposes of assigning oxidation states, the atom that is more electronegative is assigned both electrons in the bond. The electron that was brought to the bond by the less electronegative atom thus contributes -1 towards the oxidation state of the more electroctronegative atom. Of course, this type of assignment does not exactly reflect the reality that in every true bond, the electrons in the bond are shared at least somewhat by both atoms. However, it does provide a way to keep track of the electrons in a molecule in a rationale and useful way.

As an example, consider now the water molecule. When H_2O is formed from its consituent elements, two hydrogen atoms bring one electron each for bonding to the neutral oxygen. Since oxygen is more electronegative than hydrogen, the oxygen is assigned both of those electrons, and so we see now why the oxygen in water is assigned an oxidation state of −II. Since each of the hydrogen atoms has "lost" its one electron to the oxygen, we see now why each of the hydrogens in water is said to have an oxidation state of +I (or simply I). Because: 1) oxygen is usually more electronegative than the other atoms to which it bonds; and 2) it usually forms two bonds in molecules, then when both of its bonds are to elements other than oxygen, then its oxidation state will usually be −II. By the same token, because: 1) hydrogen is usually less electronegative than the other atoms to which it bonds, and 2) each hydrogen usually forms just one bond, then its oxidation state will usually be I.[6] Figure 19.11 summarizes the electronegatives of the elements.

[6]Hydrogen in the −I oxidation is also known, as in the organic synthesis reagent lithium aluminum hydride, $LiAlH_4$. The electrons in the bonds to hydrogen are assigned to the hydrogens because H is more electronegative than either lithium or aluminum. $LiAlH_4$ is very unstable since H(−I) is very much less stable than H(I); each H(−I) will readily lose the two electrons assigned to it and will be converted to H(I). $LiAlH_4$ is thus a powerful reducing agent.

In a *neutral* molecule, we see that the above convention requires that *the sum of all the oxidation states of the atoms in the molecule is zero.* For the molecule H_2O_2 which has the structure H-O-O-H, each of the hydrogens is in the I oxidation state. This leaves a charge of −2 to be distributed over the two oxygens, or an oxidation state of −I for each of the two oxygens. We see that this assignment is consistent with the above guidelines since each of the two oxygens has only one bond to a less electronegative atom.

For an *ion*, the source of the charge is attributable either to a net excess or net deficiency of electrons. Thus, for an ion, *the sum of all of the oxidation states of the atoms in the ion must equal the charge on the ion.* For OH^-, with the hydrogen in the I oxidation state and the oxygen in the −II oxidation state, we obtain the net charge for the ion of −1.

While oxygen is often in the −II oxidation state and hydrogen is often in the I oxidation state, all other elements do not behave so simply. For example, carbon can vary its oxidation state from IV down to −IV. For the HCO_3^- ion, we give the three oxygens a total charge of −6, then add +1 back in for the hydrogen for a sum of −5. Since HCO_3^- bears a charge of −1, the carbon in the HCO_3^- must be in the IV oxidation state. The carbon in both $H_2CO_3^*$ and CO_3^{2-} may be seen to also be in the IV oxidation state. In the formate ion, $HCOO^-$, we give the two oxygens a net charge of −4, add +1 back for the hydrogen for a sum of −3. Since the formate ion has a net charge of −1, the carbon in the formate ion is assigned an oxidation state of II.

Sometimes, within a molecule, a given element will have different combinations of bonds in different places in the molecule. For example, in the molecule ethanol (CH_3-CH_2-OH), one of the carbons has three bonds to hydrogen and one bond to carbon. The other has two bonds to hydrogen, one bond to carbon, and one bond to oxygen. We can compute the *average* oxidation state of the carbons in ethanol by computing a net charge of +6 for the hydrogens, subtracting out 2 for the oxygen, leaving +4 to be distributed over the two carbons for an average oxidation state of II for each carbon.

The compounds formaldehyde (CH_2O) and glucose ($C_6H_{12}O_6$) may be viewed as hydrates of elemental carbon which is C(0) (note that both molecules contain an integral number of water molecules per carbon atom. In both formaldehyde and glucose, therefore, we obtain an average oxidation state of C(0) for these molecules. Although the *assigned* average oxidation states of carbon are all zero in graphite (a form of elemental carbon), diamond (another form of elemental carbon), formaldehyde, and glucose, since the carbon atoms in these different forms have different average electronic environments, they are actually at somewhat different average energy levels.

With bare ions like Fe^{3+}, the oxidation state equals the charge, and so as we already know, the iron in Fe^{3+} is in the III oxidation state. The sodium in Na^+ is in the I oxidation state. The iron in $FeOH^{2+}$ is also in the III oxidation state. In terms of the guidelines outlined, this may be understood by assigning a charge of −2 to the oxygen and +1 to the hydrogen for a charge of −1 on the OH^- unit. Since $FeOH^{2+}$ has a charge of 2+, the iron must be assigned a charge of +3, that is, an oxidation state of III. In the species $FeOH^+$, application of the above guidelines leads to the conclusion that the iron is in the II oxidation state. A variety of other examples are summarized in Table 19.5.

Problems illustrating the principles of this chapter are presented by Pankow (1992).

TABLE 19.5. Examples of oxidation state calculations.

Molecule or Ion	Name	Oxidation States
O_2	molecular, elemental oxygen	O(0)
H_2O_2	hydrogen peroxide	O(-I), H(I)
H_2O	water	O(-II), H(I)
OH^-	hydroxide	O(-II), H(I)
H^+	proton	H(I)
HOCl	molecular hypochlorous acid	Cl(I), O(-II), H(I)
OCl^-	hypochlorite ion	Cl(I), O(-II)
Cl_2	molecular, elemental chlorine	Cl(0)
HCl	molecular hydrochloric acid	Cl(-I), H(I)
Cl^-	chloride ion	Cl(-I)
$H_2CO_3^*$	carbonic acid	C(IV), O(-II), H(I)
HCO_3^-	bicarbonate ion	C(IV), O(-II), H(I)
CO_3^{2-}	carbonate ion	C(IV), O(-II)
$HCOO^-$	formate ion	C(II), O(-II), H(I)
CH_3-CH_2-OH	ethanol	C(II), O(-II), H(I)
$C_{(s)}$	elemental carbon, e.g. graphite or diamond	C(0)
CH_2O	formaldehyde	C(0)
$C_6H_{12}O_6$	glucose	C(0)
CH_3-CH_3	ethane	C(-III)
CH_4	methane	C(-IV)
H_2SO_4	molecular sulfuric acid	S(VI), O(-II), H(I)
HSO_4^-	bisulfate	S(VI), O(-II), H(I)
SO_4^{2-}	sulfate	S(VI), O(-II)
$S_{8(s)}$	molecular, elemental sulfur	S(0)
H_2S	hydrogen sulfide	S(-II), H(I)
HS^-	bisulfide	S(-II), H(I)
S^{2-}	sulfide	S(-II)
HNO_3	molecular nitric acid	N(V), O(-II), H(I)
NO_3^-	nitrate	N(V), O(-II)
HNO_2	molecular nitrous acid	N(III), O(-II), H(I)
NO_2^-	nitrite	N(III), O(-II)
N_2	molecular, elemental nitrogen	N(0)
NH_3	ammonia	N(-III), H(I)
NH_4^+	ammonium	N(-III), H(I)

H 2.1																	He -
Li 1.0	Be 1.5											B 2.0	C 2.5	N 3.0	O 3.5	F 4.0	Ne -
Na 0.9	Mg 1.2											Al 1.5	Si 1.8	P 2.1	S 2.5	Cl 3.0	Ar -
K 0.8	Ca 1.0	Sc 1.3	Ti 1.5	V 1.6	Cr 1.6	Mn 1.5	Fe 1.8	Co 1.8	Ni 1.8	Cu 1.9	Zn 1.6	Ga 1.6	Ge 1.8	As 2.0	Se 2.4	Br 2.8	Kr -
Rb 0.8	Sr 1.0	Y 1.2	Zr 1.4	Nb 1.6	Mo 1.8	Tc 1.9	Ru 2.2	Rh 2.2	Pd 2.2	Ag 1.9	Cd 1.7	In 1.7	Sn 1.8	Sb 1.9	Te 2.1	I 2.5	Xe -
Cs 0.7	Ba 0.9	Lu 1.1	Hf 1.3	Ta 1.5	W 1.7	Re 1.9	Os 2.2	Ir 2.2	Pt 2.2	Au 2.4	Hg 1.9	Tl 1.8	Pb 1.8	Bi 1.9	Po 2.0	At 2.2	Rn -
Fr 0.7	Ra 0.9	Lr 1.1	Th 1.3	Pa 1.5	U 1.7												

Figure 19.11. Electronegativities of the elements. Electronegativity is a measure of electron affinity

19.6 REFERENCES

BARD, A. J., R. PARSONS, and J. JORDAN. 1985. *Standard Potentials in Aqueous Solution.* New York: Marcel Dekker.

PANKOW, J. F. 1992. *Aquatic Chemistry Problems.* Portland: Titan Press-OR, P.O. Box 91399, Portland, Oregon 97291-1399.

STUMM, W. and J. J. MORGAN. 1981. *Aquatic Chemistry.* New York: Wiley-Interscience.

20

Introduction to pe-pH Diagrams: The Cases of Aqueous Chlorine, Hydrogen, and Oxygen

20.1 INTRODUCTION

The chemistry of natural waters is controlled by a number of variables. pe, pH, and p_{CO_2} are three very important ones. As discussed in previous chapters, these three parameters are often referred to as "master variables." We are therefore interested in learning how to draw diagrams that will tell us how the aqueous chemistry of a given element of interest changes with pe and pH. A natural way to draw a diagram of this type would be to have various forms of the element predominate in various pe-pH domains. When this is done, the resulting diagram is referred to as a *pe-pH predominance diagram.*

Prior exposure in this text to the concept of predominance diagrams occurred in our discussion of Figures 14.1, 14.3, and 14.4. In those figures, the variables were $\log p_{CO_2}$ and pH. The species having predominance regions included Fe^{2+}, $FeOH^+$, $Fe(OH)_2^o$, $Fe(OH)_3^-$, $FeCO_{3(s)}$, and $Fe(OH)_{2(s)}$. At any given point in one of the regions labeled Fe^{2+}, the species Fe^{2+} was explained as being present at a concentration higher than that of any other Fe(II)-containing species. However, in the region for a given solid, that solid was discussed as *controlling* the aqueous chemistry of the Fe(II); it may or may not actually represent the form in which *most* of the Fe(II) is present in the total (i.e, aqueous solution + solid) system. Indeed, *just after* crossing into one of the solid regions from the left (i.e., from the dissolved Fe(II) region), that specific solid is neither present in the system to the *exclusion* of dissolved Fe(II) nor even present in amounts greater than dissolved Fe(II). Rather, that solid now just *controls* the Fe(II) chemistry.

As discussed in Chapter 14, strictly speaking, one should perhaps refer to diagrams such as Figures 14.1, 14.3, and 14.4 as "*predominance and/or solubility-controlling*" diagrams. However, since these diagrams are constructed in a *log/log format*, significant chemical change can occur when you move a fraction of a unit in either the *y* or *x* direction. Therefore, not long after crossing into a solid region, that solid may well predominate as well as control. Thus, the simpler name *predominance diagram* will suffice for Figures 14.1, 14.3, and 14.4, and also for the pe-pH diagrams to be discussed in this chapter.

With pe as a very important master variable in natural waters, we recognize that Figures 14.1, 14.3, and 14.4 must have been prepared with some type of assumption concerning the value of pe. The forms of Fe in those figures represent Fe in a reduced oxidation state, but not the most reduced state possible. That is, Fe(II) is reduced relative to Fe(III), but not as reduced as Fe(0) which would be solid, elemental $Fe_{(s)}$. Thus, Figures 14.1, 14.3, and 14.4 were prepared assuming a sufficiently low pe that Fe(II) dominates the chemistry, but not so low as to permit the formation of Fe(0).

If we add a *z*-axis to Figure 14.1 for pe, we obtain Figure 20.1. A log p_{CO_2} vs. pH diagram thus represents an *xy* slice at some constant pe. Figure 14.1 is an *xy* slice at fairly low pe. A figure which presented Fe(III) chemistry in a $\log p_{CO_2}$ vs. pH format would be an *xy* slice at high pe. A pe vs. pH diagram would be an *xz* slice at constant $\log p_{CO_2}$.

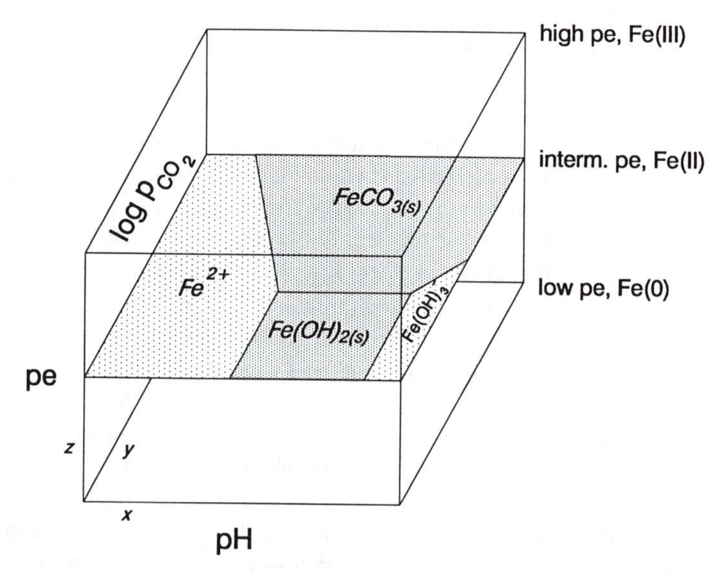

Figure 20.1 Three-dimensional pe-pH-log p_{CO_2} coordinate system for iron for $Fe_T = 10^{-4}$ *M*. At high pe, iron in the Fe(III) oxidation state will dominate the chemistry, though some Fe(II) species will also be present. At intermediate pe, iron in the Fe(II) oxidation state will dominate the chemistry, though some Fe(III) species will also be present. At very low pe, solid Fe(0) will be present, along with low levels of Fe(II) and exceedingly low levels of Fe(III). Figure 14.1 for $Fe_T = 10^{-4}$ *M* is shown as a horizontal, $\log p_{CO_2}$ vs. pH slice through the three-dimensional coordinate system.

We now discuss how pe vs. pH diagrams can be used to study the manner in which the chemistry of an element will change in response to changes in the redox and pH conditions in the solution. We will consider pe-pH diagrams for chlorine, hydrogen and oxygen. For all three elements, the value of $\log p_{CO_2}$ will be considered to be sufficiently low that CO_2 has no effect on the chemistry of these elements. For chlorine, we will follow the approach used by Stumm and Morgan (1981).

20.2 pe-pH DIAGRAM FOR AQUEOUS CHLORINE

20.2.1 Drawing a Generic pe-pH Diagram for Aqueous Chlorine

Chlorine has three oxidation states that are of interest here, I, 0, and −I.[1] Cl(0) is elemental chlorine, and exists in aqueous solution as Cl_2. The species Cl_2 is the reference point from which all other oxidation state values are measured. In particular, a chlorine atom that has lost one electron relative to the number of electrons which a Cl unit has in Cl_2 is in the I oxidation state. A chlorine atom that has gained one electron relative to the number of electrons each Cl has in Cl_2 is in the −I oxidation state. Two different Cl species are possible in the I oxidation state, HOCl and OCl^-. Two species are also possible in the −I oxidation state. This chemistry is summarized in Table 20.1.

Based on the information presented in Table 20.1, *generically*, we expect that the pe-pH diagram for chlorine will look *something* like the diagram in Figure 20.2. All of the acid/base boundarylines (i.e., the HOCl/OCl^- and HCl/Cl^- boundarylines) in that figure have been drawn as simple vertical lines. We will find that it is proper to do this. For simplicity, all of the OX/RED boundarylines (i.e., the HOCl/Cl_2, OCl^-/Cl_2, Cl_2/HCl, and Cl_2/Cl^- boundarylines) in Figure 20.2 have been drawn as simple horizontal lines. Within the context of the assumptions made in this section, we will determine that while

TABLE 20.1. Redox chemistry of aqueous chlorine.

oxidation state	species	name
I	HOCl	hypochlorous acid
	OCl^-	hypochlorite ion
0	Cl_2	elemental chlorine
−I	HCl	hydrochloric acid
	Cl^-	chloride ion

[1]Chlorine has other oxidation states besides I, 0, and −I. For example potassium perchlorate ($KClO_4$) contains Cl in the +VII oxidation state. Chlorine dioxide (ClO_2) contains Cl in the IV oxidation state. As will be seen, even the I state is far outside the stability range of water, and so we will not bother to consider oxidation states above I here. The interested reader may, however, refer to Cotton and Wilkinson (1980).

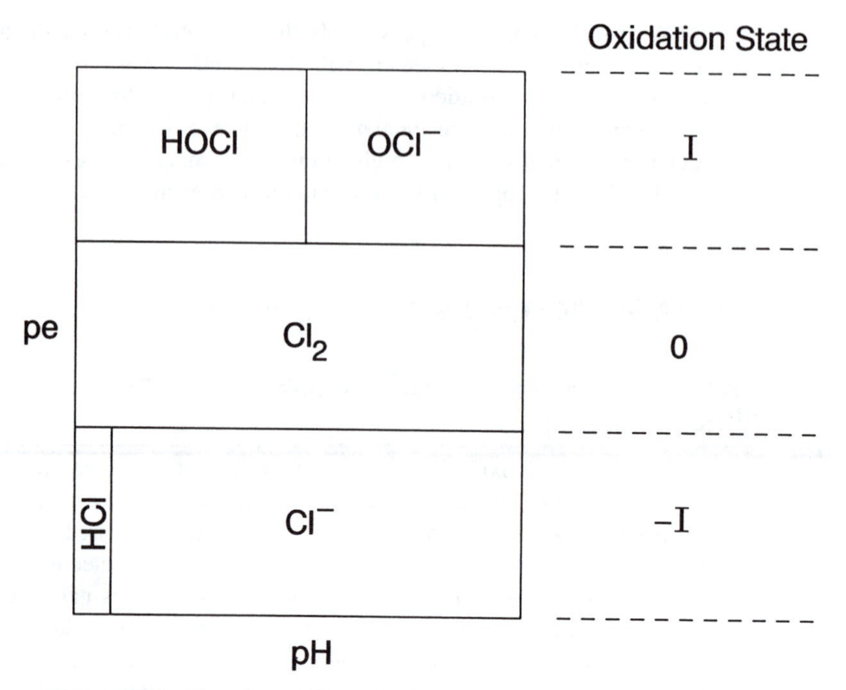

Figure 20.2 Generic pe-pH predominance diagram for aqueous chlorine at a Cl_T value large enough for a region for Cl_2 to be present. A line between predominance regions for two species in the same oxidation state (e.g., HOCl and OCl^-) will be a simple vertical line at a specific pH. For HOCl and OCl^-, this occurs when the pH equals the p*K* for HOCl. A line between predominance regions for two species not in the same oxidation state may or may not appear as a simple horizontal line; the slope of such a line will depend on the numbers of protons (or hydroxides) and electrons participating in the reduction half reaction. If no protons (or hydroxides) participate, then the boundaryline will be a straight horizontal line. Thus, the Cl_2/Cl^- boundaryline is essentially a straight line as long as one can assume that no other species contribute in a significant way to Cl_T.

this is correct for some of these OX/RED boundarylines, it will not be the case for others. In particular, we will determine that

1. Redox half-reactions that *do not* consume or produce protons or hydroxide ions will depend only on pe, and will in fact give rise to simple *horizontal* boundarylines. For example, when Cl_2 and Cl^- are assumed to be the only important chlorine-containing species, the boundaryline between the Cl_2 and Cl^- regions will be a simple horizontal line.
2. Redox half-reaction that *do* consume or produce protons or hydroxide ions will depend on both the pe and the pH, and will give rise to boundarylines with finite, non-zero slopes. The slopes can be positive or negative. When a redox half-reaction

written as a reduction (i.e., as consuming electrons) also involves the *consumption* of protons (or production of hydroxide ions), the *d*pe/*d*pH slope will be *negative*. The *production* of protons (or consumption of hydroxides) will lead to a *positive* slope.

3. Non-redox equilibria (no electrons consumed or produced) that consume or produce protons or hydroxides (e.g., acid dissociations, metal ion complexations with hydroxide, etc.) will give rise to simple *vertical* boundarylines. For the case at hand, the boundaryline between the HOCl and OCl^- regions will be given by a vertical line drawn at pH = p*K* for HOCl (= 7.3 at 25°C/1 atm); the boundaryline between the HCl and the Cl^- regions will be a vertical line drawn at pH = p*K* for HCl (= –3 at 25°C/1 atm).

20.2.2 Drawing the Actual pe-pH Diagram for Aqueous Chlorine for Specific Cl_T

Based on the above considerations, we know that the slopes of some of the OX/RED boundarylines given roughly in Figure 20.2 will not, in fact, be zero. Since some of the lines may have significant slopes, the actual pe-pH diagram might look much different from Figure 20.2. Indeed, the actual positions of some of the lines may be so far from other lines that "expected" boundarylines in a given pe-pH diagram become non-applicable. When this happens, the preparation of the pe-pH plot can be a little tricky. As it turns out, the case of aqueous chlorine is an example of this type of situation. This will be discussed in detail.

The task now is to determine the actual equations for the various boundarylines for aqueous chlorine. The following equilibrium data are available at 25°C/1 atm:

$$HOCl + H^+ + e^- = \frac{1}{2}Cl_2 + H_2O \qquad K = 10^{26.9} \tag{20.1}$$

$$\frac{1}{2}Cl_2 + e^- = Cl^- \qquad K = 10^{23.6} \tag{20.2}$$

$$HOCl = H^+ + OCl^- \qquad K = 10^{-7.3}. \tag{20.3}$$

These reaction equilibria can be used to locate the various predominance region boundarylines. For the boundaryline between the HOCl and Cl_2 regions, neglecting activity corrections for the dissolved chlorine species, the equilibrium in Eq. (20.1) gives

line	equation	depends on	boundary	
(1)	$pe = 26.9 + \log[HOCl] - \frac{1}{2}\log[Cl_2] - pH$	pe and pH	$HOCl/Cl_2$	(20.4)

Equation (20.4) cannot be used directly to draw a pe vs. pH line because log[HOCl] and $\log[Cl_2]$ remain as variables. This means that to draw the $HOCl/Cl_2$ boundaryline, we must either assume specific values for log[HOCl] and $\log[Cl_2]$, or some relationship between these two quantities that will allow both terms to be removed as variables from Eq. (20.4). Let us pick a specific value for the total amount of Cl_T in the solution. Since

HCl is such a strong acid, it can be neglected in any equation for Cl_T. Therefore, for $Cl_T = 0.04\ M$, we have

$$Cl_T = 0.04\ M = [HOCl] + [OCl^-] + 2[Cl_2] + [Cl^-]. \tag{20.5}$$

Along most of the boundaryline between HOCl and Cl_2, OCl^- and Cl^- will not contribute significantly to Cl_T. Therefore, along most of that boundaryline we have

$$Cl_T = 0.04\ M \simeq [HOCl] + 2[Cl_2]. \tag{20.6}$$

A logical way to apportion the concentrations of the two above species for drawing the $HOCl/Cl_2$ boundaryline is to have equal amounts of chlorine atoms present in the two species. That is, along the boundaryline, we set

$$[HOCl] = \left(\frac{1}{2}\right) 0.04\ M \qquad \text{and} \qquad [Cl_2] = \left(\frac{1}{4}\right) 0.04\ M. \tag{20.7}$$

Thus, by Eq. (20.4), we have

$$pe = 26.9 - 1.70 + 1.00 - pH \tag{20.8}$$

and so

$$\text{line 1}: \qquad pe = 26.2 - pH \qquad \frac{dpe}{dpH} = -1. \tag{20.9}$$

Line 1 is drawn in Figure 20.3.

We now consider the equation for the boundaryline between the regions for Cl_2 and Cl^-. By utilizing the equilibrium constant describing the redox equilibrium between these two species, neglecting activity corrections for Cl_2 and Cl^-, we have

line	equation	depends on	boundary	
(2)	$pe = 23.6 + \frac{1}{2}\log[Cl_2] - \log[Cl^-]$	pe only	Cl_2/Cl^-	(20.10)

To be consistent with the conditions used to obtain Eq. (20.9), we again let $Cl_T = 0.04\ M$, but this time we presume that Cl_2 and Cl^- are the major species. Assuming equal amounts of chlorine in the two forms along the boundaryline, we set

$$[Cl_2] = 0.01\ M \qquad [Cl^-] = 0.02\ M \tag{20.11}$$

and so

$$\text{line 2}: \quad pe = 24.3 \qquad \frac{dpe}{dpH} = 0. \tag{20.12}$$

Lines 1 and 2 will define the predominance region for Cl_2 at $Cl_T = 0.04\ M$. The two lines cross at the point pH = 1.9, pe = 24.3. What happens beyond that point? If we examine the shape of the predominance region for Cl_2 enclosed by the two lines, we see that it becomes narrower and narrower as the pH increases towards 1.9. Thus, at $Cl_T = 0.04\ M$, for pH > 1.9: 1) the Cl_2 (oxidation state 0) prodominance region must cease to exist, and we have a direct boundaryline between the HOCl and Cl^- regions; 2) Cl_2 (oxidation state 0) can no longer be a predominating, dissolved chlorine species; 3) the Cl^- region (oxidation state $= -$I) is pulled up directly against the HOCl region (oxidation state = I). The lack of an intermediate, oxidation state = 0 region beyond pH = 1.9 is a consequence of the facts that: a) the Cl_2/Cl^- line plots very high in the diagram; and b) the Cl_2/HOCl line has a negative slope.

Like $Fe_2(OH)_2^{4+}$, the species Cl_2 is a dimer. Thus, like $Fe_2(OH)_2^{4+}$, the importance of Cl_2 will depend on the total dissolved concentration of the element. As Cl_T is decreased, the area of the predominance region for Cl_2 will shrink. Since 0.04 M is a fairly high chlorine concentration for fresh waters, we can conclude that at typical total levels of chlorine, the region for Cl_2 will be even smaller than in Figure 20.3.

If Cl_2 cannot exist at Cl_T = 0.04 M as a dominant species beyond pH = 1.9, we now need an equation that describes the redox equilibrium and therefore the direct boundaryline between HOCl and Cl^-. Since the new boundaryline (line 3) must emanate from the vertex of lines 1 and 2, we can derive the equation for line 3 by adding together the chemical equilibrium equations underlying lines 1 and 2. In particular, we have

line	equation	depends on	boundary	
(1)	$pe = 26.9 + \log[HOCl] - \frac{1}{2}\log[Cl_2] - pH$	pe and pH	$\frac{HOCl}{Cl_2}$	(20.4)
+ (2)	$pe = 23.6 + \frac{1}{2}\log[Cl_2] - \log[Cl^-]$	pe	$\frac{Cl_2}{Cl^-}$	(20.10)
(3)	$2pe = 50.5 + \log[HOCl] - \log[Cl^-] - pH$	pe and pH	$\frac{HOCl}{Cl^-}$	(20.13)

In contrast to the cases encountered with lines 1 and 2, the two species appearing in the equation for line 3 both have just one chlorine per species. In this case, when equal amounts of chlorine are present in the two species along the boundaryline, the concentrations of the two species are exactly equal along the boundaryline. Thus, at 25°C/1 atm we can write

$$\text{line 3}: \quad pe = 25.25 - \frac{1}{2}\,pH \qquad \frac{dpe}{dpH} = -1/2. \tag{20.14}$$

Line 3 gives the locus of all points for which [HOCl] = $[Cl^-]$, even those lying in regions of the diagram wherein HOCl and Cl^- are present at low concentrations relative to other species.

It is important to note that line 3 (Eq. (20.14)) is independent of Cl_T. The reason for this is that both of the species in the two adjacent regions have the same number of chlorine atoms in them, that is, one. Since the positions (but not the slopes) of lines 1 and 2 depend on the value of Cl_T, this means that the specific point along line 3 at which lines 1 and 2 intersect will depend on Cl_T.

It happens that the equation for line 3 can also be derived directly by combining the equations for lines 1 and 2:

line 1 :	$pe = 26.2 - pH$	(20.9)
+ line 2 :	$pe = 24.3$	(20.12)
line 3 :	$2pe = 50.5 - pH.$	(20.14)

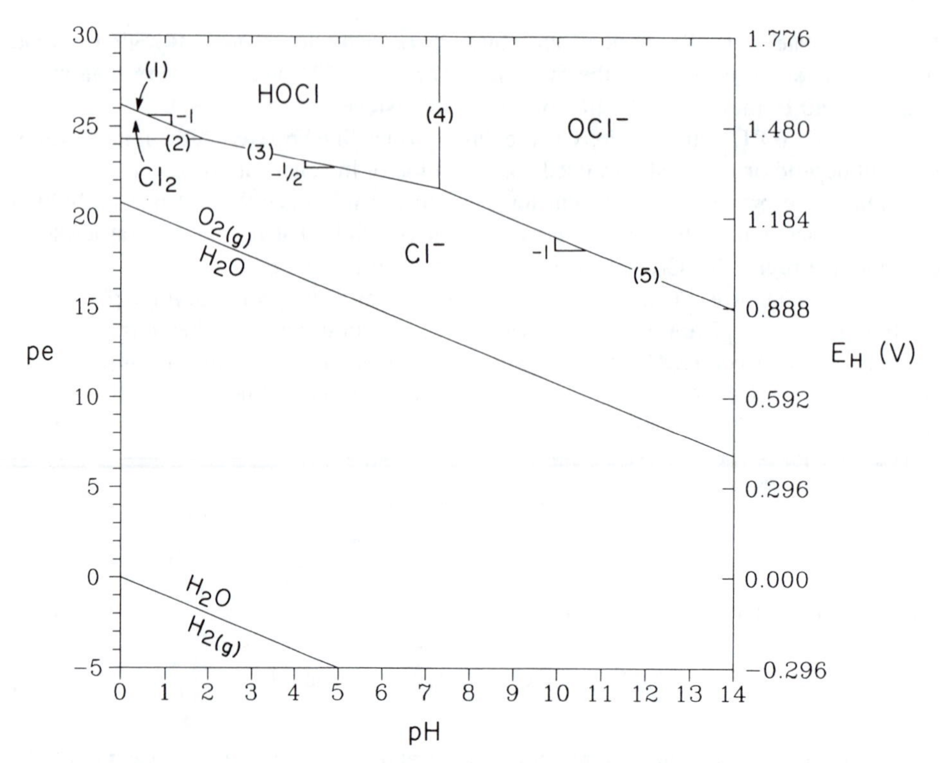

Figure 20.3 pe-pH predominance diagram for aqueous chlorine at 25°C/1 atm and $Cl_T = 0.04\ M$. The criterion for each boundary line is that there be equal mass amounts of chlorine present for both species along the line. This assumption leads to a contradiction at the intersection of lines 1 and 2. In actuality, the Cl_2 is not a sharp wedge as shown here, but a blunt wedge.

This simple combination works because: 1) the conditions along lines 1 and 2 collapse into the line 3 conditions:

$$\text{line 1 conditions}: \quad [\text{HOCl}] = 0.02M,\ [\text{Cl}_2] = 0.01M \tag{20.15}$$

$$+\ \text{line 2 conditions}: \quad [\text{Cl}_2] = 0.01M,\ [\text{Cl}^-] = 0.02M \tag{20.16}$$

$$\text{line 3 conditions}: \quad [\text{HOCl}] = [\text{Cl}^-] \tag{20.17}$$

and 2) the "*polarities*" of lines 1 and 2 as given above *happen to have been correct* for direct combination. In contrast, if line 2 had been written in the equivalent format 24.3 = pe then added to line 1, the result *would not have been correct*. We see then that the equations have to be written in the "proper direction" for direct combination to be successful.

What is behind the definition of "proper direction"? If we reexamine the foregoing discussion, we can see that this means that the log[ ] terms for the chemical species *implicitly present* in the underlying equations must cancel out during the addition. Thus, we have to be *very* careful in how simplified pe-pH equations are handled if they are just added together to yield new lines for the plot. All in all, it is probably too dangerous to try and do this; it is easier and certainly much safer to just go back and combine the equations for the lines in the non-simplified forms when they still explicitly contain the concentrations (e.g., see development of Eq. (20.13)).

For the boundaryline between the regions for HOCl and OCl^-, we use the equilibrium given in Eq. (20.3). Neglecting activity corrections for the dissolved chlorine species, at 25°C/1 atm, we obtain

line	equation	depends on	boundary	
(4)	$pH = 7.3 - \log[HOCl] + \log[OCl^-]$	pH	$\frac{HOCl}{OCl^-}$	(20.18)

Since the pK for HOCl is 7.3, it is clear that predominance between HOCl and OCl^- switches at pH = 7.3. At pH = 7.3, there will be equal amounts of the conjugate acid and conjugate base. Thus, we have

$$\text{line 4}: \quad pH = 7.3 \quad \text{independent of pe.} \tag{20.19}$$

Like line 3, line 4 is independent of Cl_T because both chlorine-containing species have the same number (one) of chlorine atoms. Also, in a manner that is analogous to line 3, line 4 gives the locus of all points for which $[HOCl] = [OCl^-]$, even those lying in regions of the diagram wherein HOCl and OCl^- are present at low concentrations relative to other species.

We now obtain the boundaryline between OCl^- and Cl^- by adding up the equations underlying lines 1, 2, and 4:

line	equation	depends on	boundary	
(1)	$pe = 26.9 + \log[HOCl] - \frac{1}{2}\log[Cl_2] - pH$	pe and pH	$\frac{HOCl}{Cl_2}$	(20.4)
(2)	$pe = 23.6 + \frac{1}{2}\log[Cl_2] - \log[Cl^-]$	pe	$\frac{Cl_2}{Cl^-}$	(20.10)
+ (4)	$pH = 7.3 - \log[HOCl] + \log[OCl^-]$	pH	$\frac{HOCl}{OCl^-}$	(20.18)
(5)	$2pe = 57.8 + \log[OCl^-] - \log[Cl^-] - 2pH$	pe and pH	$\frac{OCl^-}{Cl^-}$	(20.20)

When there are equal amounts of chlorine in the two forms OCl^- and Cl^-, we obtain

$$\text{line (5)}: \quad pe = 28.9 - pH \quad \frac{dpe}{dpH} = -1. \tag{20.21}$$

As with lines 3 and 4, line 5 is independent of Cl_T. Line 5 gives the locus of all points for which $[OCl^-] = [Cl^-]$, even those lying in regions of the diagram wherein OCl^- and Cl^- are present at very low concentrations relative to other species.

As a general note, it may be pointed out that for a pe-pH diagram that has n predominance regions, we will need a *minimum* of $(n - 1)$ equilibrium expressions for defining the boundarylines. The first expression considered will give a boundaryline between *two* regions. Every expression thereafter will allow the drawing of one additional boundaryline. Hence, the minimum total number of expressions needed is $(n-1)$. For the system described in Figure 20.3, four predominance regions were identified using $n=3$ equilibrium expressions (Eqs. (20.1)–(20.3)). If HCl were not as acidic as it is, the acid dissociation constant for HCl ($n=4$) would have allowed us to define a fifth predominance region (HCl). As will be made more clear in subsequent chapters, the quantity $(n-1)$ is given as a *minimum* above because species that do not themselves possess predominance regions will nevertheless affect the locations of the boundarylines between the regions that are present.

In addition to having the various chlorine lines present in Figure 20.3, we are also interested in drawing in two boundarylines that describe the stability limits of water. For the reduction of water to $H_{2(g)}$, at 25°C/1 atm, we have

$$\text{pe} = -\frac{1}{2}\log p_{H_2} - \text{pH}. \qquad (19.104)$$

When $p_{H_2} = 1$ atm,

$$\text{pe} = -\text{pH} \qquad (p_{H_2} = 1 \text{ atm}) \qquad (20.22)$$

Below the line given by Eq. (20.22), $H_{2(g)}$ bubbles will start forming in the water.

For the oxidation of water to $O_{2(g)}$ at high pe, at 25°C/1 atm, we have

$$\text{pe} = 20.78 - \text{pH} + \frac{1}{4}\log p_{O_2} \qquad (19.109)$$

When $p_{O_2} = 1$ atm,

$$\text{pe} = 20.78 - \text{pH} \qquad (p_{O_2} = 1 \text{ atm}) \qquad (20.23)$$

Above the line given by Eq. (20.23), $O_{2(s)}$ bubbles will start forming in the water. With the lines for Eqs. (20.22) and (20.23) included in Figure 20.3, it may be concluded that Cl^- is by far the most dominant equilibrium Cl species at the surface of the earth ($P = 1$ atm). Nevertheless, everywhere in the diagram we will be able to compute a non-zero value for the concentration of each of the different Cl species.

At the beginning of this section it was stated that drawing the pe-pH diagram for aqueous chlorine can be a bit tricky. The fact that it did not seem tricky is no accident. This may be explained as follows. The need for all this discussion about "trickiness" stems from the fact that the Cl_2/Cl^- boundaryline occurs at such a high pe value. The sequence according to which the boundarylines were drawn was chosen intentionally: it is in fact a sequence that allows us to draw the diagram without encountering any problems. What kind of "problems" do we refer to here?

Consider what would have happened if another sequence had been chosen for drawing the boundarylines. Let us say that the first line drawn was line 1. But then, say that line 4 had been drawn next. Indeed, line 4 is an easy one to draw, and it would have been tempting to draw it early on in the process. The boundaryline that we would have then been tempted to draw next would have been one that emanated from the vertex of an intersection between line 1 and line 4, producing a Cl_2/OCl^- boundaryline. All that would have remained to do then would have been to draw the Cl_2/Cl^- boundaryline. To our surprise and dismay, however, we would have drawn that line at such a high pe

(24.3) that it would have intersected the $Cl_2/HOCl$ line before pH = 7.3! This would mean that: 1) a significant portion of line 1 as drawn was not appropriate (i.e., the portion that extended to pe values lower than 24.3); and 2) that the line drawn as a boundary between Cl_2 and OCl^- was completely inappropriate. The only thing that one could then do would be to start over again, this time drawing the diagram as originally outlined above. Other sequences can also lead to difficulties, for example, any sequence starting with the drawing of a boundaryline between Cl_2 and OCl^-.

20.2.3 Interesting Aspects of the Aqueous Chlorine pe-pH Diagram

As Stumm and Morgan (1981) note, the nature of the pe-pH diagram for chlorine has some interesting implications. Firstly, significant amounts of Cl_2 cannot exist (i.e., predominate) at most pH values. If one tries to place Cl_2 in water at typical pH and pe values, most of it will be converted to other forms of chlorine. What other forms?

Because Cl_2 is the only Cl(0) form of chlorine, the only thing that can happen is the formation of another oxidation state. This can happen in one of two ways. Most simply, even if the equivalent partial pressure of O_2 in the system is as large as that given by the ambient value of 0.21 atm, water itself can be oxidized by the Cl_2, forming Cl^- and O_2. Alternatively, the conversion of Cl_2 to other oxidation states can occur if the Cl_2 reacts with itself. As seen in Figure 20.4, the Cl_2 molecule can be formed when two Cl atoms (each with seven outer-shell electrons) share a pair of electrons. The resulting single bond holds the two Cl atoms together. If the Cl_2 unit is broken up in a manner

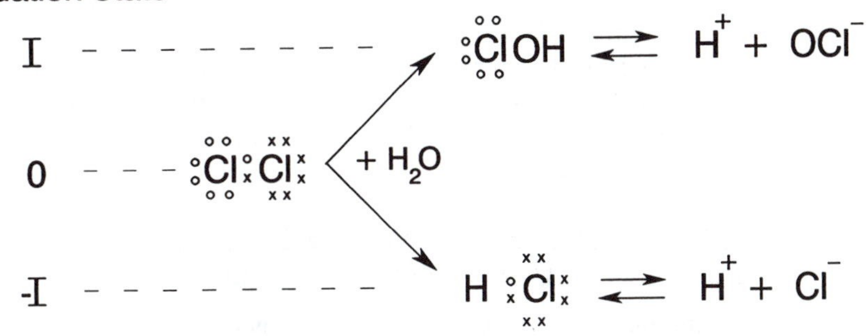

Figure 20.4 Disproportionation (autooxidation) of Cl_2 which is in the Cl(0) oxidation state to one chlorine in the Cl(I) oxidation state, and one chlorine in the Cl(−I) oxidation state. An equilibrium is established between the Cl(0) state and the Cl(I) and Cl(−I) states. Similarly, the HOCl that is formed equilibrates with H^+ and OCl^-, and the HCl that is formed equilibrates with H^+ and Cl^-. (Note: even at pH values as low as 1, there will be essentially complete conversion of HCl to H^+ and Cl^-.) Since the two oxidation states formed in the disproportionation are symmetrically distributed around the originating oxidation state, equal amounts of chlorine are formed in the I and −I states. If the disproportionation involved the conversion of Cl(0) to Cl(II) and Cl(−I), the amounts of chlorine converted to the Cl(−I) state would have been double the amount converted to the Cl(II) state.

such that *both* of the electrons in the bond go with one of the chlorines, that chlorine will be converted to the –I oxidation state species (Cl^-). The other chlorine will be converted to the I oxidation state (partly HOCl and partly OCl^-, the ratio depending on the pH). This type of reaction is termed an *autooxidation* reaction (also called a *disproportionation* reaction) because the Cl_2 molecule oxidizes a portion of itself. To accomplish this, a portion of the Cl_2 molecule must be reduced. Thus, some of the Cl(0) goes up to Cl(I), and some goes down to Cl(–I). At pH values less than 7.3, the net reaction for the process proceeds predominantly as given by

$$\frac{1}{2}Cl_2 + e^- = Cl^- \qquad K = 10^{23.6} \tag{20.2}$$

$$- \qquad HOCl + H^+ + e^- = \frac{1}{2}Cl_2 + H_2O \qquad K = 10^{26.9} \tag{20.1}$$

$$Cl_2 + H_2O = HOCl + H^+ + Cl^- \qquad K = 10^{-3.3} \tag{20.24}$$

While the K value for the disproportionation reaction given in Eq. (20.24) may look small, the value is somewhat deceiving. Indeed, consider that the numerator of the equilibrium concentration quotient will involve the product of *three* dissolved species, one of which is H^+:

$$K = \frac{[HOCl][H^+][Cl^-]}{[Cl_2]} = 10^{-3.3}. \tag{20.25}$$

The product of the concentrations of the three species in the numerator in Eq. (20.25) can be small without either [HOCl] or $[Cl^-]$ needing to be very small individually. Thus, at pH values that are basic to slightly acidic, $10^{-3.3}$ is large enough to drive the reaction far to the right.

Since the HOCl and OCl^- regions are both well above the pe-pH line for O_2/H_2O redox equilibrium for $p_{O_2} = 1$ atm, unless the in-situ partial pressure of O_2 in the system is rather high, the HOCl or OCl^- formed by the disproportionation reaction will be able to go on to oxidize water to form O_2 and Cl^-.

20.2.4 Violation of the Mass Balance Equation at the Intersection of Lines 1 and 2 in Figure 20.3

Along line 1 in Figure 20.3 we have assumed that [HOCl] = 0.02 M and $[Cl_2]$ = 0.01 M. Along line 2, we have assumed that $[Cl_2]$ = 0.01 M and $[Cl^-]$ = 0.02 M. At the intersection of these two lines (beyond which point we say that disproportion prevents Cl_2 from being the dominant species), we see that Figure 20.3 has been drawn assuming that [HOCl] = 0.02 M, $[Cl_2]$ = 0.01 M, and $[Cl^-]$ = 0.02 M, for a Cl_T of 0.06 M. This clearly violates our statement that Cl_T = 0.04 M for the diagram. A similar problem does *not* exist at the intersection of lines 3, 4 and 5 because all that we require at that intersection is that [HOCl] = $[OCl^-]$ = $[Cl^-]$ and that is possible for any value of Cl_T. It is only the lines bounding the Cl_2 region that lead to a violation of the MBE.

We can conclude that the assumptions concerning lines 1 and 2 begin to break down more and more as the intersection of lines 1 and 2 is approached. Thus, as in the $\log p_{CO_2}$ vs. pH diagrams presented in Figures 14.1, 14.3, and 14.4, we will need to take all of the solution species into consideration if we are to avoid that problem. One way to accomplish that would be to compute the line that gives $\alpha_{Cl_2} = 0.5$ where α_{Cl_2} is the fraction of dissolved chlorine present as Cl_2. All of the dissolved, Cl-containing species would thereby taken into consideration.[2]

The value of α_{Cl_2} will depend on both pH and pe. (In a manner that is analogous to what we observed in Chapter 18 for the fraction α_{22} for the dimer $Fe_2(OH)_2^{4+}$, we note that α_{Cl_2} will also depend on Cl_T.) The result will be a *blunt wedge* rather than the sharp wedge that we see in Figure 20.3. Line 3 would appear in exactly the same position as it is in Figure 20.3, except that it would extend all the way up to that wedge. The coordinates of the resulting intersection between line 3 and the wedge may be calculated to be pH = 1.72, and pe = 24.39. Lines 4 and 5 would remain unchanged.

Computation of the locus of (pH, pe) coordinates giving the blunt Cl_2 boundaryline would be obtained by plugging a range of pe values[3] into the equation $\alpha_{22} = 0.5$, and determining the corresponding pH values by trial and error for each pe value.

There are two reasons why we prepared Figure 20.3 as we did (i.e., assuming some species to be negligible along some of the lines) rather than using the approach outlined in the preceding three paragraphs. Firstly, even though doing so can lead to significant violations of the pertinent MBE (most severely at one or more line intersections), most introductory texts in aquatic chemistry take the assumptions approach in preparing pe-pH diagrams. Thus, we have desired to provide the reader with some instruction in that approach so that those other texts can be made as understandable and useful as possible to the reader. Secondly, the differences between Figure 20.3 and an "exact" diagram in which the Cl_2 boundary wedge is computed without assumptions (e.g., using $\alpha_{Cl_2} = 0.5$) are not prominent enough to be worth the trouble if all we want to do is get an overview of Cl chemistry.

From this point onwards, this text employs the exact approach when computing all pe-pH boundarylines.

20.3 pe-pH DIAGRAM FOR AQUEOUS HYDROGEN

For hydrogen, the two stable oxidation states that we need to consider in aqueous systems are H(I) and H(0). In water, most of the H(I) is present as liquid water itself. Some is also present as H^+ and OH^-, and as the protons on acids other than water. H(0) will be present as dissolved H_2. If a gas phase is present in addition to the aqueous phase, then some H(I) will be present as $H_2O_{(g)}$, and some H(0) will be present as $H_{2(g)}$.

[2]This "simple majority" boundary would be slightly different from a second alternative boundary along which the amount of chlorine in Cl_2 predominates in a simple way over the amounts of chlorine in the other species, that is, is greater than that in any other *individual* species (a case of "species plurality"). Although we usually think of predominance diagrams in a pluraility type of way, viewpont, calculating the plurality line would be even more complicated than calculating the simple majority line because we would have no clear criterion like $\alpha_{Cl_2} = 0.5$ with which to work.

[3]Doing the opposite would be more complicated since the function viewed as pe = f(pH) is not single-valued (i.e., we can get two different pe values for a single pH).

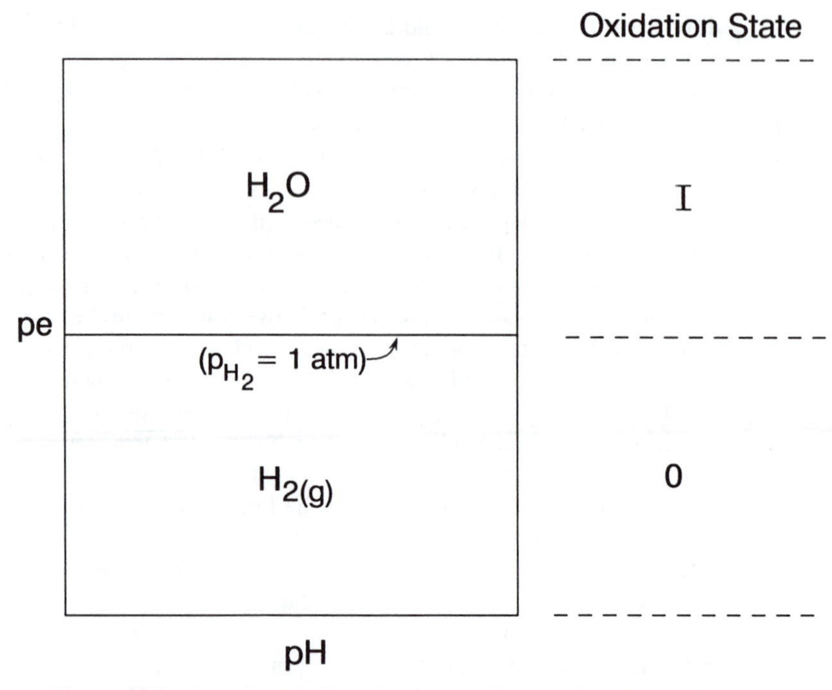

Figure 20.5 Generic pe-pH predominance diagram for aqueous hydrogen. The boundaryline between H(I) and H(0) is assumed to be that along which $p_{H_2} = 1$ atm.

Given the high molar concentration of H_2O in water, it is not very physically realistic to talk about drawing the H(I)/H(0) boundaryline along the line that gives equal amounts of H in dissolved H_2 and in H_2O. Moreover, the relative insolubility of $H_{2(g)}$ in water means that even when the molarity of H_2 in water is still fairly low compared to the molarity of H_2O, p_{H_2} will be comparatively high, and the system will be extremely reducing. For these reasons, and because H_2 is usually handled as a gas in equilibrium constant expressions, the H(I)/H(0) OX/RED boundaryline in the pe-pH diagram for hydrogen is usually drawn according to the redox equilibrium between H_2O and $H_{2(g)}$ rather than between H_2O and dissolved H_2. Accordingly, the generic pe-pH diagram for hydrogen may be found in Figure 20.5. Since a negative *d*pe/*d*pH slope will result whenever protons are consumed along with electrons in a boundaryline half reaction, according to Eq. (19.102), we can expect that the slope of the actual H(I)/H(0) boundaryline will be negative.

The H(I)/H(0) equilibrium line for $p_{H_2} = 1$ atm at 25°C/1 atm was included in Figure 20.3 as a lower stability limit for water. It is based on the equation

$$\text{pe} = -\frac{1}{2}\log p_{H_2} - \text{pH}. \qquad (19.104)$$

For $p_{H_2} = 1$ atm, we have simply that

$$pe = -pH. \tag{20.26}$$

Equation (20.26) is drawn again in Figure 20.6 as a solid line and represents a pe-pH "predominance boundaryline" for hydrogen. The quotation marks emphasize again that the conditions of $[H_2O] = 1.0$ (mole fraction scale) and $p_{H_2} = 1$ atm do not correspond to equal amounts of hydrogen present as H_2O liquid and as dissolved H_2.

Although the H(I)/H(0) boundaryline drawn most often using Eq. (19.104) corresponds to that for $p_{H_2} = 1$ atm, actually, Eq. (19.104) can be used to draw a boundaryline at 25°C/1 atm for any value of $p_{H_2} \leq 1$ atm. Strictly speaking, it cannot be used for $p_{H_2} > 1$ atm because the value of the equilibrium constant upon which it is based is for a *total* pressure of only 1 atm. Lines for a range of p_{H_2} values less than 1 atm are given as dashed lines in Figure 20.6. When the system at a given pe and pH does not contain any gas phase, the lines give the in-situ value of p_{H_2}. This applies even along the $p_{H_2} = 1$ atm line because, from Chapter 19, we recall that actual bubbles of pure $H_{2(g)}$

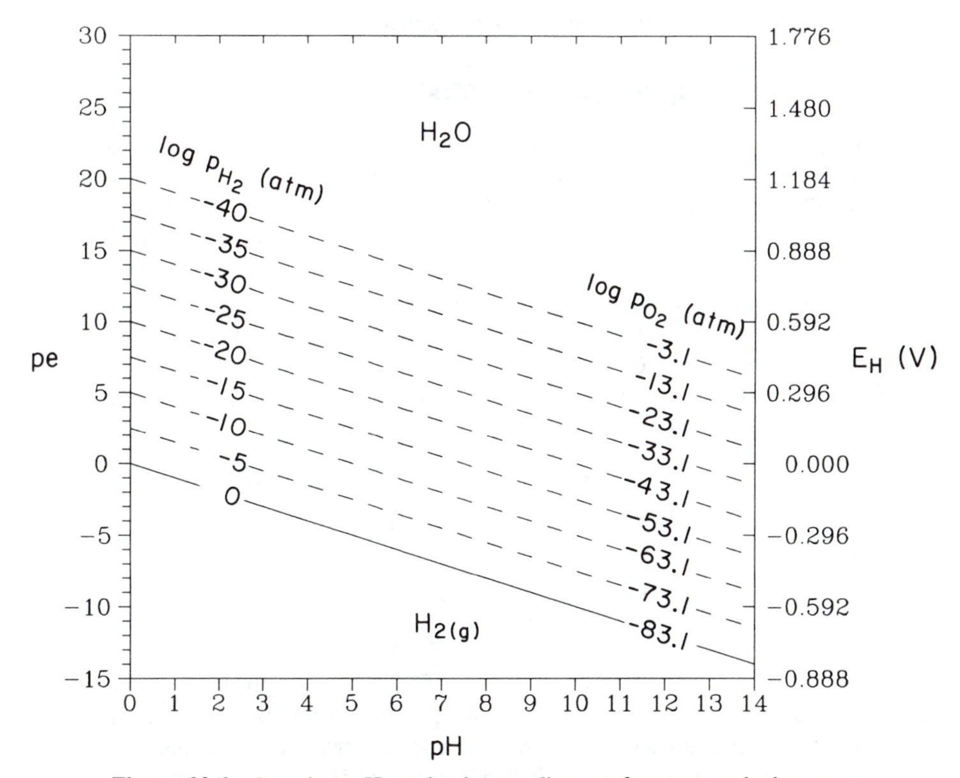

Figure 20.6 Actual pe-pH predominance diagram for aqueous hydrogen at 25°C/1 atm. The main H(I)/H(0) OX/RED boundaryline for $p_{H_2} = 1$ atm is shown with a solid line. Boundaries for p_{H_2} values other than 1 atm are shown with dashed lines. Corresponding values for p_{O_2} calculated according to Eq. (20.27) are shown on each line.

will be able to form spontaneously in water under 1 atm of total pressure only when the system is chemically pushed to yield $p_{H_2} > 1$ atm.

The values of p_{O_2} that correspond at 25°C/1 atm to the various p_{H_2} values in Figure 20.6 can be calculated by combining

$$2(2H^+ + 2e^- = H_{2(g)} \qquad \log K = 0 \qquad (19.102)$$

$$- \quad O_{2(g)} + 4H^+ + 4e^- = 2H_2O \qquad \log K = 83.1 \qquad (19.106)$$

$$2\ H_2O = 2H_{2(g)} + O_{2(g)} \qquad \log K = -83.1 \qquad (20.27)$$

With the activity of liquid water expressed as usual on the mole fraction scale, Eq. (20.27) yields

$$2 \log p_{H_2} + \log p_{O_2} = -83.1. \qquad (20.28)$$

When $p_{H_2} = 1$ atm, then at equilibrium, $p_{O_2} = 10^{-83.1}$ atm. This very low value for p_{O_2} provides a direct measure of the tremendous tendency for O_2 to be reduced in the presence of H_2 at 1 atm. The p_{O_2} values corresponding to the other p_{H_2} lines in Figure 20.6 have been calculated in a similar manner.

20.4 pe-pH DIAGRAM FOR AQUEOUS OXYGEN

For oxygen, the three stable oxidation states that we need to consider in aqueous systems are O(0), O(−I), and O(−II). O(−II) is usually present as water and hydroxide ion, and O(0) will be present as dissolved O_2. If a gas phase is present, then O(−II) will be present as gaseous H_2O, and O(0) will be present as $O_{2(g)}$. For reasons exactly analogous to those discussed for H_2, OX/RED boundarylines involving O_2 are usually drawn using $O_{2(g)}$ rather than dissolved O_2 in the redox equilibria.

The compound in which O(−I) can be found is hydrogen peroxide, H_2O_2. Located as it is between O(O) and O(−II), the oxidation state represented by H_2O_2 occupies a middle ground in the preliminary generic pe-pH diagram that is Figure 20.7*a*. At 25°C/1 atm, the redox half reaction between H_2O_2 and H_2O is

$$H_2O_2 + 2H^+ + 2e^- = 2H_2O \qquad \log K = 59.6. \qquad (20.29)$$

The ratio of the molar concentration of oxygen in H_2O to that in H_2O_2 at any pe and pH is given by

$$\frac{55.5X_{H_2O}}{2[H_2O_2]} = \frac{55.5 \times 10^{59.6} 10^{-(2pH+2pe)}}{2X_{H_2O}}. \qquad (20.30)$$

The factor of 55.5 (= 1000/18.01) on the LHS of Eq. (20.30) converts the concentration of water from units of mole fraction to molarity; the factor of 2 in the denominator accounts for the fact that each H_2O_2 molecule contains two oxygen atoms.

The ratio given by Eq. (20.30) will decrease and H_2O_2 will become increasingly favored as pe and/or pH increase. For 25°C/1 atm, when the ratio equals 1.0, Eq. (20.30) gives the boundaryline as

$$pe = 30.52 - pH. \qquad (20.31)$$

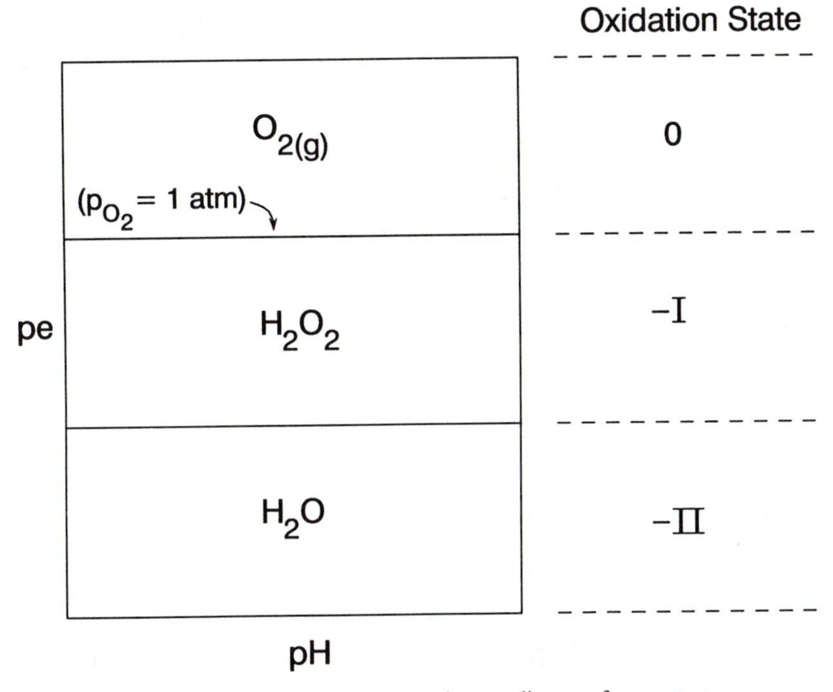

Figure 20.7 (*a*) Generic pe-pH predominance diagram for aqueous oxygen allowing the possibility of all three main oxygen oxidation states. The boundaryline between O(0) and O(−I) is assumed to be that along which $p_{O_2} = 1$ atm. It is assumed that the O(−I)/O(−II) boundaryline lies below the O(0)/O(−I) boundaryline for $p_{O_2} = 1$ atm. As discussed in the text, this assumption is incorrect, and that conclusion leads us to Figure 20.7*b*.

As the reader might expect, the line giving equimolar concentrations of oxygen in O_2 and H_2O_2 lies above Eq. (20.31). Thus, if we computed all boundarylines on an equimolar basis, there would be a predominance region for H_2O_2. However, as we will see in a moment, Eq. (20.31) plots *above* the $O_{2(g)}/H_2O$ line for 25°C/1 atm and $p_{O_2} = 1$ atm. Thus, the stability limit of water at 1 atm prevents us from being able to access the types of pe and pH values that would lie inside that predominance region for H_2O_2.

Based on these comments, we see that at 1 atm pressure, the generic pe-pH predominance diagram at 25°C/1 atm is not given by Figure 20.7*a*, but rather by Figure 20.7*b*. We emphasize that in the latter, the OX/RED line for O_2/H_2O is the $O_{2(g)}/H_2O$ line for $p_{O_2} = 1$ atm, and not the equimolar oxygen line for (dissolved O_2)/H_2O. Thus, a concentrated solution of H_2O_2 is seen to be unstable relative to oxygen production at 1 atm total pressure; the net reaction is simply $2\ H_2O_2 = 2H_2O + O_{2(g)}$. Like the conversion of Cl_2 to HOCl and Cl^-, the decomposition of H_2O_2 is a disproportionation reaction. Half of the O(−I) in H_2O_2 goes up to O(0), and the other half goes down to O(−II).

At 3 percent H_2O_2, the type of hydrogen peroxide solution that is typically used for disinfecting wounds is quite concentrated when viewed in the above context. Thus,

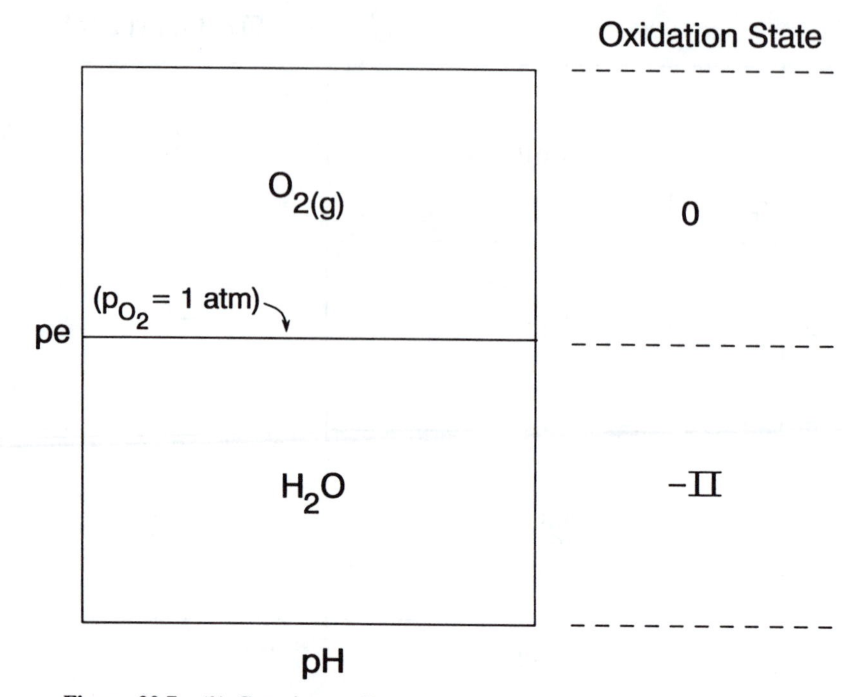

Figure 20.7 (*b*) Generic pe-pH predominance diagram for aqueous oxygen. The boundaryline between O(0) and O(–II) is assumed to be that along which $p_{O_2} = 1$ atm.

we now understand why such solutions can fizz so vigorously when placed in a wound. The gas that is released is O_2. The medicinal worth of this O_2 is that it tends to kill the types of anaerobic bacteria that can cause septic infections (e.g., the staphylococcus bacteria). When left undisturbed in a dark bottle, an H_2O_2 solution that also contains dilute phosphoric acid as a stabilizer can have a shelf life of years. Exposure to light permits the reaction $2\ H_2O_2 = 2H_2O + O_{2(g)}$ to proceed photolytically.

We now concern ourselves with transforming Figure 20.7*b* into a pe-pH diagram for oxygen. As discussed, similar to H_2 in Figure 20.6, O_2 will be handled as a gas. The O(0)/O(–II) equilibrium line for $p_{O_2} = 1$ atm at 25°C/1 atm was included in Figure 20.3 as a upper stability limit for water. It is based on the equation

$$\text{pe} = 20.78 - \text{pH} + \frac{1}{4}\log p_{O_2}. \qquad (19.109)$$

At $p_{O_2} = 1$ atm, we have

$$\text{pe} = 20.78 - \text{pH} \qquad (20.32)$$

Equation (20.32) is drawn in Figure 20.8 as a solid line representing a pe-pH boundaryline for oxygen. As anticipated, Eq. (20.32) plots below Eq. (20.31); H_2O_2 does not have a predominance region in systems at 25°C/1 atm.

Although the O(0)/O(−II) boundaryline drawn most often using Eq. (19.109) corresponds to that for $p_{O_2} = 1$ atm, Eq. (19.109) can be used to draw a boundaryline at 25°C/1 atm for any $p_{O_2} \leq 1$ atm. Strictly speaking, it cannot be used for $p_{O_2} > 1$ atm because the value of the equilibrium constant upon which it is based is for a *total* pressure of only 1 atm. Lines for a range of p_{O_2} values less than 1 atm are given as dashed lines in Figure 20.8. When the system at a given pe and pH does not contain any gas phase, the lines give the in-situ value of p_{O_2}. This applies even along the $p_{O_2} = 1$ atm line because, from Chapter 19, we recall that actual bubbles of pure $O_{2(g)}$ will be able to form spontaneously in water under 1 atm of total pressure only when the system is chemically pushed to yield $p_{O_2} > 1$ atm. The values of p_{H_2} that correspond to the various p_{O_2} values in Figure 20.8 have been calculated using Eq. (20.28). When $p_{O_2} = 1$ atm, at 25°C/1 atm, then $p_{H_2} = 10^{-41.55}$ atm. When $p_{O_2} = 0.21$ atm (the ambient atmospheric value), then the equilibrium value at 25°C/1 atm for p_{H_2} is $10^{-41.21}$ atm; it is not surprising then that there is little H_2 in the earth's atmosphere. Figure 20.9 contains the O(0)/O(−II)

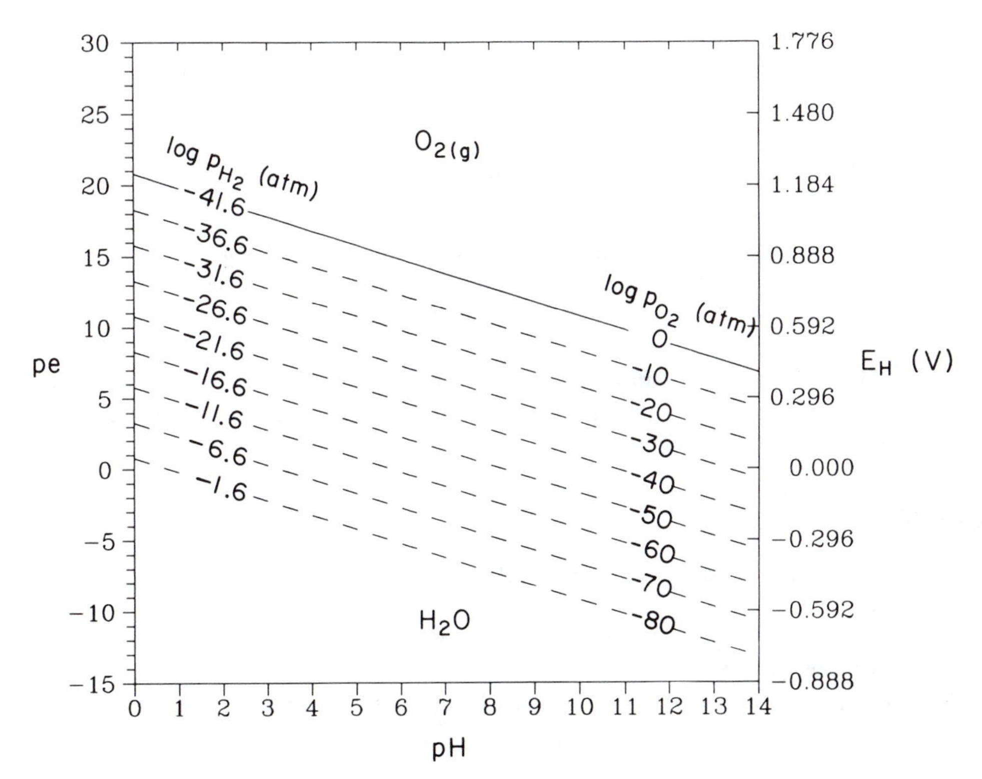

Figure 20.8 Actual pe-pH predominance diagram for aqueous oxygen at 25°C/1 atm. The main O(0)/O(−II) OX/RED boundaryline for $p_{O_2} = 1$ atm is shown with a solid line. Boundaries for p_{O_2} values other than 1 atm are shown with dashed lines. Corresponding values for p_{H_2} calculated according to Eq. (20.28) are shown on each line.

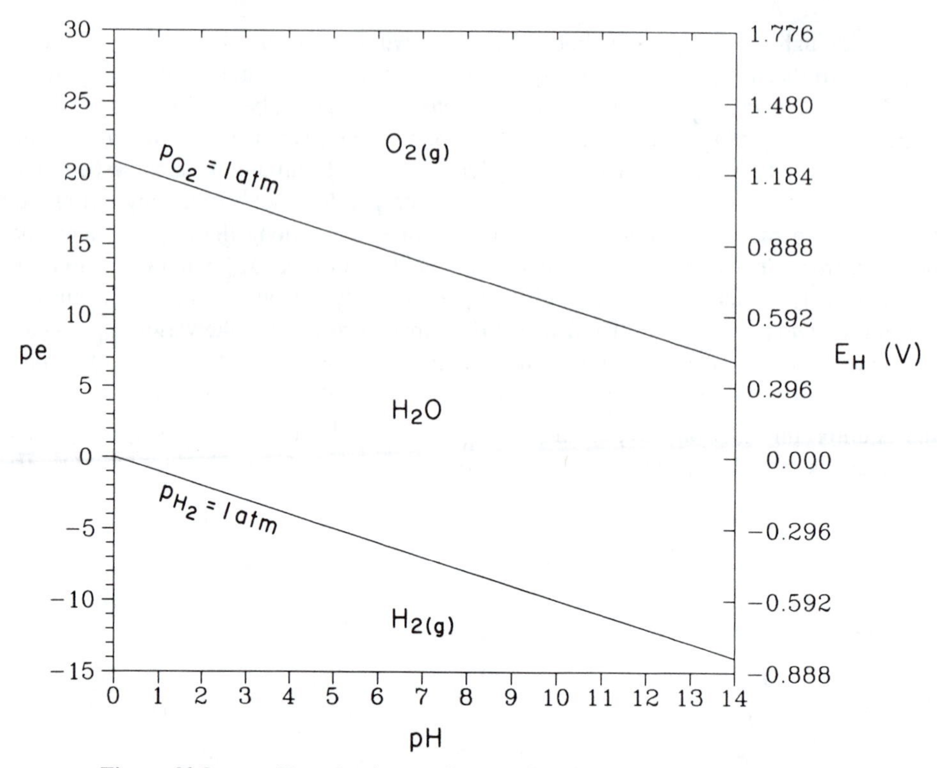

Figure 20.9 pe-pH predominance diagram showing the stability field for water at 25°C/1 atm. Water will not boil chemically as long as one remains between the lines $p_{O_2} = 1$ atm at $p_{H_2} = 1$ atm.

OX/RED line for $p_{O_2} = 1$ atm and the H(I)/H(O) OX/RED line for $p_{H_2} = 1$ atm. Both lines are drawn for 25°C/1 atm. Problems illustrating the principles of this chapter are presented by Pankow (1992).

20.5 REFERENCES

COTTON, F. A., and G. WILKINSON. 1980. *Advanced Inorganic Chemistry*. New York: Wiley-Interscience.

PANKOW, J. F. 1992. *Aquatic Chemistry Problems*. Portland: Titan Press-OR, P.O. Box 91399, Portland, Oregon 97291-1399.

STUMM, W., and J. J. MORGAN. 1981. *Aquatic Chemistry*. New York: Wiley-Interscience.

21

pe-pH Diagrams for Aqueous Lead in the Absence of CO_2

21.1 INTRODUCTION

In Chapter 20, the principles governing the construction and use of pe-pH diagrams were introduced, and the cases of aqueous chlorine, hydrogen, and oxygen were discussed in some detail. In pe-pH diagrams for chlorine (e.g., as in Figure 20.3), we usually do not consider the possibility of any chlorine-containing solids. All of the chemical forms containing chlorine are usually assumed to be *dissolved.* For example, solid salts of the chloride ion (e.g., $NaCl_{(s)}$) are just too soluble to precipitate unless the Na^+ concentration is very high. For hydrogen and oxygen, such large amounts of these two elements are present in water itself that pe-pH diagrams of these elements do not consider predominance regions for H-containing or O-containing solids (e.g., (am)$Fe(OH)_{3(s)}$).

In contrast to the situation discussed above, for pe-pH diagrams for *metallic* elements, predominance regions for one or more solids can be present. Moreover, as the level of a given metal in a system increases, the probability that solids containing that element will form will also increase. Conversely, if there is a predominance region for some solid, then as the level of the element decreases, the size of that predominance region will shrink, and will eventually disappear completely.

Lead (Pb) is a good example of a redox-active metallic element that can form solids. The case of aqueous lead in the absence of CO_2 was considered in the first edition of the text by Stumm and Morgan (1970), and also by Hem (1976). In this chapter, we will discuss how to prepare the pe-pH predominance diagrams for three different levels of total Pb, that is, $Pb_T = 10^{-2}$ *M*, $Pb_T = 10^{-4}$ *M*, and $Pb_T = 10^{-6}$ *M*. All three cases will be considered in the absence of dissolved CO_2 (i.e., taking $p_{CO_2} = 0$). This corresponds to an infinitely-low value of $\log p_{CO_2}$ when viewed in the context of a Pb-analog of Figure 20.1. Chapter 22 will consider the case of aqueous lead in the presence of CO_2, at a fixed level of C_T.

21.2 pe-pH DIAGRAM FOR $Pb_T = 10^{-2}$ *M*

21.2.1 General

Lead has three major oxidation states that must be considered. They are: IV, II, and 0. $PbO_{2(s)}$ is the important form of Pb(IV). This solid is so insoluble that Pb(IV) species like Pb^{4+}, $PbOH^{3+}$, etc. need not be considered in the aqueous phase. For Pb(II), the solid that we will consider in this chapter is "red" $PbO_{(s)}$. $PbO_{(s)}$ can dissolve to form Pb^{2+}, $PbOH^+$, $Pb(OH)_2^o$, and $Pb(OH)_3^-$. Other pe-pH diagrams could be drawn considering "yellow" $PbO_{(s)}$, or $Pb(OH)_{2(s)}$. At 25°C/1 atm, both yellow-$PbO_{(s)}$ and $Pb(OH)_{2(s)}$ are more soluble than red-$PbO_{(s)}$, and so are not thermodynamically possible under equilibrium conditions at 25°C/1 atm.[1]

In the case of Pb(0), the only solid is simply elemental lead, that is, $Pb_{(s)}$. As with the dissolved Pb(IV) species, we will assume that the concentrations of dissolved forms of Pb(0) (i.e., Pb^o, $PbOH^-$, etc.), are negligibly small.

Since 10^{-2} *M* is a fairly high value for Pb_T, let us assume for the moment that: 1) along all solution phase/solid phase boundarylines at "low pH," Pb^{2+} is by far the dominant dissolved species (i.e., $[Pb^{2+}] \simeq 10^{-2}$ *M*); and 2) along all solution phase/solid phase boundarylines at "high pH," $Pb(OH)_3^-$ is the dominant dissolved species (i.e., $[Pb(OH)_3^-] \simeq 10^{-2}$ *M*). For example, at intermediate values of pe, where Pb(II) chemistry dominates, this means that $PbO_{(s)}$ will be assumed to precipitate at a pH that is low enough that Pb^{2+} is still the dominant solution phase species, and that $PbO_{(s)}$ will redissolve at a pH that is high enough that $Pb(OH)_3^-$ is the dominant solution phase species.

Under the assumptions outlined in the preceding paragraph, we conclude that the pe-pH diagram for aqueous Pb for $Pb_T = 10^{-2}$ *M might* appear *roughly* as shown in Figure 21.1. There are several reasons why the words "might" and "roughly" are used in the previous sentence. First, as we learned in Chapter 20, disproportionation reactions can sometimes bring a predominance region for a relatively reduced form of an element (e.g., Cl(−I) in Cl^- and O(−II) in H_2O) up against the predominance region for a more oxidized form (e.g., Cl(I) in HOCl, and O(0) in O_2). In such cases, the predominance region for a species of intermediate oxidation state (e.g., Cl_2 and H_2O_2) might not be present at all, or might be present only over a limited pH range. As it turns out, Pb(II) does not tend to disproportionate to Pb(IV) and Pb(0) to any significant

[1] If one of these other solids, e.g., $Pb(OH)_{2(s)}$, forms quickly from solution and is converted to red-$PbO_{(s)}$ at a rate that is slow relative to the environmental time frame of interest, then it could become of interest to prepare a pe-pH diagram for Pb taking $Pb(OH)_{2(s)}$ to be the Pb(II) solid of interest rather than red-$PbO_{(s)}$.

degree under any conditions,[2] and so nowhere will we encounter a direct boundaryline between Pb(0) and Pb(IV). This will be made apparent in the pe-pH diagrams presented below.

Before going on, we should note though that in contrast to the stable oxidation state of Pb(II), the unstable states Pb(III) and Pb(I) *do* disproportionate strongly: virtually all of any significant amount of Pb(III) will disproportionate to Pb(IV) and Pb(II), and virtually all of any significant amount of Pb(I) will disproportionate to Pb(II) and Pb(0). The equilibrium positions of these two disproportionation reactions disfavor Pb(III) and Pb(I) to such extents that Pb(III) and Pb(I) will be limited to exceedingly low (but not zero) concentrations everywhere in the pe-pH diagram. Thus, while disproportionation

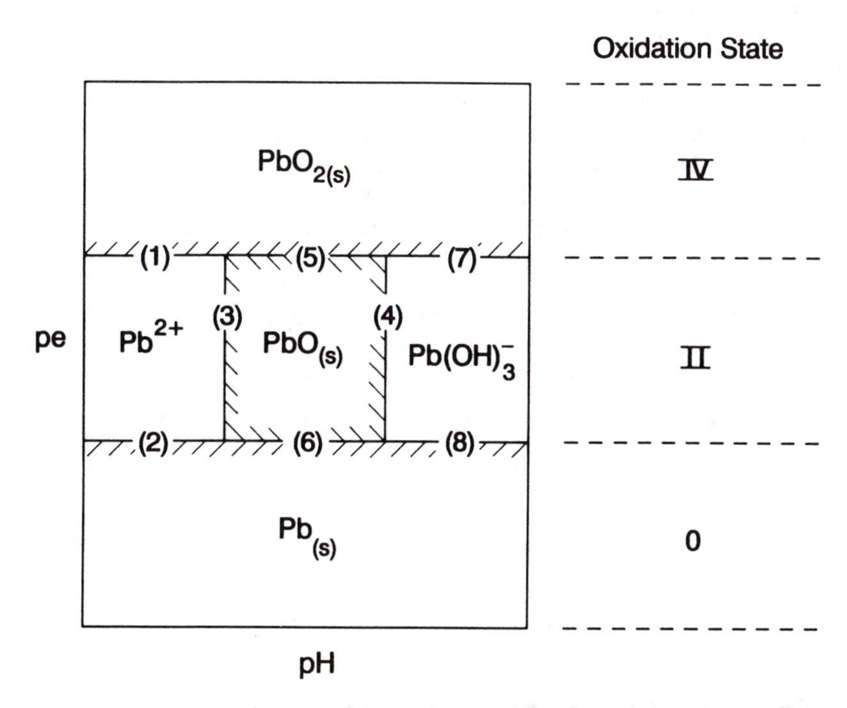

Figure 21.1 Generic pe-pH diagram for $Pb_T = 10^{-2}$ M assuming that Pb^{2+} and $Pb(OH)_3^-$ are the dominant, dissolved phase redox species at low and high pH, respectively. As will be shown, we do not in fact need to make these assumptions to draw the various solution/solid boundarylines. Making the assumptions just simplifies the initial visualization of the problem.

[2]The word "significant" is used advisedly here. For example, say that we prepare a solution of $PbCl_2$. Equilibrium between the Pb^{2+} and all other possible Pb-containing species will *require* that at least some of the Pb(II) disproportionate to form finite, though exceedingly small, dissolved concentrations of both Pb(IV) and Pb(0).

reactions are important in the Pb system, they occur in a manner that does not complicate the pe-pH diagrams.[3]

A second reason why the actual pe-pH diagram for $Pb_T = 10^{-2}\ M$ might appear only roughly like Figure 21.1 is that the actual slopes of the lines are not likely to be simple horizontal or vertical lines as in Figure 21.1; some of them might have some slope. A third reason is that one or more of the solution phase/solid phase boundarylines might be noticeably *curved*. As will be seen below, curvature in such lines will be visible when there are significant changes in the distribution of the Pb(II) among the various solution phase species. While not visible in the diagram for $Pb_T = 10^{-2}\ M$, we will find that curvature is visible at levels of Pb_T that are significantly lower than $10^{-2}\ M$.

As discussed in Chapter 20, at least $(n - 1)$ independent equilibrium equations are needed to draw a pe-pH diagram with n predominance regions. For the generic diagram depicted in Figure 21.1, since we have predominance regions for five species (four dissolved and one solid), at least four equilibrium equations interrelating those species will be required. However, $PbOH^+$ and $Pb(OH)_2^o$ are also species that need to be considered in aqueous Pb systems. Thus, even though they do not have predominance regions in Figure 21.1, we will need two more equilibrium constant equations that can be used to quantify the extent to which $PbOH^+$ and $Pb(OH)_2^o$ will form. The extent to which these species form will determine the exact locations of some of the boundarylines.

One possible set of six independent equilibrium equations is

$$PbO_{2(s)} + 4H^+ + 2e^- = 2H_2O + Pb^{2+} \qquad \log K = 49.2 \qquad (21.1)$$

$$Pb^{2+} + 2e^- = Pb_{(s)} \qquad \log K = -4.26 \qquad (21.2)$$

$$PbO_{(s)} + 2H^+ = Pb^{2+} + H_2O \qquad \log\ {}^*K_{s0} = 12.7 \qquad (21.3)$$

$$Pb^{2+} + H_2O = PbOH^+ + H^+ \qquad \log\ {}^*K_{H1} = -7.7 \qquad (21.4)$$

$$PbOH^+ + H_2O = Pb(OH)_2^o + H^+ \qquad \log\ {}^*K_{H2} = -9.4 \qquad (21.5)$$

$$Pb(OH)_2^o + H_2O = Pb(OH)_3^- + H^+ \qquad \log\ {}^*K_{H3} = -11.0. \qquad (21.6)$$

The specific equilibrium constant values given in Eqs. (21.1)–(21.6) are for 25°C/1 atm.

21.2.2 The Two Straight Solution/$PbO_{(s)}$ Boundarylines

Since we have assumed that both Pb(IV) and Pb(0) are essentially insoluble, in the solution phase, we have

$$Pb_T = [Pb^{2+}] + [PbOH^+] + [Pb(OH)_2^o] + [Pb(OH)_3^-]. \qquad (21.7)$$

[3]For another example of this type of situation, consider the case of chromium. Cr has three stable oxidation states, Cr(VI), Cr(III), and Cr(0). These three oxidation states are the ones that will appear in a pe-pH diagram for Cr. Virtually all of any significant amounts of Cr in the oxidation states Cr(V), Cr(IV), Cr(II), and Cr(I) will tend to disproportionate to yield a mix of other oxidation states.

Neglecting activity corrections, Eq. (21.7) can be expressed as

$$Pb_T = [Pb^{2+}]\left(1 + \frac{{}^*K_{H1}}{[H^+]} + \frac{{}^*K_{H1}\,{}^*K_{H2}}{[H^+]^2} + \frac{{}^*K_{H1}\,{}^*K_{H2}\,{}^*K_{H3}}{[H^+]^3}\right) \quad (21.8)$$

where the term in the parentheses on the LHS may be recognized as being equal to α_0^{-1} for the complexation of dissolved Pb(II) by OH^-. Using the ${}^*\beta$ notation, Eq. (21.8) yields

$$Pb_T = [Pb^{2+}]\left(1 + \frac{{}^*\beta_1}{[H^+]} + \frac{{}^*\beta_2}{[H^+]^2} + \frac{{}^*\beta_3}{[H^+]^3}\right). \quad (21.9)$$

From Eqs. (21.4)–(21.6), at 25°C/1 atm, we have ${}^*\beta_1 = 10^{-7.7}$, ${}^*\beta_2 = 10^{-17.1}$, and ${}^*\beta_3 = 10^{-28.1}$.

On both the LHS and RHS of line 3 in Figure 21.1, the Pb oxidation state of interest is Pb(II). Since no electrons are exchanged when converting dissolved Pb(II) species to $PbO_{(s)}$, line 3 will be a simple vertical line in the actual pe-pH diagram. A similar conclusion can be reached for line 4. In fact, what we can realize is that the pH values which locate lines 3 and 4 in Figure 21.1 are nothing more than the low and high pH roots of Eq. (21.9) where each of the species is controlled by equilibrium with $PbO_{(s)}$. The fact that both Pb(IV) and Pb(0) are essentially insoluble makes the process of locating the pH values for lines 3 and 4 exactly like the cases in Chapter 11 involving the location of low and high pH roots for dissolution/precipitation of a metal oxide or hydroxide (e.g., see Figure 11.7). Thus, when we are saturated with $PbO_{(s)}$, Eq. (21.3) may be used to convert Eq. (21.9) to

$${}^*K_{s0}[H^+]^2\left(1 + \frac{{}^*\beta_1}{[H^+]} + \frac{{}^*\beta_2}{[H^+]^2} + \frac{{}^*\beta_3}{[H^+]^3}\right) = Pb_T \quad (21.10)$$

which is equivalent to

$${}^*K_{s0}[H^+]^2 + {}^*K_{s1}[H^+] + K_{s2} + \frac{{}^*K_{s3}}{[H^+]} = Pb_T. \quad (21.11)$$

From Eq. (21.3), at 25°C/1 atm, we have ${}^*K_{s0} = 10^{12.7}$. For $Pb_T = 10^{-2}$ M, by trial and error, we obtain a low pH root of pH = 7.45, and a high pH root of 13.40.

Because Eq. (21.4) indicates that $[Pb^{2+}]/[PbOH^+] = 1.8$ at pH = 7.45, we see that Pb^{2+} is indeed the dominant dissolved Pb(II) species at the pH at which $PbO_{(s)}$ becomes saturated. Thus, at $Pb_T = 10^{-2.0}$ M, there will be no predominance region for $PbOH^+$, and the Pb^{2+} predominance region extends all the way up to the $PbO_{(s)}$ predominance region. At the high pH root of 13.40, Eq. (21.6) indicates that $[Pb(OH)_3^-]/[Pb(OH)_2^o] = 251$, and so $Pb(OH)_3^-$ certainly dominates over $Pb(OH)_2^o$ at that pH. At pH values higher than 13.40, that dominance can only increase, and so $Pb(OH)_3^-$ is the only Pb(II) species for which a predominance region is needed for pH > 13.40.

Knowing the pH values for lines 3 and 4 does not tell us everything we need to draw those lines. Indeed, although we know where they lie, we do not know where either of them start or stop. For line 3, the starting point will be given by the intersection of line 3 with line 2 (and line 6), and the stopping point will be given by the intersection with line 1 (and line 5). For line 4, the starting point will be given by the intersection of line 4 with line 6 (and line 8), and the stopping point will be given by the intersection with line 5 (and line 7). Thus, to draw lines 3 and 4, we need to know where some of the other lines lie.

21.2.3 The Curved $PbO_{2(s)}$/Solution Boundaryline

In a manner similar to the solution phase/$FeCO_{3(s)}$ boundarylines depicted in Figures 14.1, 14.3, and 14.4, lines 1 and 7 in Figure 21.1 are really portions of the *same*, curved, $PbO_{2(s)}$/solution boundaryline along which $Pb_T = 10^{-2}$ *M*. In the pH interval between lines 3 and 4, the curved $PbO_{2(s)}$/solution boundaryline that we can calculate *does not apply*. Thus, we do not plot it over that interval, but rather plot the $PbO_{2(s)}/PbO_{(s)}$ boundaryline (line 5). The reader should compare this situation with that in Figure 14.1 wherein we do not draw the portion of the $FeCO_{3(s)}$/solution boundaryline that falls below $\log p_{CO_2} = -5.43$.

When we have equilibrium between $PbO_{2(s)}$ and the solution phase, then at 25°C/1 atm, by Eq. (21.1) we can write

$$2\text{pe} = 49.2 + \log\{PbO_{2(s)}\} - \log\{Pb^{2+}\} - 2\log\{H_2O\} - 4\text{ pH}. \tag{21.12}$$

Assuming unit activity coefficients in the aqueous phase, and unit mole fraction for $PbO_{2(s)}$ and H_2O, combination of Eq. (21.9) with Eq. (21.12) yields

$$2\text{pe} = 49.2 - \log Pb_T + \log\left(1 + \frac{{}^*\beta_1}{[H^+]} + \frac{{}^*\beta_2}{[H^+]^2} + \frac{{}^*\beta_3}{[H^+]^3}\right) - 4\text{ pH}. \tag{21.13}$$

Solving for pe for a range of pH values yields the desired, curved $PbO_{2(s)}$/solution boundaryline. This solution process is carried out easily using a personal computer and a spreadsheet program such as Lotus 123. (Eq. (21.15) below can be solved in the same manner.) As discussed, the boundaryline embodied in Eq. (21.13) will apply only for the pH values for which $PbO_{(s)}$ does not form. This range extends from very low pH up to the low pH root of Eq. (21.10) for equilibrium with $PbO_{(s)}$ (pH = 7.45), then again from the high pH root of Eq. (21.10) (pH = 13.40) towards even higher pH values. The pertinent portions of Eq. (21.13) are plotted as solid lines in Figure 21.2. The portion of Eq. (21.13) that extends over the range that $PbO_{(s)}$ can form is plotted as a dashed line.

As the pH changes, Eq. (21.13) adapts to shifts among between the various dissolved Pb(II) species. For example, when Pb^{2+} dominates, the 1 in the log term on the RHS will dominate, and a *d*pe/*d*pH slope near −2 will be obtained. When $PbOH^+$ dominates, the ${}^*\beta_1/[H^+]$ term in the argument of the logarithm in Eq. (21.13) will dominate, and a slope near −3/2 will be obtained over that dominance interval. Analogous statements can be made concerning dominance by $Pb(OH)_2^o$ and $Pb(OH)_3^-$.

21.2.4 The Curved Solution/$Pb_{(s)}$ Boundaryline

As with lines 1 and 7 in Figure 21.1, lines 2 and 8 in that figure are really parts of the same, curved, solution phase/$Pb_{(s)}$ boundaryline. The explicit equation for that

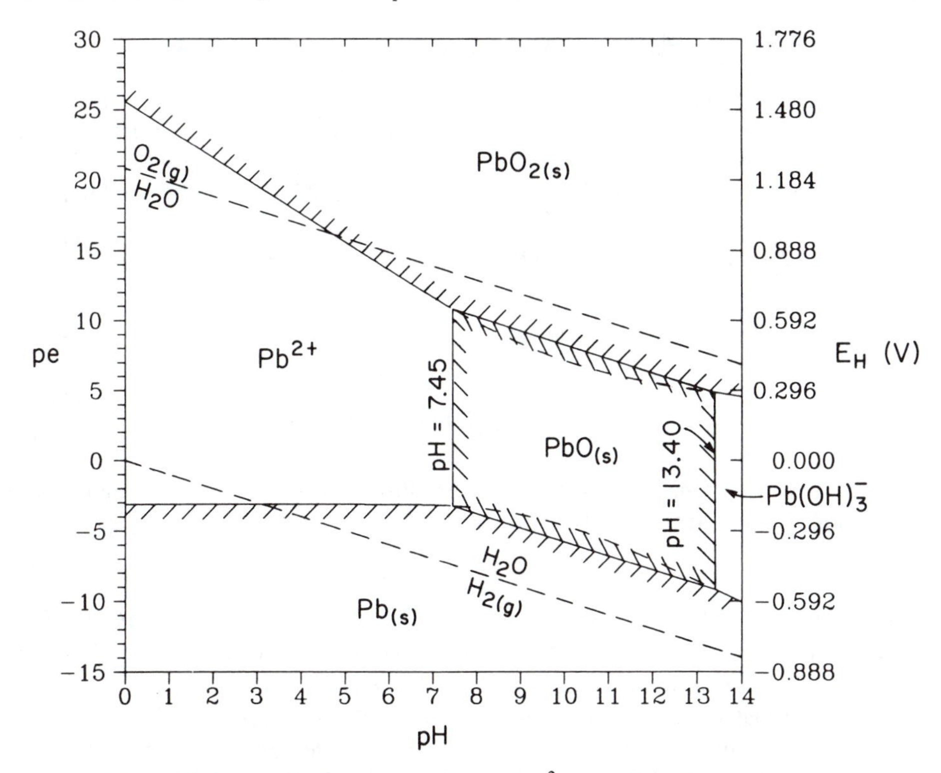

Figure 21.2 pe-pH diagram for $Pb_T = 10^{-2}$ *M* at 25°C/1 atm. The only assumptions are: 1) unit activity for all solids when present; and 2) unit activity coefficients for all solution species.

boundaryline may be developed in a manner that is highly analogous to the procedure followed in the preceding section. In particular, at 25°C/1 atm, we have

$$2pe = -4.26 + \log\{Pb^{2+}\} - \log\{Pb_{(s)}\}. \tag{21.14}$$

Assuming unit activity coefficients again in the aqueous phase, and unit activity for $Pb_{(s)}$, combination of Eq. (21.9) with Eq. (21.14) yields

$$2pe = -4.26 + \log Pb_T - \log\left(1 + \frac{{}^*\beta_1}{[H^+]} + \frac{\beta_2}{[H^+]^2} + \frac{{}^*\beta_3}{[H^+]^3}\right). \tag{21.15}$$

As with Eq. (21.13), solving for pe for a range of pH values yields the solution/$Pb_{(s)}$ boundaryline for both low and high pH. Equation (21.15) is plotted in Figure 21.2 for $Pb_T = 10^{-2}$ *M*. The low and high pH ranges over which that boundaryline will be applicable will be the same as discussed for Eq. (21.13). In other words, neither Eq. (21.13) nor Eq. (21.15) apply over the pH range in which $PbO_{(s)}$ will form. As with Eq. (21.13), over that range, Eq. (21.15) is plotted in Figure 21.2 as a dashed line.

21.2.5 The Straight $PbO_{2(s)}/PbO_{(s)}$ Boundaryline

After having determined the equations for the various solution/solid boundarylines, we now turn our attention to the two solid/solid boundarylines in Figure 21.1, namely lines 5 and 6. A consideration of the principles discussed in the previous chapter lead to the conclusion that the equation for line 5 can be obtained at 25°C/1 atm by combining

$$2\text{pe} = 49.2 + \log\{PbO_{2(s)}\} - \log\{Pb^{2+}\} - 2\log\{H_2O\} - 4\text{pH} \qquad (21.12) \qquad \frac{PbO_{2(s)}}{Pb^{2+}}$$

$$-\quad \log\{Pb^{2+}\} + \log\{H_2O\} = 12.7 - 2\text{pH} + \log\{PbO_{(s)}\} \qquad (21.16) \qquad \frac{PbO_{(s)}}{Pb^{2+}}$$

$$2\text{pe} = 36.5 - 2\text{pH} + \log\{PbO_{2(s)}\} - \log\{PbO_{(s)}\} - \log\{H_2O\} \qquad (21.17) \qquad \frac{PbO_{2(s)}}{PbO_{(s)}}$$

where Eq. (21.16) originates in Eq. (21.3). When $PbO_{2(s)}$, $PbO_{(s)}$, and H_2O are all present at unit mole fraction, then Eq. (21.17) reduces to

$$\text{pe} = 18.25 - \text{pH} \qquad (21.18)$$

which is line 5 at 25°C/1 atm, and is plotted in Figure 21.2. The slope *d*pe/*d*pH of the line is -1 because for every electron that is consumed in the conversion of $PbO_{2(s)}$ to $PbO_{(s)}$, one proton is also consumed: $PbO_{2(s)} + 2e^- + 2H^+ = PbO_{(s)} + H_2O$.

We note that it was *not necessary to assume solution phase ideality* for any dissolved Pb species in order to generate Eq. (21.18). The reason is that line 5 describes a boundaryline between two predominance regions for which both Pb species are solids (see Eq. (21.17)): no solution phase Pb species are involved. For the same reason, we note that line 5 is *independent of* Pb_T.

21.2.6 The Straight $PbO_{(s)}/Pb_{(s)}$ Boundaryline

For the equilibrium pertaining to line 6, we combine:

$$2\text{pe} = -4.26 + \log\{Pb^{2+}\} - \log\{Pb_{(s)}\} \qquad (21.14) \qquad \frac{Pb^{2+}}{Pb_{(s)}}$$

$$+\quad \log\{Pb^{2+}\} + \log\{H_2O\} = 12.7 - 2\text{pH} + \log\{PbO_{(s)}\} \qquad (21.16) \qquad \frac{PbO_{(s)}}{Pb^{2+}}$$

$$2\text{pe} + \log\{H_2O\} = 8.44 - 2\text{pH} + \log\{PbO_{(s)}\} - \log\{Pb_{(s)}\} \qquad (21.19) \qquad \frac{PbO_{(s)}}{Pb_{(s)}}$$

When $PbO_{(s)}$, $Pb_{(s)}$, and H_2O are all present at unit mole fraction, then Eq. (21.19) reduces to

$$\text{pe} = 4.22 - \text{pH}. \qquad (21.20)$$

As with Eq. (21.18), Eq. (21.20) does not depend on Pb_T. Thus, as with line 5, line 6 does not depend on Pb_T. The slope *d*pe/*d*pH of the line is -1 because for every electron that is consumed in the conversion of $PbO_{(s)}$ to $Pb_{(s)}$, one proton is also consumed: $PbO_{(s)} + 2e^- + 2H^+ = Pb_{(s)} + H_2O$. Equation (21.20) is plotted in Figure 21.2.

21.3 THE pe-pH PREDOMINANCE DIAGRAM FOR Pb—GENERAL COMMENTS

The stability lines for the oxidation and reduction of water are included in Figure 21.2. As in preceding chapters, the water redox lines at 25°C/1 atm are based on the equations

$$\text{pe} = -\frac{1}{2}\log p_{H_2} - \text{pH} \qquad (19.104)$$

and

$$\text{pe} = 20.78 - \text{pH} + \frac{1}{4}\log p_{O_2}. \qquad (19.109)$$

For $p_{H_2} = 1$ atm, Eq. (19.104) becomes

$$\text{pe} = -\text{pH} \qquad (p_{H_2} = 1 \text{ atm}) \qquad (20.26)$$

and for $p_{O_2} = 1$ atm, Eq. (19.109) becomes

$$\text{pe} = 20.78 - \text{pH} \qquad (p_{O_2} = 1 \text{ atm}). \qquad (20.32)$$

Including the lines given in Eqs. (20.26) and (20.32) in the pe-pH diagram allows us to see that when $Pb_T = 10^{-2}$ *M*, a pe-pH regime in which each of the three oxidation states is important can be found. Pb(II) enjoys the most extensive predominance domain.

The dependence of pe on p_{O_2} is very weak (see Eq. (19.109) above). As a result, the dependence of pe on pH when p_{O_2} equals that for ambient air (0.21 atm) is given to a good approximation by Eq. (20.32) as it is plotted in Figure 21.2. Thus, for water saturated with atmospheric levels of O_2, at pH values of ~6.5 or higher, we will be very much inside Figure 21.2's region for $PbO_{2(s)}$. In other words, when we have a total system level of Pb_T of 10^{-2} mols per liter of water and when pH > ~6.5, we know $PbO_{2(s)}$ will be present. $Pb_{(s)}$ will not be present. This may come as somewhat of a surprise to many who have observed that solid *elemental* lead (i.e., $Pb_{(s)}$) in various forms (e.g., fishing weights, drapery weights, scuba diving weights, etc.) seems to be perfectly stable in the presence of air and water.

The reason for the seemingly contradictory observations made above is the fact that, in the presence of oxygen, a thin layer of $PbO_{2(s)}$ rapidly forms on the surface of Pb(0), and this layer protects (the chemical term is "passivates") the underlying Pb(0) from further oxidation. When one scratches the surface, a portion of the tarnished gray $PbO_{2(s)}$ is removed, and the shiny metallic color of the underlying, unstable Pb(0) is revealed. In a relatively short period of time, the $Pb_{(s)}$ surface in the scratch is oxidized to $PbO_{2(s)}$. The presence of liquid water will hasten the oxidation because it provides a medium in which the O_2 can attack the surface and electrons can be exchanged.

Consider now the specific point at pe = 14, pH = 7 in Figure 21.2. This point is well inside the $PbO_{2(s)}$ region, and so the dissolved Pb level is well below 10^{-2} *M*. We know that the level of dissolved oxygen is high because we are close to the $p_{O_2} = 1$ atm line. Consider now what happens as we move at constant pe towards lower pH. Lowering the pH means that we are moving closer to the solution/$PbO_{2(s)}$ boundaryline. This means that the dissolved Pb level is increasing towards 10^{-2} *M*. We see then that even in highly oxygenated waters, as the pH of the water decreases, the solubility of $PbO_{2(s)}$ increases rapidly to the point that very high levels of dissolved Pb^{2+} (e.g., 10^{-2} *M*) can be present and in equilibrium with $PbO_{2(s)}$. If the $PbO_{2(s)}$ that is dissolving is only a passivating surface on Pb(0), the large amounts of the underlying elemental lead can then also begin to dissolve. This type of behavior can lead to Pb poisoning for persons (especially young children) consuming low pH drinking water that has passed through lead pipes, or even through copper pipes connected with tin/lead solder.[4] Moreover, instances unfortunately continue to be reported of persons making, consuming, and/or selling whiskey produced using distillation ("still") equipment made with lead tubing, and/or joints connected with tin/lead solder. Those who drink spirits made with such equipment are at risk for Pb poisoning.

Finally, we note that since the guts of higher animals are acidic so as to promote the digestion of proteins, wild water fowl such as ducks and geese are at great risk for Pb poisoning when they ingest the Pb(0) shot from the shotguns used to hunt them. Many environmental groups are proposing that steel shot should be substituted for the Pb(0) shot.

21.4 UNDERSTANDING THE WIDTH OF THE $PbO_{(S)}$ DOMAIN AS A FUNCTION OF Pb_T

When there is a region for $PbO_{(s)}$ in a pe-pH predominance diagram for Pb, then the width of that region will depend on the value of Pb_T. As Pb_T decreases, then the value of the low pH root of Eq. (21.10) increases, and the value of the high pH root decreases. As a result, as Pb_T decreases, then the width of the $PbO_{(s)}$ domain decreases. Much light may be shed on the nature of this situation by considering the plot governing red-$PbO_{(s)}$ solubility as a function of pH. This plot, presented in Figure 21.3, gives log concentration vs. pH lines for all of the dissolved Pb(II) species, as well as for Pb_T. It should be clear that all of the lines in Figure 21.3, including the one for total dissolved Pb, pertain only when there is saturation equilibrium with red-$PbO_{(s)}$.

As discussed in Chapter 11, if a system contains only a limited amount of some given metal (i.e., a limited M_T), then the corresponding metal oxide or metal hydroxide can only be present when the curve for total dissolved metal gives a value that is *lower* than M_T. Thus, when a horizontal line is drawn through Figure 21.3 at $\log Pb_T = -2.0$, then: 1) the two pH values at which the line $y = -2.0$ intersects the $\log Pb_T$ curve are the low pH and high pH roots of Eq. (21.10) for $Pb_T = 10^{-2}$; and 2) the width and horizontal location of the interval described by the pH intersections in Figure 21.3 correspond to the width and horizontal location of the $PbO_{(s)}$ region in Figure 21.2.

[4]It may be noted in this regard that the current U.S. EPA standards call for no more than 20 μg/L ($\sim 10^{-7}$ *M*) or less of Pb in drinking water.

As the value of Pb_T for a pe-pH diagram decreases, then the width of the $PbO_{(s)}$ region will decrease. Its width will become equal to zero (i.e., no $PbO_{(s)}$ region possible) when the line $y = \log Pb_T$ just touches the bottom of the $\log Pb_T$ curve at one pH value (both roots to Eq. (21.10) are then identical). At that pH, the solution is saturated with $PbO_{(s)}$, but no $PbO_{(s)}$ can form; raising or lowering the pH from that point causes the solution to become undersaturated again. For values of Pb_T that are less than the minimum for Pb_T in Figure 21.3, then the solution cannot become saturated with $PbO_{(s)}$ at any pH. Thus, for any value of Pb_T that is equal to or less than the minimum for Pb_T in Figure 21.3, there will be no region for $PbO_{(s)}$ in the pe-pH diagram prepared for that Pb_T.

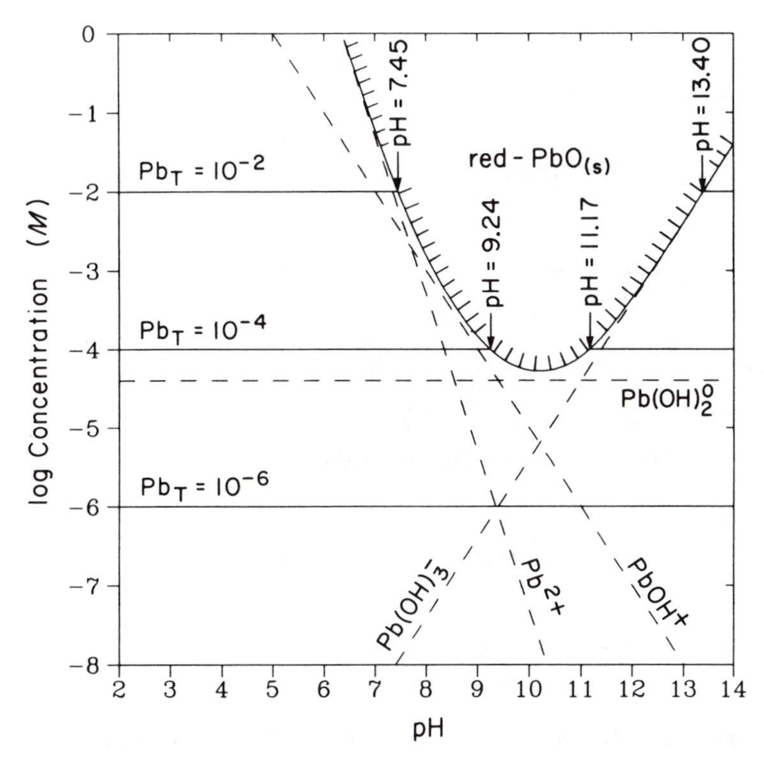

Figure 21.3 Log concentration vs. pH lines for various individual Pb(II) species and for $\log Pb_T$ for equilibrium with red-$PbO_{(s)}$. When $\log Pb_T$ is limited (and so cannot increase indefinitely at either low or high pH), the intersections of the horizontal line at that limit with the $\log Pb_T$ curve for red-$PbO_{(s)}$ will give the appropriate low and high pH roots of Eq. (21.10). At $\log Pb_T = -2$, the roots correspond to pH = 7.45 and 13.40. At $\log Pb_T = -4$, the roots correspond to pH = 9.24 and 11.17. Note that the pH range given by the roots for $Pb_T = 10^{-4}$ *M* is significantly narrower than the range for $Pb_T = 10^{-2}$ *M*. At $\log Pb_T = -6$, there are no real roots for Eq. (21.10) since $\log Pb_T$ can never be this low and still be in equilibrium with red-$PbO_{(s)}$.

21.5 pe-pH DIAGRAM FOR $Pb_T = 10^{-4}$ *M*

An examination of Figure 21.3 reveals that $PbO_{(s)}$ can become saturated when $Pb_T = 10^{-4}$ *M*. For this value of Pb_T, Eq. (21.10) gives a low pH root of 9.24, and a high pH root of 11.17. Figure 21.3, which gives the speciation when there is saturation with $PbO_{(s)}$, indicates that at the low pH root, then $PbOH^+$ is the dominant dissolved Pb(II) species; at the high pH root, $Pb(OH)_3^-$ is the dominant species. On this basis, we conclude that the pe-pH diagram for $Pb_T = 10^{-4}$ *M* will look something like what is presented in Figure 21.4. Lines 3b and 4 in Figure 21.4 are plotted in Figure 21.5 at pH values of 9.24 and 11.17, respectively.

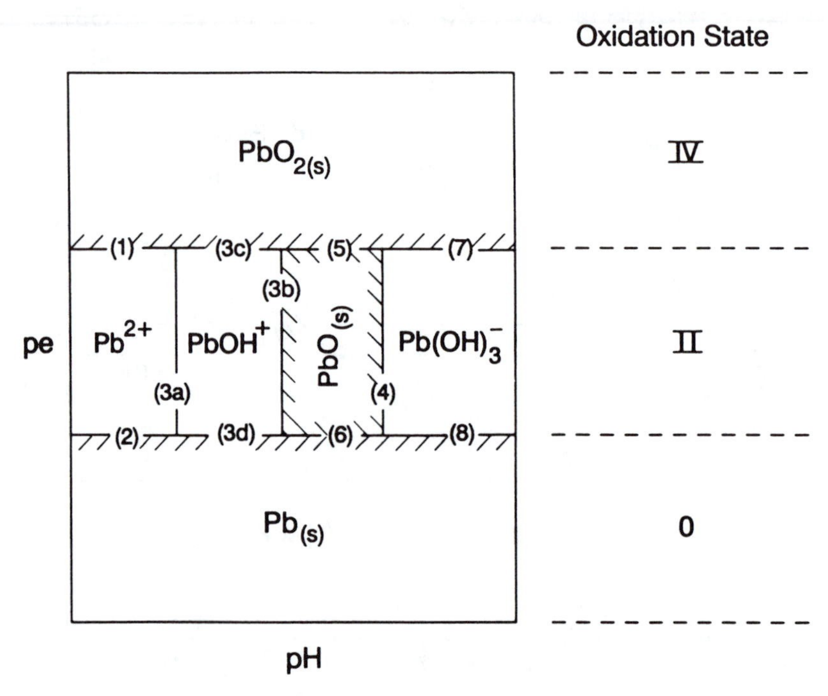

Figure 21.4 Generic pe-pH diagram for $Pb_T = 10^{-4}$ *M*. A region for $PbOH^+$ appears in this generic diagram because relative to $Pb_T = 10^{-2}$ *M*, a significantly higher pH is needed to reach the low pH root of Eq. (21.10). Thus, at the point that red-$PbO_{(s)}$ becomes saturated, $PbOH^+$ dominates the solution chemistry. We note that we do not in fact need to know that there has been this change in the Pb(II) solution phase chemistry in order to start drawing the solution/$PbO_{(s)}$ boundarylines in the actual pe-pH diagram (Figure 21.5). Indeed, we could go ahead and calculate the low and high pH roots of Eq. (21.10) (pH = 9.24 and pH 11.17), and *then* determine which dissolved Pb(II) species dominates the chemistry at the pH values of those roots. The pH of the high pH root is not sufficiently low to cause a $Pb(OH)_2^o$ region to appear.

Lines 1, 3c, and 7 in Figure 21.4 are all part of the same $PbO_{2(s)}$/solution boundaryline, and lines 2, 3d, and 8 are all part of the same solution/$Pb_{(s)}$ boundaryline. These two boundarylines are plotted in Figure 21.5 using Eqs. (21.13) and (21.15) with $Pb_T = 10^{-4}\ M$. As in Figure 21.2, these two boundarylines do not apply over the whole pH range because $PbO_{(s)}$ can form. In the interval between pH 9.24 and 11.17, then, the two boundarylines are given in dashed form in Figure 21.5. In that interval, lines 5 and 6 are used.

As noted in Section 21.2, lines 5 and 6 do not depend on Pb_T. Therefore, for $Pb_T = 10^{-4}\ M$ will be *exactly the same* as their positions for $Pb_T = 10^{-2}\ M$. Their lengths, however, will be shorter since the pH range of the $PbO_{(s)}$ dominance region is less extensive.

The only task remaining is to obtain the equation for line 3a in Figure 21.4. From Eq. (21.4), at 25°C/1 atm, we obtain:

$$\log\{PbOH^{+}\} = -7.7 + pH + \log\{Pb^{2+}\} \qquad (21.21)$$

$\frac{Pb^{2+}}{PbOH^{+}}$.

Along line 3a, we will take $[Pb^{2+}] = [PbOH^{+}]$. Thus, neglecting activity corrections, for this line we obtain

$$pH = 7.7. \qquad (21.22)$$

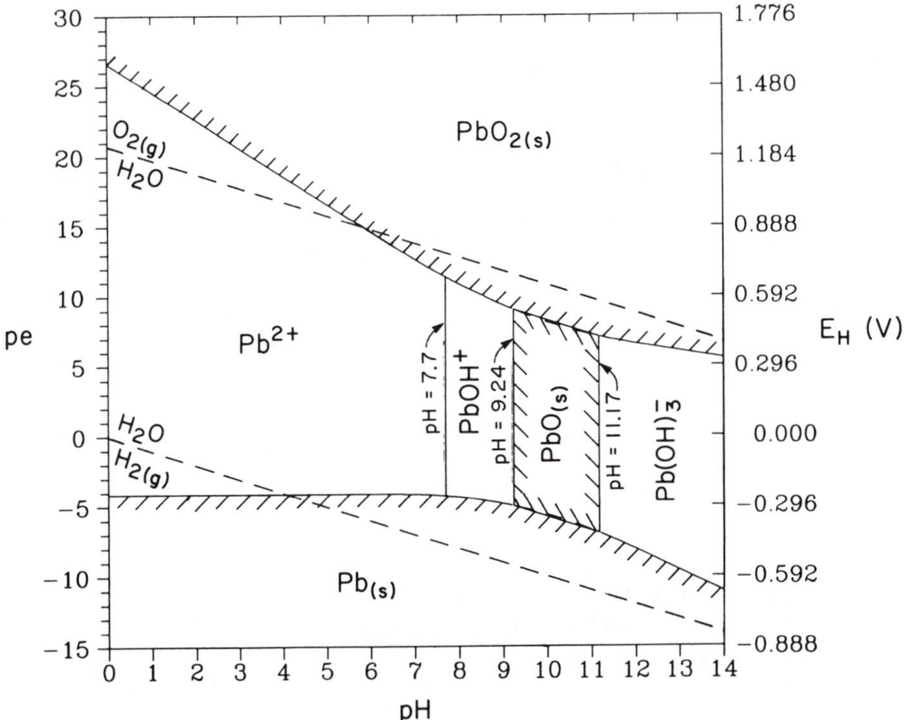

Figure 21.5 pe-pH diagram for $Pb_T = 10^{-4}\ M$ at 25°C/1 atm. The only assumptions are: 1) unit activity for all solids when present; and 2) unit activity coefficients for all solution species.

Compared to Figure 21.2, we see that reducing Pb_T from 10^{-2} M to 10^{-4} M has indeed caused the dominance region for $PbO_{(s)}$ to decrease in size. In addition, we note that the sizes of the dominance regions for $PbO_{2(s)}$ and $Pb_{(s)}$ have also retreated. Thus, as Pb_T decreases, all boundarylines between a solid form of Pb and a dissolved form will be shifted. In each case, the shifting will occur in a manner that causes the solid region to recede in favor of the solution phase. The reason for this behavior is that it becomes harder and harder, chemically speaking, to precipitate any given solid. This is true whether the species "driving" the precipitation is hydroxide, the electron, or a combination of the two.

For example, as Pb_T decreases, then when increasing the pH from an initially low value, one has to go to higher and higher pH values to precipitate $PbO_{(s)}$ from the solution phase. Thus, the size of the $PbO_{(s)}$ predominance region shrinks from the left. At the

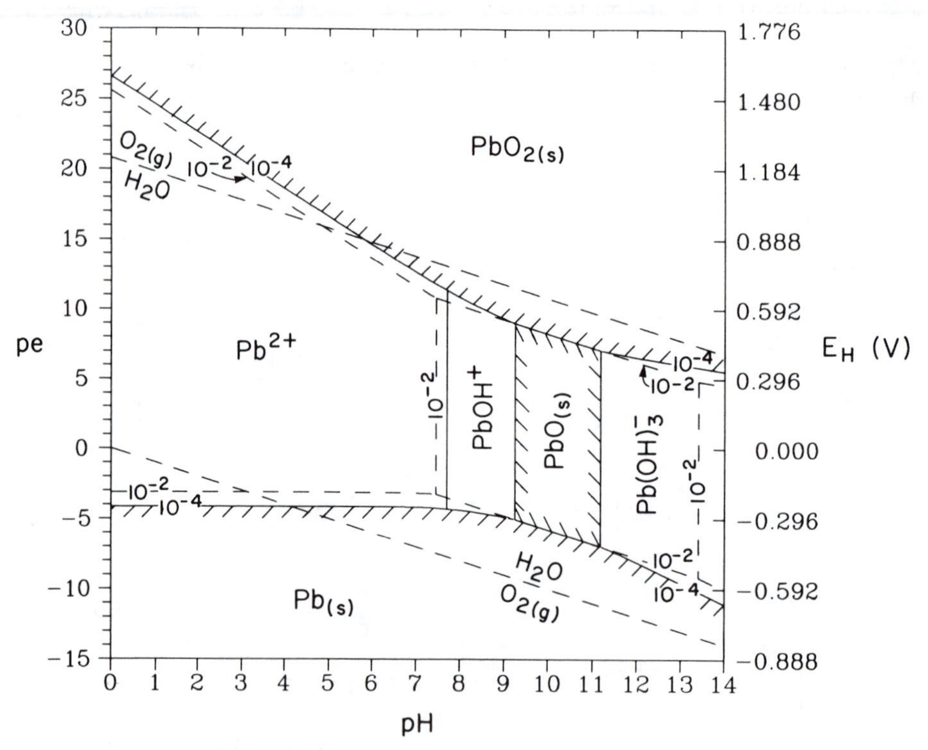

Figure 21.6 Combined pe-pH diagram for $Pb_T = 10^{-2}$ M and $Pb_T = 10^{-4}$ M, with assumptions as outlined in the captions for Figures 21.2 and 21.5. Boundarylines or portions of boundarylines applicable only to $Pb_T = 10^{-2}$ M are given as dashed lines. The $Pb^{2+}/PbOH^+$ line that appears in the $Pb_T = 10^{-4}$ M diagram is not labelled "10^{-4}" in this figure. Although it only appears as a visible boundaryline in the 10^{-4} M diagram, when activity corrections can be neglected at 25°C/1 atm, then pH = 7.70 always describes the pH at which $[Pb^{2+}]/[PbOH^+] = 1$, regardless of whether any $PbO_{(s)}$ is present.

same time, the pH value required to redissolve all of $PbO_{(s)}$ decreases as Pb_T decreases, and the $PbO_{(s)}$ region shrinks from the right. In a similar manner, as Pb_T decreases, one has to go to: 1) higher and higher pe values in order to oxidize dissolved Pb(II) to form $PbO_{2(s)}$; and 2) lower and lower pe values to reduce dissolved Pb(II) to form $Pb_{(s)}$.

In reference texts, pe-pH diagrams for a given element are often combined so that information concerning systems that contain variable levels of the element of interest can be presented in one diagram. As an example of how such diagrams might look, Figure 21.6 combines Figures 21.2 and 21.5 into a single plot. In order to make Figure 21.6 more readable, some of the lines from Figure 21.2 are presented in a dashed form.

21.6 pe-pH DIAGRAM FOR $Pb_T = 10^{-6}$ *M*

When $Pb_T = 10^{-6}$ *M*, a consideration of Figure 21.3 reveals that $PbO_{(s)}$ cannot form at any pH. It is likely then that the pe-pH diagram for $Pb_T = 10^{-6}$ *M* will look something like Figure 21.7. In contrast to Figures 21.1 and 21.4, since $PbO_{(s)}$ no longer has any predominance region, lines 5 and 6 are now absent. The trend observed in Figures 21.2 and 21.5 that the lengths of lines 5 and 6 decrease as Pb_T decreases has now led to a total disappearance of those lines.

In Figure 21.7, both the $PbO_{2(s)}$/solution boundaryline and the solution/$Pb_{(s)}$ boundaryline extend uninterrupted from low to high pH. Lines 1, 3c, 4c, and 7 are all intervals of the first boundaryline. That line can be computed using Eq. (21.13) with $Pb_T = 10^{-6}$ *M*. Lines 2, 3d, 4d, and 8 are all intervals of the solution/$Pb_{(s)}$ boundaryline. That line can be computed using Eq. (21.15) with $Pb_T = 10^{-6}$ *M*.

While line 3a in Figure 21.7 is located at the same position (pH = 7.7) as determined for $Pb_T = 10^{-4}$ *M*, lines 4a and 4b are new lines for us. For line 4a, from Eq. (21.5) we have at 25°C/1 atm that

$$\log\{Pb(OH)_2^o\} = -9.4 + pH + \log\{PbOH^+\} \qquad (21.22) \qquad \frac{PbOH^+}{Pb(OH)_2^o}.$$

Along line 4a, we will take $[PbOH^+] = [Pb(OH)_2^o]$. Thus, neglecting activity corrections, for this line we obtain

$$pH = 9.4. \qquad (21.23)$$

For the equilibrium pertaining to line 4b, from Eq. (21.6), at 25°C/1 atm, we have

$$\log\{Pb(OH)_3^-\} = -11.0 + pH + \log\{Pb(OH)_2^o\} \qquad (21.24) \qquad \frac{Pb(OH)_2^o}{Pb(OH)_3^-}.$$

Along line 4b, we will take $[Pb(OH)_2^o] = [Pb(OH)_3^-]$. Thus, neglecting activity corrections, for this line we obtain

$$pH = 11.0. \qquad (21.24)$$

Figure 21.8 gives the pe-pH diagram for $Pb_T = 10^{-6}$ *M* that is based on the equations discussed above.

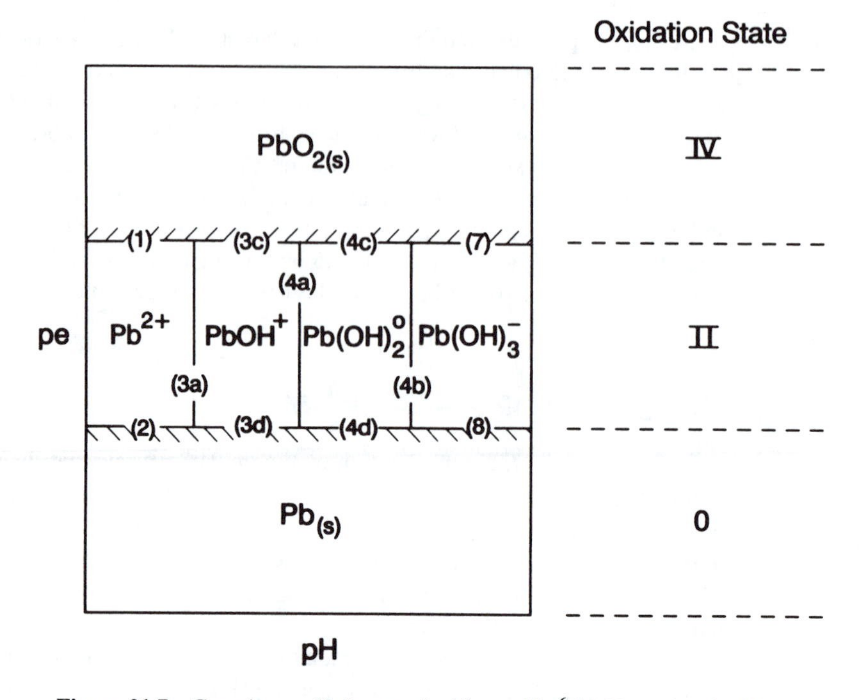

Figure 21.7 Generic pe-pH diagram for $Pb_T = 10^{-6}$ M. No region for $PbO_{(s)}$ appears in this generic diagram. We note that we do not, in fact, need to know that there is no region for $PbO_{(s)}$ in order to start drawing the actual pe-pH diagram (Figure 21.8). Indeed, if we tried to calculate the low and high pH roots of Eq. (21.10) for $Pb_T = 10^{-6}$ M, the trial and error method used would not converge on a "zero-crossing." That result would tell us that $PbO_{(s)}$ cannot be present at $Pb_T = 10^{-6}$ M (see also Figure 21.3).

21.7 USE OF "SIMPLIFYING" ASSUMPTIONS TO DRAW pe-pH DIAGRAMS

Many texts that describe the preparation of pe-pH diagrams for metals use assumptions that are claimed to simplify the task. For Figure 21.1, the approach used in those texts would call for assuming that $[Pb^{2+}] = 10^{-2}$ M all along lines 1, 2, and 3. That approach would also call for assuming that $[Pb(OH)_3^-] = 10^{-2}$ M all along lines 4, 7, and 8. Such assumptions do minimize the level of sophistication of the calculations required to draw the pe-pH diagram. However, now that sophisticated calculators and personal computers are readily available to do the calculations, it is far easier to ignore the simplifying assumptions and draw all of the lines as they really exist. Indeed, consider the following observations.

In the assumptions approach, a *completely separate line* must be developed for *each* of the lines in Figure 21.1. For line 1, we use the line obtained by ignoring all but the 1 in the logarithm term in Eq. (21.13). For line 7, we ignore all but the $^*\beta_3/[H^+]^3$ term in

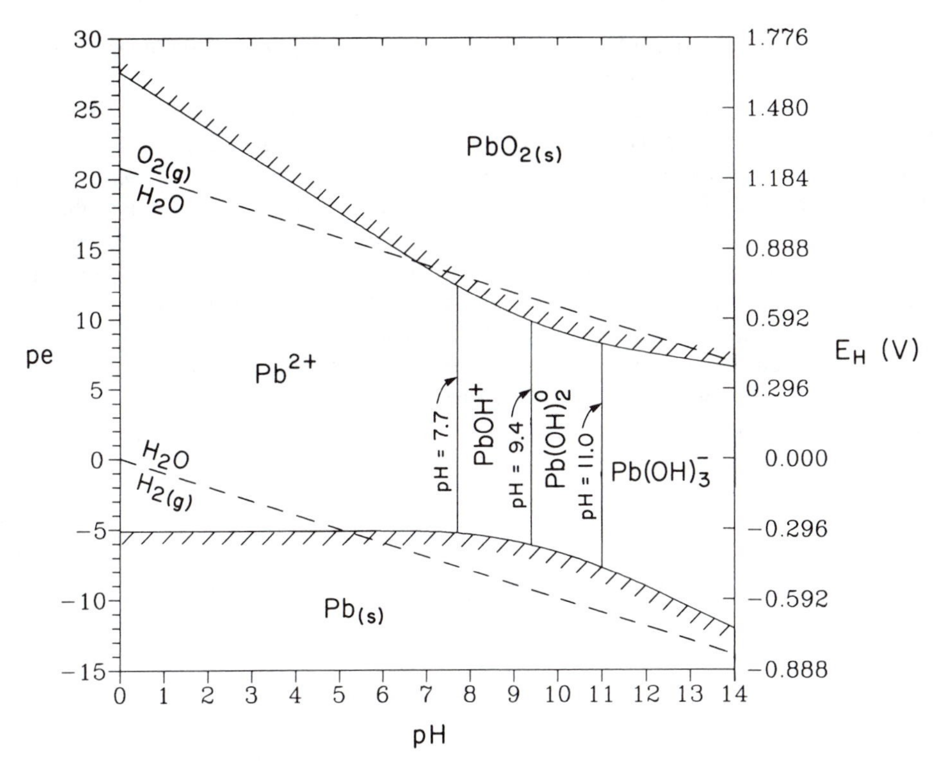

Figure 21.8 pe-pH diagram for $Pb_T = 10^{-6}$ *M* at 25°C/1 atm. The only assumptions are: 1) unit activity for all solids when present; and 2) unit activity coefficients for all solution species.

Eq. (21.13). For line 2, we ignore all but the 1 in the logarithm term in Eq. (21.15). For line 8, we ignore all but the $^*\beta_3/[H^+]^3$ term in Eq. (21.15). For line 3, we ignore all but the 1 inside the parentheses in Eq. (21.10). For line 4, we ignore all but the $^*\beta_3/[H^+]^3$ inside the parentheses in Eq. (21.10). Each of the six approximate equations would be subject to some error.

In the "exact"[5] approach that was used in Section 21.2, instead of developing six different, slightly wrong equations, we developed and used just three exact equations. These equations produced curvature in some of the lines, and automatically dealt with changing dominance among the dissolved Pb(II) species. Moreover, we found that the same three equations could be used when considering levels of Pb_T other than 10^{-2} *M*. Finally, in the exact approach, locating line 3 did not *require* that we first assume which dissolved Pb(II) species dominates when $PbO_{(s)}$ first becomes saturated. Indeed, in the exact approach, all we needed to do to locate line 3 was solve Eq. (21.10) for the low pH root. With that root in hand, we could then determine which dissolved Pb(II) species

[5]Exact to within activity corrections, that is.

dominates at the saturation pH and proceed to label the adjacent predominance region accordingly. The same comments apply to line 4.

For purposes of comparison, the pe-pH diagrams for $Pb_T = 10^{-2}$, 10^{-4}, and 10^{-6} *M* drawn using the assumptions approach are presented in Figures 21.9–21.11, respectively. In addition to the fact that lines which are actually curved are approximated as being composed of a series of straight lines with constant slopes, we note that the assumptions approach leads to solid regions that are larger than they should be. The reason for this is that ignoring some of the dissolved Pb(II) species along a given line leads to a concentration for the dominant species that is too high along all of that solid/solution boundary. (Along part of the boundary, the error might be very minor, as for lines 1 and 2 in Figure 21.9 at low pH.) Thus, in the context of that type of assumption, the diagrams imply that it is easier to form the solids than is actually the case. Since no dissolved Pb species are involved with lines 5 and 6, there are no errors in those lines in Figures 21.9–21.11.

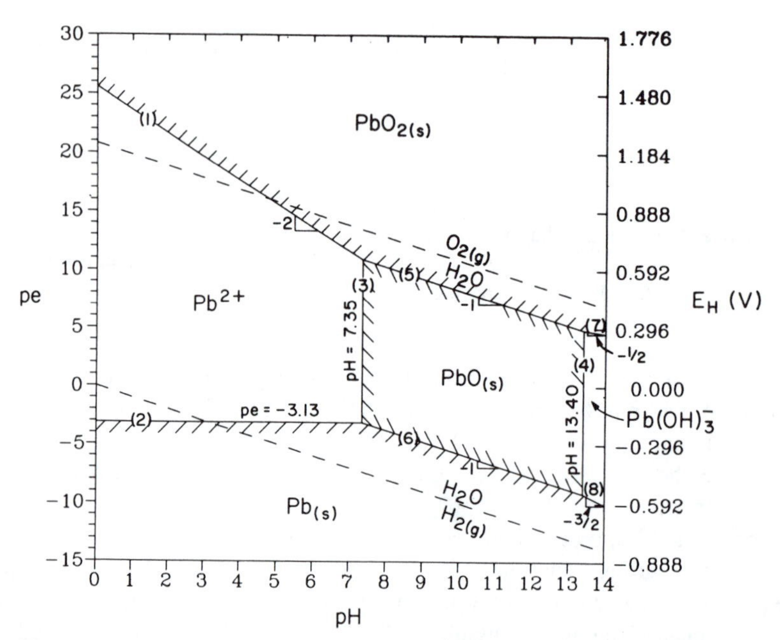

Figure 21.9 pe-pH diagram for $Pb_T = 10^{-2}$ *M* at 25°C/1 atm obtained using "simplifying" assumptions regarding dissolved Pb(II) chemistry. "Simplifying" assumptions include: 1) $[Pb^{2+}] \simeq 10^{-2}$ *M* along lines 1–3; and 2) $[Pb(OH)_3^-] \simeq 10^{-2}$ *M* along lines 4, 7, and 8. Other assumptions are that all solution species have unit activity coefficients and all solids are pure (when present). The numbers of the individual lines correspond to the numbers given in Figure 21.1. For each solution/solid boundaryline, the slope given corresponds to the value that would be obtained if the concentration of the dominant dissolved species equals Pb_T all along the line.

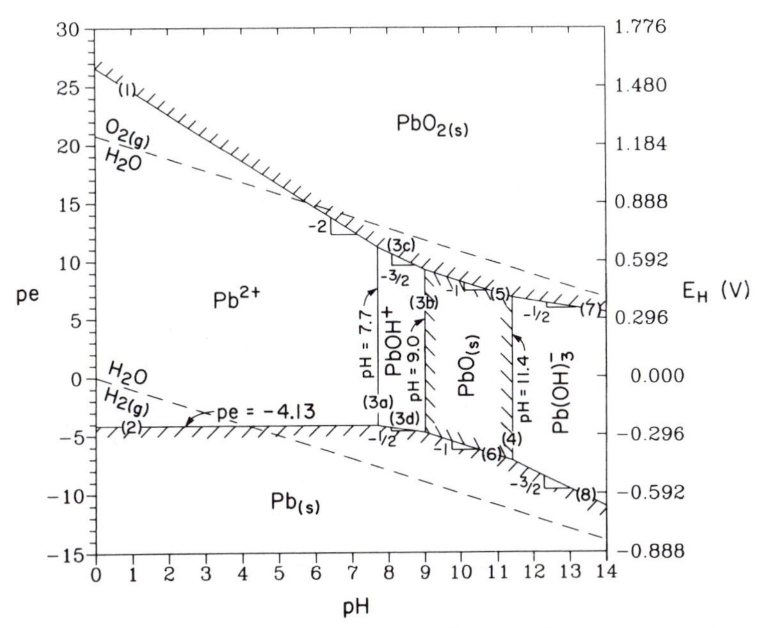

Figure 21.10 pe-pH diagram for $Pb_T = 10^{-4}$ M at 25°C/1 atm obtained using "simplifying" assumptions regarding dissolved Pb(II) chemistry. "Simplifying" assumptions include: 1) $[Pb^{2+}] \simeq 10^{-4}$ M along boundarylines 1 and 2; 2) $[PbOH^{+}] \simeq 10^{-4}$ M along lines 3b–3d; and 3) $[Pb(OH)_3^-] \simeq 10^{-4}$ M along boundarylines 4, 7, and 8. (Note that these assumptions cause the MBE on Pb_T to be violated at the following intersections: lines 1 and 3c; and, lines 2 and 3d.) Other assumptions are that all solution species have unit activity coefficients and all solids are pure (when present). The numbers of the individual lines correspond to the numbers given in Figure 21.4. For each solution/solid boundaryline, the slope given corresponds to the value that would be obtained if the concentration of the dominant dissolved species equals Pb_T all along the line.

21.8 INCLUDING SOLUTION PHASE ACTIVITY CORRECTIONS WHEN DRAWING pe-pH DIAGRAMS

In Sections 21.2, 21.5, and 21.6, we discussed how to draw pe-pH boundarylines in a manner which avoided the need to make any "simplifying" assumptions regarding dissolved Pb(II) chemistry. In those sections, however, we did not include any solution phase activity corrections. There are two ways in which such corrections might be included. In the simplest case, we could draw the diagram for a single value of the ionic strength I, assuming then that all points in the pe-pH diagram have that value of I. With this approach, all we would have to do is adjust all of the infinite dilution constants to ${}^{c}K$-type constants, then proceed as already outlined. However, such a diagram cannot extend outside of the range

$$-\log I < \text{pH} < \log I + 14. \tag{21.25}$$

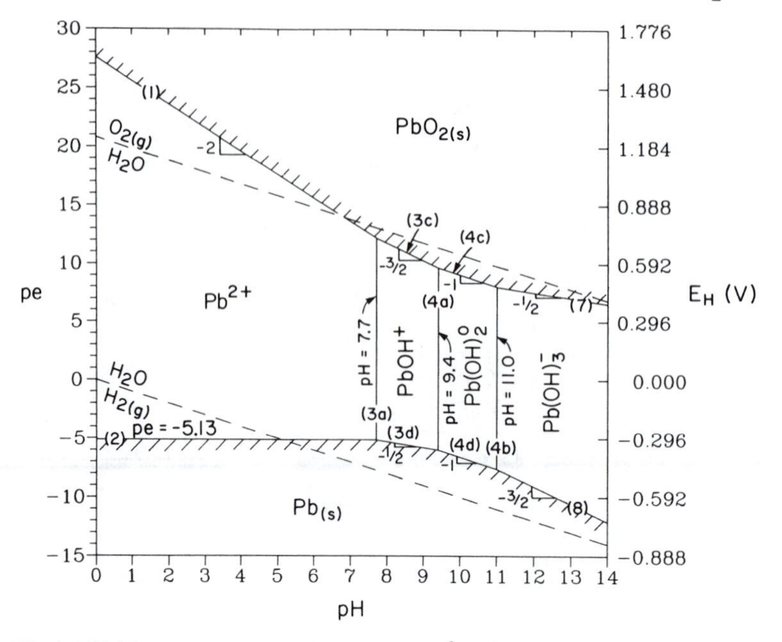

Figure 21.11 pe-pH diagram for $Pb_T = 10^{-6}$ M at 25°C/1 atm obtained using "simplifying" assumptions regarding dissolved Pb(II) chemistry. "Simplifying" assumptions include: 1) $[Pb^{2+}] \simeq 10^{-6}$ M along boundarylines 1 and 2; 2) $[PbOH^+] \simeq 10^{-6}$ M along boundarylines 3c and 3d; 3) $[Pb(OH)_2^o] = 10^{-6}$ M along lines 4c and 4d; 4) $[Pb(OH)_3^-] \simeq 10^{-6}$ M along lines 7 and 8. (Note that these assumptions cause the MBE on Pb_T to be violated at the following intersections: lines 1 and 3c; lines 3c and 4c; lines 4c and 7; lines 2 and 3d; lines 3d and 4d; and lines 4d and 8.) Other assumptions are that all solution species have unit activity coefficients and all solids are pure (when present). The numbers of the individual lines correspond to the numbers given in Figure 21.7. For each solution/solid boundaryline, the slope given corresponds to the value that would be obtained if the concentration of the dominant dissolved species equals Pb_T all along the line.

As an example, if $I = 0.1$, then our diagram cannot extend below pH = 1; at pH values lower than 1, just the contributions to I from H^+ and a typical anion like Cl^- will be greater than 0.1. On the high pH side, if $I = 0.1$, the diagram cannot extend beyond pH = 13 because at that pH the contributions to I from OH^- and a typical cation like Na^+ will be greater than 0.1.

If the value of I is not constant for the entire pe-pH domain, then one will need to carry out activity corrections using the I value expected at each given (pe,pH) point in order to determine whether that point was actually inside or outside of a given pe-pH domain of interest. Problems illustrating the principles of this chapter are presented by Pankow (1992).

21.9 REFERENCES

HEM, J. D. 1976. Inorganic chemistry of lead in water, in *Lead in the Environment*, pp. 5–11. T. G. Lovering, Ed., U.S. Geological Survey Professional Paper 957, Washington, D.C.

PANKOW, J. F. 1992. *Aquatic Chemistry Problems*, Portland: Titan Press-OR, P.O. Box 91399, Portland, Oregon 97291-1399.

STUMM, W., and J. J. MORGAN. 1970. *Aquatic Chemistry*, New York: Wiley-Interscience.

22

pe-pH Diagrams for Aqueous Lead in the Presence of CO_2, with Fixed C_T

22.1 INTRODUCTION

The case of aqueous lead (Pb) in the absence of CO_2 was discussed in Chapter 21. While some natural water systems do exist in which little or no CO_2 is present, most systems contain significant amounts of dissolved CO_2 species. In a pe-pH plot for a metallic element, if some CO_2 is present, then we must consider the possibility that there will be predominance regions for one or more metal carbonate solid. In this chapter, we will consider the pe-pH diagrams that result for C_T equal to constant values of 10^{-2} *M* and 10^{-3} *M* for two values of Pb_T, namely 10^{-4} *M* and 10^{-6} *M*. Thus, four specific cases will be examined. For C_T, the subscript *T* will refer only to total *dissolved* CO_2. The value of C_T will be taken to be constant for all portions of a given pe-pH diagram. However, as in Chapter 21, for Pb_T the subscript *T* will mean total lead (including dissolved forms *plus* solid forms) per liter of aqueous phase.

In Chapter 21, two different approaches are discussed for preparing pe-pH diagrams for aqueous Pb. In the *exact* approach (see Sections 21.2, 21.5, and 21.6), the only assumptions made are that all activity coefficients in the aqueous phase are unity, and that when present, all solid phases are pure. In the approach using additional *"simplifying" assumptions* (see Section 21.7), one assumes that along each solution/solid boundaryline, the concentration of the species in the solution predominance region equals the total number of mols of element per liter of water in the system. For example, along a $Pb^{2+}/PbO_{2(s)}$ boundaryline, we would assume that $[Pb^{2+}]$ equals the Pb_T for the diagram.

When there are numerous solution/solid boundarylines (e.g., see Figure 21.7), Section 21.7 discussed how the "simplifying" assumptions approach can be unattractive since a different equation will need to be painstakingly generated for each of the straight, approximate boundarylines. Moreover, at the intersection of two solution/solid boundarylines, the assumptions corresponding to the two lines will violate the MBE for dissolved metal (e.g., consider lines 1 and 3c in Figure 21.10).

When CO_2 is added to a Pb-containing system, the "simplifying" assumptions approach has the potential to be even more complicated to use than when CO_2 is absent. Indeed, if a given carbonate solid does form, then to be able to draw simple straight lines, we would have to allow for different boundarylines in the three pH ranges in which CO_2 chemistry behaves asymptotically. For example, if a given carbonate solid exhibits a predominance region that extends over a pH range that straddles pK_2 for $H_2CO_3^*$, then in constructing the boundarylines for that region we would need to assume that $[HCO_3^-] \simeq C_T$ for a boundaryline for $pK_1 < pH < pK_2$, and we would need to assume that $[CO_3^{2-}] \simeq C_T$ for another boundaryline for $pH > pK_2$. At the intersection of those boundarylines, the MBE on total dissolved C_T would be violated: for one line we would have $[HCO_3^-] \simeq C_T$ and for the other we would have $[CO_3^{2-}] \simeq C_T$.

When a personal computer is available to make the calculations through the application of a suitable spreadsheet program, the exact approach for drawing pe-pH diagrams is very easy to use. For example, in Chapter 21 we found that a single equation could be developed that describes all applicable portions of the $PbO_{2(s)}$/solution boundaryline over the entire pH range, that is, even when dominance is shifting from one dissolved Pb(II) species to another. We can handle changing CO_2 chemistry in the same way. For these various reasons, in this chapter, we will focus entirely on the exact approach.

22.2 pe-pH DIAGRAM FOR $Pb_T = 10^{-4}$ *M* AND $C_T = 10^{-2}$ *M*

22.2.1 General

As discussed in Chapter 21, Pb has three major oxidation states that must be considered, IV, II, and 0. As when CO_2 is absent, we will assume that: 1) Pb(IV) is essentially insoluble and is limited to the phase $PbO_{2(s)}$; 2) Pb(0) is essentially insoluble and is limited to the phase $Pb_{(s)}$; and 3) the important aqueous Pb(II) species are Pb^{2+}, $PbOH^+$, $Pb(OH)_2^o$, and $Pb(OH)_3^-$. The species $PbCO_3^o$ can also form to a very small extent, but it will be neglected here. As far as Pb(II) solids are concerned, we will consider red-$PbO_{(s)}$ as one possibility. Since C_T is not zero, two additional Pb(II) solids that we will consider to be possible are $PbCO_{3(s)}$ and $Pb_3(CO_3)_2(OH)_{2(s)}$.

One possible set of independent equilibrium relations with which we can develop the desired pe-pH diagram is

$$PbO_{2(s)} + 4H^+ + 2e^- = 2H_2O + Pb^{2+} \qquad \log K = 49.2 \qquad (22.1)$$

$$Pb^{2+} + 2e^- = Pb_{(s)} \qquad \log K = -4.26 \qquad (22.2)$$

$$PbO_{(s)} + 2H^+ = Pb^{2+} + H_2O \qquad \log K = 12.7 \qquad (22.3)$$

$$Pb^{2+} + H_2O = PbOH^+ + H^+ \qquad \log {}^*K_{H1} = -7.7 \qquad (22.4)$$

$$PbOH^+ + H_2O = Pb(OH)_2^o + H^+ \qquad \log {}^*K_{H2} = -9.4 \qquad (22.5)$$

$$Pb(OH)_2^o + H_2O = Pb(OH)_3^- + H^+ \qquad \log\ {}^*K_{H3} = -11.0 \quad (22.6)$$

$$PbCO_{3(s)} = Pb^{2+} + CO_3^{2-} \qquad \log K = -12.8 \quad (22.7)$$

$$Pb_3(CO_3)_2(OH)_{2(s)} + 2H^+ = 3Pb^{2+} + 2CO_3^{2-} + 2H_2O \qquad \log K = -18.8. \quad (22.8)$$

The values of the equilibrium constants given for Eqs. (22.1)–(22.8) are for 25°C/1 atm.

22.2.2 Identification of the Pb(II) Solids Having Predominance Regions

In Chapter 21, we concluded that as long as Pb(IV) is assumed to be totally insoluble, then no matter how low Pb_T becomes, the pe can be made high enough to form $PbO_{2(s)}$ at any pH. (Depending on the pH and on how low we make Pb_T, this may mean that p_{O_2} is raised to levels substantially greater than 1 atm.) In a similar manner, as long as Pb(0) is assumed to be totally insoluble, then no matter how low Pb_T becomes, the pe can be made low enough to form $Pb_{(s)}$ at any pH. (Depending on the pH and how low we make Pb_T, this may mean that p_{H_2} is raised to levels greater than 1 atm.)[1]

Because Pb(II) solids are much more soluble than either $PbO_{2(s)}$ or $Pb_{(s)}$, the pe-pH situation for Pb(II) is different from that for either $PbO_{2(s)}$ or $Pb_{(s)}$. Indeed, as discussed in Chapter 21, when CO_2 is absent, when Pb_T is sufficiently small, it will not be possible to form $PbO_{(s)}$ at any pH. In a similar manner, for any given nonzero value of C_T, when Pb_T is sufficiently small, it will not be possible to form any Pb(II) carbonate solid at any pH. As Pb_T increases, predominance regions for at least one of the Pb(II) solids in Eqs. (22.3), (22.7), and (22.8) will eventually appear in the pe-pH diagram.

For any given value of C_T, an important first task will be to determine: 1) which (if any) of the above three solids can exhibit predominance regions for the Pb_T and C_T values of interest; and 2) over which pH ranges those predominance regions extend. In order to complete this task, we recall from Chapter 14 that:

1. The solid that prescribes the lowest concentration of any given dissolved metal species or total dissolved metal is the *solubility limiting* solid.
2. When a given solid is solubility-limiting, then at equilibrium the solution phase does not have the ability to hold more of any given dissolved metal species (or total dissolved metal) than is prescribed by that specific solid.
3. When a given solid is solubility limiting as well as actually *solubility controlling*, then the solution is characterized by the dissolved metal species and total dissolved metal concentration values that are prescribed by equilibrium with that solid for the solution conditions of interest.

[1]Of course, neither Pb(IV) nor Pb(0) will have exactly zero solubilities at any pH. Thus, Pb_T can in fact be made low enough that neither $PbO_{2(s)}$ nor $Pb_{(s)}$ can form at *any* values of pe and pH. Such values of Pb_T are very low in comparison to values like 10^{-6} M.

4. Two different solids can potentially be present simultaneously, but only when the concentrations that they prescribe are exactly the same for each individual dissolved metal species as well as for the total dissolved metal.

Based on these observations, we can conclude that if we seek to determine which of the three Pb(II) solids have predominance regions and what the pH ranges of those regions are, we first need to determine where the different solids are solubility limiting. Whether or not a given Pb solid will be present *and* solubility controlling (i.e., have a pe-pH predominance region) under specific circumstances will then depend on the value of Pb_T for which the diagram is being constructed. As discussed in Chapter 21, a given solubility-limiting solid can only occur over the pH range for which that solid prescribes total dissolved Pb values that are *less* than that of Pb_T.

To probe the issue of solubility limitation by the different Pb(II) solids, let us compute $[Pb^{2+}]$ as it is determined by the three different solids as: 1) a function of pH; 2) for the C_T of interest; and 3) assuming that regardless of the pH, this is being done at a pe value such that Pb(II) is the only stable oxidation state. The last assumption is reasonable because: 1) we have already assumed that neither Pb(IV) nor Pb(0) have any significant solubility in water (i.e., they are not important when Pb(II) is important); and 2) the regions for $PbO_{2(s)}$ and $Pb_{(s)}$ will not overlap with a region for a Pb(II) solid.

Each of the three Pb(II) solids will give a different equation for $[Pb^{2+}]$ as a function of pH. For $PbO_{(s)}$, $PbCO_{3(s)}$, and $Pb_3(CO_3)_2(OH)_{2(s)}$, we will refer to these three functions for $[Pb^{2+}]$ as $f([H^+])$, $g([H^+])$, and $h([H^+])$, respectively. At any given pH, the function that gives the lowest value for $[Pb^{2+}]$ will correspond to the solid that is solubility limiting. When the lowest two out of the three functions cross at a given pH, then as discussed in Chapter 14, that pH value will be a stability transition pH. On one side of the transition pH, one of the two solids will be solubility limiting. On the other side, the other solid will be solubility limiting. At the transition pH, both solids could be present. We now examine the nature of the three functions for the three Pb(II) solids. Unit activity coefficients in the aqueous phase and unit activities for all solids present will be assumed.

For $PbO_{(s)}$, we have:

$$[Pb^{2+}] = f([H^+]) = 10^{12.7}[H^+]^2 \tag{22.9}$$

$$\log f([H^+]) = 12.7 - 2\text{ pH}. \tag{22.10}$$

For $PbCO_{3(s)}$, we have:

$$[Pb^{2+}] = g([H^+]) = 10^{-12.8}/[CO_3^{2-}] \tag{22.11}$$

$$= \frac{10^{-12.8}}{\alpha_2 C_T} \tag{22.12}$$

$$\log g([H^+]) = -12.8 - \log \alpha_2 - \log C_T. \tag{22.13}$$

For pH > pK_2, we know that $d\alpha_2/d\text{pH} \simeq 0$. Therefore, for constant C_T,

$$\frac{d \log g([H^+])}{d\text{pH}} \simeq 0. \tag{22.14}$$

For $pK_1 < pH < pK_2$, we know that $d\alpha_2/dpH \simeq +1$, and so for constant C_T,

$$\frac{d \log g([H^+])}{dpH} \simeq -1. \tag{22.15}$$

For $pH < pK_1$, $d\alpha_2/dpH \simeq +2$, and so for constant C_T,

$$\frac{d \log g([H^+])}{dpH} \simeq -2. \tag{22.16}$$

For $Pb_3(CO_3)_2(OH)_{2(s)}$, we have:

$$[Pb^{2+}] = h([H^+]) = \left(\frac{10^{-18.8}[H^+]^2}{[CO_3^{2-}]^2}\right)^{1/3} \tag{22.17}$$

$$= \left(\frac{10^{-18.8}[H^+]^2}{\alpha_2^2 C_T^2}\right)^{1/3} \tag{22.18}$$

$$\log h([H^+]) = -6.27 - \frac{2}{3}\text{ pH} - \frac{2}{3}\log\alpha_2 - \frac{2}{3}\log C_T. \tag{22.19}$$

For constant C_T, for $pH > pK_2$ we have

$$\frac{d \log h([H^+])}{dpH} = -\frac{2}{3} - \frac{2}{3}(\sim 0) \simeq -\frac{2}{3}. \tag{22.20}$$

For $pK_1 < pH < pK_2$, we have

$$\frac{d \log h([H^+])}{dpH} = -\frac{2}{3} - \frac{2}{3}(\sim 1) \simeq -\frac{4}{3}. \tag{22.21}$$

For $pH < pK_1$, we have

$$\frac{d \log h([H^+])}{dpH} = -\frac{2}{3} - \frac{2}{3}(\sim 2) \simeq -2. \tag{22.22}$$

The lines for $\log[Pb^{2+}]$ vs. pH as prescribed by $f([H^+])$, $g([H^+])$, and $h([H^+])$ are plotted for the three solids in Figure 22.1 for $C_T = 10^{-2}$ M. The functions $g([H^+])$ and $h([H^+])$ cross at a pH value somewhat above 7. To compute the exact transition pH between the two carbonate solids, we require that the two solids prescribe the same value for $[Pb^{2+}]$. By equating Eqs. (22.13) and (22.19), at 25°C/1 atm we obtain

$$-12.8 - \log\alpha_2 - \log C_T = -6.27 - \frac{2}{3}\text{ pH} - \frac{2}{3}\log\alpha_2 - \frac{2}{3}\log C_T \tag{22.23}$$

or

$$-6.53 - \frac{1}{3}\log\alpha_2 - \frac{1}{3}\log C_T + \frac{2}{3}\text{ pH} = 0. \tag{22.24}$$

Solving Eq. (22.24) for $C_T = 10^{-2}$ M, we obtain that transition pH between the two carbonate solids is pH = 7.20. Considering the relatively low values of $[Pb^{2+}]$ given by the functions $g([H^+])$ and $h([H^+])$ in Figure 22.1, it is probable that a Pb_T value of 10^{-4} M is sufficiently high that both of the carbonate solids will have predominance regions

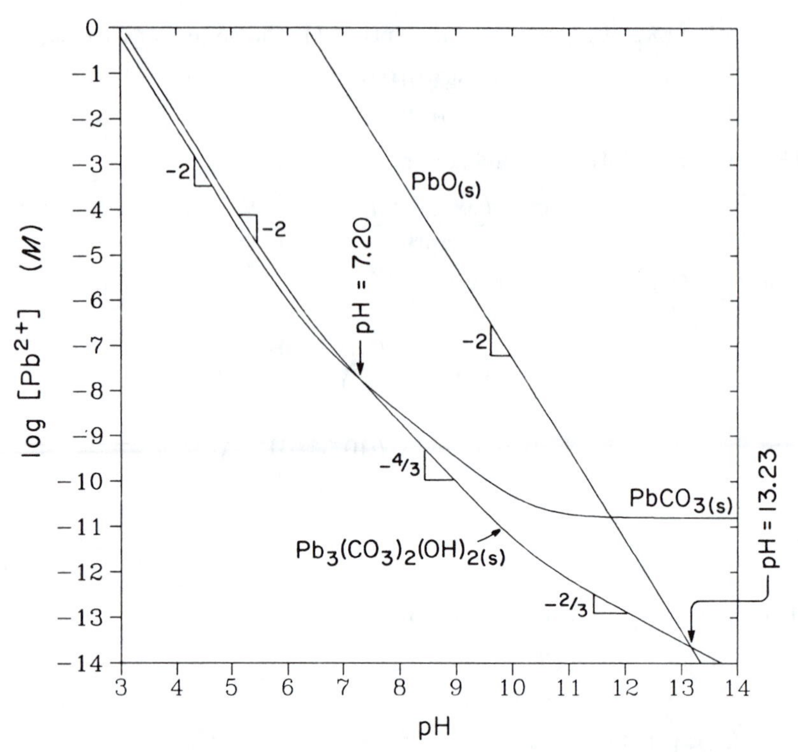

Figure 22.1 $\log[Pb^{2+}]$ vs. pH for $C_T = 10^{-2}$ *M* and 25°C/1 atm, as controlled by three different solids: $PbO_{(s)}$, $PbCO_{3(s)}$, and $Pb_3(CO_3)_2(OH)_{2(s)}$. Unit activity coefficients in the aqueous phase and unit activities for the solids are assumed.

(this will be verified shortly). For intermediate pe values, at pH values immediately to the left of 7.20, we are therefore inside the predominance region for $PbCO_{3(s)}$; at pH values immediately to the right of 7.20, we are inside the predominance region for $Pb_3(CO_3)_2(OH)_{2(s)}$.

We note that although $\log g([H^+])$ and $\log h([H^+])$ appear essentially parallel in Figure 22.1 at pH < pK_1, by Eqs. (22.13) and (22.19), at any given pH, their slopes are never *exactly* equal for any value of C_T. Thus, for pH < pK_1, if we carefully changed (decreased) the value of C_T, we could not make an extended portion of the $\log g([H^+])$ line lie on top of the $\log h([H^+])$ line. Therefore, there is no value of C_T such that both carbonate solids are thermodynamically possible at more than a single pH value.

The lines for $Pb_3(CO_3)_2(OH)_{2(s)}$ and $PbO_{(s)}$ cross at a relatively high pH in Figure 22.1. Combining Eqs. (22.10) and (22.19), at 25°C/1 atm we obtain

$$18.97 - \frac{4}{3}\,\text{pH} + \frac{2}{3}\log \alpha_2 + \frac{2}{3}\log C_T = 0. \tag{22.25}$$

Solving for $C_T = 10^{-2}$ *M*, we obtain a transition between $Pb_3(CO_3)_2(OH)_{2(s)}$ and $PbO_{(s)}$ at pH = 13.23. Since the equilibrium constants used above were infinite dilution constants, under these high pH conditions, the ionic strength is high enough to now require significant activity corrections. Therefore, a specific estimate of the ionic strength would need to be made at this point if we are to go back and include activity corrections for use in calculating the *actual* $Pb_3(CO_3)_2(OH)_{2(s)}/PbO_{(s)}$ transition pH. However, there is no reason to actually try and make these activity corrections since $PbO_{(s)}$ cannot have a predominance region in this $Pb_T = 10^{-4}$ M system. Indeed, Figure 21.3 illustrates that for pH $\simeq$ 13, total dissolved Pb as limited by $PbO_{(s)}$ is about 10^{-2} M, that is, much above 10^{-4} M. Thus, if Pb_T is only 10^{-4} M, then $PbO_{(s)}$ cannot form in this system.

Considering that: 1) at $Pb_T = 10^{-4}$ M the high pH root for $PbO_{(s)}$ is quite high (11.17); and 2) $h([H^+])$ in Figure 22.1 lies far below $f([H^+])$ for much of the pH range (i.e., $Pb_3(CO_3)_2(OH)_{2(s)}$ is much less soluble than $PbO_{(s)}$ for much of the pH range), we can conclude that a $Pb(OH)_3^-$ predominance region must bound the $Pb_3(CO_3)_2(OH)_{2(s)}$ region on the right. This conclusion is validated by an examination of Figure 22.2 which gives the concentrations of the various dissolved Pb(II) species at equilibrium with $Pb_3(CO_3)_2(OH)_{2(s)}$ when $C_T = 10^{-2}$ M. In particular, when $Pb_T = 10^{-4}$ M, the dominant dissolved Pb(II) species in equilibrium with this solid on the high pH side is in fact $Pb(OH)_3^-$. The location of the boundaryline between the regions for $Pb_3(CO_3)_2(OH)_{2(s)}$ and $Pb(OH)_3^-$ will be given by the high pH root value for $Pb_3(CO_3)_2(OH)_{2(s)}$ and a total dissolved Pb value of 10^{-4} M.

In a manner that is similar to that outlined in the preceding paragraph, through a comparison between $g([H^+])$ and $f([H^+])$, we can expect that the $PbCO_{3(s)}$ that forms at pH < 7.20 will not become solubilized until the pH becomes rather low. The predominance region that bounds the $PbCO_{3(s)}$ region on the left will therefore be a region dominated by Pb^{2+}. In other words, if $PbO_{(s)}$ at $Pb_T = 10^{-4}$ M does not become solubilized on the low pH end until such point that Pb^{2+} dominates the dissolved Pb(II) chemistry, and if $PbCO_{3(s)}$ is less soluble at $C_T = 10^{-2}$ M than is $PbO_{(s)}$, then $PbCO_{(s)}$ at $Pb_T = 10^{-4}$ M will surely not become solubilized on the low pH end until such point that Pb^{2+} dominates the dissolved Pb(II) chemistry. This conclusion is validated by an examination of Figure 22.3 which gives the concentrations of the various dissolved Pb(II) species at equilibrium with $PbCO_{3(s)}$ when $C_T = 10^{-2}$ M. In particular, when $Pb_T = 10^{-4}$ M, the dominant dissolved Pb(II) species in equilibrium with this solid on the low pH side is in fact Pb^{2+}. In generic terms, we may therefore expect that the pe-pH diagram desired will look something like Figure 22.4.

22.2.3 The Two Solution/Pb(II) Solid Boundarylines

For the two solution/Pb(II) solid boundarylines, we must solve Eq. (21.7)

$$[Pb^{2+}] + [PbOH^+] + [Pb(OH)_2^o] + [Pb(OH)_3^-] = Pb_T \tag{21.7}$$

for the appropriate value of Pb_T where, each of the dissolved Pb(II) species is controlled at low pH by $PbCO_{3(s)}$, and at high pH by $Pb_3(CO_3)_2(OH)_{2(s)}$. We note now that Eq. (21.9) contains the factor that relates Pb_T to $[Pb^{2+}]$. In particular, we have

$$Pb_T = [Pb^{2+}]\left(1 + \frac{{}^*\beta_1}{[H^+]} + \frac{{}^*\beta_2}{[H^+]^2} + \frac{{}^*\beta_3}{[H^+]^3}\right). \qquad (21.9)$$

At 25°C/1 atm, ${}^*\beta_1 = 10^{-7.7}$, ${}^*\beta_2 = 10^{-17.1}$, and ${}^*\beta_3 = 10^{-28.1}$.

For the $Pb_3(CO_3)_2(OH)_{2(s)}$/solution boundaryline (line 15 in Figure 22.4), Eq. (22.19) gives the relationship between $\log[Pb^{2+}]$ and pH when $Pb_3(CO_3)_2(OH)_{2(s)}$ controls the solubility. Substituting into Eq. (22.19) for $[Pb^{2+}]$ using Eq. (21.9), we obtain

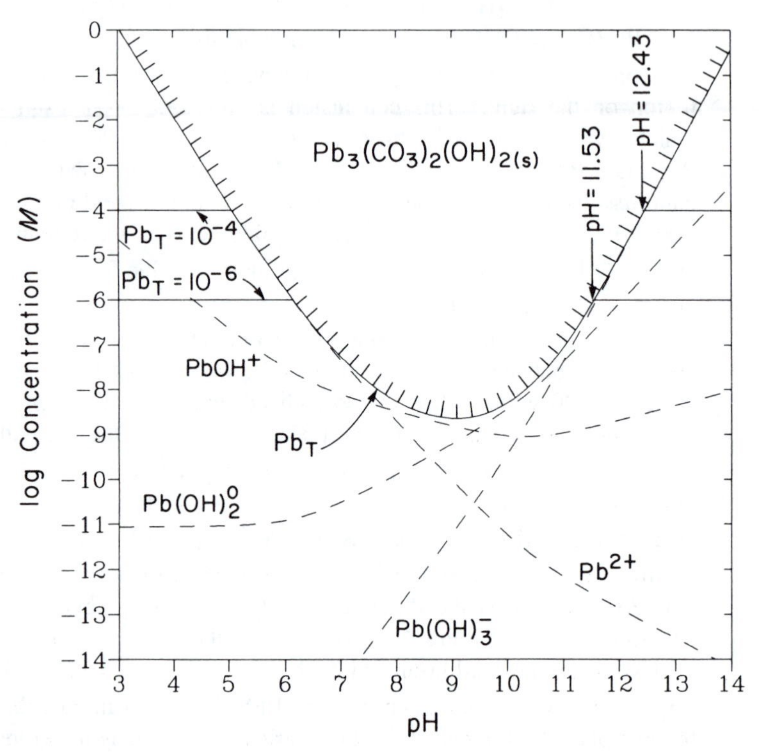

Figure 22.2 Log concentration vs. pH when $C_T = 10^{-2}$ *M* at 25°C/1 atm for various dissolved Pb(II) species in equilibrium with $Pb_3(CO_3)_2(OH)_{2(s)}$. The various lines have been drawn with $C_T = 10^{-2}$ *M* using the following equilibrium information:

$$\frac{1}{3}Pb_3(CO_3)_2(OH)_{2(s)} + \frac{2}{3}H^+ = Pb^{2+} + \frac{2}{3}CO_3^{2-} + \frac{2}{3}H_2O, \qquad \log K = -6.27$$

$$\frac{1}{3}Pb_3(CO_3)_2(OH)_{2(s)} + \frac{1}{3}H_2O = PbOH^+ + \frac{2}{3}CO_3^{2-} + \frac{1}{3}H^+, \qquad \log K = -13.97$$

$$\frac{1}{3}Pb_3(CO_3)_2(OH)_{2(s)} + \frac{4}{3}H_2O = Pb(OH)_2^o + \frac{2}{3}CO_3^{2-} + \frac{4}{3}H^+, \qquad \log K = -23.37$$

$$\frac{1}{3}Pb_3(CO_3)_2(OH)_{2(s)} + \frac{7}{3}H_2O = Pb(OH)_3^- + \frac{2}{3}CO_3^{2-} + \frac{7}{3}H^+, \qquad \log K = -34.37.$$

$$\log \mathrm{Pb_T} - \log\left(1 + \frac{{}^*\beta_1}{[\mathrm{H^+}]} + \frac{{}^*\beta_2}{[\mathrm{H^+}]^2} + \frac{{}^*\beta_3}{[\mathrm{H^+}]^3}\right) = -6.27 - \frac{2}{3}\,\mathrm{pH} - \frac{2}{3}\log\alpha_2 - \frac{2}{3}\log C_\mathrm{T}. \tag{22.26}$$

Solving Eq. (22.26) for $Pb_T = 10^{-4}$ M and $C_T = 10^{-2}$ M at 25°C/1 atm, we obtain:

$$\mathrm{pH} = 12.43. \tag{22.27}$$

This pH value is sufficiently high that Eq. (22.27) can only be considered as an approximation for this boundaryline; activity corrections would be required if an accurate estimate of the boundaryline is needed.

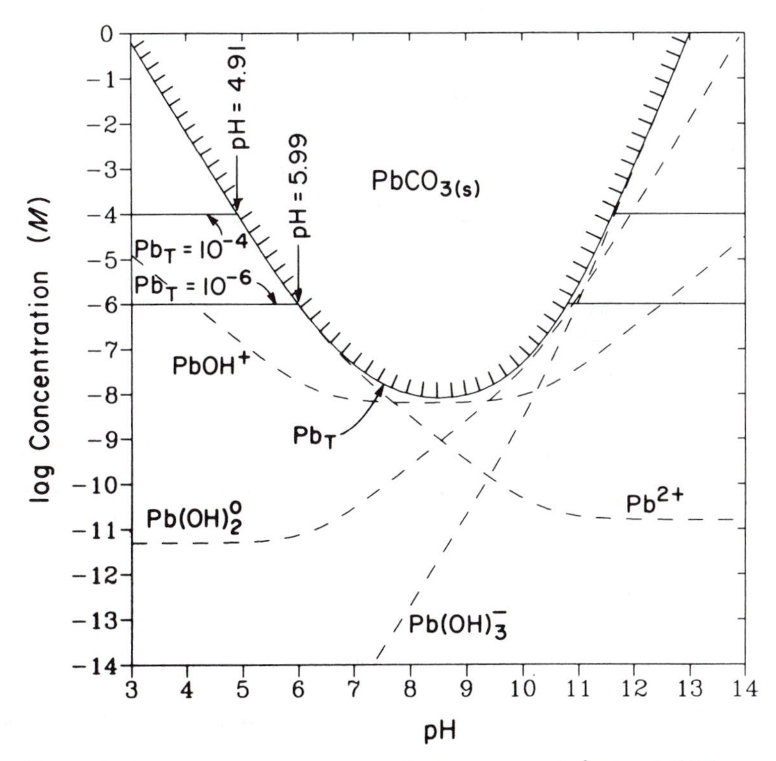

Figure 22.3 Log concentration vs. pH when $C_T = 10^{-2}$ M at 25°C/1 atm for various dissolved Pb(II) species in equilibrium with $PbCO_{3(s)}$. The various lines have been drawn with $C_T = 10^{-2}$ M using the following equilibrium information:

$PbCO_{3(s)} = Pb^{2+} + CO_3^{2-}$, $\log K = -12.8$

$PbCO_{3(s)} + H_2O = PbOH^+ + H^+ + CO_3^{2-}$, $\log K = -20.5$

$PbCO_{3(s)} + 2H_2O = Pb(OH)_2^o + 2H^+ + CO_3^{2-}$, $\log K = -29.9$

$PbCO_{3(s)} + 3H_2O = Pb(OH)_3^- + 3H^+ + CO_3^{2-}$, $\log K = -41.0$.

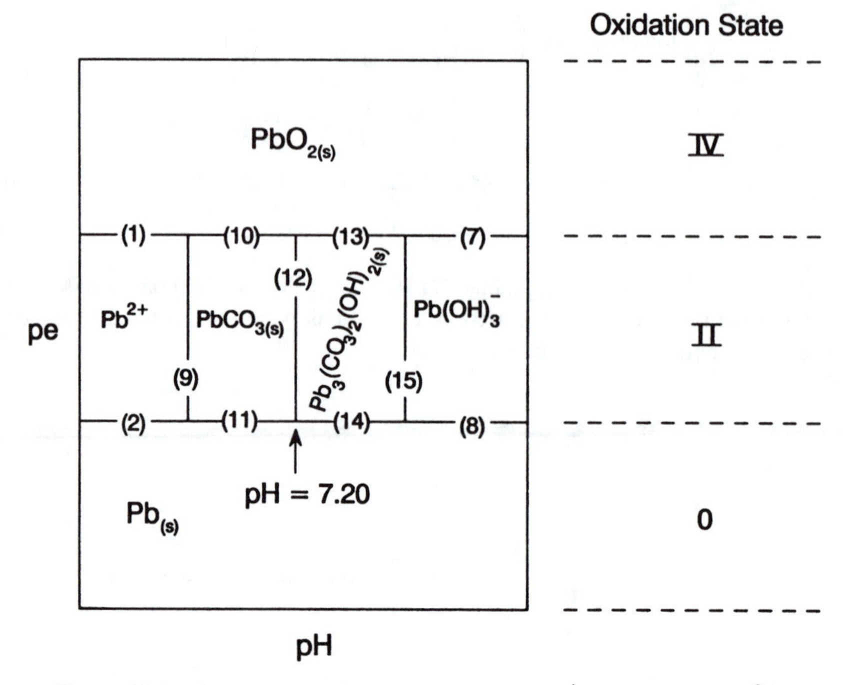

Figure 22.4 Generic pe-pH diagram for $Pb_T = 10^{-4}$ M and $C_T = 10^{-2}$ M. Also applies for $Pb_T = 10^{-6}$ M and $C_T = 10^{-2}$ M. Reduction of C(IV) carbon to methane at low pe is neglected.

For the solution/$PbCO_{3(s)}$ boundaryline (line 9), Eq. (22.13) gives the relationship between log[Pb^{2+}] and pH when $PbCO_{3(s)}$ controls the solubility. Substituting into Eq. (22.13) for [Pb^{2+}] using Eq. (21.9), we obtain

$$\log \mathrm{Pb_T} - \log\left(1 + \frac{{}^*\beta_1}{[\mathrm{H^+}]} + \frac{{}^*\beta_2}{[\mathrm{H^+}]^2} + \frac{{}^*\beta_3}{[\mathrm{H^+}]^3}\right) = -12.8 - \log \alpha_2 - \log C_\mathrm{T}. \tag{22.28}$$

Solving Eq. (22.28) for $Pb_T = 10^{-4}$ M and $C_T = 10^{-2}$ M at 25°C/1 atm, we obtain:

$$\mathrm{pH} = 4.91. \tag{22.29}$$

22.2.4 The $PbO_{2(s)}$/Solution Boundaryline, and the Solution/$Pb_{(s)}$ Boundaryline

We note that according to the exact approach for preparing pe-pH diagrams, lines 1 and 7 in Figure 22.4 are parts of the same, curved $PbO_{2(s)}$/solution phase boundaryline, and lines 2 and 8 are parts of the same solution phase/$Pb_{(s)}$ boundaryline.

From Chapter 21, at 25°C/1 atm, we have for the $PbO_{2(s)}$/solution phase boundaryline (lines 1 and 7) that

$$2\ \text{pe} = 49.2 - \log \text{Pb}_T + \log\left(1 + \frac{{}^*\beta_1}{[\text{H}^+]} + \frac{{}^*\beta_2}{[\text{H}^+]^2} + \frac{{}^*\beta_3}{[\text{H}^+]^3}\right) - 4\ \text{pH}. \quad (21.13)$$

For the solution phase/$Pb_{(s)}$ boundaryline (lines 2 and 8) we have

$$2\ \text{pe} = -4.26 + \log \text{Pb}_T - \log\left(1 + \frac{{}^*\beta_1}{[\text{H}^+]} + \frac{{}^*\beta_2}{[\text{H}^+]^2} + \frac{{}^*\beta_3}{[\text{H}^+]^3}\right). \quad (21.15)$$

22.2.5 The $PbO_{2(s)}$/Pb(II) and Pb(II)/$Pb_{(s)}$ Solid/Solid Boundarylines

For the various Pb(IV)/Pb(II) and Pb(II)/Pb(0) boundaries of the solid/solid type, we combine the needed equations as follows. For the $PbO_{2(s)}/PbCO_{3(s)}$ boundaryline (line 10), we combine

$$\text{PbO}_{2(s)} + 4\text{H}^+ + 2\text{e}^- = 2\text{H}_2\text{O} + \text{Pb}^{2+} \qquad \log K = 49.2 \quad (22.1)$$

$$-\quad \text{PbCO}_{3(s)} = \text{Pb}^{2+} + \text{CO}_3^{2-} \qquad \log K = -12.8 \quad (22.7)$$

$$\text{PbO}_{2(s)} + 4\text{H}^+ + 2\text{e}^- + \text{CO}_3^{2-} = \text{PbCO}_{3(s)} + 2\text{H}_2\text{O} \qquad \log K = 62.0. \quad (22.30)$$

Assuming unit activity for the solids and unit activity coefficients in the solution phase, at 25°C/1 atm, Eq. (22.30) yields

$$2\ \text{pe} = 62.0 + \log \alpha_2 C_T - 4\ \text{pH}. \quad (22.31)$$

For the $PbO_{2(s)}/Pb_3(CO_3)_2(OH)_{2(s)}$ boundaryline (line 13), we combine

$$\text{PbO}_{2(s)} + 4\text{H}^+ + 2\text{e}^- = 2\text{H}_2\text{O} + \text{Pb}^{2+} \qquad \log K = 49.2 \quad (22.1)$$

$$-\quad \frac{1}{3}\Big(\text{Pb}_3(\text{CO}_3)_2(\text{OH})_{2(s)} + 2\text{H}^+ = 3\text{Pb}^{2+} + 2\text{CO}_3^{2-} + 2\text{H}_2\text{O} \qquad \log K = -18.8\Big) \quad (22.8)$$

$$\text{PbO}_{2(s)} + \frac{10}{3}\text{H}^+ + \frac{2}{3}\text{CO}_3^{2-} + 2\text{e}^- = \frac{1}{3}\text{Pb}_3(\text{CO}_3)_2(\text{OH})_{2(s)} + \frac{4}{3}\text{H}_2\text{O} \qquad \log K = 55.47. \quad (22.32)$$

Assuming unit activity for the solids and unit activity coefficients in the solution phase, at 25°C/1 atm, Eq. (22.32) yields

$$2\ \text{pe} = 55.47 + \frac{2}{3}\log \alpha_2 C_T - \frac{10}{3}\ \text{pH}. \quad (22.33)$$

For the $PbCO_{3(s)}/Pb_{(s)}$ boundaryline (line 11), we combine

$$PbCO_{3(s)} = Pb^{2+} + CO_3^{2-} \qquad \log K = -12.8 \tag{22.7}$$

$$+ \quad Pb^{2+} + 2e^- = Pb_{(s)} \qquad \log K = -4.26 \tag{22.2}$$

$$PbCO_{3(s)} + 2e^- = Pb_{(s)} + CO_3^{2-} \qquad \log K = -17.06. \tag{22.34}$$

Assuming unit activity for the solids and unit activity coefficients in the solution phase, at 25°C/1 atm, Eq. (22.34) yields

$$2\ \text{pe} = -17.06 - \log \alpha_2 C_T. \tag{22.35}$$

For the $Pb_3(CO_3)_2(OH)_{2(s)}/Pb_{(s)}$ boundaryline (line 14), we combine

$$\frac{1}{3}\big(Pb_3(CO_3)_2(OH)_{2(s)} + 2H^+ = 3Pb^{2+} + 2CO_3^{2-} + 2H_2O \qquad \log K = -18.8\big) \tag{22.8}$$

$$+ \quad Pb^{2+} + 2e^- = Pb_{(s)} \qquad \log K = -4.26 \tag{22.2}$$

$$\frac{1}{3}Pb_3(CO_3)_2(OH)_{2(s)} + \frac{2}{3}H^+ + 2e^- = Pb_{(s)} + \frac{2}{3}CO_3^{2-} + \frac{2}{3}H_2O \qquad \log K = -10.53. \tag{22.36}$$

Assuming unit activity for the solids and unit activity coefficients in the solution phase, at 25°C/1 atm, Eq. (22.36) yields

$$2\ \text{pe} = -10.53 - \frac{2}{3}\log \alpha_2 C_T - \frac{2}{3}\ \text{pH}. \tag{22.37}$$

22.2.6 The $PbCO_{3(s)}/Pb_3(CO_3)_2(OH)_{2(s)}$ Solid/Solid Boundaryline

From above, for equilibrium between $PbCO_{3(s)}$ and $Pb_3(CO_3)_2(OH)_{2(s)}$ (line 12), at 25°C/1 atm, we have

$$-6.53 - \frac{1}{3}\log \alpha_2 - \frac{1}{3}\log C_T + \frac{2}{3}\text{pH} = 0. \tag{22.24}$$

As stated above, for $C_T = 10^{-2}\ M$, Eq. (22.24) gives

$$\text{pH} = 7.20. \tag{22.38}$$

The boundarylines represented by Eqs. (21.13), (21.15), (22.27), (22.29), (22.31), (22.33), (22.35), (22.37), and (22.38) are plotted in Figure 22.5.

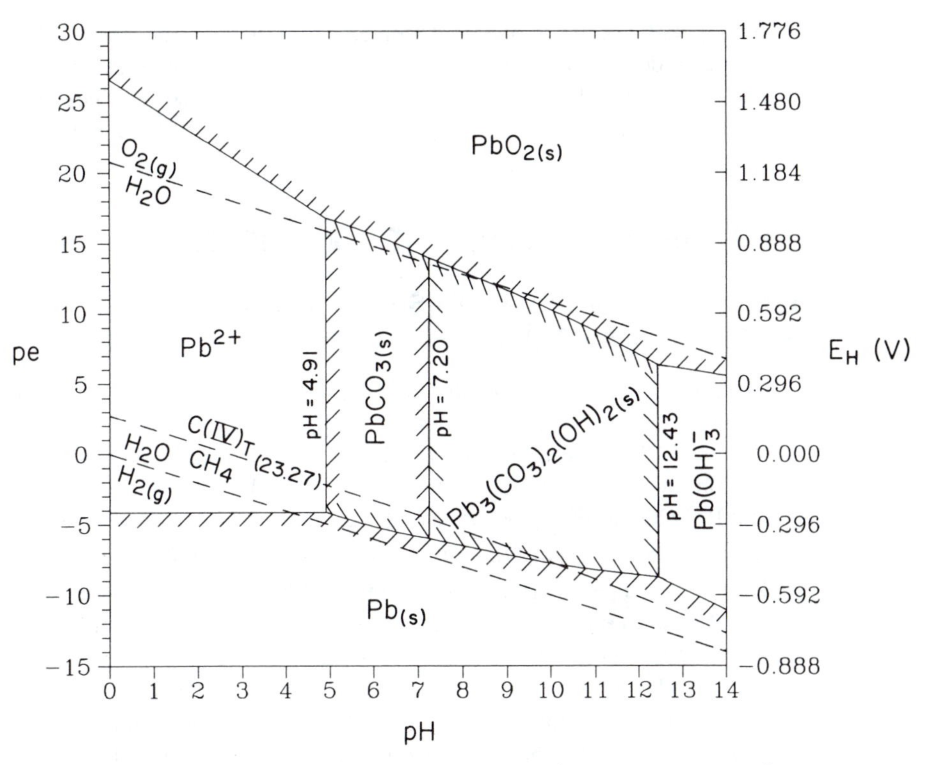

Figure 22.5 pe-pH diagram for $Pb_T = 10^{-4}$ M and $C_T = 10^{-2}$ M at 25°C/1 atm. Unit activity coefficients in the aqueous phase, and unit activities for the solids (when present) are assumed. Reduction of C(IV) carbon to methane at low pe is neglected. The predominance boundaryline between total dissolved C(IV) and CH_4 is given by Eq. (23.27,) which is pe = $2.70 + \frac{1}{8}\alpha_0 -$ pH where the α_0 corresponds to the fraction of C(IV) present as $H_2CO_3^*$. Near as well as anywhere below that line, at full equilibrium, most of the dissolved C_T will be present as CH_4, and so the lines drawn assuming that $[CO_3^{2-}] = \alpha_2 C_T$ will be in error.

22.3 C_T SUBJECT TO CONVERSION TO CH_4 UNDER VERY REDUCING CONDITIONS

Like Pb, CO_2 species themselves are subject to redox reactions. As we will see in Chapter 23, the two main carbon oxidation states of interest at full redox equilibrium are C(IV), as in $H_2CO_3^*$, HCO_3^-, and CO_3^{2-}, and C(-IV), as in methane. At 25°C/1 atm, Eq. (23.27) gives the predominance boundaryline between total C(IV) and CH_4 as

$$\text{pe} = 2.70 + \frac{1}{8}\alpha_0 - \text{pH} \qquad (23.27)$$

where the α_0 equals the fraction of C(IV) present as $H_2CO_3^*$. Above the line given by Eq. (23.27), most of the dissolved C_T will be present as C(IV), and the pe-pH diagrams developed in this chapter will be valid. Below that line, at full equilibrium, most of the dissolved C_T will be present as CH_4, and so the lines drawn assuming that $[CO_3^{2-}] = \alpha_2 C_T$ will be in significant error. Line 23.27 is included in all of the pe-pH diagrams for this chapter.

22.4 pe-pH DIAGRAM FOR $Pb_T = 10^{-6}$ *M* AND $C_T = 10^{-2}$ *M*

22.4.1 Identification of the Pb(II) Solids Having Predominance Regions

Since $C_T = 10^{-2}$ *M* as in the previous case, Figure 22.1 remains applicable. In order to determine whether or not both solids will still have predominance regions for $Pb_T = 10^{-6}$ *M*, we turn again to Figures 22.2 and 22.3. As discussed in the previous section, at 25°C/1 atm and when $C_T = 10^{-2}$ *M*, the transition pH between $PbCO_{3(s)}$ and $Pb_3(CO_3)_2(OH)_{2(s)}$ occurs at pH 7.20 (Eq. (22.38)). A consideration of Figure 22.3 indicates that: 1) $PbCO_{3(s)}$ limits the total dissolved Pb to less than 10^{-6} *M* for at least some range of pH values less than 7.20; and 2) Pb^{2+} is still the dominant species for pH values equal to and less than the pH for which equilibrium with $PbCO_3$ gives a total dissolved lead concentration of 10^{-6} *M*. Similarly, Figure 22.2 indicates that: 1) $Pb_3(CO_3)_2(OH)_{2(s)}$ limits the total dissolved Pb to less than 10^{-6} *M* for some range of pH values greater than 7.20; and 2) $Pb(OH)_3^-$ is still the dominant species for pH values equal to and greater than the pH for which equilibrium with $Pb_3(CO_3)_2(OH)_{2(s)}$ gives a total dissolved lead concentration of 10^{-6} *M*.

On the basis of the preceding paragraph, we conclude that: 1) both Pb(II) carbonate solids will retain predominance regions for $Pb_T = 10^{-6}$ *M* and $C_T = 10^{-2}$ *M*; and 2) the generic pe-pH diagram given in Figure 22.2 still applies, as do Eqs. (21.13) (lines 1 and 7), (21.15) (lines 2 and 8), (22.27) (line 15), (22.29) (line 9), (22.31) (line 10), (22.33) (line 13), (22.35) (line 11), (22.37) (line 14), and (22.38) (line 12). Application of these equations to the problem at hand yields Figure 22.6. The two solution/Pb(II) solid boundarylines (lines 9 and 15) occur at pH = 5.99 and 11.53. All solid/solid boundarylines are independent of Pb_T, and therefore are in the same positions in Figure 22.6 as they are in Figure 22.5. Below the line given by Eq. (23.27), at full equilibrium, most of the dissolved C_T will be present as CH_4. Thus, the lines drawn assuming that $[CO_3^{2-}] = \alpha_2 C_T$ are in error below that line.

22.4.2 Effects of Changing Pb_T on the pe-pH Diagram at $C_T = 10^{-2}$ *M*

Because the widths of the regions for $PbCO_{3(s)}$ and $Pb_3(CO_3)_2(OH)_{2(s)}$ are narrower for $Pb_T = 10^{-6}$ *M* than for $Pb_T = 10^{-4}$ *M*, the lengths of the Pb(IV)/Pb(II) solid/solid boundarylines are shorter in Figure 22.6 than they are in Figure 22.5. However, the vertical length of the $PbCO_{3(s)}/Pb_3(CO_3)_2(OH)_{2(s)}$ boundaryline is the same in the two figures.

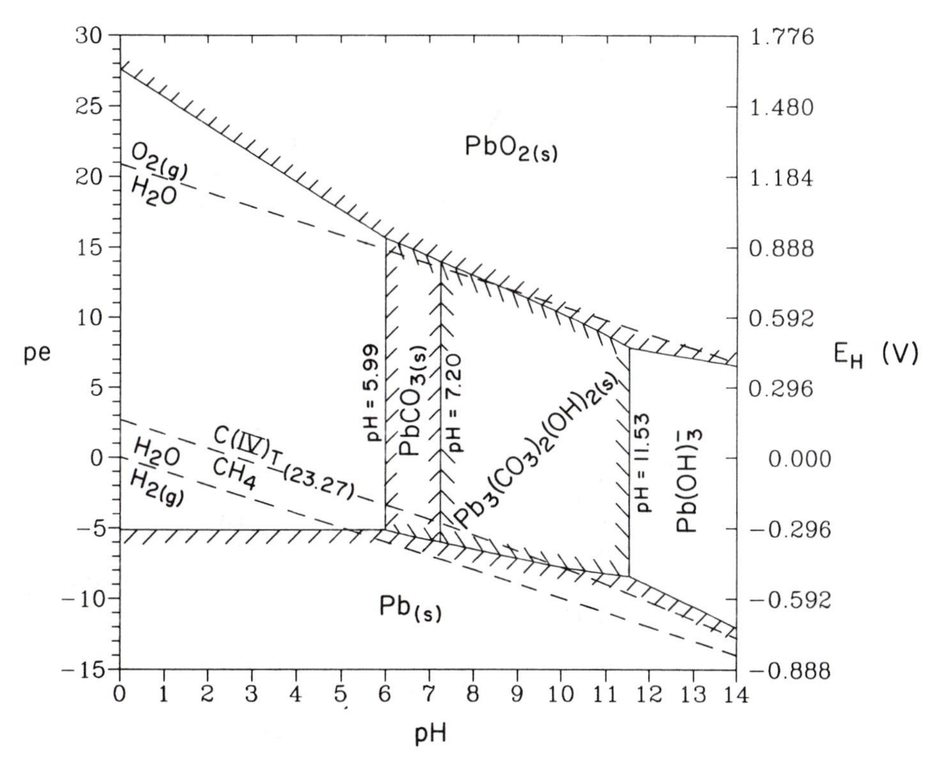

Figure 22.6 pe-pH diagram for $Pb_T = 10^{-6}$ M and $C_T = 10^{-2}$ M at 25°C/1 atm. Unit activity coefficients in the aqueous phase, and unit activities for the solids (when present) are assumed. The predominance boundaryline between total dissolved C(IV) and CH_4 (i.e., Eq. (23.27)) is included. Near as well as anywhere below that line, at full equilibrium, the lines drawn assuming that $[CO_3^{2-}] = \alpha_2 C_T$ will be in error.

All solution/solid boundarylines depend on Pb_T, and therefore are in different positions in Figures 22.5 and 22.6. These lines are longer in Figure 22.6 as a result of the expansions of the solution phase predominance regions that can occur when Pb_T decreases.

22.5 pe-pH DIAGRAM FOR $Pb_T = 10^{-4}$ M AND $C_T = 10^{-3}$ M

When C_T is lowered from 10^{-2} to 10^{-3} M, the lines for log$[Pb^{2+}]$ vs. pH as controlled by $PbCO_{3(s)}$ and $Pb_3(CO_3)_2(OH)_{2(s)}$ shift from the positions given in Figure 22.1 to those given in Figure 22.7; the position of the line for $PbO_{(s)}$ remains unchanged since that solid does not involve CO_3^{2-}.

As illustrated in Figure 22.7, lowering C_T to 10^{-3} M causes the log$[Pb^{2+}]$ line for $PbCO_{3(s)}$ to be shifted above the corresponding line for $Pb_3(CO_3)_2(OH)_{2(s)}$ for all pH. What this means is that there is no longer any pH range for which $PbCO_{3(s)}$ is stable, for

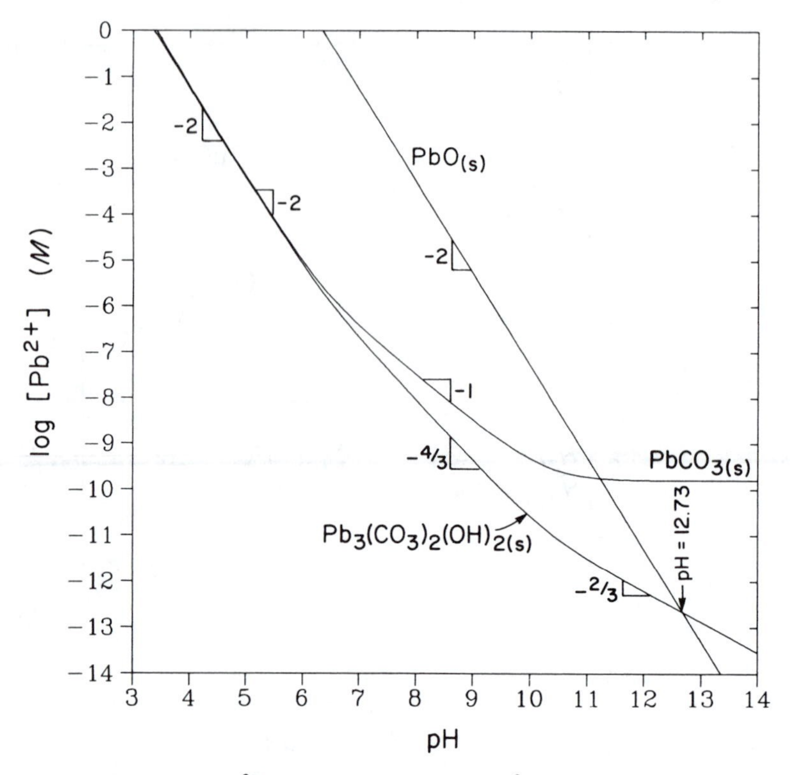

Figure 22.7 Log $[Pb^{2+}]$ vs. pH for $C_T = 10^{-3}$ M and 25°C/1 atm, as controlled by three different solids: $PbO_{2(s)}$, $PbCO_{3(s)}$, and $Pb_3(CO_3)_2(OH)_{2(s)}$. Unit activity coefficients in the aqueous phase and unit activities for the solids are assumed.

any value of Pb_T. Above a pH of 12.73, however, $PbO_{(s)}$ is possible, providing of course that the Pb_T for the system of interest is larger than the value prescribed in Figure 21.3 at pH = 12.73. Since $Pb_T = 10^{-4}$ M does not satisfy that criterion, $PbO_{(s)}$ cannot form in this system. (The pH value of 12.73 may be obtained by solving Eq. (22.25) for $C_T = 10^{-3}$ M.)

Whether or not $Pb_3(CO_3)_2(OH)_{2(s)}$ can form when $Pb_T = 10^{-4}$ M and $C_T = 10^{-3}$ M is not answered by Figure 22.2, but by an analogous diagram constructed for $C_T = 10^{-3}$ M. The lines for all of the dissolved species Pb(II) in that analogous diagram will be shifted upwards for $C_T = 10^{-3}$ M relative to their positions in Figure 22.2. Since there are two carbonates per Pb^{2+} in $Pb_3(CO_3)_2(OH)_{2(s)}$, the magnitude of that upward shift will be given by $-(2/3)\Delta \log C_T$, which in this case is 0.67. Since an upwards shift of this magnitude will not bring the minimum in the log of total dissolved Pb(II) in Figure 22.2 above −4, we can conclude that a region for $Pb_3(CO_3)_2(OH)_{2(s)}$ will be present for $Pb_T = 10^{-4}$ and $C_T = 10^{-3}$ M.

Based on the above results, we conclude that the pe-pH diagram for this case (Figure 22.8) will be similar to Figure 22.5, but the region for $PbCO_{3(s)}$ will be missing.

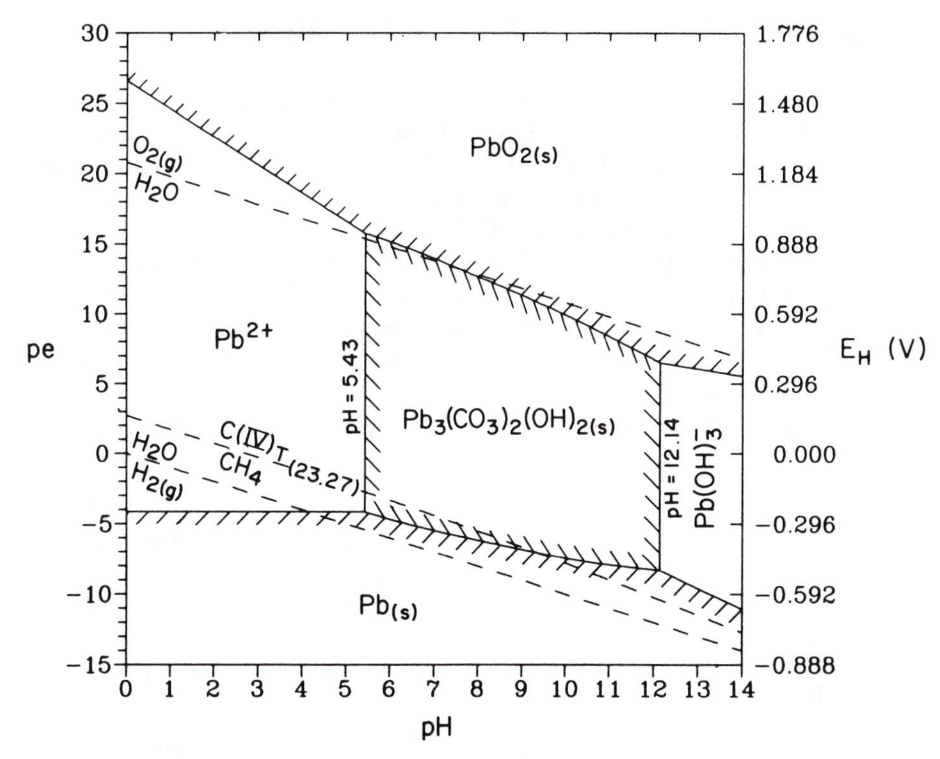

Figure 22.8 pe-pH diagram for $Pb_T = 10^{-4}$ M and $C_T = 10^{-3}$ M at 25°C/1 atm. Unit activity coefficients in the aqueous phase, and unit activities for the solids (when present) are assumed. The predominance boundaryline between total dissolved C(IV) and CH_4 (i.e., Eq. (23.27)) is included. Near as well as anywhere below that line, at full equilibrium, the lines drawn assuming that $[CO_3^{2-}] = \alpha_2 C_T$ will be in error.

Now, *both* the low pH and the high pH roots of Eq. (22.26) apply, and we obtain pH = 5.43 and 12.14 for the two solution/$Pb_3(CO_3)_2(OH)_{2(s)}$ boundarylines. (At the low pH root, Pb^{2+} is dominant, and at the high pH root, $Pb(OH)_3^-$ is dominant.) The remaining boundarylines are given by Eqs. (21.13), (21.15), (22.33), and (22.37) using the appropriate values of Pb_T and C_T. Below the line given by Eq. (23.27), at full equilibrium, most of the dissolved C_T will be present as CH_4. Thus, the lines drawn assuming that $[CO_3^{2-}] = \alpha_2 C_T$ are in error below that line.

Although the region for $Pb_3(CO_3)_2(OH)_{2(s)}$ is larger in Figure 22.8 than it is in Figure 22.5, overall, the area covered by all Pb(II) solids is smaller in Figure 22.8 than it is in Figure 22.5. The reason for this is that both of these Pb(II) solids contain carbonate, and C_T is lower in Figure 22.8 than it is in Figure 22.5. Thus, lowering C_T allows the solution phase regions of the pe-pH diagram to grow at the expense of the regions for solids containing carbonate.

22.6 pe-pH DIAGRAM FOR $Pb_T = 10^{-6}$ *M* AND $C_T = 10^{-3}$ *M*

For reasons that are analogous to those presented in the previous section, even at $Pb_T = 10^{-6}$ *M*, for $C_T = 10^{-3}$ *M* there will be a predominance region for $Pb_3(CO_3)_2(OH)_{2(s)}$. In this case, the low and high pH roots of Eq. (22.26) are 6.57 and 11.19, respectively. As in the previous case, these pH values are such that Pb^{2+} is the only dissolved species having a predominance region to the left of the solid, and $Pb(OH)_3^-$ is the only species having a predominance region to the right of the solid. The resulting pe-pH diagram is presented in Figure 22.9. As in Figure 22.8, Eqs. (21.13), (21.15), (22.33), and (22.37) (with the appropriate values of Pb_T and C_T) give the remaining boundarylines. Below the line given by Eq. (23.27), at full equilibrium, most of the dissolved C_T will be present as CH_4. Thus, the lines drawn assuming that $[CO_3^{2-}] = \alpha_2 C_T$ are in error below that line. Problems illustrating the principles of this chapter are discussed by Pankow (1992).

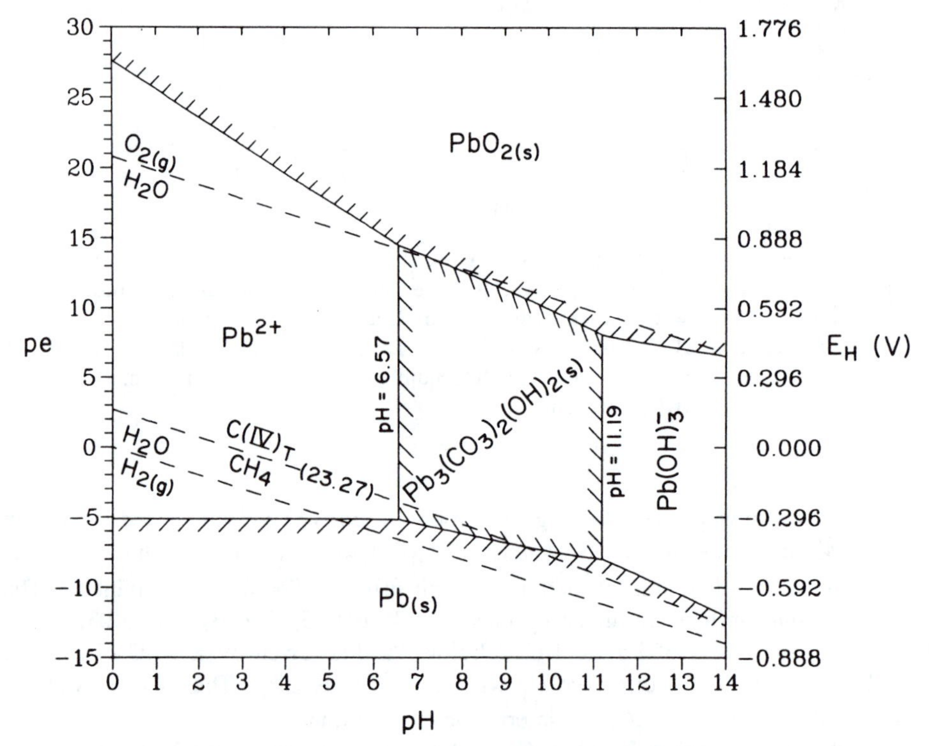

Figure 22.9 pe-pH diagram for $Pb_T = 10^{-6}$ *M* and $C_T = 10^{-3}$ *M* at 25°C/1 atm. Unit activity coefficients in the aqueous phase, and unit activities for the solids (when present) are assumed. The predominance boundaryline between total dissolved C(IV) and CH_4 (i.e., Eq. (23.27)) is included. Near as well as anywhere below that line, at full equilibrium, the lines drawn assuming that $[CO_3^{2-}] = \alpha_2 C_T$ will be in error.

22.7 REFERENCES

PANKOW, J. F. 1992. *Aquatic Chemistry Problems*. Portland: Titan Press-OR, P.O. Box 91399, Portland, Oregon 97291-1399.

23

pe and Natural Systems

23.1 INTRODUCTION

23.1.1 pe°, pe°(W) and pe°(pH)

When a given half-reaction is written as a reduction, we recall from Chapter 19 that the term $\{OX\}$ represents the product of the activities of all of the reactants (except e^-) raised to their respective stoichiometric powers, and $\{RED\}$ represents the product of all of the activities of all of the products similarly exponentiated. As discussed in Section 19.2.3, pe° values provide a useful reference scale for comparing the relative oxidation and reduction strengths of different redox half-reactions. This comparison requires a *specific set of conditions* for each half-reaction. In particular, it requires a set of conditions in which each species is present at unit activity, or at least gives $\{OX\}/\{RED\} = 1.0$. Conversely, since $pe = pe^\circ$ when $\{OX\} = \{RED\}$ for a given half-reaction, pe° values give us information on what types of pe values are required to obtain $\{OX\}/\{RED\} \simeq 1$ for that half-reaction.

In simple cases like the reduction of Fe^{3+} to Fe^{2+} where the only species involved are the ones that are gaining or losing electrons, knowing pe° translates directly into a knowledge of the type of pe that is obtained when just the two species exchanging the electrons are present at similar activities. As is evident from Table 19.2, however, a great variety of redox half-reactions involve more than just the two chemical species exchanging the electron(s). Examples of other species that can be involved in redox reactions include H^+ and HCO_3^-. Consider for example the reduction of nitrate to nitrite:

$$NO_3^- + 2H^+ + 2e^- = NO_2^- + H_2O \qquad K = \frac{\{NO_2^-\}}{\{NO_3^-\}\{H^+\}^2\{e^-\}^2} = 10^{28.3}. \qquad (23.1)$$

The value of the constant is the one for 25°C/1 atm. By Eq. (19.61), we have

$$pe^{o} = \left(\frac{1}{2}\right) 28.3 = 14.15. \tag{23.2}$$

If we assume unit activity for the water, then $\{RED\} = \{NO_2^-\}$, and $\{OX\} = \{NO_3^-\}\{H^+\}^2$. At 25°C/1 atm then, we know that $\{NO_3^-\}/\{NO_2^-\}$ will equal 1.0 when $pe = pe^o = 14.15$, *and* $\{H^+\} = 1.0$ (i.e., pH = 0). A pH of 0 is not, however, very environmentally relevant.

Let us now suppose that we are interested in the question as to whether NO_3^- will tend to oxidize Fe^{2+} to Fe^{3+} under conditions when $\{NO_3^-\} = \{NO_2^-\}$, and when $\{Fe^{3+}\} = \{Fe^{2+}\}$. If $\{H^+\} = 1.0$, then the answer is yes, because, as shown in Table 19.2, the pe^o for the nitrate/nitrite couple (14.15) is 1.15 units higher than the pe^o for the ferric ion/ferrous ion couple (13.0). In other words, under the above conditions, the nitrate/nitrate couple is prescribing a lower electron activity than the ferric ion/ferrous ion couple, and so some of the ferrous ions will yield up their electrons to the nitrate, and be oxidized to ferric ions in the process.[1] However, if the pH is different from zero, that is, at a more environmentally relevant value such as 7, then the pe^o values given above for the two half-reactions do not provide a direct answer to the question; the pe^o values do provide the *basis* for the calculations that will provide the answer.

For pH = 7.0, Eq. (23.1) indicates that $\{NO_3^-\}/\{NO_2^-\}$ will be 1.0 when pe = 7.15. As noted briefly in Section 19.3.3, this special pe value is referred to as the $pe^o(W)$ value for this half-reaction. The W is intended invoke the idea of "water," although that is a bit misleading since aqueous solutions can obviously take on a whole range of pH values. Since the Fe^{3+}/Fe^{2+} redox couple does not involve H^+, $\{H^+\}$ does not play a *direct* role in setting the pe for this couple,[2] and so for this couple we have $pe^o(W) = pe^o = 13.0$. At pH = 7, for $\{Fe^{3+}\}/\{Fe^{2+}\} = 1$, this couple will create a much more oxidizing environment (pe = 13.0) than is created by $\{NO_3^-\}/\{NO_2^-\} = 1.0$ (pe = 7.15). Thus, the reaction that will occur at pH = 7 for these unit activity ratios is the oxidation of some of the NO_2^- by some of the Fe^{3+}.

As noted in Section 19.3.3, the $pe^o(W)$ scale is not the only pH-adjusted pe^o scale that is possible. Adjustments to other pH values can also be carried out. In fact, we can discuss $pe^o(pH)$ values for any pH. We note that for any given half-reaction, $pe^o = pe^o(0)$, and $pe^o(7) \equiv pe^o(W)$. A given $pe^o(pH)$ scale allows us to factor out H^+ effects for all reactions for that pH. That is, it allows us to directly compare the oxidizing capabilities of different half-reactions at that pH. For example, as should now be clear, $pe^o(W)$ values may be compared directly when examining the relative oxidizing capabilities of various half-reactions at pH = 7.

[1]Note that a positive value for a difference in pe^o values corresponds to a positive and proportional value for the corresponding E^o value for an overall cell.

[2]The value of $\{H^+\}$ *will* play an *indirect* role in setting the pe for the ferric/ferrous couple inasmuch as the pH will control the fraction of dissolved Fe(III) present as Fe^{3+}, as well as the fraction of dissolved Fe(II) present as Fe^{2+}.

We can develop a formal method for converting pe° values to pe°(pH) values for any half-reaction and any pH. Consider the general redox half-reaction

$$aA + bB + n_H H^+ + n_e e^- = cC + dD. \tag{23.3}$$

Species A will be considered to be the species that contains the element that is being reduced. Species C will be considered to be the species that contains the element that has been reduced. B and D represent other possible participants in the half-reaction. The value of n_H will be considered to be positive if protons are *consumed* in the process of the reduction; n_H is negative if protons are *produced*. For the reaction in Eq. (23.3), we have

$$K = \frac{\{C\}^c\{D\}^d}{\{A\}^a\{B\}^b\{H^+\}^{n_H}\{e^-\}^{n_e}}. \tag{23.4}$$

By definition,

$$pe = \frac{1}{n_e}\log K + \frac{1}{n_e}\log\frac{\{A\}^a\{B\}^b\{H^+\}^{n_H}}{\{C\}^c\{D\}^d}. \tag{23.5}$$

At any given pH, we have $\{H^+\} = 10^{-pH}$, and so

$$pe = \frac{1}{n_e}\log K + \frac{1}{n_e}\log\frac{\{A\}^a\{B\}^b}{\{C\}^c\{D\}^d} - \frac{n_H}{n_e}pH. \tag{23.6}$$

Equation (23.6) allows us to construct pe°(pH) scales for any desired pH value. On any of these scales, each $\{A\}^a\{B\}^b/\{C\}^c\{D\}^d$ quotient will equal 1.0. Thus, we have

$$pe^o(pH) = pe^o - \frac{n_H}{n_e}pH. \tag{23.7}$$

The parameter given here as n_H/n_e is given simply as n_H in Stumm and Morgan (1981, p. 445).

Applying Eq. (23.7) to the case of the reduction of nitrate to nitrite, at 25°C/1 atm, we obtain that

$$pe^o(W) = 14.15 - \frac{2}{2}7.0 = 7.15 \tag{23.8}$$

as was noted previously. Table 19.2 contains pe° as well as pe°(W) values for a large number of redox half-reactions. For all half-reactions like $Fe^{3+} + e^- = Fe^{2+}$ in which there is no direct participation of protons in the reaction (i.e., $n_H = 0$), then pe°(pH) = pe° for all pH, and in particular, pe°(W) = pe°.

At any given pH, a pe°(pH) "redox ladder" can be constructed. Each different half-reaction comprises a rung on the ladder. At the pe level corresponding to a given rung, the activity quotient on the RHS of Eq. (23.6) will equal 1. If the stoichiometric coefficients b and d in Eq. (23.3) are zero (no participants in the reaction besides A, C, and possibly H^+), and if $a = c$ (i.e., A and C contain the same number of atoms of the redox-active element), then on each rung the activity of the electron-accepting species will equal the activity of the electron-donating species.

"Climbing" a redox ladder means increasing the pe by some means, for example, by slowly adding the oxidized form of the redox pair occupying the highest rung in the ladder. A redox ladder constructed for pH = 7 places the rungs at the levels corresponding to the pe°(W) values for the various half-reactions. Figure 23.1 presents redox ladders for a series of pH values. The height of the rung for any half-reaction for which $n_H \neq 0$ will depend on pH; when $n_H = 0$, the height of the rung is independent of pH.

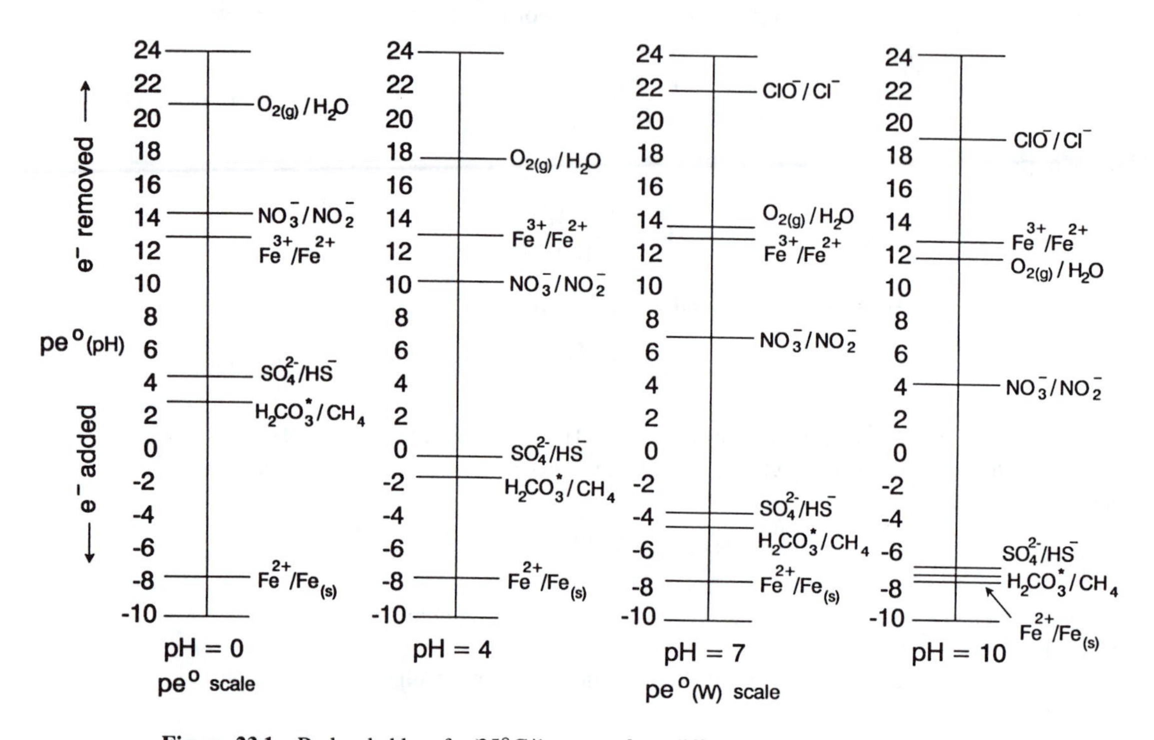

Figure 23.1 Redox ladders for 25°C/1 atm at four different pH values. Each rung in each ladder corresponds to the pe°(pH) value for that half-reaction for that pH. The ladder at pH = 0 corresponds to the pe° scale. The ladder at pH = 7 corresponds to the pe°(W) scale. A half-reaction that does not involve H^+ (e.g., Fe^{3+}/Fe^{2+}) is located at the same elevation on all of the ladders. For a half-reaction involving two dissolved species each of which contain the same number of atoms of the redox-active element (e.g., NO_3^-/NO_2^- but not NO_3^-/N_2), the two species are present at equal activities at the pe of the ladder rung. For a ladder rung involving a gaseous species, at the pe of the rung, the in-situ pressure of the gas is assumed to be 1 atm, and the dissolved species is assumed to be present at unit activity. For a ladder rung involving a solid, at the pe of the rung, the solid is assumed to be present at a mole fraction of 1, and the dissolved species is assumed to be present at unit activity. Climbing up a redox ladder corresponds to removing electrons from the system, as through the addition of a strong oxidizing agent. Moving down a redox ladder corresponds to adding electrons to the system, as through the addition of a strong reducing agent.

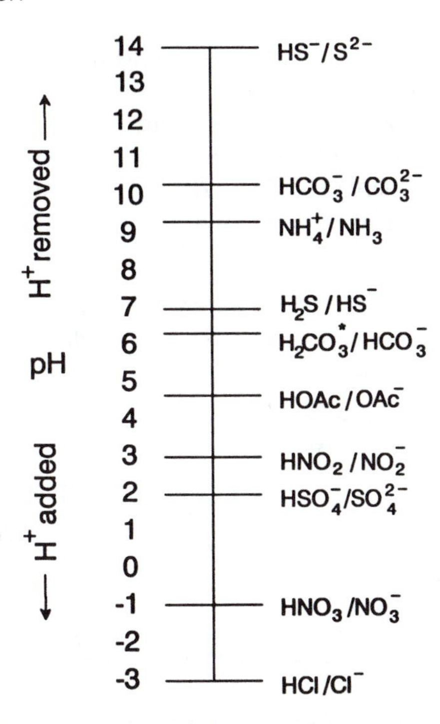

Figure 23.2 A pH ladder for 25°C/1 atm. Each rung in each ladder corresponds to the pK for the acid for that rung. At the level of a given rung, the activities of the conjugate acid and conjugate base are equal. Climbing up the pH ladder corresponds to removing protons from the system, as through the addition of a strong base. Moving down the pH ladder corresponds to adding protons to the system, as through the addition of a strong acid.

A redox ladder is perfectly analogous to what might be referred to as an "acid/base ladder." Indeed, in Figure 23.2, we see that if we are to have $\{A^-\}/\{HA\} \simeq 1$ for a strong acid, then the pH must be very low. In particular, it must be near the pK of the acid. At the lowest rung, the pH will equal the pK for the strongest acid. To remove protons from a system containing multiple acids HA, HB, HC, etc., we must add strong base. In so doing, we climb the acid/base ladder, and move to successively higher pH rungs. To summarize this analogy between a redox ladder and an acid/base ladder, we see that the corresponding parameters are: $e^- \Longleftrightarrow H^+$, pe $\Longleftrightarrow$ pH; generic $\{OX\}/\{RED\} \simeq 1 \Longleftrightarrow$ generic $\{A^-\}/\{HA\} \simeq 1$; and pe°(pH) $\Longleftrightarrow$ pK_{HA}.

23.1.2 pe°, pe°(W,3), and pe°(pH, pHCO$_3^-$)

Like the proton, the HCO_3^- ion is: 1) a participant in some redox half-reactions of interest; and 2) not present at unit activity in environmental systems. Therefore, even when one has gone to the trouble of creating a special pe°(pH) scale for reactions of interest, if one of those reactions involves HCO_3^-, then a direct comparison of the pe°(pH) values may lead to incorrect conclusions regarding the relative oxidizing capabilities of the various half-reactions.

Based on the discussion provided in the preceding paragraph, we infer that we can do the same thing for HCO_3^- that was done for H^+. That is, we can create a special pe° scale for any desired value of $\{HCO_3^-\}$. Consider a general redox half-reaction involving HCO_3^-

$$aA + n_{HCO_3^-}HCO_3^- + n_H H^+ + n_e e^- = cC + dD. \tag{23.9}$$

As before, A and C represent the species exchanging the electron(s). The species B in Eq. (23.3) is now taken to be HCO_3^-; the species D represents some other possible participant in the half-reaction. $n_{HCO_3^-}$ is the number of HCO_3^- ions involved in the half-reaction; $n_{HCO_3^-}$ will be considered to be positive if HCO_3^- are consumed in the process of the reduction. For the reaction in Eq. (23.9), we have

$$K = \frac{\{C\}^c\{D\}^d}{\{A\}^a\{HCO_3^-\}^{n_{HCO_3^-}}\{H^+\}^{n_H}\{e^-\}^{n_e}}. \tag{23.10}$$

By definition,

$$pe = \frac{1}{n_e}\log K + \frac{1}{n_e}\log\frac{\{A\}^a\{HCO_3^-\}^{n_{HCO_3^-}}\{H^+\}^{n_H}}{\{C\}^c\{D\}^d}. \tag{23.11}$$

With $pHCO_3^- = -\log\{HCO_3^-\}$ and $\{H^+\} = 10^{-pH}$, we have

$$pe = \frac{1}{n_e}\log K + \frac{1}{n_e}\log\frac{\{A\}^a}{\{C\}^c\{D\}^d} - \frac{n_H}{n_e}pH - \frac{n_{HCO_3^-}}{n_e}pHCO_3^-. \tag{23.12}$$

If $pe^o(pH, pHCO_3^-)$ is defined as the value of the pe that obtains when $\{H^+\} = 10^{-pH}$, $\{HCO_3^-\} = 10^{-pHCO_3^-}$, and $\{A\}^a/\{C\}^c\{D\}^d = 1.0$, then

$$pe^o(pH, pHCO_3^-) = pe^o - \frac{n_H}{n_e}pH - \frac{n_{HCO_3^-}}{n_e}pHCO_3^-. \tag{23.13}$$

For any half-reaction that *does not* involve HCO_3^-, we have $pe^o(pH, pHCO_3^-) = pe^o(pH)$ for any value of $pHCO_3^-$. For any half-reaction that does not involve either H^+ or HCO_3^-, then $pe^o(pH, pHCO_3^-) = pe^o$ for any values of pH and $pHCO_3^-$. Since a typical value for $pHCO_3^-$ in natural waters is ~3, Table 23.1 presents pe°(W, 3) values at 25°C/1 atm for two important half-reactions involving HCO_3^-.

TABLE 23.1 Data for two redox half-reactions involving HCO_3^- at 25°C/1 atm.

Reduction Half-Reaction	log K	pe°	pe°(W)	pe°(W, 3)	E_H^o(= 0.05916 pe°)
$MnO_{2(s)} + HCO_3^- + 3H^+ + 2e^- = MnCO_{3(s)} + 2H_2O$	25.8	15.9	5.4	3.90	0.94
$FeOOH_{(s)} + HCO_3^- + 2H^+ + e^- = FeCO_{3(s)} + 2H_2O$	16.2	16.2	2.20	−0.80	0.96

23.1.3 pe°, pe°(pH), and pe°(pH, $pHCO_3^-$) Values for Reactions Involving Solids

Sections 23.1.1 and 23.1.2 have discussed how pe° values for redox half-reactions involving H^+ and HCO_3^- can be scaled so that comparisons of relative oxidizing and reducing strength can be more made more easily under specific conditions of interest. In particular, we discussed how the comparisons can be made more directly applicable to environmentally-relevant values of pH and $pHCO_3^-$. Unfortunately, because some OX and RED species of interest are solids, the pe°(pH) and pe°(pH, $pHCO_3^-$) scales will not allow us to directly compare all environmentally-relevant redox half-reactions.

As an example, consider the reduction $SO_4^{2-} + 8H^+ + 8e^- = S_{(s)} + 4H_2O$. A pe°(pH) scale will tell us the value of the pe when $\{SO_4^{2-}\}/\{S_{(s)}\} = 1$ and when the pH takes on the value of interest. However, since $\{S_{(s)}\}$ is expressed on the mole fraction scale, in order to obtain $\{SO_4^{2-}\}/\{S_{(s)}\} = 1$, we need $\{SO_4^{2-}\} = 1\ m$, and that is certainly not typical in the environment. Thus, in this case, we conclude that to obtain a rough estimate of the oxidizing or reducing strength of the $SO_4^{2-}/S_{(s)}$ half-reaction, not only do we need an estimate of the pH, but we also need a specific estimate of $\{SO_4^{2-}\}$: we can't just ask what the pe is when there are roughly equal activities for both SO_4^{2-} and $S_{(s)}$. This situation contrasts with that of NO_3^-/NO_2^- wherein both the oxidized and reduced forms of the element are dissolved, and so we can easily talk about environmentally-relevant situations in which $\{NO_3^-\}/\{NO_2^-\} = 1$.

By way of summary of this and the preceding two sections, we can say that pe°(pH) and pe°(pH, $pHCO_3^-$) scales can be used to estimate relative oxidizing and reducing strengths of half-reactions, providing that neither the oxidized form nor the reduced form of the element in a reaction of interest is a solid. However, for reactions involving one redox-active solid and one redox-active solution phase species, we will not be able to make *general* statements about relative oxidizing and reducing strengths. As in any *specific* situation involving an *overall* redox reaction made up of two redox half-reactions, to compare the relative tendencies of the two half-reactions to proceed as reductions (as in computing an E value for an overall cell), we will require specific knowledge of all pertinent reactant and product activities.

23.2 REDOX-CONTROLLING ELEMENTS IN NATURAL WATERS

23.2.1 Major and Minor Redox Elements

In natural waters, the group of elements that can play a *controlling* role in determining the redox conditions is rather small. This group includes carbon, oxygen, hydrogen, nitrogen, iron, sulfur, and manganese (COHNISM). These elements play important redox roles in natural systems because in addition to simply being redox active, several of them are virtually always present in amounts that are sufficiently large to help control the redox chemistry; certainly the presence of water allows forms of oxygen and hydrogen to always be present in large amounts.

In addition to the COHNISM elements, many comparatively minor elements (e.g., arsenic, copper, lead, chromium, mercury, selenium, etc.) can also exhibit redox behavior. However, the positions of their redox couples are usually reflective of what the redox-controlling elements are doing in establishing the pe. Often, only a few of the COHNISM group are actually present in significant amounts in any given system; when a given COHNISM element is present at low levels, it will tend to have its redox chemistry controlled by those COHNISM elements that are: 1) present at high levels; and 2) capable of reacting in a redox fashion on the time scale of interest.

The redox chemistry situation described above is highly analogous to what happens with acid/base chemistry in natural waters. In particular, the group of acidic and basic species that play a controlling role in setting the pH of natural waters is fairly small. The group includes: water; the various CO_2 species; the strong acids; the strong bases; boric acid; organic acids; ammonia; the organic amines; and various carbonate, oxide, hydroxide, and clay solids. In any given situation, usually only a few from this group play a major role in setting the pH. Any minor acid/base pairs will have their equilibrium positions essentially controlled by the most abundant acid/base pairs. As an example, we know that in natural waters the pH is often almost completely determined by what is going on in the CO_2 system. While the presence of any minor acid/base pair(s) will always have *some* effect(s) on the pH, the magnitude of the effect(s) will be small. In like manner, the presence of all redox couples in a given system will have *some* effect on the equilibrium pe, but that pe will tend to be set by the most abundant couples.

It is worthwhile to note the use of the plural "couples" at the end of the last paragraph. As discussed in Chapter 19, in aqueous systems the electron cannot achieve very high concentrations, and we always need two redox couples to: 1) write an overall reaction that is free of the explicit presence of electrons; and therefore 2) establish the balance that sets the pe. Thus, in order to set the pe in an aqueous system, at least two redox couples must be present; one of the redox couples donates electrons, and the other accepts those electrons until an equilibrium position is reached. Again, the situation is similar with aqueous acid/base chemistry. For example, if an acid HA partially dissociates to form H^+ and A^-, then as discussed in Chapter 3, very little truly-free H^+ is present (most is solvated by water molecules). Thus, the HA donates protons, and the solvating water accepts the protons: two acid/base pairs are involved, HA/A^-, and solvated proton/H_2O.

23.2.2 The pe of Pure Water

As noted above, water is composed of two of the COHNISM elements. Water can therefore provide two redox couples, the H^+/H_2 couple and the O_2/H_2O couple. In the absence of any other redox active elements, at equilibrium the pe of water will be well defined due to the natural tendency of water to dissociate to a very small extent into O_2 and H_2. This is analogous to the fact that in the absence of any other acids or bases, the acid/base properties of water will set the pH. The value of the pe for pure water may be calculated as follows. At 25°C/1 atm, the pH of pure water will be 7.00. At 25°C/1 atm, from Eq. (20.27) we also have

$$2\ H_2O = 2H_{2(g)} + O_{2(g)} \qquad \log K = -83.1. \tag{23.14}$$

By mass balance, in absolutely pure water with no head space, we will have

$$[H_2] = 2[O_2]. \qquad (23.15)$$

The Henry's Law constant for O_2 at 25°C/1 atm is $10^{-2.90}$ M/atm, and that for H_2 is $10^{-3.10}$ M/atm. Combining these Henry's Law constants with the equilibrium constant in Eq. (23.14), we obtain

$$\begin{array}{llll} & 2\,H_2O = 2H_{2(g)} + O_{2(g)} & \log K = -83.1 & (23.14) \\ 2\times & (\;H_{2(g)} = H_2 & \log K = -3.10\;) & (23.16) \\ + & O_{2(g)} = O_2 & \log K = -2.90 & (23.17) \\ \hline & 2\,H_2O = 2H_2 + O_2 & \log K = -92.20. & (23.18) \end{array}$$

With the activity of pure liquid water given by $X_{H_2O} = 1$, by Eqs. (23.15) and (23.18) we have

$$4[O_2]^3 = 10^{-92.20} \qquad [O_2] = 1.25 \times 10^{-31}\ M. \qquad (23.19)$$

By the Henry's Law constant for O_2, we then obtain an in-situ p_{O_2} of 9.23×10^{-29} atm. By Eq. (19.109), at pH = 7.00 this value of p_{O_2} yields a pe value of 6.78. This pe falls very near the middle of the pe-pH stability field for H_2O as it is demarcated at pH = 7.0 by the lines $p_{O2} = 1$ atm and $p_{H_2} = 1$ atm. This symmetry is a direct result of the approximate equality: 1) between $[H_2]$ and $[O_2]$ in Eq. (23.15); and 2) between the two K_H values for H_2 and O_2. It is analogous to the fact that because $[H^+] = [OH^-]$ in pure water, at 25°C/1 atm, there is a symmetrical division between the acidic and basic regimes at pH = 7.0.

When another redox couple is added to pure water, say Fe(II)/Fe(III), then the pe will be set by the new balance that is achieved. If the $\{OX\}/\{RED\}$ activity ratio and the pe° value of the new couple have an oxidizing effect, then the balance will be shifted in favor of O_2, and the pe will go up. Conversely, if adding another redox couple causes a reducing effect, then the balance will be shifted in favor of H_2, and the pe will go down. It is useful to consider these conclusions together with a re-examination of Figures 20.6, 20.8, and 20.9.

23.2.3 Full Redox Equilibrium Often Not Obtained in Natural Waters

The exchange of protons among the different dissolved acids and bases in water may in many circumstances be considered to proceed to equilibrium within a minute or less.[3] As discussed in Section 19.3.5, in natural waters, one may essentially always assume that all conjugate acid/base pairs like HA/A^-, HB/B^-, HC/C^-, etc. are describing the same pH according to

$$pH = pK_{HA} + \log\frac{\{A^-\}}{\{HA\}} = pK_{HB} + \log\frac{\{B^-\}}{\{HB\}} = pK_{HC} + \log\frac{\{C^-\}}{\{HC\}} = \text{etc.} \qquad (19.99)$$

[3]An important exception to this general rule pertains to cases involving the hydrolysis of metal ions where the release of protons as caused by the equilibrium formation of complex hydrolysis products such as large polynuclear complexes can take months or longer.

For redox reactions, the analog of Eq. (19.99) is Eq. (19.100)

$$\mathrm{pe} = (1/n_1)\left(\log K_1 + \log\frac{\{\mathrm{OX}_1\}}{\{\mathrm{RED}_1\}}\right) = (1/n_2)\left(\log K_2 + \log\frac{\{\mathrm{OX}_2\}}{\{\mathrm{RED}_2\}}\right)$$
$$= (1/n_3)\left(\log K_3 + \log\frac{\{\mathrm{OX}_3\}}{\{\mathrm{RED}_3\}}\right) = \text{etc.} \tag{19.100}$$

As discussed in Section 19.3.5, unlike proton exchange reactions, certain redox reactions can be slow. This is particularly true for reactions involving the exchange of multiple electrons. For example, while the addition of a single electron to Fe^{3+} to from Fe^{2+} can proceed relatively quickly under many circumstances, the abiotic reduction of SO_4^{2-} to S^{2-} involves the addition of *eight* electrons and is by comparison very slow. Similarly, the abiotic multi-electron oxidation of carbohydrates to CO_2 is very slow. Even in the presence of suitable catalyzing bacteria, if the proper conditions for bacterial growth are not present, then the reactions involving SO_4^{2-} and carbohydrates can still be slow.

A large part of the reason why multiple electron reactions are slow is that they usually entail complex mechanisms involving many steps. The fact that the concentration of the electron in aqueous systems is always very low is also a contributing factor. In the absence of bacteria, many multiple electron redox reactions are so slow that it would be difficult to even think about the application of equilibrium principles on time scales shorter than many thousands to hundreds of millions of years. In exchange for catalyzing a wide variety of redox reactions and moving nature towards equilibrium, bacteria can obtain the chemical potential energy that is stored in a nonequilibrium redox situation. Bacteria use this energy to continue living, to build cell mass, and to reproduce. It is useful to note that in nature, virtually any redox reaction that can be out of equilibrium will have one or more strains of bacteria that can move the system towards equilibrium: if there is a ecological niche that can be occupied, nature will fill it. Many of these bacteria have been given names that reflect what they do. A good example of this is *Thiobacillus ferrooxidans*. This bacterium can oxidize both H_2S and $S_{(s)}$ to SO_4^{2-}, and can also oxidize Fe(II) to Fe(III). (The prefix *thio* means sulfur.)

If the constant input of solar energy were to cease, in time, natural waters at the earth's surface would gradually attain full redox equilibrium. However, such a system would be totally dead: life itself represents a nonequilibrium condition. For example, in an oxic environment, the carbon compounds making up a bacterial cell (or a human) are not at equilibrium with respect to oxidation by ambient levels of oxygen. It is therefore the daily input of solar energy that continously frustrates the global approach to total redox equilibrium. This solar energy provides the energy stream for photosynthesis in which most forms of life in our solar system live.[4]

The general lack of total redox equilibrium in natural systems means that Eq. (19.100) is not nearly as generally applicable in natural systems as is Eq. (19.99).

[4]Relatively recently, oceanographers have discovered deep ocean, *chemosynthetic* ecosystems built upon bacteria that live by using dissolved O_2 to oxidize H_2S. The H_2S emanates from deep within the earth along the lines separating large tectonic plates.

For example, if one were to examine the acid/base activity ratios in a given aqueous system together with the appropriate pK data, the same pH would very likely be indicated by all of the acid/base pairs. If one of the acid/base pairs did not indicate the pH given by the others, an inaccurate pK value for that acid is more probable than is a lack of pH equilibrium. In contrast, if one were to examine the $\{OX\}/\{RED\}$ values for a series of redox half-reactions pertinent to a given aqueous system together with the appropriate pe° values, it is very likely that all of the resulting calculated pe values would be at least somewhat different. The actual pe value that a given water will have will be influenced to a large degree by the two redox half-reactions that exchange electrons with the greatest facility. The main purpose of equilibrium redox calculations for natural waters is therefore not so much to give us a means to predict exactly what the actual pe will be for a given water sample, but rather to indicate: 1) the general redox conditions in that system (e.g., very oxidizing, somewhat oxidizing, somewhat reducing, very reducing, etc.); and 2) which redox reactions are in fact possible in that system and therefore the overall species composition towards which that system is being driven thermodynamically.

23.3 pe-pH DIAGRAMS FOR CARBON, NITROGEN, SULFUR, AND IRON

23.3.1 pe-pH Diagram for Carbon

23.3.1.1 Redox Equilibria Governing Carbon Species. The pe-pH diagrams for two of the COHNISM elements, oxygen and hydrogen, were considered in Chapter 20. This chapter develops pe-pH diagrams for most of the remaining five COHNISM elements. In the case of carbon, we note that there are a number of possible oxidation states. The forms of carbon and their oxidation states that will be considered in this section are summarized in Table 23.2. They range from CO_2-type species, which are in the IV oxidation state, down to CH_4 which is in the −IV oxidation state. Carbon can also exist in the elemental forms of graphite and diamond (both oxidation state 0), but these two solids are so slow to form at 25°C/1 atm that they will be ignored within the context of this chapter.

Some of the possible redox half-reactions involving the Table 23.2 carbon compounds are given in Table 23.3. The question that we now face is whether or not all of

TABLE 23.2 Some forms of carbon in natural waters.

oxidation state	species
IV	$H_2CO_3^*$, HCO_3^-, CO_3^{2-}
II	HCOOH (formic acid), $HCOO^-$ (formate ion)
0	$C_6H_{12}O_6$ (glucose), and CH_2O (formaldehyde)
−II	CH_3OH (methanol)
−IV	CH_4 (methane)

TABLE 23.3 Data for selected redox reactions involving carbon at 25°C/1 atm in order of increasing pe° and E^o_H.

Reduction Half-Reaction	log K	pe°	pe°(W)	E^o_H (= 0.05916 pe°)
$CO_{2(g)} + H^+ + 2e^- = HCOO^-$	−9.66	−4.83	−8.33	−0.29
$CO_{2(g)} + 4H^+ + 4e^- = CH_2O + H_2O$	−4.8	−1.2	−8.2	−0.071
$CO_{2(g)} + 4H^+ + 4e^- = \frac{1}{6}C_6H_{12}O_6$ (glucose) $+ H_2O$	−0.8	−0.2	−7.2	−0.012
$CO_{2(g)} + 6H^+ + 6e^- = CH_3OH + H_2O$	3.2	0.53	−6.47	0.032
$HCOO^- + 3H^+ + 2e^- = CH_2O + H_2O$	5.64	2.82	−7.68	0.17
$CO_{2(g)} + 8H^+ + 8e^- = CH_{4(g)} + 2H_2O$	23.0	2.87	−4.13	0.17
$CH_2O + 2H^+ + 2e^- = CH_3OH$	8.0	4.0	−3.0	0.24
$HCOO^- + 7H^+ + 6e^- = CH_{4(g)} + 2H_2O$	32.66	5.44	−2.72	0.32
$CH_2O + 4H^+ + 4e^- = CH_{4(g)} + H_2O$	27.8	6.94	−0.06	0.41
$CH_3OH + 2H^+ + 2e^- = CH_{4(g)} + H_2O$	19.8	9.88	2.88	0.58

the different carbon oxidation states and species present in Table 23.2 can have pe-pH predominance regions at 25°C/1 atm. As we shall soon see, the answer is no because disproportionation plays an important role in the carbon system. We recall that the first favorable disproportionation reaction that we encountered was in Chapter 20, that is,

$$Cl_2 + H_2O = HOCl + H^+ + Cl^-. \qquad (23.20)$$

23.3.1.2 Identification of the pe-pH Predominance Regions for Aqueous Carbon. The equilibrium reaction that we consider first is the very important one between CO_2 and CH_4. Combining the $CO_{2(g)}/CH_{4(g)}$ redox half-reaction given in Table 23.2 with the appropriate Henry's Law equilibria (i.e., Eqs. (23.22) and (23.23)) at 25°C/1 atm, we obtain

$$\begin{array}{llr} CO_{2(g)} + 8H^+ + 8\,e^- = CH_{4(g)} + 2H_2O & \log K = 23.0 & (23.21) \\ H_2CO_3^* = CO_{2(g)} + H_2O & \log K = 1.47 & (23.22) \\ +\ \ CH_{4(g)} = CH_4 & \log K = -2.87 & (23.23) \\ \hline H_2CO_3^* + 8H^+ + 8e^- = CH_4 + 3H_2O & \log K = 21.60. & (23.24) \end{array}$$

Equation (3.24) yields

$$pe = 2.70 + \frac{1}{8}\log\frac{\{H_2CO_3^*\}}{\{CH_4\}} - pH. \qquad (23.25)$$

CH_4 has no acid/base chemistry. However, although $H_2CO_3^*$ will dominate the other C(IV) species for pH < pK_1, this will not be true for pH ≥ pK_1. Let us therefore substitute $[H_2CO_3^*] = \alpha_0 C(IV)_T$ where $C(IV)_T$ represents *only* the C(IV) portion of the total dissolved carbon. Neglecting activity corrections in Eq. (3.25), the result is

$$pe = 2.70 + \frac{1}{8}\log\frac{C(IV)_T}{[CH_4]} + \frac{1}{8}\log\alpha_0 - pH. \qquad (23.26)$$

When there are equal amounts in the C(IV) and C(−IV) forms, we have $C(IV)_T = [CH_4]$, and Eq. (23.26) yields

$$pe = 2.70 + \frac{1}{8}\log\alpha_0 - pH \qquad C(IV)_T/CH_4. \qquad (23.27)$$

When solid, elemental carbon (i.e., either as graphite or diamond) is excluded from consideration, disproportionation in the carbon system makes $C(IV)_T$ and CH_4 the only two stable forms at 25°C/1 atm. This means that at 25°C/1 atm, any system that is initially predominately formate, glucose, formaldehyde, or methanol will tend to disproportionate to some combination of $C(IV)_T$ *and* CH_4. The carbons in glucose ($C_6H_{12}O_6$), formaldehyde (CH_2O), and any other carbohydrate with a 1:2:1 stoichiometry for C:H:O are (on average) oxidatively midway between $C(IV)_T$ and CH_4. Their disproportionation will thus lead to equal amounts of $C(IV)_T$ and CH_4. For glucose, the reaction is

$$C_6H_{12}O_6 = 3\ CH_4 + 3\ CO_2. \qquad (23.28)$$

The fact that total formic (= formic_T = $[HCOOH]+[HCOO^-]$), glucose, formaldehyde, and methanol cannot have predominance regions at 25°C/1 atm may be shown by developing a pe-pH equation between each of these forms and $C(IV)_T$. (The same objective may be accomplished by developing each of the pe-pH equations with CH_4.) Using data in Table 23.3 and proceeding in a manner similar to that used to develop Eq. (23.27), we obtain the results presented in Table 23.4. It is important to note that all of the equilibria in Table 23.4 involving CO_2 and methane have been converted so as to involve *dissolved* $C(IV)_T$ and *dissolved* CH_4. Equations (23.27), (23.34), (23.38), (23.42), and (23.46) are plotted in Figure 23.3. We consider now what the positions of the various lines in Figure 23.3 indicate about formic_T, glucose, formaldehyde, and methanol.

$C(II)_T$ (i.e., formic_T, i.e., $HCOOH + HCOO^-$): (a) Above line (23.42), formic_T cannot predominate since $C(IV)_T > \text{formic}_T$ in that region. (b) Since line (23.42) is below the $C(IV)_T/CH_4$ line, along line (23.42) we know that CH_4 will predominate over $C(IV)_T$ and therefore also over formic_T. Since CH_4 is more reduced than formic_T, decreasing the pe by moving below line (23.42) will favor CH_4 even more. Thus, formic_T cannot predominate below line (23.42).

C(0), glucose: One may use line (23.34) and arguments that are exactly analogous to those given for formic_T to prove that glucose cannot predominate at 25°C/1 atm.

C(0), formaldehyde: One may use line (23.46) and arguments that are exactly analogous to those given for formic_T and glucose to prove that formaldehyde cannot predominate at 25°C/1 atm.

C(-II), methanol: One may use line (23.38) and arguments that are exactly analogous to those given for formic_T, glucose, and formaldehyde to prove that methanol cannot predominate at 25°C/1 atm.

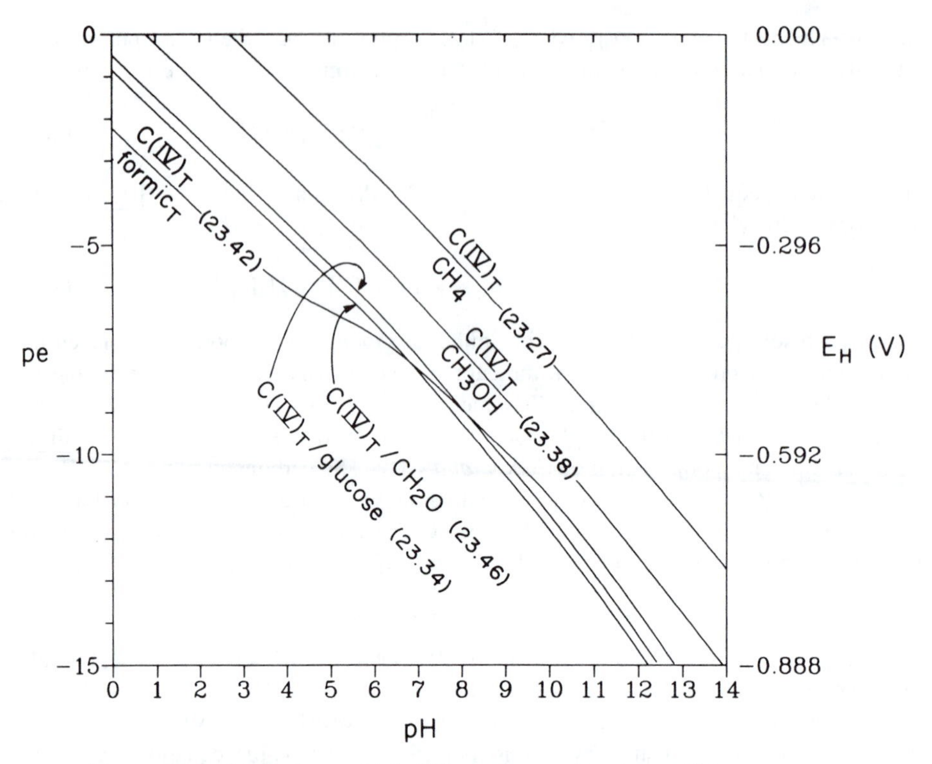

Figure 23.3 Lines along which different forms of carbon will be present at concentrations equal to the total concentration of CO_2-type carbon (i.e., $C(IV)_T$) at 25°C/1 atm. Activity corrections are neglected. For the line involving glucose, it is assumed that the sum of all forms of dissolved carbon equals 10^{-3} *M*. The relative positions of the lines indicate that as long as elemental carbon in the forms of graphite and carbon are excluded from consideration (due to slow kinetics of formation), then $C(IV)_T$ and CH_4 are the only forms of carbon that can have predominance regions. The lines drawn represent only five of the 15 possible combinations among the six different forms of carbon considered.

There is a pattern in the positions of the lines in Figure 23.3. In particular, for all species that are less oxidized than $C(IV)_T$, the lines of equivalent concentration with $C(IV)_T$ plot *below* the $C(IV)_T/CH_4$ line. What this means is that when we exclude both graphitic and diamond carbon from consideration, there cannot be a sequential transition in oxidation state predominance from C(IV) to C(−IV) in the carbon system. For such a transition to occur, we would need the following highest-to-lowest order for the lines: $C(IV)_T$/formic$_T$, $C(IV)_T/CH_2O$ (or $C(IV)_T$/glucose), $C(IV)_T$/methanol, and *then* $C(IV)_T/CH_4$. (The reader may wish to show that if the various lines are formed with CH_4 rather than with $C(IV)_T$, all of the lines will plot *above* the $C(IV)_T/CH_4$ line, a result which leads to the same conclusion regarding predominance in the carbon system.)

TABLE 23.4 Selected equations governing various carbon redox half-reactions in aqueous systems with equilibrium constant data for 25°C/1 atm.

couple	equations		
$C(IV)_T/CH_4$	$H_2CO_3^* + 8H^+ + 8e^- = CH_4 + 3H_2O$	$\log K = 21.60$	(23.24)
	$pe = 2.70 + \frac{1}{8}\log\frac{\{H_2CO_3^*\}}{\{CH_4\}} - pH$		(23.25)
For unit γ values, then	$pe = 2.70 + \frac{1}{8}\log\frac{\alpha_0 C(IV)_T}{[CH_4]} - pH.$		(23.29)
When $C(IV)_T = [CH_4]$, then	$pe = 2.70 + \frac{1}{8}\log\alpha_0 - pH.$		(23.27)
$C(IV)_T$/glucose	$H_2CO_3^* + 4H^+ + 4e^- = \frac{1}{6}C_6H_{12}O_6(\text{glucose}) + 2H_2O$	$\log K = 0.66$	(23.30)
	$pe = 0.17 + \frac{1}{4}\log\frac{\{H_2CO_3^*\}}{\{C_6H_{12}O_6\}^{1/6}} - pH$		(23.31)
For unit γ values,	$pe = 0.17 + \frac{1}{4}\log\frac{\alpha_0 C(IV)_T}{[C_6H_{12}O_6]^{1/6}} - pH.$		(23.32)
Unlike Eq. (23.25) obtaining a pe-pH line from Eq. (23.32) requires specification of the total amount of dissolved carbon. For a value of 10^{-3} *M*, there are equal amounts of carbon in the two species when			
	$C(IV)_T = 6[C_6H_{12}O_6] = 5.0 \times 10^{-4}\ M.$		(23.33)
When Eq. (23.33) applies, then	$pe = -0.49 + \frac{1}{4}\log\alpha_0 - pH.$		(23.34)
$C(IV)_T/CH_3OH$	$H_2CO_3^* + 6H^+ + 6e^- = CH_3OH + 2H_2O$	$\log K = 4.66$	(23.35)
	$pe = 0.77 + \frac{1}{6}\log\frac{\{H_2CO_3^*\}}{\{CH_3OH\}} - pH.$		(23.36)
For unit γ values,	$pe = 0.77 + \frac{1}{6}\log\frac{\alpha_0 C(IV)_T}{[CH_3OH]} - pH.$		(23.37)
When $C(IV)_T = [CH_3OH]$, then	$pe = 0.77 + \frac{1}{6}\log\alpha_0 - pH.$		(23.38)
$C(IV)_T/\text{formic}_T$	$H_2CO_3^* + H^+ + 2e^- = HCOO^- + H_2O$	$\log K = -8.20$	(23.39)
	$pe = -4.10 + \frac{1}{2}\log\frac{\{H_2CO_3^*\}}{\{HCOO^-\}} - \frac{1}{2}pH.$		(23.40)
For unit γ values,	$pe = -4.10 + \frac{1}{2}\log\frac{\alpha_0 C(IV)_T}{\alpha_{1,\text{formic}}\text{formic}_T} - \frac{1}{2}pH.$		(23.41)
When $C(IV)_T = \text{formic}_T$, then	$pe = -4.10 + \frac{1}{2}\log\alpha_0 - \frac{1}{2}\log\alpha_{1,\text{formic}} - \frac{1}{2}pH.$		(23.42)
$C(IV)_T/CH_2O$	$H_2CO_3^* + 4H^+ + 4e^- = CH_2O + 2H_2O$	$\log K = -3.34$	(23.43)
	$pe = -0.84 + \frac{1}{4}\log\frac{\{H_2CO_3^*\}}{\{CH_2O\}} - pH.$		(23.44)
For unit γ values,	$pe = -0.84 + \frac{1}{4}\log\frac{\alpha_0 C(IV)_T}{[CH_2O]} - pH.$		(23.45)
When $C(IV)_T = [CH_2O]$, then	$pe = -0.84 + \frac{1}{4}\log\alpha_0 - pH.$		(23.46)

The fact that Eq. (23.27) plots *well above* all of the other lines in Figure 23.3 means that not only are C(IV) and C(−IV) the only predominant carbon oxidation states, but also that everywhere in the diagram, the concentrations of the other species will be *extremely* small in comparison to the concentrations of either $C(IV)_T$ or CH_4. The species that does come the closest to dominating is CH_3OH since the line closest to Eq. (23.27) is the line for $C(IV)_T/CH_3OH$ (Eq. (23.35).) However, even CH_3OH will be a very minor species everywhere in the diagram. The interested reader who desires more proof of this can compute values of the logarithm of the redox α for CH_3OH as a function of pe and pH, then plot those values in a gridded format over the pe and pH domain used for Figure 23.3.

Formic acid, formate ion, glucose, formaldehyde, and methanol are only a few representatives of the many organic compounds that are possible in natural systems. Nevertheless, the stability of all organic compounds is sufficiently similar that the basic conclusions obtained above concerning carbon systems remain unchanged. In particular, we conclude that when solid carbon in the forms of graphite and diamond is excluded from consideration, then at 25°C/1 atm: 1) $C(IV)_T$ and CH_4 are the only forms of carbon that will have predominance regions; 2) the pe-pH diagram will appear as drawn generically in Figure 23.4; 3) and all other species (formic acid, formate ion, glucose,

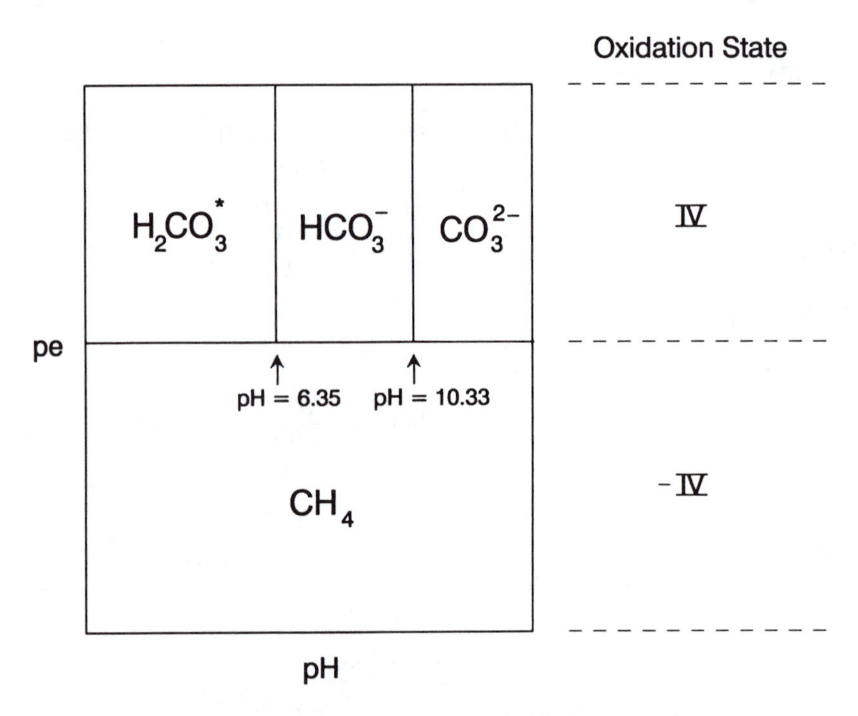

Figure 23.4 Generic pe-pH diagram for aqueous carbon at 25°C/1 atm and when solid elemental forms of carbon (i.e., graphite and diamond) are assumed to be too slow to form for the time frame of interest.

etc.) will be present at very low concentrations everywhere in the pe-pH domain; and 4) at any given point in that domain, a knowledge of the activity of any one C(IV) species (or of CH_4) together with the redox equilibrium constant for conversion to any other carbon species y will allow the calculation of the equilibrium value of $\{y\}$.

23.3.1.3 Finalizing the pe-pH Diagram for Aqueous Carbon. The curved $C(IV)_T/CH_4$ boundaryline is given by Eq. (23.27). Within the $C(IV)_T$ region, the boundarylines between the three C(IV) species are simply pe-independent lines drawn at pH = pK_1 and pH = pK_2. The result is presented in Figure 23.5. Given the fact that $d \log \alpha_0/d\text{pH} \simeq 0, -1$, and -2 for pH < pK_1, pK_1 < pH < pK_2, and pH > pK_2, respectively, in those three regions line (23.27) will take on approximate of slopes of -1, $-9/8$, and $-10/8$, respectively.

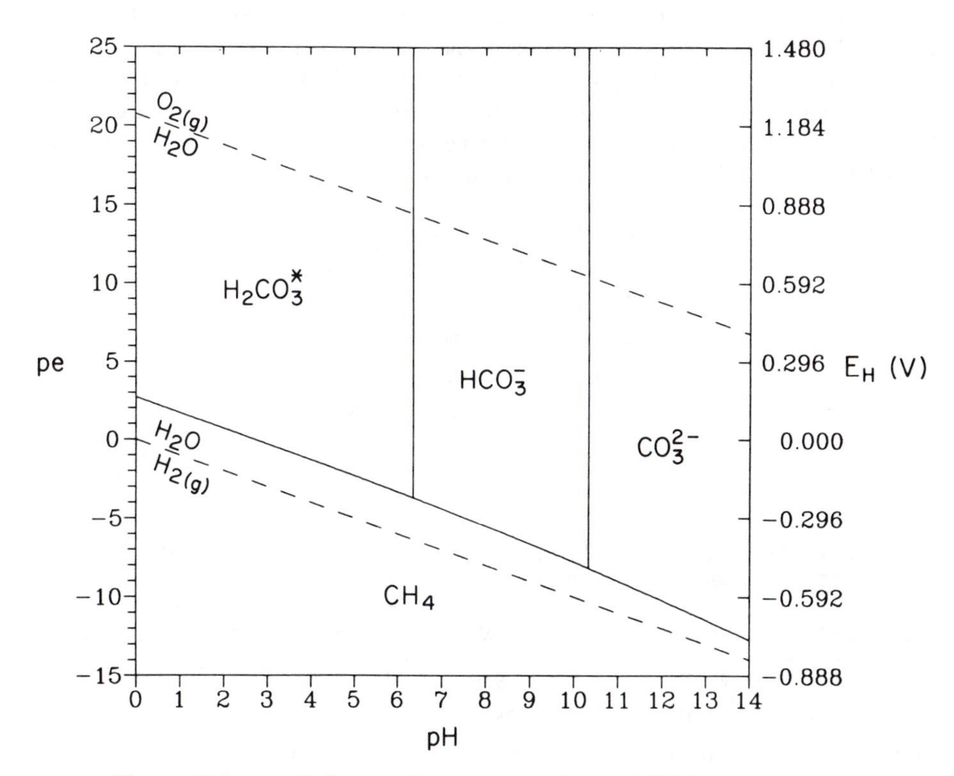

Figure 23.5 pe-pH diagram for aqueous carbon at 25°C/1 atm, and when solid elemental forms of carbon are not considered. Activity corrections are neglected. Since $H_2CO_3^*$, HCO_3^-, CO_3^{2-}, and CH_4 all contain one atom of carbon, and since no solid forms of carbon are considered, the diagram is independent of C_T. As the pe is lowered at any given pH value, the diagram indicates that CH_4 can become the dominant form of carbon before the in-situ p_{H_2} can reach 1 atm.

23.3.2 pe-pH Diagrams for Nitrogen

23.3.2.1 Redox Equilibria Governing Nitrogen Species. Like carbon, nitrogen can exist as a number of different species and in more than one oxidation state. As shown in Table 23.5, nitrogen can range from the N(V) species NO_3^- down to the N(−III) species NH_4^+ and NH_3. As these oxidation states indicate, eight electrons must be added to NO_3^- to convert it to either NH_4^+ or NH_3. This difference in oxidation states is the same as the overall difference of eight between C(IV) and C(−IV). However, since nitrogen is in the Vb column of elements in the periodic table rather than in the IVb column with carbon, nitrogen can exist in the V oxidation state. (An expanded discussion of this matter is given in Section 23.3.3.1.) When NO_3^- is protonated, we obtain HNO_3. This species is such a strong acid that it will not have a predominance region in the pH range of interest here. All of the redox half-reactions in Table 19.2 that involve the Table 23.5 species are collected in Table 23.6.

As with aqueous carbon, disproportionation reactions are possible with aqueous nitrogen. When we started our discussion of the carbon system, we explicitly identified the oxidation states that have predominance regions at 25°C/1 atm. We then used a specific group of lines derived from the various redox half-reaction equilibria to support that statement. For this case, we will assume no such a priori knowledge, and will proceed to interpret the nitrogen case without any such crutch. In this case, then, we

TABLE 23.5 Important inorganic forms of nitrogen in natural waters

oxidation state	species
V	NO_3^-
III	HNO_2 and NO_2^-
0	N_2
−III	NH_4^+ and NH_3

TABLE 23.6 Data for selected redox reactions involving nitrogen at 25°C/1 atm in order of increasing pe° and E_H^o.

Reduction Half-Reaction	log K	pe°	pe°(W)	E_H^o (= 0.05916 pe°)
$N_{2(g)} + 6H^+ + 6e^- = 2NH_3$	9.5	1.58	−5.42	0.093
$N_{2(g)} + 8H^+ + 6e^- = 2NH_4^+$	28.1	4.68	−4.65	0.28
$NO_2^- + 7H^+ + 6e^- = NH_3 + 2H_2O$	81.5	13.58	5.41	0.80
$NO_3^- + 2H^+ + 2e^- = NO_2^- + H_2O$	28.3	14.15	7.15	0.84
$NO_3^- + 10H^+ + 8e^- = NH_4^+ + 3H_2O$	119.2	14.9	6.15	0.88
$NO_2^- + 8H^+ + 6e^- = NH_4^+ + 2H_2O$	90.8	15.14	5.82	0.90
$NO_3^- + 6H^+ + 5e^- = 1/2\ N_{2(g)} + 3H_2O$	105.3	21.05	12.65	1.25
$2NO_2^- + 8H^+ + 6e^- = N_{2(g)} + 4H_2O$	153.5	25.58	16.25	1.51

will plot all six of the pe-pH lines that are possible among the four oxidation states of nitrogen.[5] We will include lines in the diagram that are between forms of nitrogen that are not likely to have adjacent predominance regions. We will then use that diagram to try and determine which of the various species actually have predominance regions.

An alternative way to identify the predominance regions would be to prepare redox α vs. pe plots for each of the nitrogen forms of interest. We would then learn how the α values vary with pe and therefore how predominance changes with pe. This method is utilized in Stumm and Morgan (1981, p. 454). *However*, as in Figures 19.7*a* and 19.7*b* in this text, since each such plot is only valid at *one* pH, we conclude that α plots will need to be prepared for numerous pH values over the pH range of interest. The great advantage of the approach introduced in this section is that we can deduce which species predominate (as a function of pe) over the entire pH range by using just one plot containing all of the possible pe-pH boundarylines.

The equilibria in Table 23.6 have been converted in Table 23.7 into the pe-pH lines represented by Eqs. (23.51), (23.55), (23.59), (23.63), (23.68), and (23.73); Figure 23.6 plots those lines in a pe vs. pH format. For the reduction of any nitrogen-containing species y to N(−III), when half-reaction data for the reduction of y to both NH_4^+ and NH_3 were available, just one of the half-reactions was needed in Table 23.7 to obtain the $y/N(-III)_T$ pe-pH line. It is important to note that both half-reactions will yield the same $y/N(-III)_T$ line, though some manipulation will be needed to make them look exactly the same. The lines involving $N(-III)_T$ have been drawn in Figure 23.6 taking the pK for NH_4^+ to be 9.3 at 25°C/1 atm. The lines involving $N(III)_T$ have been drawn taking the pK for HNO_2 to be 3.0 at 25°C/1 atm. Those equilibria involving $N_{2(g)}$ have been converted into equilibria involving the dissolved species N_2 using the equilibrium (25°C/1 atm)

$$N_{2(g)} = N_2 \qquad K = 10^{-3.20}. \tag{23.47}$$

Since all of the other nitrogen-containing species in Table 23.6 are dissolved, for each half-reaction involving N_2 and some other nitrogen-containing species y, these conversions allow the application of the boundaryline criterion of equal concentrations of nitrogen in the two species along each N_2/y or y/N_2 line in Table 23.7 and Figure 23.6; a value of $N_T = 10^{-3}$ M was chosen for those lines (see Table 23.7).

23.3.2.2 Identification of the pe-pH Predominance Regions for Aqueous Nitrogen When $N_T = 10^{-3}$ *M* and N_2 is Redox Active. The triple bond that binds the two nitrogen atoms in N_2 is exceedingly strong. For $N_{2(g)}$ at 25°C/1 atm, the bond strength is 948.9 kJ/mol. By comparison, the bond strength of a typical carbon-carbon single bond is about 350 kJ/mol. A very important consequence of the strength of the nitrogen-nitrogen bond is the fact that N_2 often behaves as if it is inert from a redox point of view (i.e., as if it is not redox active). For example, when water at typical pH values is in equilibrium with the atmosphere, we will see that thermodynamics tells us that N_2

[5]If there are n items, then combinatorial theory says that the number of different sets of 2 items will be given by the expression $n!/((n-2)!2!)$. When $n = 4$, we obtain six possible combinations of 2. For the carbon case considered in Figure 23.3 where six different types of carbon were considered, there are 15 possible lines; only five were plotted in Figure 23.3.

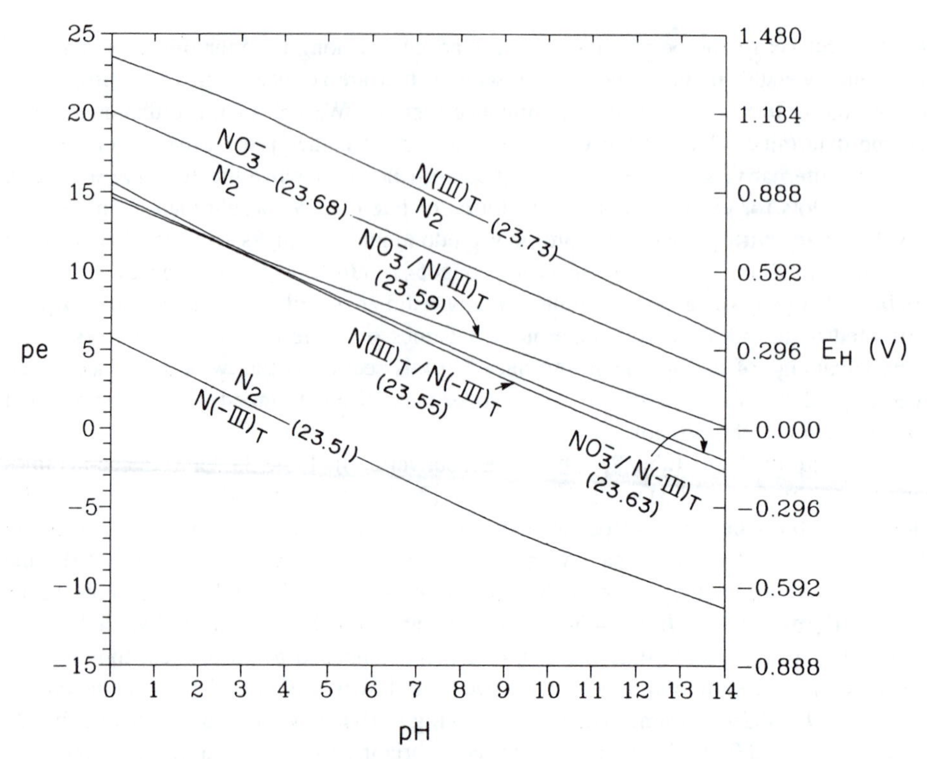

Figure 23.6 Lines along which different pairs of nitrogen forms will contain equal concentrations of nitrogen. The conditions are $N_T = 10^{-3}$ *M* and 25°C/1 atm. Activity corrections are neglected. N_2 is considered to be redox active. Since a total of four different forms of nitrogen are considered (i.e., NO_3^-, $N(III)_T$, N_2, and $N(-III)_T$), a total of six lines can be drawn. The relative positions of the lines indicate that at equilibrium, $N(III)_T$ cannot predominate at any pH; the other three forms of nitrogen will have predominance regions.

should be virtually completely oxidized by atmospheric oxygen to NO_3^-. However, this reaction is exceedingly slow. Indeed, the slowness of this reaction allows substantial amounts of N_2 to accumulate in the atmosphere, and often makes N_2 appear non-redox active. This section will examine the pe-pH diagram for nitrogen when N_2 is in fact considered redox active. The next section will examine the pe-pH diagram that results when N_2 is considered non-redox active.

We can use Figure 23.6 to elucidate which species will have predominance regions in the pe-pH diagram when $N_T = 10^{-3}$ *M*. The need to specify N_T stems from the fact that N_2 contains two atoms of nitrogen, and the other dissolved forms contain only one atom of nitrogen.[6] We focus on each of the various species in turn. We begin with

[6]This situation is analogous to what we encountered in Chapter 20 with aqueous chlorine: Cl_2 contains two atoms of chlorine, but HClO, ClO^-, and Cl^- only contain one atom of chlorine, and so the pe-pH diagram for aqueous chlorine depends on Cl_T.

$N(III)_T$ because we can show that it will not have a predominance region in this diagram, and doing so will be convenient for subsequent considerations.

$N(III)_T$ (i.e., HNO_2 + NO_2^-): According to Figure 23.6, N_2 will predominate over $N(III)_T$ everywhere below line (23.73). $N(III)_T$ cannot therefore have a predominance region below line (23.73). We now note that line (23.73) is above line (23.68) (i.e., above the NO_3^-/N_2 line), and so everywhere along line (23.73), we have $[NO_3^-] > N(III)_T$. At any pH, as we increase the pe by moving above line (23.73), we will further increase the $[NO_3^-]/N(III)_T$ ratio, and so $N(III)_T$ cannot predominate above line (23.73). With $N(III)_T$ predominance now ruled out everywhere in the pe-pH diagram, we can ignore any line in Figure 23.6 involving $N(III)_T$.

N(V) (i.e., NO_3^-): As the most oxidized species under consideration, we know that the predominance region for NO_3^- will occupy all of the upper portion of the pe-pH diagram. As the pH changes, its lower boundary could be with first one, then another less oxidized species. However, line (23.68) tells us that this lower boundary is with just one species, N_2. We note that this boundaryline is well below the O_2/H_2O line for $p_{O2} = 1$ atm. Since the line for $p_{O2} = 0.21$ atm will be very close to the $p_{O2} = 1$ atm line, it is apparent that the predominance region for NO_3^- extends well into the pe-pH region for typical oxygenated waters. The facts that little NO_3^- is actually found in most surface waters but that much N_2 is present are testaments to the inertness of N_2.

N(0) (i.e., N_2): We know that N_2 will predominate over NO_3^- below line (23.68). Since N_2 is not the most reduced form of nitrogen under consideration, we know that its predominance region cannot extend forever as pe is lowered. Indeed, line (23.51) shows that $N(-III)_T$ will predominate over N_2 below line (23.51). Since we need only consider the forms NO_3^-, N_2, and $N(-III)_T$ (recall that $N(III)_T$ has been ruled out for the diagram), we can conclude that the predominance region for N_2 will extend from line (23.68) to line (23.51).

$N(-III)_T$ (i.e., NH_3 + NH_4^+): Since $N(-III)_T$ is the most reduced form of nitrogen under consideration, below some boundaryline, the predominance region for $N(-III)_T$ will occupy all of the lower portion of the pe-pH diagram. For reasons discussed, we can identify that boundary as line (23.51). (Although line (23.63) accurately describes the location of the line along which $[NO_3^-] = [N(-III)_T]$, this line occurs in a region in which neither NO_3^- nor $N(-III)_T$ can predominate, and so it will not appear in the final pe-pH diagram.) There will be a vertical line within the $N(-III)_T$ region that separates the NH_4^+ and NH_3 portions of $N(-III)_T$ predominance region.

23.3.2.3 Finalizing the pe-pH Diagram for Aqueous Nitrogen When N_2 is Redox Active and $N_T = 10^{-3}$ *M*. Based on these considerations, we conclude that the pe-pH diagram for nitrogen at 25°C/1 atm and $N_T = 10^{-3}$ *M* appears as given in Figure 23.7. This figure is based on lines (23.51) and (23.68) plus the boundaryline equation that separates NH_4^+ and NH_3, namely pH = 9.3. Given that $d\log\alpha_1/d$pH for NH_4^+ is approximately +1 and 0 for pH < pK and pH > pK, respectively, the slope of line (23.51) in those two regions will be approximately $-4/3$, and -1, respectively. In the transition pH range between those two regions, the slope gradually shifts from $-4/3$ to -1. Although N(III) species will be of very minor importance everywhere in

TABLE 23.7 Equations governing various nitrogen redox half-reactions in aqueous systems with equilibrium constant data for 25°C/1 atm.

couple	equations	
$N_2/N(-III)_T$	$N_2 + 6H^+ + 6e^- = 2NH_3 \quad \log K = 12.7$	(23.48)
	$pe = 2.12 + \frac{1}{6}\log\frac{\{N_2\}}{\{NH_3\}^2} - pH$	(23.49)

Since the numbers of nitrogens in the species N_2 and NH_3 are different, obtaining a pe-pH line from Eq. (23.49) requires specification of the total amount of dissolved nitrogen. For water saturated with atmospheric levels of N_2, at 25°C/1 atm, $[N_2] = 4.9 \times 10^{-4}$ M. At this level, excluding other nitrogen species, we then have $N_T = 9.8 \times 10^{-4}$ M. Therefore, selecting $N_T = 10^{-3}$ M for this pe-pH diagram there are equal amounts of nitrogen present as N_2 and N(−III) when

$$2[N_2] = N(-III)_T = [NH_3] + [NH_4^+] = 5 \times 10^{-4}\ M \qquad (23.50)$$

Substituting $[NH_3] = \alpha_{1,N(-III)}N(-III)_T$, and for unit γ values, Eqs. (23.49) and (23.50) yield

$$pe = 2.62 - \frac{1}{3}\log\alpha_{1,N(-III)} - pH \qquad (23.51)$$

couple	equations	
$N(III)_T/N(-III)_T$	$NO_2^- + 7H^+ + 6e^- = NH_3 + 2H_2O \quad \log K = 81.5$	(23.52)
	$pe = 13.58 + \frac{1}{6}\log\frac{\{NO_2^-\}}{\{NH_3\}} - \frac{7}{6}pH$	(23.53)
For unit γ values, then	$pe = 13.58 + \frac{1}{6}\log\frac{\alpha_{1,N(III)}N(III)_T}{\alpha_{1,N(-III)}N(-III)_T} - \frac{7}{6}pH$	(23.54)
When $N(III)_T = N(-III)_T$, then	$pe = 13.58 + \frac{1}{6}\log\alpha_{1,N(III)} - \frac{1}{6}\log\alpha_{1,N(-III)} - \frac{7}{6}pH$	(23.55)
$NO_3^-/N(III)_T$	$NO_3^- + 2H^+ + 2e^- = NO_2^- + H_2O \quad \log K = 28.3$	(23.56)
	$pe = 14.15 + \frac{1}{2}\log\frac{\{NO_3^-\}}{\{NO_2^-\}} - pH$	(23.57)
For unit γ values, then	$pe = 14.15 + \frac{1}{2}\log\frac{[NO_3^-]}{\alpha_{1,N(III)}N(III)_T} - pH$	(23.58)
When $[NO_3^-] = N(III)_T$, then	$pe = 14.15 - \frac{1}{2}\log\alpha_{1,N(III)} - pH$	(23.59)

the pe-pH diagram, the concentrations of HNO_2 and NO_2^- will not be zero at any point in the diagram.

23.3.2.4 Identification of the pe-pH Predominance Regions for Aqueous Nitrogen When $N_T = 10^{-3}$ *M* and N_2 is Not Redox-Active. When all of the lines involving N_2 are removed from Figure 23.6, just three remain, the $NO_3^-/N(III)_T$ line, the $NO_3^-/N(-III)_T$ line, and the $N(III)_T/N(-III)_T$ line. These three lines are plotted in Figure 23.8 over the range pH = 0 to 6. We note that while the three lines come rather close to one another near pH = 2.94, they do not actually intersect. Above line (23.59), NO_3^- predominates. Between lines (23.59) and (23.55), $N(III)_T$ predominates. Below line (23.55), $N(-III)_T$ predominates. Therefore, line (23.63) will not appear in the pe-pH diagram as a boundaryline because: 1) between it and line (23.59), $N(III)_T$

TABLE 23.7 continued

couple	equations		
$NO_3^-/N(-III)_T$	$NO_3^- + 10H^+ + 8e^- = NH_4^+ + 3H_2O$	$\log K = 119.2$	(23.60)
	$pe = 14.9 + \frac{1}{8}\log\frac{\{NO_3^-\}}{\{NH_4^+\}} - \frac{5}{4}pH$		(23.61)
For unit γ values,	$pe = 14.9 + \frac{1}{8}\log\frac{[NO_3^-]}{\alpha_{0,N(-III)}N(-III)_T} - \frac{5}{4}pH$		(23.62)
When $[NO_3^-] = N(-III)_T$, then	$pe = 14.9 - \frac{1}{8}\log\alpha_{0,N(-III)} - \frac{5}{4}pH$		(23.63)
NO_3^-/N_2	$NO_3^- + 6H^+ + 5e^- = \frac{1}{2}N_2 + 3H_2O$	$\log K = 102.1$	(23.64)
	$pe = 20.42 + \frac{1}{5}\log\frac{\{NO_3^-\}}{\{N_2\}^{1/2}} - \frac{6}{5}pH$		(23.65)
For unit γ values,	$pe = 20.42 + \frac{1}{5}\log\frac{[NO_3^-]}{[N_2]^{1/2}} - \frac{6}{5}pH$		(23.66)
For $N_T = 10^{-3}$ *M*, there are equal amounts of nitrogen present as NO_3^- and N_2 when			
	$2[N_2] = [NO_3^-] = 5 \times 10^{-4}$ *M*		(23.67)
then	$pe = 20.12 - \frac{6}{5}pH$		(23.68)
$N(III)_T/N_2$	$2NO_2^- + 8H^+ + 6e^- = N_2 + 4H_2O$	$\log K = 150.3$	(23.69)
	$pe = 25.05 + \frac{1}{6}\log\frac{\{NO_2^-\}^2}{\{N_2\}} - \frac{4}{3}pH$		(23.70)
For unit γ values,	$pe = 25.05 + \frac{1}{6}\log\frac{\alpha_{1,N(III)}^2 N(III)_T^2}{[N_2]} - \frac{4}{3}pH$		(23.71)
For $N_T = 10^{-3}$ *M*, there are equal amounts of nitrogen present as $N(III)_T$ and N_2 when			
	$2[N_2] = N(III)_T = 5 \times 10^{-4}$ *M*		(23.72)
Thus,	$pe = 24.55 + \frac{1}{3}\log\alpha_{1,N(III)} - \frac{4}{3}pH$		(23.73)

predominates over NO_3^-; and 2) between it and line (23.55), $N(III)_T$ predominates over $N(-III)_T$.

If NO_2^- could not be protonated, the three lines in Figure 23.7 would all intersect at a single point at pH = 2.94. Line (23.63) would remain the center line, but line (23.63) would become the lowest line and line (23.55) would become the highest line. For reasons considered above in discussions of various disproportionation cases, such a reversal in positions would lead to a disappearance of the $N(III)_T$ predominance region for pH < 2.94, and a direct adjacency of the NO_3^- and $N(-III)_T$ predominance regions. It is the inclusion of the $\log\alpha_{1,N(III)}$ terms in Eqs. (23.55) and (23.59) that causes the former line to be pulled down, and the latter line to be pulled up, thus preventing the disappearance of the $N(III)_T$ predominance region for pH < 2.94.

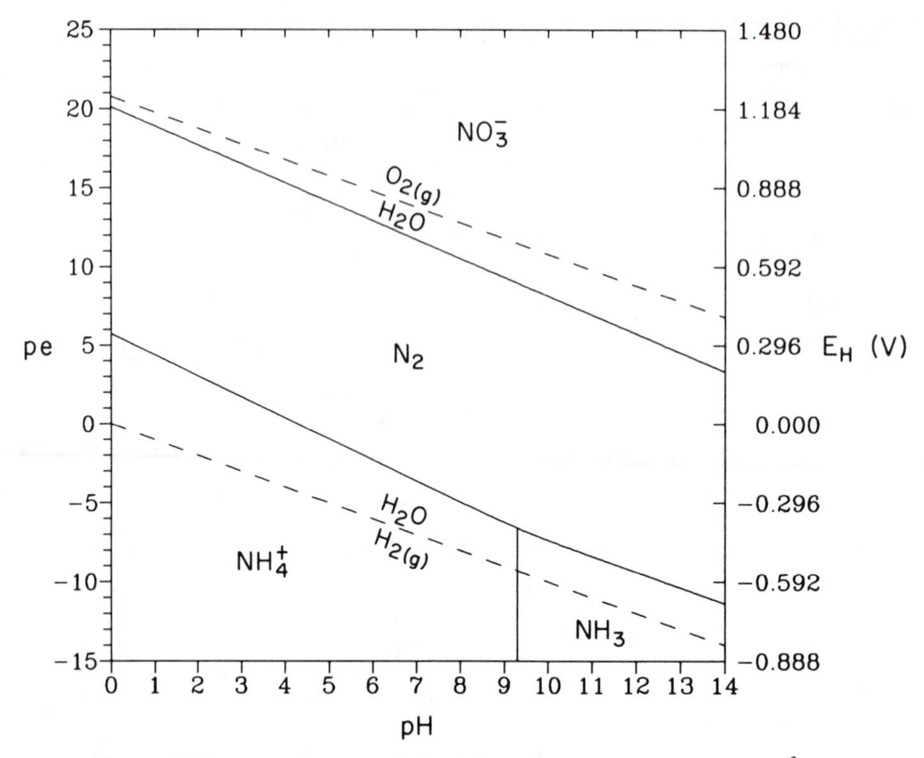

Figure 23.7 pe-pH diagram for aqueous nitrogen when $N_T = 10^{-3}$ *M*, 25°C/1 atm, and when N_2 is considered to be redox active. Activity corrections are neglected. As the pe is lowered at any given pH value, the diagram indicates that N(−III) (either as NH_3 or as NH_4^+) can become the dominant form of nitrogen before the in-situ p_{H_2} can reach 1 atm.

23.3.2.5 Finalizing the pe-pH Diagram for Aqueous Nitrogen When $N_T = 10^{-3}$ *M* and N_2 is Not Redox Active. Based on the above considerations, the pe-pH predominance diagram for nitrogen when N_2 is not redox active is composed of the lines given by Eqs. (23.59) and (23.55), plus boundarylines at pH = pK_{HNO_2} = 3.0 and pH = $pK_{NH_4^+}$ = 9.3. For pH < 3.0, the $N(III)_T$ predominance region will be labelled as HNO_2; for pH > 3.0 it will be labeled as NO_2^-. For pH < 9.3, the $N(-III)_T$ predominance region will be labeled as NH_4^+, and for pH > 9.3 as NH_3. The overall result is given in Figure 23.9.

23.3.3 pe-pH Diagram for Sulfur

23.3.3.1 Redox Equilibria Governing Sulfur Species. Several of the most important of the sulfur species and their oxidation states are listed in Table 23.8. The important redox equilibria relating the Table 23.8 species are presented in Table 23.9. Using nomenclature that is analogous to that employed with $H_2CO_3^*$, the species $H_2SO_3^*$

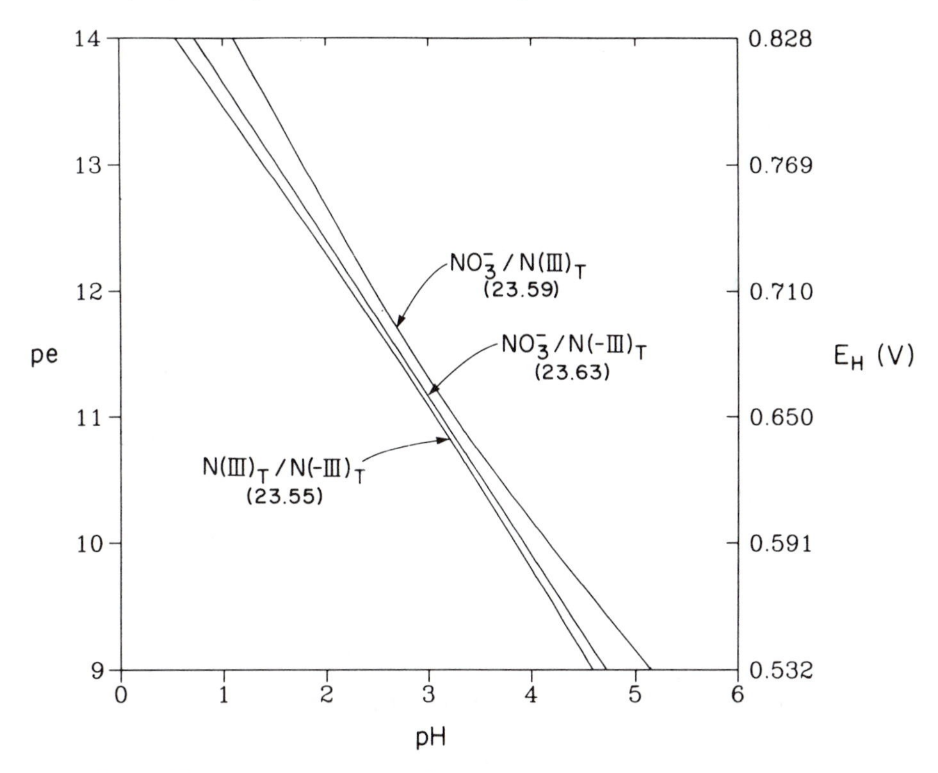

Figure 23.8 Lines along which different pairs of nitrogen forms (except N_2) will contain equal concentrations of nitrogen at 25°C/1 atm. Activity corrections are neglected. N_2 is viewed as being inert from a redox point of view. Since a total of three different forms of nitrogen are considered (i.e., NO_3^-, $N(III)_T$, and $N(-III)_T$), a total of three lines can be drawn. The relative positions of the lines indicate that when N_2 is not redox active, NO_3^-, $N(III)_T$, and $N(-III)_T$ will all have predominance regions that extend over the full pH range.

TABLE 23.8 Important inorganic forms of sulfur in natural waters.

oxidation state	species
VI	HSO_4^-, SO_4^{2-}
IV	$H_2SO_3^*$, HSO_3^-, SO_3^{2-}
0	$S_{(s)}$
−II	H_2S, HS^-, S^{2-}

refers to the sum of SO_2 and the true hydrate H_2SO_3. Based on Table 23.8, we see that sulfur ranges from the two S(VI) species HSO_4^- and SO_4^{2-} down to the three S(−II) species H_2S, HS^-, and S^{2-}. As with the conversion of NO_3^- to NH_4^+ or NH_3, and the conversion of $H_2CO_3^*$ or HCO_3^- or CO_3^{2-} to CH_4, eight electrons must be added to

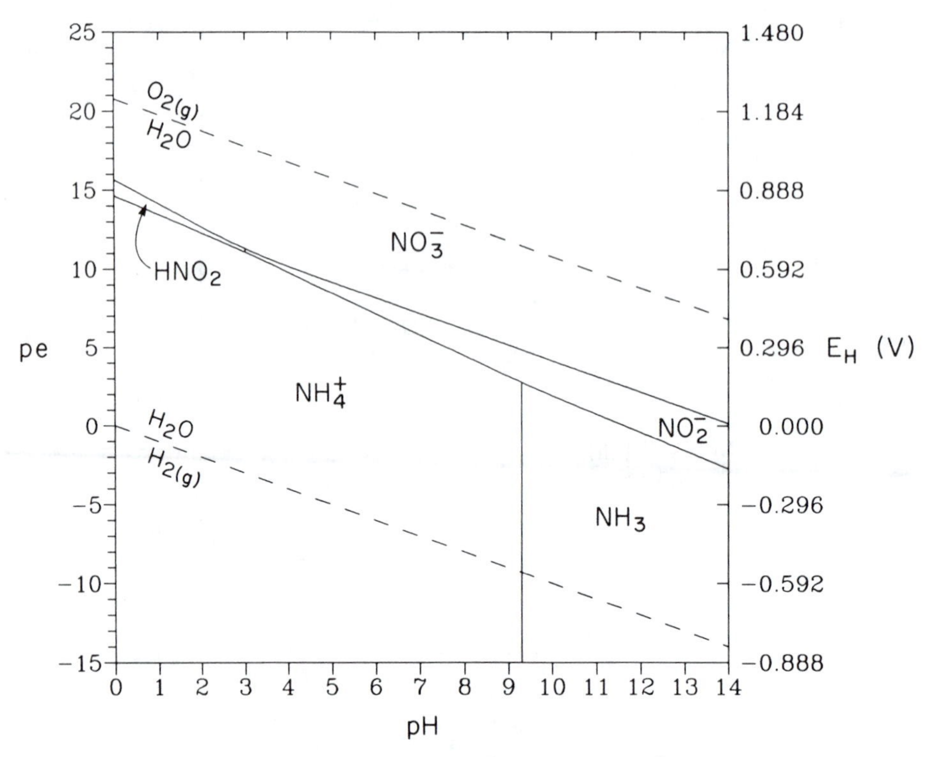

Figure 23.9 pe-pH diagram for aqueous nitrogen at 25°C/1 atm when N_2 is not redox active. Activity corrections are neglected. A line at pH = 3.0 separates the HNO_2 and NO_2^- regions. Since all of the forms of nitrogen considered contain the same number of nitrogen atoms (i.e., one), and since no solid forms of nitrogen are considered, the diagram is independent of N_T. The $N(III)_T$ region becomes quite narrow at pH $\simeq$ 3.

convert SO_4^{2-} to H_2S, HS^-, or S^{2-}. The reason why eight electrons plays this recurring role with all three of these elements may be understood as follows.

Sulfur is in the VIb column of the periodic table, and as discussed in preceding sections, carbon and nitrogen are in the IVb and Vb columns, respectively. In each case, the column number gives the number of electrons in the outer shell of the elemental form (oxidation state zero) of the element. When all of those outer shell electrons are removed, the highest oxidation state is attained, and so the highest oxidation state for these elements is equal to its column number. A total of eight electrons are needed to completely fill the outer electron shells of these elements. Therefore, the lowest oxidation state may be attained by adding eight electrons to the highest oxidation state. The lowest oxidation state may also be attained by adding (8 – c) electrons to the elemental (zero oxidation state) form of the element, where c is the column number.

At 25°C/1 atm, for H_2SO_4 we have $pK_1 \simeq -3$, and $pK_2 = 2.0$. For $H_2SO_3^*$, we have $pK_1 = 1.91$ and $pK_2 = 7.18$. For H_2S, we have $pK_1 = 7.0$. The pK_2 for H_2S is not well known, but is believed to be close to 14, though some estimates have placed it as high as 15.[7] We will take pK_2 for H_2S to be 14.0. The fact that pK_2 for H_2S is at least 14 means that S^{2-} will not have a predominance region in a pe-pH diagram that extends from pH = 0 to pH = 14.

The very low pK_1 for H_2SO_4 means that H_2SO_4 will not have a predominance region in a pe-pH diagram for sulfur. The species S^{2-} has been included in Table 23.8 but H_2SO_4 has not because: 1) pH values near the pK_2 for H_2S are closer to the range pH = 0 to 14 than are pH values near the pK_1 for H_2SO_4; and 2) the species S^{2-} plays an important role in low pe waters in controlling the solubility of metals like Fe(II) and Pb(II) through the formation of highly insoluble metal sulfides like $FeS_{(s)}$ and $PbS_{(s)}$.

As far as sulfur in the S(0) state is concerned, depending on the pe and the pH, *dissolved* S(0) will only be present at very low concentrations, and so can be neglected. In contrast, solid elemental sulfur can be important in certain regions of the diagram. As with any solid in a pe-pH diagram, the presence and the extent of a predominance region for solid sulfur will depend on the value of S_T in the system.

Solid elemental sulfur is composed predominantly of eight-membered, ringed molecules, that is, S_8. However, both smaller and larger rings can also be found in any given sample of solid sulfur. Thus, solid elemental sulfur is really a *solid solution* of ringed S molecules of different sizes. The fact that the chemical potentials of sulfur in the various rings are all very similar means that we may consider solid elemental sulfur to be simply $S_{(s)}$.

With the four oxidation states which it includes, Table 23.8 presents a picture that is similar in many ways to the case of aqueous nitrogen. In particular: the acid H_2SO_4 is analogous to the acid HNO_3; the intermediate oxidation state acid $H_2SO_3^*$ is analogous to HNO_2; and $S_{(s)}$ is analogous to N_2. Finally, H_2S is analogous to NH_3, though the very available free electron pair on NH_3 and the very low basicity of the hydrogens on NH_3 make NH_3 a base rather than an acid like H_2S.

The equilibria in Table 23.9 yield the six pe-pH equations given in Table 23.10. These pe-pH lines represent all six possible combinations among the four forms $S(VI)_T$, $S(IV)_T$, S(0), and $S(-II)_T$. For those equations involving S(0) (i.e., those involving $S_{(s)}$), an S_T value of 10^{-3} *M* was chosen. Equations (23.77), (23.81), (23.85), (23.89), (23.93), and (23.97) from Table 23.10 are plotted in Figure 23.10. This figure provides for sulfur what Figure 23.6 provides for nitrogen.

23.3.3.2 Identification of the pe-pH Predominance Regions for Aqueous Sulfur when $S_T = 10^{-3}$ *M*. We can use Figure 23.10 to determine which forms of sulfur will have predominance regions in the pe-pH diagram for $S_T = 10^{-3}$ *M* at 25°C/1 atm. We focus on each form in turn. We will start with $S(IV)_T$ because we will be able to eliminate it, and it will simplify things to do so.

$S(IV)_T$ (i.e., $H_2SO_3^* + HSO_3^- + SO_3^{2-}$): $S(IV)_T$ cannot have a predominance region in this system. Indeed, the $S(IV)_T/S(-II)_T$ line (23.93) plots above the $S(VI)_T/S(-II)_T$ line (23.89). Above line (23.89), including the points along line (23.93), $S(VI)_T$ will

[7] It is the fact that pK_2 for H_2S is so high that makes it so difficult to measure. Indeed, we recall from Chapters 7 and 8 that OH^- buffering makes it difficult to study the deprotonation of an acid that is very weak.

TABLE 23.9 Data for selected redox half-reactions involving sulfur at 25°C/1 atm in order of increasing pe° and E_H^o.

Reduction Half-Reaction	log K	pe°	pe°(W)	E_H^o(= 0.05916 pe°)
$S_{(s)} + H^+ + 2e^- = HS^-$	−2.2	−1.1	−4.6	−0.065
$S_{(s)} + 2H^+ + 2e^- = H_2S$	4.8	2.4	−4.6	0.14
$SO_4^{2-} + 9H^+ + 8e^- = HS^- + 4H_2O$	34.0	4.25	−3.63	0.25
$SO_4^{2-} + 10H^+ + 8e^- = H_2S + 4H_2O$	41.0	5.13	−3.62	0.30
$HSO_4^- + 7H^+ + 6e^- = S_{(s)} + 4H_2O$	34.2	5.7	−2.47	0.34
$SO_4^{2-} + 8H^+ + 6e^- = S_{(s)} + 4H_2O$	36.2	6.03	−3.3	0.36
$H_2SO_3^* + 4e^- + 4H^+ = S_{(s)} + 3H_2O$	33.8	8.45	1.45	0.50

predominate over $S(-II)_T$. Thus, $S(VI)_T$ will predominate over S(IV) along line (23.93). Increasing the pe by going above line (23.93) will only further increase the $S(VI)_T/S(IV)_T$ ratio. $S(IV)_T$ cannot predominate anywhere below line (23.93) since in that region, $S(-II)_T > S(IV)_T$.

$S(VI)_T$ (i.e., HSO_4^- + SO_4^{2-}): Since we have ruled out a predominance region for $S(IV)_T$, we now focus on Figure 23.11, which is the same as Figure 23.10, except that all of the lines involving $S(IV)_T$ have been removed. As the most oxidized form of sulfur under consideration, we know that the predominance region for $S(VI)_T$ will occupy all of the upper portion of the pe-pH diagram. As the pH changes, its lower boundary may be with first one, then another less oxidized form of sulfur. Figure 23.11 shows that for pH values less than ~5, the lower boundary of the $S(VI)_T$ region is with S(0) (line (23.77)). As the pH is increased beyond ~5, the lower boundary is with $S(-II)_T$ (line (23.89)). The exact pH of the intersection of lines (23.77) and (23.89) is pH = 5.05.

S(0) (i.e., $S_{(s)}$): Based on Figure 23.11, when pH < 5.05, and when $S_T = 10^{-3}$ M, $S_{(s)}$ will begin to precipitate from a solution predominated by $S(VI)_T$ at pe values *somewhat below* those given by line (23.77). (Actually, we will see shortly that a more exact requirement for $S_{(s)}$ being possible when $S_T = 10^{-3}$ M is not pH < 5.05, but rather pH < 4.40.) The reason for the qualifier "somewhat below" is the fact that line (23.77) assumes that $S(VI)_T = 10^{-3}$ M. In fact, if $S_T = 10^{-3}$ M, then $S(VI)_T$ will always be at least a little less than 10^{-3} M. Near line (23.77), then, the next most important form of sulfur in solution will be $S(-II)_T$.

As we drop down further in pe, the same SO_4^{2-} that precipitated as $S_{(s)}$ will eventually begin to re-dissolve as $S(-II)_T$. The dissolution will be complete at pe values *somewhat above* line (23.97). The reason for the qualifier "somewhat above" is the fact that line (23.97) assumes that $S(-II)_T = 10^{-3}$ M. In fact, if $S_T = 10^{-3}$ M, then $S(-II)_T$ will always be at least a little less than 10^{-3} M. Near line (23.97), the next most important form of sulfur in solution will be $S(VI)_T$.

Based on the above comments, we conclude that for pH < 4.40, $S_{(s)}$ can exist in a narrow wedge of pe-pH values between lines (23.77) and (23.97). For pH > ~4.40 and

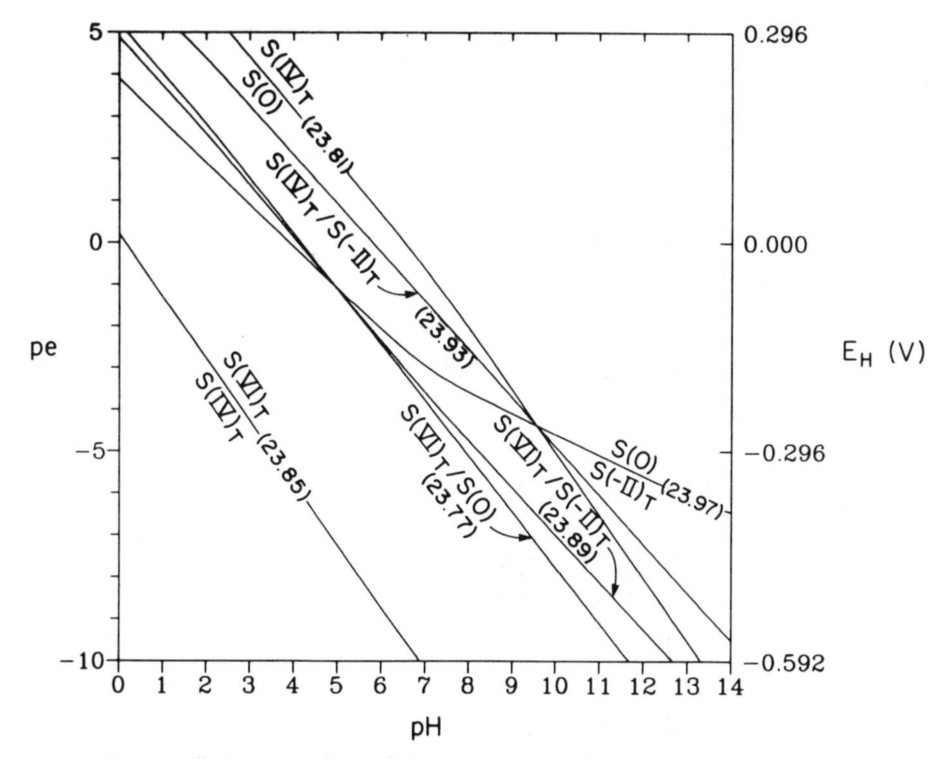

Figure 23.10 Lines which can be used to deduce the nature of the pe-pH diagram for aqueous sulfur when $S_T = 10^{-3}$ M and 25°C/1 atm. Activity corrections are neglected. Along a line involving $S_{(s)}$, the single dissolved species is taken to be present at 10^{-3} M and an infinitesimal amount of $S_{(s)}$ is taken to be present at a mole fraction of 1. Since a total of four different forms of sulfur are considered (i.e., $S(VI)_T$, $S(IV)_T$, $S_{(s)}$, and $S(-II)_T$), a total of six lines can be drawn. The relative positions of the lines indicate that at equilibrium, $S(IV)_T$ cannot predominate at any pH; the other three forms have predominance regions, though $S_{(s)}$ can only form at relatively low pH.

when $S_T = 10^{-3}$ M, $S_{(s)}$ becomes unstable with respect to a dissolution disproportionation to $S(VI)_T$ and $S(-II)_T$ at all pe values. Line (23.89) then describes the direct boundaryline between $S(VI)_T$ and $S(-II)_T$. The wedge *will not* be perfectly pointed as would be obtained by taking lines (23.77) and (23.97) as the upper and lower boundaries of the wedge. Rather, for the same reasons that required the use of the qualifiers "somewhat below" and "somewhat above," the wedge is a *curved, blunt wedge*.

The single boundaryline defining the $S_{(s)}$ wedge may be computed by calculating the set of pe and pH values for which $S_{(s)}$ specifies that the dissolved $S_T = 10^{-3}$ M. Neglecting S(IV) species as well as S^{2-}, then everywhere except actually inside the boundaryline for the wedge we have for the solution phase that

$$[HSO_4^-] + [SO_4^{2-}] + [H_2S] + [HS^-] = S_T. \tag{23.98}$$

TABLE 23.10 Equations governing various sulfur redox half-reactions in aqueous systems with equilibrium constant data for 25°C/1 atm.

couple	equations
$\frac{S(VI)_T}{S(0)}$	$SO_4^{2-} + 8H^+ + 6e^- = S_{(s)} + 4H_2O \qquad \log K = 36.2 \qquad (23.74)$ $pe = 6.03 + \frac{1}{6}\log\frac{\{SO_4^{2-}\}}{\{S_{(s)}\}} - \frac{4}{3}pH \qquad (23.75)$ For unit γ values and $\{S_{(s)}\} = 1$, we have $pe = 6.03 + \frac{1}{6}\log \alpha_{2,S(VI)} S(VI)_T - \frac{4}{3}pH \qquad (23.76)$ For $S_T = 10^{-3}$ *M*, and assuming that $S(VI)_T = S_T$, Eq. (23.76) becomes $pe = 5.53 + \frac{1}{6}\log \alpha_{2,S(VI)} - \frac{4}{3}pH \qquad (23.77)$
$\frac{S(IV)_T}{S(0)}$	$H_2SO_3^* + 4e^- + 4H^+ = S_{(s)} + 3H_2O \qquad \log K = 33.8 \qquad (23.78)$ $pe = 8.45 + \frac{1}{4}\log\frac{\{H_2SO_3^*\}}{\{S_{(s)}\}} - pH \qquad (23.79)$ For unit γ values and $\{S_{(s)}\} = 1$, we have $pe = 8.45 + \frac{1}{4}\log \alpha_{0,S(IV)} S(IV)_T - pH \qquad (23.80)$ For $S_T = 10^{-3}$ *M*, and assuming that $S(IV)_T = S_T$, Eq. (23.80) becomes $pe = 7.70 + \frac{1}{4}\log \alpha_{0,S(IV)} - pH \qquad (23.81)$
$\frac{S(VI)_T}{S(IV)_T}$	By subtracting Eq. (23.78) from Eq. (23.74), we obtain $SO_4^{2-} + 4H^+ + 2e^- = H_2SO_3^* + H_2O \qquad \log K = 2.4 \qquad (23.82)$ $pe = 1.2 + \frac{1}{2}\log\frac{\{SO_4^{2-}\}}{\{H_2SO_3^*\}} - 2pH \qquad (23.83)$ For unit γ values, we have $pe = 1.2 + \frac{1}{2}\log\frac{\alpha_{2,S(VI)} S(VI)_T}{\alpha_{0,S(IV)} S(IV)_T} - 2pH \qquad (23.84)$ When $S(VI)_T = S(IV)_T$, then $pe = 1.2 + \frac{1}{2}\log\frac{\alpha_{2,S(VI)}}{\alpha_{0,S(IV)}} - 2pH \qquad (23.85)$

At 25°C/1 atm, K_2 for H_2SO_4 is $10^{-2.0}$, and K_1 for H_2S is $10^{-7.0}$. Therefore, neglecting activity corrections, Eq. (23.98) becomes

$$[HSO_4^-] + \frac{10^{-2.0}}{[H^+]}[HSO_4^-] + [H_2S] + \frac{10^{-7.0}}{[H^+]}[H_2S] = S_T \tag{23.99}$$

or

$$[HSO_4^-]\left(1 + \frac{10^{-2.0}}{[H^+]}\right) + [H_2S]\left(1 + \frac{10^{-7.0}}{[H^+]}\right) = S_T. \tag{23.100}$$

TABLE 23.10 continued

couple	equations	
$\frac{S(VI)_T}{S(-II)_T}$	$SO_4^{2-} + 9H^+ + 8e^- = HS^- + 4H_2O \quad \log K = 34.0$	(23.86)
	$pe = 4.25 + \frac{1}{8}\log\frac{\{SO_4^{2-}\}}{\{HS^-\}} - \frac{9}{8}pH$	(23.87)
	For unit γ values, we have	
	$pe = 4.25 + \frac{1}{8}\log\frac{\alpha_{2,S(VI)}S(VI)_T}{\alpha_{1,S(-II)}S(-II)_T} - \frac{9}{8}pH$	(23.88)
	When $S(VI)_T = S(-II)_T$, then	
	$pe = 4.25 + \frac{1}{8}\log\frac{\alpha_{2,S(VI)}}{\alpha_{1,S(-II)}} - \frac{9}{8}pH$	(23.89)
$\frac{S(IV)_T}{S(-II)_T}$	By subtracting Eq. (23.82) from Eq. (23.86), we obtain	
	$H_2SO_3^* + 5H^+ + 6e^- = HS^- + 3H_2O \quad \log K = 31.6$	(23.90)
	$pe = 5.27 + \frac{1}{6}\log\frac{\{H_2SO_3^*\}}{\{HS^-\}} - \frac{5}{6}pH$	(23.91)
	For unit γ values, we have	
	$pe = 5.27 + \frac{1}{6}\log\frac{\alpha_{0,S(IV)}S(IV)_T}{\alpha_{1,S(-II)}S(-II)_T} - \frac{5}{6}pH$	(23.92)
	When $S(VI)_T = S(-II)_T$, then	
	$pe = 5.27 + \frac{1}{6}\log\frac{\alpha_{0,S(IV)}}{\alpha_{1,S(-II)}} - \frac{5}{6}pH$	(23.93)
$\frac{S(0)}{S(-II)_T}$	$S_{(s)} + 2H^+ + 2e^- = H_2S \quad \log K = 4.8$	(23.94)
	$pe = 2.4 + \frac{1}{2}\log\frac{\{S_{(s)}\}}{\{H_2S\}} - pH$	(23.95)
	For unit γ values and $\{S_{(s)}\} = 1$, we have	
	$pe = 2.4 - \frac{1}{2}\log\alpha_{0,S(-II)}S(-II)_T - pH$	(23.96)
	For $S_T = 10^{-3}$ M, and assuming that $S(-II)_T = S_T$, Eq. (23.86) becomes	
	$pe = 3.9 - \frac{1}{2}\log\alpha_{0,S(-II)} - pH$	(23.97)

Note that as required for consistency of the thermodynamic data used for these lines, Eqs. (23.77), (23.89), and (23.97) all intersect at a single point, and Eqs. (23.81), (23.93), and (23.97) in all intersect at a single point.

Along the $S_{(s)}$ wedge boundaryline, the concentrations of each of the species in Eq. (23.98) will be governed by equilibrium with $S_{(s)}$. Eq. (23.100) simplifies the computations to the point that we need to substitute in for just $[HSO_4^-]$ and $[H_2S]$ as controlled by $S_{(s)}$. Therefore, from Table 23.8, at 25°C/1 atm, we have

$$\frac{1}{10^{34.2}[H^+]^7\{e^-\}^6}\left(1 + \frac{10^{-2.0}}{[H^+]}\right) + 10^{4.8}[H^+]^2\{e^-\}^2\left(1 + \frac{10^{-7.0}}{[H^+]}\right) = S_T. \qquad (23.101)$$

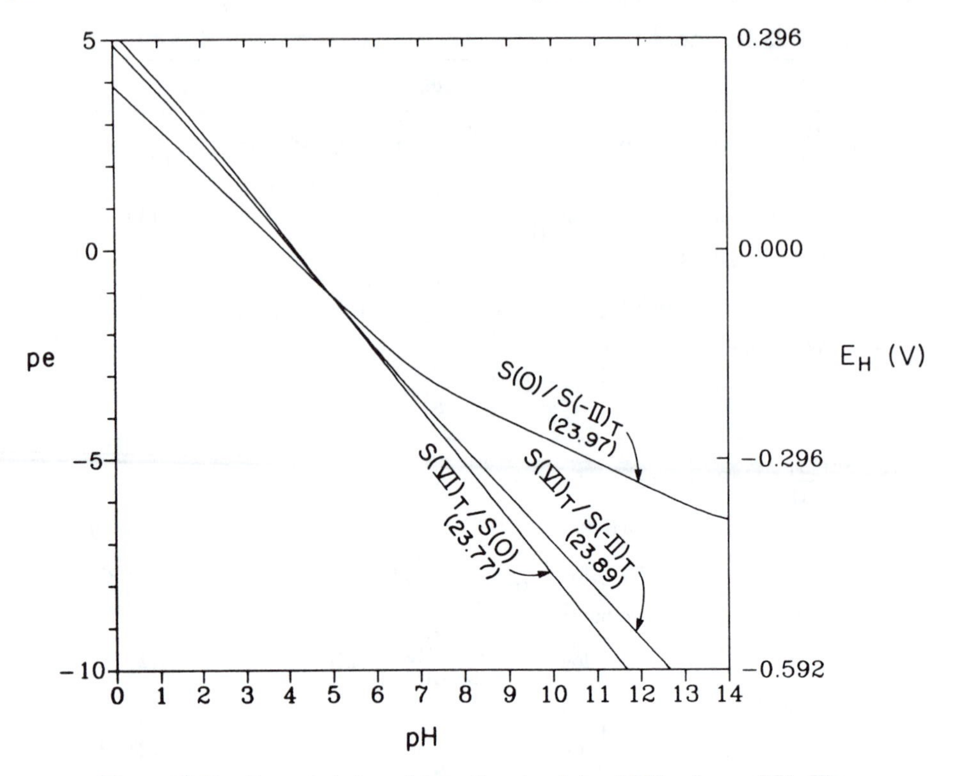

Figure 23.11 Expanded view of three lines involving $S(VI)_T$, $S_{(s)}$, and $S(-II)_T$ for $S_T = 10^{-3}$ *M* and 25°C/1 atm. Activity corrections are neglected. The solution/$S_{(s)}$ lines (i.e., lines (23.77) and (23.97)) were drawn assuming that: 1) $S(VI)_T$ and $S(-II)_T$, respectively, are present at 10^{-3} *M*; and 2) $S_{(s)}$ is present at a mole fraction of 1. This diagram indicates that $S_{(s)}$ has a wedge-shaped predominance region. The wedge will not be given exactly by the portions of lines (23.77) and (23.97) that are left of the intersection due to the fact that those lines were drawn using assumptions. The actual $S_{(s)}$ wedge will be somewhat thinner as well as blunted. To the right of the $S_{(s)}$ wedge, there will be a $S(VI)_T/S(-II)_T$ boundaryline.

We note now that the predominance wedge for $S_{(s)}$ is angled downward. This means that for the pH range over which the wedge extends, for each value of $[H^+]$, Eq. (23.101) will have two solutions for $\{e^-\}$. One of these $\{e^-\}$ values represents a high pe root to Eq. (23.101), and the other will be a low pe root. As hinted above, at $S_T = 10^{-3}$ *M*, the maximum pH to which the wedge extends is pH = 4.40.

$S(-II)_T$ (i.e., $H_2S + HS^- + S^{2-}$): Since $S(-II)_T$ is the most reduced form of sulfur under consideration, its predominance region will occupy all of the lower portion of the pe-pH diagram. For reasons given earlier, the upper boundary of that region will be given at low pH by the boundary with the $S_{(s)}$ wedge, then by line (23.89) for pH $\geq$ 4.40.

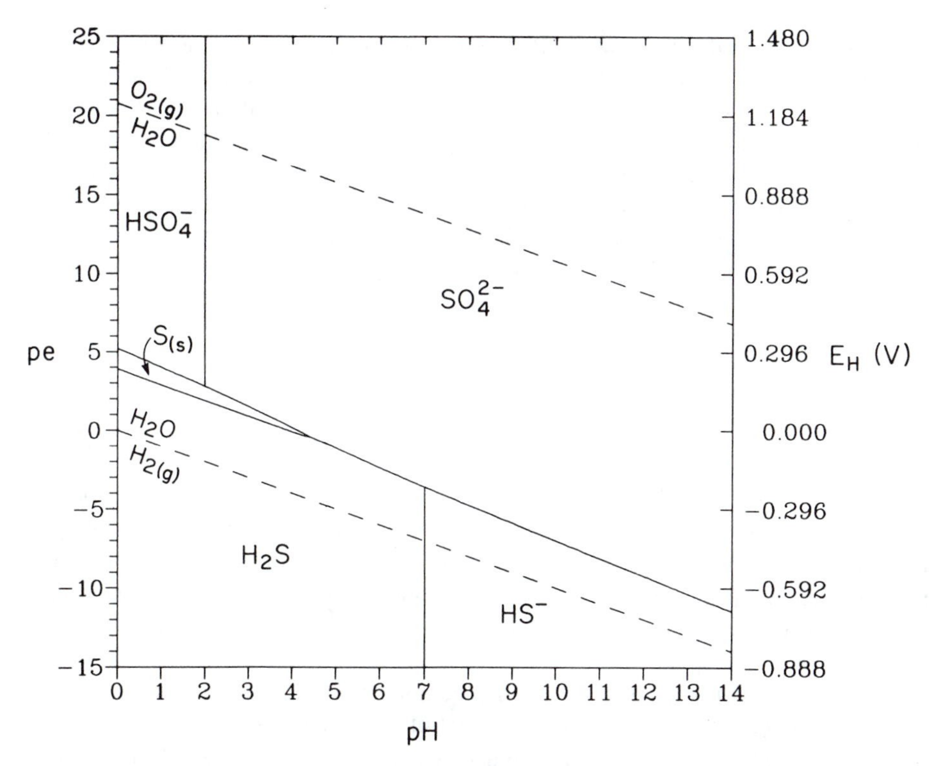

Figure 23.12 pe-pH diagram for aqueous sulfur when $S_T = 10^{-3}$ *M* and 25°C/1 atm. Activity corrections are neglected. $S_{(s)}$ is only possible at relatively low pH. As the pe is lowered at any given pH value, the diagram indicates that S(−II) (either as H_2S or as HS^-) can become the dominant form of sulfur before the in-situ p_{H_2} can reach 1 atm.

23.3.3.3 Finalizing the pe-pH Diagram for Aqueous Sulfur When $S_T = 10^{-3}$ *M*.

Figure 23.12 gives the final pe-pH diagram for $S_T = 10^{-3}$ *M* and 25°C/1 atm. At equilibrium, the concentrations of S(IV) species are extremely low (though never zero) everywhere in the pe-pH domain. Equation (23.101) is plotted in Figure 23.12 as the blunt wedge defining the $S_{(s)}$ region. While it does not actually seem very blunt, inspection at a finer scale would reveal its rounded nature. As discussed above, the degree of blunting makes the maximum pH of the wedge be pH = 4.40 rather than the pH of 5.05 that is obtained as the intersection of lines (23.77) and (23.89). Lines at pH = 2.0 and pH = 7.0 provide the HSO_4^-/SO_4^{2-} and H_2S/HS^- boundarylines, respectively. The $S(VI)_T/S(-II)_T$ line given by Eq. (23.89) in Figures 23.10 and 23.11 is valid over its entire length. However, in a pe-pH diagram like Figure 23.12, the convention is to plot only the portion that is outside of the $S_{(s)}$ wedge. This convention is analogous to that governing the HSO_4^-/SO_4^{2-} and H_2S/HS^- boundarylines. In particular, at 25°C/1 atm, while pH = 2.0 and pH = 7.0 define $[HSO_4^-]/[SO_4^{2-}] = 1$ and $[H_2S]/[HS^-] = 1$ for all pe, we only plot those lines as far as they are needed to define the corresponding predominance regions.

23.3.3.4 Other Sulfur Species. Before proceeding, it is worthwhile pointing out that besides the species in Table 23.10, a whole host of additional dissolved sulfur species can form in natural waters. However, at equilibrium, those additional dissolved species will not predominate anywhere in the typical pe-pH diagram at 25°C/1 atm. Nevertheless, since redox equilibrium is more the exception than the rule in the environment, just as a great many carbon-containing compounds can be found in natural waters, the reader should be aware that many different sulfur species can also be found. As a result, the natural water chemistry of sulfur can be *exceedingly* complex. For example, a sulfur atom can add to S^{2-} to form the species S_2^{2-}:

$$S + S^{2-} = S_2^{2-}. \tag{23.102}$$

Similarly, a second S can add to the S_2^{2-} according to

$$S + S_2^{2-} = S_3^{2-}. \tag{23.103}$$

In general, species with the formula S_n^{2-} and $n \geq 2$ are referred to as polysulfides. The specific species S_2^{2-} is referred to as disulfide. A very well known solid containing S_2^{2-} is pyrite ("fool's gold") which has the formula $FeS_{2(s)}$. The average oxidation state of the sulfur atoms in the species S_n^{2-} is $-2/n$. Each polysulfide species is dibasic, and can pick up two protons leading to H_2S_n. Each such species will be characterized by a pK_1 and a pK_2.

A few of the many other species that complicate sulfur chemistry include species like dithionate ($S_2O_4^{2-}$) which may be thought of as: 1) being the result of the addition of an S atom to SO_4^{2-}; and 2) having an average sulfur oxidation state of +6/2 = III. A large group of anions called the polythionites, are also possible. These species have the general formula $S_nO_6^{2-}$, where n can range up to 22 or higher (Cotton and Wilkinson, 1988). The average sulfur oxidation state in the polythionites is $+10/n$.

23.3.4 pe-pH Diagrams for Iron

23.3.4.1 $Fe_T = 10^{-5}$ *M*, No Carbonate Carbon.

23.3.4.1.1. General Comments. The manner in which pe-pH diagrams for aqueous iron are prepared is similar to that discussed in Chapters 21 and 22 for aqueous lead. In particular, as with aqueous lead, with aqueous iron we: 1) need to consider the possibility of hydroxide, oxide, and carbonate solids; 2) do not need to consider disproportionation reactions between oxidation states that have predominance regions;[8]

[8]With regard to this second point, in the case of aqueous lead, Pb(III) is an unstable oxidation state that will disproportionate to Pb(IV) and Pb(II). However, since Pb(III) cannot predominate over any portion of the pH range, as far as the mechanics of preparing a Pb pe-pH diagram are concerned, there is a uniform transition from oxidation state IV to II to 0 for all pH. Likewise, with aqueous iron, since Fe(I) cannot predominate over any portion of the pH range, we will see that there is a uniform transition from oxidation state III to II to 0 over all pH. In contrast, with sulfur for example, we need to consider disproportionation between oxidation states that have predominance regions. Indeed, when the level of S_T is high enough to permit a region for $S_{(s)}$, the pe-pH diagram for sulfur is complicated by a transition from oxidation state VI to 0 to −II at low pH, then from VI directly to −II at higher pH; the disproportionation of oxidation state 0 is favored at high pH.

and 3) may neglect the presence of any dissolved metal in elemental (i.e., zero oxidation state) form.

In the absence of any carbonate carbon, and over the time frame of interest, let us assume that $(am)Fe(OH)_{3(s)}$ and $Fe(OH)_{2(s)}$ are the only possible Fe(III) and Fe(II) solids, respectively. Under these conditions, the various forms of Fe that may have predominance regions in a pe-pH diagram for iron are given in Table 23.11. The dimer $Fe_2(OH)_2^{4+}$ has not been included because it will not be present at significant concentrations when Fe_T is low (e.g., $Fe_T = 10^{-3.5}$ *M* or less; see Chapter 18). Table 23.11 indicates that in contrast to aqueous lead, the solubility of the most oxidized form of iron cannot be neglected. That is, while we could neglect the presence of dissolved Pb(IV) species over the pH range 0 to 14, dissolved Fe(III) can be very significant over that range. Table 23.12 summarizes the redox reactions for iron at 25°C/1 atm, and Table 23.13 summarizes other important equilibrium constants for both Fe(III) and Fe(II).

Figure 23.13 is a generic pe-pH diagram for aqueous iron that has been drawn based on the information in Table 23.11 and the knowledge that disproportionation reactions among the oxidation states III, II, and 0 are not significant. It has been assumed that Fe_T is high enough that both $(am)Fe(OH)_{3(s)}$ and $Fe(OH)_{2(s)}$ can have predominance regions.

TABLE 23.11 Important forms of iron in natural waters.

oxidation state	species
III	$(am)Fe(OH)_{3(s)}$, Fe^{3+}, $FeOH^{2+}$, $Fe(OH)_2^+$, $Fe(OH)_4^-$
II	$Fe(OH)_{2(s)}$, Fe^{2+}, $FeOH^+$, $Fe(OH)_2^o$, $Fe(OH)_3^-$
0	$Fe_{(s)}$

TABLE 23.12 Data for selected redox reactions involving iron at 25°C/1 atm and infinite dilution in order of increasing pe° and E_H^o.

Reduction Half-Reaction	log *K*	pe°	pe°(W)	E_H^o (= 0.05916 pe°)
$FeCO_{3(s)} + 2e^- = Fe_{(s)} + CO_3^{2-}$	−25.58	−12.79	−12.79	−0.76
$Fe^{2+} + 2e^- = Fe_{(s)}$	−14.9	−7.45	−7.45	−0.44
$Fe^{3+} + e^- = Fe^{2+}$	13.0	13.0	13.0	0.77
$\alpha-FeOOH_{(s)} + HCO_3^- + 2H^+ + e^- = FeCO_{3(s)} + 2H_2O$	13.15	13.15	−0.85	0.78
$\alpha-FeOOH_{(s)} + 3H^+ + e^- = Fe^{2+} + 3H_2O$	13.5	13.5	−7.5	0.80
$(am)Fe(OH)_{3(s)} + 3H^+ + e^- = Fe^{2+} + 3H_2O$	16.2	16.2	−4.8	0.96
$Fe^{3+} + e^- + CO_3^{2-} = FeCO_{3(s)}$	23.68	23.68	23.68	1.40

TABLE 23.13 Important equilibrium constants for Fe(II) and Fe(III) at 25°C/1 atm and infinite dilution.

Complexation or Dissolution Reaction	log *K*
$Fe^{2+} + H_2O = FeOH^+ + H^+$	$\log {}^*\beta_1 = -9.5$
$Fe^{2+} + 2H_2O = Fe(OH)_2^o + 2H^+$	$\log {}^*\beta_2 = -20.6$
$Fe^{2+} + 3H_2O = Fe(OH)_3^- + 3H^+$	$\log {}^*\beta_3 = -32.0$
$Fe(OH)_{2(s)} + 2H^+ = Fe^{2+} + 2H_2O$	$\log {}^*K_{s0} = 12.9$
$Fe(OH)_{2(s)} + H^+ = FeOH^+ + H_2O$	$\log {}^*K_{s1} = 3.4$
$Fe(OH)_{2(s)} = Fe(OH)_2^o$	$\log K_{s2} = -7.7$
$Fe(OH)_{2(s)} + H_2O = Fe(OH)_3^- + H^+$	$\log {}^*K_{s3} = -19.1$
$FeCO_{3(s)} = Fe^{2+} + CO_3^{2-}$	$\log K_{s0} = -10.68$
$Fe^{3+} + H_2O = FeOH^{2+} + H^+$	$\log {}^*\beta_1 = -2.2$
$Fe^{3+} + 2H_2O = Fe(OH)_2^+ + 2H^+$	$\log {}^*\beta_2 = -5.7$
$Fe^{3+} + 4H_2O = Fe(OH)_4^- + 4H^+$	$\log {}^*\beta_4 = -21.6$
$(am)Fe(OH)_{3(s)} + 3H^+ = Fe^{3+} + 3H_2O$	$\log {}^*K_{s0} = 3.2$
$(am)Fe(OH)_{3(s)} + 2H^+ = FeOH^{2+} + 2H_2O$	$\log {}^*K_{s1} = 1.0$
$(am)Fe(OH)_{3(s)} + H^+ = Fe(OH)_2^+ + H_2O$	$\log {}^*K_{s2} = -2.5$
$(am)Fe(OH)_{3(s)} + H_2O = Fe(OH)_4^- + H^+$	$\log {}^*K_{s4} = -18.4$

On both the left- and right-hand sides of the diagram, the solution phase has been divided into two regions labeled simply as $Fe(III)_T$ and $Fe(II)_T$. Once we decide what level of Fe_T is of interest, and once we then locate the various solution/solid boundarylines, we will be able to subdivide those regions into regions labeled as Fe^{3+}, $FeOH^{2+}$, etc., and Fe^{2+}, $FeOH^+$, etc., as necessary. As usual, for the purposes of this generic diagram, all of the boundarylines have been drawn as simple straight lines with either zero or infinite slope.

As indicated in Figure 23.13, lines 1, 3, and 5 will intersect at a single point since it is the same solution phase region that is located to the left of line 1, line 3, and the leftmost point of line 5. However, there is no a priori reason why the rightmost end of line 9 should intersect exactly at the 1/3/5 intersection; thus, it is likely that line 9 will end either above or below that intersection. We have anticipated reality in Figure 23.13 by drawing line 9 above the 1/3/5 intersection. For analogous reasons, we can conclude: 1) that lines 4, 5, and 7 will intersect at a single point; but 2) that there is no a priori reason why the leftmost end of line 10 should intersect exactly at the line 4/5/7 intersection; thus, it is likely that the leftmost end of line 10 will be located either above or below that intersection. We have anticipated reality in Figure 23.13 by drawing line 10 below the 4/5/7 intersection.

23.3.4.1.2. Solution/(am)Fe(OH)$_{3(s)}$ Boundaryline. We first consider the solution/(am)$Fe(OH)_{3(s)}$ boundaryline. Lines 1 and 7 in the generic diagram Figure 23.13 represent different portions of a single, curved line. When we developed the equation for the solution/$PbO_{2(s)}$ boundaryline in Chapter 21, we only included Pb(II) in the equation for Pb_T along that boundaryline. In this case, since Fe(III) has significant solubility at

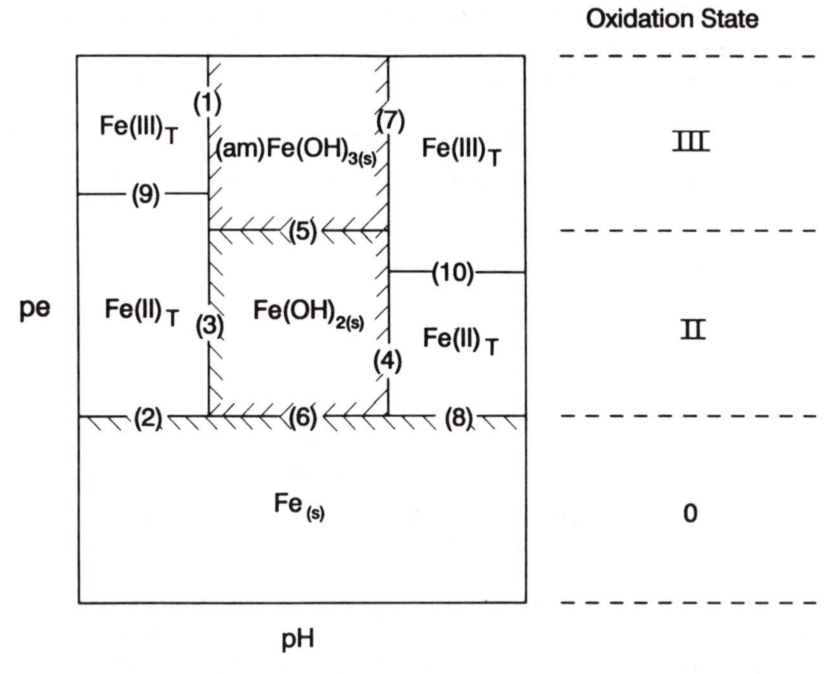

Figure 23.13 Generic pe-pH diagram for aqueous iron for $Fe_T = 10^{-5}$ M, no carbonate carbon, and 25°C/1 atm. The transition in dominance between any two adjacent oxidation states occurs over a range of pe values, and not at a single pe as might be interpreted based on the nature of the right hand side of the figure.

both low and high pH, we will need to include both Fe(II) *and* Fe(III) species when developing the equation for the solution/(am)$Fe(OH)_{3(s)}$ boundaryline. Thus, for that boundaryline, we have

$$\mathrm{Fe(III)_T} + \mathrm{Fe(II)_T} = \mathrm{Fe_T}. \tag{23.104}$$

Neglecting the Fe(III) species $Fe(OH)_3^o$ as well as the dimer $Fe_2(OH)_2^{4+}$, Eq. (23.104) becomes

$$[\mathrm{Fe^{3+}}] + [\mathrm{FeOH^{2+}}] + [\mathrm{Fe(OH)_2^+}] + [\mathrm{Fe(OH)_4^-}]$$
$$+[\mathrm{Fe^{2+}}] + [\mathrm{FeOH^+}] + [\mathrm{Fe(OH)_2^o}] + [\mathrm{Fe(OH)_3^-}] = \mathrm{Fe_T}. \tag{23.105}$$

Based on the information in Table 23.12, neglecting activity corrections, Eq. (23.105) can be rewritten

$$[\mathrm{Fe^{3+}}]\left(1 + \frac{{}^*\beta_1}{[\mathrm{H^+}]} + \frac{{}^*\beta_2}{[\mathrm{H^+}]^2} + \frac{{}^*\beta_4}{[\mathrm{H^+}]^4}\right) + [\mathrm{Fe^{2+}}]\left(1 + \frac{{}^*\beta_1}{[\mathrm{H^+}]} + \frac{{}^*\beta_2}{[\mathrm{H^+}]^2} + \frac{{}^*\beta_3}{[\mathrm{H^+}]^3}\right) = \mathrm{Fe_T} \tag{23.106}$$

where the values of the various $^*\beta$ values will depend on whether we are referring to

Fe(III) or Fe(II); the $^*\beta$ values in the first half of Eq. (23.106) refer to Fe(III), and the $^*\beta$ values in the second half refer to Fe(II).

Since we have equilibrium with the solid (am)$Fe(OH)_{3(s)}$ along the line of interest,

$$[Fe^{3+}] = {}^*K_{s0}[H^+]^3. \qquad (23.107)$$

For redox equilibrium between Fe^{3+} and Fe^{2+}, neglecting activity corrections, we have

$$[Fe^{2+}] = 10^{13.0}10^{-pe}[Fe^{3+}]. \qquad (23.108)$$

Combining Eqs. (23.107) and (23.108) with Eq. (23.106), we obtain

$$^*K_{s0}[H^+]^3\left(1 + \frac{^*\beta_1}{[H^+]} + \frac{^*\beta_2}{[H^+]^2} + \frac{^*\beta_4}{[H^+]^4}\right)$$
$$+ 10^{13.0}10^{-pe}\,{}^*K_{s0}[H^+]^3\left(1 + \frac{^*\beta_1}{[H^+]} + \frac{^*\beta_2}{[H^+]^2} + \frac{^*\beta_3}{[H^+]^3}\right) = Fe_T. \qquad (23.109)$$

Rearranging Eq. (23.109) to give pe as a function of pH, we obtain

$$pe = -\log\frac{Fe_T - {}^*K_{s0}[H^+]^3(1 + {}^*\beta_1/[H^+] + {}^*\beta_2/[H^+]^2 + {}^*\beta_4/[H^+]^4)}{10^{13.0}\,{}^*K_{s0}[H^+]^3(1 + {}^*\beta_1/[H^+] + {}^*\beta_2/[H^+]^2 + {}^*\beta_3/[H^+]^3)}. \qquad (23.110)$$

The $^*K_{s0}$ in Eq. (23.110) refers to (am)$Fe(OH)_{3(s)}$; the $^*\beta$ values in the numerator refer to Fe(III); and the $^*\beta$ values in the denominator refer to Fe(II). Equation (23.110) can be plotted for any value of Fe_T larger than the minimum solubility of (am)$Fe(OH)_{3(s)}$. It is plotted in Figure 23.14 for $Fe_T = 10^{-5}$ *M*. The pH range for Figure 23.14 extends to pH = 16 so that the behavior of some of the other lines in the figure can be examined fully.

We note that for any value of Fe_T larger than the minimum solubility of (am) $Fe(OH)_{3(s)}$, the argument of the logarithm in Eq. (23.110) will become negative for some low pH value and for some high pH value. When this happens, the log term becomes undefined. This means that at these two pH values, the slopes of the *upper portions* of lines 1 and 7 become infinite and the lines become vertical; the slope of line 1 becomes infinitely negative, and the slope of line 7 becomes infinitely positive. For $Fe_T = 10^{-5}$ *M*, the two pH values are 3.09 and 13.40. These two pH values correspond to the low and high pH roots of the equation $Fe(III)_T = Fe_T$. Thus, at very high pe values where the slopes are infinite, the various Fe(II) species are negligible, and the two solution/(am)$Fe(OH)_{3(s)}$ lines are given by the simple lines pH = 3.09 and pH = 13.40.

23.3.4.1.3. Solution $Fe(III)_T$/Solution $Fe(II)_T$ Boundaryline. We now consider the boundaryline between the $Fe(III)_T$ and $Fe(II)_T$ regions. Lines 9 and 10 in the generic diagram Figure 23.13 represent different portions of this same line. For this boundaryline, we have

$$Fe(III)_T = Fe(II)_T. \qquad (23.111)$$

Neglecting the Fe(III) species $Fe(OH)_3^o$ as well as the dimer $Fe_2(OH)_2^{4+}$, Eq. (23.111) becomes

$$[Fe^{3+}] + [FeOH^{2+}] + [Fe(OH)_2^+] + [Fe(OH)_4^-]$$
$$= [Fe^{2+}] + [FeOH^+] + [Fe(OH)_2^o] + [Fe(OH)_3^-]. \qquad (23.112)$$

Neglecting activity corrections, Eq. (23.112) can be rewritten

$$[Fe^{3+}]\left(1+\frac{{}^*\beta_1}{[H^+]}+\frac{{}^*\beta_2}{[H^+]^2}+\frac{{}^*\beta_4}{[H^+]^4}\right) = [Fe^{2+}]\left(1+\frac{{}^*\beta_1}{[H^+]}+\frac{{}^*\beta_2}{[H^+]^2}+\frac{{}^*\beta_3}{[H^+]^3}\right). \quad (23.113)$$

The ${}^*\beta$ values on the LHS of Eq. (23.113) refer to Fe(III); the ${}^*\beta$ values on the RHS refer to Fe(II).

Since we will always have redox equilibrium between Fe^{3+} and Fe^{2+} throughout the pe-pH diagram, neglecting activity corrections, Eq. (23.108) again applies. Combining Eq. (23.108) with (23.113), we obtain

$$1+\frac{{}^*\beta_1}{[H^+]}+\frac{{}^*\beta_2}{[H^+]^2}+\frac{{}^*\beta_4}{[H^+]^4} = 10^{13.0}10^{-pe}\left(1+\frac{{}^*\beta_1}{[H^+]}+\frac{{}^*\beta_2}{[H^+]^2}+\frac{{}^*\beta_3}{[H^+]^3}\right). \quad (23.114)$$

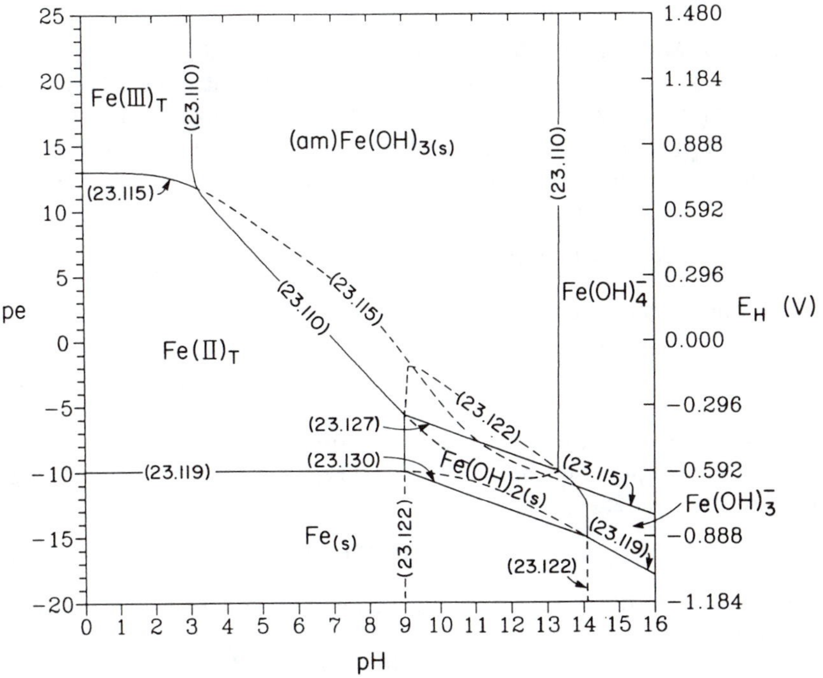

Figure 23.14 pe-pH diagram for aqueous iron for $Fe_T = 10^{-5}$ *M*, no carbonate carbon, and 25°C/1 atm. Activity corrections are neglected. The various lines are drawn as solid lines over the portions which will appear in an actual pe-pH diagram, and as dashed lines everywhere else. Although the $Fe(III)_T/Fe(II)_T$ line is dashed when it passes through the $(am)Fe(OH)_{3(s)}$ region, that dashed portion is still physically meaningful. The same is not true for the other dashed lines. Though not environmentally relevant, the diagram is extended to pH = 16 so that a portion of all of the mathematically-possible predominance regions can be seen.

Rearranging Eq. (23.114) to give pe as a function of pH, we obtain

$$pe = 13.0 + \log \frac{1 + {}^*\beta_1/[H^+] + {}^*\beta_2/[H^+]^2 + {}^*\beta_3/[H^+]^3}{1 + {}^*\beta_1/[H^+] + {}^*\beta_2/[H^+]^2 + {}^*\beta_4/[H^+]^4}. \quad (23.115)$$

The ${}^*\beta$ values in the numerator refer to Fe(II); and the ${}^*\beta$ values in the denominator refer to Fe(III).

Equation (23.115), which is plotted in Figure 23.14, is independent of Fe_T. If we had included the dimer $Fe_2(OH)_2^{4+}$ in our considerations, then this boundaryline would not have been independent of Fe_T. At low pH, where $Fe(III)_T \simeq [Fe^{3+}]$ and $Fe(II)_T \simeq [Fe^{2+}]$, the log term in Eq. (23.115) is zero, and the $Fe(III)_T/Fe(II)_T$ boundaryline occurs at pe $\simeq$ 13.0. At very high pH, where $Fe(III)_T \simeq [Fe(OH)_4^-]$ and $Fe(II)_T \simeq [Fe(OH)_3^-]$, the log term will be given approximately by $\log({}^*\beta_3[H^+]/{}^*\beta_4)$ and the boundaryline will be given by $pe \simeq 13.0 + \log({}^*\beta_3/{}^*\beta_4) - pH$; the species $Fe(OH)_4^-$ is favored over $Fe(OH)_3^-$ as the pH increases because of the difference in the number of complexing hydroxides between the two species.

23.3.4.1.4. Solution/$Fe_{(s)}$ Boundaryline. Lines 2 and 8 in the generic pe-pH diagram in Figure 23.13 are parts of the same solution/$Fe_{(s)}$ boundaryline. Strictly speaking, for this boundaryline, we need to again apply Eqs. (23.104)–(23.106). However, elemental iron ($Fe_{(s)}$) is an extremely reduced material. Therefore, along the solution/$Fe_{(s)}$ boundaryline, there will be essentially no error in neglecting the Fe(III) species in Eqs. (23.104)–(21.106). Thus, along this boundaryline, we let

$$[Fe^{2+}] + [FeOH^+] + [Fe(OH)_2^o] + [Fe(OH)_3^-] = Fe_T. \quad (23.116)$$

Neglecting activity corrections, we have

$$[Fe^{2+}]\left(1 + \frac{{}^*\beta_1}{[H^+]} + \frac{{}^*\beta_2}{[H^+]^2} + \frac{{}^*\beta_3}{[H^+]^3}\right) = Fe_T \quad (23.117)$$

where the values of the various ${}^*\beta$ values refer to Fe(II).

Since there is equilibrium between Fe^{2+} and $Fe_{(s)}$ along the boundaryline, neglecting activity corrections, at 25°C/1 atm, we have

$$[Fe^{2+}] = 10^{14.9}10^{2pe}. \quad (23.118)$$

Substituting Eq. (23.118) into Eq. (23.117) and rearranging to obtain pe as a function of pH, we have

$$pe = \frac{1}{2}\log \frac{10^{-14.9}Fe_T}{1 + {}^*\beta_1/[H^+] + {}^*\beta_2/[H^+]^2 + {}^*\beta_3/[H^+]^3}. \quad (23.119)$$

Equation (23.119) will give pe vs. pH for any value of Fe_T. Equation (23.119) is plotted in Figure 23.14 for $Fe_T = 10^{-5}$ *M*. At low pH where the denominator is essentially 1.0, $pe \simeq \frac{1}{2}\log 10^{-14.9}Fe_T$; for $Fe_T = 10^{-5}$, this gives pe = −9.95. Thus, as noted above, the solution/$Fe_{(s)}$ line occurs at very low pe values, and so we can indeed expect to be able to neglect dissolved Fe(III) species under such conditions. As the pH increases, the pe trends downward along line (23.119) because Fe^{2+} begins to represent a smaller and smaller fraction of Fe_T, and so a higher electron activity is required to precipitate $Fe_{(s)}$.

23.3.4.1.5. Solution/$Fe(OH)_{2(s)}$ Boundaryline. In Section 23.3.4.1.2, we discussed that lines 1 and 7 were parts of the same solution phase/(am)$Fe(OH)_{3(s)}$ boundaryline. For that line, equilibrium with (am)$Fe(OH)_{3(s)}$ was used to develop the pe-pH relationship between the dissolved Fe_T and the solid phase. For completely analogous reasons, lines 3 and 4 are parts of the same, solution phase/$Fe(OH)_{2(s)}$ boundaryline. In this case, equilibrium with $Fe(OH)_{2(s)}$ will be used to develop the pe-pH relationship between the dissolved Fe_T and the solid phase.

When we developed the boundarylines between the solution and the intermediate oxidation state solid $PbO_{(s)}$ in Chapter 21, we solved an expression (Eq. (21.9)) that equated the total dissolved Pb(II) with the Pb_T for the diagram. The two boundarylines represented the low and high pH roots of a single equation. In discussions in this chapter, the same approach was found to apply to the solution/(am)$Fe(OH)_{3(s)}$ boundaries at high pe where the Fe(II) species can be neglected. However, since both Fe(II) and Fe(III) species need to be considered at intermediate pe when developing the solution/(am)$Fe(OH)_{3(s)}$ boundaryline, we can expect that same will be true for intermediate pe for the solution/$Fe(OH)_{2(s)}$ boundaryline. Indeed, the upper portions of lines 3 and 4 extend to pe-pH domains where the solid (am)$Fe(OH)_{3(s)}$ begins to be present and $Fe(OH)_{2(s)}$ disappears. Thus, we will again utilize Eqs. (23.104)–(23.106).

Since we now have equilibrium with $Fe(OH)_{2(s)}$, then along the solution/$Fe(OH)_{2(s)}$ line we have

$$[Fe^{2+}] = {}^*K_{s0}[H^+]^2. \qquad (23.120)$$

For redox equilibrium between Fe^{3+} and Fe^{2+}, neglecting activity corrections, Eq. (23.108) once again applies. Combining Eqs. (23.108) and (23.120) with Eq. (23.106), we obtain

$$10^{-13}10^{pe}\,{}^*K_{s0}[H^+]^2\left(1+\frac{{}^*\beta_1}{[H^+]}+\frac{{}^*\beta_2}{[H^+]^2}+\frac{{}^*\beta_4}{[H^+]^4}\right) + {}^*K_{s0}[H^+]^2\left(1+\frac{{}^*\beta_1}{[H^+]}+\frac{{}^*\beta_2}{[H^+]^2}+\frac{{}^*\beta_3}{[H^+]^3}\right) = Fe_T. \qquad (23.121)$$

Rearranging Eq. (23.121) to give pe as a function of pH, we obtain

$$pe = \log\frac{(Fe_T - {}^*K_{s0}[H^+]^2(1 + {}^*\beta_1/[H^+] + {}^*\beta_2/[H^+]^2 + {}^*\beta_3/[H^+]^3))}{10^{-13.0}\,{}^*K_{s0}[H^+]^2(1 + {}^*\beta_1/[H^+] + {}^*\beta_2/[H^+]^2 + {}^*\beta_4/[H^+]^4)}. \qquad (23.122)$$

The ${}^*K_{s0}$ in Eq. (23.122) refers to $Fe(OH)_{2(s)}$; the ${}^*\beta$ values in the numerator refer to Fe(II); and the ${}^*\beta$ values in the denominator refer to Fe(III). There is much similarity between Eqs. (23.110) and (23.122).

Equation (23.122) can be plotted for any value of Fe_T larger than the minimum solubility of $Fe(OH)_{2(s)}$. It is plotted for $Fe_T = 10^{-5}$ *M* in Figure 23.14. As with Eq. (23.110), for any such value of Fe_T, the argument of the logarithm in Eq. (23.122) will become negative for some low pH and some high pH, and the log term will become undefined. At these two pH values, the slopes of lines 3 and 4 become infinite; the slope of line 3 becomes infinitely positive, and the slope of line 4 becomes infinitely negative. For $Fe_T = 10^{-5}$ *M*, the two pH values are 9.01 and 14.10. These two values correspond to the low and high pH roots of the equation $Fe(II)_T = Fe_T$. Thus, at very low pe where the slopes are infinite, the various Fe(III) species are negligible, and the two solution/$Fe(OH)_{2(s)}$ lines are approximated by the simple lines pH = 9.01 and pH = 13.40.

23.3.4.1.6.* (am)$\mathbf{Fe(OH)_{3(s)}/Fe(OH)_{2(s)}}$ *Boundaryline. Since the three solids (am)$Fe(OH)_{3(s)}$, $Fe(OH)_{2(s)}$, and $Fe_{(s)}$ will all be present in the pe-pH diagram, we need to develop the equations for the (am)$Fe(OH)_{3(s)}/Fe(OH)_{2(s)}$ and $Fe(OH)_{2(s)}/Fe_{(s)}$ boundarylines. For the former, we combine

$$\begin{array}{llll} & (am)Fe(OH)_{3(s)} + 3H^+ = Fe^{3+} + 3H_2O & \log K = 3.2 & (23.123) \\ + & Fe^{3+} + e^- = Fe^{2+} & \log K = 13.0 & (23.124) \\ - & Fe(OH)_{2(s)} + 2H^+ = Fe^{2+} + 2H_2O & \log K = 12.9 & (23.125) \\ \hline & (am)Fe(OH)_{3(s)} + H^+ + e^- = Fe(OH)_{2(s)} + H_2O & \log K = 3.3. & (23.126) \end{array}$$

Equation (23.126) gives

$$pe = 3.3 - pH \tag{23.127}$$

as the (am)$Fe(OH)_{3(s)}/Fe(OH)_{2(s)}$ boundaryline at 25°C/1 atm. Equation (23.127) is plotted in Figure 23.14, but only for the pH range over which the (am)$Fe(OH)_{3(s)}$ is present. That range is given by the two intersections of line (23.127) with line (23.110).

23.3.4.1.7.* $\mathbf{Fe(OH)_{2(s)}/Fe_{(s)}}$ *Boundaryline. For the $Fe(OH)_{2(s)}/Fe_{(s)}$ boundaryline, we combine

$$\begin{array}{llll} & Fe(OH)_{2(s)} + 2H^+ = Fe^{2+} + 2H_2O & \log K = 12.9 & (23.125) \\ + & Fe^{2+} + 2e^- = Fe_{(s)} & \log K = -14.9 & (23.128) \\ \hline & Fe(OH)_{2(s)} + 2H^+ + 2e^- = Fe_{(s)} + 2H_2O & \log K = -2.0. & (23.129) \end{array}$$

Equation (23.129) gives

$$pe = -1.0 - pH \tag{23.130}$$

as the $Fe(OH)_{2(s)}/Fe_{(s)}$ boundaryline at 25°C/1 atm. Equation (23.130) is plotted in Figure 23.14, but only for the pH range over which $Fe(OH)_{2(s)}$ is present. That range is given by the two intersections of line (23.130) with line (23.122).

23.3.4.1.8. Converting Figure 23.14 into the pe-pH Diagram for* $\mathbf{Fe_T = 10^{-5}}$ *M. Figure 23.14 contains all of the solution/solid boundarylines, the $Fe(III)_T/Fe(II)_T$ boundaryline, and the two solid/solid boundarylines. All of the lines describing a boundary with a solution have been drawn as solid lines in those regions where they will appear in the diagram, and as dashed lines everywhere else. Wherever a solution/solid line is dashed, the line has no physical meaning. However, along the portion that the $Fe(III)_T/Fe(II)_T$ line is dashed, that line still describes a physically-meaningful set of pe-pH points along which the dissolved $Fe(III)_T$ equals the dissolved $Fe(II)_T$.

In Figure 23.15, the dashed portions of the lines in Figure 23.14 have been deleted, and the various regions have been labelled. While a pH of 16 will not be physically

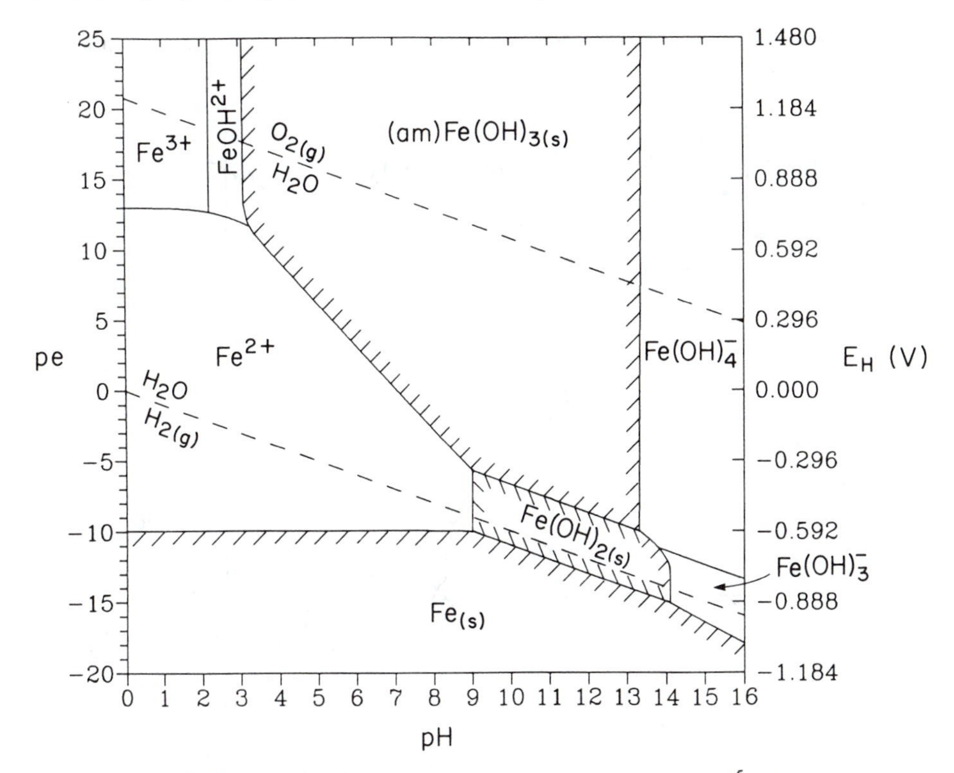

Figure 23.15 pe-pH diagram for aqueous iron for $Fe_T = 10^{-5}$ *M*, no carbonate carbon, and 25°C/1 atm. Activity corrections are neglected. Only the pertinent portions of the various lines are shown. As the pe is lowered at any given pH, the diagram indicates that when $Fe_T = 10^{-5}$ *M*, then the in-situ p_{H_2} will reach 1 atm before $Fe_{(s)}$ can form.

attainable,[9] we have retained this full range to allow us to better understand the chemical behavior in this system. On the LHS of the diagram, the $Fe(III)_T$ has been divided along the line pH = 2.2 because based on data from Table 23.13, that is the pH at which $[Fe^{3+}]/[FeOH^{2+}] = 1$ (assuming that activity corrections can be neglected). No boundary between $FeOH^{2+}$ and $Fe(OH)_2^+$ has been drawn because $[FeOH^{2+}]/[Fe(OH)_2^+] = 1$ at pH = 3.5, and this pH is to the right of the left side of the solution/(am)$Fe(OH)_{3(s)}$ line.

On the RHS of Figure 23.14, the $Fe(III)_T$ region has not been subdivided, but has been labelled simply as $Fe(OH)_4^-$. Indeed, since pH $\simeq$ 13.4 describes the right-hand portion of the solution/(am)$Fe(OH)_{3(s)}$ line, $Fe(OH)_4^-$ is by far the dominant Fe(III) species in this region. Similar arguments can be used to explain why the LHS and RHS $Fe(II)_T$ regions have been labelled simply as Fe^{2+} and $Fe(OH)_3^-$, respectively. If we return to our usual pH = 0 to 14 format, Figure 23.16 provides a partially obscured, but more realistic pe-pH diagram. The $Fe(OH)_3^-$ region is very small in that diagram.

[9]The assumption that activity corrections can be neglected even at pH = 14 is, moreover, extremely poor.

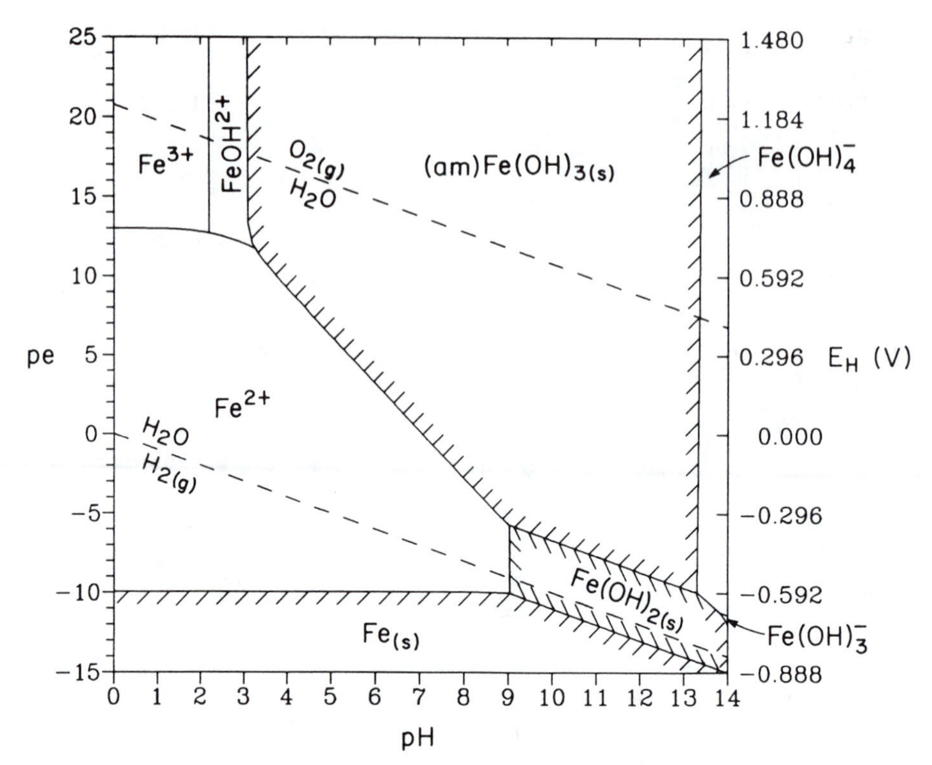

Figure 23.16 pe-pH diagram for aqueous iron for $Fe_T = 10^{-5}$ *M*, no carbonate carbon, and 25°C/1 atm. Activity corrections are neglected. The pH extends only to pH = 14; the species $Fe(OH)_3^-$ has only a very small predominance region in this diagram.

It is clear from Figure 23.16 that Fe(III) will be the dominant form of Fe in waters containing even small amounts of oxygen. Fe(II) can only be formed in significant amounts under rather reducing conditions. When $Fe_{(s)}$ is present, then unless $\{Fe^{2+}\}$ or the pH are rather high, the $Fe_{(s)}$ will be capable of reducing water to form hydrogen gas at p_{H_2} values greater than 1 atm.

23.3.4.2 $Fe_T = 10^{-5}M$ and $C_T = 10^{-3}M$.

23.3.4.2.1. General Comments. In the absence of any carbonate carbon, the preceding section revealed that $Fe(OH)_{2(s)}$ can form in a system in which $Fe_T = 10^{-5}$ *M*. When the amount of carbonate carbon is non-zero but still very low, then we can expect that $Fe(OH)_{2(s)}$ will remain the only possible Fe(II) solid. However, since Fe(II) can form a relatively insoluble metal carbonate, then as C_T increases, we can expect for some intermediate range of C_T values that both $Fe(OH)_{2(s)}$ and $FeCO_{3(s)}$ will have predominance regions. When C_T is sufficiently high, then $FeCO_{3(s)}$ will be the only Fe(II) solid that can possess a predominance region for $Fe_T = 10^{-5}$ *M*. As in the treatment given Pb in Chapter 22, we will assume that no reduction of CO_2-type carbon to CH_4 occurs, no matter how low the pe becomes. However, below the line given by Eq. (23.27), at full equilibrium, most of the dissolved C_T will be present at CH_4; the lines drawn assuming that $[CO_3^{2-}] = \alpha_2 C_T$ will be in error below that line. Adjustments for simultaneous changes in the redox chemistries of both iron and carbon are possible (e.g., see Garrels and Christ, 1965), and will be considered in a subsequent edition of this text.

Since we desire to prepare the pe-pH diagram for $C_T = 10^{-3}$ M, we need to determine whether there is a pH range over which $FeCO_{3(s)}$ is more stable than $Fe(OH)_{2(s)}$. If there is, we will then need to determine if $FeCO_{3(s)}$ prescribes a total dissolved iron value that is lower than 10^{-5} M. If it does, then $FeCO_{3(s)}$ will have a predominance region in the pe-pH diagram for $Fe_T = 10^{-5}$ M when $C_T = 10^{-3}$ M. This situation is analogous to that discussed in Section 22.2.2 where the relative stabilities of $PbO_{(s)}$, $PbCO_{3(s)}$, and $Pb_3(CO_3)_2(OH)_{2(s)}$ were considered as a function of pH for $C_T = 10^{-3}$ M. We proceed now in a similar manner here.

To probe the issue of solubility limitation by the two different Fe(II) solids, let us compute $[Fe^{2+}]$ as it is determined by the two different solids as: 1) a function of pH; 2) for the C_T of interest; and 3) assuming that regardless of the pH, this is being done at a pe value such that Fe(II) is the only stable oxidation state. Neglecting the presence of Fe(III) species will not change our conclusions regarding relative stability since the Fe(III) equilibria will not affect the values of $[Fe^{2+}]$ as prescribed by $Fe(OH)_{2(s)}$ and $FeCO_{3(s)}$.

Each of the two Fe(II) solids of interest will give a different function for $[Fe^{2+}]$ as determined by pH. For $Fe(OH)_{2(s)}$ and $FeCO_{3(s)}$, we will refer to these functions for $[Fe^{2+}]$ as $f([H^+])$ and $g([H^+])$, respectively. At any given pH, the function that gives the lowest value will be the one corresponding to the solubility-limiting solid. When two such functions cross at a given pH, then as discussed in Chapter 14, that pH value will be a stability transition pH. On one side of the transition pH, one of the two solids will be solubility limiting. On the other side, the other solid will be solubility limiting. At the transition pH, both solids could be present and solubility controlling, but only if the value of Fe_T that the solids prescribe is less than the value of Fe_T for which the pe-pH diagram is being constructed.[10]

Assuming unit activity coefficients in the aqueous phase and unit activities for all solids when present, we obtain the following results. At 25°C/1 atm, for $Fe(OH)_{2(s)}$, we have

$$[Fe^{2+}] = f([H^+]) = 10^{12.9}[H^+]^2 \tag{23.131}$$

$$\log f([H^+]) = 12.9 - 2\text{pH}. \tag{23.132}$$

At 25°C/1 atm, for $FeCO_{3(s)}$, we have

$$[Fe^{2+}] = g([H^+]) = \frac{10^{-10.68}}{[CO_3^{2-}]} \tag{23.133}$$

$$= \frac{10^{-10.68}}{\alpha_2 C_T} \tag{23.134}$$

[10]Although both solids *could* be present under these circumstances, all that these circumstances actually require is that one of the solids be present.

$$\log g([H^+]) = -10.68 - \log \alpha_2 - \log C_T. \tag{23.135}$$

Equations (23.132) and (23.135) are plotted in Figure 23.17. The functions $f([H^+])$ and $g([H^+])$ seen to intersect at pH = 10.06, a result which can be obtained by solving $f([H^+]) = g([H^+])$ by trial and error. $g([H^+])$ is lower than $f([H^+])$ for pH < 10.06. In the absence of any carbonate carbon, we recall from Section 23.3.4.1 that at $Fe_T = 10^{-5}$ M, $Fe(OH)_{2(s)}$ can be present at intermediate pe for some pH values greater than ~9. Since $FeCO_{3(s)}$ prescribes an even lower $[Fe^{2+}]$ than does $Fe(OH)_{2(s)}$ over the range ~9 to 10.06, we automatically know that at $Fe_T = 10^{-5}$ M, $FeCO_{3(s)}$ will have a predominance region for some pH values less than 10.06. The region will likely extend to pH values lower than 9. Since a pH of 10.06 is less than the upper pH limit of the $Fe(OH)_{2(s)}$ region in the absence of any carbonate carbon (pH $\simeq$ 14.1), for $C_T = 10^{-3}$ M, we know that $Fe(OH)_{2(s)}$ will retain its predominance in the range pH = 10.06 to ~14.1.

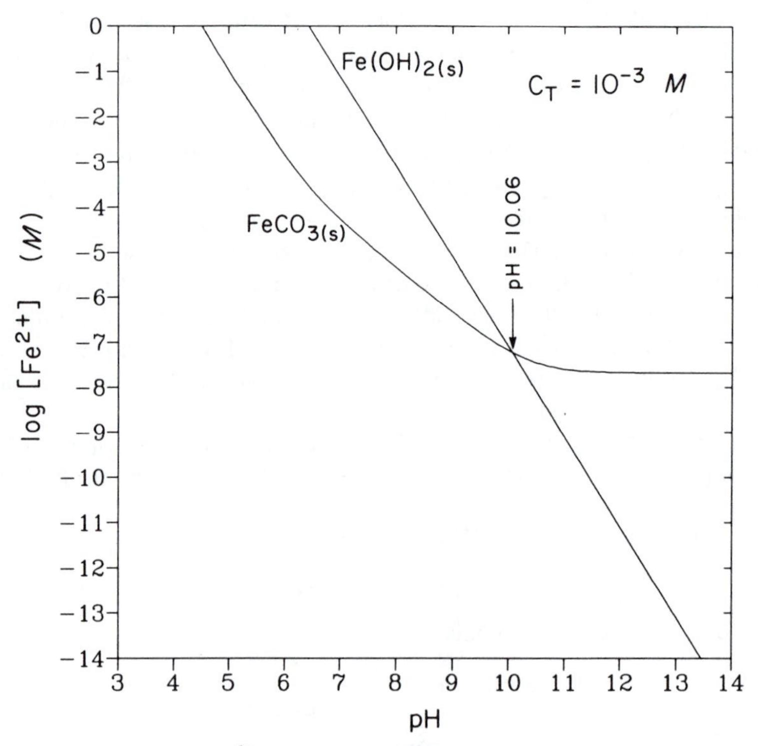

Figure 23.17 Log $[Fe^{2+}]$ with solubility control by $Fe(OH)_{2(s)}$, and with solubility control by $FeCO_{3(s)}$. Conditions are $C_T = 10^{-3}$ M and 25°C/1 atm. Activity corrections are neglected. $FeCO_{3(s)}$ has the lowest solubility for pH values less than 10.06, and $Fe(OH)_{2(s)}$ has the lowest solubility for pH values greater than 10.06.

On the basis of the above considerations, we conclude that a generic version of the pe-pH diagram for this case will appear as given in Figure 23.18. Lines 1, 2, 4, 5, 6, 7, 8, and 10 are all the same as developed for $C_T = 0$. Lines 3′, 5′, and 6′ are new, and equations need to be developed for them.

23.3.4.2.2. (am)Fe(OH)$_{3(s)}$/FeCO$_{3(s)}$ Boundaryline. For the (am)Fe(OH)$_{3(s)}$/FeCO$_{3(s)}$ boundaryline, we combine

	$(am)Fe(OH)_{3(s)} + 3H^+ = Fe^{3+} + 3H_2O$	$\log K = 3.2$	(23.123)
+	$Fe^{3+} + e^- = Fe^{2+}$	$\log K = 13.0$	(23.124)
−	$FeCO_{3(s)} = Fe^{2+} + CO_3^{2-}$	$\log K = -10.68$	(23.136)
	$(am)Fe(OH)_{3(s)} + 3H^+ + CO_3^{2-} + e^- = FeCO_{3(s)} + 3H_2O$	$\log K = 26.88.$	(23.137)

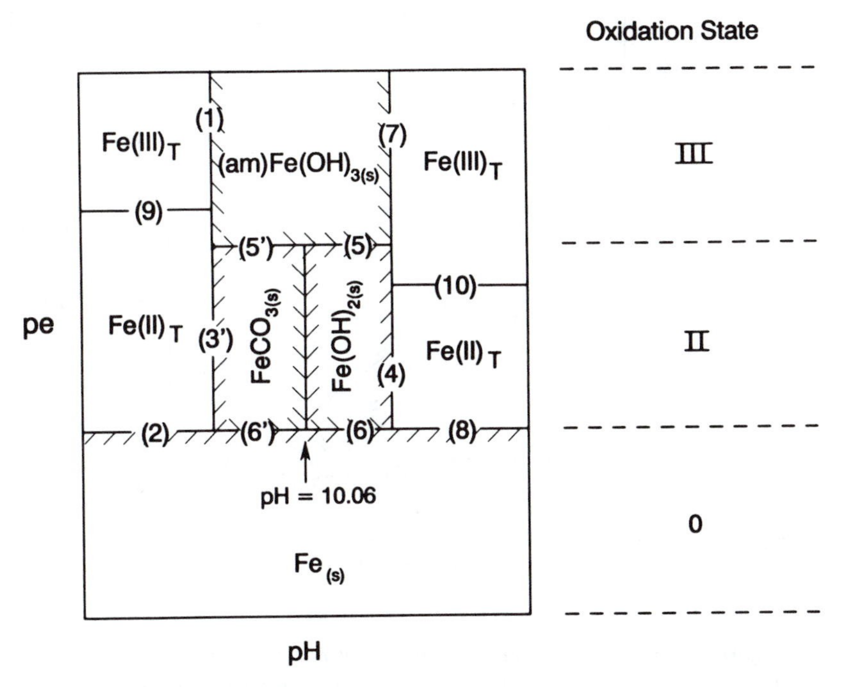

Figure 23.18 Generic pe-pH diagram for aqueous iron for $Fe_T = 10^{-5}$ *M*, $C_T = 10^{-3}$ *M* carbonate carbon, and 25°C/1 atm. The transition in dominance between any two adjacent oxidation states occurs over a range of pe values, and not at a single pe as might be interpreted based on the nature of the right hand side of the figure.

Equation (23.137) gives

$$pe = 26.88 + \log \alpha_2 + \log C_T - 3pH \tag{23.138}$$

as the (am)$Fe(OH)_{3(s)}/FeCO_{3(s)}$ boundaryline at 25°C/1 atm. Equation (23.138) is plotted in Figure 23.19, but only for the pH range over which both (am)$Fe(OH)_{3(s)}$ and $FeCO_{3(s)}$ are present. That range is given by the intersection of line (23.138) with line (23.127), and the intersection of line (23.138) with line (23.110).

23.3.4.2.3. $FeCO_{3(s)}/Fe_{(s)}$ Boundaryline. For the $FeCO_{3(s)}/Fe_{(s)}$ boundaryline, from Table 23.12 we obtain

$$FeCO_{3(s)} + 2e^- = Fe_{(s)} + CO_3^{2-} \qquad \log K = -25.58. \tag{23.139}$$

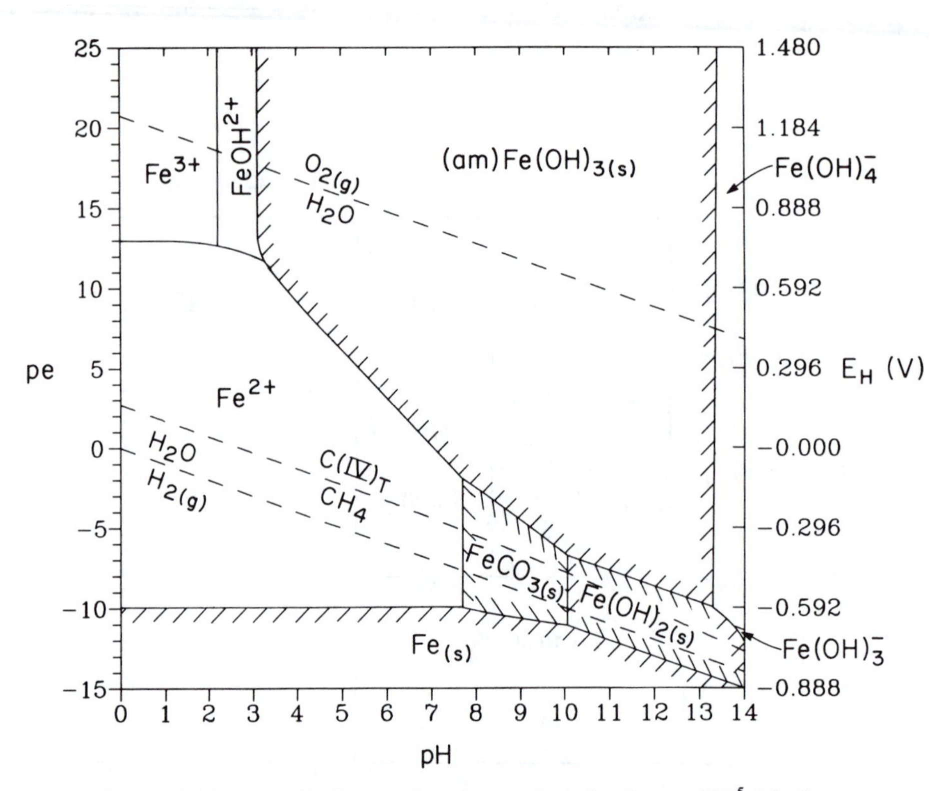

Figure 23.19 pe-pH diagram for aqueous iron for $Fe_T = 10^{-5}$ M, $C_T = 10^{-3}$ M, and 25°C/1 atm. Activity corrections are neglected. Only the pertinent portions of the various lines are shown. The $C(IV)_T/CH_4$ line (Eq. (23.27)) is included in the diagram. Below that line, at full equilibrium, there will be significant error in all of the pe-pH lines involving $FeCO_{3(s)}$. Indeed, as drawn here, the lines involving $FeCO_{3(s)}$ assume that C(IV) carbon (i.e., carbonate carbon) is present at a constant level of 10^{-3} M.

Equation (23.139) gives

$$pe = -12.79 - \frac{1}{2}\log\alpha_2 - \frac{1}{2}\log C_T \qquad (23.140)$$

as the $FeCO_{3(s)}/Fe_{(s)}$ boundaryline at 25°C/1 atm. $FeCO_{3(s)}$ becomes increasingly favored as α_2 increases because increasing α_2 promotes the precipitation of $FeCO_{3(s)}$. The same may be said about the effects of increasing C_T. Equation (23.140) is plotted in Figure 23.19, but only for the pH range over which both $FeCO_{3(s)}$ and $Fe_{(s)}$ are both present. That range is given by the intersection of line (23.140) with line (23.130), and the intersection of line (23.140) with line (23.119).

23.3.4.2.4. Solution/$FeCO_{3(s)}$ Boundaryline. In Section 23.3.4.1.5, we developed the equation for the solution phase/$Fe(OH)_{2(s)}$ boundaryline. When $Fe_T = 10^{-5}$ M and $C_T = 10^{-3}$ M, the low pH portion of that boundary is not present in the pe-pH diagram. In its place we have a solution/$FeCO_{3(s)}$ boundaryline (line 3′), and equilibrium with $FeCO_{3(s)}$ will be used to develop the pe-pH relationship between the dissolved Fe_T and the solid phase. As with line (23.122), we base our derivation on Eqs. (23.104)–(23.106).

Since we have equilibrium with $FeCO_{3(s)}$, neglecting activity corrections, along the solution/$FeCO_{3(s)}$ line, we have

$$[Fe^{2+}] = \frac{K_{s0}}{\alpha_2 C_T}. \qquad (23.141)$$

For redox equilibrium between Fe^{3+} and Fe^{2+}, neglecting activity corrections, Eq. (23.108) once again applies. Combining Eqs. (23.108) and (23.141) with Eq. (23.106), we obtain

$$\frac{10^{-13}10^{pe}K_{s0}}{\alpha_2 C_T}\left(1 + \frac{^*\beta_1}{[H^+]} + \frac{^*\beta_2}{[H^+]^2} + \frac{^*\beta_4}{[H^+]^4}\right) + \frac{K_{s0}}{\alpha_2 C_T}\left(1 + \frac{^*\beta_1}{[H^+]} + \frac{^*\beta_2}{[H^+]^2} + \frac{^*\beta_3}{[H^+]^3}\right) = Fe_T. \qquad (23.142)$$

Rearranging Eq. (23.142) to give pe as a function of pH, we obtain

$$pe = \log\frac{Fe_T - \dfrac{K_{s0}}{\alpha_2 C_T}\left(1 + \dfrac{^*\beta_1}{[H^+]} + \dfrac{^*\beta_2}{[H^+]^2} + \dfrac{^*\beta_3}{[H^+]^3}\right)}{10^{-13.0}\dfrac{K_{s0}}{\alpha_2 C_T}\left(1 + \dfrac{^*\beta_1}{[H^+]} + \dfrac{^*\beta_2}{[H^+]^2} + \dfrac{^*\beta_4}{[H^+]^4}\right)}. \qquad (23.143)$$

The K_{s0} in Eq. (23.143) refers to $FeCO_{3(s)}$; the $^*\beta$ values in the numerator refer to Fe(II); and the $^*\beta$ values in the denominator refer to Fe(III). Equation (23.143) can be plotted for any value of Fe_T larger than the minimum solubility of $FeCO_{3(s)}$ at $C_T = 10^{-3}$ M. For $Fe_T = 10^{-5}$ M, only the pertinent, low pH portion is plotted in Figure 23.19.

Equations (23.143), (23.110) and (23.122) are all fundamentally very similar. As with Eqs. (23.110) and (23.122), for any value of Fe_T, the numerator portion of the argument of the logarithm in Eq. (23.143) will become negative for some low pH value and for some high pH value, and the log term will become undefined. At these two pH values, the slope of the line given by Eq. (23.143) becomes infinite. For $Fe_T = 10^{-5}$ M, the slope becomes infinite near pH = 7.68 and pH = 11.42. These two values correspond to the low and high pH roots of the equation $Fe(II)_T = Fe_T$ when $FeCO_{3(s)}$ controls the solubility and the various Fe(III) species are negligible. As the pH is increased, the fact that $FeCO_{3(s)}$ is capable of completely redissolving at a pH (~11.4) that is lower than the pH at which $Fe(OH)_{2(s)}$ redissolves (~14.1) is consistent with the fact that $FeCO_{3(s)}$ cannot exist beyond pH = 10.06, but $Fe(OH)_{2(s)}$ can exist at such pH values.

23.4 ORDER OF REDUCTION WHEN pe IS REDUCED

This and previous chapters have discussed how the order in which species are oxidized or reduced will follow a sequence that is established by the pe° values for the various half-reactions. We have discussed how the levels of H^+, carbonate species, etc., can affect the reaction sequence. The effects of pH and $pHCO_3^-$ can be factored out using an appropriate $pe^o(pH, pHCO_3^-)$ scale. When a redox-active solid can form, the effects of the total level of element (e.g., Fe_T, S_T, etc.) on the reaction sequence can be determined through substitution into the equation for pe based on the half-reaction of interest.

We now note that a set of pe-pH *diagrams* for the elements of interest will also allow us to deduce a redox reaction sequence. Consider for example a system at 25°C/1 atm and pH = 8.0 that initially contains the following levels of redox active elements: 2.63×10^{-4} *M* of dissolved O_2 (in-situ $p_{O_2} = 0.21$ atm); 10^{-3} *M* each of dissolved C_T, N_T, and S_T; and $Fe_T = 10^{-5}$ *M*. By Eq. (19.109), the initial pe will be 12.6. Using Figure 23.20, which is a composite of Figures 23.5, 23.7, 23.12, and 23.19, we now consider what happens as the pe is lowered. This reduction in pe could be carried out through the continued incremental addition of a highly reducing chemical species. A selected series of pe-pH positions that would result are identified in Figure 23.20 as points 1 through 8. Assuming that N_2 is redox active, these points have the following characteristics:

1. **pe = 12.6**. In-situ $p_{O_2} = 0.21$ atm. At equilibrium, the dissolved oxygen will be present as O_2 (and a great deal more of H_2O). Since significant dissolved O_2 will be present, the initial pe is fairly high. The carbon, nitrogen, and sulfur will initially be present primarily as HCO_3^-, NO_3^-, and SO_4^{2-}, respectively. (am)$Fe(OH)_{3(s)}$ will be present. The in-situ p_{H_2} will be exceedingly low.
2. **pe = 12.1**. As the pe is reduced, the first electron acceptor (oxidant) to be reduced in large amounts (but not exclusively) will be the dissolved O_2, which will be converted to more H_2O. By the time pe = 12.1, the in-situ p_{O_2} will have been reduced to 0.0021 atm.
3. **pe = 10.0**. When essentially all of the O_2 is exhausted, the next electron acceptor to be reduced in large amounts will be the NO_3^-, which will be converted primarily to N_2. Once the pe drops to 10, the dominant nitrogen species will be N_2.
4. **pe = −2.8**. After most of the NO_3^- has been exhausted, the next electron acceptor to be reduced in large amounts will be the Fe(III). Indeed, once the pe drops to −2.8, (am)$Fe(OH)_{3(s)}$ is no longer present, and we are inside the $FeCO_{3(s)}$ region. Since $Fe_T \ll C_T$, most of the carbon is still present in solution; in solution, we still have $C(IV)_T \simeq 10^{-3}$ *M*.

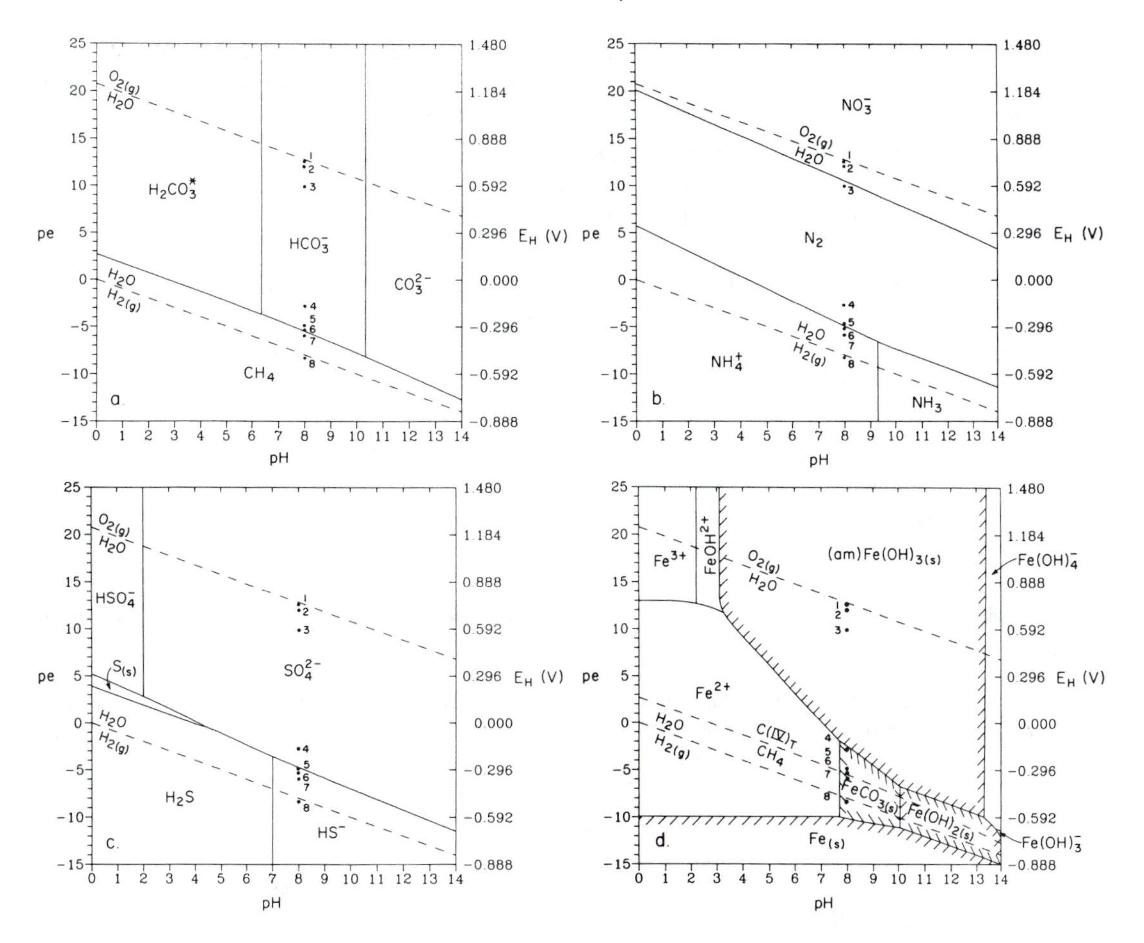

Figure 23.20 Collected pe-pH diagrams for carbon, nitrogen, sulfur, and iron at 25°C/1 atm showing eight increasingly reducing points at pH = 8. For the sulfur diagram, $S_T = 10^{-3}$ M, and for the iron diagram, $Fe_T = 10^{-5}$ M and $C_T = 10^{-3}$ M. Activity corrections are neglected.

5. **pe = −4.9**. After most of the Fe(III) has been exhausted, the next electron acceptor to be reduced in large amounts will be the SO_4^{2-} which will be converted primarily to HS^-. However, since the N_2/NH_4^+ and HCO_3^-/CH_4 boundaries are only slightly below the SO_4^{2-}/HS^- boundary, as SO_4^{2-} begins to accept electrons in significant amounts, N_2 and HCO_3^- will also begin to accept some electrons. Once the pe reaches −4.9, the dominant sulfur species will be HS^-; N_2 and HCO_3^- will still be the dominant nitrogen and carbon species, respectively, but significant conversion to NH_4^+ and CH_4 will also have taken place. With $C(IV)_T$ still close to 10^{-3} M, $FeCO_{3(s)}$ would still be present if it were not for the fact that there was so much sulfur in this system. Indeed, since $FeS_{(s)}$ is very insoluble, at pe = −4.9 the $FeCO_{3(s)}$ will have been converted to $FeS_{(s)}$.

6. **pe = −5.3.** Once the pe reaches −5.3, we drop below the N_2/NH_4^+ line, and NH_4^+ will become the dominant nitrogen species. HCO_3^- will still be the dominant carbon species, but significant amounts of CH_4 will by now also be present.
7. **pe = −6.0.** Once the pe reaches −6.0, we are below the HCO_3^-/CH_4 line, and CH_4 is the dominant carbon species. The bulk of the Fe is present as $FeS_{(s)}$. The level of dissolved Fe(II) is very low and the level of dissolved Fe(III) is exceedingly low.
8. **pe = −8.4.** If the pe continues to be lowered, the last major electron acceptor in the system will be H_2O. Once the pe drops below 8.4, p_{H_2} will be greater than 1 atm. Since we are still above the line given by Eq. (23.119), $Fe_{(s)}$ is not present.

Additional examples illustrating the principles of this chapter are presented by Pankow (1992).

TABLE 23.14 Change in dominance, presence, or absence of redox species at pH = 8.0 as pe is reduced. $C_T = 10^{-3}$ *M* and $Fe_T = 10^{-5}$ *M*.

point	pe	in-situ pressure (atm) of $O_{2(g)}$ and $H_{2(g)}$	Dominant C-Containing Species	Dominant N-containing Species	Dominant S-containing Species	$Fe(OH)_{3(s)}$ present?	$FeCO_{3(s)}$ present?	$FeS_{(s)}$ present?
1	12.6	$p_{O_2} = 0.21$ $p_{H_2} = 6.2 \times 10^{-42}$	HCO_3^-	NO_3^-	SO_4^{2-}	Yes	No	No
2	12.1	$p_{O_2} = 0.0021$ $p_{H_2} = 6.2 \times 10^{-41}$	HCO_3^-	NO_3^-	SO_4^{2-}	Yes	No	No
3	10.0	$p_{O_2} = 7.6 \times 10^{-12}$ $p_{H_2} = 1.0 \times 10^{-36}$	HCO_3^-	N_2	SO_4^{2-}	Yes	No	No
4	−2.8	$p_{O_2} = 4.8 \times 10^{-63}$ $p_{H_2} = 1.6 \times 10^{-11}$	HCO_3^-	N_2	SO_4^{2-}	No	Yes	No
5	−4.9	$p_{O_2} = 1.9 \times 10^{-71}$ $p_{H_2} = 6.5 \times 10^{-7}$	HCO_3^-	N_2	HS^-	No	No	Yes
6	−5.3	$p_{O_2} = 4.8 \times 10^{-73}$ $p_{H_2} = 4.1 \times 10^{-6}$	HCO_3^-	NH_4^+	HS^-	No	No	Yes
7	−6.0	$p_{O_2} = 7.6 \times 10^{-76}$ $p_{H_2} = 1.0 \times 10^{-4}$	CH_4	NH_4^+	HS^-	No	No	Yes
8	−8.4	$p_{O_2} = 1.9 \times 10^{-85}$ $p_{H_2} = 6.3$	CH_4	NH_4^+	HS^-	No	No	Yes

23.5 REFERENCES

COTTON, F. A. and G. WILKINSON. 1980. *Advanced Inorganic Chemistry*. New York: Wiley-Interscience.

GARRELLS, R. M., and C. L. CHRIST. 1965. *Solutions, Minerals, and Equilibria*. New York: Harper and Row.

PANKOW, J. F. 1992. *Aquatic Chemistry Problems*. Portland: Titan Press-OR, P.O. Box 91399, Portland, Oregon 97291–1399.

24

The pe Changes in a Stratified Lake During a Period of Summer Stagnation: An Example of a Redox Titration

24.1 INTRODUCTION

24.1.1 Lake Dynamics

Over the course of a year, a large lake (or reservoir) can provide the stage for some very dramatic and predictable changes in redox chemistry. This is because the bottom portion of such a lake can become quite isolated during the summer, allowing the oxygen (as well as other oxidants present) to be exhausted. The details of how this might occur in one specific case is discussed in this chapter. We first consider some basic principles and processes governing lake dynamics.

Figure 24.1 shows how the density of pure water depends on temperature. In lakes of moderate depth in the middle latitudes of the temperate zones, the interplay of temperature, density, and wind during the different seasons of the year produces a cycle of characteristic patterns of thermal stratification. Figure 24.2 illustrates a typical series of temperature and density gradients for such lakes. The next four paragraphs paraphrase the discussion of this cycle by Fair, Geyer, and Okun (1968).

During the winter, the water immediately below the ice is essentially at 0°C. The ice itself can be much colder than 0°C at and near the air surface.[1] The diffusion of heat upwards through the ice will allow the thickness of the ice to increase over the course of the winter. The temperature of the water at the bottom of the lake is not far from 4°C, which is the temperature at which water reaches its maximum density. Under these

[1]Fortunately, the ice is less dense than the underlying water. Thus, the ice floats, and remains as a protective cover for the lake during the winter. If ice were more dense than water, it would sink, the lake could fill up with ice, and many forms of life would die.

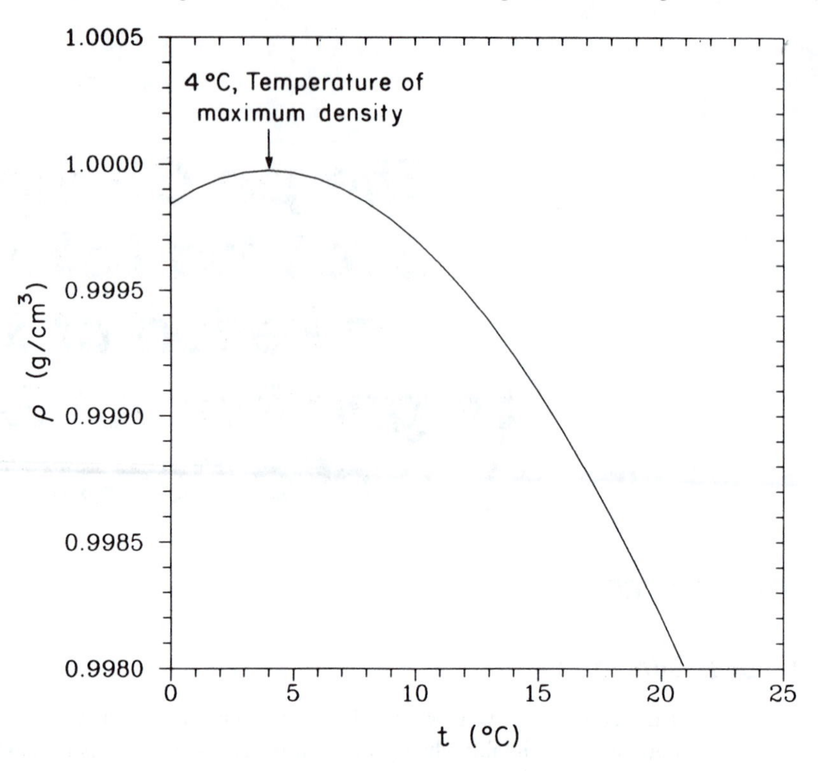

Figure 24.1 Density (g/cm^3) of pure water as a function of temperature (°C).

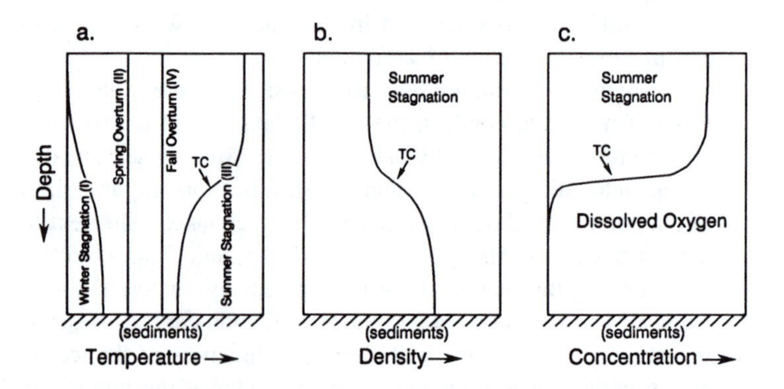

Figure 24.2 (*a*) Temperature as a function of depth during the four seasons of the year in a lake or reservoir in the middle latitudes. It is assumed that the lake or reservoir mixes completely during both the spring and fall overturning. (Note: TC = thermocline.) Adapted from Fair, Geyer, and Okun (1968). (*b*) Density as a function of depth during the period of summer stagnation. (*c*) Dissolved oxygen as a function of depth during the period of summer stagnation. Adapted from Fair, Geyer, and Okun (1968).

conditions, the water column is in a state of stable physical equilibrium. It is *inversely* stratified in terms of temperature, but *directly* and *stably* stratified in terms of density. The ice cover shuts out the shear forces of the wind, and so there is little movement of the surface waters; "roll cells" that might provide vertical mixing by moving surface water down and deep water up cannot develop. This is the period of *winter stagnation.*

In the spring, the ice begins to break up, and the water near the surface begins to warm. As the temperature of water's maximum density is approached, the surface waters also become denser and tend to sink. Equilibrium is further upset by the wind, as well as by diurnal fluctuations in the temperature. Once the water temperature is nearly uniform at all depths and close to the temperature of maximum density, mixing becomes especially pronounced. This is the period of the *spring overturn.* It varies in length during different years, and can last several weeks.

With the onset of summer, the surface waters become progressively warmer. Soon, lighter water overlies denser water once again. As the temperature differences between the upper and lower waters increase, circulation becomes confined more and more to the upper waters. A second period of stable equilibrium is established. The water column is now *directly* stratified in terms of both temperature and density. In comparatively deep lakes, the water below ~8 m is often nearly stagnant, and the temperature at the bottom remains almost constant and near ~4°C. This is the period of *summer stagnation.* For lakes in northern latitudes in the northern hemisphere, it can extend from April to November.

With the arrival of autumn, surface layers cool and sink, and the summer equilibrium is upset. The water is mixed to greater and greater depths, and the temperature gradient eventually becomes essentially vertical. At this point, the waters are easily put into circulation by autumn winds, and the *great overturn* or *fall overturn* occurs. At this time, many lakes become fully mixed over the entire water column. When winter arrives, the surface waters once again freeze, and the condition of winter stagnation begins to establish itself once again. (In very deep lakes, winter stagnation can begin before full mixing occurs, and the deepest waters remain stagnant and never mix directly with the overlying waters.)

The conditions in a lake at one point during the summer are represented in Figure 24.3. As discussed, the "epilimnion" or upper portion of the lake is warmer than the "hypolimnion," or lower portion of the lake. Since the temperature of the epilimnion increases steadily as the summer progresses, the stratification in the lake becomes increasingly more stable. The bulk of the temperature change occurs over a narrow vertical interval referred to as the "thermocline." Those persons who have dived into a lake in the summer may have been surprised to encounter some rather cold water below a certain depth: *that* was an encounter with a thermocline.[2] The stability of the summer

[2]On such occasions, if there is enough light, it is even possible to "see" the thermocline with the help of a diving mask. Indeed, the changing index of refraction that accompanies changing density causes a distortion of visual images. This is similar to what occurs in warm air atop asphalt or atop an automobile roof, except that in those cases the stratification is not stable and so an image viewed through that stratification is not only distorted, but also unstable (i.e., it shimmers spontaneously as the lighter and warmer surface air continually rises into the cooler and more dense overlying air).

stratification is rooted in the fact that any downward movement of a "parcel" of water from the epilimnion towards the hypolimnion will be resisted by buoyancy effects, and most of that parcel will remain in the epilimnion. The greater the temperature difference between the two layers, the more stable is the stratification.

As the summer progresses, the position of the thermocline gradually deepens as heat is transported downward by the parcels of water "impacting" upon the thermocline. Small portions of these parcels succeed in advecting heat into the hypolimnion (the simple diffusion of heat by molecular motions plays a relatively minor role). Unless the lake is comparatively shallow, the fall overturn usually occurs before the thermocline can reach the bottom.

24.1.2 The Effects of Lake Dynamics on pe

Stratification in the summer can cause a major problem in many lakes and reservoirs. In particular, since advective mixing across the thermocline is very limited during that

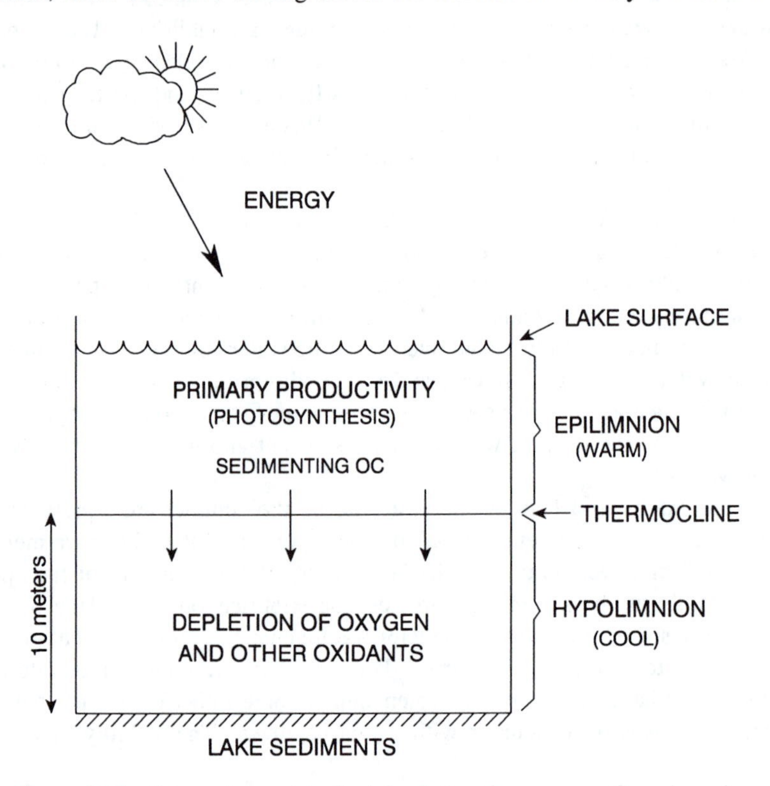

Figure 24.3 Schematic model of a lake during the summer. Organic carbon (OC) as a result of primary biological production in the epilimnion is sedimenting through the thermocline into the 10 m deep hypolimnion. Oxygen and other oxidants in the hypolimnion are reduced sequentially as the summer progresses. Limited mixing across the thermocline prevents replenishment of the dissolved oxygen.

period, and since simple molecular diffusion is comparatively slow, there is little ability for surface oxygen to move from the epilimnion into the hypolimnion. The oxygen supply in the hypolimnion is thus essentially limited to what was present at the beginning of the summer, prior to the onset of stratification. That oxygen supply might not last until the fall overturn remixes the lake. Indeed, over the course of the summer, oxygen will be consumed due to the biological degradation of: 1) organic matter initially in the water column in the hypolimnion; 2) dead and/or live organic matter (e.g., algal and diatom cells) in the epilimnion that subsequently sediments through the thermocline and into the hypolimnetic waters; and 3) organic matter in the sediments. We note that although advective transport across the thermocline is inhibited, organic matter as well as other particles (e.g., $CaCO_{3(s)}$ from "whiting" events[3]) can be more dense than the hypolimnetic waters, and can sediment through the thermocline. The production of organic matter in the epilimnion is more likely to occur in lakes that are nutrient rich (eutrophic) than in lakes that are nutrient poor (oligotrophic).

The consumption of oxygen in the hypolimnion can proceed to the point that the hypolimnion becomes completely anoxic; this will lead to the death of fish and other aerobic organisms. The consumption of O_2 and the concomitant production of CO_2-type carbon will lead to concentration gradients of these and other species across the thermocline.

The pe of the hypolimnion can be lowered far below what is required to remove essentially all of the oxygen. Indeed, after virtually all of the O_2 is gone, the pe can be lowered sufficiently for the sequential reduction of nitrate to N_2, ferric iron (Fe(III)) to ferrous iron (Fe(II)), sulfate (S(VI)) to sulfide (S(−II)), carbonate carbon C(IV) to CH_4 (C-IV), and even some H^+ and H_2O (H(I)) to H_2 (H(0)). As described below, the combined presence of Fe(II) and S(−II) can lead to the precipitation of $FeS_{(s)}$. Figure 9.9 in Stumm and Morgan (1981) is a photograph showing black, $FeS_{(s)}$-containing sediment layers ("varves") in a sediment core from Lake Zurich. (In that lake, the waters at the very bottom never become mixed and re-oxygenated; the $FeS_{(s)}$ is deposited at the sediment/water interface during the winter portion of a permanent stagnation.)

During the course of the summer, the pe and the pH will be changing along a track that can be plotted on a pe-pH diagram. In effect, we will be taking a pe-pH train ride along that track. However, like a train trip through France, we won't be able to get an *overview* of what the whole pe-pH countryside looks like; all that we will be able to see is the types of pe-pH predominance regions that lie along the track.

24.2 MODEL CONSIDERATIONS

24.2.1 Initial Conditions

Figure 24.3 gives a view of our lake in schematic terms. The example of changing hypolimnion redox chemistry considered here will be discussed in terms of an equilibrium

[3]When algal blooms occur in the surface waters, the rapid consumption of $H_2CO_3^*$ for photosynthesis can lead to a rise in pH that causes $CaCO_{3(s)}$ to precipitate. This effect can be sufficiently pronounced for the surface waters to appear whitened by the white $CaCO_{3(s)}$.

model. Thus, as the chemistry of the system changes with time due to the input of organic matter, we will be computing a series of equilibrium states for a series of discrete points in time. When we calculate the shape of an acid/base titration curve, we are also computing a series of equilibrium states. In that case, we compute the pH as a function of $(C_B - C_A)$. We could also compute an acid/base titration curve as a function of time if there was a case when $(C_B - C_A)$ was being added according to a known function of time.

Table 24.1 gives the parameters that will be assumed to remain constant in our lake model. Table 24.2 summarizes the assumed initial concentrations of various chemical species in the hypolimnion. Many significant figures will be retained in the calculations that will be made involving the numbers in those tables. Although the accuracy with which we can know equilibrium constants and initial conditions will not justify a great number of significant figures in any particular calculated result, once we pick our best values for those constants and conditions, we need to carry many significant figures in the calculations because at certain points during the summer, premature rounding to a reasonable number of significant figures will have a big effect on the calculated pe and pH values and therefore on the concentrations of pe- and pH-dependent species. In other words, given specific values for the equilibrium constants and initial conditions, we wish to obtain the one true-mathematical solution to the model for each point in time. *Then*, although we will not do it, one could go back and place confidence intervals around the equilibrium constants and the initial conditions so that errors can be propagated and confidence intervals placed around the pe, pH, and speciation results for any given point in time.

Since an equilibrium model assumes that there are no kinetic limitations on the rates of the reactions of interest, such a model will predict that changes will occur *instantaneously* when chemical materials are added or removed from the system. While it is very unlikely that perfect redox equilibrium will ever exist at any point in any real lake, as discussed in Chapter 1, results obtained using an equilibrium lake model can nevertheless provide great insight. The manner in which certain kinetic considerations can be incorporated into our lake model will be pointed out where appropriate. Since the main objective here is to illustrate some redox chemistry, and not to develop a detailed and perfect lake model, we will also assume that:

TABLE 24.1. Conditions in hypolimnion assumed to remain constant

thickness of hypolimnion (i.e., distance from thermocline down to sediment/water interface)	10 m
loading rate (flux) of organic carbon† from the epilimnion to the hypolimnion (as carbon)	$\begin{cases} 450 \text{ mg C/m}^2\text{-day} = \\ 3.746566 \times 10^{-2} \text{ mol/m}^2\text{-day} = \\ 1.498626 \times 10^{-1} \text{ eq/m}^2\text{-day} = \\ 3.746566 \times 10^{-6} \text{ mol/L-day} = \\ 1.498626 \times 10^{-5} \text{ eq/L-day} \end{cases}$

†The organic carbon (OC) is assumed to have the empirical formula of the carbohydrate glucose ($C_6H_{12}O_6$). In this form, the average carbon oxidation state is zero. Like formaldehyde (CH_2O), glucose contains the equivalent of one water molecule per atom of zero-oxidation-state carbon.

TABLE 24.2. Initial chemical conditions in the hypolimnion

pH = 8.00, $C(IV)_T = 4.915840 \times 10^{-4}$ *M*, *Alk* = 4.840975×10^{-4} eq/L,
$N_T = 9.985723 \times 10^{-4}$ *M*

other species	mg/L	*M*	*n* (equivalents/mol)	eq/L
O_2	8.459662	2.643943×10^{-4}	4 (if reduced to water)	1.057497×10^{-3}
$Fe(OH)_{3(s)}$ (suspended)	0.50 (as Fe)	8.953032×10^{-6} (FW per liter)	1 (if reduced to Fe(II))	8.953032×10^{-6}
NO_3^-	0.20 (as N)	1.427888×10^{-5}	2 (if reduced to NO_2^-)	2.855776×10^{-5}
			5 (if reduced to $\frac{1}{2}N_2$)	7.139440×10^{-5}
			8 (if reduced to NH_4^+)	1.142310×10^{-4}
N_2	13.786703	4.921467×10^{-4}	6 (if reduced to 2 NH_4^+)	2.952880×10^{-3}
SO_4^{2-}	3.00 (as S)	9.357455×10^{-5}	6 (if reduced to $S_{(s)}$)	5.614473×10^{-4}
			8 (if reduced to H_2S/HS^-)	7.485964×10^{-4}
OC†	1.0 (as C)	8.325701×10^{-5}	4 (if oxidized to CO_2)	3.330281×10^{-4}

†The organic carbon (OC) is assumed to have the average formula of the carbohydrate $C_6H_{12}O_6$ (glucose). In this form, the average carbon oxidation state is zero. Like formaldehyde (CH_2O), glucose contains the equivalent of one water molecule per atom of zero-oxidation-state carbon.

1. A simple one-box model is appropriate for the hypolimnion (more complicated multi-box models can be constructed [e.g., see Imboden and Lerman, 1978], but we do not need one here).
2. The reducing role of organic matter in the sediments may be neglected.
3. The depth of the thermocline and therefore the thickness of the hypolimnion may be assumed to remain constant.
4. Tabulated equilibrium constants available at 25°C/1 atm apply even though most hypolimnions will be significantly cooler than 25°C and under higher pressure than 1 atm (the corrections for pressure effects would be very minor).

The approach that we will use here will set up and solve all of the pertinent details of our problem. This will help to build chemical intuition in a manner that is difficult to match by simply looking at output listings from a large, general equilibrium computer program like MINEQL, MINTEC, WATEQ2, or PHREEQE. However, for the *routine* solution of problems which are as complicated as the one considered in this chapter, the practicing aquatic chemist will probably want to use one of those programs.

Prior to the summer stratification, the water in the lake will be assumed to be well mixed. Therefore, the initial levels of the dissolved gases O_2, N_2, and $H_2CO_3^*$ in Table 24.2 for the hypolimnion after stratification have been taken to be those for direct equilibrium with the atmosphere at 25°C/1 atm. We will take that: 1) in the atmosphere,

$p_{N_2} = 0.78$ atm, $p_{O_2} = 0.21$ atm, and $p_{CO_2} = 10^{-3.5}$ atm; and 2) at 25°C/1 atm, for N_2 we have $K_H = 10^{-3.20}$ M/atm, for O_2 we have $K_H = 10^{-2.90}$ M/atm, for H_2 we have $K_H = 10^{-3.10}$ M/atm, and for CO_2 we have $K_H = 10^{-1.47}$ M/atm.

As indicated in Table 24.2, essentially all of the initial amounts of sulfur, iron, and non-N_2 nitrogen in the hypolimnion have been taken to be in their highest-possible oxidation states. This is consistent with the redox behavior of these elements as discussed in Chapter 23. The initial organic carbon (OC) has been taken to be in an average oxidation state of zero. Thus, for the oxidation of that OC to CO_2, the value of n in Table 24.2 is 4, and we have

$$\frac{1}{6}C_6H_{12}O_6 + H_2O = CO_2 + 4H^+ + 4e^-. \tag{24.1}$$

Assuming a different average initial oxidation state for the OC would have given a different value of n.

Levels of dissolved or suspended OC of a few mg/L are typical for natural waters. However, as discussed in Chapter 23, when any appreciable amount of oxygen is present in water at natural pH values, C(IV) will be the stable carbon oxidation state. Thus, the presence of OC in the hypolimnion at the beginning of the stratification does not represent an equilibrium situation. If we are to adopt an equilibrium model for the hypolimnion, we will need to allow for the oxidation of the initial OC before considering any changes caused by the subsequent sedimentation of additional OC into the hypolimnion.

24.2.2 Assumptions Governing Nitrogen Redox Chemistry

Given the nature of Figure 23.7, the formation of the species NO_2^- will not be considered in the model. Since the kinetics of N_2 oxidation (e.g. to NO_3^-) as well as the kinetics of N_2 reduction (e.g. to NH_4^+) are very slow, we will also assume that any N_2 present or formed will be stable. Thus, we will consider N_2 to be non-redox active: 1) at high pe when oxidation of N_2 to NO_3^- or NO_2^- is thermodynamically possible; and 2) at low pe when reduction of N_2 to NH_3 and NH_4^+ is thermodynamically possible. At intermediate pe, where NO_3^- can be converted to N_2, we will allow N_2 to be redox active so that $\{N_2\}$ can govern the extent of that conversion. Handling N_2 in this fashion means that our model will not be a full equilibrium model. However, doing so builds some important kinetic considerations into the model.

24.2.3 "eq/L" as Useful Units in Situations Involving Redox Reactions

If the hypolimnion is assumed to be well mixed (this assumption is the essence of a one-box model), dividing the flux of sedimenting OC by the thickness of the box will yield the input rate of OC in units of M/time:

$$\frac{3.746566 \times 10^{-2}\ \text{mol/m}^2\text{-day}}{10\ \text{m}} \times \frac{1\ \text{m}^3}{1000\ \text{L}} = 3.746566 \times 10^{-6}\ M/\text{day}. \tag{24.2}$$

Units of M are not, however, the most convenient when determining the effects of this reduced carbon on the hypolimnion because of the different numbers of electrons involved in the various possible redox reactions.

In Chapter 7, for acids and bases, we found it convenient to use units of eq and eq/L to describe mass quantities and solution concentrations, respectively. For example, 1.0 eq of strong acid will exactly neutralize 1.0 eq of strong base. For the acid HCl, 1.0 FW equals 1 eq, and a 1.0 F solution is a 1.0 eq/L solution. For the acid H_2SO_4, 0.5 FW equals 1.0 eq, and a 0.5 F solution is a 1.0 eq/L solution. For the base NaOH, 1.0 FW equals 1 eq, and a 1.0 F solution is a 1.0 eq/L solution. For the base $Mg(OH)_2$, 0.5 FW equals 1.0 eq, and a 0.5 F solution is a 1.0 eq/L solution. In summary, multiplying the FW by the number of acidic or basic units per FW will convert units of FW to units of eq, and units of F to units of eq/L.

Units of eq and eq/L are also very useful for redox-active species. For such species, the conversion factor between FW (or mols) and eq, and between F (or M) and eq/L is the number of electrons n being exchanged in the redox reaction of interest. As noted, for the oxidation of the OC considered here to CO_2, $n = 4$. Therefore, multiplying by the associated conversion factor of (4 eq C/mol C) leads to

$$\begin{array}{l}\text{input rate of reduced}\\ \text{carbon to hypolimnion}\end{array} = 3.746566 \times 10^{-6}\ M/\text{day} \times \frac{4\text{ eq C}}{\text{mol C}} = 1.498626 \times 10^{-5}\text{ eq/L-day.} \qquad (24.3)$$

24.2.4 Calculating the "Initial" pe

We can calculate the pe given by the initial oxygen concentration and the initial pH in the hypolimnion. At 25°C/1 atm,

$$\text{pe} = 20.78 - \text{pH} + \frac{1}{4}\log p_{O_2}. \qquad (24.4)$$

Since we know that the initial in-situ p_{O_2} of the water is 0.21 atm, we could use that value in Eq. (24.4) to calculate an initial pe. However, during the course of the summer, we will be keeping track of how much oxygen is present in terms of its concentration in units of eq/L. Thus, it will be useful to recast Eq. (24.4) in a form where the O_2 concentration is expressed in eq/L.

For O_2 at 25°C/1 atm, we have $K_H = 10^{-2.90}\ M/\text{atm} = 10^{-2.30}$ eq/L-atm. Therefore, if [O_2, eq/L] represents the O_2 concentration in units of eq/L, then for the in-situ p_{O_2}, we have

$$p_{O_2} = \frac{[O_2, \text{eq/L}]}{10^{-2.30}\ \text{eq/L- atm}}. \qquad (24.5)$$

Substitution of Eq. (24.5) into Eq. (24.4) yields for 25°C/1 atm that

$$\begin{aligned}\text{pe} &= 20.78 - \text{pH} + \frac{1}{4}\log[O_2, \text{eq/L}] - \frac{1}{4}\log 10^{-2.30}\\ &= 21.35 - \text{pH} + \frac{1}{4}\log[O_2, \text{eq/L}]. \qquad (24.6)\end{aligned}$$

Thus, for $[O_2, \text{eq/L}] = 1.057497 \times 10^{-3}$ eq/L and pH = 8.00, we obtain

$$pe = 21.35 - 8.00 + \frac{1}{4}\log(1.057497 \times 10^{-3}\text{eq/L}) = 12.606. \quad (24.7)$$

As anticipated above, the big problem with 12.606 as an initial pe value is that it does not represent an *equilibrium* pe. Indeed, for an initial value of p_{CO_2} of $10^{-3.5}$ atm, and extending the assumption that the initial OC not only has a chemical makeup equivalent to $C_6H_{12}O_6$ but is also actually present as glucose[4] at a concentration of $(8.33 \times 10^{-5}\ M)/6 = 1.39 \times 10^{-5}\ M$, then by redox reaction 9 in Table 24.3, we obtain an initial pe of only −7.86. Thus, as is also discussed in Sections 19.3.5 and 23.2.3, when different redox reactions are not in equilibrium with one another, the different $\{OX_1\}/\{RED_1\}$, $\{OX_2\}/\{RED_2\}$, $\{OX_3\}/\{RED_3\}$, etc. ratios will prescribe different pe values. None of those values might be equal to the actual system pe which, moreover, would be changing as the various redox pairs exchange electrons and the system moves toward overall redox equilibrium. In such cases, if one desires to know the actual, nonequilibrium pe, it will be necessary to try and identify the *pair* of redox reactions that is exchanging electrons the most rapidly, and therefore tending to set the electron activity. Usually, this is a very difficult task.

24.2.5 Factors Affecting the Sequential Reduction of Oxidants in the Hypolimnion

The initial presence as well as the slow addition of organic carbon (OC) to the hypolimnion will cause a sequential reduction of the various oxidants (i.e., electron acceptors) present in the water. Actually, just as the titration of a solution containing more than one acid/base pair does not occur in a *perfectly* step-wise manner, all of the various possible redox couples will be undergoing *some* reduction as the OC is added. Nevertheless, given the fact that the redox properties of the initial oxidants are significantly different, the acceptance of electrons from the OC will likely occur in a *nearly* step-wise manner.

The order of reduction will be determined by the pe° values of the various appropriate half-reactions together with the concentrations of the various participating species (including H^+). Various relevant redox half-reactions have been collected in Table 24.3 together with their pe°(pH) values for a pH of 7. Although the initial pH of the hypolimnion is 8.00, a pH of 7 was selected for the tabulated pe°(pH) values because the oxidation of OC to CO_2 quickly brings the lake to near-neutral pH (see below). For the reduction of (am)$Fe(OH)_{3(s)}$ to $FeCO_{3(s)}$, the value of pe°(7, $pHCO_3^-$) is given assuming that $pHCO_3^- \simeq pAlk_o$. Since the pH and $pHCO_3^-$ will be *changing* as a result of the chemical changes occurring during the summer, as a rule, the tabulated pe°(7) and pe°(7, $pAlk_o$) values will not be scaled exactly correctly to the conditions in the hypolimnion at any given point in time during the summer.

[4]If we had an estimate of the half-reaction K value for the natural organic carbon so that did not need to make this assumption, we would still get a rather low pe for the $H_2CO_3^*/OC$ redox equilibrium.

TABLE 24.3. Redox half-reactions of interest in the hypolimnion. Those reactions that we will actually need to consider in the model are marked with an asterisk. Reactions involving the reduction of N_2 to NH_3 (or NH_4^+) and the reduction of NO_3^- to NO_2^- are not included for reasons discussed in Section 24.2.2.

Reaction	log *K*	pe°	pe°(7.00)[†]
*(1) $O_{2(g)} + 4H^+ + 4e^- = 2H_2O$	83.1	20.78	20.78 – (1)(7) = 13.78
*(2) $NO_3^- + 6H^+ + 5e^- = \frac{1}{2}N_2 + 3H_2O$	103.7	20.74	20.74 – (6/5)(7) = 12.34
(3) $(am)Fe(OH)_{3(s)} + 2H^+ + HCO_3^- + e^- = FeCO_{3(s)} + 3H_2O$	16.55	16.55	16.55 – (2)(7) = 2.55 (pe°(7,3.32)[†] = 2.55 – 3.32 = –0.77)
*(4) $SO_4^{2-} + 9H^+ + 8e^- = HS^- + 4H_2O$	34.0	4.25	4.25 – (9/8)(7) = –3.63
(5) $SO_4^{2-} + 8H^+ + 6e^- = S_{(s)} + 4H_2O$	36.2	6.03	6.03 – (8/6)(7) = –3.30
*(6) $H_2CO_3^* + 8H^+ + 8e^- = CH_4 + 3H_2O$	21.6	2.7	2.7 – (1)(7) = –4.3
(7) $S_{(s)} + 2H^+ + 2e^- = H_2S$	4.8	2.4	2.4 – (1)(7) = –4.6
*(8) $(am)Fe(OH)_{3(s)} + 3H^+ + e^- = Fe^{2+} + 3H_2O$	16.2	16.2	16.2 – (3)(7) = –4.8
(9) $H_2CO_3^* + 4H^+ + 4e^- = \frac{1}{6}C_6H_{12}O_6 + 2H_2O$	0.67	0.17	0.17 – (1)(7) = –6.83
*(10) $2H^+ + 2e^- = H_2$	–3.10	–1.55	– 1.55 – (2)(7.00) = –15.55

[†] pe°(7) is the pe°(pH) value for pH = 7. pe°(7,3.32) is the pe(pH,pHCO_3^-) value for pH = 7 and pHCO_3^- = 3.32.

24.3 CONSIDERATION OF THE POSSIBLE COMPLETE DISSOLUTION OF THE INITIAL (am)Fe(OH)$_{3(s)}$, AND THE POSSIBLE FORMATION OF FeCO$_{3(s)}$, S$_{(s)}$, AND FeS$_{(s)}$ DURING THE SUMMER

24.3.1 General Comments

During the course of the summer, the reduction of (am)Fe(OH)$_{3(s)}$ to Fe(II) might lead to the complete dissolution of that solid, as well as the precipitation of some FeCO$_{3(s)}$. We note in this regard that for the specific conditions considered in Figure 23.19 ($Fe_T = 10^{-5}$ *M* and $C_T = 10^{-3}$ *M*), a region for FeCO$_{3(s)}$ lies below the region for (am)Fe(OH)$_{3(s)}$. Thus, under the Fe_T and C_T conditions for that diagram, our pe-pH train ride could pass out of the (am)Fe(OH)$_{3(s)}$ region and into (and through) the FeCO$_{3(s)}$ region.

At any given point in time during the summer, whether or not solids like (am)Fe(OH)$_{3(s)}$ and FeCO$_{3(s)}$ are present will affect how we compute solution concentrations, as well as the details of how the model handles mass balance calculations. Thus, when constructing the model, it will be very helpful if we can determine beforehand which solids are likely to come and go in the system. Solids other than $FeCO_{3(s)}$ that might form as the system becomes reducing are $S_{(s)}$ and $FeS_{(s)}$.[5]

24.3.2 Will All of the Initial (am)$Fe(OH)_{3(s)}$ Disappear? Yes.

As seen in Figure 23.20, Fe(III) is only stable under intermediate to high pe conditions. Thus, if as in many cases the addition of the OC occurs to the point that rather reducing conditions are achieved in the hypolimnion, then we can be certain that at some point, there will be complete conversion of all of the (am)$Fe(OH)_{3(s)}$ to Fe(II) plus trace levels of dissolved Fe(III)). Indeed, conversion to Fe(II) can occur at pe values that are not highly reducing, and are moreover much higher than the $C(IV)_T/CH_4$ line (23.27). The position of the $C(IV)_T/CH_4$ line is relevant in this context because the continual addition of OC will allow us to reduce the pe to that point that we will start approaching that line. We know this because once all of the oxidants like O_2, (am)$Fe(OH)_{3(s)}$, etc. are essentially exhausted, then the fact that OC is unstable with respect to disproportionation to $C(IV)_T$ and CH_4 will cause a build up in $[CH_4]$ and therefore a significant decrease in the $C(IV)_T/[CH_4]$ ratio. For example, large amounts of CH_4 can form in landfills.

24.3.3 Will $FeCO_{3(s)}$ Form in This System? No.

Initially, when large amounts of oxygen are present, the level of dissolved Fe(II) will be extremely low. Thus, we know that $FeCO_{3(s)}$ will not be able to precipitate during the initial stages of the reduction of the (am)$Fe(OH)_{3(s)}$. However, as the dissolved Fe(II) builds up, at some point we may need to consider the effects of the formation of $FeCO_{3(s)}$ on the evolution of the pe-pH track. Let us assume for the moment that the pH and $C(IV)_T$ values remain constant at 8.00 and 4.915840×10^{-4} M, respectively.

If $Fe(II)_T$ is the total *dissolved* Fe(II), and if $\alpha_{0,Fe(II)}$ is the fraction that is present as Fe^{2+}, then

$$[Fe^{2+}] = \alpha_{0,Fe(II)} Fe(II)_T \tag{24.8}$$

where

$$\alpha_{0,Fe(II)} = \frac{[Fe^{2+}]}{[Fe^{2+}] + [FeOH^+] + [Fe(OH)_2^o] + [Fe(OH)_3^-]}. \tag{24.9}$$

At the type of approximately neutral pH values found in the hypolimnion, $Fe(OH)_2^o$ and $Fe(OH)_3^-$ will be present at negligible concentrations. Thus, to a very good approximation we will have

$$\alpha_{0,Fe(II)} \simeq \frac{1}{1 + K_{H1}[OH^-]}. \tag{24.10}$$

Based on data from Table 12.1, at pH = 8.00, $\alpha_{0,Fe(II)} = 0.969$.

[5]We do not need to consider the possibility of the formation of $Fe(OH)_{2(s)}$ in this system. Initially, the in-situ p_{CO_2} value equals $10^{-3.5}$ atm. From Chapter 14, the p_{CO_2} for equilibrium between $FeCO_{3(s)}$ and $Fe(OH)_{2(s)}$ at 25°C/1 atm is $10^{-5.43}$ atm. Thus, unless the in-situ p_{CO_2} drops by two orders of magnitude during the course of the summer (an unlikely event), $Fe(OH)_{2(s)}$ will never be an equilibrium component of this system.

The K_{s0} for $FeCO_{3(s)}$ is $10^{-10.68}$. Thus, for $FeCO_{3(s)}$ to become saturated, when activity corrections can be neglected, we will need

$$[Fe^{2+}][CO_3^{2-}] = \alpha_{0,Fe(II)}Fe(II)_T\alpha_2C(IV)_T = 10^{-10.68}. \quad (24.11)$$

As discussed above, at pH = 8.00, $\alpha_{0,Fe(II)} = 0.969$. For the CO_2 system, at pH = 8.00, $\alpha_2 = 4.55 \times 10^{-3}$. Thus, for $C(IV)_T = 4.915840 \times 10^{-4}\ M$, for $FeCO_{3(s)}$ to become saturated at this pH, we will need

$$Fe(II)_T = 9.63 \times 10^{-6}\ M. \quad (24.12)$$

Since this represents more than 100 percent (in particular, 108%) of the Fe initially present per liter of water in the hypolimnion, unless the factor $\alpha_2C(IV)_T$ increases during the course of the summer, we know that $FeCO_{3(s)}$ will not be able to form in this system. Given that the various oxidants will convert the OC to C(IV) (note that oxidation states such as C(II) will be of very minor importance at equilibrium), we know that $C(IV)_T$ will tend to increase during the course of the summer. However, the formation of C(IV) as CO_2 according to

$$\frac{1}{6}C_6H_{12}O_6 + O_2 = CO_2 + H_2O \quad (24.13)$$

will tend to reduce the pH, and so α_2 will tend to decrease. We need to know what the net effect on the product $\alpha_2C(IV)_T$ will be.

A quick estimate of the factor α_2C_T later in the summer can be obtained as follows. We know that significant reduction of the Fe(III) will not take place until some time after essentially all of the O_2 is exhausted (note that Figure 23.19 shows that Fe(III) is stable well below the $p_{O_2} = 1$ atm line). To exhaust nearly 100 percent of the O_2 (we will never be able to exhaust 100 percent), an equal number of eq/L of OC, that is, 1.057497×10^{-3} eq/L, will have been oxidized to CO_2. This will increase $C(IV)_T$ by $(1.057497 \times 10^{-3}\text{eq/L})/(4\text{eq/L-}\ M) = 2.64 \times 10^{-4}\ M$, bringing $C(IV)_T$ to $7.56 \times 10^{-4}\ M$. The alkalinity will not have changed significantly since the vast majority of the OC oxidation to that point in time will have taken place according to reaction (24.13), that is, without any production or consumption of H^+ or OH^-. Other OC oxidation reactions can change *Alk*.[6]

With *Alk* still essentially equal to its initial value of 4.840975×10^{-4} eq/L at the point that the most of the O_2 has been reduced, using the new $C(IV)_T$ we can calculate that the pH at that point will be about 6.60. At this pH, for the new $C(IV)_T$, we obtain $\alpha_2C(IV)_T = 8.98 \times 10^{-8}\ M$. Since this value is even smaller than the initial

[6]As an example, OC oxidation by the electron acceptor NO_3^- will tend to increase *Alk* because H^+ is consumed during the net oxidation reaction

$$\frac{5}{12}C_6H_{12}O_6 + 2NO_3^- + 2H^+ = \frac{5}{2}H_2CO_3^* + N_2 + H_2O.$$

$\alpha_2 C(IV)_T$, we can be confident that $FeCO_{3(s)}$ will not be able to form in this particular system. Thus, the pe-pH track that will be followed in this problem will drop down to the *left* of the $FeCO_{3(s)}$ region that could be drawn for $Fe_T = 8.95 \times 10^{-6}\ M$ and $C_T \simeq 10^{-3}\ M$.

24.3.4 Will $S_{(s)}$ Form in This System? No.

An examination of Figure 23.20 reveals that acidic pH values favor the formation of $S_{(s)}$. As the pH increases, the vertical thickness of the sulfur predominance wedge will decrease for any given value of S_T. Thus, what we need to determine is the maximum pH value to which the $S_{(s)}$ wedge extends for our value of S_T. At that point, the predominance region for $S_{(s)}$ will end for that value of S_T, and solid elemental sulfur will no longer be thermodynamically possible at equilibrium. If the pe-pH track followed in the hypolimnion drops down at a pH higher than that maximum, then $S_{(s)}$ will not be possible at equilibrium in the hypolimnion.

We can seek the maximum pH of the $S_{(s)}$ wedge using a number of different mathematical approaches. For example, we can find the pH at which the *d*pH/*d*pe derivative of the $S_{(s)}$ wedge boundaryline (Eq. (23.101)) equals zero. Or, we can find the pH of the intersection of the $S_{(s)}$ wedge boundaryline with the $S(VI)_T/S(-II)_T$ boundaryline (Eq. (23.89).) Or, we can find the pH at which the same pe is specified by both: a) Eq. (23.76) for $S(VI)_T/S(0)$ equilibrium, with $S(VI)_T = S_T/2$; and b) Eq. (23.96) for $S(-II)_T/S(0)$ equilibrium, with $S(-II)_T = S_T/2$. The last approach is the easiest.

We recall from Chapter 23 that at 25°C/1 atm, the acidity constant for HSO_4^- is $10^{-2.0}$. Under the same conditions, we take K_1 and K_2 for H_2S to be $10^{-7.0}$ and $10^{-14.0}$, respectively. For $S_T = 9.357455 \times 10^{-5}\ M$, we then obtain pH = 2.12 as the maximum pH of the $S_{(s)}$ predominance wedge. Thus, for this S_T, the $S_{(s)}$ wedge certainly does not extend out to approximately neutral pH values, and so the only S(VI) reduction process that we will need to consider will be the one that leads to the formation of S(−II): $S_{(s)}$ will not be able to form. In other words, for this level of S_T, as conditions become more reducing, the pe-pH track will miss the $S_{(s)}$ predominance region by passing down to the right of it. The track will pass directly from one of the $S(VI)_T$ subregions (in particular the SO_4^{2-} subregion) down into one of the $S(-II)_T$ subregions (the H_2S subregion for pH < 7, or the HS^- subregion for pH > 7).

24.3.5 Will $FeS_{(s)}$ Form in This System? Yes.

Since the production of the dissolved $S(-II)_T$ will take place after essentially all of the Fe(III) is reduced to Fe(II), we finally ask if the $S(-II)_T$ will build up enough to allow $FeS_{(s)}$ to form. The common mineral form of $FeS_{(s)}$ is mackinawite. At 25°C/1 atm, the K_{s0} for this metal sulfide is $10^{-18.0}$. Once the system has become sufficiently reducing for significant formation of $S(-II)_T$ to occur, all of the $(am)Fe(OH)_{3(s)}$ will have been reduced to $Fe(II)_T$ (with only trace levels of dissolved $Fe(III)_T$ remaining), and so $Fe(II)_T$ will equal $8.953032 \times 10^{-6}\ M$. Taking pH = 6.60 as an estimate for that point, $\alpha_{0,Fe(II)}$ will equal ~1.00, and α_2 for the H_2S system will equal 1.13×10^{-8}. Thus, since $S(-II)_T$

represents the total concentration of dissolved S(−II), $FeS_{(s)}$ will start to form once

$$S(-II)_T = \frac{10^{-18.0}}{(1.13 \times 10^{-8})(1.00)(8.953032 \times 10^{-6})} = 9.92 \times 10^{-6}\ M. \qquad (23.14)$$

This is only ~10% of the total sulfur present per liter, and so we can expect that $FeS_{(s)}$ will become saturated and will precipitate at some point during the summer. The reader will note that we did not consider complexation of the Fe(II) by HS^- in the above calculations; available thermodynamic data suggests that such complexation is indeed weak.

24.4 A REDOX TITRATION MODEL FOR WATER IN THE HYPOLIMNION

24.4.1 The Redox Titration Equation (RTE)

24.4.1.1 General Considerations. As OC continually sediments into the hypolimnion, the reducing effects of the OC will slowly accumulate as various groups of bacteria catalyze the various redox reactions. We can develop a titration equation that keeps track of exactly how many equivalents of electrons have been added to the system. Including both the initial OC present, and the amount of OC added during the course of the summer, we have

$$\text{initial OC} + \text{added OC (eq/L)} = 3.330281 \times 10^{-4}\ \text{eq/L} + 1.489626 \times 10^{-5}\ (\text{eq/L-day})t$$
$$= \text{total eq/L of all reduced species formed.} \qquad (24.15)$$

The time t has units of days.

The reducing equivalents of the OC will accumulate in the various reduced species that are formed. Specifically, we have

$$\begin{aligned}\text{initial OC} + \text{added OC (eq/L)} &= \text{eq/L of } O_2 \text{ reduced (to water or hydroxide)}\\ &+ \text{eq/L of Fe(II) formed} + \text{eq/L of } \textit{new}\ N_2 \text{ formed}\\ &+ \text{eq/L of S(−II) formed} + \text{eq/L of } CH_4 \text{ formed}\\ &+ \text{eq/L of } H_2 \text{ formed.} \qquad (24.16)\end{aligned}$$

At any given point along the titration, there will be one pair of equilibrium pe and pH values, and that pair will satisfy Eq. (24.16).

24.4.1.2 The Chemistry Underlying the Terms in the RTE. The O_2 will be reduced to just one oxidation state, O(−II) (as water or hydroxide). This requires four electrons per O_2 molecule. The term "eq/L of O_2 reduced" in Eq. (24.16) is a direct measure of the OC that is consumed by reaction with O_2; with so much water present, it is much more direct (but not more correct) to do the bookkeeping in terms of the amount of O_2 that is *reduced*, rather than in terms of an increase in the concentration total for H_2O plus OH^-.

The term "eq/L of Fe(II) formed" in Eq. (24.16) includes all Fe(II) produced (both dissolved and solid), where the concentration in eq/L is computed for the one electron reduction of Fe(III) to Fe(II). With the possibility of $S_{(s)}$ ruled out (at least in a model based on the assumption of equilibrium), the only term for sulfur in Eq. (24.16) is that for the "eq/L of S(−II) formed". This quantity includes both the dissolved $S(-II)_T$ and the S(−II) in any $FeS_{(s)}$ that forms.

A term for "eq/L of NO_2^- formed" has not been included in Eq. (24.16) because, as discussed in Chapter 23, NO_2^- will only be a very minor species provided that N_2 is considered redox active at intermediate pe. However, as discussed in Section 24.2.2, if we allow N_2 to be fully redox active, then at the high pe values that will be present early in the summer, our model will convert N_2 to NO_3^-, and we know that this happens only *very slowly* in oxic waters. Thus, the way we will have to handle our equilibrium model is to exclude the possibility of N_2 being redox active as long as the pe and pH conditions would allow any of the initial N_2 to be oxidized to form more NO_3^- than is initially present. Then, once the pe is sufficiently low that the initial NO_3^- can become unstable relative to conversion to N_2, we can allow N_2 to be redox active in the NO_3^-/N_2 half-reaction (but not in the N_2/NH_3 half-reaction). At that point, the N_2 can play its role in regulating, in an equilibrium manner, the extent of production of new N_2.[7]

A term for "eq/L of NH_4^+/NH_3 formed" has not been included in Eq. (24.16) because ammonia-type nitrogen does not usually form in anoxic waters by the reduction of either NO_3^- or N_2 (see also Section 24.2.2). High levels of ammonia can be present in low pe waters, but the source is mainly the "de-amination" breakdown of biogenic proteins rather than the thermodynamically possible reduction of N_2 (Berner, 1980).

The term "eq/L of CH_4 formed" has been included to allow for the autooxidation (i.e., disproportionation) of the $C_6H_{12}O_6$. Since the $C(IV)_T/CH_4$ boundaryline in Figure 23.4 falls at rather low pe values, appreciable amounts of CH_4 will only begin to form towards the very end of the titration. The formation of any measurable H_2 will also be limited to the very end of the titration.

When $C_6H_{12}O_6$ disproportionates, equal amounts of CH_4 and CO_2 are formed:[8]

$$C_6H_{12}O_6 = (CO_2)_3(CH_4)_3 = 3CO_2 + 3CH_4. \qquad (24.17)$$

Some of the zero-oxidation-state carbon in the $C_6H_{12}O_6$ is oxidized up to C(IV), and an equal amount is reduced down to C(−IV). Alternately, but *completely equivalently*, we can think of each $C_6H_{12}O_6$ unit as reducing three molecules of CO_2 (C(IV)) in the solution to methane (C(−IV)), with all of the OC being oxidized to CO_2, that is,

$$C_6H_{12}O_6 + 3CO_2 = 6CO_2 + 3CH_4. \qquad (24.18)$$

Subtracting 3 CO_2 from both sides of Eq. (24.18) yields Eq. (24.17).

[7]We note that when bacteria have a fuel like OC available, they can harness some of the energy that results from the biochemical conversion of the fuel in order to drive a reaction in a direction that is opposite to that which would be indicated by equilibrium calculations. Both lower and higher organisms do this all the time in order to obtain needed nutrients and to build cell mass. Thus, just allowing thermodynamics to decide when any reaction will proceed does not always exactly simulate what happens in the real world.

[8]One can write a disproportionation reaction like Eq. (24.17) for any oxygen-containing organic compound. For CH_3-CH_2-COOH (propanoic acid), we obtain

$$4CH_3\text{-}CH_2\text{-}COOH + 2H_2O = 5CO_2 + 7CH_4$$

Based on this discussion, we now reexpress Eq. (24.16) in more specific terms. Continuing to use $Fe(II)_T$ and $S(-II)_T$ to represent the molar concentrations of *dissolved* Fe(II) and S(−II), respectively, we have

$$\begin{aligned} \text{eq/L of OC added} &= \text{eq/L of } O_2 \text{ reduced (to water)} + Fe(II)_T + [FeS_{(s)}] \\ &\quad + \text{eq/L of } NO_3^- \text{ reduced (to } N_2) + 8\ S(-II)_T \\ &\quad + 8\ [FeS_{(s)}] + 8[CH_4] + 2[H_2] \end{aligned} \qquad (24.19)$$

where $[FeS_{(s)}]$ has units of FW/(L of water). $[FeS_{(s)}]$ appears twice in Eq. (24.19) because it must be counted for the Fe(II) it contains, and also for the S(−II) it contains. The term "eq/L of NO_3^- reduced (to N_2)" has been substituted for "eq/L of *new* N_2 formed" for simplicity in the electron bookkeeping. Prior to the start of the model, it will be assumed that the RHS of Eq. (24.19) is zero. Although one can expect that some $Fe(II)_T$, CH_4, and H_2 will be present in the initial water, those levels will be exceedingly low.

When counting the S(−II) that has been formed from S(VI), we include a factor of 8 in Eq. (24.19) because eight electrons are required for that reduction. When counting the H(0) in H_2 that is formed from H(I), we include a factor of 2 because two electrons are required for that reduction. When counting the CH_4 that has been formed, we include a factor of 8. In this case, the formation of CH_4 from the disproportionation of the OC effectively stores *all* of the reducing equivalents of the OC in dissolved CH_4. As indicated by the stoichiometry given in reaction (24.17), each carbon in the OC that ends up as a CH_4 molecule not only gets the four electrons that it had by starting out as C(0), but it also gets four electrons yielded by one C(0) carbon that is raised to C(IV). Thus, each mol of CH_4 contains 8 electrons worth of reducing equivalents. When the reaction of the OC is viewed in terms of reaction 24.18, the OC reduces C(IV) to C(−IV), and so the eq/L of CH_4 formed is again seen to be $8[CH_4]$.

24.4.1.3 Expressing the Terms in the RTE as Functions of pe and pH.

We now seek to express each of the terms in the RHS of Eq. (24.19) as a function of pe and pH. We focus first on the oxygen, nitrogen, carbon, and hydrogen because none of these totally-dissolved elements will be participating in any solid-forming reactions. Thus, the functions for these elements will remain the same throughout the titration process.

As noted in the previous section, the initial oxygen level will be $[O_2, \text{eq/L}] = 1.057497 \times 10^{-3}$ eq/L. Thus, at any given time during the summer, *for the oxygen*, we have

$$\text{eq/L of } O_2 \text{ reduced} = \text{OXR} = 1.057497\times10^{-3}\ \text{eq/L} - \text{eq/L } O_2 \text{ remaining.} \qquad (24.20)$$

By Eq. (24.6), at 25°C/1 atm,

$$\text{eq/L of } O_2 \text{ reduced} = \text{OXR} = 1.057497 \times 10^{-3}\ \text{eq/L} - 10^{4(\text{pe}-21.35+\text{pH})}. \qquad (24.21)$$

For the nitrogen, we have

$$\text{eq/L of } NO_3^- \text{ reduced} = \text{NO3R} = 7.139440\times10^{-5}\ \text{eq/L} - 5[NO_3^-] \text{ remaining} \qquad (24.22)$$

where the factor of 5 in Eq. (24.22) stems from the fact that each NO_3^- is being reduced to 1/2 N_2. From Table 24.3, for equilibrium between NO_3^- and N_2 at 25°C/1 atm, neglecting activity corrections, we have

$$[NO_3^-] = \frac{[N_2]^{1/2}}{10^{103.7}10^{-6pH}10^{-5pe}}. \quad (24.23)$$

Neglecting both $[NO_2^-]$ and ammonia-type nitrogen, for the nitrogen MBE we have that

$$N_T = 2[N_2] + [NO_3^-]. \quad (24.24)$$

Substituting Eq. (24.23), we obtain

$$2[N_2] + \frac{[N_2]^{1/2}}{10^{103.7}10^{-6pH}10^{-5pe}} - N_T = 0. \quad (24.25)$$

At given values of pe, pH, and N_T, Eq. (24.25) can be solved for $x = [N_2]^{1/2}$ by the quadratic equation. In our system, $N_T = 9.985723 \times 10^{-4}$ M. When we compare the equation $ax^2 + bx + c = 0$ to Eq. (24.25) we see that b in Eq. (24.25) is positive for all pe and pH. Therefore, to get a positive root to Eq. (24.25), we know that we want the root $(-b + \sqrt{(b^2 - 4ac)})/2a$ for all pe and all pH.

With $[N_2]$ determined from Eq. (24.25), NO3R can be calculated utilizing Eqs. (24.23) and (24.22), in that order. (Use of Eq. (24.24) to calculate $[NO_3^-]$ might lead to roundoff error when $2[N_2]$ approaches N_T.) As discussed above, we will take the NO3R term in the RHS of Eq. (24.19) to be zero until the pe and pH conditions are brought to the point that the initial $[NO_3^-]$ is thermodynamically consistent with the initial $[N_2]$. By the conditions given in Table 24.2, those conditions must satisfy

$$10^{-5pe}10^{-6pH} = \frac{(4.921467 \times 10^{-4})^{1/2}}{(1.427888 \times 10^{-5})10^{103.7}} \quad (24.26)$$

or

$$5pe + 6pH = 100.51. \quad (24.27)$$

Prior to as well as at the point at which Eq. (24.27) is satisfied, we will assume that NO3R = 0. Before any redox chemistry occurs, the "initial" pe and pH values of 12.606 and 8.00 give (5pe + 6pH) = 111.03. Reductions in pe and/or pH will be able to lead to satisfaction of Eq. (24.27). Once that occurs, we can consider the NO_3^-/N_2 couple to be redox active, and we can use Eq. (24.25) (with $N_T = 9.985723 \times 10^{-4}$ M) and Eqs. (24.23) and (24.22) (in that order) to compute NO3R as a function of pe and pH.

For the carbon, we have

$$\text{eq/L of } CH_4 \text{ formed} = (8 \text{ eq/mol})[CH_4] = 8\text{MEF}. \quad (24.28)$$

From Table 24.3, for equilibrium between CO_2 and CH_4 at 25°C/1 atm, neglecting activity corrections, we have

$$[CH_4] = 10^{21.6}[H_2CO_3^*]10^{-8pH}10^{-8pe}. \quad (24.29)$$

Neglecting $C_6H_{12}O_6$ in the carbon MBE (see Section 23.3.1.2), we have that

$$C_T = C(IV)_T + [CH_4] = [H_2CO_3^*]/\alpha_0 + [CH_4] \quad (24.30)$$

where as usual $C(IV)_T$ represents the concentration sum for all CO_2-type carbon species. Substituting Eq. (24.29), we obtain

$$C_T = [CH_4]10^{8pH}10^{8pe}10^{-21.6}/\alpha_0 + [CH_4] \tag{24.31}$$

or

$$[CH_4] = \frac{C_T}{10^{8pH}10^{8pe}10^{-21.6}/\alpha_0 + 1}. \tag{24.32}$$

The value of C_T will depend on time (t) in days according to

$$C_T = 5.748410 \times 10^{-4}\ M + (3.746566 \times 10^{-6}\ M/\text{day})t \tag{24.33}$$

where the two terms on the RHS represent: a) the sum of the initial $C(IV)_T$ and the initial OC, and b) the sedimented OC, respectively. At any given values of pe, pH, we have

$$8MEF = (8\text{eq/mol})\frac{5.748410 \times 10^{-4}\ M + (3.746566 \times 10^{-6}\ M/\text{day})t}{10^{8pH}10^{8pe}10^{-21.6}/\alpha_0 + 1}. \tag{24.34}$$

For the hydrogen, from Table 24.3, for equilibrium between H^+ and H_2 at 25°C/1 atm, neglecting activity corrections, we have

$$[H_2] = 10^{-3.10}10^{-2pH}10^{-2pe} \tag{24.35}$$

$$\text{eq/L of } H_2 \text{ formed} = (2\text{eq/mol})[H_2] = 2 \times 10^{-3.10}10^{-2pH}10^{-2pe}. \tag{24.36}$$

For the iron, as long as $(am)Fe(OH)_{3(s)}$ is present, neglecting activity corrections, we have

$$Fe(II)_T = 10^{16.2}10^{-3pH}10^{-pe}/\alpha_{0,Fe(II)} \tag{24.37}$$

where as before, $\alpha_{0,Fe(II)} = [Fe^{2+}]/Fe(II)_T$. We will assume that $(am)Fe(OH)_{3(s)}$ is sufficiently insoluble at neutral pH (see Figure 11.3) that it will remain present until $Fe(II)_T = Fe_T = 8.953032 \times 10^{-6}\ M$. By Eq. (24.37) then, $(am)Fe(OH)_{3(s)}$ will remain present until

$$3\ pH + pe + \log \alpha_{0,Fe(II)} = 21.25. \tag{24.38}$$

The initial pe, pH, and $\alpha_{0,Fe(II)}$ values of 12.606, 8.00, and 0.969 give ($3pH + pe + \log \alpha_{0,Fe(II)}) = 36.59$.

For the sulfur, from Table 24.3, we have

$$pe = 4.25 + \frac{1}{8}\log\frac{\alpha_{2,S(VI)}S(VI)_T[H^+]^9}{\alpha_{1,S(-II)}S(-II)_T}. \tag{24.39}$$

Before any $FeS_{(s)}$ precipitates, assuming that only very small amounts of S(IV) can form (see Section 23.3.3.2), the sulfur MBE will be

$$S(VI)_T + S(-II)_T = S_T. \tag{24.40}$$

Elemental sulfur is not included in Eq. (24.40) because: 1) it was shown that $S_{(s)}$ cannot form in this system; and 2) we can assume that the concentration of dissolved S(0) will be negligible. Assuming that $\alpha_{2,S(VI)} \simeq 1$, by Eq. (24.39) we have

$$S(VI)_T = \frac{10^{8(pe-4.25)}}{10^{-9pH}}\alpha_{1,S(-II)}S(-II)_T. \tag{24.41}$$

Substitution of Eq. (24.41) into Eq. (24.40) yields

$$S(-II)_T = S1 = \frac{S_T}{1 + \frac{10^{8(pe-4.25)}}{10^{-9pH}}\alpha_{1,S(-II)}}. \quad (24.42)$$

Up to the point that $FeS_{(s)}$ just becomes saturated, $S(-II)_T$ as given by Eq. (24.42) may be used in Eq. (24.19) as the total measure of all S(−II) produced.

Since the K_{s0} for $FeS_{(s)}$ is $10^{-18.0}$, neglecting activity corrections at 25°C/1 atm, that solid will just become saturated when

$$[Fe^{2+}][S^{2-}] = \alpha_{0,Fe(II)}Fe(II)_T\alpha_{2,S(-II)}S(-II)_T = 10^{-18.0}. \quad (24.43)$$

Since all of the $(am)Fe(OH)_{3(s)}$ will surely be reduced by the time that the $FeS_{(s)}$ just starts to precipitate, at that point $Fe(II)_T$ will essentially equal Fe_T (there will also be trace levels of dissolved Fe(III)), and so the equation that will need to be satisfied is

$$\alpha_{2,S(-II)}S(-II)_T\alpha_{0,Fe(II)}8.95 \times 10^{-6} = 10^{-18.0}. \quad (24.44)$$

Since no $FeS_{(s)}$ will yet be present when the $FeS_{(s)}$ first becomes saturated, Eq. (24.42) can be used to provide values of $S(-II)_T$ when using Eq. (24.44) to test for the onset of $FeS_{(s)}$ saturation.

When expressed in units of (FW of solid)/(L of water), once it starts to form, $[FeS_{(s)}]$ will be given by

$$[FeS_{(s)}] = FES = S_T - S(-II)_T - S(VI)_T. \quad (24.45)$$

Substitution of Eq. (24.41) yields

$$[FeS_{(s)}] = S_T - S(-II)_T - \frac{10^{8(pe-4.25)}}{10^{-9pH}}\alpha_{1,S(-II)}S(-II)_T. \quad (24.46)$$

When $FeS_{(s)}$ is present, computing $S(-II)_T$ and $Fe(II)_T$ as a function of pe and pH requires consideration of the coupled effects of the iron and the sulfur on the time evolution of the pe and the pH. This coupling is most easily dealt with using the K_{s0} for $FeS_{(s)}$, that is, Eq. (24.43). With no $(am)Fe(OH)_{3(s)}$ present and only trace levels of dissolved Fe(III) in solution, $Fe(II)_T = Fe_T - [FeS_{(s)}]$. With $[FeS_{(s)}]$ given by Eq. (24.46), we obtain

$$\alpha_{0,Fe(II)}\left[Fe_T - S_T + S(-II)_T + \frac{10^{8(pe-4.25)}}{10^{-9pH}}\alpha_{1,S(-II)}S(-II)_T\right]\alpha_{2,S(-II)}S(-II)_T = 10^{-18.0} \quad (24.47)$$

or

$$\alpha_{0,Fe(II)}\left[1 + \frac{10^{8(pe-4.25)}}{10^{-9pH}}\alpha_{1,S(-II)}\right]\alpha_{2,S(-II)}S(-II)_T^2 + \alpha_{0,Fe(II)}(Fe_T - S_T)\alpha_{2,S(-II)}S(-II)_T - 10^{-18.0} = 0. \quad (24.48)$$

which, for specified values of pe and pH, can be solved by the quadratic equation for $S(-II)_T$. The root of interest here is $(-b + \sqrt{(b^2 - 4ac)})/2a$ for all pe and all pH. Since $FeS_{(s)}$ will be forming during the *fourth* period of interest during the summer (see

Tables 24.6 and 24.7), we will refer to this root as S4. With values for $S(-II)_T$ and pH, a corresponding value for $Fe(II)_T$ can be calculated using Eq. (24.43), and $[FeS_{(s)}]$ (= FES) can be calculated using Eq. (24.46).

We now have expressions for each of the terms on the RHS of Eq. (24.19) as functions of pe and pH. The way we will use them will depend on whether: 1) we are taking NO_3^-/N_2 to be redox active; 2) $(am)Fe(OH)_{3(s)}$ is still present; and/or 3) $FeS_{(s)}$ has precipitated yet. However, even for a specific value of time t, we are still left with two unknowns in the equation, namely pe and pH. We need another equation to solve for pe and pH as a function of the eq/L of OC added. The ENE can serve as that equation.

24.4.2 The ENE

Because we need a second equation linking time with pe and pH, we now establish the ENE in that role. Prior to the onset of any redox chemistry (i.e., at time zero where the pH is 8.00), for the ENE we have

$$[H^+]_o+[Na^+] = (\alpha_1 C(IV)_T)_o+2(\alpha_2 C(IV)_T)_o+[NO_3^-]_o+2[SO_4^{2-}]_o+[OH^-]. \qquad (24.49)$$

where it has been assumed that the initial levels of $Fe(II)_T$ and $S(-II)_T$ are negligibly low, that $[HSO_4^{2-}]$ is unimportant, and that Na^+ makes up all of the *net* positive charge needed to obtain initial electroneutrality.[9] The small subscript zero signifies an initial value.

Based on Eq. (24.49), we have

$$10^{-8}+[Na^+] = (0.973649)(4.915840\times10^{-4})+2(4.554097\times10^{-3})(4.915840\times10^{-4})$$
$$+\ 1.427888\times10^{-5}+2(9.357455\times10^{-5})+10^{-6}. \qquad (24.50)$$

Since Na^+ will neither be produced nor removed from solution during the summer, a subscript zero on the $[Na^+]$ is not needed, and during the entire summer, we have

$$[Na^+] = 6.855255\times10^{-4}\ M. \qquad (24.51)$$

Once some oxidation of OC begins, various reduced species will begin to form and the ENE will be

$$[H^+]+[Na^+]+2[Fe^{2+}]+[FeOH^+] = \alpha_1 C(IV)_T + 2\alpha_2 C(IV)_T + [NO_3^-]$$
$$+\ 2[SO_4^{2-}] + a_{1,S(-II)} S(-II)_T + [OH^-]. \qquad (24.52)$$

The values of the concentrations of S^{2-} and all Fe(III) species have been assumed to be negligible (see also the development of Eq. (24.38)). We also note that the sum $[Fe^{2+}]+[FeOH^+]$ equals $(2\alpha_{0,Fe(II)}+\alpha_{1,Fe(II)})Fe(II)_T \simeq (1+\alpha_{0,Fe(II)})Fe(II)_T$.

[9] Even if there were some other ions such as Cl^- present in the solution, then Eq. (24.49) would still be valid since then $[Na^+]$ as given would just represent the value of $([Na^+]-[Cl^-]) = (C_B - C_A)$.

With the exception of $[Na^+]$ which is a constant, all of the terms in Eq. (24.52) can be expressed as functions of pe, pH, and time. By Eqs. (24.30) and (24.32), we have that

$$\begin{aligned} C(IV)_T &= C_T - \frac{C_T}{10^{8pH}10^{8pe}10^{-21.6}/\alpha_0 + 1} = C_T - \text{MEF} \\ &= [5.748410 \times 10^{-4}\ M + (3.746566 \times 10^{-6}\ M/\text{day})t] \\ &\quad \times \left[1 - \frac{1}{10^{8pH}10^{8pe}10^{-21.6}/\alpha_0 + 1}\right]. \end{aligned} \tag{24.53}$$

Equation (24.52) can now be used together with Eq. (24.19) to solve by trial and error for pe and pH as functions of (eq/L of OC added). The solution can be obtained using the Gauss-Seidel method. With that method, for a given value of (eq/L of OC), a guess for pH is made, and Eq. (24.19) is then solved by trial and error for pe. The resulting pe is then used to solve Eq. (24.52) for pH. That pH is then used to solve Eq. (24.19) again for pe, and so on, back and forth until convergence for both pe and pH is attained for the time (i.e., eq/L of OC) of interest.[10] The values obtained at convergence are the equilibrium pe and pH for that point in time. The process is then repeated for a new larger value of time (i.e., eq/L of OC added). After carrying out this process for a whole range of times, we obtain the equilibrium pe and pH as functions of time. The concentrations of the individual species can then be calculated as functions of time.

24.4.3 Switching Between Different Versions of the RTE and Between the Different Versions of the ENE

As has been discussed in the preceding sections, there are four distinct periods of interest during the summer. During *Period 1*, the initial N_2 is unstable with respect to oxidation to NO_3^-, and so we consider the NO_3^-/N_2 redox couple to be frozen; during the same period, (am)$Fe(OH)_{3(s)}$ is still present, but no $FeS_{(s)}$ has yet formed. During *Period 2*, we reach pe and pH values at which NO_3^- can be reduced to N_2, and so we unfreeze the NO_3^-/N_2 redox couple; the solid (am)$Fe(OH)_{3(s)}$ is still present, but $FeS_{(s)}$ is not yet present. During *Period 3*, we have reached conditions under which all of the (am)$Fe(OH)_{3(s)}$ has dissolved to form $Fe(II)_T$. The solid $FeS_{(s)}$ is still not yet present. During *Period 4*, $FeS_{(s)}$ is present. The versions of the RTE that apply during the various periods are summarized in Tables 24.4 and 24.5. The versions of the ENE that apply during the same periods are summarized in Tables 24.4 and 24.6.

Within each period, and for any given value of time t, the appropriate version of the RTE from Table 24.5 would be used with the corresponding ENE from Table 24.6 to solve by Gauss-Seidel iteration for the equilibrium pe and pH values; the concentrations of all of the species for that time may then be calculated using those values.

[10] Alternatively, an initial guess for pe could be made and Eq. (24.19) could be solved for pH, then that pH could be used with Eq. (24.53) to solve for pe, and so on, back and forth until convergence is obtained.

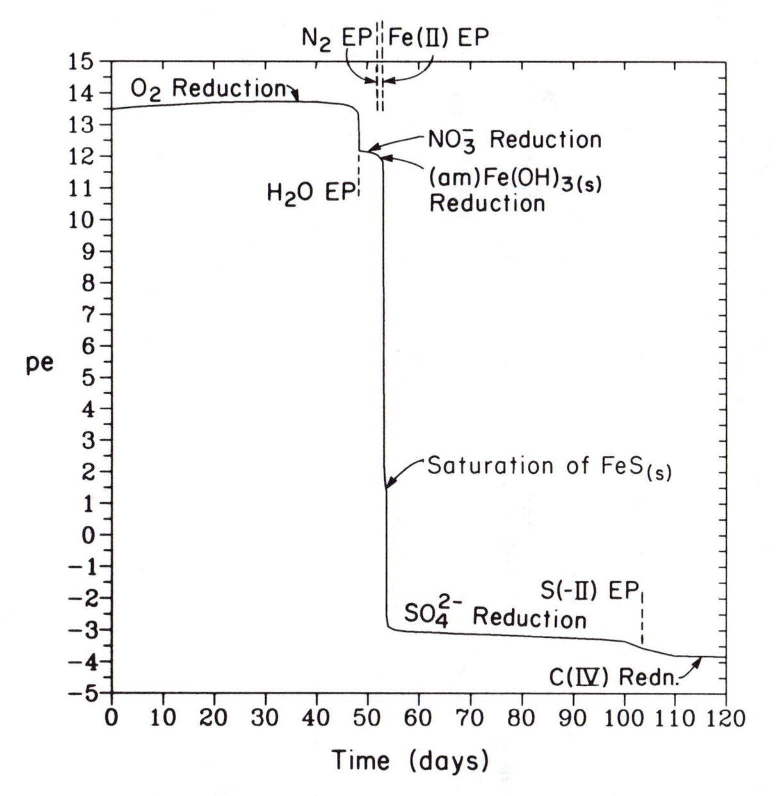

Figure 24.4 pe vs. time in the hypolimnion.

TABLE 24.4. Summary of the four periods during the summer in the specific hypolimnion case of interest. Names are given to the four versions of the redox titration equation (RTE), and to the four versions of the electroneutrality equation (ENE).

Period	NO_3^-/N_2 Redox Active?	$N_2/N(-III)$ Redox Active?	$(am)Fe(OH)_{3(s)}$ Present?	$FeS_{(s)}$ Present?	Version of RTE	Version of ENE
1	No	No	Yes	No	RTE1	ENE1
2	Yes	No	Yes	No	RTE2	ENE2
3	Yes	No	No	No	RTE3	ENE3
4	Yes	No	No	Yes	RTE4	ENE4

Table 24.7 shows how the pe, the pH, and the concentrations of the various chemical species change with time in the model. Figures 24.4 and 24.5 show how the pe and pH vary with time. Figure 24.6 shows the pe-pH track on a pe vs. pH plot. Figure 24.7 shows the time-dependence of the *M* (or FW per liter of solution) concentrations of each of the nine important redox species.

TABLE 24.5. Versions of the redox titration equation (RTE) for four different periods during the summer.

initial plus added OC = (eq/L)		O_2 reduced + (eq/L)	$Fe(II)_T$ (eq/L)	+ new N_2 formed + (eq/L)	$8S(-II)_T$ (dissolved) + (eq/L)	$9[FeS_{(s)}]$ + (eq solid/L of water)	8MEF + (eq/L)	$2[H_2]$ (eq/L)
Period 1. RTE1. NO_3^-/N_2 is considered frozen, $Fe(OH)_{3(s)}$ is still present, $FeS_{(s)}$ not yet present.								
$(3.330281 \times 10^{-4} + 1.498626 \times 10^{-5})t$	=	OXR	$+ \dfrac{10^{16.2}10^{-3pH}10^{-pe}}{\alpha_{0,Fe(II)}}$	+ 0	+ 8S1	+ 0	+ 8MEF	$+ 2 \times 10^{-3.10}10^{-2pe}10^{-2pH}$
Period 2. RTE2. NO_3^- begins to be reduced, $Fe(OH)_{3(s)}$ is still present, $FeS_{(s)}$ not yet present.								
$(3.330281 \times 10^{-4} + 1.498626 \times 10^{-5})t$	=	OXR	$+ \dfrac{10^{16.2}10^{-3pH}10^{-pe}}{\alpha_{0,Fe(II)}}$	+ NO3R	+ 8S1	+ 0	+ 8MEF	$+ 2 \times 10^{-3.10}10^{-2pe}10^{-2pH}$
Period 3. RTE3. $Fe(OH)_{3(s)}$ is no longer present, $FeS_{(s)}$ not yet present.								
$(3.330281 \times 10^{-4} + 1.498626 \times 10^{-5})t$	=	OXR	$+ 8.953032 \times 10^{-6}$	+ NO3R	+ 8S1	+ 0	+ 8MEF	$+ 2 \times 10^{-3.10}10^{-2pe}10^{-2pH}$
Period 4. RTE4. $FeS_{(s)}$ is present.								
$(3.330281 \times 10^{-4} + 1.498626 \times 10^{-5})t$	=	OXR	$+ \dfrac{10^{-18.0}}{\alpha_{0,Fe(II)}\alpha_{2,S(-II)}S4}$	+ NO3R	+ 8S4	+ 9FES	+ 8MEF	$+ 2 \times 10^{-3.10}10^{-2pe}10^{-2pH}$

Note :

$$\text{OXR (eq/L)} = (1.057497 \times 10^{-4} - 10^{4(pe-21.35+pH)}) \quad (24.21)$$

$$\text{8MEF (eq/L)} = 8\frac{5.748410 \times 10^{-4} + (3.746566 \times 10^{-6})t}{10^{8pH}10^{8pe}10^{-21.6}/\alpha_0 + 1} \quad (24.34)$$

$$\text{S1 } (M) = \frac{S_T}{1 + \dfrac{10^{8(pe-4.25)}}{10^{-9pH}}\alpha_{1,S(-II)}}$$

$$\text{S4 } (M) = (-\text{BSUL} + (\text{BSUL}^2 - 4 * \text{ASUL} * \text{CSUL})^{1/2})/(2 * \text{ASUL})$$

$$\text{FES (FW/L)} = S_T - \text{S4} - \frac{10^{8(pe-4.25)}}{10^{-9pH}}\alpha_{1,S(-II)}\text{S4}$$

$$\text{NO3R (eq/L)} = 7.139440 \times 10^{-5} - 5[(-\text{BNIT} + (\text{BNIT}^2 - 4 * \text{ANIT} * \text{CNIT})^{1/2})/(2 * \text{ANIT})] * \text{BNIT}$$

$$\text{ANIT} = 2$$

$$\text{BNIT} = (10^{103.7}10^{-6pH}10^{-5pe})^{-1}$$

$$\text{CNIT} = -N_T$$

$$\text{ASUL} = \alpha_{0,Fe(II)}\left[1 + \frac{10^{8(pe-4.25)}}{10^{-9pH}}\alpha_{1,S(-II)}\right]\alpha_{2,S(-II)}$$

$$\text{BSUL} = \alpha_{0,Fe(II)}(Fe_T - S_T)\alpha_{2,S(-II)}$$

$$\text{CSUL} = -10^{-18}$$

TABLE 24.6. Versions of the ENE for four different periods during the summer.

$[H^+]$	+	$[Na^+]$	+	$(2[Fe^{2+}]+[FeOH^+])$	=	$[HCO_3^-]+2[CO_3^{2-}]$	+	$[NO_3^-]$	+	$2[SO_4^{2-}]$	+	$[HS^-]$	+	$[OH^-]$
Period 1. ENE1. NO_3^-/N_2 is considered frozen, $Fe(OH)_{3(s)}$ is still present, $FeS_{(s)}$ not yet present.														
10^{-pH}	+	6.855255×10^{-4}	+	$(1+\alpha_{0,Fe(II)})\dfrac{10^{16.2}10^{-3pH}10^{-pe}}{\alpha_{0,Fe(II)}}$	=	$\alpha_1 C(IV)_T + 2\alpha_2 C(IV)_T$	+	1.427888×10^{-5}	+	$2(S_T - S1)$	+	$\alpha_{1,S(-II)}S1$	+	$K_w 10^{pH}$
Period 2. ENE2. NO_3^- begins to be reduced, $Fe(OH)_{3(s)}$ is still present, $FeS_{(s)}$ not yet present.														
10^{-pH}	+	6.855255×10^{-4}	+	$(1+\alpha_{0,Fe(II)})\dfrac{10^{16.2}10^{-3pH}10^{-pe}}{\alpha_{0,Fe(II)}}$	=	$\alpha_1 C(IV)_T + 2\alpha_2 C(IV)_T$	+	NO3	+	$2(S_T - S1)$	+	$\alpha_{1,S(-II)}S1$	+	$K_w 10^{pH}$
Period 3. ENE3. $Fe(OH)_{3(s)}$ is no longer present, $FeS_{(s)}$ not yet present.														
10^{-pH}	+	6.855255×10^{-4}	+	$(1+\alpha_{0,Fe(II)})(8.95 \times 10^{-6})$	=	$\alpha_1 C(IV)_T + 2\alpha_2 C(IV)_T$	+	NO3	+	$2(S_T - S1)$	+	$\alpha_{1,S(-II)}S1$	+	$K_w 10^{pH}$
Period 4. ENE4. $FeS_{(s)}$ is present.														
10^{-pH}	+	6.855255×10^{-4}	+	$(1+\alpha_{0,Fe(II)})\dfrac{10^{-18.0}}{\alpha_{0,Fe(II)}\alpha_{2,S(-II)}S4}$	=	$\alpha_1 C(IV)_T + 2\alpha_2 C(IV)_T$	+	NO3	+	$2(S_T - S4 - FES)$	+	$\alpha_{1,S(-II)}S4$	+	$K_w 10^{pH}$

Note :

$$S1\ (M) = \frac{S_T}{1 + \dfrac{10^{8(pe-4.25)}}{10^{-9pH}}\alpha_{1,S(-II)}}$$

$$S4\ (M) = (-BSUL + (BSUL^2 - 4 * ASUL * CSUL)^{1/2})/(2 * ASUL)$$

$$FES\ (FW/L) = S_T - S4 - \frac{10^{8(pe-4.25)}}{10^{-9pH}}\alpha_{1,S(-II)}S4$$

$$C(IV)_T\ (M) = 5.748410 \times 10^{-4} + (3.746566 \times 10^{-6})t - MEF$$

$$NO3\ (M) = [(-BNIT + (BNIT^2 + 8N_T)^{1/2})/4] * BNIT$$

$$ANIT = 2$$

$$BNIT = (10^{103.7}10^{-6pH}10^{-5pe})^{-1}$$

$$CNIT = -N_T$$

$$ASUL = \alpha_{0,Fe(II)}\left[1 + \frac{10^{8(pe-4.25)}}{10^{-9pH}}\alpha_{1,S(-II)}\right]\alpha_{2,S(-II)}$$

$$BSUL = \alpha_{0,Fe(II)}(Fe_T - S_T)\alpha_{2,S(-II)}$$

$$CSUL = -10^{-18}$$

TABLE 24.7. pe, pH, and Concentrations of Various Species of Interest During the Course of the Summer.

	time (days)	OC oxidized (eq/L)	pH	pe	TEST	log O_2 (M)	log O_2 (eq/L)	log NO_3^- (M)	log NO_3^- (eq/L)	log $Fe(OH)_{3(s)}$ (eq/L, FW/L)	log $Fe(II)_T$ (eq/L)	$\alpha_{0,Fe(II)}$
	Period 0. Redox equilibrium not yet established.											
	0	0.00E + 00	8.00000	12.60560	111.0280	– 3.58	– 2.97	– 4.85	– 4.15	– 5.05	–	0.9693
	Period 1. NO_3^-/N_2 couple is considered frozen, $(am)Fe(OH)_{3(s)}$ is still present, $FeS_{(s)}$ not yet present. The parameter TEST examines whether the initial NO_3^- can be reduced to N_2; the critical value is 100.51.											
Mostly O_2 Reduction	0	3.33 E–04	7.075153	13.489852	109.90	– 3.74	– 3.14	– 4.85	– 4.15	– 5.05	– 18.51	0.9963
	5	4.08 E–04	6.994340	13.558811	109.76	– 3.79	– 3.19	– 4.85	– 4.15	– 5.05	– 18.34	0.9969
	10	4.83 E–04	6.926171	13.613671	109.63	– 3.84	– 3.24	– 4.85	– 4.15	– 5.05	– 18.19	0.9973
	15	5.58 E–04	6.867235	13.657437	109.49	– 3.90	– 3.30	– 4.85	– 4.15	– 5.05	– 18.05	0.9977
	20	6.33 E–04	6.815331	13.691700	109.35	– 3.97	– 3.37	– 4.85	– 4.15	– 5.05	– 17.93	0.9979
	25	7.08 E–04	6.768965	13.716994	109.20	– 4.06	– 3.46	– 4.85	– 4.15	– 5.05	– 17.82	0.9981
	30	7.83 E–04	6.727069	13.732718	109.03	– 4.16	– 3.56	– 4.85	– 4.15	– 5.05	– 17.71	0.9983
	35	8.58 E–04	6.688858	13.736372	108.82	– 4.30	– 3.70	– 4.85	– 4.15	– 5.05	– 17.60	0.9985
	40	9.32 E–04	6.653737	13.720507	108.52	– 4.51	– 3.90	– 4.85	– 4.15	– 5.05	– 17.48	0.9986
	45	1.01 E–03	6.621244	13.653687	108.00	– 4.90	– 4.30	– 4.85	– 4.15	– 5.05	– 17.31	0.9987
	46	1.02 E–03	6.615028	13.621302	107.80	– 5.06	– 4.45	– 4.85	– 4.15	– 5.05	– 17.26	0.9987
	47	1.04 E–03	6.608899	13.566980	107.49	– 5.30	– 4.70	– 4.85	– 4.15	– 5.05	– 17.19	0.9987
	48	1.05 E–03	6.602855	13.424641	106.74	– 5.89	– 5.29	– 4.85	– 4.15	– 5.05	– 17.03	0.9987
	48.2	1.06 E–03	6.601656	13.330498	106.26	– 6.27	– 5.67	– 4.85	– 4.15	– 5.05	– 16.93	0.9987
H_2O EP →	48.342	1.06 E–03	6.600807	12.622699	102.72	– 9.11	– 8.51	– 4.85	– 4.15	– 5.05	– 16.22	0.9987
	Period 2. NO_3^- begins to be reduced, $(am)Fe(OH)_{3(s)}$ is still present, $FeS_{(s)}$ not yet present. The parameter TEST examines whether $(am)Fe(OH)_{3(s)}$ is still present; the critical value is 21.25. Under the assumption that $(am)Fe(OH)_{3(s)}$ is largely insoluble, and when $FeS_{(s)}$ is not yet present, complete dissolution of $(am)Fe(OH)_{3(s)}$ is also observable when $\log Fe(II)_T = \log Fe_T = -5.05$.											
Mostly NO_3^- reduction	48.343	1.06 E–03	6.600807	12.180746	31.98	– 10.88	– 10.27	– 4.85	– 4.15	– 5.05	– 15.78	0.9987
	49	1.07 E–03	6.601780	12.166600	31.97	– 10.93	– 10.33	– 4.91	– 4.21	– 5.05	– 15.77	0.9987
	50	1.08 E–03	6.603251	12.140461	31.95	– 11.03	– 10.43	– 5.03	– 4.33	– 5.05	– 15.74	0.9987
	51	1.10 E–03	6.604709	12.104835	31.92	– 11.15	– 10.56	– 5.20	– 4.50	– 5.05	– 15.71	0.9987
	52	1.11 E–03	6.606153	12.047037	31.86	– 11.39	– 10.79	– 5.48	– 4.78	– 5.05	– 15.66	0.9987
	53	1.13 E–03	6.607585	11.841645	31.66	– 12.21	– 11.60	– 6.50	– 5.80	– 5.05	– 15.46	0.9987
N_2 EP →	53.106	1.13 E–03	6.607736	11.295826	31.12	– 14.39	– 13.79	– 9.23	– 8.53	– 5.05	– 14.91	0.9987
Mostly $(am)Fe(OH)_{3(s)}$ reduction	53.2	1.13 E–03	6.614057	2.210492	22.05	– 50.70	– 50.10	– 54.61	– 53.92	– 5.12	– 5.852	0.9987
	53.4	1.13 E–03	6.627604	1.674024	21.56	– 52.80	– 52.19	– 57.22	– 56.52	– 5.34	– 5.356	0.9987
	53.65	1.14 E–03	6.644684	1.355421	21.29	– 54.00	– 53.40	– 58.71	– 58.01	– 6.10	– 5.088	0.9986
	53.7	1.14 E–03	6.648123	1.306908	21.25	– 54.18	– 53.58	– 58.93	– 58.23	– 7.27	– 5.050	0.9986
Fe(II) EP →	53.703	1.14 E–03	6.648329	1.304100	21.25	– 54.19	– 53.59	– 58.94	– 58.24	– 8.04	– 5.048	0.9986
	Period 3. $Fe(OH)_{3(s)}$ is no longer present, $FeS_{(s)}$ not yet present. The parameter TEST examines the value of $[Fe^{2+}][S^{2-}]$; the critical value for saturation of $FeS_{(s)}$ is $K_{s0} = 1.00 \times 10^{-18}$.											
Mostly SO_4^{2-} reduction	53.704	1.14 E–03	6.648372	– 2.526310	8.83 E–23	– 69.51	– 68.91	– 78.09	– 77.39	–	– 5.048	0.9986
	54	1.14 E–03	6.648873	– 2.888073	6.82 E–20	– 70.96	– 70.36	– 79.90	– 79.20	–	– 5.048	0.9986
	55	1.16 E–03	6.650551	– 2.971320	3.00 E–19	– 71.29	– 70.68	– 80.30	– 79.61	–	– 5.048	0.9986
	56	1.17 E–03	6.652208	– 3.005493	5.35 E–19	– 71.42	– 70.81	– 80.47	– 79.77	–	– 5.048	0.9986
	57	1.19 E–03	6.653843	– 3.028250	7.73 E–19	– 71.50	– 70.90	– 80.57	– 79.87	–	– 5.048	0.9986
$FeS_{(s)}$ → saturation	57.943	1.20 E–03	6.655367	– 3.044861	1.00 E–18	– 71.56	– 70.96	– 80.64	– 79.94	–	– 5.048	0.9986
	Period 4. $FeS_{(s)}$ is present.											
Mostly SO_4^{2-} reduction	57.944	1.20 E–03	6.655362	– 3.044867	–	– 71.56	– 70.96	– 80.64	– 79.94	–	– 5.048	0.9986
	58	1.20 E–03	6.655232	– 3.045128	–	– 71.56	– 70.96	– 80.65	– 79.95	–	– 5.050	0.9986
	60	1.23 E–03	6.651459	– 3.055334	–	– 71.62	– 71.02	– 80.72	– 80.02	–	– 5.143	0.9986
	65	1.31 E–03	6.648240	– 3.085389	–	– 71.75	– 71.15	– 80.89	– 80.19	–	– 5.356	0.9986
	70	1.38 E–03	6.650365	– 3.116650	–	– 71.87	– 71.27	– 81.03	– 80.33	–	– 5.529	0.9986
	75	1.46 E–03	6.654759	– 3.146960	–	– 71.97	– 71.37	– 81.16	– 80.46	–	– 5.666	0.9986
	80	1.53 E–03	6.660035	– 3.176657	–	– 72.07	– 71.47	– 81.27	– 80.57	–	– 5.778	0.9986
	85	1.61 E–03	6.665605	– 3.206883	–	– 72.17	– 71.57	– 81.39	– 80.69	–	– 5.873	0.9985
	90	1.68 E–03	6.671209	– 3.239547	–	– 72.28	– 71.67	– 81.52	– 80.82	–	– 5.953	0.9985
	95	1.76 E–03	6.676722	– 3.278691	–	– 72.41	– 71.81	– 81.68	– 80.98	–	– 6.025	0.9985
	100	1.83 E–03	6.682083	– 3.338741	–	– 72.63	– 72.03	– 81.95	– 81.25	–	– 6.088	0.9985
S(–II) EP →	103.665	1.89 E–03	6.685662	– 3.563268	–	– 73.51	– 72.91	– 83.05	– 82.35	–	– 6.129	0.9985
Mostly C(IV) Reduction (OC Dispro-portionation)	110	1.98 E–03	6.670750	– 3.792801	–	– 74.49	– 73.89	– 84.29	– 83.59	–	– 6.105	0.9985
	115	2.06 E–03	6.659162	– 3.811264	–	– 74.61	– 74.01	– 84.45	– 83.75	–	– 6.086	0.9986
	120	2.13 E–03	6.647864	– 3.818331	–	– 74.68	– 74.08	– 84.56	– 83.86	–	– 6.067	0.9986
	150	2.58 E–03	6.585456	– 3.804453	–	– 74.88	– 74.28	– 84.86	– 84.16	–	– 5.961	0.9988
	200	3.33 E–03	6.497683	– 3.745094	–	– 74.99	– 74.39	– 85.09	– 84.39	–	– 5.811	0.9990

TABLE 24.7. (continued)

	log S(−II)T (*M*)	log $S(-II)_T$ (eq/L)	$\alpha_{1,S(-II)}$	$\alpha_{2,S(-II)}$	log $FeS_{(s)}$ (FW/L)	log $FeS_{(s)}$ (eq/L)	log SO_4^{2-} (*M*)	log CH_4 (*M*)	log CH_4 (eq/L)	log $C(IV)_T$ (*M*)	α_0	log H_2 (*M*)
Period 0. Redox equilibrium not yet established.												
	–	–	0.9091	9.09 E–07	–	–	– 4.03	–	–	– 3.31	0.0218	–
Period 1. NO_3^-/N_2 couple is considered frozen, (am)$Fe(OH)_{3(s)}$ is still present, $FeS_{(s)}$ not yet present. The parameter TEST examines whether the initial NO_3^- can be reduced to N_2; the critical value is 100.51.												
Mostly O_2 Reduction	– 141.36	– 140.46	0.5431	6.46 E–08	–	–	– 4.03	– 146.96	– 146.06	– 3.24	0.1583	– 44.23
	– 141.14	– 140.24	0.4967	4.90 E–08	–	–	– 4.03	– 146.79	– 145.88	– 3.23	0.1848	– 44.21
	– 140.93	– 140.03	0.4576	3.86 E–08	–	–	– 4.03	– 146.61	– 145.71	– 3.21	0.2096	– 44.18
	– 140.72	– 139.82	0.4241	3.12 E–08	–	–	– 4.03	– 146.43	– 145.53	– 3.20	0.2330	– 44.15
	– 140.50	– 139.59	0.3952	2.58 E–08	–	–	– 4.03	– 146.24	– 145.33	– 3.19	0.2550	– 44.11
	– 140.25	– 139.35	0.3700	2.17 E–08	–	–	– 4.03	– 146.02	– 145.12	– 3.17	0.2758	– 44.07
	– 139.98	– 139.07	0.3478	1.86 E–08	–	–	– 4.03	– 145.77	– 144.87	– 3.16	0.2955	– 44.02
	– 139.64	– 138.73	0.3281	1.60 E–08	–	–	– 4.03	– 145.46	– 144.55	– 3.15	0.3142	– 43.95
	– 139.17	– 138.27	0.3106	1.40 E–08	–	–	– 4.03	– 145.01	– 144.11	– 3.14	0.3319	– 43.85
	– 138.32	– 137.42	0.2948	1.23 E–08	–	–	– 4.03	– 144.19	– 143.28	– 3.13	0.3487	– 43.65
	– 138.00	– 137.10	0.2918	1.20 E–08	–	–	– 4.03	– 143.87	– 142.97	– 3.13	0.3519	– 43.57
	– 137.51	– 136.60	0.2889	1.17 E–08	–	–	– 4.03	– 143.38	– 142.48	– 3.12	0.3551	– 43.45
	– 136.31	– 135.41	0.2860	1.15 E–08	–	–	– 4.03	– 142.19	– 141.28	– 3.12	0.3583	– 43.15
	– 135.54	– 134.64	0.2855	1.14 E–08	–	–	– 4.03	– 141.42	– 140.52	– 3.12	0.3590	– 42.96
H_2O EP →	– 129.87	– 128.97	0.2851	1.14 E–08	–	–	– 4.03	– 135.75	– 134.85	– 3.12	0.3594	– 41.55
Period 2. NO_3^- begins to be reduced, (am)$Fe(OH)_{3(s)}$ is still present, $FeS_{(s)}$ not yet present. The parameter TEST examines whether (am)$Fe(OH)_{3(s)}$ is still present; the critical value is 21.25. Under the assumption that (am)$Fe(OH)_{3(s)}$ is largely insoluble, and when $FeS_{(s)}$ is not yet present, complete dissolution of (am)$Fe(OH)_{3(s)}$ is also observable when $\log Fe(II)_T = \log Fe_T = -5.05$.												
Mostly NO_3^- reduction	– 126.34	– 125.43	0.2851	1.14 E–08	–	–	– 4.03	– 132.22	– 131.32	– 3.12	0.3594	– 40.66
	– 126.23	– 125.33	0.2855	1.14 E–08	–	–	– 4.03	– 132.11	– 131.21	– 3.12	0.3589	– 40.64
	– 126.04	– 125.14	0.2862	1.15 E–08	–	–	– 4.03	– 131.91	– 131.01	– 3.12	0.3581	– 40.59
	– 125.77	– 124.86	0.2869	1.15 E–08	–	–	– 4.03	– 131.64	– 130.74	– 3.12	0.3573	– 40.52
	– 125.32	– 124.42	0.2876	1.16 E–08	–	–	– 4.03	– 131.19	– 130.28	– 3.11	0.3566	– 40.41
	– 123.69	– 122.79	0.2883	1.17 E–08	–	–	– 4.03	– 129.55	– 128.65	– 3.11	0.3558	– 40.00
N_2 EP →	– 119.33	– 118.42	0.2883	1.17 E–08	–	–	– 4.03	– 125.19	– 124.29	– 3.11	0.3557	– 38.91
Mostly (am)Fe $(OH)_{3(s)}$ reduction	– 46.70	– 45.80	0.2913	1.20 E–08	–	–	– 4.03	– 52.56	– 51.66	– 3.11	0.3524	– 20.75
	– 42.54	– 41.64	0.2978	1.26 E–08	–	–	– 4.03	– 48.39	– 47.48	– 3.11	0.3453	– 19.70
	– 40.16	– 39.26	0.3061	1.35 E–08	–	–	– 4.03	– 45.98	– 45.08	– 3.11	0.3365	– 19.10
	– 39.81	– 38.90	0.3078	1.37 E–08	–	–	– 4.03	– 45.63	– 44.72	– 3.11	0.3347	– 19.01
Fe(II) EP →	– 39.79	– 38.88	0.3079	1.37 E–08	–	–	– 4.03	– 45.60	– 44.70	– 3.11	0.3346	– 19.00
Period 3. $Fe(OH)_{3(s)}$ is no longer present, $FeS_{(s)}$ not yet present. The parameter TEST examines the value of $[Fe^{2+}][S^{2-}]$; the critical value for saturation of $FeS_{(s)}$ is $K_{s0} = 1.00 \times 10^{-18}$.												
Mostly SO_4^{2-} reduction	– 9.14	– 8.24	0.3079	1.37 E–08	–	–	– 4.03	– 14.96	– 14.06	– 3.11	0.3346	– 11.34
	– 6.26	– 5.35	0.3082	1.37 E–08	–	–	– 4.03	– 12.07	– 11.17	– 3.11	0.3343	– 10.62
	– 5.61	– 4.71	0.3090	1.38 E–08	–	–	– 4.04	– 11.42	– 10.52	– 3.11	0.3335	– 10.46
	– 5.37	– 4.46	0.3098	1.39 E–08	–	–	– 4.05	– 11.16	– 10.25	– 3.11	0.3326	– 10.39
	– 5.21	– 4.31	0.3106	1.40 E–08	–	–	– 4.06	– 10.99	– 10.08	– 3.10	0.3318	– 10.35
$FeS_{(s)}$ → saturation	– 5.10	– 4.20	0.3114	1.41 E–08	–	–	– 4.07	– 10.87	– 9.96	– 3.10	0.3310	– 10.32
Period 4. $FeS_{(s)}$ is present.												
Mostly SO_4^{2-} reduction	– 5.10	– 4.20	0.3114	1.41 E–08	– 9.41	– 8.45	– 4.07	– 10.87	– 9.96	– 3.10	0.3310	– 10.32
	– 5.10	– 4.19	0.3113	1.41 E–08	– 7.27	– 6.32	– 4.07	– 10.86	– 9.96	– 3.10	0.3311	– 10.32
	– 5.00	– 4.10	0.3094	1.39 E–08	– 5.75	– 4.80	– 4.09	– 10.74	– 9.84	– 3.10	0.3330	– 10.29
	– 4.78	– 3.88	0.3078	1.37 E–08	– 5.34	– 4.39	– 4.14	– 10.47	– 9.56	– 3.09	0.3347	– 10.23
	– 4.61	– 3.71	0.3089	1.38 E–08	– 5.22	– 4.27	– 4.20	– 10.22	– 9.32	– 3.08	0.3336	– 10.17
	– 4.48	– 3.58	0.3111	1.40 E–08	– 5.17	– 4.21	– 4.27	– 10.01	– 9.11	– 3.07	0.3313	– 10.12
	– 4.38	– 3.47	0.3137	1.43 E–08	– 5.14	– 4.18	– 4.35	– 9.81	– 8.91	– 3.06	0.3286	– 10.07
	– 4.29	– 3.39	0.3164	1.47 E–08	– 5.12	– 4.16	– 4.46	– 9.61	– 8.70	– 3.05	0.3258	– 10.02
	– 4.22	– 3.32	0.3192	1.50 E–08	– 5.11	– 4.15	– 4.59	– 9.38	– 8.48	– 3.04	0.3230	– 9.96
	– 4.16	– 3.26	0.3220	1.53 E–08	– 5.10	– 4.14	– 4.79	– 9.11	– 8.21	– 3.03	0.3202	– 9.90
	– 4.10	– 3.20	0.3247	1.56 E–08	– 5.09	– 4.14	– 5.16	– 8.67	– 7.76	– 3.02	0.3175	– 9.79
S(−II) EP →	– 4.07	– 3.17	0.3265	1.58 E–08	– 5.09	– 4.13	– 6.89	– 6.90	– 5.99	– 3.02	0.3158	– 9.34
Mostly C(IV) Reduction (OC Disproportionation)	– 4.07	– 3.17	0.3190	1.49 E–08	– 5.09	– 4.13	– 8.85	– 4.92	– 4.02	– 3.01	0.3232	– 8.86
	– 4.07	– 3.17	0.3132	1.43 E–08	– 5.09	– 4.14	– 9.10	– 4.67	– 3.77	– 3.01	0.3291	– 8.80
	– 4.07	– 3.17	0.3077	1.37 E–08	– 5.09	– 4.14	– 9.25	– 4.51	– 3.61	– 3.00	0.3349	– 8.76
	– 4.07	– 3.16	0.2779	1.07 E–08	– 5.10	– 4.15	– 9.67	– 4.06	– 3.16	– 2.98	0.3676	– 8.66
	– 4.06	– 3.16	0.2392	7.53 E–09	– 5.13	– 4.18	– 9.93	– 3.74	– 2.84	– 2.94	0.4157	– 8.61

The fact that the pe goes up during the first 35 days or so is a result of the fact that the functional dependence of the pe on the pH is stronger than on p_{O_2} (see Eq. (24.4).) Thus, as long as O_2 remains present at some non-negligible concentration, the initial lowering of the pH due to the production of CO_2 according to Eq. (24.13) makes the addition of reduced OC cause the pe to *rise*! Only after the value of p_{O_2} is driven to low levels and the lowering of the pH slows does the pe finally go down. As in an acid/base titration, where the change in pH at an **equivalence point** (EP) can be very dramatic, for the case under consideration here, the pe finally starts to drop as the amount of OC that has been added becomes close to being equivalent to the initial O_2. We will call this the H_2O redox EP since it is H_2O that is formed from the O_2. By way of comparison, we remember that for the titration of the acid HA, we used the term "NaA EP" to refer to the $f = 1$ alkalimetric EP.

The length of time required to reach the H_2O EP may be calculated according to

$$\text{time to reach } H_2O\text{ EP} = \frac{(1.057497 \times 10^{-3} - 3.330281 \times 10^{-4})\text{ eq/L}}{1.498626 \times 10^{-5}\text{ eq/L-day}} = 48.342 \text{ days.} \qquad (24.54)$$

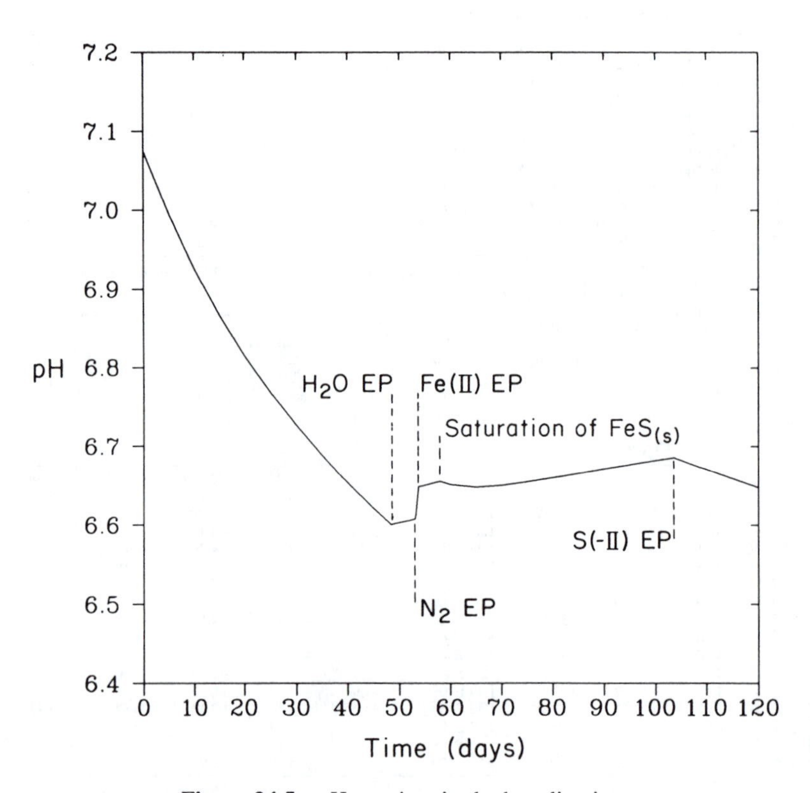

Figure 24.5 pH vs. time in the hypolimnion.

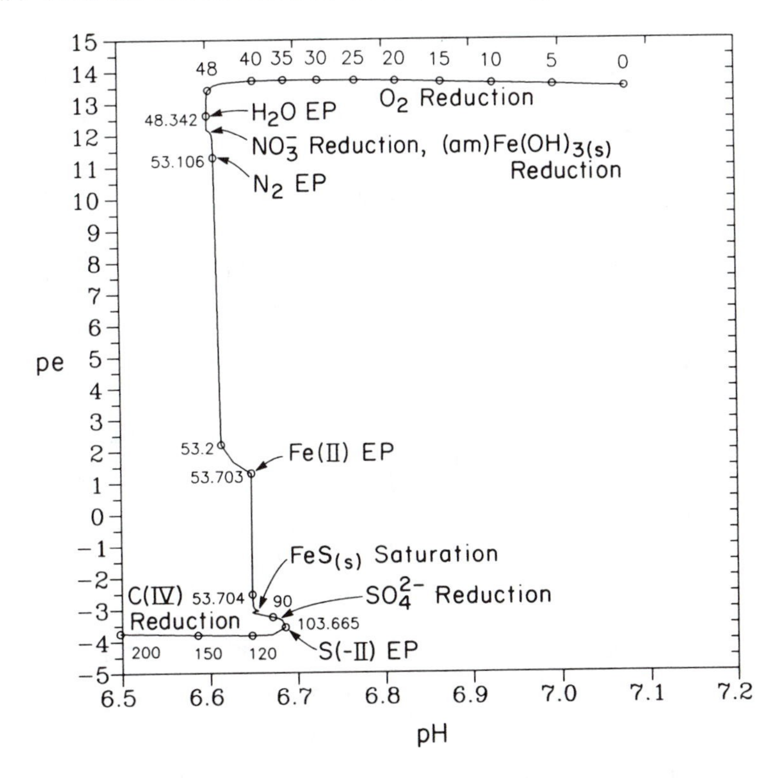

Figure 24.6 pe vs. pH in the hypolimnion as a function of time.

The numerator in Eq. (24.54) is the difference between the initial O_2 level and the initial OC level. It is, therefore, the initial amount of O_2 minus the amount that is titrated by the initial OC. As seen in Table 24.7, at t $\simeq$ 48.342 days, the pe does drop significantly. The $[O_2]$ is not zero at the H_2O EP. As an analogy, in the alkalimetric titration of the acid HA, [HA] remains nonzero at the NaA EP.

While $t = 48.342$ days corresponds to the H_2O EP, to within 0.001 days it is also the time at which we must consider the nitrogen to be redox active. This is no coincidence. Indeed, once the O_2 is essentially gone, the pe can drop down to the types of values that make NO_3^- unstable relative to reduction to N_2: we now enter Period 2.

For reduction of NO_3^- to N_2, the number of equivalents of NO_3^- available at the beginning of the summer is 7.139440×10^{-5} eq/L. Thus,

$$\begin{aligned}\text{time to reach } N_2 \text{ EP} &= 48.342 \text{ days} + \frac{7.139440 \times 10^{-5} \text{eq/L}}{1.498626 \times 10^{-5} \text{ eq/L-day}} \\ &= 48.342 \text{ days} + 4.764 \text{ days} \\ &= 53.106 \text{ days.}\end{aligned} \quad (24.55)$$

Between $t = 48.342$ and $t = 53.106$ days, the majority of the reduction going on involves the conversion of NO_3^- to N_2. The net balanced redox reaction for NO_3^- reduction by

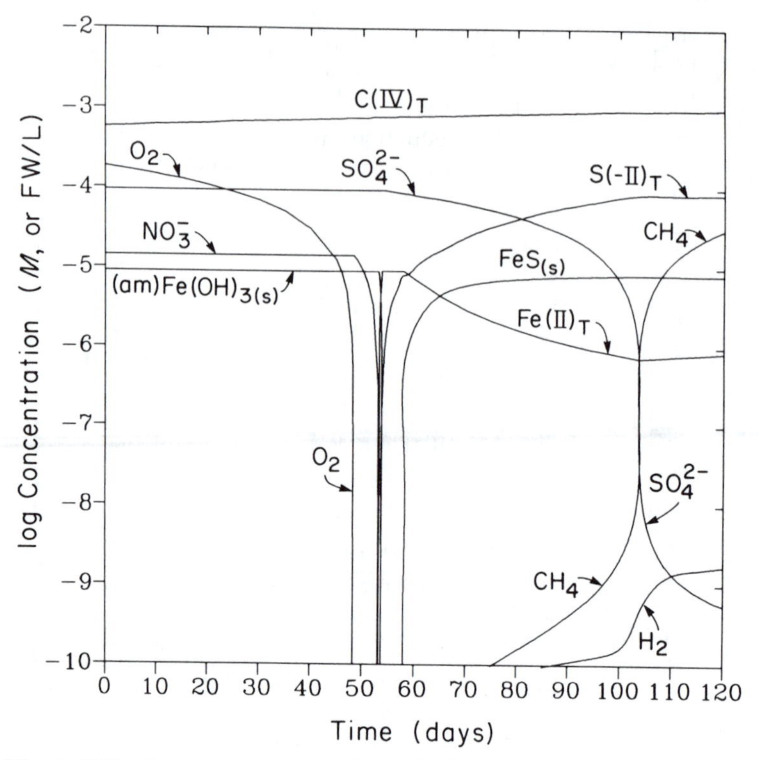

Figure 24.7 Log concentration (*M*, or FW/liter of water) vs. time in the hypolimnion for various important redox species.

the OC is

$$\frac{5}{6}C_6H_{12}O_6 + 4NO_3^- + 4H^+ = 5H_2CO_3^* + 2N_2 + 2H_2O. \tag{24.56}$$

Since removing 4 H^+ from solution tends to raise the pH more than adding 5 $H_2CO_3^*$ tends to lower it, the pH rises between 48.342 and 53.106 days.

At t = 53.106 days, significant amounts of the (am)$Fe(OH)_{3(s)}$ will begin to be reduced. The amount of (am)$Fe(OH)_{3(s)}$ initially available for reduction is 8.953032×10^{-6} eq/L. Thus, the (am)$Fe(OH)_{3(s)}$ will last until

$$\begin{aligned}\text{time to reach Fe(II) EP} &= 53.106 \text{ days} + \frac{8.953032 \times 10^{-6}\text{eq/L}}{1.498626 \times 10^{-5}\text{ eq/L} - \text{day}} \\ &= 53.106 \text{ days} + 0.597 \text{ days} \\ &= 53.703 \text{ days.}\end{aligned} \tag{24.57}$$

Between t = 53.106 and 53.703 days, the majority of the reduction going on involves the conversion of (am)$Fe(OH)_{3(s)}$ to $Fe(II)_T$. The net balanced redox reaction for (am)$Fe(OH)_{3(s)}$ reduction by the OC is

$$\frac{1}{6}C_6H_{12}O_6 + 4(am)Fe(OH)_{3(s)} + 8H^+ = H_2CO_3^* + 4Fe^{2+} + 10H_2O. \quad (24.58)$$

Since removing 8 H^+ from solution tends to raise the pH much more than adding a single $H_2CO_3^*$ tends to lower it, the pH rises strongly between 53.106 and 53.703 days. Even though the number of equivalents of $(am)Fe(OH)_{3(s)}$ that can be reduced is much smaller than the number of NO_3^- equivalents, because of the numbers of H^+ and $H_2CO_3^*$ involved, the pH rise is much larger than the rise during the ~5 day long interval during which NO_3^- is the dominant electron acceptor.

At t = 53.703 days, Period 2 ends, and Period 3 begins. During Period 3, the $S(-II)_T$ continues to build up until enough of it is present to saturate the system with respect to $FeS_{(s)}$ ($K_{s0} = 10^{-18.0}$). With $[Fe^{2+}]$ now roughly constant at $\sim 10^{-5.048}$ M, Period 3 will last until the product $\alpha_{2,S(-II)}S(-II)_T$ can build up to $10^{-18.0}/(\sim 10^{-5.048}) = \sim 10^{-12.952}$. With $\alpha_{2,S(-II)}$ constant at $\sim 1.37 \times 10^{-8}$ during this period, $FeS_{(s)}$ saturation will therefore occur once $S(-II)_T$ builds up to $\sim 8.15 \times 10^{-6}$ M. Since very little $S(-II)_T$ has accumulated in the solution by the time that Period 2 ends, Period 3 will last until (8 eq/mol)($\sim 8.15 \times 10^{-6}$ M) eq/L = $\sim 6.52 \times 10^{-5}$ eq/L of SO_4^{2-} have been reduced to $S(-II)_T$. Thus,

$$\begin{aligned}\text{time to saturate } FeS_{(s)} &= 53.703 \text{ days} + \frac{\sim 6.52 \times 10^{-5} \text{ eq/L}}{1.498626 \times 10^{-5} \text{ eq/L-day}} \\ &= 53.703 \text{ days} + \sim 4.351 \text{ days} \\ &= \sim 58.054 \text{ days.}\end{aligned} \quad (24.59)$$

Given the estimates involved in the above calculations, t = 58.054 is only an approximation of the time required to reach $FeS_{(s)}$ saturation. By carefully checking the product $[Fe^{2+}][S^{2-}]$ during this period (see Table 24.7), it may be seen that $FeS_{(s)}$ actually becomes saturated at t = 57.943 days.

Between t = 53.703 and 57.943 days, the majority of the reduction going on involves the conversion of SO_4^{2-} to HS^-. The net balanced redox reaction for SO_4^{2-} reduction by the OC is

$$\frac{2}{6}C_6H_{12}O_6 + H^+ + SO_4^{2-} = 2H_2CO_3^* + HS^-. \quad (24.60)$$

Since removing H^+ from solution tends to raise the pH more than adding two $H_2CO_3^*$ tends to lower it, the solution pH continues to rise between 53.703 and 57.943 days.

At t = 57.944 days, Period 4 begins. $FeS_{(s)}$ is now saturated, and starts to accumulate in the system. Since S_T for the system is much greater than Fe_T, the $S(-II)_T$ continues to increase during Period 4. This drives the $Fe(II)_T$ down due to the continued precipitation of $FeS_{(s)}$.[11] However, during the portion of this period that $FeS_{(s)}$ is precipitating in significant amounts, the pH *falls*. Now, much of the the S(−II) that

[11]If Fe(II) exhibited greater tendency to form complexes with HS^- than it does, then the increase in $S(-II)_T$ could eventually cause a resolubilization of the $FeS_{(s)}$ much like increasing the pH can first precipitate, then resolubilize the solid $Fe(OH)_{2(s)}$.

forms combines with Fe^{2+} rather than with H^+, and so rather than occurring according to Eq. (24.60), much of the reduction of SO_4^{2-} occurs according to

$$\frac{2}{6}C_6H_{12}O_6 + Fe^{2+} + SO_4^{2-} = 2H_2CO_3^* + FeS_{(s)}. \tag{24.61}$$

Thus, due to the absence of any pH-increasing effects due to the consumption of H^+, the production of $H_2CO_3^*$ tends to lower the pH. Once essentially 100 percent of the Fe(II) has been precipitated as $FeS_{(s)}$, then the pH can reverse its course and begin to rise again by virtue of the process in Eq. (24.60).

If reduced to S(−II), the amount of SO_4^{2-} initially in the system is 7.485964×10^{-4} eq/L. Thus,

$$\begin{aligned}\text{time to reach S(−II)EP} &= 53.703 \text{ days} + \frac{7.485964 \times 10^{-4} \text{ eq/L}}{1.498626 \times 10^{-5} \text{ eq/L-day}} \\ &= 53.703 \text{ days} + 49.952 \text{ days} \\ &= 103.655 \text{ days}.\end{aligned} \tag{24.62}$$

Once we reach 103.655 days, a significant amount of the SO_4^{2-} has been reduced. However, since the SO_4^{2-} is a weak oxidant and the OC is not a strong reducing agent, the drop in pe at the S(−II) EP is not nearly as large as at the preceding EPs. By way of comparison, we recall that only very slight changes in pH, HA, and A^- concentrations will be observed at the NaA EP when HA is a very weak acid; this becomes even more evident if the titrant used is not a strong base. For the case at hand, OC is not a strong reducing agent.

Once the pe has been reduced sufficiently that little SO_4^{2-} remains, the continued addition of OC begins to lead to significant reduction of existing C(IV) to CH_4. Equivalently, the $C_6H_{12}O_6$ begins to disproportionate to CO_2 and CH_4. From either viewpoint, the net result is Eq. (24.17), and the pH will begin to fall again. There will be no redox EP for the carbon because additional $C_6H_{12}O_6$ added will just continue to disproportionate. If we do not totally reduce essentially all of the water to H_2, we will not reach the H_2 EP.

24.5 GENERAL COMMENTS ON CONCENTRATION CHANGES DURING THE SUMMER

The interrelated concentration changes illustrated in Figure 24.7 summarize very well the processes occurring in the hypolimnion during the summer. The first oxidant to be reduced in significant amounts is O_2, followed by NO_3^-. Once the NO_3^- is mostly gone, (am)$Fe(OH)_{3(s)}$ begins to be reduced in large amounts, and so the $Fe(II)_T$ concentration comes up very rapidly. Meanwhile, some reduction of the SO_4^{2-} to S(−II) has been proceeding all along, and this reduction starts to take place in significant amounts once the (am)$Fe(OH)_{3(s)}$ is gone. The $Fe(II)_T$ concentration holds steady until $FeS_{(s)}$ becomes saturated, then $Fe(II)_T$ starts to go down. Significant reduction of C(IV) (i.e., disproportionation of the OC) to CH_4 begins once the SO_4^{2-} is nearly exhausted. At the same time, the reduction of H^+ to H_2 increases, but the concentration of H_2 does not rise very high because the drop in pe with time becomes essentially stalled. The OC is only a weak reducing agent. It cannot effect much reduction of H(I) to H(0), and so disproportionation of the OC is the only redox reaction remaining that can take place in significant measure.

The value of $[C(IV)_T]$ increases steadily throughout the summer since $C(VI)_T$ is always being produced whether by direct oxidation of the OC, or by disproportionation of the OC. Initially, each $C_6H_{12}O_6$ unit gives rise to close to six CO_2. Very late in the summer where OC disproportionation is important, the number is close to three. Thus, once we are well past the $S(-II)_T$ EP and prior to significant H^+ reduction, for any given time increment, the molar change in CH_4 concentration will be essentially equal to the molar change in $C(IV)_T$ concentration. The two changes will never be exactly equal, however, because a small amount of the reducing effect of the $C_6H_{12}O_6$ will always be spent on reducing O_2, NO_3^-, SO_4^{2-}, and H(I). Thus, for any given Δt, $\Delta[CH_4]$ will always be less than $\Delta[C(IV)_t]$. Between t = 120 and 115 days, $\Delta[CH_4]$ is 9.3360×10^{-6} *M*, and $\Delta[C(IV)_t]$ is 9.3370×10^{-6} *M*.

Calculated results are given in Table 24.7 for times that are longer than most summer periods of hypolimnetic stagnation (e.g., t = 150 and 200 days). They are included to show what will happen after essentially 100 percent of all of the "more available" electron acceptors have been consumed. Since its pe°(7) value is so low, H(I) is not very "available." One case where essentially all of the readily-available oxidants would indeed be gone would involve the bottom waters of a lake that never completely overturns (e.g., Lake Zürich). As seen in Table 24.7, at long times the pe starts to increase again. At t = 120, 150, and 200 days, we have pe = −3.81833, −3.804453, and −3.745094. It is true that the reduced species CH_4 is building up during this period. However, the fall in pH caused by the continued addition of CO_2 causes the pe to increase as it did prior to reaching the H_2O EP. Thus, while $[CH_4]$ might be increasing, $[H^+]^8$ is increasing faster (see reaction 6 in Table 24.3).

24.6 COMPARING THE OBSERVED ORDER OF TITRATION WITH PREDICTIONS MADE USING pe°(7) VALUES

As described in great detail in the preceding section, the observed order according to which the oxidants initially present in the hypolimnion are titrated is: mostly O_2, then mostly NO_3^-, then mostly (am)$Fe(OH)_{3(s)}$, then mostly SO_4^{2-}, then mostly C(IV) with some H^+. Since none of these oxidants are involved with HCO_3^- during their reductions, we can examine the corresponding pe°(7) values as opposed to the pe°(7,pHCO_3^-) values when comparing the observed order of titration with what might be expected based on the data in Table 24.3. The fact that the pH remains reasonably close to 7 during the summer justifies the use of pe°(7) values as opposed to pe°(pH) values at some other pH.

From Table 24.3, the order of pe°(7) values is 13.78 for O_2 to H_2O, 12.34 for NO_3^-, −3.63 for SO_4^{2-} to HS^-, −4.3 for $H_2CO_3^*$ to CH_4, −4.8 for (am)$Fe(OH)_{3(s)}$ to Fe^{2+}, and −15.55 for H^+ to H_2. Although we know that the (am)$Fe(OH)_{3(s)}$ is reduced primarily between the periods that NO_3^- and SO_4^{2-} are reduced, *taken by themselves*, the pe°(pH) values would predict that significant reduction of the (am)$Fe(OH)_{3(s)}$ would not occur until after significant reduction of C(IV) (i.e., disproportionation of the OC) begins.

The problem here is that the the reactant species (am)$Fe(OH)_{3(s)}$ is a *solid*, and the reduced species Fe^{2+} is *dissolved* (see also the discussion in Section 23.1.3). For this iron redox couple, while pe°(7) values factor out the effects of pH when pH = 7, the assumption of unit activity for the solid makes the pe°(7) value give the pe when $\{Fe^{2+}\}$ is *1 molal*. Such a condition is obviously exceedingly more reducing than when $\{Fe^{2+}\}$ is limited to $\sim 10^{-5}$ *M* or less, as in the case at hand. If we were to consider the pe value that would obtain for this one electron process even when $\{Fe^{2+}\}$ is the highest that it can be in this problem, that is, $\sim 10^{-5}$ *M*, then at pH 7 and when (am)$Fe(OH)_{3(s)}$ is present, we would calculate that pe $= -4.8 - \log\{Fe^{2+}\} = -4.8 - (-5) = +0.2$. This is in fact well above the pe°(pH) value for the reduction of SO_4^{2-}, and so it is not surprising that the (am)$Fe(OH)_{3(s)}$ is totally reduced to $Fe(II)_T$ before significant SO_4^{2-} reduction begins. For redox couples like those involving NO_3^- and SO_4^{2-}, both the OX and RED halves of the process are *dissolved*, and so the pe°(pH) values give pe values when the concentration of the OX species is approximately equal to the concentration[12] of the RED species.

Problems similar to those for the iron pe°(7) value may be encountered when either the OX or RED species is a *gas*, and the other is dissolved. In such cases, the pe°(7) values will correspond to unit gas pressures and unit molalities. This complication can be removed as it has been in Table 24.3 for H_2 by application of the appropriate Henry's Law constant, thereby converting the species of interest in the reaction from the gaseous form to the dissolved form.[13]

Simply put, this section provides a practical example for the discussion in Section 23.1.3. Thus, pe°(pH) values are most useful for the direct prediction of reaction sequences when both the OX and RED species are *dissolved*, and contain the *same* number of the redox-active element. When either the OX or the RED species is a *solid* (or a gas) and the other is dissolved, pe°(pH) values cannot be compared directly, and actual concentrations (and possibly in-situ gas pressures) must be taken into consideration.

24.7 KINETICS AND LABILE VS. NON-LABILE ORGANIC CARBON

It has been assumed in the preceding discussion that 100 percent of the OC that is input to the hypolimnion can react *immediately*, with large pe changes occurring over a matter little more than a minute (see Table 24.7). These calculated pe changes are probably

[12]The concentrations will be *exactly* equal when the both the OX and RED species contain the same number of the element of interest, as in the conversion of SO_4^{2-} to HS^-, but not in the conversion of NO_3^- to N_2.

[13]For the very special case of the reduction of O_2 to H_2O of interest in this chapter, the initial p_{O_2} was in fact near 1 atm. Also, the concentration of the corresponding redox partner (water) is typically expressed on the mole fraction scale, and X_{H_2O} remains essentially unchanged during any reduction of O_2. Thus, the pe°(7) value used in this section to predict when the O_2 would be reduced (relative to the other oxidants in the hypolimnion) was useful without further consideration of the concentrations involved.

significantly faster than are achieved in natural systems. In real systems, of course, *all* of the OC will not react immediately, and there will be a whole range of reactivities. This range of reactivities will be due to the facts that: 1) natural OC is made up of a range of different compounds; and 2) as a given molecular portion of the OC begins to react, the remaining portion can become more highly reactive, or less highly reactive. Thus, some of the OC introduced to the hypolimnion will react quickly, and some of the OC may be stable over very, very long periods of time. With regard to the stable type of OC, we note that most oxygenated waters contain a least some dissolved organic humic material.

For modelling purposes, one can subdivide the OC that is added into a *labile* fraction and a *non-labile* fraction. The labile fraction would be assumed to react immediately, and the non-labile fraction would be assumed to be essentially inert over the time frame of interest. This type of approach can help incorporate real world kinetic considerations into an equilibrium-based model.

In a simple application of the labile/non-labile approach, the fraction of the OC equivalents that are labile would be assumed to be *fixed* during all periods of the summer. In a more advanced application, the labile fraction could be different during each of the different periods.[14] Although all of the analysis of pe-pH vs. time in this chapter has been framed in the context of 100 percent of the OC being reactive during all periods of the summer, our results could also be viewed as corresponding to a case in which the labile fraction is fixed, and our input rate for OC just corresponds to that for the labile fraction. When considering any real system, assignment of values to the fractions of labile and non-labile OC would require field study, laboratory study, and/or theoretical considerations.

24.8 OTHER REDOX TITRATION CASES

There are many, many ways in which redox titrations occur in the environment. The case considered in this chapter is just one very specific example. Other initial conditions would have led to different results. For example, had the initial level of (am)$Fe(OH)_{3(s)}$ been higher, $FeCO_{3(s)}$ might have precipitated at some point during the summer. Also, we took the composition of the OC to be $C_6H_{12}O_6$. Actually, aquatic biogenic carbon is more accurately described as $C_{106}H_{263}O_{110}N_{16}P_1$ (Redfield et al. 1966), which is equivalent to $C_{6.0}H_{14.9}O_{6.2}N_{0.9}P_{0.06}$. A model involving the oxidation of such OC would need to consider the effects of the nitrogen and phosphorus on the time evolution of the pe and the pH.

Environmental redox titrations can just as easily involve the addition of an oxidant to a mixture of reduced species as they can the addition of a reductant (like OC) to a mixture of oxidants. Examples of the former would be the diffusion of oxygen into anoxic lake or river sediments, and the transport of oxygenated groundwater into an anoxic portion of an aquifer. Pankow (1992) discusses additional such problems.

[14]As a rationale for assigning different values to the labile fraction during the different periods we can note that different bacteria will tend to differ in their ability to attack the OC.

24.9 REFERENCES

BERNER, R. A. 1980. *Early Diagenesis: A Theoretical Approach.* Princeton: Princeton University Press.

FAIR, G. M., J. C. GEYER, and D. A. OKUN. 1968. *Water and Wastewater Engineering*, vol 2. New York: John Wiley and Sons.

IMBODEN, D. M., and A. LERMAN. 1978. Chemical Models of Lakes. *Lakes: Chemistry, Geology, and Physics*, A. Lerman, Ed. New York: Springer, p. 363.

PANKOW, J. F. 1992. *Aquatic Chemistry Problems*, Titan Press-OR, P.O. Box 91399, Portland, Oregon.

REDFIELD, A. C., B. H. KETCHUM, and F. A. RICHARDS. 1966. in *The Sea*, Vol. II, M. N. Hill, Ed. New York: Wiley-Interscience.

STUMM, W., and J. J. MORGAN. 1981. *Aquatic Chemistry.* New York: Wiley-Interscience.

PART VI

Effects of Electrical Charges on Solution Chemistry

25

The Debye-Hückel Law and Related Equations for the Activity Coefficients of Aqueous Ions

25.1 INTRODUCTION

In previous chapters, we have mentioned that it is often necessary to make corrections for the *nonideal* behavior of ions in aqueous solutions. As we now know, such corrections require a knowledge of the pertinent *activity coefficients*. The Debye-Hückel Law and the Extended Debye-Hückel Law allow us to predict ion activity coefficients over certain ranges of ionic strength. These two equations as well as other related activity coefficient equations are summarized in Table 2.3.

The original derivation of the Debye-Hückel Law in 1923 was considered one of the greatest accomplishments in physical chemistry during the early part of the 20th century. In this chapter, we will present the derivation of this and related equations. This will provide a basic understanding of the origins and meanings of activity coefficients for ionic species, and will also lay the groundwork for a firm understanding of Chapter 26 which discusses the principles governing electrical double layers. Other treatments of ion activity coefficients may be found in Bockris and Reddy (1970), Moore (1972), and Robinson and Stokes (1959).

As discussed in Chapter 2, on the infinite dilution scale, the activity coefficient γ of a dissolved species is equal to 1.0 when the solution is in fact infinitely dilute. As the solution becomes more concentrated, the resultant interactions between the various dissolved species will lead to γ values that deviate from unity. For species which are ionic, by far the most important interactions are coulombic in nature (i.e., electrostatic). On this basis, Debye and Hückel (1923) reasoned that if one could understand how the extent of these coulombic interactions depend on the nature and concentrations of the dissolved ions in solution, then it would be possible to predict γ values for ions as a function of solution composition.

Coulomb's Law states that

$$F = \frac{Kq_1q_2}{\epsilon r^2} \tag{25.1}$$

where F is the electrostatic force between the two charges, q_1 and q_2 are the charges in units of coulombs (C), K is a constant ($= 8.99 \times 10^9\ N - m^2/C^2$), ϵ is the dielectric constant of the medium (dimensionless), and r is the distance separating the charges. Equation (25.1) is consistent with what we know about activity coefficients of ionic species. Firstly, by charge balance, an ion of a particular sign will on the average be surrounded by more charge of the opposite sign than by charge of the same sign. Thus, there will be more attractive forces acting on the ion than repulsive forces. This will tend to *lower* the activity of the ion, and that lowering will be manifested in an activity coefficient that is less than 1.0. Secondly, we note that coulombic forces depend inversely on distance. Thus, as an ion becomes closer (on average) to other ions (i.e., more concentrated), the greater will be the strength of the coulombic forces, and therefore the greater the deviation of γ from 1.0.

An important equation for us here is the one that relates to how the chemical potential of species i depends on its activity, that is,

$$\mu_i = \mu_i^{\text{o}} + RT\ln a_i \tag{25.2}$$

$$a_i = m_i\gamma_i \tag{25.3}$$

where for species i, μ_i^{o} is the standard state chemical potential, a_i is the activity, m_i is the molality in the *bulk solution* (essentially the same as M for the solutions of interest here), and γ_i is the activity coefficient. Substituting Eq. (25.3) into (25.2) leads to

$$\mu_i = \mu_i^{\text{o}} + RT\ln m_i + RT\ln \gamma_i \tag{25.4}$$

Assuming that ion-ion interactions are the only types of interactions that are affecting the activity coefficient of species i, then the activity coefficient term in Eq. (25.4) provides a direct measure of the intensity of those ion-ion interactions for species i. That is, $RT\ln\gamma_i$ represents the portion of μ_i that is due to ion-ion interactions. Since the units of R are energy/mol-K, and since $\ln\gamma_i$ is dimensionless, the units of $RT\ln\gamma_i$ are energy/mol. It is important to note that these interactions include those with other ions *as well as* those between ions of type i. In other words, the activity coefficient of Na^+ in a solution of NaCl will be affected by the Cl^- ions as well as by other Na^+ ions.

25.2 THE DEBYE-HÜCKEL LAW

25.2.1 The Poisson Equation and the Local Charge Density

If we could develop a relationship between $RT\ln\gamma_i$ and the solution composition, then we could predict how γ_i varies as a function of solution composition. The way to determine the needed expression is to examine the magnitude of the free energy change that results when the ions in solution are initially in a hypothetical, uncharged state, and then are slowly brought up at constant T and P to their full levels of charge.

In a constant T and P system, the change in free energy for a given change in state will equal the amount of work that is done in a reversible manner to cause that change in state. Therefore, we are interested in the work required to carry out the process depicted in Figure 25.1. That work is composed of two distinct parts. The first part involves the energy needed simply to charge each of the core ions. The second part involves the ion-ion *interaction* energy, i.e., the part responsible for the term $RT \ln \gamma_i$.

A parameter that is very important in the context of the charging process is the electrical potential, ψ (volts). For a given point, its value equals the amount of work required to bring a unit positive charge from infinity (where ψ will be assumed to be zero) to that point. (For a unit positive charge of 1 C, the amount of work would be 1 C–V = 1 Joule (J).) Since the charge distribution around a given ion will, on the average, be spherically symmetrical, ψ will also on average be spherically symmetrical, and only a function of distance r from the center of the ion (see Figure 25.2). For the derivation of the Debye-Hückel Law, all ions will be assumed to have a radius of zero. For any such point charge ion, any value of r greater than zero will be assumed to be *outside* of that ion.

If we are to compute the work done in charging up a given ion surrounded by other ions, we need an expression for $\psi(r)$. The Poisson Equation from basic electrostatic theory relates ψ to the local charge density $q(r)$ (C/volume). In spherical coordinates, we have

$$\frac{1}{r^2}\frac{d}{dr}\left(r^2\frac{d\psi}{dr}\right) = \frac{-1}{\epsilon\epsilon_o}q(r) \quad \text{Poisson Equation for Spherical Symmetry} \tag{25.5}$$

where ϵ_o is the so-called "permittivity of free space." This parameter represents the inherent ability of electrical forces to be transmitted through a perfect vacuum. (It may be noted that the K in Coulomb's Law equals $1/(4\pi\epsilon_o)$.)

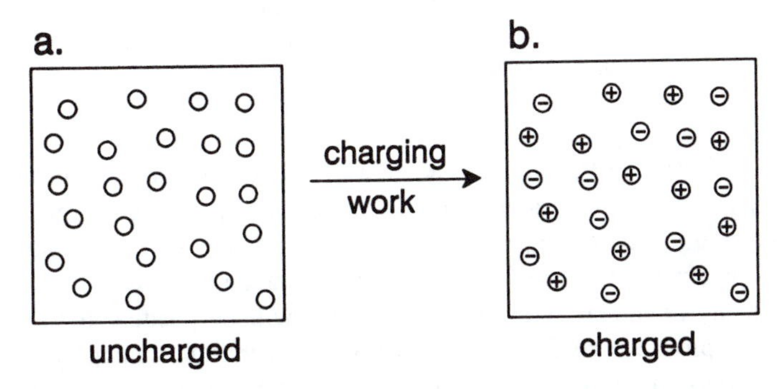

Figure 25.1 Work is done when ions are brought from a hypothetical uncharged state (a) to the fully charged and interacting state (b). Only a portion of the ions being charged are of type i.

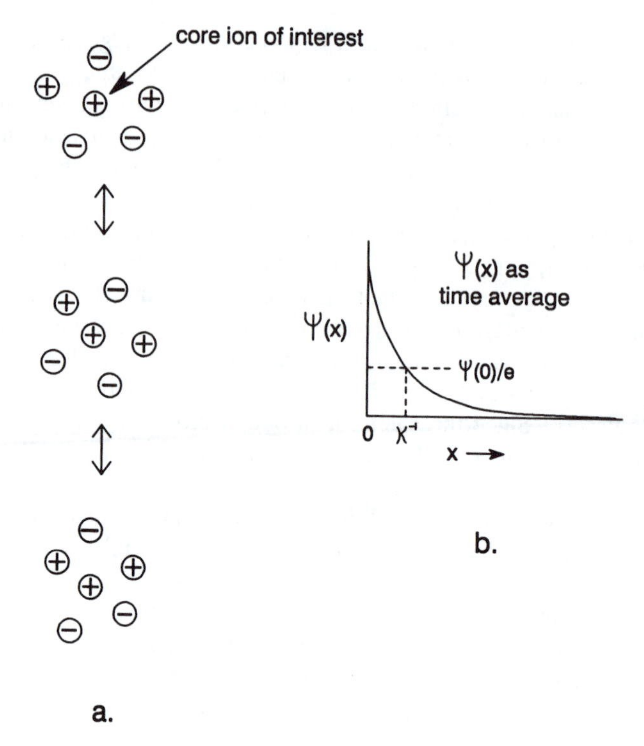

Figure 25.2 (*a*) On average, a given positive core ion will have more close neighbors of negative charge than of positive charge. The identity and positions of the ions in close proximity to the core ion are constantly changing. Three different possibilities for a cross-section through the center of the core ion are shown. (*b*) The time-average electrical potential ψ surrounding a given core ion will vary as a function of distance. For a positive core ion, ψ will decrease with distance towards a value of zero. The decrease is due to the neutralization that occurs from the relative excess of negative charge in the surrounding ion cloud. For a negative core ion, ψ will increase with distance towards a value of zero due to the neutralization that occurs from the relative excess of positive charge in the surrounding ion cloud.

The units of $q(r)$ are C/volume. The sign of $q(r)$ can be positive or negative, and will depend on whether there is a net excess of positive or negative charge, respectively, at point r. In a solution containing both positive and negative ions, if r is measured from the center of a positive ion, then on *average* $q(r)$ will be negative for all values of r, and will asymptotically approach zero for large r; $\psi(r)$ will be positive for all r, and will asymptotically approach zero for large r. Thus, to balance the charge of any positive core ion, there must be a negative cloud of charge surrounding that core ion. If r is measured from the center of a negative ion, then on *average*, $q(r)$ will be positive and $\psi(r)$ will be negative for all values of r, but both will again asymptotically approach zero for large r; there will be a positive cloud of charge surrounding any negative core ion.

Theoretically, for r measured from the center of any given ion, one must go to infinity to reach $q(r) = 0$ and $d\psi(r)/dr = 0$. The reason is that the core ion occupies one point in space, and the entire remainder of the solution is available for the cloud of opposite charge to provide overall electroneutrality for the whole system. For all intents and purposes, however, both $q(r)$ and $d\psi(r)/dr$ approach zero very rapidly, usually within a few hundred angstroms. As we will see later, the actual rate of decay is a strong function of the ionic strength.

Figure 25.3 shows how an ion cloud is relatively limited in size. As a result, once $q(r)$ diminishes to essentially zero, the solution surrounding the ion and its cloud is electrically neutral. Although that medium contains a myriad of other core ions and their ion clouds, each of those ion/(ion cloud) pairs is electrically neutral when viewed from a distance. The charge on any particular ion cloud is equal and opposite to the charge on its core ion. For example, for an ion like Na^+ which has a charge of $+e_o$ (note that $e_o = 1.60 \times 10^{-19}$ C), the charge on the cloud will be exactly $-e_o$. For an ion like SO_4^{2-}, the charge on the cloud will be $+2e_o$.

The charge present in the cloud around a given core ion is made up of portions of charge from a number of different, surrounding neighboring ions. Continual Brownian motion makes those ions flit in and out of the cloud. On *average*, when one such ion enters, another leaves, and vice versa. Each of the ions in the cloud contributes only a portion of the charge of the cloud since each of those ions (both positive and negative) is also surrounded by an ionic cloud, of which the core ion in Figure 25.3 is a partial member. Throughout this discussion then, what we are considering is the *average* nature of the ion cloud.

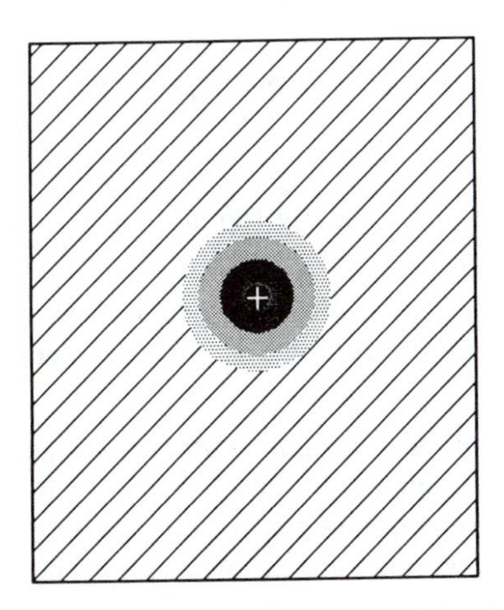

Figure 25.3 A charged ion cloud surrounds every ion in solution. The negative cloud surrounding a positive ion is depicted in this figure. The charge on the ion is equal and opposite to the total charge in the ion cloud. Surrounding the ion/(ion cloud) system is the bulk aqueous solution (cross-hatching). The bulk solution appears electrically neutral to the ion/(ion cloud) system.

25.2.2 $\psi_{\text{ion}}(r)$ and $\psi_{\text{cloud}}(r)$

Two things contribute to the electrical potential $\psi(r)$: 1) the charge on the ion itself; and 2) the charge distribution in the ion cloud. That is,

$$\psi(r) = \psi_{\text{ion}}(r) + \psi_{\text{cloud}}(r). \tag{25.6}$$

The interaction that is causing a non-unity activity coefficient is that between the ion and the ion cloud.

Let us first consider what happens when we charge all of the solution ions individually and in isolation from one another. Charging each of those ions involves adding small amounts of either positive or negative charge and building up each $\psi_{\text{ion}}(r)$ until each full charge is reached. For every ion, whether charged positively or negatively, positive work will be required to do this since more and more charge of the same type is being piled onto each core ion, and according to Coulomb's Law, charges of the same type repel one another. For a given ion, none of this work is related to its $RT \ln \gamma_i$ term since all of this work is required for the *very existence* of the ion in the aqueous medium. This remains true no matter what else is in the solution, or how concentrated the solution is.

In actuality, charging up all of the ions in a solution cannot occur with the ions in isolation from one-another. Doing the charging will in fact simultaneously build up an ion cloud and a $\psi_{\text{cloud}}(r)$ electrical potential distribution around each ion. The charging work for a given ion will thereby be *reduced* from that of charging an isolated ion. The reason is that the ion cloud will always have a charge that is opposite to that of its core ion. Thus, the sign of $\psi_{\text{cloud}}(r)$ will be opposite to that of $\psi_{\text{ion}}(r)$ for all values of r, and it becomes easier to bring in charge from infinity to charge the core ion. The magnitude of the effect will be determined by the magnitude of $\psi_{\text{cloud}}(r)$ evaluated at $r = 0$, that is, by $\psi_{\text{cloud}}(0)$.

To summarize, the work required to charge a core ion will be composed of two parts:

1. *Positive work* done in incrementally forcing one sign of charge onto the core ion and building up $\psi_{\text{ion}}(r)$.
2. *Negative work* done in placing the increments of core ion charge at the center of the oppositely-charged ion cloud, that is, at the center of $\psi_{\text{cloud}}(r)$.

It is the second component of the work that contributes to an ion activity coefficient that is less than 1.0. In particular, since the sign of that work is negative, we know that the resulting value of the activity coefficient will be less than 1.0. Equation (25.4) tells us this since doing negative work on a chemical species must lead to a reduction in the chemical potential of that species.

We need to solve the Poisson Equation so that we can get $\psi_{\text{cloud}}(r)$ as a function of the charging. We will then mathematically carry out the charging, and determine the second component of the work done as described above. This negative work will be equated with the interaction between the ion and its ion cloud. That is, the work will be equated to $(RT/N_a)\ln\gamma_i$ where we divide by Avogadro's Number N_a to get the work on a *per ion* basis.

In order to solve the Poisson Equation, we need $q(r)$, the local charge density in the cloud. Since $q(r) \neq 0$ in the cloud, we realize that the ENE is not satisfied on the very small scale of ions and their clouds. In the aggregate, however, since all ions and their surrounding clouds are of equal and opposite charge, the overall solution satisfies the ENE. The sign and magnitude of $q(r)$ at any given radius r will be determined by the extent to which the ENE is not satisfied.

25.2.3 The Linearized Poisson-Boltzmann Equation

When a solution satisfies the electroneutrality equation (ENE), we have

$$\sum z_{\text{cat}} n_{\text{cat}} + \sum z_{\text{an}} n_{\text{an}} = 0 \tag{25.7}$$

where n is the concentration in units of ions/cm^3. The subscript "cat" is used for cations, and the subscript "an" is used for anions. For dilute solutions at 25°C/atm where concentrations in units of molality and molarity are very similar, we note that for species i

$$n_i \simeq \frac{m_i N_a}{1000 \text{ cm}^3/\text{kg}} \tag{25.8}$$

where N_a is Avogadro's number. All z_{cat} are positive, and all z_{an} are negative.

When the two sums in Eq. (25.7) do indeed combine to equal zero, then the ENE is satisfied, $q = 0$, and the solution is electrically neutral. In the local zone surrounding the core ion, however, we know that Eq. (25.7) will not be satisfied. The amount of charge per unit volume in that region is given by

$$q(r) = e_o \sum z_{\text{cat}} n_{\text{cat}}(r) + e_o \sum z_{\text{an}} n_{\text{an}}(r). \tag{25.9}$$

Each of the two sums is multiplied by e_o, the elementary unit of charge. $q(r)$ has units of C/cm^3; the parameters z_{cat} and z_{an} may be thought of as having units of (units of elementary charge)/ion.

In order to determine the extent to which the ENE is satisfied, we need the concentrations of the various species as a function of r. The average concentration of any ion j (including ions of type i) in the ion cloud around an ion of type i can be determined as a function of r using a form of the Boltzmann Equation:

$$n_j(r) = n_{j,o} \exp[-U(r)/kT] \qquad \text{Boltzmann Equation} \tag{25.10}$$

where $U(r)$ = potential energy of an individual ion of type j when located at a distance r from the center of core ion i. The potential energy is assumed to be zero when the distance from the ion is so great that the ion and its cloud have no measurable effects on the concentration. That is, in the bulk solution where $r = \infty$ and $\psi = 0$, we have $U(r) = 0$ and so $n_j(r)$ equals the bulk solution value denoted in Eq. (25.10) as $n_{j,o}$. (Although it was not included in Eq. (25.4), we now use the subscript "o" to be especially clear when we are referring to the average, bulk solution concentration.)

For any given ionic species j of charge z_j, $U(r)$ will depend on $\psi(r)$. With z_j being positive for cations and negative for anions, we have[1]

$$U(r) = z_j e_o \psi(r). \tag{25.11}$$

Thus, for ion j, its potential energy at point r is equal to the product of its charge and $\psi(r)$. We recall that $\psi(r)$ gives the amount of work done to bring a unit charge from ∞ to point r. Thus, $z_j e_o \psi(r)$ represents the amount of work required to bring a charge of $z_j e_o$ from ∞ to point r.

Depending on the signs of z_j and $\psi(r)$, $z_j e_o \psi(r)$ may be either positive or negative. $z_j e_o \psi(r)$ will be positive when the core ion and ion j have the same sign. In that case, the concentration of ion j will tend on average to be *depleted* in the cloud relative to the bulk solution. Conversely, $z_j e_o \psi(r)$ will be negative when the core ion and ion j have the opposite signs. In that case, the concentration of ion j will tend on average to be *enriched* in the cloud relative to the bulk solution.

Combining Equations (25.10) and (25.11), we obtain

$$n_j(r) = n_{j,o} \exp[-z_j e_o \psi(r)/kT]. \tag{25.12}$$

Since Eq. (25.12) will be obeyed for all of the ions in the solution (*including those that are of the type of the core ion*), we can determine $q(r)$ using a single summation, that is,

$$q(r) = \sum_j z_j e_o n_{j,o} \exp[-z_j e_o \psi(r)/kT]. \tag{25.13}$$

When the magnitude of the electrical effects (measured by $z_j e_o \psi(r)$) are smaller than the thermal effects of the moving water molecules (measured by kT), that is, when the absolute value

$$\left| \frac{z_j e_o \psi(r)}{kT} \right| \ll 1 \tag{25.14}$$

then the arguments of the exponentials in Eq. (25.13) will be small and the exponentials can be *linearized*. In particular, a Taylor series expansion of the function e^y yields

$$e^y = 1 + y + y^2/2! + y^3/3! + ... \tag{25.15}$$

When y is small, then only a few terms of the above expansion are required to obtain a good approximation of e^y. In particular, if y is small, then $y^2/2!$ as well as all of the higher order terms will be much smaller in magnitude than y. Therefore, for small y, $e^y \simeq 1 + y$. Under these conditions,[2]

$$\exp[-z_j e_o \psi(r)/kT] \simeq 1 - z_j e_o \psi(r)/kT. \tag{25.16}$$

[1]We note that Eq. (25.11) is very similar to the relation $\Delta G = -nFE$ (Eq. (19.10)) wherein ΔG is the change in free energy for a redox reaction, n is the number of mols of electrons exchanged, and E is the potential difference in volts between the two half cells.

[2]As y gets smaller and smaller, fewer and fewer of the terms in Eq. (25.15) are required to obtain a good estimate of e^y. For example, for $y = 1.0$, we have $e^y = 2.7183 = 1 + 1.0 + 0.500 + 0.1667 + ...$ and only a few of the terms of the expansion are required. For $y = 0.1$, fewer terms are required to get a good approximation, since now $e^y = 1.1052 = 1 + 0.1 + 0.0050 + ...$

We can expect that we will be able to linearize the exponential in this manner when the electric field near the core ion is small.

Substituting Eq. (25.16) into Eq. (25.13), we obtain

$$q(r) = \sum_j z_j e_o n_{j,o}(1 - z_j e_o \psi(r)/kT) \tag{25.17}$$

or

$$q(r) = \sum_j z_j e_o n_{j,o} - \sum_j n_{j,o} z_j^2 e_o^2 \psi(r)/kT. \tag{25.18}$$

Since the bulk solution will be electrically neutral, the first term in Eq. (25.18) will be identically equal to zero. Therefore, combining Eq. (25.18) with Eq. (25.5) yields the so-called "Linearized Poisson-Boltzmann Equation"

$$\frac{1}{r^2}\frac{d}{dr}\left(r^2 \frac{d\psi(r)}{dr}\right) = \frac{1}{\epsilon\epsilon_o}\sum_j n_{j,o} z_j^2 e_o^2 \psi(r)/kT \qquad \text{Linearized Poisson-Boltzmann Equation} \tag{25.19}$$

$$= \frac{e_o^2 \psi(r)}{\epsilon\epsilon_o kT}\sum_j z_j^2 n_{j,o}. \tag{25.20}$$

25.2.4 Integrating the Linearized Poisson-Boltzmann Equation

Since we seek $\psi(r)$ for examination relative to Eq. (25.6), we need to integrate the linearized Poisson-Boltzmann equation. If we define

$$\kappa^2 = \frac{e_o^2}{\epsilon\epsilon_o kT}\sum_j z_j^2 n_{j,o} \tag{25.21}$$

then the Eq. (25.20) version of the linearized Poisson-Boltzmann equation becomes

$$\frac{d}{dr}\left(r^2 \frac{d\psi(r)}{dr}\right) = \kappa^2 r^2 \psi(r). \tag{25.22}$$

The parameter κ is most often discussed in terms of its inverse, κ^{-1}, which is called the "Debye length." As we will see later, κ^{-1} does in fact have dimensions of length. It provides a measure of the *thickness of the ion cloud* surrounding an ion.

We desire to solve the differential equation given in Eq. (25.22). To accomplish this, we use a change of variable. Let

$$u(r) = r\psi(r), \qquad \text{i.e.,} \qquad \psi(r) = u(r)/r. \tag{25.23}$$

Therefore, since $d(u/v) = (vdu - udv)/v^2$,

$$\frac{d\psi(r)}{dr} = \frac{r(du(r)/dr) - u(r)}{r^2}. \tag{25.24}$$

Combining Eqs. (25.22) and (25.24) yields

$$\frac{d}{dr}\left(r(du(r)/dr) - u(r)\right) = \kappa^2 r^2 u(r)/r \tag{25.25}$$

$$du(r)/dr + r(d^2u(r)/dr^2) - du(r)/dr = \kappa^2 r^2 u(r)/r. \tag{25.26}$$

The first and third terms on the LHS of Eq. (25.26) cancel, leaving

$$d^2u(r)/dr^2 = \kappa^2 u(r). \tag{25.27}$$

Differential equations like Eq. (25.27) have solutions that are exponential in nature. In particular, $u(r)$ could equal $Ae^{-\kappa r}$, $Be^{+\kappa r}$, or a combination of the two, where A and B are constants to be determined from the boundary conditions of the problem. Therefore, let us try a solution of the type

$$u(r) = Ae^{-\kappa r} + Be^{+\kappa r}. \tag{25.28}$$

For boundary conditions, we first use the fact that $\psi(r) \to 0$ as $r \to \infty$. When $r = \infty$, since κ is positive, $Ae^{-\kappa r} = 0$ no matter what the value of A. However, when $r = \infty$, the value of $Be^{+\kappa r}$, is unbounded for any finite value of B. Indeed, the value of κ will never be zero since even in pure water the concentrations of H^+ and OH^- are nonzero. This requires that $B = 0$.

We can now substitute

$$\psi(r) = u(r)/r = Ae^{-\kappa r}/r \tag{25.29}$$

into the basic form of the Poisson Equation, that is, into

$$\frac{1}{r^2}\frac{d}{dr}\left(r^2\frac{d\psi(r)}{dr}\right) = \frac{-q(r)}{\epsilon\epsilon_o}. \tag{25.5}$$

We first note that

$$d\psi(r)/dr = -e^{-\kappa r}\frac{A}{r^2} - \frac{A\kappa}{r}e^{-\kappa r} \tag{25.30}$$

$$r^2(d\psi(r)/dr) = -Ae^{-\kappa r} - rA\kappa e^{-\kappa r}. \tag{25.31}$$

Substituting Eqs. (25.30) and (25.31) into Eq. (25.5), we have

$$\frac{d}{dr}\left(r^2\frac{d\psi(r)}{dr}\right) = A\kappa e^{-\kappa r} - A\kappa e^{-\kappa r} + rA\kappa^2 e^{-\kappa r} \tag{25.32}$$

and so

$$\frac{1}{r^2}\frac{d}{dr}\left(r^2\frac{d\psi(r)}{dr}\right) = \frac{A\kappa^2}{r}e^{-\kappa r} = \frac{-q(r)}{\epsilon\epsilon_o} \tag{25.33}$$

or

$$q(r) = -\frac{A\kappa^2\epsilon\epsilon_o}{r}e^{-\kappa r}. \tag{25.34}$$

As discussed in Section 25.2.1, the charge on the ion cloud is equal in magnitude, but opposite in sign to the charge on the ion itself. Therefore, if the charge on the ion is $z_i e_o$ (we cannot just say z_i here since we are at too detailed a level for that), then the charge on the cloud is $-z_i e_o$. If $q(r)$ is integrated from $r = 0$ to ∞ (see Figure 25.4), then $-z_i e_o$ must be the value of the integral. We start the integration at 0 since we have assumed that the core ion itself is a *point charge*. Therefore, we have

$$\int_0^\infty 4\pi r^2 q(r) dr = -z_i e_o. \tag{25.35}$$

Substituting Eq. (25.34) into Eq. (25.35), we obtain

$$4\pi A \kappa^2 \epsilon\epsilon_o \int_0^\infty r e^{-\kappa r} dr = z_i e_o. \tag{25.36}$$

With the help of integral tables, we obtain

$$\int_0^\infty r e^{-\kappa r} dr = \frac{e^{-\kappa r}}{\kappa^2}(\kappa r - 1)\bigg|_0^\infty = 1/\kappa^2. \tag{25.37}$$

Combining Eqs. (25.36) and (25.37), we have

$$A = \frac{z_i e_o}{4\pi\epsilon\epsilon_o} \tag{25.38}$$

and so for point charge ions,

$$\psi(r) = \frac{z_i e_o}{4\pi\epsilon\epsilon_o r} e^{-\kappa r}. \tag{25.39}$$

The pre-exponential term in Eq. (25.39) is the value of ψ at $r = 0$. We therefore see that under the assumptions utilized above (small potentials (see Eq. (25.14)) and point charges), κ is the distance to which one has to go for ψ to drop to $1/e$ of its $r = 0$ value (see Figure 25.2).

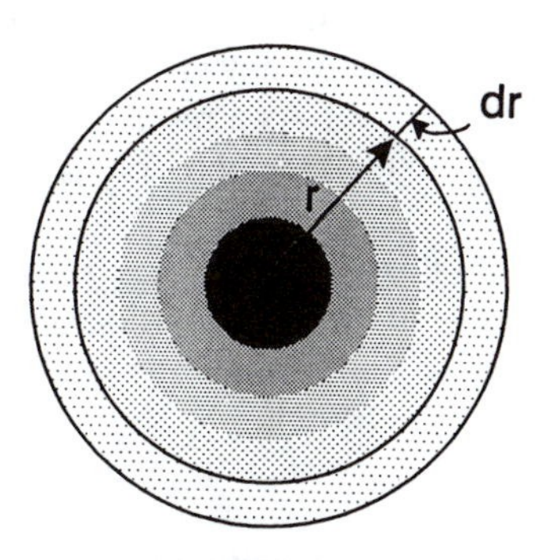

Figure 25.4 Integration of the charge density $q(r)$ in an ion cloud in spherical coordinates. The charge contained in each successive shell of volume $4\pi r^2 dr$ is equal to $4\pi r^2 q(r) dr$.

25.2.5 Determining $\psi_{\text{cloud}}(r)$

Equation (25.39) was developed just by considering the charge density in the cloud as a function of distance. By Eq. (25.6), we have that $\psi(r) = \psi_{\text{ion}}(r) + \psi_{\text{cloud}}(r)$. Thus, we note at this point that the *total* electrical potential function (including the contribution from the ion) can be deduced from the charge density in the cloud alone. This is a result of the facts that: 1) the cloud has been assumed to be continuous all the way up to the surface of the ion; and 2) both $\psi_{\text{cloud}}(r)$ *and* $\psi_{\text{ion}}(r)$ affect $q(r)$ in the cloud, and so evaluating the properties of the cloud reveal information about $\psi_{\text{cloud}}(r)$ as well as about $\psi_{\text{ion}}(r)$.

Very near the "surface" of the ion (if that term can be used for a point charge), r will be very small. Therefore, we can linearize the exponential in Eq. (25.39), with the result that for small r,

$$\psi(r) = \frac{z_i e_o}{4\pi\epsilon\epsilon_o r}(1 - \kappa r) \tag{25.40}$$

or,

$$\psi(r) = \frac{z_i e_o}{4\pi\epsilon\epsilon_o r} - \frac{z_i e_o \kappa}{4\pi\epsilon\epsilon_o}. \tag{25.41}$$

From basic electrostatic theory, we recall that the first term on the RHS of Eq. (24.41) is exactly the potential as a function of r for an ion of charge $z_i e_o$ in a medium with dielectric constant ϵ. That is, the first term on the RHS is $\psi_{\text{ion}}(r)$. Therefore, *near the ion* (i.e., for small r),

$$\psi_{\text{cloud}} = -\frac{z_i e_o \kappa}{4\pi\epsilon\epsilon_o}. \tag{25.42}$$

25.2.6 Use of $\psi_{\text{cloud}}(r)$ Near the Core Ion to Derive the Debye-Hückel Law

It is important to note that Eq. (25.42) is valid for any value of the charge on the core ion. That is, while we know that the parameter z_i is in practice a whole number, there is nothing in the original derivation of Eq. (25.42) that actually requires it to be a whole number. Therefore, Eq. (25.42) is valid for any charge on the core ion, including charges that are only a very small fraction of e_o. As a result, we can use Eq. (25.42) to give us the value of the potential of the cloud at the core ion as the charging process discussed in Section 25.2.1 is in progress.

The portion of the change in free energy (i.e., work done) while placing charge onto a *single* core ion i that is due to the ψ_{cloud} portion of the potential will be given by the integral

$$\Delta G_{\text{ion-cloud}} = \int_0^{z_i e_o} \psi_{\text{cloud}} d(z_i e_o). \tag{25.43}$$

The parentheses around the variable of integration $(z_i e_o)$ are used to set it apart from the final, actual value for the charge on the core ion, which is $z_i e_o$. Only ψ_{cloud} is included in Eq. (25.43) since, as discussed in Section 25.2.2, what we are interested in here are the effects that concern γ values, and the ψ_{ion} portion of the potential does not affect γ values. Substituting Eq. (25.42) into Eq. (25.43), we obtain

$$\Delta G_{\text{ion-cloud}} = \frac{-\kappa}{4\pi\epsilon\epsilon_o}\int_0^{z_i e_o}(z_i e_o)d(z_i e_o). \tag{25.44}$$

Integration yields

$$\Delta G_{\text{ion-cloud}} = \frac{-\kappa}{8\pi\epsilon\epsilon_o} z_i^2 e_o^2. \tag{25.45}$$

As expected, regardless of the sign of z_i, the sign of $\Delta G_{\text{ion-cloud}}$ is negative; the ionic cloud surrounding each ion lowers that ion's free energy, or equivalently, it lowers that ion's chemical potential μ_i. We recall that

$$\mu_i = \mu_i^{\text{o}} + RT\ln m_i + RT\ln\gamma_i \tag{25.4}$$

where we recall that $c_i (\simeq 1000 n_i(\infty)/\text{N}_a)$ is the molal concentration of ion i in the *bulk solution*. As we know, μ_i is the free energy per mol of ion i. Since there have been no changes in the concentration of an ion in the *bulk solution* due to any ion-cloud interactions, if μ_i is to be lowered, this means that the cloud around each such ion must be causing the activity coefficient γ_i to be less than 1.0. Since concentrations on the molal scale will be virtually directly proportional to concentration on the ion/cm^3 scale, the activity coefficients on these two scales will be virtually identical.

Dividing each of the terms on the RHS of Eq. (25.4) by Avogadro's Number (N_a) yields the components of the free energy on a *per ion basis*. Therefore, the per ion free energy due to activity coefficient effects is given by

$$\begin{array}{l}\text{per ion } G\\ \text{due to activity}\\ \text{corrections}\end{array} = \frac{RT}{\text{N}_a}\ln\gamma_i = kT\ln\gamma_i. \tag{25.46}$$

Combining Eqs. (25.45) and (25.46), we obtain

$$kT\ln\gamma_i = \frac{-\kappa}{8\pi\epsilon\epsilon_o} z_i^2 e_o^2 \tag{25.47}$$

or,

$$\ln\gamma_i = \frac{-\kappa}{8\pi\epsilon\epsilon_o kT} z_i^2 e_o^2. \tag{25.48}$$

It would be convenient if Eq. (25.48) could be simplified and placed into a more usable form. We recall that

$$\kappa^2 = \frac{e_o^2}{\epsilon\epsilon_o kT}\sum_j n_{j,o} z_j^2. \tag{25.49}$$

The sum in Eq. (25.49) is a simple function of the ionic strength. We recall that the units selected for all of the bulk phase concentrations $n_{j,o}$ are ions/cm^3. Since the typical units for I are mol/L, and since

$$I = \frac{1}{2}\sum_j (n_{j,o}(1000 \text{ cm}^3/\text{L})/\text{N}_a)z_j^2 \tag{25.50}$$

we have that

$$\sum_j n_{j,o}z_j^2 = \frac{2\,I\text{N}_a}{1000 \text{ cm}^3/\text{L}}. \tag{25.51}$$

Therefore,

$$\kappa = \left(\frac{2e_o^2\,I\text{N}_a}{\epsilon\epsilon_o kT\,1000 \text{ cm}^3/\text{L}}\right)^{1/2} \tag{25.52}$$

and

$$\log_{10}\gamma_i = -\frac{e_o^3(2\text{N}_a)^{1/2}}{2.303\;8\pi(\epsilon\epsilon_o kT)^{3/2}(1000 \text{ cm}^3)^{1/2}}\,z_i^2 I^{1/2} \tag{25.53}$$

where it may be noted that the dimensions of $I^{1/2}$ are mols$^{1/2}$-L$^{-1/2}$.

The expression multiplying $z_i^2 I^{1/2}$ in Eq. (25.53) is usually collected into a single constant **A**, which is different from the non-boldface italic A in Eq. (25.28).

$$\mathbf{A} = \frac{e_o^3(2\text{N}_a)^{1/2}}{2.303\;8\pi(\epsilon\epsilon_o kT)^{3/2}(1000 \text{ cm}^3)^{1/2}} \tag{25.54}$$

$$\log_{10}\gamma_i = -\mathbf{A}z_i^2 I^{1/2} \qquad \text{Debye-Hückel Law.} \tag{25.55}$$

As required, for $I = 0$, $\gamma_i = 1.0$. For the evaluation of the parameter **A**, we summarize the values of the various constants:

$$e_o = 1.60 \times 10^{-19}\text{C} \qquad \epsilon_o = 8.854 \times 10^{-14}\text{C/V-cm}$$

$$\text{N}_a = 6.02 \times 10^{23}/\text{mol} \qquad k = 1.381 \times 10^{-23}\text{V-C}/K.$$

At $T = 298°$K, $\epsilon = 78.3$ for water. Therefore, at 298°K, **A** $= 0.51$. At temperatures other than 298°K, neglecting the effects of temperature on the density of water,

$$\mathbf{A} = (1.82 \times 10^6)(\epsilon \text{T})^{-3/2}. \tag{25.56}$$

Chemical handbooks such as the one published yearly by CRC Press tabulate ϵ for water as a function of temperature. Several pertinent values are: 278°K, 85.8; 288°K, 82.0; 293°K, 80.1; 298°K, 78.3; and 308°K, 74.8.

25.3 DERIVATION OF THE EXTENDED DEBYE-HÜCKEL LAW

The pre-exponential factor A in Eq. (25.28) was evaluated by making use of the fact that the charge on the ion cloud is equal in magnitude (but opposite in sign) to the charge on the core ion. In particular, $q(r)$ was integrated from $r = 0$ to ∞, and the result was equated to $-z_i e_o$. In the derivation of the *Extended* Debye-Hückel Law, the assumption that the core ion itself is a point charge is not made, and it is acknowledged that the core ion has a *finite radius* equal to a. Values of a for various ions of interest are collected in Table 2.4.

For finite a, Eq. (25.36) becomes

$$4\pi A\kappa^2\epsilon\epsilon_o \int_a^\infty re^{-\kappa r}\,dr = z_i e_o. \tag{25.57}$$

With the help of integral tables, we obtain

$$\int_a^\infty re^{-\kappa r}\,dr = \frac{e^{-\kappa r}}{\kappa^2}(\kappa r - 1)\Big|_a^\infty = \frac{e^{-a\kappa}(1 + a\kappa)}{\kappa^2}. \tag{25.58}$$

Combining Eqs. (25.57) and (25.58), we obtain

$$A = \frac{z_i e_o e^{a\kappa}}{4\pi\epsilon\epsilon_o(1 + a\kappa)} \tag{25.59}$$

and so for non-point charge ions,

$$\psi(r) = \frac{z_i e_o e^{a\kappa}}{4\pi\epsilon\epsilon_o(1 + a\kappa)r}e^{-\kappa r}. \tag{25.60}$$

It may be noted that as the parameter a becomes increasingly small, Eq. (25.60) reduces as expected to Eq. (25.39).

At the surface of the ion, $r = a$, and so

$$\psi(a) = \frac{z_i e_o}{4\pi\epsilon\epsilon_o(1 + a\kappa)a}. \tag{25.61}$$

Since

$$\psi_{\text{ion}}(a) = \frac{z_i e_o}{4\pi\epsilon\epsilon_o a} \tag{25.62}$$

and since $\psi_{\text{cloud}}(r) = \psi(r) - \psi_{\text{ion}}(r)$, combining Eqs. (25.61) and (25.62) yields

$$\psi_{\text{cloud}}(a) = \frac{z_i e_o}{4\pi\epsilon\epsilon_o a}\left(\frac{1}{(1 + a\kappa)} - 1\right) = -\frac{\kappa z_i e_o}{4\pi\epsilon\epsilon_o(1 + \kappa a)}. \tag{25.63}$$

Substituting Eq. (25.63) into Eq. (25.43), we obtain

$$\Delta G_{\text{ion-cloud}} = \frac{-\kappa}{4\pi\epsilon\epsilon_o(1 + \kappa a)}\int_0^{z_i e_o}(z_i e_o)d(z_i e_o). \tag{25.64}$$

Integrating, Eq. (25.64) becomes

$$\Delta G_{\text{ion-cloud}} = \frac{-\kappa}{8\pi\epsilon\epsilon_o(1 + \kappa a)}z_i^2 e_o^2. \tag{25.65}$$

The only difference between Eq. (25.65) and Eq. (25.45) is the presence of the $(1 + \kappa a)$ term in the denominator in Eq. (25.65). As the parameter a becomes increasingly small, Eq. (25.65) reduces to Eq. (25.45). The $(1 + \kappa a)$ term propagates through all of the subsequent mathematics. Consequently, the Extended Debye-Hückel analog of Eq. (25.48) is

$$\ln\gamma_i = \frac{-\kappa}{8\pi\epsilon\epsilon_o kT(1 + \kappa a)}z_i^2 e_o^2 \tag{25.66}$$

and so we obtain

$$\log_{10} \gamma_i = \frac{-\mathbf{A} z_i^2 I^{1/2}}{(1 + \kappa a)} \qquad \text{Extended Debye-Hückel Law} \tag{25.67}$$

where **A** is the same as defined in Eq. (25.54). As required, for the limit of $I = 0$, we have $\gamma_i = 1.0$. Equation (25.67) reduces to the simple Debye-Hückel Law (Eq. (25.55)) for very small a, or very small κ.

For water at 298°K, Eq. (25.52) yields

$$\kappa = 3.258 \times 10^7 \; I^{1/2} (\text{cm}^{-1}) \tag{25.68}$$

In all portions of the above derivation, the ion size parameter a was assumed to be given in cgs units, that is, in cm. If we switch to the more relevant scale of angstroms ($1\text{Å} = 10^{-8}$ cm), for water at 298°K, the denominator in Eq. (25.67) becomes $(1+\mathbf{B}aI^{1/2})$ where from Eq. (25.68) we see that

$$\mathbf{B} = (3.258 \times 10^7)(1 \times 10^8) = 0.33. \tag{25.69}$$

For temperatures other than 298°K, if we neglect the effects of temperature on the density of water, Eq. (25.52) can be used to obtain

$$\mathbf{B} = 50.2(\epsilon \text{T})^{-1/2}. \tag{25.70}$$

To summarize then, when the ion size parameter a is expressed in angstroms, then we have

$$\log_{10} \gamma_i = \frac{-\mathbf{A} z_i^2 I^{1/2}}{(1 + \mathbf{B}aI^{1/2})} \qquad \text{Extended Debye-Hückel Law.} \tag{25.71}$$

25.4 GENERAL COMMENTS ON THE DEBYE-HÜCKEL AND EXTENDED DEBYE-HÜCKEL LAWS

As long as the assumptions underlying their derivations apply, the Debye-Hückel and Extended Debye-Hückel Laws are applicable for all ions in a given solution. For example, say that there are three different types of ions in a given solution, let us call them ions w, x, and y. One would first calculate I as usual by including all of the different ions. To compute γ_w, one would then use z_w and either Eq. (25.55) or (25.71). To compute γ_x, one would use the same I, and z_x. To compute γ_y, one would use the same I, and z_y.

A number of assumptions were made in the derivation of the Debye-Hückel Law. One important one that was not mentioned explicitly was that the *bulk* dielectric constant ϵ is applicable even at the very small scale of individual ions. The two assumptions that were mentioned explicitly were: 1) the exponential in Eq. (25.39) can be linearized as in Eq. (25.40); and 2) $a = 0$. Despite these non-trivial assumptions, as discussed on pages 212–218 in Bockris and Reddy, (1970), the Debye-Hückel Law performs very well for I values lower than about 0.01.

As I increases, the ions in solution become on the average closer and closer, and so each individual ion acts less and less like a point charge. This is due to the fact that the Debye length (κ^{-1}), which provides a measure of the physical extent of a given ionic cloud (see Eq. (25.60)), decreases as I increases. The volume of the core ion thus begins to take on a significant fraction of the volume inside of the ionic cloud, and so the integration of the ionic cloud charge must begin at $r = a$ rather than $r = 0$ (see Eq. (25.57)).

The effect of the term $(1 + \mathbf{B}aI^{1/2})$ in the Extended Debye-Hückel Law is to reduce the deviation of the value of γ from 1.0 relative to what would be predicted by the Debye-Hückel Law. The reason for this is that giving some dimension to the ion allows a reduction of $\psi(r)$ very near the ion from what it would be for a virtual point charge. This reduces the interaction between the charge on the ion and the charge in the surrounding ion cloud.

25.5 THE GÜNTLBERG AND DAVIES EQUATIONS

The four most commonly used equations to *estimate* γ values for ions are collected in Table 2.3. A comparison of the Güntlberg Equation with the Extended Debye-Hückel Law reveals that the former is a simplified version of the latter, with a taken to be ~3Å. Both of the Debye-Hückel Laws as well as the Güntlberg Equation predict that ion activity coefficients will tend to decrease monotonically as the ionic strength decreases. In contrast, the Davies Equation with its $-0.2\ I$ term predicts that for high values of I, ion activity coefficients will eventually start to *increase* as I increases. The Davies Equation must therefore take into account some type(s) of interactions that are not of the simple electrostatic type considered by the two Debye-Hückel Laws.

As discussed in Section 2.8.1, the Davies Equation considers the fact that as the total ion concentration increases, the various ions begin to have to *compete* for solvating water molecules. For example, for a constant concentration of NaCl, as the concentration of some other salt (say KBr) increases to the 0.5 F range, the competition gets more difficult, the Na^+ and Cl^- ions start to become less well solvated and therefore more "exposed." Their chemical activities thus begin to increase even though their concentrations are remaining constant. The concentration ranges over which the four major ion activity coefficient equations can be used are summarized in Table 2.3. Problems illustrating the principles of this chapter are presented by Pankow (1992).

25.6 REFERENCES

BOCKRIS, J. O'M., and A. K. N. REDDY. 1970. *Modern Electrochemistry*, Volume 1, Chapter 3. New York: Plenum Press.

DEBYE, P., and E. HÜCKEL. *Z. Physik.* 1923. *The Theory of Electrolytes. Pt. 1. Lowering of Freezing Points and Related Phenomena*, vol. 24, 185–206.

MOORE, W. J. 1972. *Physical Chemistry*, Fourth Edition. Englewood Cliffs: Prentice-Hall.

PANKOW, J. F. 1992. *Aquatic Chemistry Problems*. Portland: Titan Press-OR, P.O. Box 91399, Portland, Oregon 97291-1399.

ROBINSON, R. A., and R. H. STOKES. 1959. *Electrolyte Solutions*. London: Butterworths.

26

Electrical Double Layers in Aqueous Systems

26.1 INTRODUCTION

When a particle of a mineral solid, a living cell, some organic detrital matter, or even a purely organic phase like mineral oil is equilibrated with water, electrical charges almost always develop at the interface with the aqueous solution. Such charged interfaces will be found in the natural environment wherever water is found, that is, in aerosols in the atmosphere, in rain, in fresh surface waters of all types, in the oceans, in soil pore waters both above and below the water table, and in sediment pore waters of all types.

There are several reasons why the presence of charges at particle/water interfaces are important. Firstly, for particles suspended in the water column, the nature of the charged interface and the properties of the surrounding aqueous medium will determine whether or not the particles will coagulate (i.e., agglomerate) into larger masses that will tend to be removed by sedimentation (density greater than 1), or flotation (density less than 1). Coagulation is of great interest because it is a removal mechanism for: 1) natural sediment materials suspended in the waters of lakes, rivers, estuaries, and the oceans; 2) organic[1] and inorganic material in drinking water; and 3) organic matter in domestic sewage (about 50 percent of the organic carbon in the effluent from secondary sewage treatment plants is in particulate form).

Secondly, if the charges on the surfaces of particles affect the fate of suspended particles, they must necessarily also affect the fate of any inorganic or organic contaminants that are *sorbed* to and carried along with those particles. Indeed: 1) hydrophobic

[1]Removing organic matter from drinking water prior to chlorination is now practiced widely so as to minimize the formation of halomethanes like chloroform by the reaction of Cl_2, HOCl and/or OCl^- with the organic matter.

organic compounds like PCBs, PAHs, and certain pesticides like DDT and dieldrin will be strongly sorbed to particles high in natural organic matter; and 2) many metals (e.g., Pb, Cd, etc.) that are of great interest from an environmental contamination point of view can be strongly adsorbed to inorganic particle surfaces as well as to particles that are mostly organic.

Thirdly, since the presence of charges will affect ion behavior at a particle/water interface, the very tendency of ions like Pb^{2+} to sorb to an interface will be affected by the nature of the surface charge. Thus, sorption of Pb^{2+} onto a negatively-charged (am)$Fe(OH)_{3(s)}$ surface will be stronger than sorption onto a positively-charged (am)$Fe(OH)_{3(s)}$ surface.

26.2 THE ORIGINS OF CHARGE AT SOLID/WATER INTERFACES

26.2.1 Surface Charge Resulting from the Effects of a Potential-Determining Ion (pdi)

26.2.1.1 A Constituent Ion of the Solid is the pdi. When solids are equilibrated with initially-pure water, charges at the solid/water interface will usually arise because the various species constituting the solid will exhibit different tendencies to dissolve into the aqueous phase. For example, consider a solid $AB_{(s)}$ composed of the species A^+ and B^-. Let us assume that A^+ has a greater affinity for the aqueous phase than does B^-. Under these conditions, if $AB_{(s)}$ is equilibrated with initially-pure water, then the surface of the solid will become negatively-charged since more A^+ will enter the aqueous phase than will B^-. Net *charge separation* from the solid occurs because of the preferential dissolution of A^+ over B^-. A schematic depiction of this process is given in Figure 26.1.

In absolute terms, the amount of surface charge buildup that can occur when $AB_{(s)}$ dissolves into initially-pure water will always be relatively small. The reason is that the magnitude of the elementary charge e_o carried by both A^+ and B^- is sufficiently large that even small surface charge densities (in units of C/cm^2) can result in large electrical potentials. As will be discussed in considerable detail below, these potentials act to limit the amount of charge separation that occurs.

The negative charge at the interface in Figure 26.1 attracts the dissolved A^+ and repels the dissolved B^-. With H^+ and OH^- from the dissociation of water also in the solution, the negative surface charge also attracts the H^+ and repels the OH^-, but we will reserve discussion of that for a bit later. Because more A^+ dissolves than does B^-, there is excess positive charge in the solution. That positive charge is distributed diffusely in the solution. As discussed in Chapter 25 for the ion cloud surrounding a single ion in solution, strictly speaking, the aqueous charge layer extends very far out into the bulk solution. However, the vast majority of the net aqueous charge is located near the solid/water interface.

The distribution of charge near the solid/water interface in Figure 26.1 is often referred to as an "electrical double layer." There are two layers of charge. One charge

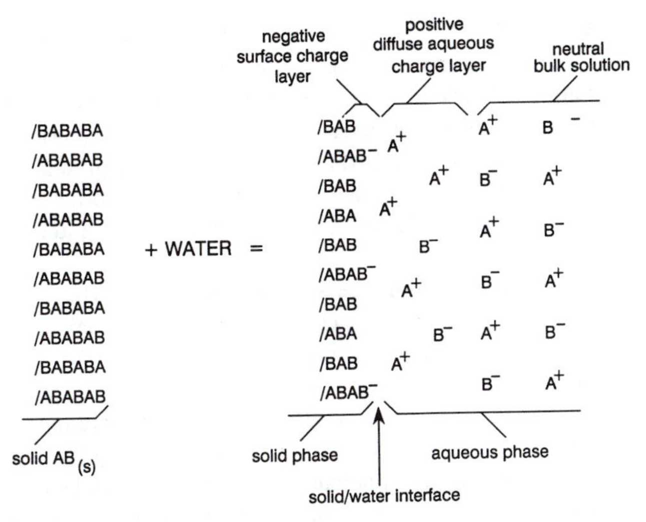

Figure 26.1 Dissolution of $AB_{(s)}$ into initially-pure water. The solid/aqueous system is presented in cross section. In this case, as with $AgI_{(s)}$, it is assumed that the A^+ ion has a greater inherent affinity for the aqueous phase than does B^-. These different affinities cause slightly more A^+ than B^- to enter into the initially-pure water. Near the surface, $[A^+] > [B^-]$. Out in the bulk solution, $[A^+] = [B^-]$. It is the excess A^+ that dissolves that causes the aqueous charge to be positively charged, and the surface to be negatively charged. The behaviors of H^+ and OH^- are not addressed in this figure.

layer is on the surface of the solid, and the other is in the adjacent aqueous charge layer. Like the ion/(ion cloud) system considered in Chapter 25, the aqueous layer bears a charge that is equal in magnitude but opposite in sign to the surface charge layer.

In the bulk solution, that is, at essentially infinite distance from the surface, the charge effects of the surface are no longer felt. Thus, when $AB_{(s)}$ dissolves into initially-pure water, in the bulk solution we have $[A^+] = [B^-]$. Although we do not have $[A^+] = [B^-]$ in the aqueous charge layer near the surface, that charge layer is so thin that $[A^+] = [B^-]$ makes a perfectly adequate MBE[2] for solving for the properties of most[3] systems.

$AgCl_{(s)}$, $AgBr_{(s)}$, and $AgI_{(s)}$ are examples of the many solids that exhibit the type of behavior depicted in Figure 26.1. In addition to the particular case involving dissolution into initially-pure water, double layers will usually also be present when these solids are equilibrated with solutions containing initial amounts of Ag^+ and/or Cl^-, Ag^+ and/or

[2]Since the concentrations of $[A^+]$ and $[B^-]$ are not equal everywhere in the solution, we also note that for this case, the *total amounts* of A^+ and B^- that dissolve are not equal. Again, however, since the aqueous charge layer is so thin, these two amounts are usually exceedingly close in magnitude.

[3]One exception that can arise involves systems containing a great many small particles per unit volume, and very little aqueous phase. In the aqueous phase of such a system, it is not possible to get far from a solid/water interface, and there is no electrically-neutral bulk solution.

Br^-, and Ag^+ and/or I^-, respectively. For the general solid $AB_{(s)}$, when $\{A^+\}$ is high and $\{B^-\}$ is low, A^+ can be driven onto the surface (onto B^- surface sites), and a positive surface charge will result. This will remain true *regardless* of which of the two ions (A^+ or B^-) is inherently more attracted to the aqueous phase. The magnitude of the positive charge density on the surface as well as the magnitude of the electrical potential at the surface will be determined in part by the equilibrium value of $\{A^+\}$ in the bulk solution. Thus, for the solid $AB_{(s)}$, A^+ is referred to as a **potential-determining ion** (pdi).

The pdi controls both the magnitude and the sign of the surface charge. If the pdi is positively charged, then increasing the activity of the pdi will make the surface charge more positive. If the pdi is negatively charged, then increasing the activity of the pdi will make the surface charge more negative.

When $\{B^-\}$ is high and $\{A^+\}$ is low, B^- will be driven onto the surface (onto A^+ surface sites), and negative charge will develop there. The magnitude of the negative charge density as well as the electrical potential at the surface will be determined in part by the equilibrium value of $\{B^-\}$ in the bulk solution. Thus, like A^+, we see that B^- can be viewed as a pdi for the $AB_{(s)}$ surface. Since the bulk solution values of $\{A^+\}$ and $\{B^-\}$ are relatable through the K_{s0} of the solid, we will be able to parameterize the sign of the surface charge as well as the sign and magnitude of the electrical potential at the surface using the activity of *either* of the two species. As we will see later, if the ionic strength in the solution is known, then the magnitude of the surface charge density may also be determined.

26.2.1.2 H^+ and OH^- as Potential-Determining Ions.

Surface charging can also occur with solids that can be hydrated, protonated, and/or deprotonated. For example, an $-SiO$ unit on the surface of solid silica $SiO_{2(s)}$ can be protonated directly to form $-SiOH^+$. In addition, an $-SiO$ unit on the surface can be hydrated by H_2O to form surface $-Si(OH)_2$ groups. Protons can then be removed to yield $-Si{<}^{O^-}_{OH}$, and a negative surface charge. (Alternatively, the formation of $-Si{<}^{O^-}_{OH}$ can be viewed as the direct adsorption of an OH^- ion to a surface $-SiO$ unit.) The proton (or OH^-) can thus be considered the pdi for the $SiO_{2(s)}$ surface. It is the pdi role that H^+ plays on the $SiO_{2(s)}$ surface of a glass pH electrode that is the source of the pH-dependent electrical potential that is measured with a pH meter.

A solid like $SiO_{2(s)}$ can be viewed in the $AB_{(s)}$ context discussed in the previous section. Thus, in addition to H^+ (or OH^-), the aqueous species Si^{4+} and O^{2-} can also be thought of as being pdi's for the $SiO_{2(s)}$ surface. Indeed, the solution-phase values of $\{Si^{4+}\}$ and $\{O^{2-}\}$ will be directly related through equilibrium constants to $\{H^+\}$. However, the actual mechanism by which charge is generated at the $SiO_{2(s)}$ surface is most easily visualized when viewing H^+ (or OH^-) as the pdi. Since most environmentally-important oxides, hydroxides, and oxyhydroxides (e.g., $Fe_2O_{3(s)}$, $MnO_{2(s)}$, $Fe(OH)_{3(s)}$, $FeOOH_{(s)}$, etc.) follow the pattern of behavior exhibited by the oxide $SiO_{2(s)}$, H^+ is usually used as the pdi in double-layer computations involving oxides, hydroxides, and oxyhydroxides.

Another class of surfaces for which H^+ is the pdi includes organic biomembranes and organic detrital matter. With the former, $\{H^+\}$ controls the magnitude of the negative surface charge by regulating the fraction (α_1) of surface carboxyl groups that are ionized. With the latter, $\{H^+\}$ controls the magnitude of the negative surface charge by regulating the fraction of the surface carboxyl, phenol, and other acid groups that are ionized.

26.2.2 Particles with Fixed Surface Charge—Clays

For some types of particles, for example, the true clays, the surface charge is not controlled by a pdi. We note that the term "true clay" is used here to refer to small particles that have a crystal structure involving two-dimensional arrays of silicon-oxygen tetrahedra and two-dimensional arrays of aluminum- or magnesium-oxygen-hydroxyl octahedra (van Olphen 1977).[4] These sheets can be layered one on top of one another in different ways to achieve different overall classes of clay crystal structures.

In many clay minerals, an atom of lower oxidation state replaces an atom of a higher oxidation state in the basic lattice structure. For example, an Al^{3+} ion can replace an Si^{4+} ion in a tetrahedral sheet. The substituting Al^{3+} ion completely fills the lattice site of the Si^{4+} ion. This type of complete substitution is called *isomorphous substitution.* The lower charge of the Al^{3+} relative to the Si^{4+} causes a deficit of positive charge, that is, a negative charge on the lattice. Negative-lattice charge can also result from the isomorphous substitution of Mg^{2+} or Fe^{2+} for Al^{3+} in the octahedral sheet. Given the many different ways in which the sheets can be layered and in which isomorphous substitution can occur (e.g., one can also have substitution of Cr^{3+} for Al^{3+} in the octahedral sheets), clay chemistry and mineralogy are complicated and richly diverse topic areas.

When isomorphous substitution leads to net negative charge on the lattice, overall charge balance is achieved through the presence of other, usually +1 cations, adsorbed on the surfaces of the layers and at the edges. These ions might be Na^+, K^+, NH_4^+, or even an organic cation. When placed in water, some clays like the montmorillonites take in enough water and swell (expand) sufficiently that virtually all of the interlayer cations are free to exchange with other cations in the aqueous solution. This ability to exchange cations is quantified in terms of a **cation exchange capacity** (CEC) for a clay. The CEC is usually expressed in units of millequivalents of positive charge per 100 g of air-dried clay. The CEC for some montmorillonite-type[5]clays can be as high as 100 meq/100 g. In this context, the unit meq refers to meq of charge; 100 meq of Na^+ charge corresponds to 0.1 mols of Na^+.

All clays are not swelling clays. Important non-swelling clays include the illites, kaolinites, and the chlorites. For non-swelling clays, the layers are held together *tightly*, and water cannot penetrate and expand the layers. Thus, the interlayer cations balancing the negative-lattice charge caused by isomorphous substitution are unable to exchange

[4]In soil science, the term "clay" is also used to refer to all mineral particles with diameters less than about 2 μm. While most of this "clay fraction" is indeed composed of true clays, it is also likely to also include at least some small particles of quartz and other non-clay minerals.

[5]The clays in the montmorillonite class include montmorillonite itself as well as hectorite, nontronite, and several other clays.

TABLE 26.1 Cation exchange capacities of different clay and soil types

Material	CEC (meq/100 g)
Clays	
Montmorillonites	60 to 100
Illites	10 to 40
Kaolinites	1 to 10
Chlorites	< 10
Soils[†]	
Sands	1 to 5
Fine sandy loams	5 to 10
Loams and silt loams	5 to 15
Clay loams	15 to 30
Clays	1 to 150

[†] Note that each soil class is a mixture of sands, clays, and organic humic material.

with cations in the aqueous solution. Only those balancing cations on the exteriors and edges of the layers will contribute to the CEC values of such clays. The CEC ranges for illites, kaolinites, and chlorites are given in Table 26.1. For some kaolinites, the CEC can be as low as 1 meq/100 g. Since soils are mixtures of different clays, sands, as well as organic humic material, the CEC of a soil will represent the mass-weighted average CEC for the mixture. Table 26.1 summarizes the CEC values for a variety of different soil types.

When a clay particle is equilibrated with liquid water, the exchangeable charge will tend to leave the particle and enter the solution. As seen in Figure 26.2, all of the charge that departs from a given clay particle will leave that particle with a net negative charge. The negative charge may be viewed as being distributed over the surface with an average charge density (C/cm^2). We note that the magnitude of the net negative charge in/on a

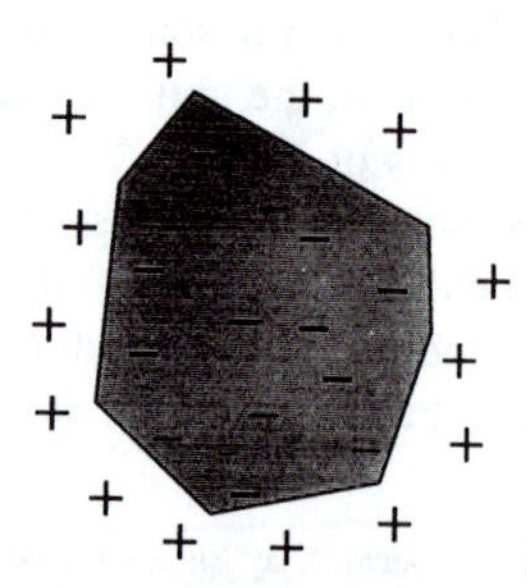

Figure 26.2 Cross-sectional view of a negatively-charged clay particle together with a surrounding, positive aqueous charge layer. Some of the positive charge associated with the clay particle may reside deep within the clay lattice, and some may be associated with lattice sites directly on the surface.

given clay particle is usually *fixed*; the amount of charge that is sufficiently mobile that it can leave the lattice is usually fixed within the context of time scales of years or longer.[6]

The *aqueous* charge layer that develops around a clay particle is always positive in sign, and the amount of charge that it contains equals the amount of charge that departed from the clay particle. The magnitude of the charge will vary from particle to particle (even for the same type of clay), and will depend on the particle size, among other things.

26.3 SIMILARITIES BETWEEN A DOUBLE-LAYER SYSTEM AND AN ION/(ION CLOUD) SYSTEM

As has been hinted above, the case of the aqueous double layer is a very close analog of the *ion/(ion cloud)* case discussed in Chapter 25. In particular, the charge that develops on the surface of the solid is analogous to the charge on a core ion. As seen in Figure 26.3, the diffuse layer of counterbalancing charge present in the aqueous solution near a charged solid surface is analogous to the diffuse sphere of charge present in the ion cloud surrounding a core ion. Moreover, just as the charge on a core ion is exactly balanced by the charge in its surrounding ion cloud, so too is the charge on the solid surface exactly balanced by the charge in its adjacent aqueous charge layer.

In the case of clays, the analogy discussed here may be extended even further. Indeed, the magnitude of the charge on a core ion is fixed, and as stated earlier, that is usually also the case with clay particles. When the surface charge is controlled by a pdi, however, the magnitude of the surface charge as well as its sign will change in response to changes in the aqueous chemistry.

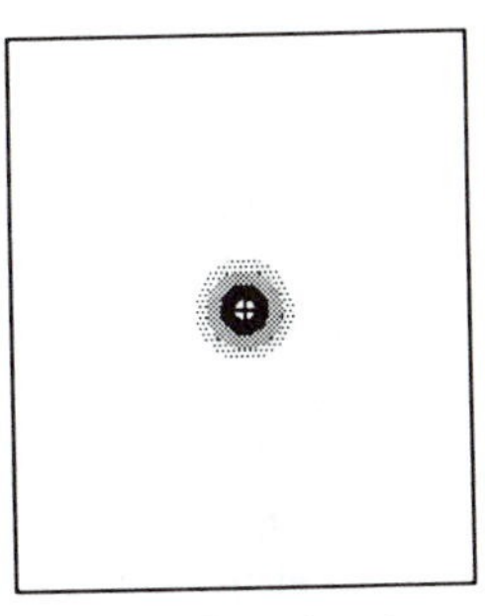

a. ion/ion cloud system

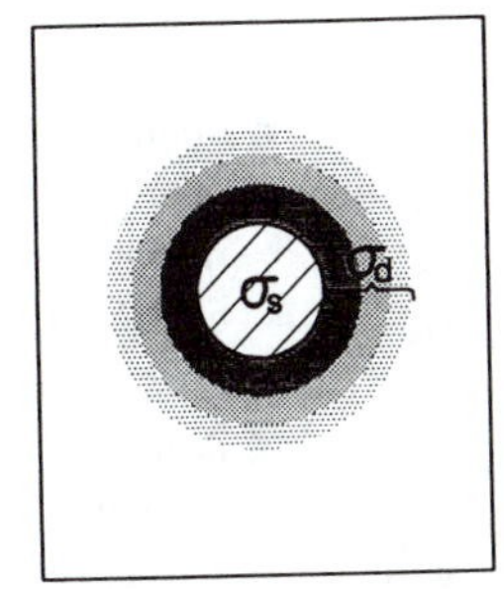

b. particle/aqueous charge layer system

Figure 26.3 Comparison of two systems involving central charge surrounded by a diffuse aqueous layer of balancing, opposite charge. (*a*) An ion surrounded by its diffuse ion cloud; (*b*) A spherical particle with charge σ_s is surrounded by a diffuse, aqueous charge layer with charge σ_d.

[6] Over the types of timescales required to weather and transform clays, it may be noted that the CEC of a given clay material may be subject to change.

26.4 THE ELECTROCHEMICAL POTENTIAL

26.4.1 The Electrochemical Potential as the Criterion of Equilibrium in the Presence of an Electrical Field

A useful introductory treatment of aqueous double-layer theory was provided some years ago by de Bruyn and Agar (1962). Certain aspects of the theoretical discussion provided here will follow their approach, though a different specific example will be used to illustrate the basic principles under consideration.

Consider again the case of a piece of $AB_{(s)}$ in equilibrium with initially-pure water. We will again assume that A^+ has a greater inherent affinity for the aqueous phase than does B^-, and so at equilibrium the surface will be negatively charged, and the aqueous charge layer will be positively charged as in Figure 26.1. If the movement of A^+ between the solid and the solution was influenced solely by chemical forces, then the dissolution of A^+ would continue until μ_{A^+} was equal in both phases. However, the electrical potential that is present near a charged surface introduces *powerful electrical forces*. Because of them, the dissolution of A^+ stops (i.e., equilibrium is reached) *before* μ_{A^+} is equal in both phases.

Equilibrium will be attained when an exact counterbalancing is achieved between: 1) the residual *chemical* force tending to drive A^+ from the solid into the bulk solution, and 2) the attractive *electrical* force involving negative surface charge that is tending to hold A^+ in the solid phase. Thus, at equilibrium we have that

$$\mu^{\beta}_{A^+}(0) - \mu^{\alpha}_{A^+}(\infty) = z_+ e_o N_a \left(\psi^{\alpha}(\infty) - \psi^{\beta}(0)\right)$$
$$\text{Chemical Imbalance} = \text{Electrical Imbalance} \tag{26.1}$$

where:

$\mu^{\beta}_{A^+}(0)$ = chemical potential of A^+ at the surface ($x = 0$) of the solid (β) phase
$\mu^{\alpha}_{A^+}(\infty)$ = chemical potential of A^+ in the bulk ($x = \infty$) aqueous (α) phase
z_+ = nominal charge on a positive potential-determining ion = +1 for A^+
e_o = elementary charge = 1.60×10^{-19} C = F/N_a
F = Faraday Constant = 96,485 C/mol
N_a = Avogadro's Number = 6.02×10^{23} mol^{-1}
$\psi^{\alpha}(\infty)$ = electrical potential (V) in the bulk ($x = \infty$), aqueous (α) phase
$\psi^{\beta}(0)$ = electrical potential (V) at the surface ($x = 0$) in the solid (β) phase

Both sides of Eq. (26.1) have units of energy. The same may be said about the basic thermodynamic equation $\Delta G = -nFE$ which gives the change in free energy when n mols of elementary electrical charge move across a potential difference of E volts. In Eq. (26.1), $(\mu^{\beta}_{A^+}(0) - \mu^{\alpha}_{A^+}(\infty))$ plays the role of ΔG, z_+ plays the role of n, $e_o N_a$ equals the Faraday constant F, and $(\psi^{\alpha}(\infty) - \psi^{\beta}(0))$ plays the role of the potential difference E. From Chapter 19, we recall that E can be positive or negative.

For the case considered in Figure 26.1, both sides of Eq. (26.1) are positive. However, Eq. (26.1) is equally valid when a high $\{A^+\}$ in the solution drives positive charge *onto* the solid surface. In this case, the phase transfer of positive charge again stops before the chemical potential of A^+ can become equal in both phases. Now, both sides of Eq. (26.1) are negative, and the equilibrium position that is attained is the one that is established when an exact counterbalancing is achieved between: 1) the residual chemical force tending to drive A^+ onto the surface; and 2) the repulsive electrical force between the positive surface charge and the positive A^+ ions.

Equation (26.1) may be rearranged to obtain

$$\mu^{\beta}_{A^+}(0) + z_+ F \psi^{\beta}(0) = \mu^{\alpha}_{A^+}(\infty) + z_+ F \psi^{\alpha}(\infty) \tag{26.2}$$

where F has been substituted for $e_o N_a$. The terms on the LHS of Eq. (26.2) pertain only to the solid (β) phase, and the terms on the RHS pertain only to aqueous (α) phase. Thus, we may recognize the parameter $\mu_i(x) + z_i F \psi(x)$ as the pertinent, equilibrium-controlling potential for species i in the presence of electrical forces. This potential is referred to as the *electrochemical potential*, and is given the symbol $\bar{\mu}_i$. It should be obvious that the use of the adjective "electrochemical" is meant to indicate that $\bar{\mu}_i$ contains both electrical and chemical components. Thus, as the criterion for equilibrium across a charged interface, we have

$$\bar{\mu}^{\beta}_i = \bar{\mu}^{\alpha}_i. \tag{26.3}$$

Equality of Electrochemical Potentials

When ψ in both phases is either zero or at least equal, the criterion for equilibrium reduces to the equality of the chemical potentials μ^{α}_i and μ^{β}_i.

For the case of a negatively-charged surface, the electrical forces attract positive ions towards the surface. This creates a gradient in the concentration of those ions (see Figure 26.1). This concentration gradient creates a gradient in the chemical potential. The positive ions tending to diffuse away from the surface, by virtue of the nature of the concentration gradient, are at the same time pulled towards the surface by the electrical forces. The dynamic equilibrium that is established will be the one that satisfies Eq. (26.3). In the case of negatively-charged ions, there will be a gradient in concentration that leads towards a negatively-charged surface. Diffusion of those negative ions towards the surface will at the same time be counterbalanced by the repulsive electrical forces associated with moving towards the surface. Again, the dynamic equilibrium that is established will be the one that satisfies Eq. (26.3). We will be seeking a more explicit representation of the nature of that equilibrium.

26.4.2 The Distribution of Gases in the Atmosphere: An Analogy for the Electrochemical Potential

The fact that a simple equality of chemical potentials is not the criterion of equilibrium in the presence of an electrical field may seem a little strange at first. However, an analogous case that goes a long way towards helping us to understand this phenomenon may be found in the distribution of the gases making up the atmosphere. Indeed, we realize that the gases in the earth's atmosphere are not uniformly distributed. They are most concentrated at the surface of the earth where the pressure is nearly 1 atm. As one moves upwards, the pressure decreases exponentially upward according to the equation (Seinfeld 1986)

$$p(z) = p^{\circ} \exp\left[-M_a z g / RT\right] \tag{26.4}$$

where z is the elevation measured relative to sea level, p° is the pressure at sea level, M_a is the average molecular weight of air, and g is the gravitation constant (i.e., the strength of the earth's gravitational field).

Gravitation attracts gases towards the earth, and thereby creates a gradient in the concentration. This concentration gradient creates a gradient in the chemical potential, and so as in the case of an aqueous double layer, equality of the chemical potential is not the criterion for chemical equilibrium. Rather, the criterion for equilibrium may be thought of as being a gravochemical potential.[7] Gases tending to diffuse upward by virtue of the nature of the concentration gradient are at the same time pulled downwards by gravitation. The dynamic equilibrium that is reached is the one represented by Eq. (26.4).

26.5 DOUBLE-LAYER PROPERTIES AS A FUNCTION OF THE ACTIVITY OF A POTENTIAL-DETERMINING ION

26.5.1 ψ° as a Function of the Activity of the Potential-Determining Ion

Often, when surface charges are present on particles suspended in aqueous systems, the particles in question may be viewed more or less either as spheres or as long cylinders. When bodies of these geometries bear charges that are uniformly distributed over the surface, basic electrostatic theory states that the electrical potential ψ inside of such a body will be constant. (The same will be true for spherical or long cylindrical clay particles if the interior sites vacated by the charge-balancing cations are uniformly distributed throughout the interior of the particle.) Therefore, for the piece of $AB_{(s)}$ under consideration here, we will assume that

$$\psi^{\beta}(x) = \psi^{\beta}(0) = \psi^{\beta} \qquad \text{for all } x. \tag{26.5}$$

Since the composition of the solid changes very little as a result of the creation of a small amount of charge (i.e., excess A^+ or excess B^-) on the surface, $\mu^{\beta}_{A^+}(0)$ will essentially equal the chemical potential of A^+ in the pure solid $AB_{(s)}$. We will refer to this chemical potential as $\mu^{\beta,\text{pure}}_{A^+}$. (We don't call this the *standard* chemical potential of pure A^+ itself in the solid because even the neutral solid is not pure A^+.) Therefore, by substituting the usual equation $\mu_i = \mu_i^{\circ} + RT \ln a_i$ for the chemical potential of A^+ in the aqueous (α) phase in Eq. (26.1), and by using 26.5, we obtain

$$\mu^{\beta,\text{pure}}_{A^+} - (\mu^{\text{o},\alpha}_{A^+} + RT \ln a^{\alpha}_{A^+}) = z_+ F(\psi^{\alpha}(\infty) - \psi^{\beta}) \tag{26.6}$$

where $a^{\alpha}_{A^+}$ is the activity of A^+ in the bulk aqueous solution.

[7] The expression for the gravochemical potential of a gaseous species i is the sum of the terms μ_i°, $RT \ln p_i$, and the gravitational potential energy term $M_i g z$.

Equation (26.6) may be rearranged to obtain

$$\left(\psi^{\beta} - \psi^{\alpha}(\infty)\right) = \frac{\mu_{A^+}^{o,\alpha} - \mu_{A^+}^{\beta,\text{pure}}}{z_+F} + \frac{RT}{z_+F}\ln a_{A^+}^{\alpha}. \tag{26.7}$$

Since we will now only be concerned with activities in the aqueous phase, we will drop the α superscript in $a_{A^+}^{\alpha}$. Since the chemical potentials $\mu_{A^+}^{o,\alpha}$ and $\mu_{A^+}^{\beta,\text{pure}}$ are constants, taking the derivative of Eq. (26.7) yields

$$d(\psi^{\beta} - \psi^{\alpha}(\infty)) = \frac{RT}{z_+F} d\ln a_{A^+} = \frac{2.303RT}{z_+F} d\log a_{A^+}. \tag{26.8}$$

If the term $(\psi^{\beta} - \psi^{\alpha}(\infty))$ is abbreviated as $\Delta\psi$, Eq. (26.8) becomes

$$d(\Delta\psi) = \frac{2.303RT}{z_+F} d\log a_{A^+}. \tag{26.9}$$

As noted in Chapter 19, at 298°K, the term $2.303RT/F = 0.05916$ V = 59.16 mV.

Equation (26.9) states that the difference in electrical potential between the solid and the aqueous phases is determined by $\log a_{A^+}$. Thus, we now see explicitly why A^+ is referred to as a pdi for $AB_{(s)}$. We point out again that clays do not have a pdi.

Re-integrating Eq. (26.9), we obtain

$$\Delta\psi = \frac{2.303RT}{z_+F}\log a_{A^+} + C \tag{26.10}$$

where C is a constant of integration. The value of C will depend on the nature of the system, that is, it will depend on the properties of the solvent (in this case water), and on those of the solid.

As discussed above, when a_{A^+} is very low, then the surface will be negatively charged. This corresponds to a negative value of $\Delta\psi$. When a_{A^+} is high, then the surface will be positively charged, and the value of $\Delta\psi$ will be positive. Based on this, we realize that there must be a value of a_{A^+} for which there is *no net charge* on the surface. This activity will be referred to as the **zero point of charge** (zpc) concentration of the pdi A^+, and will be given the symbol $a_{A^+}^{o}$.

Since $a_{A^+}^{o}$ will be a constant for a given solid, without any loss of generality, we can let

$$C = \chi - \frac{2.303RT}{z_+F}\log a_{A^+}^{o} \tag{26.11}$$

where χ is another constant. Substituting Eq. (26.11) into Eq. (26.10), we obtain

$$\Delta\psi = \frac{2.303RT}{z_+F}\log\frac{a_{A^+}}{a_{A^+}^{o}} + \chi. \tag{26.12}$$

$\Delta\psi$ is plotted as a function of $\log a_{A^+}$ for a generic case in Figure 26.4. When $a_{A^+} = a_{A^+}^{o}$, the pdi is at its zpc, and there is no net charge on the surface. However, $\Delta\psi$ is not zero at that point since χ is not in general zero. As de Bruyn and Agar (1962) point out, there could easily be some polarity to the surface that will cause a net alignment at the solid/water interface of the oxygen in the water molecules either towards

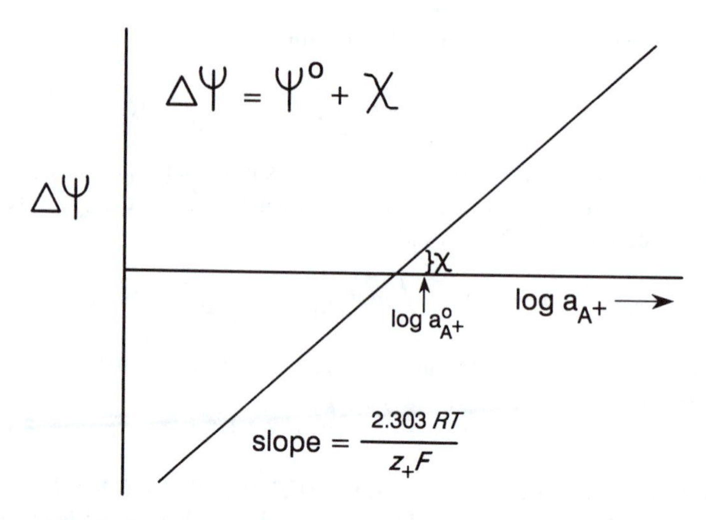

Figure 26.4 $\Delta\psi$ vs. $\log a_{A^+}$ for a case when χ is positive. The slope is positive because z_+ is positive. $\Delta\psi$ is not zero when $\log a_{A^+} = \log a^o_{A^+}$ because $\chi \neq$ zero.

or away from the surface. Since water is a dipole, this type of net charge orientation will lead to a certain amount of potential difference. This potential difference is equal to χ in Eq. (26.12). Dropping the subscript A^+, the expression that can be developed for χ based on Eqs. (26.7) and (26.12) is $(\mu^\alpha_{a=a^o} - \mu^{\beta,\text{pure}})/z_+F$ where $\mu^\alpha_{a=a^o}$ is the chemical potential of A^+ in the aqueous phase when $a_{A^+} = a^o_{A^+}$. For our purposes, we will not need to know the value of χ.

Based on the above discussion, we see that $\Delta\psi$ is always composed of two parts: 1) the portion that is due to the double layer charges; and 2) χ. The first term in Eq. (26.12) is usually abbreviated as ψ^o. That is,

$$\Delta\psi = \psi^o + \chi \tag{26.13}$$

$$\psi^o = \frac{2.303RT}{z_+F} \log \frac{a_{A^+}}{a^o_{A^+}}. \tag{26.14}$$

A plot of ψ^o as a function of $\log a_{Ag^+}$ is provided for $AgI_{(s)}$ in Figure 26.5. The manner in which ψ varies as a function of distance into the solution is shown generically in Figure 26.6 for a case when $\Delta\psi$ is positive. The manners in which $\bar{\mu}$ and its component terms vary as a function of distance are depicted in Figure 26.7 for a case when $\mu^\beta > \mu^\alpha$.

26.5.2 General Equation for ψ^o as a Function of the Activity of the Potential-Determining Ion

The behavior of the cation A^+ was used to derive Eq. (26.14). However, there is no reason why we could not have based our derivation of an equation for ψ^o on the behavior of the anion B^-. Therefore, it should not come as a surprise that an equation very similar to Eq. (26.14) may be obtained by considering B^- to be the pdi. We first note that the

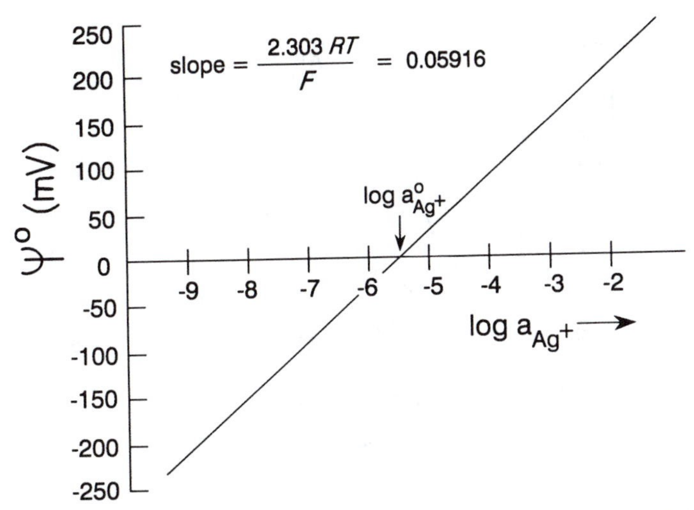

Figure 26.5 $\psi^{\rm o}$ vs. $\log a_{\rm Ag^+}$ for $\rm AgI_{(s)}$ at 25°C/1 atm. $K_{\rm s0} = 10^{-16.1}$, $a^{\rm o}_{\rm Ag^+} = 10^{-5.6}$, and $a^{\rm o}_{\rm I^-} = 10^{-10.5}$. Note that when $\rm AgI_{(s)}$ is equilibrated with initially-pure water, the AgI surface will be strongly negatively-charged because $\log a_{\rm Ag^+} = \log a_{\rm I^-} = 10^{-8.05}$ in that system.

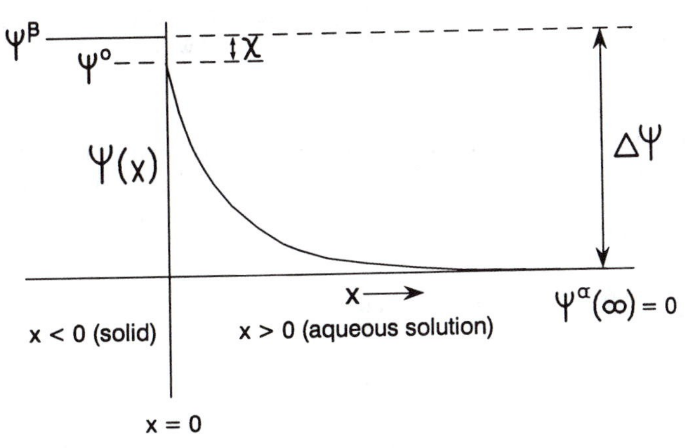

Figure 26.6 ψ as a function of distance when $\Delta\psi$ is positive (i.e, the surface is positively charged), and χ is also positive. $x = 0$ at the solid/water interface. The value of $\psi^{\alpha}(\infty)$ is taken to be zero.

$K_{\rm s0}$ for the solid will *always* be satisfied whenever there is equilibrium between the solid and the aqueous phase, that is,

$$a_{\rm A^+}a_{\rm B^-} = K_{\rm s0}. \tag{26.15}$$

The ion activities are those in the bulk solution. Equation (26.15) applies when the surface is negatively charged, when it is positively charged, and when the conditions pertain to the zpc and there is no charge at all.

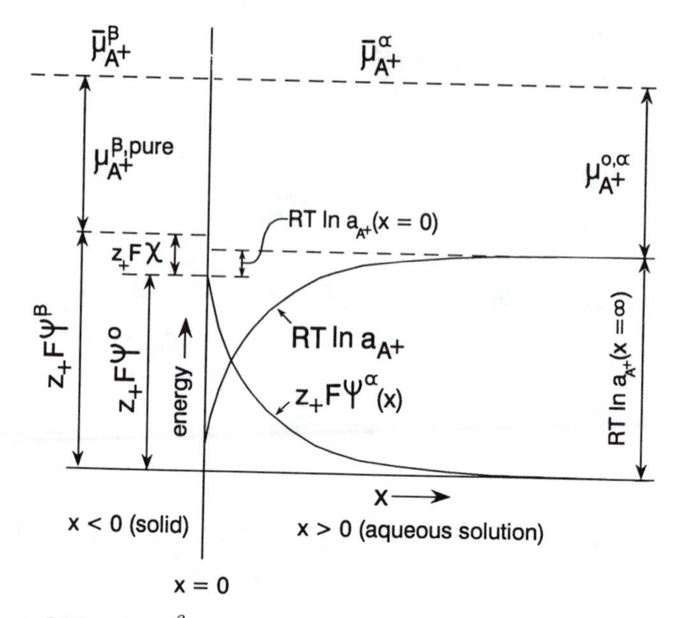

Figure 26.7 $\bar{\mu}^{\alpha}_{A^+}$, $\bar{\mu}^{\beta}_{A^+}$, $z_+F\psi^{\alpha}(x)$, and $z_+F\psi^{\beta}(x)$ as a function of distance when $\mu^{\beta}_{A^+} = \mu^{\beta,\text{pure}}_{A^+} < \mu^{\alpha}_{A^+} = \mu^{o,\alpha}_{A^+} + RT \ln a_{A^+}(x = \infty)$, and the value of $\Delta\psi$ is positive (i.e., the surface is positively charged). The value of $\psi^{\alpha}(\infty)$ in the bulk solution) is taken to be zero. $x = 0$ at the solid/water interface.

Since there can be only one zpc condition, when $a_{A^+} = a^{o}_{A^+}$, then at the same time we must also have that $a_{B^-} = a^{o}_{B^-}$. Therefore, as a special case of Eq. (26.15), we have

$$a^{o}_{A^+}a^{o}_{B^-} = K_{s0}. \tag{26.16}$$

Combining Eqs. (26.14)–(26.16), we obtain

$$\psi^{o} = \frac{2.303RT}{z_+F} \log \frac{K_{s0}/a_{B^-}}{K_{s0}/a^{o}_{B^-}} \tag{26.17}$$

$$= \frac{2.303RT}{z_+F} \log \frac{a^{o}_{B^-}}{a_{B^-}}. \tag{26.18}$$

Since $z_+ = +1 = -z_- = -(-1)$, substituting for z_+ in terms of z_- in Eq. (26.18) yields

$$\psi^{o} = \frac{2.303RT}{z_-F} \log \frac{a_{B^-}}{a^{o}_{B^-}} \tag{26.19}$$

which is analogous to Eq. (26.14).

Figures 26.8*a–d* illustrate in a schematic manner how taking the pdi to be A^+ is equivalent to taking the pdi to be B^-. This remains true regardless of the sign of the

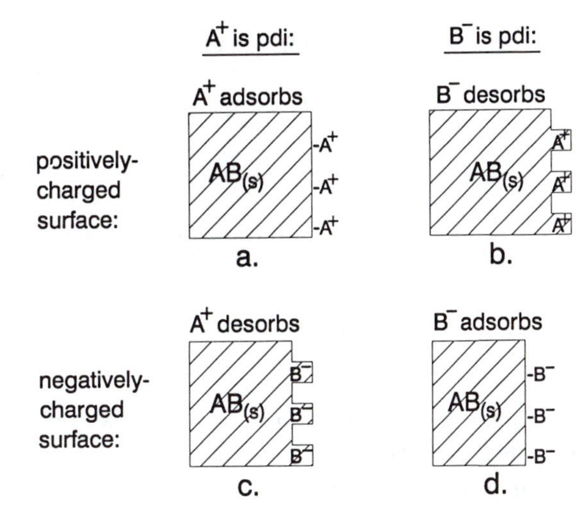

Figure 26.8 When A^+ is taken to be the pdi at the $AB_{(s)}$/water interface, a positively-charged surface is viewed as being the result of the adsorption of A^+ ions on the $AB_{(s)}$ surface (part *a*). A negatively-charged surface is viewed as being the result of the desorption of A^+ ions from the $AB_{(s)}$ surface (part *c*). When B^- is taken to be the pdi at the $AB_{(s)}$/water interface, a positively-charged surface is viewed as being the result of the desorption of B^- ions from the $AB_{(s)}$ surface (part *b*), and a negatively-charged surface is viewed as being the result of the adsorption of B^- ions on the $AB_{(s)}$ surface (part *d*). One can take either A^+ or B^- to be the pdi. Thus, parts *a* and *b* of this figure are equivalent views of a positively-charged surface, and parts *c* and *d* are equivalent views of a negatively-charged surface. For ψ^o, we may use either

$$\psi^o = \frac{2.303RT}{z_+F}\log\frac{a_{A^+}}{a^o_{A^+}} \qquad \text{or} \qquad \psi^o = \frac{2.303RT}{z_-F}\log\frac{a_{B^-}}{a^o_{B^-}}$$

A^+ is viewed as the pdi — B^- is viewed as the pdi

charge on the surface. In general then, for any potential determining ion i of charge z_i (z_i could be either positive or negative), we have

$$\psi^o = \frac{2.303RT}{z_iF}\log\frac{a_i}{a^o_i}. \tag{26.20}$$

26.5.3 Characteristics Affecting the Magnitude of the Zero Point of Charge Activity of a Potential-Determining Ion

As is noted by de Bruyn and Agar (1962), for a solid like $AB_{(s)}$, it is the relative characteristics of the constituent ions that determine their zpc activities. For $AgI_{(s)}$, at 25°C/1 atm, $K_{s0} = 10^{-16.1}$. Neglecting activity corrections, if both Ag^+ and I^- were equally well attracted to the aqueous phase, both $a^o_{Ag^+}$ and $a^o_{I^-}$ would equal $10^{-8.05}$ *M*. However, at 25°C/1 atm, $a^o_{Ag^+} = 10^{-5.5}$ *M* and $a^o_{I^-} = 10^{-10.6}$ *M*; the Ag^+ ion is much

TABLE 26.2 Zero point of charge (zpc) concentrations for three silver halides $AgX_{(s)}$ at 25°C/1 atm.

Silver Halide $AgX_{(s)}$	K_{s0}	$a^o_{Ag^+}$	$a^o_{X^-}$	$a^o_{Ag^+}/a^o_{X^-}$
$AgCl_{(s)}$	$10^{-9.7}$	$10^{-4.0}$	$10^{-5.7}$	50
$AgBr_{(s)}$	$10^{-12.3}$	$10^{-5.4}$	$10^{-6.9}$	32
$AgI_{(s)}$	$10^{-16.1}$	$10^{-5.6}$	$10^{-10.5}$	79,000

more strongly attracted to the aqueous phase than is the I^- ion. Thus, when equilibrated with initially-pure water, $AgI_{(s)}$ will take on a strongly negative charge since a_{Ag^+} in the bulk solution will equal 10^{-8} *M*.

Compared to Ag^+, the I^- ion is bulky, polarizable, and capable of interacting with other positive ions. Thus, I^- "prefers" to remain in the ionic solid phase. Ag^+, on the other hand, is relatively small in size, and it exhibits a strongly negative ΔH of hydration. Significantly more energy is released when a mol of Ag^+ ions becomes solvated with water molecules than is released when a mol of I^- ions is solvated. These properties are responsible for the fact that $a^o_{Ag^+} = 10^{-5.6} \gg a^o_{I^-} = 10^{-10.4}$. As is noted by de Bruyn and Agar (1962), one might expect that the zpc values for the solids $AgBr_{(s)}$ and $AgCl_{(s)}$ will be more symmetrical with respect to their K_{s0} values than is the case for $AgI_{(s)}$; Br^- and Cl^- are much less polarizable than is the I^- ion. These expectations are confirmed by the experimental data summarized in Table 26.2. A measure of the symmetry of the zpc values is obtained by ratioing the $a^o_{Ag^+}$ to the corresponding $a^o_{X^-}$ where X^- is Cl^-, Br^-, or I^-. For $AgI_{(s)}$, $a^o_{Ag^+}/a^o_{X^-} = 79{,}000$. For $AgBr_{(s)}$ and $AgCl_{(s)}$, it is only 32 and 50, respectively.

26.6 CONCENTRATIONS AS A FUNCTION OF DISTANCE IN THE AQUEOUS CHARGE LAYER

26.6.1 Charge Distribution in the Aqueous Charge Layer: σ_+ and σ_-

As discussed previously, the electrical charge on the surface is exactly counterbalanced by the charge in the diffuse aqueous charge layer. Symbolically, we can write this as

$$\sigma_s + \sigma_d = 0 \tag{26.21}$$

where σ_s is the *surface* charge density (C/cm²), and σ_d is the *diffuse* aqueous charge density (C/cm²). It is the need for consistent units for both σ_s and σ_d that causes us to choose C/cm² as the units for σ_d. While this selection might seem a little strange at first for a quantity that is spread over a diffuse aqueous layer, we will see very shortly how the mathematical definition of σ_d makes C/cm² the appropriate units.

In order to better understand Eq. (26.21) and its implications, consider a specific case. Let us say that charge on the surface is governed by a pdi (i.e., the surface is not a clay surface), and that the pdi has a charge of -1. In Figure 26.9*a*, the initial, pre-equilibrium state is considered to be an aqueous solution containing the ions of interest at concentrations equal to their bulk phase concentrations all the way up to the solid/water interface. In Figure 26.9*b* the solid/water system is taken to be at equilibrium under the assumption that the bulk solution activity of the negatively-charged pdi exceeds the a^o value for that ion. That is, it is assumed that the activity of the pdi in the bulk solution phase is large enough to cause the surface to be negatively charged.

As shown in Figure 26.9*b*, the positive ions in the aqueous solution are attracted to the negative surface, and the negative ions are repelled. In the aqueous charge layer, then, *electroneutrality is not obeyed.* Out in the bulk solution, however, things get back to normal, and the electroneutrality equation applies once again. Equation (26.21) states that if the surface plus the solution is taken as a whole, then electroneutrality is maintained in that *overall* system.

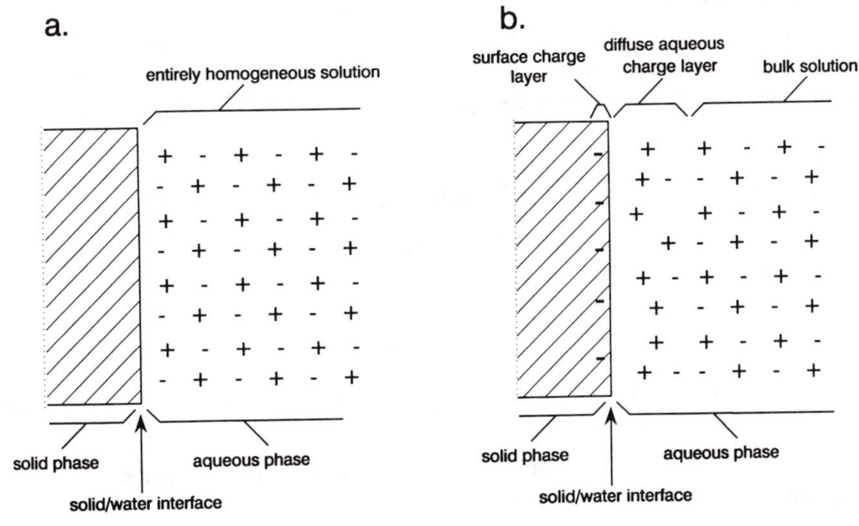

Figure 26.9 (*a*) A surface and the aqueous solution shown in cross section *before* any charge moves from the solution onto the surface. The pdi is assumed to have a charge of -1. Note that on average, cations and anions are evenly balanced throughout the solution. (*b*) Surface and aqueous solution shown in cross section *after* some negative charge moves from the solution onto the surface. Overall charge balance requires that the amount of negative charge that moves onto the surface be equal to the deficiency of negative charge left in the solution. Thus, $\sigma_s = -\sigma_d$. Near the surface, there is an excess of positive charge relative to the bulk solution, and a deficiency of negative charge relative to the bulk solution. In the bulk solution itself, cations and anions remain evenly balanced.

In an aqueous charge layer, the cation(s) and the anion(s) undergo transitions in concentration as one moves from near the charged surface towards the bulk solution. For the specific case of a surface that is negatively-charged, the transition for a given cation is from a state of enrichment back down to bulk solution concentration; for a given anion, the transition is from a state of depletion back up to the bulk solution concentration. The story is exactly reversed for a positively-charged surface.

Thus, regardless of the sign of the charge on the surface, the aqueous charge layer is always composed of two parts. One part is borne by the cation(s) and is given the symbol σ_+. The second part is borne by the anion(s), and is given the symbol σ_-. That is,

$$\sigma_d = \sigma_+ + \sigma_- . \tag{26.22}$$

Both σ_+ and σ_- always bear the same sign, regardless of the sign of σ_s. For the specific case of a negatively-charged surface, both σ_+ and σ_- are positive. In that case, the *enrichment* of *positive* charge in the aqueous charge layer makes σ_+ positive. The *depletion* of *negative* charge in that layer also makes σ_- positive for a negatively-charged surface. For a positively-charged surface, on the other hand, both σ_+ and σ_- are negative since in that case there is a depletion of positive charge and an enrichment of negative charge in the aqueous charge layer.

It is clear that if we are to develop explicit expressions for σ_+ and σ_-, we will need to know how the concentrations of the cations and anions in the aqueous solution change as a function of distance from the surface. That is, we will need explicit expressions for the concentration deficiencies and the concentration excesses to know how they combine to translate into the net charge density as given by σ_d.

26.6.2 The Boltzmann Distribution in the Aqueous Charge Layer

The constancy of the electrochemical potential of a given ion i throughout the aqueous solution means that $\bar{\mu}_i^\alpha(x)$ for any value of x is equal to the electrochemical potential for that species in the bulk solution. That is,

$$\bar{\mu}_i^\alpha(x) = \bar{\mu}_i^\alpha(\infty).$$

$$\underset{\text{at any point } x}{\text{electrochemical potential}} = \underset{\text{in bulk solution}}{\text{electrochemical potential}} \tag{26.23}$$

As before, $x = 0$ at the surface and $x = \infty$ in the bulk solution. Equation (26.23) may be expanded to yield

$$\mu_i(x) + z_i F\psi(x) = \mu_i(\infty) + z_i F\psi(\infty). \tag{26.24}$$

The α has been dropped as a superscript on both μ and ψ since we will now only be concerned with μ and ψ values in the aqueous phase; there will be no ambiguities as to which phase is under discussion.

Both of the μ_i terms may be expanded using $\mu_i = \mu_i^\text{o} + RT\ln a_i$. When this is done and the μ_i^o values are cancelled, one obtains

$$RT\ln a_i(x) + z_i F\psi(x) = RT\ln a_i(\infty) + z_i F\psi(\infty). \tag{26.25}$$

We now neglect activity corrections in the aqueous phase. Since the density of water is very close to unity at 25°C/1 atm, concentrations expressed in molality and molarity will be essentially identical. Moreover, the molarity of i will be proportional to n_i, the concentration in units of ions/cm^3. Eq. (26.25) can then be rearranged to obtain

$$\ln \frac{n_i(x)}{n_i(\infty)} = \frac{-z_i F(\psi(x) - \psi(\infty))}{RT} \tag{26.26}$$

or

$$n_i(x) = n_i(\infty) \exp\left[\frac{-z_i F(\psi(x) - \psi(\infty))}{RT}\right]. \tag{26.27}$$

Equation (26.27) may be recognized as a Boltzmann distribution.[8] If $(\psi(x) - \psi(\infty))$ is negative (i.e., negatively-charged surface), then for a negatively-charged ion ($z_i < 0$), we have $\ln[n_i(x)/n_i(\infty)] < 0$ for all finite values of x. As one approaches $x = 0$, the value of $(\psi(x) - \psi(\infty))$ becomes increasingly negative, and so $\ln[n_i(x)/n_i(\infty)]$ becomes increasingly negative. Therefore, relative to the bulk solution, the concentration of a negatively-charged ion becomes increasingly depleted as one approaches $x = 0$. Conversely, a positively-charged ion becomes increasingly enriched as one approaches $x = 0$. The situation for cations and anions is reversed for a surface that is positively charged.

Equation (26.27) illustrates that, strictly speaking, the aqueous charge layer extends infinitely far into the solution. That is, strict equality of $n_i(x)$ and $n_i(\infty)$ occurs only at $x = \infty$. For all intents and purposes, however, once x reaches values of a few hundred angstroms, $n_i(x)$ and $n_i(\infty)$ will usually be essentially identical.

26.6.3 Selecting the Reference Value for $\psi(\infty)$

The value of the quantity $(\psi(x) - \psi(\infty))$ does not depend on the absolute electrical potential at the point x, but only on the *difference* between $\psi(x)$ and $\psi(\infty)$. Therefore, to obtain the correct value of $(\psi(x) - \psi(\infty))$, all we need to do is measure both quantities relative to the same reference potential. As an analogy, consider the situation depicted in Figure 26.10 where the quantity of interest is the difference in physical height between two particular steps on a staircase.

A convenient value for the reference potential $\psi(\infty)$ is zero. Under these conditions, Eq. (26.27) can be rewritten

$$n_i(x) = n_i(\infty) \exp\left[\frac{-z_i F \psi(x)}{RT}\right]. \tag{26.28}$$

If we can determine an equation for ψ as a function of x, we will be able to calculate n_i as a function of x for any ion i.

A relation that is useful in conjunction with Eq. (26.28) is

$$F/R = 11{,}605 \text{ K/V}. \tag{26.29}$$

At 298°K, then,

$$F/RT = 38.941 \text{ V}^{-1}. \tag{26.30}$$

[8]Equation (26.4) is also a Boltzmann distribution. For that gravitational case, however, the strength of the field is assumed to remain constant. That is, g is assumed to remain essentially constant with height. In contrast, in Eq. (26.27), $w(x)$ varies with x.

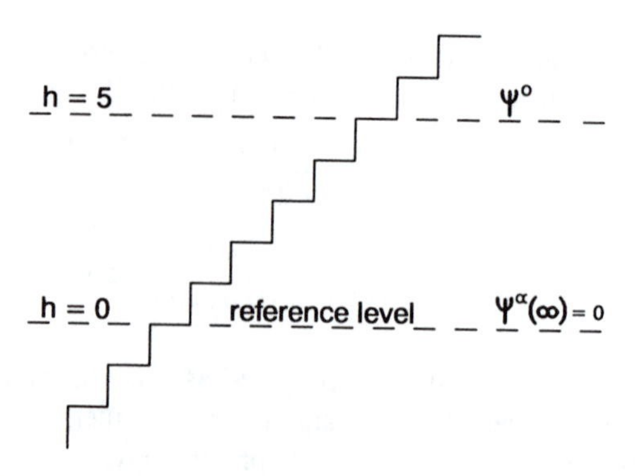

Figure 26.10 Analogy between measurements of electrical potential ψ and height h. Setting a specific reference level for ψ° is analogous to setting a specific reference level for h. Thus, the *difference* in potential between ψ° and $\psi(\infty)$ does not depend on the value assigned to $\psi(\infty)$, and the difference in height between any given stairstep and a reference stairstep does not depend on the height assigned to the reference stairstep.

26.6.4 σ_+ and σ_- for $AB_{(s)}$ in Equilibrium with Initially-Pure Water

We consider for the moment the case when only A^+ and B^- are present in the solution that is in equilibrium with $AB_{(s)}$. The expressions for σ_+ and σ_- are then

$$\sigma_+ = z_+ e_o \int_0^{\infty} (n_+(x) - n_+(\infty))\, dx = \text{charge (in C per cm}^2\text{) borne by positive ions} \tag{26.31}$$

$$\sigma_- = z_- e_o \int_0^{\infty} (n_-(x) - n_-(\infty))\, dx = \text{charge (in C per cm}^2\text{) borne by negative ions.} \tag{26.32}$$

Figures 26.11*a–b* illustrate the nature of these integrals for a case involving a negatively-charged surface.

When the value of the parameter z is small, then as discussed for Eq. (25.15), we have $e^y \simeq 1 + y$. Thus, when the absolute value $|\psi(x)|$ is small, $\exp[-z_i F \psi(x)/RT] \simeq 1 - z_i F \psi(x)/RT$. The value of $|\psi(x)|$ will be small *for all* x when $|\psi^{\circ}|$ is small (what is meant by "small" will be discussed shortly). Thus, when $|\psi^{\circ}|$ is in fact small, the charge in the aqueous charge layer due to the excess of one type of ion (cation or anion) *throughout* the aqueous charge layer will be nearly equal to the charge due to the deficiency of the other type of ion *throughout* that layer (see Eqs. (26.31) and (26.32)). That is,

$$\sigma_+ \simeq \sigma_- \tag{26.33}$$

since $z_+(n_+(x) - n_+(\infty))$ will be very nearly equal to $z_-(n_-(x) - n_-(\infty))$. (Remember

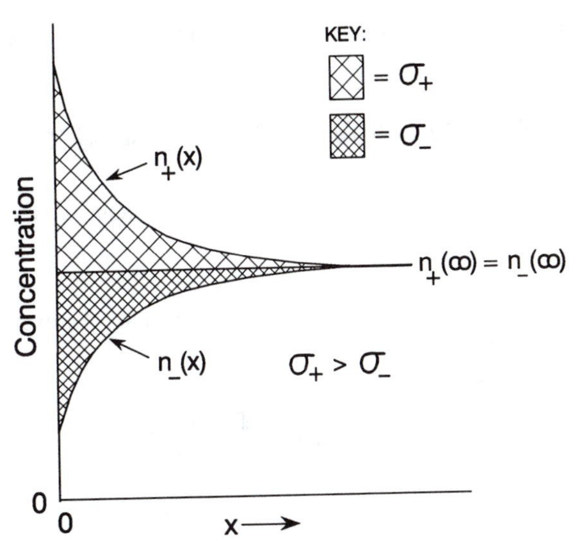

Figure 26.11 Concentration as a function of distance for negative ψ^o, and assuming $z_+ = -z_- = 1$. For negative ψ^o, the area under the $n_+(x)$ curve (i.e., σ_+) is always greater than the area between the x-axis and the $n_-(x)$ curve (i.e., σ_-). Thus, σ_+ will always be greater than σ_- for a negatively-charged surface. (σ_- will always be greater than σ_+ for a positively-charged surface.)

When $|\psi^o|$ is small, then for a negatively-charged surface σ_+ will be only slightly greater than σ_-. The difference will increase as $|\psi^o|$ increases because $n_-(x)$ has a mathematical lower bound of zero, and there is no mathematical upper bound on $n_+(x)$. (There will, however, be a physical upper limit on how concentrated $n_+(x)$ can become.) When multiple cations of the same charge z_+ are present, and when multiple anions of the same charge z_- are present, then: 1) $n_+(x)$ refers to the concentration sum over all cations; 2) $n_-(x)$ refers to the concentration sum over all anions; 3) σ_+ refers to the area sum over all cations; and 4) σ_- refers to the area sum over all anions.)

that z_+ and z_- have sign.) For the case of a negatively-charged surface, the excess of positive ions will nearly equal the deficiency of negative ions.

How small is "small" for $|\psi^o|$? The answer is determined by Eq. (26.28) in conjunction with the values of z_+, z_-, and the nature of the exponential function. If $z_+ = -z_- = 1$, "small" turns out to be about 0.008 V (8 mV). Indeed, for $\psi^o = +0.005$ V, $\exp[-z_+F\psi^o/RT] = 0.82$, and $\exp[-z_-F\psi^o/RT] = 1.22$. That is, very near the surface, there is about a 20 percent depletion of the cation, and about a 20 percent enrichment of the anion. The degree of symmetry present always improves as x increases since the value of $|\psi(x)|$ always decreases with distance.

For values of $|\psi^o|$ as large as 0.025 V (25 mV), the symmetry is lost except at large x. E.g., for $z_+ = -z_- = 1$ and $\psi^o = 25$ mV, $\exp[-z_+F\psi^o/RT] = 0.38$, and $\exp[-z_-F\psi^o/RT] = 2.65$. We can see for large $|\psi^o|$, the cause of the asymmetry is just the nature of the exponential: e^y is unbounded for large positive y, but bounded by 0 for large negative y.

26.7 THE DIFFERENTIAL EQUATION GOVERNING THE AQUEOUS CHARGE LAYER

If the electrolyte in the aqueous bulk solution and in the aqueous charge layer is symmetrical, then we have that

$$z_+ = -z_- = z. \tag{26.34}$$

Salts that obey Eq. (26.34) include AgCl, KCl, $CaSO_4$, etc. An example of a salt that does not obey Eq. (26.34) is $MgCl_2$.

From Eqs. (26.27) and (27.34), for a symmetrical electrolyte solution we have

$$n_+(x) = n_+(\infty) \exp\left[\frac{-zF\psi(x)}{RT}\right] \tag{26.35}$$

$$n_-(x) = n_-(\infty) \exp\left[\frac{zF\psi(x)}{RT}\right]. \tag{26.36}$$

When the aqueous solution satisfies the electroneutrality equation at some point x,

$$z_+n_+(x) + z_-n_-(x) = 0. \tag{26.37}$$

The net amount of charge per unit volume at any distance x from the surface is given by $q(x)$ according to

$$q(x) = z_+e_o n_+(x) + z_-e_o n_-(x). \tag{26.38}$$

By Eq. (26.34,) then,

$$q(x) = ze_o(n_+(x) - n_-(x)). \tag{26.39}$$

A value of x that will satisfy Eq. (26.37) is $x = \infty$. In the bulk solution then, we have that $q(\infty) = 0$. That is, the bulk solution is electrically neutral. In the aqueous charge layer, however, we know that Eq. (26.37) will not be satisfied, and in fact will be less and less satisfied as one approaches the surface. Equation (26.38) is similar to Eq. (25.9) except that x is now the coordinate of interest rather than the r used in the spherically symmetrical case of charge density around an ion. Again, as in Eq. (25.9), the concentrations in Eqs. (26.35)–(26.39) are in units of ions per cm^3. The factor of e_o (the elementary charge on a species with +1 charge) is included in Eqs. (26.38) and (26.39) so as to obtain the required units of C/cm^3.

The reader may now recall that the electrical potential ψ at a given point is related to the charge density through the Poisson Equation. Since the dimensions of the particles upon which double layers form are generally very large relative to the thicknesses of their corresponding aqueous charge layers, we can usually assume that the aqueous charge layer is essentially a flat plate of charge. That is, we can assume that properties of the aqueous charge layer vary only with one dimension, x. In this one-dimensional problem in *Cartesian* coordinates then, the Poisson Equation is

$$\nabla^2\psi(x) = \frac{d^2\psi(x)}{dx^2} = \frac{-q(x)}{\epsilon\epsilon_o} \tag{26.40}$$

where, as before, ϵ is the dielectric constant of water and ϵ_o is the permittivity of free space (= 8.85×10^{-14} C–V/cm). Substituting Eq. (26.39) into Eq. (26.40), we obtain

$$\frac{d^2\psi(x)}{dx^2} = \frac{-ze_o}{\epsilon\epsilon_o}(n_+(x) - n_-(x)). \tag{26.41}$$

Substituting Eqs. (26.35) and (26.36) into Eq. (26.41), we obtain

$$\frac{d^2\psi(x)}{dx^2} = \frac{-ze_o}{\epsilon\epsilon_o}\left[n_+(\infty)\exp\left[\frac{-zF\psi(x)}{RT}\right] - n_-(\infty)\exp\left[\frac{zF\psi(x)}{RT}\right]\right]. \tag{26.42}$$

For this symmetrical electrolyte, since $z_+ = -z_-$, by Eq. (26.37), we know that

$$n_+(\infty) = n_-(\infty) = n(\infty) \tag{26.43}$$

and so *both* bulk phase concentrations may be referred simply as $n(\infty)$. We now note that the hyperbolic sine function ($\sinh u$) is given by

$$\sinh u = \frac{e^u - e^{-u}}{2}. \tag{26.44}$$

As is the case for the trigonometric sine function, the sinh function is dimensionless. In addition, the sinh function bears a number of other more significant similarities to the sine function. One important example is that for small arguments, both $\sin u$ and $\sinh u$ are approximately equal to u.

By Eqs. (26.42)–(26.44), we obtain

$$\frac{d^2\psi(x)}{dx^2} = \frac{ze_o n(\infty)}{\epsilon\epsilon_o} 2\sinh(zF\psi(x)/RT). \tag{26.45}$$

As with $n_+(x)$ and $n_-(x)$, the dimensions for $n(\infty)$ are ions/cm^3, and we recall that z has dimensions of units of charge per ion. Using these facts, a quick check of the units of Eq. (26.45) reveals V/cm^2 for the LHS. For the RHS, we have C–cm^{-3}/(C–V^{-1}–cm^{-1}), which reduces to V/cm^2 (if it did not, we would know something was wrong). If we multiply and divide the RHS of Eq. (26.45) by zF/RT, we obtain

$$\frac{d^2\psi(x)}{dx^2} = \left(\frac{2ze_o n(\infty)zF}{\epsilon\epsilon_o RT}\right)\left(\frac{\sinh(zF\psi(x)/RT)}{zF/RT}\right). \tag{26.46}$$

Let us focus now for the moment on the first factor on the RHS of Eq. (26.46). From Eq. (25.21), we have that

$$\kappa^2 = \frac{e_o^2}{\epsilon\epsilon_o kT}\sum_j z_j^2 n_j(\infty). \tag{26.47}$$

For Eq. (25.68), for water at 298°K, we have

$$\kappa = 3.258 \times 10^7\ I^{1/2}\ \text{cm}^{-1} \tag{26.48}$$

$$\kappa^{-1} = 3.069 \times 10^{-8}\ I^{-1/2}\ \text{cm}. \tag{26.49}$$

For the case at hand, $2z^2n(\infty) = \sum_j z_j^2 n_j(\infty)$. Also, we know that $F/R = (F/N_a)/(R/N_a) = e_o/k$. Therefore, the first factor in Eq. (26.46) is simply κ^2, and so Eq. (26.46) simplifies to

$$\frac{d^2\psi(x)}{dx^2} = \kappa^2 \frac{\sinh(zF\psi(x)/RT)}{zF/RT} \tag{26.50}$$

which is the differential equation that we must solve in order to obtain $\psi(x)$.

26.8 REAL AQUEOUS SOLUTIONS GENERALLY CONTAIN MORE THAN ONLY ONE CATION AND MORE THAN ONLY ONE ANION

The implication up to now in our discussions in this chapter has been that there is only one cation/anion pair present in the aqueous solution. For example, in the case of $AB_{(s)}$ dissolving into initially-pure water, we did not pay much attention to the presence of H^+ and OH^- in the solution. Since multiple cations and anions will generally be present in most aqueous solutions of interest, we need to include a discussion of their behavior in our treatment.

Firstly, we note that the fact that κ could be substituted into Eq. (26.46) was based on the implication that we were limiting our discussion to the case when only one cation/anion pair is present in the aqueous phase, and that pair moreover satisfies Eq. (26.34). However, even for the very simple case of $AgI_{(s)}$ equilibrated with initially-pure water, we will at least have Ag^+, I^-, H^+ and OH^- all present in the solution. Our approach to this complication is to note that all of these anions and cations are singly charged, so perhaps $n_+(x)$ as we have used it can refer to the concentration *sum* for all of the cations at point x, and $n_-(x)$ can refer to the concentration *sum* for all of the anions at point x. The value of κ can then be computed as usual using the concentrations of *all* of the ions in the bulk solution. This approach would also apply to solutions containing other salts of the same z value, as might occur in a system containing some $AgI_{(s)}$, some initially-pure water, plus some dissolved $AgNO_3$ as a means to alter the activity of the pdi Ag^+.

It is this *sums approach* that in fact underlies the various double layer equations developed above and also derived below. However, strictly speaking, the only way that this approach can work is if we consider all of the symmetrically-charged ions in solution as *simple point charges*. For example, consider the system involving some $AgI_{(s)}$ in equilibrium with some initially-pure water. In the bulk solution, at 25°C/1 atm, we know that we will have $[H^+] = [OH^-] = 10^{-7}$ M, and $[Ag^+] = [I^-] = 10^{-8.05}$ M. As we approach the surface from out in the bulk solution, $\psi(x)$ will become more and more negative, and so according to Eq. (26.27) there will be an increase in $[Ag^+]$ from $10^{-8.05}$ M, and a decrease in $[I^-]$ from $10^{-8.05}$ M. If one could measure the *true area* under the $[Ag^+]$ vs. x curve, we would obtain a measure of the amount of excess Ag^+ that dissolved relative to the line $n_{Ag^+}(x) = n_{Ag^+}(\infty)$ for all x. Similarly, the true area between the x-axis and the $[I^-]$ curve will give a measure of the amount of I^- that stayed on the surface relative to the line $n_{I^-}(x) = n_{I^-}(\infty)$ for all x. The sum of these true areas will give σ_d.

Continuing our discussion of the above example, for H^+ we know that near the surface there will be an enrichment relative to the bulk solution, and for OH^- there will be a depletion relative to the bulk solution. However, unlike Ag^+ and I^-, the integrals from zero to ∞ for H^+ and OH^- must be exactly equal: neither H^+ nor OH^- sorbs on the surface or is released to the solution from the surface. Thus, the enrichment of H^+ that must occur at small x must in reality be compensated for by a depletion in H^+ relative to 10^{-7} M at larger x (see Figure 26.12), and the depletion of OH^- that must occur at small x must in reality compensated for by an enrichment in OH^- relative to 10^{-7} M at larger x (see Figure 26.12). In the bulk solution, the intermediate depletion of H^+ and the intermediate enrichment of OH^- relax back to the bulk concentrations of 10^{-7} M.

What is happening with H^+ and OH^- in this example is exactly the same as would occur if a single electrode was placed in pure water and the electrode was given a negative charge. Some H^+ ions at intermediate distance would be drawn in towards the electrode, and some OH^- ions would be repelled from the electrode and take up positions at intermediate x. The bulk solution concentrations, that is, those very far from

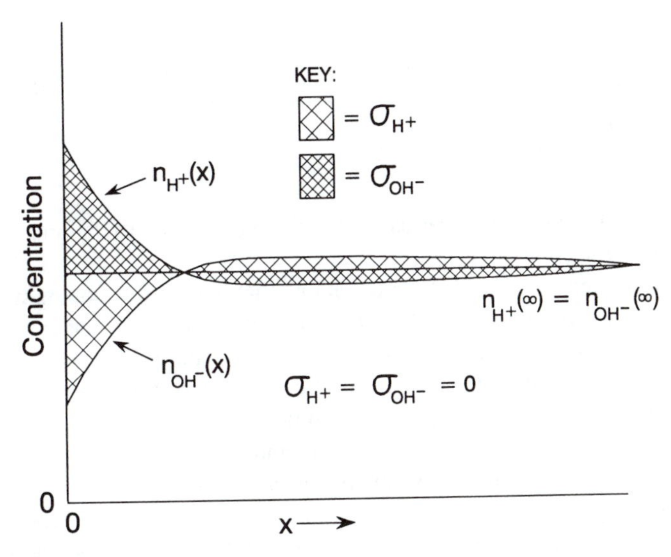

Figure 26.12 True concentrations of H^+ and OH^- vs. distance when $AgI_{(s)}$ dissolves into initially-pure water, with the resulting solid/water interface taking on a negative charge. Some H^+ ions at intermediate distance are drawn in towards the surface, and some OH^- ions are repelled and take up positions at intermediate x. The bulk solution concentrations, that is, those very far from the surface and out of range of its effects, remain at 10^{-7} M for both H^+ and OH^-. The migration of some H^+ towards the surface allows an equal amount of the Ag^+ providing a positive σ_+ to move *out* into the well in the H^+ concentration plot. Similarly, the repulsion of OH^- from the surface allows an equal amount of I^- at moderate distances to move *in* closer to the surface and "neutralize" the hump in the OH^- concentration plot.

the surface and out of range of its effects, would remain at 10^{-7} M for both H^+ and OH^-. The net effect would be distributions like those in Figure 26.12. When the charged surface is being created by $AgI_{(s)}$, the migration of some H^+ towards the surface allows an equal amount of the Ag^+ making for a positive σ_+ to move *out* into the long, shallow well in the H^+ concentration plot. Similarly, the repulsion of OH^- from the surface allows an equal amount of I^- at moderate distances to move *in* closer to the surface and "neutralize" the long, low hump in the OH^- concentration plot. *Strictly speaking,* however, in this example we still have that $(+1)e_o \int_0^\infty (n_{Ag^+}(x) - n_{Ag^+}(\infty))\,dx = \sigma_+$ and $(-1)e_o \int_0^\infty (n_{I^-}(x) - n_{I^-}(\infty))\,dx = \sigma_-$, and we have that $\int_0^\infty (n_{H^+}(x) - n_{H^+}(\infty))\,dx = 0$ and $\int_0^\infty (n_{OH^-}(x) - n_{OH^-}(\infty))\,dx = 0$. When considering all ions to be simple point charges that are distinguished only by the values of their charges (i.e., taking the sums approach), then as discussed above we obtain in this case that

$$\sigma_+ = (+1)e_o \int_0^\infty (n_{Ag^+}(x) - n_{Ag^+}(\infty))\,dx \; + \; (+1)e_o \int_0^\infty (n_{H^+}(x) - n_{H^+}(\infty))\,dx$$

$$= (+1)e_o \int_0^\infty (n_+(x) - n_+(\infty))\,dx \tag{26.51}$$

$$\sigma_- = (-1)e_o \int_0^\infty (n_{I^-}(x) - n_{I^-}(\infty))\,dx \; + \; (-1)e_o \int_0^\infty (n_{OH^-}(x) - n_{OH^-}(\infty))\,dx$$

$$= (-1)e_o \int_0^\infty (n_-(x) - n_-(\infty))\,dx \tag{26.52}$$

and the properties of the double layer can determined according to the approach outlined here and also in the sections that follow.

In summary, then, in order for the equations that we are developing here to apply, the criterion that must be satisfied is that all of the cations have the same charge $z_+ = z$, and all of the anions have the same charge, $z_- = -z$. The quantities $n_-(x)$ and $n_+(x)$ then correspond to the sums of the concentrations of all of the cations and all of the anions, respectively, at point x. Similarly, the quantities $n_-(\infty)$ and $n_+(\infty)$ become the sums of the concentrations of all of the cations and all of the anions, respectively, in the bulk solution. In this context, when the surface charge is created by virtue of a potential-determining ion (pdi), we note that while it is the pdi that is responsible for the *presence* of that charge and the magnitude of ψ^o, *all* of the ions in the solution play a role in determining the behavior of $\psi(x)$. As we will see later, for a clay surface for which there is no pdi, both $\psi(x)$ *and* ψ^o are affected by the concentrations of all of the ions.

For all intents and purposes, most natural waters satisfy the criterion that $z_+ = z = -z_-$ for nearly all ions. In particular, in many natural waters, the vast majority of the cations have a +1 charge, and the majority of the anions have a -1 charge. Thus, Eq. (26.50) remains: 1) correct for mixtures of ions with the same charge magnitude with the value of κ determined by *all* of the cations and anions in the solution; and 2) a very good approximation for most natural waters (including seawater), with z usually equal to 1. In the specific case of seawater, the value of the ionic strength that is used to calculate κ in Eqs. (26.48) and (26.49) is $I = 0.65$.

26.9 INTEGRATING THE DIFFERENTIAL EQUATION GOVERNING THE AQUEOUS CHARGE LAYER SO AS TO OBTAIN $\psi(x)$

26.9.1 $|\psi^o|$ Is Small

When the absolute value of the argument $zF\psi(x)/RT$ is small, the exponential functions comprising the sinh function in Eq. (26.50) can be simplified to the point that integıation of the differential equation is very straightforward. To be able to carry out this integration, however, it is important that $|zF\psi(x)/RT|$ be small over the *entire aqueous charge layer* so that the simplification applies over the entire layer. This will happen only when $|zF\psi^o/RT|$ is small. Since: 1) z is usually 1; 2) F and R are constants; and 3) T does not vary much over ambient temperatures, the criterion that $|zF\psi^o/RT|$ is small usually reduces simply to a requirement that $|\psi^o|$ is small.

As discussed in Section 25.2.3, a Taylor series expansion of the function e^u yields

$$e^u = 1 + u + u^2/2! + u^3/3! + \cdots. \tag{26.53}$$

When u is smaller than about 1, only a few of the terms of the expansion will be required. It may be noted that since $k = 1.381 \times 10^{-23}$ V–C/K and $e_o = 1.602 \times 10^{-19}$ C,

$$\left|\frac{zF\psi^o}{RT}\right| = \left|\frac{ze_o\psi^o}{kT}\right| = 1 \text{ when } z|\psi^o| = 0.0253 \text{ V, at } T = 293°\text{K}. \tag{26.54}$$

Thus, in the context of the above discussion of approximating exponentials, a very useful "smallness" criterion would be that $z|\psi^o|$ should be equal to or less than about 0.025 V. For $z = 1$, the criterion reduces to $|\psi^o| \leq 0.025$ V. In the remainder of this chapter we will consider the two cases that $|\psi^o|$ is: a) small; and 2) not necessarily small. When $z \neq 1$, the reader should remember that actually it is the magnitude of $z|\psi^o|$ that should be examined.

For small $|\psi^o|$ then, expanding both of the exponentials as far as three terms in the sinh function in Eq. (26.50) yields

$$\frac{d^2\psi(x)}{dx^2} = \frac{\kappa^2}{\frac{zF}{RT}} \frac{\left[\left(1 + \frac{zF\psi(x)}{RT} + \frac{1}{2}\left(\frac{zF\psi(x)}{RT}\right)^2\right) - \left(1 - \frac{zF\psi(x)}{RT} + \frac{1}{2}\left(\frac{zF\psi(x)}{RT}\right)^2\right)\right]}{2}. \tag{26.55}$$

The first terms of the expansions cancel with one another, and interestingly, the third terms (and all other odd terms) also cancel each other indicating that no improvement in accuracy is attained for the trouble of expanding to three terms. (The even terms will not cancel.) Therefore, Eq. (26.55) reduces to

$$\frac{d^2\psi(x)}{dx^2} = \frac{\kappa^2}{zF/RT} \frac{2zF\psi(x)/RT}{2} \tag{26.56}$$

$$= \kappa^2\psi(x). \tag{26.57}$$

If we compare Eqs. (26.50) and (26.57), we see that we have just proved the approximation that $\sinh u \simeq u$ for the small u that was mentioned earlier.

We need to solve the differential equation that is Eq. (26.57). In a manner that is virtually identical to the approach used in Eq. (25.28) when developing the Debye-Hückel Equation, we try a solution of the type

$$\psi(x) = \mathrm{A}e^{-\kappa x} + \mathrm{B}e^{+\kappa x}. \tag{26.58}$$

Taking the first and second derivatives, we obtain

$$d\psi(x)/dx = -\mathrm{A}\kappa e^{-\kappa x} + \mathrm{B}\kappa e^{+\kappa x} \tag{26.59}$$

$$d^2\psi(x)/dx^2 = \mathrm{A}\kappa^2 e^{-\kappa x} + \mathrm{B}\kappa^2 e^{+\kappa x} = \kappa^2\psi(x). \tag{26.60}$$

Since we know that κ is always finite and positive, the boundary condition that $d\psi(x)/dx = 0$ at $x = \infty$ can be used together with Eq. (26.59) to deduce that B must equal zero. The second boundary condition that we now use is that $\psi(x) = \psi^{\mathrm{o}}$ when $x = 0$. This fact combines with Eq. (26.58) to lead us to the conclusion that A must equal ψ^{o}. Thus, for small $|\psi^{\mathrm{o}}|$, we obtain

$$\psi(x) = \psi^{\mathrm{o}}e^{-\kappa x}. \tag{26.61}$$

Equation (26.61) is consistent with the convention chosen earlier that $\psi(\infty) = 0$.

26.9.2 $|\psi^{\mathrm{o}}|$ Is Not Necessarily Small (Optional)

Although Eq. (26.50) is considerably more difficult to integrate when $|\psi^{\mathrm{o}}|$ is not necessarily small, as shown by Sparnaay (1972), it can be done with some work. We follow the outline provided there but add considerably more detail for the benefit of the student.

The integration of Eq. (26.50) without simplifications requires the use of a mathematical trick. In particular, we note that for any function $y = y(x)$ and any constant C,

$$\frac{d}{dx}\left[\left(\frac{dy}{dx}\right)^2 + C\right] = 2\left(\frac{dy}{dx}\right)\left(\frac{d^2y}{dx^2}\right). \tag{26.62}$$

Multiplying by dx,

$$d\left[\left(\frac{dy}{dx}\right)^2 + C\right] = 2\left(\frac{d^2y}{dx^2}\right)dy. \tag{26.63}$$

Integrating,

$$\int d\left[\left(\frac{dy}{dx}\right)^2 + C\right] = 2\int\left(\frac{d^2y}{dx^2}\right)dy \tag{26.64}$$

$$\left(\frac{dy}{dx}\right)^2 + C = 2\int\left(\frac{d^2y}{dx^2}\right)dy. \tag{26.65}$$

If $y(x) = zF\psi(x)/RT$, then differentiating $y(x)$ yields

$$\frac{RT}{zF}\frac{d^2y}{dx^2} = \frac{d^2\psi(x)}{dx^2}. \tag{26.66}$$

Substituting Eq. (26.66) into Eq. (26.50) yields

$$\frac{RT}{zF}\frac{d^2y}{dx^2} = \kappa^2\frac{[\sinh(zF\psi(x)/RT)]}{zF/RT} \tag{26.67}$$

or, simply

$$\frac{d^2y}{\mathrm{dx}^2} = \kappa^2[\sinh \mathrm{y}]. \tag{26.68}$$

Combining Eq. (26.68) with Eq. (26.65), we obtain

$$\left(\frac{dy}{dx}\right)^2 + C = 2\kappa^2 \int \sinh y \, dy. \tag{26.69}$$

We note now that $\int \sinh y \, dy = \cosh y$ where $\cosh y$ is the hyperbolic cosine of y (cosh $y = (e^y + e^{-y})/2$). Therefore,

$$\left(\frac{dy}{dx}\right)^2 + C' = 2\kappa^2 \cosh y \tag{26.70}$$

where C' is another constant of integration that equals C plus the additional constant of integration that results in going from Eq. (26.69) to 26.70.

The hyperbolic trigonometric functions have "half-angle" relations just like the regular trigonometric functions. One such relation that is useful here is

$$\cosh y = 1 + 2\sinh^2(y/2). \tag{26.71}$$

Substituting Eq. (26.71) into Eq. (26.70) we obtain

$$\left(\frac{dy}{dx}\right)^2 + C'' = 4\kappa^2 \sinh^2(y/2) \tag{26.72}$$

where C'' is another constant ($C'' = C' - 2\kappa^2$).

We are now ready to apply some boundary conditions. Specifically, at $x = \infty$, $d\psi(x)/dx = 0$ and $\psi(x) = 0$, and the derivative on the LHS of Eq. (26.72) is thus equal to zero. Also, $y(\infty) = zF\psi(\infty)/RT = 0$ and so $\sinh(y/2) = 0$ at ∞. Therefore, C'' must equal 0. Therefore, taking the square root of Eq. (26.72),

$$\left(\frac{dy}{dx}\right) = \pm 2\kappa \sinh(y/2). \tag{26.73}$$

We need to pick which is the correct root. That is, do we need a positive or negative sign on the RHS of Eq. (26.73)? We need to examine the sign properties of the sinh function to answer this question. Sinh is an "odd function." That is, $\sinh(-y/2) = -\sinh(y/2)$. (For an even function $f(u)$, $f(-u) = f(u)$.) Thus, as seen in Figure 26.13, for $\psi^o > 0$, $\sinh(y/2) > 0$ for all x; for $\psi^o < 0$, $\sinh(y/2) < 0$ for all x. To complete

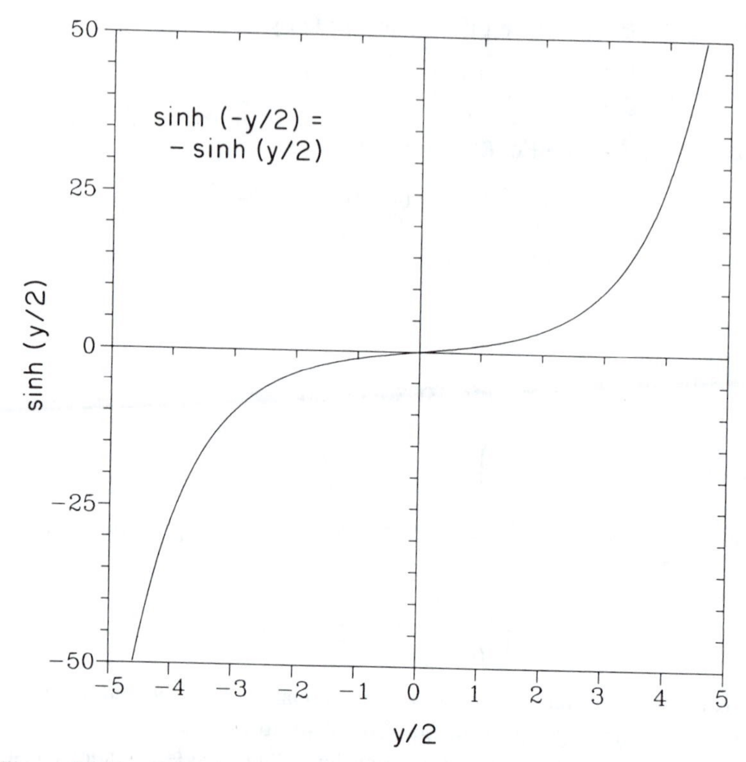

Figure 26.13 Sinh$(y/2)$ as a function of $y/2$. Sinh is an odd function. Thus, $\sinh(-y/2) = -\sinh(y/2)$.

the picture, we examine the sign behavior of dy/dx. As depicted in Figure 26.14, for $\psi^o > 0$, $dy/dx < 0$ for all values of x. For $\psi^o < 0$, $dy/dx > 0$ for all values of x.

The results of the above considerations are summarized in Table 26.3. The only way that Eq. (26.73) can satisfy the conditions in that table is for the applicable root to be the *negative* one. Thus, we conclude

$$\left(\frac{dy}{dx}\right) = -2\kappa \sinh(y/2) \tag{26.74}$$

or

$$\int \frac{dy}{\sinh(y/2)} = -2\kappa \int dx. \tag{26.75}$$

Standard integral tables (e.g., Gradshteyn and Ryzhik, 1962) show that

$$\int \frac{du}{\sinh u} = \ln \tanh(u/2) \tag{26.76}$$

where $\tanh u$ is the hyperbolic tangent of $u(= (\sinh u)/(\cosh u))$. (Since $(\sinh u)^{-1} = \operatorname{csch} u$ (hyperbolic cosecant of u), Eq. (26.76) may be listed in some integral tables as $\int \operatorname{csch} u \, du = \ln \tanh(u/2)$.) To use Eq. (26.76) to integrate (26.75), we let $u = y/2$.

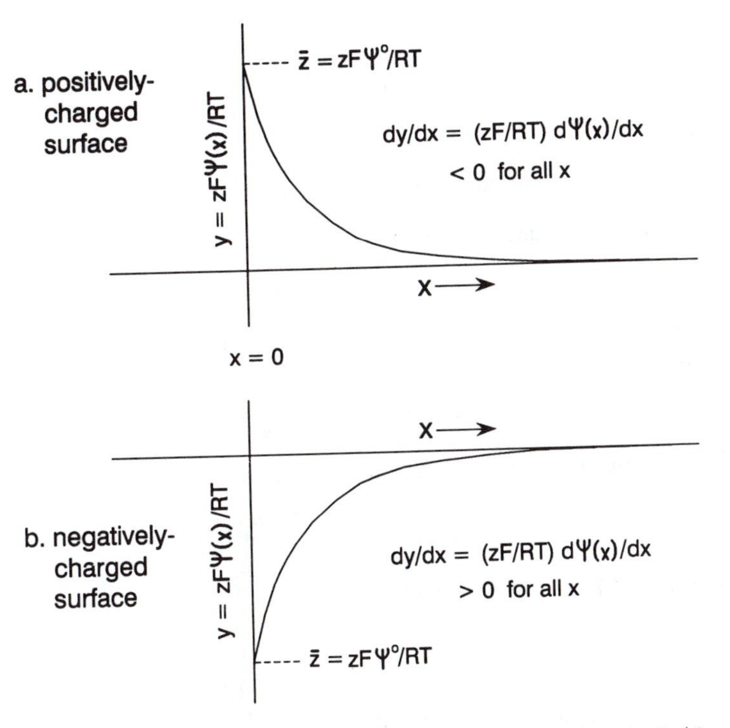

Figure 26.14 (*a*) $y = zF\psi(x)/RT$ as a function of x for $\psi^o > 0$. $dy/dx < 0$ for all values of x. (*b*) $y = zF\psi(x)/RT$ as a function of x for $\psi^o < 0$. $dy/dx > 0$ for all values of x.

TABLE 26.3 Sign properties governing Eq. (26.73) for all values of x. Note: $y = zF\psi(x)/RT$; $dy/dx = (zF/RT)d\psi(x)/dx$.

Positively-charged Surface	Negatively-charged Surface
$\psi^o > 0$	$\psi^o < 0$
$dy/dx < 0$	$dy/dx > 0$
$\sinh(y/2) > 0$	$\sinh(y/2) < 0$

Since $du = dy/2$, we substitute $dy = 2du$ into Eq. (26.75). With the help of Eq. (26.76), then, we obtain

$$\ln \tanh(y/4) + C''' = -\kappa x \qquad (26.77)$$

where C''' is some constant.

We now apply the second boundary condition to evaluate C'''. For $x = 0$, we know that $y = zF\psi^o/RT$. We will abbreviate $zF\psi^o/RT$ as $\bar{z}$. Thus,

$$C''' = -\ln \tanh(\bar{z}/4) \qquad (26.78)$$

and so

$$\kappa x = \ln \tanh(\bar{z}/4) - \ln \tanh(y/4). \tag{26.79}$$

Rearranging,

$$\ln \tanh(y/4) = -\kappa x + \ln \tanh(\bar{z}/4). \tag{26.80}$$

By exponentiating both sides, we obtain

$$\tanh(y/4) = e^{-\kappa x} \tanh(\bar{z}/4). \tag{26.81}$$

The hyperbolic trigonometric functions have corresponding "arc" functions just as do the regular trigonometric functions. Applying this to Eq. (26.82), we obtain

$$y/4 = \tanh^{-1}(e^{-\kappa x} \tanh(\bar{z}/4)). \tag{26.82}$$

One of the properties of $\tanh^{-1} u$ (= arctanh u) is that

$$\tanh^{-1} u = \frac{1}{2}\ln\frac{1+u}{1-u}. \tag{26.83}$$

Combining Eqs. (26.82) and (26.83) yields

$$y/2 = \ln\frac{1 + e^{-\kappa x} \tanh(\bar{z}/4)}{1 - e^{-\kappa x} \tanh(\bar{z}/4)} \tag{26.84}$$

where again we note that $y = zF\psi(x)/RT$ and $\bar{z} = zF\psi^{o}/RT$. Equation (26.84) allows us to calculate $\psi(x)$ for all values of $|\psi^{o}|$.

26.10 $\psi(x)$ FOR ALL VALUES OF $|\psi^{o}|$

Equation (26.84) may be used to compute $\psi(x)$ for known values of ψ^{o}, κ, and x. Eq. (26.84) is the general solution, and may be used when $|\psi^{o}|$ is not necessarily small. Many modern scientific hand calculators can compute the tanh function directly. Computers can certainly be programmed to compute it. However, when such resources are not available, Eq. (26.84) may be placed into a form that involves only exponentials. In particular, we first note that since $\tanh u = (\sinh u)/(\cosh u)$, we have

$$\tanh(\bar{z}/4) = \frac{e^{\bar{z}/4} - e^{-\bar{z}/4}}{e^{\bar{z}/4} + e^{-\bar{z}/4}}. \tag{26.85}$$

We now carry out the following operations on Eq. (26.84): 1) substitute using Eq. (26.85); 2) multiply both the top and the bottom of the RHS by $e^{\bar{z}/4}$; 3) multiply both the top and bottom of the RHS by $(e^{\bar{z}/4} + e^{-\bar{z}/4})$; and 4) exponentiate both sides. The result is

$$e^{y/2} = \frac{e^{\bar{z}/2} + 1 + (e^{\bar{z}/2} - 1)e^{-\kappa x}}{e^{\bar{z}/2} + 1 - (e^{\bar{z}/2} - 1)e^{-\kappa x}}. \tag{26.86}$$

Equation (26.86) is a form of the double layer equation that is found in many texts on solution and soil chemistry.

26.11 σ_s AND σ_d

Regardless of whether or not $|\psi^o|$ is small, by combining Eqs. (26.22), (26.31), (26.32), and (26.38), we obtain

$$-\sigma_s = \sigma_d = \int_0^\infty q(x)\,dx. \tag{26.87}$$

In words, Eq. (26.87) states that σ_d is given by the integral over x (cm) of the pointwise solution charge density (C/cm^3). The integral takes place from the surface out to the bulk solution. Using Poisson's Equation (Eq. (26.40)) to substitute for $q(x)$, Eq. (26.87) becomes

$$-\sigma_s = \sigma_d = -\epsilon\epsilon_o \int_0^\infty (d^2\psi(x)/dx^2)dx. \tag{26.88}$$

Equation (26.88) is the type of differential equation we like to see since it is so easy to integrate. In particular, we obtain

$$-\sigma_s = \sigma_d = -\epsilon\epsilon_o \frac{d\psi(x)}{dx}\bigg|_{x=0}^{x=\infty}. \tag{26.89}$$

At $x = \infty$, we know that $d\psi(x)/dx = 0$. Therefore, we obtain

$$-\sigma_s = \sigma_d = \epsilon\epsilon_o \frac{d\psi(x)}{dx}\bigg|_{x=0}. \tag{26.90}$$

For a positively charged surface, $d\psi(x)/dx|_{x=0}$ will be negative, and so Eq. (26.90) states that $\sigma_s > 0$ and $\sigma_d < 0$, as expected. For a negatively-charged surface, we have that $d\psi(x)/dx|_{x=0}$ will be positive, and so $\sigma_s < 0$ and $\sigma_d > 0$, as expected.

In more quantitative terms, we first restate Eq. (26.74) as

$$\frac{zF}{RT}\left(\frac{d\psi(x)}{dx}\right) = -2\kappa \sinh(zF\psi(x)/2RT). \tag{26.91}$$

Combining Eq. (26.90) and (26.91), we therefore obtain the important result that

$$-\sigma_s = \sigma_d = -2\epsilon\epsilon_o \frac{RT}{zF}\kappa \sinh(zF\psi^o/2RT). \tag{26.92}$$

At 298°K, the factor $2\epsilon\epsilon_o RT/F$ equals 3.56×10^{-13} C/cm.

It is important to remember that Eq. (26.92) is valid for both large and small $z|\psi^o|$. For small $z|\psi^o|$, by the approximation that $\sinh u \simeq u$ for small u, Eq. (26.92) reduces to

$$-\sigma_s = \sigma_d = -\epsilon\epsilon_o \kappa\psi^o. \tag{26.93}$$

26.12 THE PHYSICAL SIGNIFICANCE OF κ^{-1}

In Chapter 25, we discussed the fact that the electrostatic interaction that exists between a core ion and its surrounding ion *cloud* would be the same if the charge on the ion cloud was concentrated at a point that is at a distance of κ^{-1} ("the Debye length") from the core ion.

In the case of the aqueous charge layer of a double layer, the Debye length is often referred to as the "thickness of the double layer." In actuality, the aqueous charge layer is significantly thicker than κ^{-1}. In particular, as can be deduced from Eq. (26.86,) even for large $z|\psi^{o}|$ we conclude that when $x \simeq \kappa^{-1}$, $|\psi(x)|$ is still *approximately* equal to $|\psi^{o}|/e = |\psi^{o}|/2.718$. Indeed, the value of $|\psi(x)|$ is not an order of magnitude less than $|\psi^{o}|$ until $x > 2\kappa^{-1}$. Nevertheless, $x = \kappa^{-1}$ does provide a *rough* estimate of the thickness of the aqueous charge layer.

In addition to the properties of κ^{-1} discussed above, when $|\psi^{o}|$ is "small," then it turns out that the plane at $x = \kappa^{-1}$ identifies the x-coordinate for the "center of mass" of the aqueous charge layer. For a one-dimensional physical object like a thin rod of length L, the center of physical mass will be that distance x (from the end of the rod) at which the rod will be perfectly balanced if placed on a fulcrum. The density of the rod might vary along its length, just as the charge density in the aqueous charge layer varies with x. The mathematical definition of the center of physical mass (com) for the rod is

$$x_{\text{com}} = \frac{1}{M}\int_{0}^{L} x\,\rho(x)\,dx \tag{26.94}$$

where M is the total mass of the rod, and $\rho(x)$ describes the mass density of the rod as a function of x. (The reader who is unfamiliar with the center of mass concept may wish to consult an introductory text on multivariate calculus such as Buck and Willcox (1971).)

An equation for the "center of mass" of the aqueous charge layer that is analogous to Eq. (26.94) can be written. It is

$$x_{\text{com}} = \frac{1}{\sigma_d}\int_{0}^{\infty} x\,q(x)\,dx. \tag{26.95}$$

Using the Poisson Equation and Eq. (26.61) so that we make an approximate substitution for $q(x)$, we obtain

$$x_{\text{com}} = \frac{-\epsilon\epsilon_{o}\kappa^{2}\psi^{o}}{\sigma_d}\int_{0}^{\infty} xe^{-\kappa x}dx. \tag{26.96}$$

Standard integral tables reveal that

$$\int xe^{-\kappa x}dx = \frac{e^{-\kappa x}}{\kappa^{2}}(-\kappa x - 1). \tag{26.97}$$

The integral in Eq. (26.96) therefore equals $1/\kappa^{2}$. Substituting this result together with σ_d from Eq. (26.93) into Eq. (26.92,) we finally obtain the result for small $|\psi^{o}|$ that

$$x_{\text{com}} = \kappa^{-1}. \tag{26.98}$$

As discussed by Verwey and Overbeek (1948), when $z|\psi^{o}|$ is not small so that (26.61) cannot be used to help substitute in for $q(x)$ in Eq. (26.95,) then the center of mass of the aqueous space charge is nearer to the surface than is given by κ^{-1}.

26.13 EFFECTS OF κ^{-1} ON THE PROPERTIES OF THE AQUEOUS CHARGE LAYER

Using Eq. (26.49), one can easily compute κ^{-1} as a function of I at 25°C/1 atm. A range of typical values are given in Table 26.4. A corresponding plot is given in Figure 26.15. For any value of ψ^o, $\psi(x)$ approaches zero increasingly quickly as the ionic strength increases. In addition to such effects on the *thickness* of the aqueous charge layer, the value of κ^{-1} has important implications for the values of ψ^o and σ_s. In order to investigate these implications, for simplicity we consider the case of a surface for which $|\psi^o|$ is small enough that Eq. (26.93) applies. (The same conclusions would result if we were to base the following discussion on Eq. (26.92).)

Let us assume then that Eq. (26.93) is valid. For a surface for which ψ^o is controlled by a potential-determining ion (pdi), then regardless of the value of the ionic strength (i.e., regardless of the value of κ^{-1}), ψ^o will remain unchanged so long as the activity of the pdi remains unchanged. (It may be useful to recall here that the ionic strength (and thus κ^{-1}) may be changed independently of the activity of the pdi. For example, for equilibrium with the solid $AgI_{(s)}$, some $AgNO_3$ could be added to change $\{Ag^+\}$,

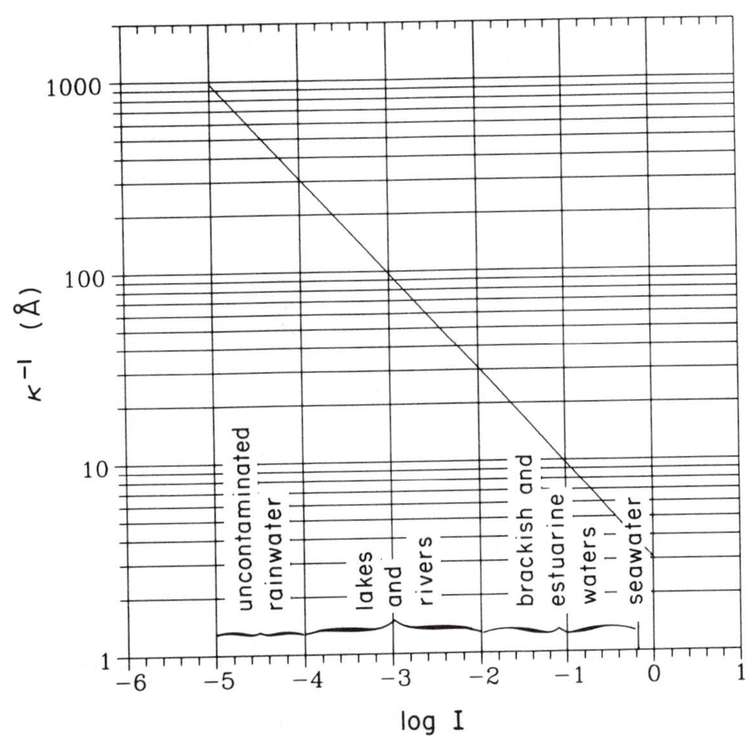

Figure 26.15 κ^{-1} in Angstroms vs. log I at 25°C/1 atm. In dilute solutions as in uncontaminated rain, κ^{-1} will be quite large, from 300 to 1000 Å. In seawater, κ^{-1} will be just ~4 Å.

TABLE 26.4 Values of κ^{-1} (Angstroms) at 25°C/1 atm for various values of ionic strength I (M).

	I	κ^{-1} (Å)
rain waters	0.00001	971
	0.00005	434
lakes and rivers	0.0001	307
	0.0005	137
	0.001	97
	0.005	43
brackish and estuarine waters	0.01	31
	0.05	14
	0.10	9.7
sea water	0.65	3.8

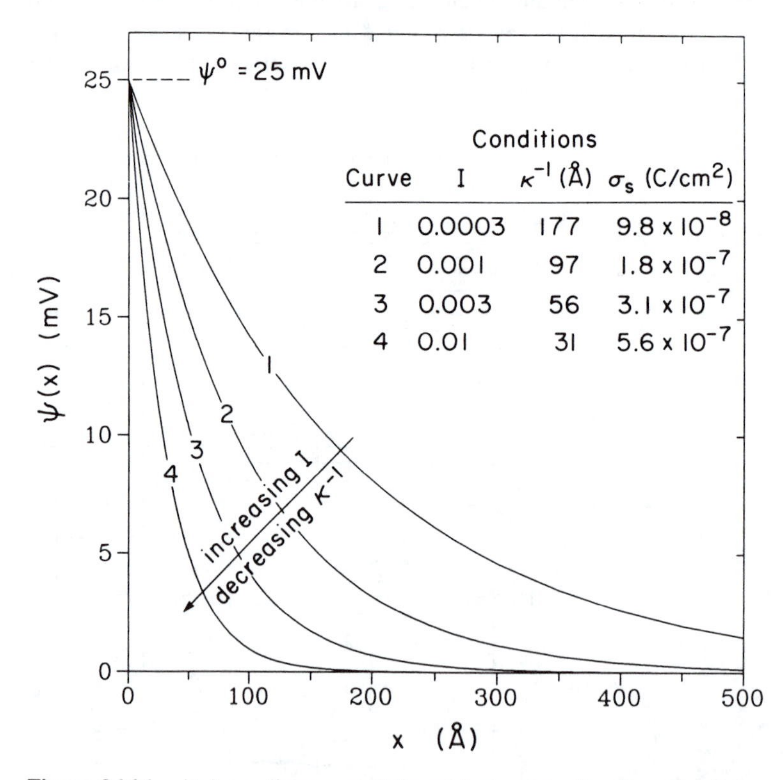

Figure 26.16 $\psi(x)$ vs. distance at different ionic strength values. In all four cases, $\psi^o = +25$ mV as controlled by the activity of a pdi. As I increases and κ^{-1} decreases, the increasing electrical shielding in the aqueous solution means that σ_s must increase if ψ^o is to remain constant. The calculations have been made assuming that the equations for small $|\psi^o|$ apply.

and some $KClO_4$ solution could be added to independently vary the value of I. Now, if ψ^o is fixed by the activity of the pdi, and if the ionic strength increases, then κ^{-1} decreases. Based on Eq. (26.93) we therefore conclude that σ_s must increase to maintain a constant ψ^o. In other words, if the ionic strength does indeed increase, the amount of physical charge required to maintain ψ^o goes up: more of the pdi must sorb on the surface. What is happening is that as the concentrations of the ions increase, the ions with a charge opposite to the surface charge tend to crowd near the solid/water interface. If the surface is positively charged, many anions crowd near the interface. If the surface is negatively charged, many cations crowd near the interface. In both cases, the surface charge is *screened*, and more charge is needed on the surface to maintain a constant value of ψ^o. These principles are summarized in Figure 26.16 for a surface for which ψ^o is maintained at +25 mV.

In contrast to the above situation, for a surface such as a clay particle where it is σ_s that is fixed, not ψ^o, then based on Eq. (26.93) we conclude that increasing the ionic strength will decrease $|\psi^o|$. Now, since there is no pdi which can serve to add surface charge when the ionic strength increases, the screening of charge that takes place at the solid/water interface succeeds in decreasing $|\psi^o|$. These principles are summarized in Figure 26.17.

26.14 STRATEGY FOR USING THE VARIOUS DOUBLE-LAYER EQUATIONS IN SOLVING PROBLEMS

As has been discussed, double-layer charges can result with or without a pdi being present. When a pdi is present, the steps for solving a double-layer problem are:

1. Calculate ψ^o based on the activity of the pdi.
2. Calculate κ^{-1} based on the ionic strength I.
3. For $|\psi^o| \leq 25$ mV, calculate $\psi(x)$ based on Eq. (26.61), and for $|\psi^o| > 25$ mV, calculate $\psi(x)$ from Eq. (26.84).

When σ_s is set at a constant value (as with a clay particle), the steps for solving the corresponding double layer problem are:

1. Calculate σ_s based on a knowledge of the amount of charge present per unit surface area of the solid.
2. Calculate κ^{-1} based on the ionic strength I.
3. Calculate ψ^o based on Eq. (26.92).
4. For $z|\psi^o| \leq 25$ mV, calculate $\psi(x)$ based on Eq. (26.61), and for $z|\psi^o| > 25$ mV, calculate $\psi(x)$ from Eq. (26.84).

Problems illustrating the principles of this chapter are presented by Pankow (1992).

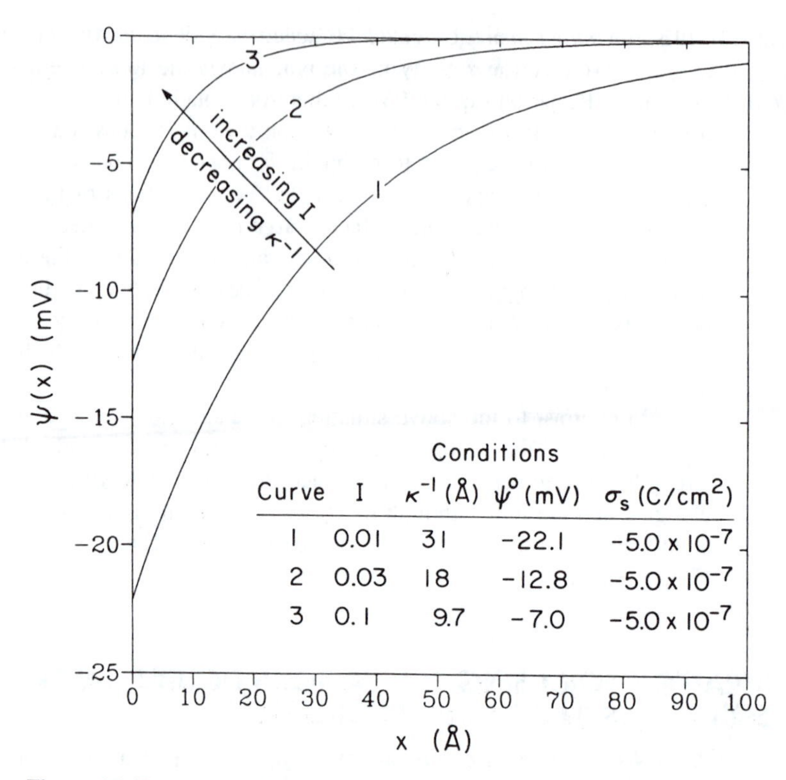

Figure 26.17 ψ(x) vs. distance at different ionic strength values. In all four cases, the value of σ_s is constant at 5×10^{-7} C/cm^2 as might be the case for a clay with a low cation exchange capacity (CEC). As I increases and κ^{-1} decreases, the increasing electrical shielding in the aqueous solution means that $|\psi^o|$ must decrease since σ_s is constant. The calculations have been made assuming that the equations for small $|\psi^o|$ apply.

26.15 REFERENCES

BUCK, R. C., and A. B. WILLCOX. 1971. *Calculus of Several Variables*, Boston: Houghton Mifflin Company.

DE BRUYN, P. L., and G. E. AGAR. 1962. *Froth Flotation*, D. W. FUERSTENAU, Ed., New York: American Institute of Mining, Metallurgical, and Petroleum Engineers.

GRADSHTEYN, I. S., and I. M. RYZHIK. 1962. *Tables of Integrals, Series, and Products*, Fourth Edition, Prepared by Yu. V. Geronimus and M. Yu. Tseytlin. New York: Academic Press.

PANKOW, J. F. 1992. *Aquatic Chemistry Problems*. Portland: Titan Press-OR, P.O. Box 91399, Portland, Oregon 97291-1399.

SEINFELD, J. H. 1985. *Air Pollution*. New York: Wiley-Interscience.

SPARNAAY, M. J. 1972. *The Electrical Double Layer*, Volume 4 of Topic 14: The Properties of Interfaces, D. H. Everett (Ed.), in *The Encyclopedia of Physical Chemistry and Chemical Physics*, Oxford: Pergamon Press (see in particular pages 57–62).

VAN OLPHEN, H. 1977. *An Introduction to Clay Colloid Chemistry*. New York: Wiley-Interscience.

VERWEY, E. J. W. and J. TH. G. OVERBEEK. 1948. *Theory of the Stability of Lyophobic Colloids*. New York/Amsterdam: Elsevier Publishing Co.

27

Stability and Coagulation of Colloidal Suspensions

27.1 INTRODUCTION

In 1861, Thomas Graham applied the word *colloid* to refer to any suspension of material in a liquid where the suspension does not appear to separate even if allowed a large amount of time to do so.[1] Colloidal suspensions in water thus contain particles of one sort or another that are dispersed within the aqueous phase. The particles may be solid or liquid; an emulsion is an example of a colloidal suspension in which the particles are liquid.

Opinions vary somewhat, but the colloidal particle size range is generally defined as 30 to 10,000 Å (0.003 to 1 μm). Particles smaller than 30 Å are of the same size as some molecules, and so may be thought of as being essentially "dissolved." It is the constant thermal (i.e., Brownian) bombardment by the relatively small, surrounding water molecules that helps to keep colloidal particles in suspension. Particles that are larger than 10,000 Å and are more dense than water are difficult to keep suspended because Brownian motion is less able to keep them from settling downwards. Particles larger than 10,000 Å that are less dense than water will tend to float upwards, as in the formation of cream on the top of a bottle of whole milk.

There are many reasons why colloids are important in aquatic systems. Firstly, a significant amount of the suspended particulate matter in lakes and rivers is colloidal in nature. This material is of interest because of: 1) the great inherent importance of sediment transport in lakes and rivers; and 2) the high surface area per unit volume of colloidal matter, and therefore the high sorption and transport capacity of such matter

[1] Prior to Graham, the terms *colloid* and *colloidal* had been used primarily in a medical context to refer to gelatinous or mucinous tissues in disease, or, normally, as in the thyroid gland.

for both organic and inorganic contaminants.[2] Secondly, as discussed in Chapter 26, the removal of colloidal matter in drinking water plants makes the water more pleasant to drink, removes viruses and bacteria, and minimizes the formation of halomethanes such as chloroform during subsequent chlorination.[3] Thirdly, many anthropogenic wastes are disposed of as aqueous suspensions. For example, about 80 percent of the biological oxygen demand (BOD) of untreated domestic sewage and 50 percent of the organic carbon in secondary sewage treatment effluent is in suspended particulate form.

Colloids are often classified as being either *lyophobic* or *lyophilic*. Lyophobic colloids are comprised of particles that are solvophobic relative to the solvent in which they are dispersed. Colloidal particles that are solvophobic when dispersed in water are hydrophobic. Lyophilic colloids are comprised of particles that are solvophilic relative to the solvent in which they are dispersed. Colloidal particles that are solvophilic when dispersed in water are hydrophilic. The close relationships between these various adjectives are summarized in Table 27.1.

Examples of lyophobic colloids in water include suspensions of clays and other minerals, metal halides like $AgI_{(s)}$ and $AgCl_{(s)}$, and emulsions of fat (as in whole milk). Examples of lyophilic colloids in water include suspensions of silica gel, proteins, and humic materials; many brands of toothpaste contain significant amounts of hydrated silica gel.

27.2 THE COAGULATION PROCESS

27.2.1 General

When some of the particles in a suspension stick together when they collide, the suspension is said to be undergoing the process of *coagulation*. First, pairs of particles form, then triplets, and so on. The continued accretion of more and more particles to each particle agglomerate will ultimately allow growth to such a size that random Brownian bombardment by the water molecules will no longer be able to prevent bouyancy effects from causing the agglomerate from either sinking downwards or floating upwards.

TABLE 27.1. Adjectives used to describe the properties of colloidal particles when the suspending solvent is water. Words in a given row are synonyms; words in a given column are antonyms.

lyophobic	$\Longleftrightarrow$	solvophobic	$\Longleftrightarrow$	hydrophobic
lyophilic	$\Longleftrightarrow$	solvophilic	$\Longleftrightarrow$	hydrophilic

[2] A 1 cm cube of material has a surface area of 6 cm^2. When cut up into 10^{12} cubes that are 1,000 Å (= 1 μm = 10^{-4} cm) on a side, the surface area increases by factor of 10,000 to 6×10^4 cm^2 = 6 m^2. While the importance of very small particles in surface waters has long been recognized, the fact that colloids can be significant in groundwater transport is a fairly recent discovery (e.g., Gschwend and Reynolds 1987).

[3] Halomethanes like chloroform are produced when organic humic materials react with Cl_2, HOCl and/or OCl^-.

Direct hydrogen bonding between lyophilic particles and water makes colloidal suspensions of such materials permanently *stable*, and the rate of particle coagulation will be *zero*. Colloidal suspensions of lyophobic particles are, on the other hand inherently *unstable*, and the rate of coagulation will always be *greater than zero*. Indeed, if brought sufficiently close together by a mechanism such as Brownian diffusion (see Figure 27.1*a*), lyophobic particles will stick together and squeeze out much of the intervening aqueous phase. If one desires to coagulate lyophilic particles, it will first be necessary to convert them to lyophobic particles. In the case of proteins, this can be accomplished by denaturing the protein with heat. In the case of humic materials, it can be accomplished by adding either some strong acid, or a solution of a di- or trivalent metal ion. Strong acid protonates the carboxylate and phenolate groups that provide humic material with its hydrophilicity. This makes the large molecules of humic material much more hydrophobic and much less soluble. Indeed, the standard method for isolating "humic acid" involves lowering the pH. A di- or trivalent metal ion will also react with carboxylate and phenolate groups, again making the humic material more hydrophobic by virtue of conversion to a rather insoluble metal-humate complex. This precipitation of humic acid by metal ions is exactly analogous to the precipitation of soaps (and the formation of soap scum) by the Ca^{2+} and Mg^{2+} that is found in hard water.

While the rate of coagulation of a lyophobic colloid will always be greater than zero, such suspensions can appear to be stable for months. This is almost always the result of stabilizing double layer charges on the particles. Indeed, since similar particles suspended in the same aqueous medium will have ψ_o values of the same sign, the various particles will be surrounded by electric fields that are mutually repulsive. As depicted in Figure 27.2, particles with $\psi_o < 0$ are surrounded by negative electric fields, and particles with $\psi_o > 0$ are surrounded by positive electric fields. For a given pair of particles, the mutually repulsive effects of the two double-layers will tend to prevent the two particles from approaching one another very closely.

The net potential ψ_{net} between two charged particles separated by distance d will be determined in part by the manner in which the two ion clouds of the two double-layers interact. As long as the two particles are not too close together, then at the midpoint of the interaction (i.e., at $d/2$) it can be assumed that the individual $\psi(x)$ values behave independently. Thus, at $d/2$ we have

$$\psi_{net}(d/2) \simeq \psi_1(d/2) + \psi_2(d/2). \tag{27.1}$$

If the two values $\psi_1(x)$ and $\psi_2(x)$ do behave independently, they may be calculated using Eq. (26.61) or Eq. (26.84), as appropriate.

27.2.2 van der Waals Forces

When $|\psi|$ is both large and also extends far out into the solution, only a very low percentage of the collisions will be of high enough energy (i.e., of high enough velocity) to be able to overcome the far-reaching repulsive electrical fields that can be present on charged particles. Thus, under these conditions, only a very low percentage of the collisions will be able to bring two particles close together. Once lyophobic particles do manage to come close together, then attractive, short range *van der Waals* forces

a. Brownian Diffusion

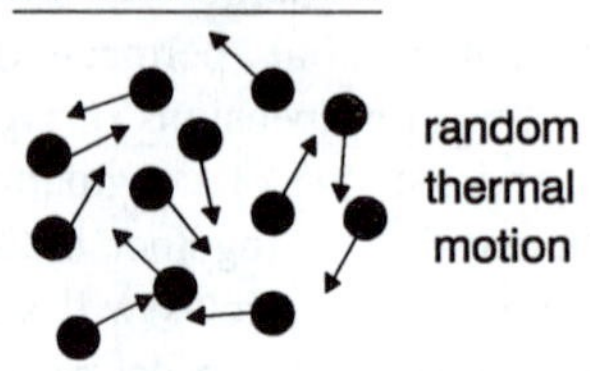

b. Velocity Gradients

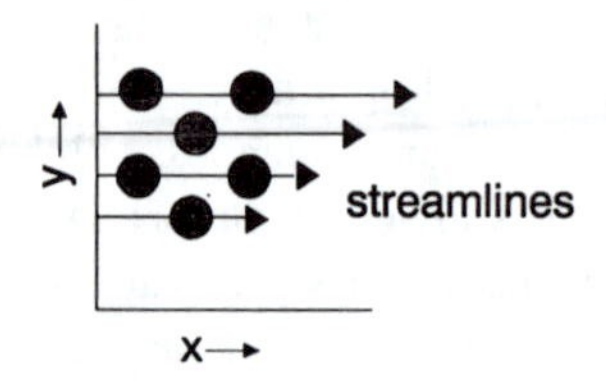

c. Differential Sedimentation

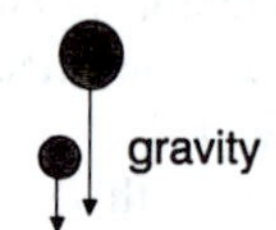

Figure 27.1 Different mechanisms by which colloidal particles can collide with one another and therefore coagulate and form agglomerates. (*a*) Brownian diffusion. (*b*) Velocity gradients. (*c*) Differential sedimentation.

can begin to be felt by the particles. These attractive forces can pull the particles closer together despite the electrostatic repulsion. The three types of interaction that are typically included in the van der Waals class are: dipole-dipole, dipole-induced dipole, and "London dispersion" interactions. Each of these types of interaction are described schematically in Figure 27.3.

When a molecule is composed of more than one type of atom, the different atoms will always exhibit at least slightly different affinities for electrons. For a given bond between two different atoms then, the more electron-attracting (i.e., "electronegative") end will be somewhat negative, and the less electronegative end will be somewhat positive. This type of permanent charge separation is called a permanent dipole, and the magnitude of the dipole is expressed as the "dipole moment."[4] Relative to many other molecules, water has a very large dipole moment. A single atom will never have a dipole moment, nor will a diatomic molecule of two identical atoms. However, a permanent

[4] A "dipole moment" is given by the product of the magnitude of the charge separation and the separation distance.

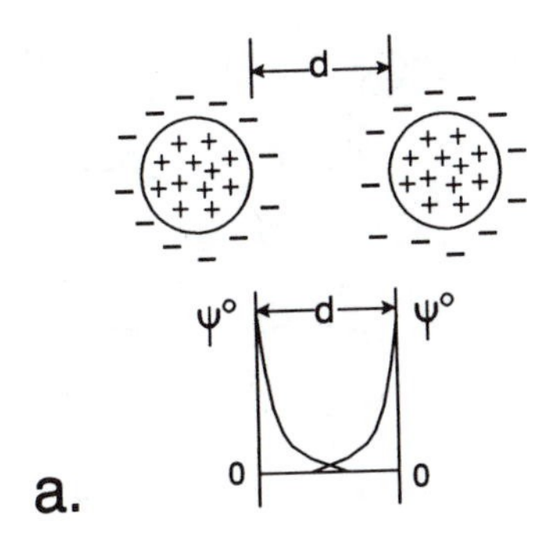

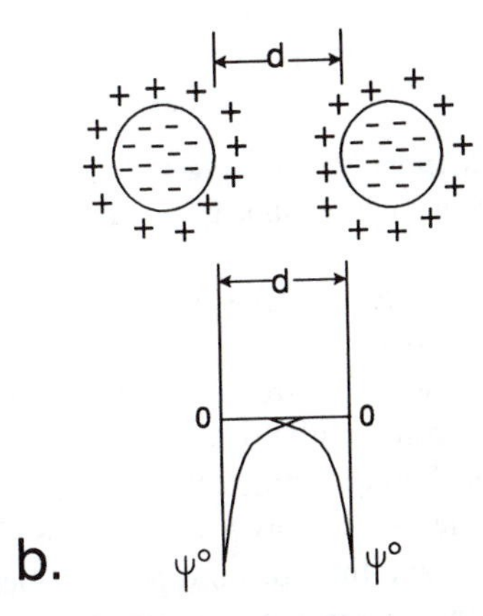

Figure 27.2 (*a*) Particles with $\psi_o < 0$ are surrounded by negative electric fields. (*b*) Particles with $\psi_o > 0$ are surrounded by positive electric fields. For any given pair of similar particles, the mutually-repulsive effects of the two double-layers will act to resist the reduction of the interparticle distance *d*. That is, they will resist the particle collisions that lead to coagulation.

dipole will always exist in a diatomic molecule composed of two *different* elements. Permanent local dipoles can also exist at various places within a large molecule.

Two permanent dipoles will tend to attract one another when they line up in a manner that has the negative end of one interacting with the positive end of the other. This type of interaction is of the *dipole-dipole* type. Hydrogen bonding is not usually included as a dipole-dipole interaction, but is put in a class by itself because of the special strength of this type of interaction. Like van der Waals forces, though, hydrogen bonding can help to hold lyophobic particles together.

A *dipole-induced dipole* interaction will occur whenever a permanent dipole from one molecule approaches a non-dipole molecule or atom. In this case, the electrical field caused by the permanent dipole is able to distort the electron distribution in the

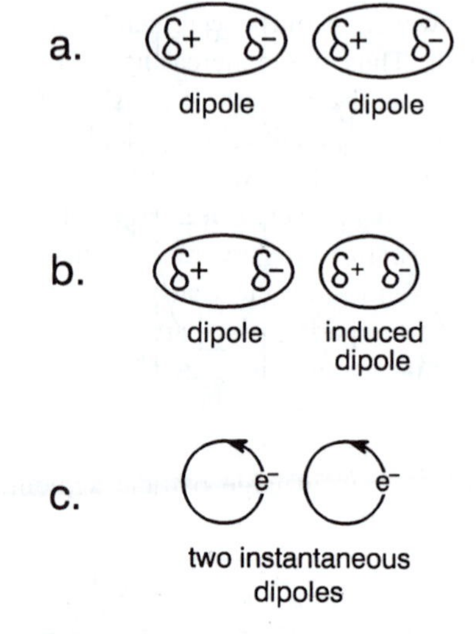

Figure 27.3 Three different types of attractive *van der Waals* interactions. (*a*) Dipole-dipole. (*b*) Dipole-induced dipole. (*c*) "London dispersion."

non-dipole to *induce* a dipole with the same orientation. The permanent dipole and the induced dipole then interact as two dipoles.

When neither of two species of interest has a permanent dipole, the species can still interact because the local electron circulations in the orbitals in the two different species will create *instantaneous*, oscillating dipoles that can get in sync, and then interact as two adjacent dipoles. For materials like hydrocarbon oils that are only weakly polar or completely non-polar (and so have little or no dipole moment), these so-called "London dispersion" interactions are by far the most important source of intermolecular attraction. For materials that are very polar and so have strong dipoles, then dipole-dipole interactions are the most important. The nature and strength of the van der Waals forces between two interacting colloidal particles will thus depend on the nature of the material comprising the particles.

27.2.3 Coagulation Mechanisms

The type of random thermal motion that causes Brownian diffusion is not the only mechanism by which particles can collide with one another. In addition, as depicted in Figure 27.1*b*, if there are velocity gradients in the fluid, then particles will be moving at different speeds along adjacent streamlines. If the repulsive electrical forces can be overcome, then particles on the faster streamlines will be able to collide with particles on adjacent, slower streamlines. In Figure 27.1*b*, a particle on one streamline will be able to collide with any particle that is ahead of it on an adjacent, slower streamline. (Because the volume that it sweeps out is larger, a large particle on any given streamline will have more collisions than a small particle on the same streamline.)

Taken to the extreme, one might think that very vigorous mixing will lead to increased coagulation rates because such mixing would create intense velocity gradients in a fluid. However, it must be remembered that the van der Waals attractions are of limited strength. Thus, while increasing the mixing intensity will increase collision rates, it will also increase the strength of the shear forces operating across the gradients in the turbulent three-dimensional velocity field. These shear forces will tear large particle agglomerates apart, and so will limit the size to which the agglomerates can grow. Agglomerates that can grow to one size with gentle mixing will be torn apart at higher mixing intensities. In water treatment plants, where particle removal is the desired end result, care is taken to mix the water at an intermediate, gentle rate.

If some of the particles or particle agglomerates in the suspension are sufficiently large that they are undergoing gravitational settling, they will be able to catch up with and collide with other particles that are smaller therefore settling at slower velocities. This mechanism, referred to as differential sedimentation, is depicted in Figure 27.1*c*. It is quite effective when large particle agglomerates or "flocs" are settling out of the solution. Differential sedimentation is often explicitly taken advantage of when treating water with chemicals that promote coagulation ("coagulants"). In particular, a mixing (velocity gradient) phase is followed with a quiescent phase during which flocs can settle out of the system and at the same time sweep out other, smaller particles.

27.3 ELECTROSTATIC REPULSION VS. VAN DER WAALS ATTRACTION IN THE COAGULATION OF LYOPHOBIC COLLOIDAL PARTICLES

27.3.1 DLVO Theory

In order to examine the net energetics of how two electrically charged particles will interact as they approach one another, Derjaguin, Landau, Verwey, and Overbeek ("DLVO") examined expressions for both the repulsive electrostatic (double-layer) and the attractive (van der Waals) contributions to the potential energy of the interaction. When the colliding particles approach one another as two flat plates, as might occur during a plate-to-plate encounter of two plate-like particles of a clay like kaolinite, the repulsive contribution is often approximated in DLVO theory as

$$\Phi_R(d) \simeq \frac{64n(\infty)kT}{\kappa}\Upsilon_0^2 e^{-\kappa d} \tag{27.2}$$

where $n(\infty)$ is the *bulk solution* concentration of the electrolyte(s) making up the ionic strength (I). The units of $n(\infty)$ are ion pairs per cm^3. For example, $n(\infty) = 6.023 \times 10^{+20}$ ion pairs per cm^3 for a $10^{-3}F$ solution of NaCl. The constant Υ_o is given by

$$\Upsilon_o = \frac{\exp[zF\psi_o/(2RT)] - 1}{\exp[zF\psi_o/(2RT)] + 1}. \tag{27.3}$$

As in Chapter 26, it is assumed that the electrolytes making up I are symmetrical. The value of $n(\infty)$ is determined as the concentration sum for those electrolytes.

At large d, the strong, exponential dependence on κ in Eq. (27.2) will cause a decrease in the strength of the repulsive forces with increasing κ (i.e, increasing I). Since the derivation underlying Eq. (27.2) involves the assumption that the plates are not very close together so that Eq. (27.1) applies, we can have confidence in this result.

Conversely, the fact that Eq. (27.1) indicates that $\Phi_R(d)$ will tend to increase with increasing I at small d (where the exponential term will be near 1) must be viewed as being unreliable.[5] The inapplicability of Eq. (27.2) at small d is not viewed as a problem since two colliding particles first interact at long range. Then, at short range, attractive forces can become stronger than the repulsive forces (as will be shown).

A second assumption underlying the derivation of Eq. (27.2) is that the value of $|z\psi_o|$ is large enough to make $|\Upsilon_o| \simeq 1$ (Hiemenez 1986). At 25°C, $F/(2RT) = 19.471\ \text{V}^{-1}$. Thus, at 25°C and when $z = 1$ for the symmetrical, ionic strength-dominating electrolyte in the bulk solution, $|\psi_o|$ must be at least $0.15V = 150$ mV for Υ_o to be 0.9. For many lyophobic colloids of interest, when $z = 1$, it can be marginal at best to assume that $|\psi_o|$ is as large as 150 mV. Nevertheless, while DLVO theory cannot therefore provide precise estimates of interaction energies and coagulation rates, it does provide predictability of the *general* manner in which changing solution properties affect those parameters and the stability of lyophobic colloids.

The attractive, van der Waals contribution to the potential energy of interaction between two flat plates is (Hamaker 1937)

$$\Phi_A(d) = -\frac{A}{12\pi d^2} \tag{27.4}$$

where A is a so-called "Hamaker constant." The specific value of A will depend on the identity of the material in the interacting plates, but a typical range is 10^{-20} to 10^{-19} J.

Both $\Phi_R(d)$ and $\Phi_A(d)$ have units of J/cm^2 which corresponds to energy per unit area. $\Phi_A(d)$ is negative because the van der Waals interactions tend to *lower* the potential energy of the system. Figures 27.4*a* and 27.4*b* show how $\Phi_R(d)$ varies as a function of d for a specific value of $|\psi_o|$ (= +150 mV) and a range of κ values, and as a function of d for constant κ and a range of ψ_o values. In Figure 27.4*a*, we see how $\Phi_R(d)$ tends to decrease with increasing I at large d, but that Eq. (27.2) incorrectly predicts that $\Phi_R(d)$ tends to increase with increasing I at small d.

The total, net potential energy of interaction $\Phi_{\text{net}}(d)$ is given by the repulsive contribution plus the attractive contribution. Thus,

$$\Phi_{\text{net}}(d) \simeq \frac{64n(\infty)kT}{\kappa}\Upsilon_o^2 e^{-\kappa d} - \frac{A}{12\pi d^2} \tag{27.5}$$

$\Phi_{\text{net}}(d)$ gives the net potential energy of interaction between opposite, *unit areas* on flat plates separated by distance d. When $d = \infty$ and the plates are infinitely far apart, $\Phi_{\text{net}}(d)$ will be zero. The value of $\Phi_{\text{net}}(d)$ will be positive when the interacting plates are at an energy level that is higher than when $d = \infty$, and $\Phi_{\text{net}}(d)$ will be negative when the interacting plates are at an energy level that is lower than when $d = \infty$. Figure 27.5 illustrates how $\Phi_R(d)$, $\Phi_A(d)$, and $\Phi_{\text{net}}(d)$ vary as a function of d for one particular case, that is, when $|\psi_o| = 150$ mV, $A = 10^{-19}$ J, $z = 1$, and $I = 0.001$ (by Eq. (25.68), $\kappa = 1.03 \times 10^6\ \text{cm}^{-1}$ at 25°C/1 atm). Since the sign of ψ_o does not affect the magnitude of $\Phi_R(d)$, Figure 27.5 applies equally well to both $\psi_o = +150$ mV and $\psi_o = -150$ mV.

[5]Note that since $c(\infty)$ is proportional to I, and since κ is proportional to $I^{1/2}$, the ratio $c(\infty)/\kappa$ will be proportional to $I^{1/2}$. Thus, at small d where the exponential term in Eq. (27.2) will be near 1, $\Phi_R(d)$ will tend to increase with increasing I.

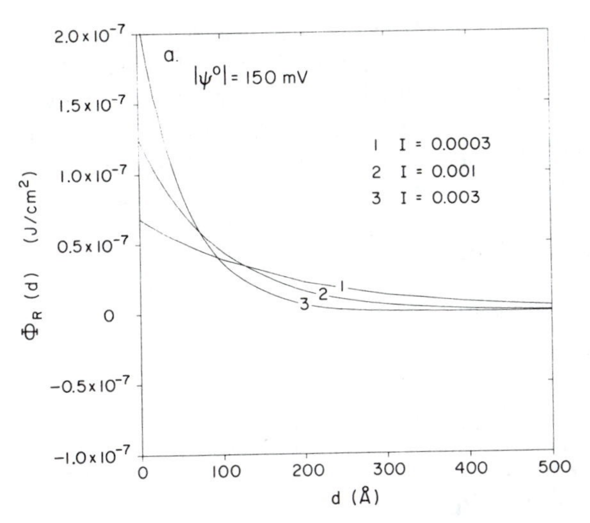

Figure 27.4 (*a*) $\Phi_R(d)$ according to Eq. (27.2) for $|\psi_o| = 150$ mV and a range of I and κ values. As given by Eq. (27.2), at large d, $\Phi_R(d)$ is correctly predicted to decrease with increasing I. At small d, however, Eq. (27.2) incorrectly predicts that $\Phi_R(d)$ tends to increase with increasing I at small d.

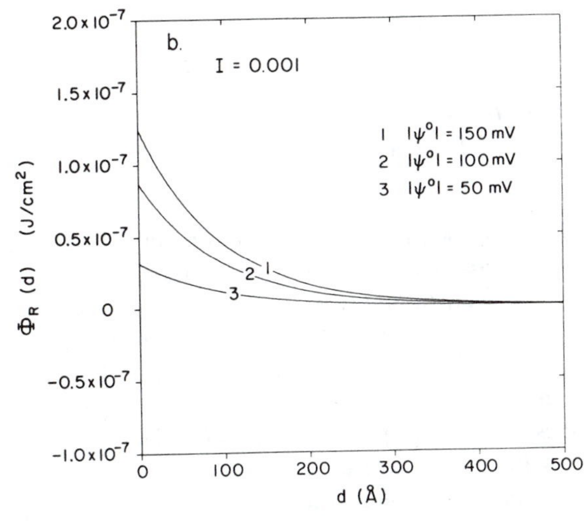

Figure 27.4 (*b*) $\Phi_R(d)$ as a function of d for constant κ and a range of ψ_o values.

Equation (27.5) neglects the fact that as the plates are pushed (even compressed) together to very close (even negative) values of d, the plates will begin to *repel* one another again. Thus, in actuality, the net potential energy curve to the left of the maximum for $\Phi_{net}(d)$ in Figure 27.5 does not go infinitely negative, but will in every case ultimately rise up again strongly as the plates are pushed into one another. Thus, the plates will

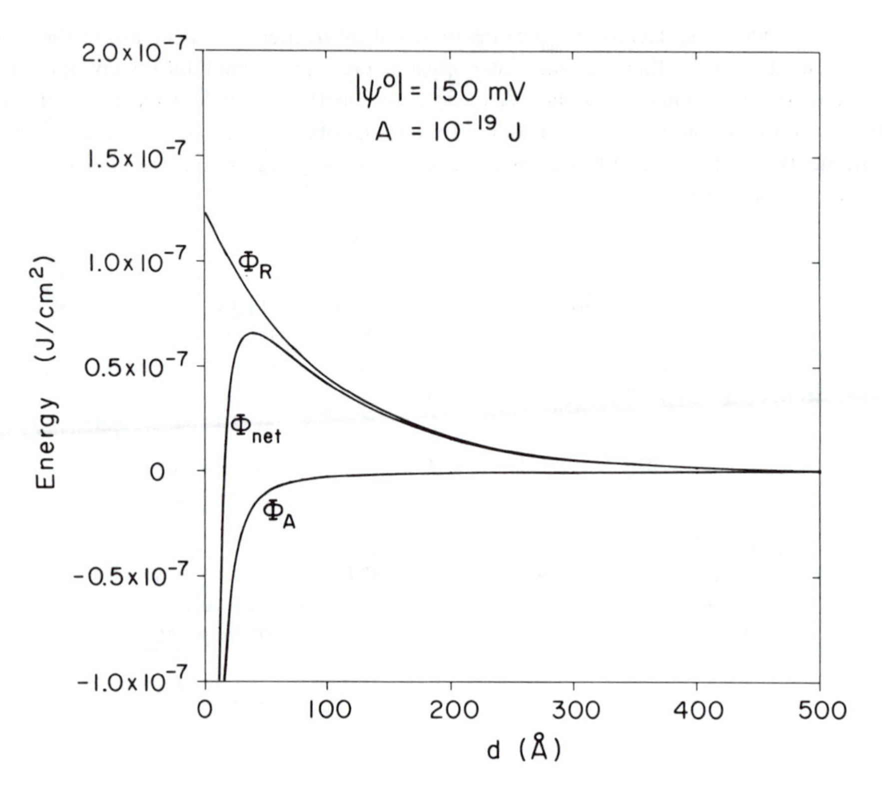

Figure 27.5 $\Phi_R(d)$, $\Phi_A(d)$, and $\Phi_{net}(d)$ as a function of d when $|\psi_o| =$ 150 mV, $A = 10^{-19}$ J, $z = 1$, and $I = 0.001$ ($\kappa = 1.03 \times 10^6$ cm^{-1}).

ultimately stop approaching one another. Since those very short range repulsive forces will be encountered only *after* the plates have passed through a negative $\Phi_{net}(d)$ regime, and since we are only concerned here with how to get the plates close enough to be in that negative $\Phi_{net}(d)$ regime, we can neglect short range repulsion.

As I increases and κ^{-1} decreases, the dependence of $\Phi_A(d)$ on distance remains constant. However, the double layer is compressed, and so $\Phi_R(d)$ as given by Eq. (27.2) decreases at all values of d except possibly at $d = 0$.[6] Thus, increasing I will decrease $\Phi_{net}(d)$, and so increasing I will favor the coagulation process. (As $d \to 0$, the value of $\Phi_A(d)$ becomes exceedingly low, and so it does not matter if $\Phi_R(d)$ does not decrease at $d = 0$.)

[6]When the surface like that of a clay exhibits a constant σ_s, then $|\psi(x)|$ will decrease for all x as I is increased (see Section 26.12). "All x" includes $x = 0$. Thus, since $|\psi_o|$ for a clay will decrease as I is increased, $\Phi_R(d)$ as given by Eq. (27.2) will decrease for all values of d including $d = 0$. However, when ψ_o is set by a potential determining ion (pdi), and if the activity of that pdi remains constant while I is increased, then $|\psi(x)|$ will decrease for all x except $x = 0$ since ψ_o will remain constant. For a pdi-type surface, then, $\Phi_R(d)$ will decrease for all values of d except $d = 0$.

As I increases, because $\Phi_{net}(d)$ decreases at all d, there is a decrease in the height of the energy barrier that the flat plates must overcome to enter into a favorable (i.e., negative) potential energy regime. A basic equation from chemical kinetics states that the rate of a reaction will be proportional to the quantity $\exp[-E_a/RT]$ where E_a (J/mol) is the height of the "activation energy barrier" separating the initial (reactant) and final (product) states. That is,

$$\text{reaction rate} \propto \exp[-E_a/RT]. \tag{27.6}$$

For the case at hand, the initial and final states correspond to d values of ∞ and 0, respectively. Thus, E_a here is the maximum value of $\Phi_{net}(d)$.

The effects of I on $\Phi_{net}(d)$ are summarized in Figure 27.6 for the case when $\psi_o = 150$ mV, A $= 10^{-19}$ J, and $z = 1$. Since increasing I decreases the maximum value of $\Phi_{net}(d)$, the height of the activation energy barrier is decreased, Thus, increasing I will increase the coagulation reaction rate by increasing the fraction of the particle collisions that can overcome that barrier. This collision efficiency fraction is sometimes called a "sticking factor."

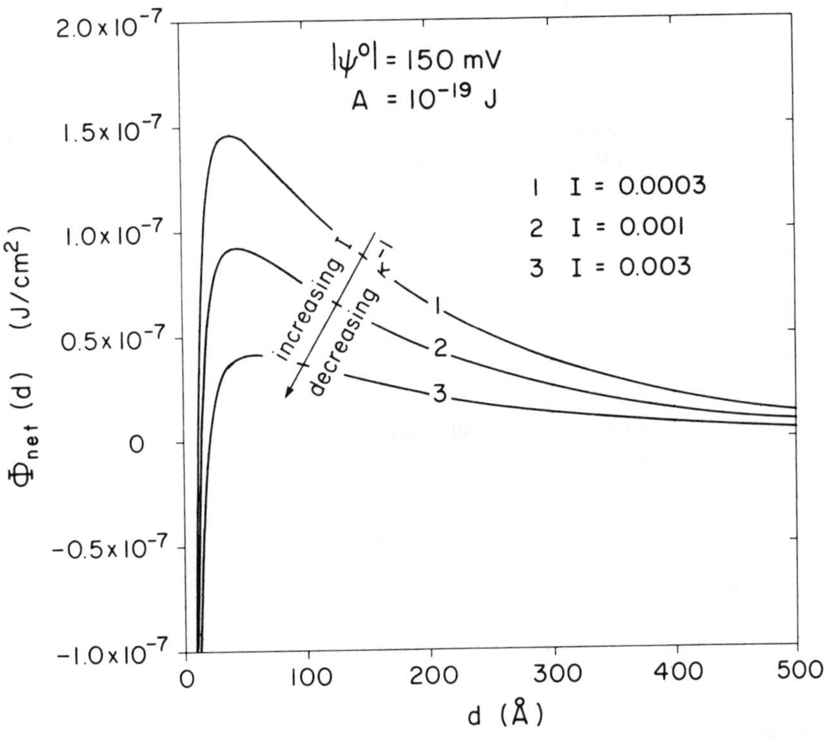

Figure 27.6 Effects of I on $\Phi_{net}(d)$ when $|\psi_o| = 150$ mV, A $= 10^{-19}$ J, and $z = 1$. Increasing I decreases the maximum value of $\Phi_{net}(d)$. Thus, increasing I decreases the height of the activation energy for coagulation.

Since the collision efficiency fraction cannot be made larger than 1.0, we cannot increase the coagulation rate beyond what will be observed when $E_a = 0$. At that point, the reaction becomes transport controlled (e.g., diffusion controlled). Thus, for given values of ψ_o, A, and z, as I is increased, one might expect that the rate of coagulation will first achieve its maximum possible value when E_a first becomes zero. This will happen when $\Phi_{net}(d)$ and $d\Phi_{net}(d)/dd$ are simultaneously equal to zero. Applying these two criteria to Eq. (27.5), we obtain

$$\frac{64n(\infty)kT}{\kappa}\Upsilon_o^2 e^{-\kappa d} = \frac{A}{12\pi d^2} \tag{27.7}$$

and

$$64n(\infty)kT\Upsilon_o^2 e^{-\kappa d} = \frac{A}{6\pi d^3}. \tag{27.8}$$

Using Eq. (27.7) to substitute for $64n(\infty)kT\Upsilon_o^2 e^{-\kappa d}$ into Eq. (27.6), we obtain that the energy barrier will first be zero and the rate of coagulation will first be maximized when

$$d = 2\kappa^{-1}. \tag{27.9}$$

When Eq. (27.9) is substituted into Eq. (27.7), we obtain that

$$\frac{64ccc(\infty)kT}{\kappa}\Upsilon_o^2 e^{-2} = \frac{A}{12\pi 4\kappa^{-2}} \tag{27.10}$$

where $ccc(\infty)$ is the "critical coagulation concentration" value of $n(\infty)$ for the z-z electrolyte determining the value of κ. Rearranging, we have

$$\kappa^3 = \frac{3072\pi ccc(\infty)kT}{A}\Upsilon_o^2 e^{-2}. \tag{27.11}$$

Recalling that

$$\kappa^2 = \frac{e_o^2 \sum(z^2 n_i(\infty))}{\varepsilon\varepsilon_o kT}. \tag{27.12}$$

For a symmetrical z-z electrolyte, the summation in Eq. (27.12) equals $2z^2 n(\infty)$. Therefore when $n(\infty) = ccc(\infty)$,

$$\kappa^3 = \frac{e_o^3 2^{3/2} z^3 (ccc(\infty))^{3/2}}{(\varepsilon\varepsilon_o kT)^{3/2}}. \tag{27.13}$$

Substituting into Eq. (27.11), we obtain

$$\frac{e_o^3 2^{3/2} z^3 (ccc(\infty))^{3/2}}{(\varepsilon\varepsilon_o kT)^{3/2}} = \frac{3072\pi ccc(\infty)kT}{A}\Upsilon_o^2 e^{-2}. \tag{27.14}$$

Rearranging,

$$ccc(\infty) = \frac{(3072\pi kT)^2(\epsilon\epsilon_o kT)^3}{A^2 e_o^6 2^3 z^6}\Upsilon_o^4 e^{-4}. \tag{27.15}$$

27.3.2 The Schulze-Hardy Rule

According to Eq. (27.15), for flat plates, the concentration of background electrolyte required to achieve the maximum rate of coagulation is proportional to z^{-6}. That is,

$$ccc(\infty) \propto \frac{1}{z^6}. \tag{27.16}$$

Thus, by Eq. (27.16), to obtain the maximum possible rate of coagulation, the $ccc(\infty)$ values of $z = 1$, 2, and 3 salts will vary as (1/1) to (1/64) to (1/729), or roughly 1 to 0.016 to 0.0014. Coagulation is thus much more effective when salts of high z are used. That is, the $ccc(\infty)$ for a solution of $MgSO_4$ should be 64 times lower than the $ccc(\infty)$ for a solution of NaCl. When carrying out coagulation in water treatment plants then, very often the high charge salts like alum (aluminum sulfate) and $FeCl_3$ are used to remove colloidal particulate matter.

Prior to the development of the DLVO theory for flat plates, an equation very similar to Eq. (27.16) had been discovered empirically for a wide variety of lyophobic colloids by Schulze (1882, 1883) and Hardy (1900). In particular, the so-called Shulze-Hardy Rule states that

$$ccc(\infty) \propto \frac{1}{z^6_{\text{counter-ion}}} \tag{27.17}$$

where $z_{\text{counter-ion}}$ is the magnitude of the charge on the ion of the I-dominating background electrolyte whose charge is *counter* (i.e., opposite) to that of the lyophobic surface itself.[7] Thus, when coagulating a negatively-charged colloid using a solution of $MgCl_2$ to compress the double layer and coagulate the suspension, $z_{\text{counter-ion}} = 2$. According to the Schulze-Hardy Rule, then, the $ccc(\infty)$ values in coagulating a given negatively-charged colloid will in general be about the same for both $MgSO_4$ and $MgCl_2$.

Given the reliability of Eq. (27.17), it was heralded as a great triumph that DLVO theory for flat plates should be able to lead to Eq. (27.16). However, since the *magnitudes* of the charges on both the co- and counter-ion will greatly affect the value of κ, DLVO theory for flat plates really predicts that *both* magnitudes should be important in determining the value of $ccc(\infty)$. The fact that the counter-ion plays a vastly more important role than the co-ion in coagulating lyophobic colloids is usually rationalized by saying something like "it is the counter-ion that is pulled into the diffuse, aqueous portion of the double-layer, and therefore it is the counter-ion that neutralizes the surface charge." In reality, however, we know from Chapter 26 that the charge density q in the aqueous portion of the double-layer is due not only to a relative excess of counter-ions, but also to a relative deficiency of co-ions.

One possibility that could account for the tremendous relative importance of the counter-ion in hastening coagulation might be the *asymmetry* in the variation of concentration with distance from the surface for counter-ions vs. co-ions (see Section 26.6.4 and Figure 26.12). This asymmetry will increase as the value of $z|\psi^o|$ for the particles increases. However, in view of some of the questionable assumptions involved in coming up with Eq. (27.16), it is not worthwhile spending a lot of time trying to discuss all

[7] An ion whose sign is the same as the charge on the colloid surface is known as a "co-ion."

the possibilities. Indeed, DLVO theory was originally developed for flat plates, and not for the oftentimes *much more realistic* (but electrostatically much more complicated) geometry of two colliding spheres. Secondly, DLVO theory neglects the possible effects of counter-ions adsorbed directly on the surface in a layer called the "Stern layer." Thirdly, we recall that it was assumed that the net potential at the midpoint between the flat plates was just the sum of the individual, non-interacting potentials. Fourthly, it is assumed in the DLVO derivation that $z|\psi_o|$ is greater than about 150 mV. Since it is certainly known that colloids with low $|\psi_o|$ will coagulate (and in fact are easy to coagulate given the low level of electrostatic repulsion), even when just this fourth assumption is relaxed and the details are worked out for two approaching flat plates, the modified the DLVO theory predicts that $ccc(\infty) \propto z^{-2}$, and not z^{-6}.

As far as the Schulze-Hardy rule itself is concerned, a significant number of the empirical observations supporting Eq. (27.17) have been based on situations in which a highly charged counter-ion is in a solution whose pH is such that only an exceedingly small percentage is uncomplexed with OH^-. For example, based in Chapter 11, we know that adding $FeCl_3$ and $Al_2(SO_4)_2$ to water at approximately neutral pH to coagulate colloidal matter in drinking water is not going to lead to the presence of much Fe^{3+} or Al^{3+}. In such cases, it will be far more important that a particle-enmeshing floc of $(am)Fe(OH)_{3(s)}$ or $Al(OH)_{3(s)}$ can form as it is that very low concentrations of ions like Fe^{3+} and Al^{3+} can compress the double-layer and thereby destabilize the colloid. Problems illustrating the principles of this chapter are presented by Pankow (1992).

27.4 REFERENCES

GSCHWEND, P. M. and M. D. REYNOLDS. 1987. Monodisperse ferrous phosphate colloids in an anoxic groundwater plume. *J. Contam. Hydrology*, **1**: 309–327.

HARDY, W. B. 1900. *Proc. Royal Soc. London*, **66**: 110, and *Z. physik. Chem.*, **33**: 385.

HAMAKER, H. C. 1937. *Physica*, **4**: 1058.

HIEMENEZ, P. C. 1986. *Principles of Colloid and Surface Chemistry*. New York: Marcel Dekker.

PANKOW, J. F. 1992. *Aquatic Chemistry Problems*, Portland: Titan Press-OR, P.O. Box 91399, Portland, Oregon 97291-1399.

SCHULZE, H. 1882. *J. prakt. Chem.*, **25**: 2.

SCHULZE, H. 1883. *J. prakt. Chem.*, **27**: 320.

Supplementary Thermodynamic Information

Standard Free Energies of Formation and *Standard Enthalpies of Formation* for selected species at 25°C/1 atm. ΔG_f^o and ΔH_f^o data are for formation of the species in their standard states from the elements in their most stable standard states. Data for species in the "aq" state have a standard state of 1.0 *m* (1.0 molal). Data for species in the "liq" state are pure liquid with a standard state of $X = 1.0$ (mole fraction = 1). Data for species in the "c" state are pure and crystalline with a standard state of $X = 1.0$ (mole fraction = 1). Data for species in the "amorph." state are amorphous (i.e., not crystalline) with a standard state of $X = 1.0$ (mole fraction = 1). Primary source: 1991–92 *CRC Handbook of Chemistry and Physics*. Other ΔG_f^o data deduced based on tabulated values of equilibrium constants at 25°C/1 atm. Abbreviations: c = crystalline solid; aq = aqueous; g = gaseous; amorph. = amorphous solid.

	State	Formula Weight (g/mol)	ΔG_f^o (kJ/mol)	ΔH_f^o (kJ/mol)
ALUMINUM (Al)				
$Al_{(s)}$	c	26.982	0	0
Al^{3+}	aq	26.982	−489.4	−531.4
$AlOH^{2+}$	aq	42.981	−698.	
$Al(OH)_2^+$	aq	60.996	−911.	
$Al(OH)_3^o$	aq	78.003	−1115.	
$Al(OH)_4^-$	aq	95.011	−1325.	
$Al_2O_{3(s)}$ (α, corundum)	c	101.961	−1582.4	−1675.7

	State	Formula Weight (g/mol)	ΔG_f^o (kJ/mol)	ΔH_f^o (kJ/mol)
ALUMINUM (Al) *(continued)*				
$AlOOH_{(s)}$ (boehmite)	c	59.988	−912.7	−987.4
$Al_2O_3 \cdot H_2O_{(s)}$ (diaspore)	c	119.977	−1841.0	−1999.95
$Al(OH)_{3(s)}$ (gibbsite)	c	78.004	−1143.7	−1281.4
$Al(OH)_{3(s)}$	amorph.	78.004	−1139.	
$Al_2Si_2O_7 \cdot 2H_2O_{(s)}$ (kaolinite)	c	258.162	−3778.15	−4098.65
$Al_2Si_2O_7 \cdot 2H_2O_{(s)}$ (halloysite)	c	258.162	−3759.32	−4079.40
ARSENIC (As)				
$As_{(s)}$ (α, gray)	c	74.922	0	0
AsO^+ As(III)	aq	90.921	−163.80	
$HAsO_2$ As(III)	aq	107.928	−402.71	−456.47
$H_2AsO_3^-$ As(III)	aq	124.936	−587.22	−714.79
H_3AsO_4 As(V)	aq	141.943	−766.0	−898.7
$H_2AsO_4^-$ As(V)	aq	140.935	−753.29	−909.56
$HAsO_4^{2-}$ As(V)	aq	139.927	−714.71	−906.34
AsO_4^{3-} As(V)	aq	138.919	−636.0	−879.3
BARIUM (Ba)				
$Ba_{(s)}$	c	137.33	0	0
Ba^{2+}	aq	137.33	−560.74	−537.64
$BaSO_{4(s)}$ (barite)	c	233.402	−1362.31	−1473.19
$BaCO_{3(s)}$ (witherite)	c	197.349	−1137.63	−1216.29
BORON (B)				
$B_{(s)}$ (β)	c	10.811	0	0
$B_2O_{3(s)}$	c	69.620	−1193.70	−1272.77
$H_3BO_{3(s)}$	c	61.833	−969.01	−1094.33
H_3BO_3	aq	61.833	−968.85	−1072.32
$B(OH)_4^-$	aq	78.841	−1153.32	−1344.03
$H_2B_4O_7$	aq	157.256	−2720.02	
BROMINE (Br)				
Br^-	aq	79.909	−103.97	−121.55
$Br_{2(l)}$	liq	159.818	0	0
Br_2	aq	159.818	3.93	−2.59
HOBr	aq	96.911	−82.2	−113.0
BrO^-	aq	95.903	−33.5	−94.1

	State	Formula Weight (g/mol)	ΔG_f^o (kJ/mol)	ΔH_f^o (kJ/mol)
CADMIUM (Cd)				
$Cd_{(s)}(\gamma)$	c	112.410	0	0
Cd^{2+}	aq	112.410	−77.58	−75.90
$CdOH^+$	aq	129.417	−284.5	
$Cd(OH)_2^o$	aq	146.425	−392.2	
$Cd(OH)_3^-$	aq	163.432	−600.8	
$Cd(OH)_4^{2-}$	aq	180.439	−758.5	
$CdO_{(s)}$	c	128.409	−228.4	−258.1
$Cd(OH)_{2(s)}$ (fresh ppt.)	c	146.417	−473.63	−560.66
$CdCl^+$	aq	147.863	−224.4	−240.6
$CdCl_2^o$	aq	183.316	−359.32	−405.01
$CdCl_3^-$	aq	218.769	−487.0	−561.0
$CdS_{(s)}$	c	144.464	−156.48	−161.92
$Cd(NH_3)^{2+}$	aq	129.431	−118.83	
$Cd(NH_3)_2^{2+}$	aq	146.461	−158.99	−266.10
$Cd(NH_3)_4^{2+}$	aq	180.522	−226.35	−450.20
$CdCO_{3(s)}$	c	172.409	−669.44	−750.61
$CdC_2O_4^o$	aq	200.430	−774.46	
$CdCN^+$	aq	138.418	64.43	
$Cd(CN)_2^o$	aq	164.436	207.94	
CALCIUM (Ca)				
Ca^{2+}	aq	40.08	−553.54	−542.83
$CaOH^+$	aq	50.087	−718.39	
$Ca(OH)_{2(s)}$	c	74.095	−898.4	−986.0
$CaCO_{3(s)}$ (calcite)	c	100.089	−1128.84	−1206.92
$CaCO_{3(s)}$ (aragonite)	c	100.089	−1127.80	−1207.13
$CaMg(CO_3)_{2(s)}$ (dolomite)	c	184.411	−2163.55	−2326.30
$Ca_5(PO_4)_3OH_{(s)}$ (hydroxyapatite)	c	502.321	−6338.4	−6721.6
$CaSO_{4(s)}$	c	136.138	−1321.7	−1434.1
$CaSO_4 \cdot 2H_2O_{(s)}$ (gypsum)	c	172.168	−1797.2	−2022.6
$CaSiO_{3(s)}$ (wollastonite)	c	116.164	−1549.9	−1635.2
CARBON (C)				
$C_{(s)}$ (graphite)	c	12.011	0	0
$C_{(s)}$ (diamond)	c	12.011	2.90	1.90
$CO_{2(g)}$	g	44.010	−394.36	−393.51
CO_2	aq	44.010	−386.02	−413.80

	State	Formula Weight (g/mol)	ΔG_f^o (kJ/mol)	ΔH_f^o (kJ/mol)
CARBON (C) *(continued)*				
$CH_{4(g)}$	g	16.043	−50.75	−74.81
CH_4	aq	16.043	−34.39	−89.04
H_2CO_3 (not $H_2CO_3^*$)	aq	62.025	−607.1	
$H_2CO_3^*$	aq	62.025	−623.16	−699.65
HCO_3^-	aq	61.017	−586.85	−691.99
CO_3^{2-}	aq	60.009	−527.90	−677.14
HCOOH (formic acid)	aq	46.026	−372.38	−425.43
$HCOO^-$ (formate ion)	aq	45.018	−351.04	−425.55
CH_3OH (methanol)	aq	32.042	−175.39	−245.93
CNO^-	aq	42.017	−97.49	−146.02
HCN (hydrogen cyanide)	aq	27.026	119.7	107.1
CN^- (cyanide)	aq	26.018	172.4	150.6
CH_3COOH (acetic acid)	aq	60.053	−396.64	−485.76
CH_3COO^- (acetate ion)	aq	59.045	−369.41	−486.01
C_2H_5OH (ethanol)	aq	46.070	−181.75	−288.28
CHLORINE (Cl)				
Cl^-	aq	35.453	−131.26	−167.16
$Cl_{2(g)}$	g	70.906	0	0
Cl_2	aq	70.906	6.90	−23.43
HClO	aq	52.460	−79.91	−120.92
ClO^-	aq	51.452	−36.82	−107.11
CHROMIUM (Cr)				
Cr^{3+}	aq	51.996	−215.5	−256.0
$CrOH^{2+}$	aq	69.003	−429.	
$Cr(OH)_2^+$	aq	86.011	−634.	
$Cr(OH)_3^o$	aq	103.018	−824.	
$Cr(OH)_4^-$	aq	120.025	−1007.	
$Cr_2O_4^{2-}$	aq	115.994	−727.85	−881.15
$HCrO_4^-$	aq	117.002	−764.84	−878.22
$Cr_2O_7^{2-}$	aq	215.988	−1301.22	−1490.34
$Cr(OH)_{3(s)}$ (fresh ppt.)	c	103.018		−1063.99
COPPER (Cu)				
$Cu_{(s)}$	c	63.546	0	0
Cu^{2+}	aq	63.546	65.52	64.77

	State	Formula Weight (g/mol)	ΔG_f^o (kJ/mol)	ΔH_f^o (kJ/mol)
COPPER (Cu) *(continued)*				
CuO (tenorite)	c	79.539	−129.70	−157.32
$CuCl^+$	aq	98.993	−68.20	
$CuS_{(s)}$	c	95.604	−53.55	−53.14
$Cu_2S_{(s)}$	c	159.152	−86.2	−79.5
$CuCO_3 \cdot Cu(OH)_{2(s)}$ (malachite)	c	221.116	−893.7	−1051.4
FLUORINE (F)				
F^-	aq	18.998	−278.82	−332.63
HYDROGEN (H)				
H^+	aq	1.008	0	0
$H_{2(g)}$	g	2.016	0	0
H_2	aq	2.016	−17.7	
OH^-	aq	17.007	−157.29	−229.99
$H_2O_{(l)}$ $(X = 1)$	liq	18.015	−237.18	−285.83
H_2O_2	aq	34.015	−134.10	−191.17
IRON (Fe)				
$Fe_{(s)}$	c	55.847	0	0
Fe^{2+}	aq	55.847	−78.87	−89.12
$FeOH^+$	aq	72.854	−261.8	
$Fe(OH)_2^o$	aq	89.862	−435.7	
$Fe(OH)_3^-$	aq	106.869	−607.8	
Fe^{3+}	aq	55.847	−4.60	−48.53
$FeOH^{2+}$	aq	72.854	−229.	
$Fe(OH)_2^+$	aq	89.862	−446.	
$Fe(OH)_3^o$	aq	106.869	~ −659.4	
$Fe(OH)_4^-$	aq	123.876	−830.	
$(am)Fe(OH)_{3(s)}$	amorph.	106.869	−698.	
$Fe_2O_{3(s)}$ (hematite)	c	159.6922	−742.24	−824.25
$Fe_3O_{4(s)}$ (magnetite)	c	231.5386	−1015.46	−1118.38
$FeOOH_{(s)}$ (goethite)	c	88.8538	−488.6	−558.98
$Fe(OH)_{3(s)}$ (fresh ppt.)	c	106.8691	−696.64	−822.99
$FeS_{(s)}$ (pyrrhotite)	c	87.911	−100.42	−100.00
$FeS_{2(s)}$ (pyrite)	c	119.975	−166.94	−178.24
$FeCO_{3(s)}$ (siderite)	c	115.8564	−666.72	−740.57

	State	Formula Weight (g/mol)	ΔG_f^o (kJ/mol)	ΔH_f^o (kJ/mol)
LEAD (Pb)				
$Pb_{(s)}$	c	207.19	0	0
Pb^{2+}	aq	207.19	−24.39	−1.67
$PbOH^+$	aq	224.197	−217.6	
$Pb(OH)_2^o$	aq	241.205	−401.	
$Pb(OH)_3^-$	aq	258.212	−574.	
$PbO_{(s)}$ (yellow)	c	223.189	−187.90	−215.33
$PbO_{(s)}$ (red)	c	223.189	−188.95	−218.99
$PbO_{2(s)}$	c	239.189	−217.36	−277.40
$PbS_{(s)}$	c	239.254	−98.74	−100.42
$PbSO_{4(s)}$	c	303.252	−813.20	−919.94
$PbCO_{3(s)}$	c	267.199	−625.51	−699.15
MAGNESIUM (Mg)				
$Mg_{(s)}$	c	24.305	0	0
Mg^{2+}	aq	24.305	−454.80	−466.85
$MgOH^+$	aq	41.312	−626.8	
$Mg(OH)_2^o$	aq	58.320	−769.4	−926.8
$Mg(OH)_{2(s)}$ (brucite)	c	58.320	−833.5	−924.5
$MgS_{(s)}$	c	56.376	−341.83	−346.02
$MgSO_4^o$	aq	120.3736	−1212.27	−1356.03
$MgCO_{3(s)}$	c	84.3214	−1012.11	−1095.79
$MgCO_3 \cdot 3H_2O_{(s)}$ (nesquehonite)	c	138.3674	−1726.32	
$MgCO_3 \cdot 5H_2O_{(s)}$ (lanfordite)	c	174.398	−2199.53	
MANGANESE (Mn)				
$Mn_{(s)}(\alpha)$	c	54.938	0	0
Mn^{2+}	aq	54.938	−228.03	−220.75
$MnO_{2(s)}$	c	86.937	−465.18	−520.03
$Mn(OH)_{2(s)}$ (fresh ppt.)	amorph.	88.926	−615.05	−695.38
$MnOOH_{(s)}$ (manganite)	c	87.945	−557.7	
$Mn_3O_{4(s)}$ (hausmannite)	c	228.812	−1281.	
$MnS_{(s)}$ (green)	c	87.002	−218.40	−214.22
$MnSO_{4(s)}$	c	150.9996	−957.42	−1065.25
$MnCO_{3(s)}$ (natural)	c	114.9474	−816.72	−894.12
MERCURY (Hg)				
$Hg_{(l)}$	liq	200.59	0	0
Hg^{2+}	aq	200.59	164.43	171.13

	State	Formula Weight (g/mol)	ΔG_f^o (kJ/mol)	ΔH_f^o (kJ/mol)
MERCURY (Hg) *(continued)*				
Hg_2^{2+}	aq	401.18	153.55	172.38
$Hg_2Cl_{2(s)}$ (calomel)	c	472.086	−210.78	−265.22
$HgOH^+$	aq	217.597	−52.3	−84.5
$HgOH_2^o$	aq	234.605	−274.9	−355.2
$HgCl^+$	aq	236.043	−5.44	−18.8
$HgCl_2^o$	aq	271.496	−173.2	−216.3
$HgCl_3^-$	aq	306.949	−309.2	−388.7
$HgCl_4^{2-}$	aq	342.402	−446.8	−554.0
$HgS_{(s)}$	c	232.650	−43.3	−46.7
NITROGEN (N)				
$N_{2(g)}$	g	28.013	0	0
N_2	aq	28.013	−18.26	
HNO_2	aq	47.013	−42.97	−119.2
NO_2^-	aq	46.006	−32.22	−104.60
NO_3^-	aq	62.005	−111.34	−207.36
NH_3	aq	17.031	−26.57	−80.29
NH_4^+	aq	18.039	−79.37	−132.51
$NH_{3(g)}$	g	17.031	−16.48	−46.1
OXYGEN (O)				
$O_{2(g)}$	g	31.9998	0	0
O_2	aq	31.9998	16.32	−11.72
$O_{3(g)}$	g	47.9982	163.18	142.67
OH^-	aq	17.0074	−157.29	−229.99
H_2O ($X = 1$)	liq	18.0153	−237.18	−285.83
H_2O_2	aq	34.0147	−134.10	−191.17
SILICON (Si)				
$Si_{(s)}$	c	28.086	0	0
$SiO_{2(s)}$ (quartz)	c	60.085	−856.67	−910.94
$SiO_{2(s)}$ (cristobalite)	c	60.085	−855.46	−909.48
$SiO_{2(s)}$ (fresh ppt.)	amorph.	60.085	−850.7	−909.5
H_4SiO_4 (silicic acid)	aq	96.115	−1316.7	−1468.6
SILVER (Ag)				
$Ag_{(s)}$	c	107.868	0	0
Ag^+	aq	107.868	77.12	105.58

		State	Formula Weight (g/mol)	ΔG_f^o (kJ/mol)	ΔH_f^o (kJ/mol)
SILVER (Ag) *(continued)*					
$AgOH^o$		aq	124.875	−92.	
$Ag(OH)_2^-$		aq	141.883	−260.2	
$AgCl^o$		aq	143.321	−72.8	
$AgCl_2^-$		aq	178.774	−215.5	
$AgCl_{(s)}$		c	143.321	−109.80	−127.07
$AgBr_{(s)}$		c	187.779	−96.90	−100.37
$AgI_{(s)}$		c	234.774	−66.19	−61.84
$Ag_2S_{(s)}$ (α orthorh.)		c	247.804	−40.67	−32.59
STRONTIUM (Sr)					
$Sr_{(s)}$		c	87.62	0	0
Sr^{2+}		aq	87.62	−559.44	−545.80
$SrOH^+$		aq	104.627	−721.	
$SrCO_{3(s)}$ (strontianite)		c	147.629	−1140.14	−1220.05
SULFUR (S)					
$S_{(s)}$ (rhombic)	S(0)	c	32.064	0	0
S_2^{2-}	S(-I)	aq	64.128	79.50	30.12
H_2S	S(-II)	aq	34.080	−27.87	−39.75
HS^-	S(-II)	aq	33.072	12.05	−17.6
S^{2-}	S(-II)	aq	32.064	85.8	33.0
H_2SO_3 (not $H_2SO_3^*$)	S(IV)	aq	82.078	−534.5	
$H_2SO_3^*$	S(IV)	aq	82.078	−537.90	−608.81
HSO_3^-	S(IV)	aq	81.070	−527.8	−626.2
SO_3^{2-}	S(IV)	aq	80.062	−486.60	−635.55
HSO_4^-	S(VI)	aq	97.070	−756.01	−887.34
SO_4^{2-}	S(VI)	aq	96.062	−744.63	−909.27
ZINC (Zn)					
$Zn_{(s)}$		c	65.38	0	0
Zn^{2+}		aq	65.38	−147.03	−153.89
$ZnOH^+$		aq	82.387	−330.1	
$Zn(OH)_2^o$		aq	99.395	−522.3	
$Zn(OH)_3^-$		aq	116.402	−694.3	
$Zn(OH)_4^{2-}$		aq	133.409	−858.7	
$ZnCl^+$		aq	100.833	−275.3	
$ZnCl_2^o$		aq	136.286	−403.8	
$ZnCl_3^-$		aq	171.739	−540.6	

	State	Formula Weight (g/mol)	ΔG_f^o (kJ/mol)	ΔH_f^o (kJ/mol)
ZINC (Zn) *(continued)*				
$ZnCl_4^{2-}$	aq	207.192	−666.1	
$Zn(OH)_{2(s)}$	c	99.395	−553.2	−641.9
$ZnS_{(s)}$ (sphalerite)	c	97.444	−201.29	−205.98
$ZnSO_{4(s)}$	c	161.442	−874.46	−982.82
$ZnCO_{3(s)}$	c	125.389	−731.57	−812.78
$ZnCO_3 \cdot H_2O_{(s)}$	c	143.404	−970.69	

L

M

Index

A

B

C

D

E

F

G

H

I

K

L

M

N

T

U

V

W

Z